Lecture Notes in Computer Science　13398

More information about this series at https://link.springer.com/bookseries/558

Günter Rudolph · Anna V. Kononova ·
Hernán Aguirre · Pascal Kerschke ·
Gabriela Ochoa · Tea Tušar (Eds.)

Parallel Problem Solving from Nature – PPSN XVII

17th International Conference, PPSN 2022
Dortmund, Germany, September 10–14, 2022
Proceedings, Part I

Springer

Editors
Günter Rudolph ⓘ
TU Dortmund
Dortmund, Germany

Hernán Aguirre ⓘ
Shinshu University
Nagano, Japan

Gabriela Ochoa ⓘ
University of Stirling
Stirling, UK

Anna V. Kononova ⓘ
Leiden University
Leiden, The Netherlands

Pascal Kerschke ⓘ
Technische Universität Dresden
Dresden, Germany

Tea Tušar ⓘ
Jožef Stefan Institute
Ljubljana, Slovenia

ISSN 0302-9743   ISSN 1611-3349 (electronic)
Lecture Notes in Computer Science
ISBN 978-3-031-14713-5   ISBN 978-3-031-14714-2 (eBook)
https://doi.org/10.1007/978-3-031-14714-2

This Springer imprint is published by the registered company Springer Nature Switzerland AG
The registered company address is: Gewerbestrasse 11, 6330 Cham, Switzerland

Preface

The first major gathering of people interested in discussing natural paradigms and their application to solve real-world problems in Europe took place in Dortmund, Germany, in 1990. What was planned originally as a small workshop with about 30 participants finally grew into an international conference named Parallel Problem Solving from Nature (PPSN) with more than 100 participants. The interest in the topics of the conference has increased steadily ever since leading to the pleasant necessity of organizing PPSN conferences biennially within the European region.

In times of a pandemic, it is difficult to find a host for a conference that should be held locally if possible. To ensure the continuation of the conference series, the 17th edition, PPSN 2022, returned to its birthplace in Dortmund. But even at the time of writing this text, it is unclear whether the conference can be held on-site or whether we shall have to switch to virtual mode at short notice.

Therefore, we are pleased that many researchers shared our optimism by submitting their papers for review. We received 185 submissions from which the program chairs have selected the top 85 after an extensive peer-review process. Not all decisions were easy to make but in all cases we benefited greatly from the careful reviews provided by the international Program Committee consisting of 223 scientists. Most of the submissions received four reviews, but all of them got at least three reviews. This led to a total of 693 reviews. Thanks to these reviews we were able to decide about acceptance on a solid basis.

The papers included in these proceedings have been assigned to 12 fuzzy clusters, entitled *Automated Algorithm Selection and Configuration, Bayesian- and Surrogate-Assisted Optimization, Benchmarking and Performance Measures, Combinatorial Optimization, (Evolutionary) Machine Learning and Neuroevolution, Evolvable Hardware and Evolutionary Robotics, Fitness Landscape Modeling and Analysis, Genetic Programming, Multi-Objective Optimization, Numerical Optimizaiton, Real-World Applications, and Theoretical Aspects of Nature-Inspired Optimization,* that can hardly reflect the true variety of research topics presented in the proceedings at hand. Following the tradition and spirit of PPSN, all papers were presented as posters. The 7 poster sessions consisting of about 12 papers each were compiled orthogonally to the fuzzy clusters mentioned above to cover the range of topics as widely as possible. As a consequence, participants with different interests would find some relevant papers in every session and poster presenters were able to discuss related work in sessions other than their own. As usual, the conference also included one day with workshops (Saturday), one day with tutorials (Sunday), and three invited plenary talks (Monday to Wednesday) for free.

Needless to say, the success of such a conference depends on the authors, reviewers, and organizers. We are grateful to all authors for submitting their best and latest work, to all the reviewers for the generous way they spent their time and provided their valuable expertise in preparing the reviews, to the workshop organizers and tutorial presenters for their contributions enhancing the value of the conference, and to the local organizers who helped to make PPSN 2022 happen.

Last but not least, we would like to thank for the donations of the *Gesellschaft der Freunde der Technischen Universität Dortmund e.V. (GdF)* and the *Alumni der Informatik Dortmund e.V. (aido)*. We are grateful for Springer's long-standing support of this conference series. Finally, we thank the *Deutsche Forschungsgemeinschaft (DFG)* for providing financial backing.

July 2022

Günter Rudolph
Anna V. Kononova
Hernán Aguirre
Pascal Kerschke
Gabriela Ochoa
Tea Tušar

Organization

General Chair

Günter Rudolph   TU Dortmund University, Germany

Honorary Chair

Hans-Paul Schwefel   TU Dortmund University, Germany

Program Committee Chairs

Hernán Aguirre   Shinshu University, Japan
Pascal Kerschke   TU Dresden, Germany
Gabriela Ochoa   University of Stirling, UK
Tea Tušar   Jožef Stefan Institute, Slovenia

Proceedings Chair

Anna V. Kononova   Leiden University, The Netherlands

Tutorial Chair

Heike Trautmann   University of Münster, Germany

Workshop Chair

Christian Grimme   University of Münster, Germany

Publicity Chairs

Nicolas Fischöder   TU Dortmund University, Germany
Peter Svoboda   TU Dortmund University, Germany

Social Media Chair

Roman Kalkreuth   TU Dortmund University, Germany

Digital Fallback Chair

Hestia Tamboer   Leiden University, The Netherlands

Steering Committee

Thomas Bäck   Leiden University, The Netherlands
David W. Corne   Heriot-Watt University, UK
Carlos Cotta   Universidad de Málaga, Spain
Kenneth De Jong   George Mason University, USA
Gusz E. Eiben   Vrije Universiteit Amsterdam, The Netherlands
Bogdan Filipič   Jožef Stefan Institute, Slovenia
Emma Hart   Edinburgh Napier University, UK
Juan Julián Merelo Guervós   Universida de Granada, Spain
Günter Rudolph   TU Dortmund University, Germany
Thomas P. Runarsson   University of Iceland, Iceland
Robert Schaefer   University of Krakow, Poland
Marc Schoenauer   Inria, France
Xin Yao   University of Birmingham, UK

Program Committee

Jason Adair   University of Stirling, UK
Michael Affenzeller   University of Applied Sciences Upper Austria,
                                           Austria
Hernán Aguirre   Shinshu University, Japan
Brad Alexander   University of Adelaide, Australia
Richard Allmendinger   University of Manchester, UK
Marie Anastacio   Leiden University, The Netherlands
Denis Antipov   ITMO University, Russia
Claus Aranha   University of Tsukuba, Japan
Rolando Armas   Yachay Tech University, Ecuador
Dirk Arnold   Dalhousie University, Canada
Anne Auger   Inria, France
Dogan Aydin   Dumlupinar University, Turkey
Jaume Bacardit   Newcastle University, UK
Thomas Bäck   Leiden University, The Netherlands
Helio Barbosa   Laboratório Nacional de Computação Científica,
                                           Brazil
Andreas Beham   University of Applied Sciences Upper Austria,
                                           Austria
Heder Bernardino   Universidade Federal de Juiz de Fora, Brazil
Hans-Georg Beyer   Vorarlberg University of Applied Sciences,
                                           Austria

Julian Blank	Michigan State University, USA
Aymeric Blot	University College London, UK
Christian Blum	Spanish National Research Council, Spain
Peter Bosman	Centrum Wiskunde & Informatica, The Netherlands
Jakob Bossek	University of Münster, Germany
Jürgen Branke	University of Warwick, UK
Dimo Brockhoff	Inria, France
Alexander Brownlee	University of Stirling, UK
Larry Bull	University of the West of England, UK
Maxim Buzdalov	ITMO University, Russia
Arina Buzdalova	ITMO University, Russia
Stefano Cagnoni	University of Parma, Italy
Fabio Caraffini	De Montfort University, UK
Ying-Ping Chen	National Chiao Tung University, Taiwan
Francisco Chicano	University of Málaga, Spain
Miroslav Chlebik	University of Sussex, UK
Sung-Bae Cho	Yonsei University, South Korea
Tinkle Chugh	University of Exeter, UK
Carlos Coello Coello	CINVESTAV-IPN, Mexico
Ernesto Costa	University of Coimbra, Portugal
Carlos Cotta	Universidad de Málaga, Spain
Nguyen Dang	St Andrews University, UK
Kenneth De Jong	George Mason University, USA
Bilel Derbel	University of Lille, France
André Deutz	Leiden University, The Netherlands
Benjamin Doerr	Ecole Polytechnique, France
Carola Doerr	Sorbonne University, France
John Drake	University of Leicester, UK
Rafal Drezewski	AGH University of Science and Technology, Poland
Paul Dufossé	Inria, France
Gusz Eiben	Vrije Universiteit Amsterdam, The Netherlands
Mohamed El Yafrani	Aalborg University, Denmark
Michael Emmerich	Leiden University, The Netherlands
Andries Engelbrecht	University of Stellenbosch, South Africa
Anton Eremeev	Omsk Branch of Sobolev Institute of Mathematics, Russia
Richard Everson	University of Exeter, UK
Pedro Ferreira	Universidade de Lisboa, Portugal
Jonathan Fieldsend	University of Exeter, UK
Bogdan Filipič	Jožef Stefan Institute, Slovenia

Ke Li	University of Exeter, UK
Arnaud Liefooghe	University of Lille, France
Giosuè Lo Bosco	Università di Palermo, Italy
Fernando Lobo	University of Algarve, Portugal
Daniele Loiacono	Politecnico di Milano, Italy
Nuno Lourenço	University of Coimbra, Portugal
Jose A. Lozano	University of the Basque Country, Spain
Rodica Ioana Lung	Babes-Bolyai University, Romania
Chuan Luo	Peking University, China
Gabriel Luque	University of Málaga, Spain
Evelyne Lutton	INRAE, France
Manuel López-Ibáñez	University of Málaga, Spain
Penousal Machado	University of Coimbra, Portugal
Kaitlin Maile	ISAE-SUPAERO, France
Katherine Malan	University of South Africa, South Africa
Vittorio Maniezzo	University of Bologna, Italy
Elena Marchiori	Radboud University, The Netherlands
Asep Maulana	Tilburg University, The Netherlands
Giancarlo Mauri	University of Milano-Bicocca, Italy
Jacek Mańdziuk	Warsaw University of Technology, Poland
James McDermott	National University of Ireland, Galway, Ireland
Jörn Mehnen	University of Strathclyde, UK
Marjan Mernik	University of Maribor, Slovenia
Olaf Mersmann	TH Köln, Germany
Silja Meyer-Nieberg	Bundeswehr University Munich, Germany
Efrén Mezura-Montes	University of Veracruz, Mexico
Krzysztof Michalak	Wroclaw University of Economics, Poland
Kaisa Miettinen	University of Jyväskylä, Finland
Edmondo Minisci	University of Strathclyde, UK
Gara Miranda	University of La Laguna, Spain
Mustafa Misir	Istinye University, Turkey
Hugo Monzón	RIKEN, Japan
Sanaz Mostaghim	Fraunhofer IWS, Germany
Mario Andres Muñoz Acosta	University of Melbourne, Australia
Boris Naujoks	TH Köln, Germany
Antonio J. Nebro	University of Málaga, Spain
Aneta Neumann	University of Adelaide, Australia
Frank Neumann	University of Adelaide, Australia
Michael O'Neill	University College Dublin, Ireland
Pietro S. Oliveto	University of Sheffield, UK
Una-May O'Reilly	MIT, USA
José Carlos Ortiz-Bayliss	Tecnológico de Monterrey, Mexico

Marc Sevaux	Université de Bretagne Sud, France
Shinichi Shirakawa	Yokohama National University, Japan
Moshe Sipper	Ben-Gurion University of the Negev, Israel
Jim Smith	University of the West of England, UK
Jorge Alberto Soria-Alcaraz	Universidad de Guanajuato, Mexico
Patrick Spettel	FH Vorarlberg, Austria
Giovanni Squillero	Politecnico di Torino, Italy
Catalin Stoean	University of Craiova, Romania
Thomas Stützle	Université Libre de Bruxelles, Belgium
Mihai Suciu	Babes-Bolyai University, Romania
Dirk Sudholt	University of Sheffield, UK
Andrew Sutton	University of Minnesota, USA
Ricardo H. C. Takahashi	Universidade Federal de Minas Gerais, Brazil
Sara Tari	Université du Littoral Côte d'Opale, France
Daniel Tauritz	Auburn University, USA
Dirk Thierens	Utrecht University, The Netherlands
Sarah Thomson	University of Stirling, UK
Kevin Tierney	Bielefeld University, Germany
Renato Tinós	University of São Paulo, Brazil
Alberto Tonda	INRAE, France
Leonardo Trujillo	Instituto Tecnológico de Tijuana, Mexico
Tea Tušar	Jožef Stefan Institute, Slovenia
Ryan J. Urbanowicz	University of Pennsylvania, USA
Koen van der Blom	Leiden University, The Netherlands
Bas van Stein	Leiden University, The Netherlands
Nadarajen Veerapen	University of Lille, France
Sébastien Verel	Université du Littoral Côte d'Opale, France
Diederick Vermetten	Leiden University, The Netherlands
Marco Virgolin	Centrum Wiskunde & Informatica, The Netherlands
Aljoša Vodopija	Jožef Stefan Institute, Slovenia
Markus Wagner	University of Adelaide, Australia
Stefan Wagner	University of Applied Sciences Upper Austria, Austria
Hao Wang	Leiden University, The Netherlands
Hui Wang	Leiden University, The Netherlands
Elizabeth Wanner	CEFET, Brazil
Marcel Wever	LMU Munich, Germany
Dennis Wilson	ISAE-SUPAERO, France
Carsten Witt	Technical University of Denmark, Denmark
Man Leung Wong	Lingnan University, Hong Kong
Bing Xue	Victoria University of Wellington, New Zealand

Kaifeng Yang	University of Applied Sciences Upper Austria, Austria
Shengxiang Yang	De Montfort University, UK
Estefania Yap	University of Melbourne, Australia
Furong Ye	Leiden University, The Netherlands
Martin Zaefferer	TH Köln, Germany
Aleš Zamuda	University of Maribor, Slovenia
Saúl Zapotecas	Instituto Nacional de Astrofísica, Óptica y Electrónica, Mexico
Christine Zarges	Aberystwyth University, UK
Mengjie Zhang	Victoria University of Wellington, New Zealand

Keynote Speakers

Doina Bucur	University of Twente, The Netherlands
Claudio Semini	IIT, Genoa, Italy
Travis Waller	TU Dresden, Germany

Contents – Part I

Benchmarking and Performance Measures

Combinatorial Optimization

(Evolutionary) Machine Learning and Neuroevolution

Evolvable Hardware and Evolutionary Robotics

Fitness Landscape Modeling and Analysis

Contents – Part II

Numerical Optimizaiton

Automated Algorithm Selection
and Configuration

Automated Algorithm Selection in Single-Objective Continuous Optimization: A Comparative Study of Deep Learning and Landscape Analysis Methods

Raphael Patrick Prager[1](✉)(iD), Moritz Vinzent Seiler[1](iD),
Heike Trautmann[1,2](iD), and Pascal Kerschke[3](iD)

[1] Data Science: Statistics and Optimization, University of Münster,
Münster, Germany
{raphael.prager,moritz.seiler,heike.trautmann}@uni-muenster.de
[2] Data Management and Biometrics, University of Twente,
Enschede, The Netherlands
[3] Big Data Analytics in Transportation, TU Dresden, Dresden, Germany
pascal.kerschke@tu-dresden.de

Abstract. In recent years, feature-based automated algorithm selection using exploratory landscape analysis has demonstrated its great potential in single-objective continuous black-box optimization. However, feature computation is problem-specific and can be costly in terms of computational resources. This paper investigates feature-free approaches that rely on state-of-the-art deep learning techniques operating on either images or point clouds. We show that point-cloud-based strategies, in particular, are highly competitive and also substantially reduce the size of the required solver portfolio. Moreover, we highlight the effect and importance of cost-sensitive learning in automated algorithm selection models.

Keywords: Automated algorithm selection · Exploratory landscape analysis · Deep learning · Continuous optimization

1 Introduction

The algorithm selection problem (ASP), nowadays, is a well-studied topic [12,31]. Essentially, it boils down to the identification of a mechanism $m : \mathcal{I} \rightarrow \mathcal{A}$, which selects an optimal algorithm a out of a collection of algorithms $\mathcal{A}$ for any given optimization problem $i \in \mathcal{I}$. Typically, this mechanism m is in need of a numerical representation of a problem instance i to make an informed decision. In the domain of single-objective continuous optimization, this role has largely been filled by exploratory landscape analysis (ELA) [21].

Various research endeavours have improved and evaluated ELA for automated algorithm selection (AAS) [3,14,29]. In this paper, we offer an alternative

G. Rudolph et al. (Eds.): PPSN 2022, LNCS 13398, pp. 3–17, 2022.
https://doi.org/10.1007/978-3-031-14714-2_1

and conceptually very different means to characterize a problem instance i in the domain of AAS. We call this collection of methods 'feature-free'. While these methods still require a sample of the search space, they do not require any computation of numerical features. Instead, the initial sample of a search space is used to construct a $2D$ image (fitness map) or a cloud of points (fitness cloud), which assists in the algorithm selection process. In a recent publication, we evaluated a few of these methods for AAS limited to $2D$ single-objective continuous problems with promising results [28]. In a subsequent publication, we successfully tested these methods for predicting a landscape's general characteristics, e.g., the degree of multi-modality, without any dimensionality restrictions [33].

This paper combines the gained insights from both works by investigating the methods' potential for AAS while simultaneously bypassing the limiting factor that $2D$ fitness maps are inherently restricted to $2D$ problems. To compare our results with existing research, we mimic the algorithm selection scenario of Kerschke and Trautmann [14]. Thereby, we will demonstrate the potential of our feature-free and deep learning (DL) based approaches for AAS and promote further research within this domain.

This paper is structured as follows. First, we briefly introduce ELA as well as the construction of the fitness maps and fitness clouds in Sect. 2. Thereafter, we describe our underlying data sets and the experimental setup in Sect. 3. This is followed by a discussion of the results in Sect. 4, as well as a summary of our findings and an outlook on further research potential in Sect. 5.

2   Background

2.1   Exploratory Landscape Analysis

ELA features [15,20] numerically characterize the landscape of continuous, single-objective optimization problems which is crucial for problem understanding and as input to automated algorithm selection models [12,23,24]. ELA features are required to be descriptive, fast to compute, and reasonably cheap, i.e., the size of the initial sample on which they are most commonly based, has to be small [4,13]. Over the years, a large variety of feature sets emerged, of which we specifically consider the following:

Classical ELA comprises 22 features of six categories. While the meta model, level set and y-distribution features are commonly used, the remaining ones require additional function evaluations [21]. Six **Fitness Distance Correlation** features provide metrics describing the distances in decision and objective space, and the relation of these distances in-between those two spaces [11]. The 16 **Dispersion** features divide the initial sample into different subsets w.r.t. sample quantiles and contrast homogeneity of both groups [18]. Five **Information Content** features rely on smoothness, ruggedness and neutrality measures of random walks over the initial sample [22]. **Nearest Better Clustering** (5 features) derives several metrics and ratios on nearest (better) neighbour distances of the initial sample [13]. Moreover, the **Miscellaneous** feature group summarizes 10 features of different concepts, e.g. principal component analysis based features or the problem dimensionality [15].

2.2   Fitness Map

a) 5D: reduced MC b) 10D: reduced MC   c) 5D: Conv-rMC   d) 10D: Conv-rMC

Fig. 1. Examples of different fitness maps. In particular, visualization of the rMC (a–b) and Conv-rMC (c–d) image-based approaches for different dimensionalities. Note, that we colored and inverted the color-scale for illustrative purposes.

Several works [2,28,32,33] have recently proposed measures to alleviate problems with feature-based AAS as it comes with three major drawbacks: instance features (1) are manually designed in an elaborative process, (2) require additional calculations which may increase selection costs, and (3) are limited to a specific domain and, thus, cannot be adapted to other problems easily (see [32]). These three drawbacks are alleviated by applying DL-based approaches directly on the raw instance information and, thereby, avoiding the need for computing instance features as well as circumnavigating the three drawbacks.

In our previous work, we have proposed several feature-free approaches [28, 33]. These approaches range from simple convolutional neural networks (CNN) [16] to point cloud transformers (PCT) [6]. The former approach is based on images while the latter one is based on point clouds.

In [33], we proposed four different techniques to create fitness maps based on an initial sample $\mathcal{X}$ of the search space with $\mathcal{X} \in \mathbb{R}^{m \times (D+1)}$. Here, m denotes the sample size and $D + 1$ represents the size of the decision space in addition to the fitness value. We formally define a fitness map as $\mathcal{F}_{map} \in \mathbb{R}^{l \times l \times 1}$ with a width and height of $l \times l$ pixels plus a singular color channel. Each technique consists of two steps. First, the D-dimensional samples and their fitness values are normalized and, then, the samples are projected into 2D by using one of four proposed dimensionality reduction techniques (cf. [33] for details about the normalization and choice of parameters like l). Afterwards, the fitness values are mapped into the respective 2D-Cartesian plane at their corresponding transformed location. This 2D-plane can then be interpreted as a gray scaled image in which unknown or bad fitness values have a brighter hue and respectively good values a darker one (see Fig. 1). These four proposed dimensionality reduction techniques are: (1) classical principal component analysis (PCA) [26], (2) PCA including the fitness-value (PCA-Func), (3) multi channel (MC), and (4) reduced multi channel (rMC). The MC approaches create an individual 2D feature map for each possible pairwise combination of search space variables which amounts

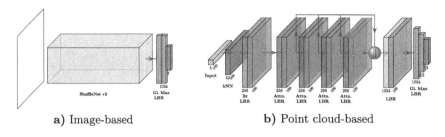

a) Image-based   **b)** Point cloud-based

Fig. 2. Visualizations of the image- (left) and point-based approach (right), adapted from [33]. a) A ShuffleNet V2 [19], extended by the following layers: Global Max. Pooling, Linear, Batch Normalization [10], ReLU [25](LBR), and Linear with Softmax activation. b) kNN-Embedding, followed by four Attention Layers (Attn.), a Global Max. Pooling, LBR, and Linear Layer with Softmax activation as proposed in [33].

to $n = \binom{D}{2}$ total feature maps. These are consolidated into a single fitness map of shape $\mathcal{F}_{map} \in \mathbb{R}^{l \times l \times n}$ with n color channels. The rMC approach projects these n feature maps into a single one by mean aggregation for each pixel of the n fitness maps instead of adding additional color channels. Thereby, rMC fitness maps retain their original shape which make them invariant to the problem dimensionality while also decreasing the number of channels and weights. For more details on the methods, we refer to [33]. Extending our previous work, we introduce an additional image-based approach below.

Convolutional Reduced Multi Channel (Conv-rMC). This approach works in a similar manner as the rMC approach. First, an individual 2D-plane is created for every pairwise combination of decision variables, resulting in n planes. As explained in [33], the major drawback of the previously considered MC approach is the exponential growth of the number $n = \binom{D}{2}$ of 2D-planes (and thus channels) w.r.t. the number of dimensions D. This has also been adressed by rMC, which reduces the amount of images n to a single 2D-plane. Yet, the aggregation may cause the individual 2D-planes to become indistinguishable from one-another [33]. Now, the herein proposed Conv-rMC approach is supposed to solve that issue. Conv-rMC projects the n images into a single, gray-scale image using a 1×1-convolutional layer, followed by a batch normalization layer [10] and a ReLU activation [25]. We choose this setup because this is identical to the main building blocks of the ShuffleNet v2 [19] architecture which we used for our experimental study. The additional weights of the convolutional layer and the batch normalization layer are trained by backpropagation. Representative fitness maps of the rMC and Conv-rMC can be found in Fig. 1.

2.3 Fitness Cloud

There are two significant issues related to image-based DL: (1) the resolution of the 2D-planes is limited by the number of pixels as well as the upper and lower bound of the optimization problem's decision space; and (2) for $D > 2$, some (not all) image-based approaches lose information due to our employed methods to reduce dimensionality [33].

Therefore, in [33], we explored the potential of a novel DL approach, called point cloud analysis. The advantage is that for a point cloud, DL can be directly applied to the individual observations of a sample without the need to project them into a $2D$ Cartesian plane first. Further, as most DL approaches for point clouds can handle any finite number of dimensions, there is no information loss for $D > 2$ (cf. Fig. 2). We formally define a point cloud as $\mathcal{F}_{cloud} \in \mathbb{R}^{m \times (D+1)}$ where m is the size of our initial sample and $D + 1$ are the dimensions plus the fitness value. Next, as the resolution is not limited by the number of the image's pixels, local neighborhoods can be processed accurately in every detail. Yet, DL for point clouds must respect the point-order isomorphic of the input data which increases the complexity of DNNs, substantially. For details, we refer to [33].

Although there are several different point cloud approaches available, in [33] we chose point cloud transformers (PCT) because of their transformer background. Transformers are part of DL and were introduced by [36] for neural language processing. They are in-particular great in capturing global relationships which is important for tasks like *high-level property prediction* or *automated algorithm selection*. In addition, [33] proposed a novel embedding technique which is based on the nodes in a k-nearest neighbor (kNN) graph. This embedding technique is supposed to embed every observation of $\mathcal{X}$ into the context of its local neighborhood. The embedding layer is represented as the first layer in Fig. 2 b) in dark blue. This is necessary as transformers are good in capturing global relations but may lose attention to local neighborhood [34]. The embedding layer comes with three additional hyperparameters: (1) k for the kNN-search, (2) p for the L_p-Norm which is used as distance measure, and (3) Δ_{max} to limit the local neighborhood. To avoid poorly chosen parameter values, we performed hyperparameter optimization for the three parameters during our experimental study (see Sect. 3.3).

3 Experiments

3.1 Algorithm Performance Features

Commonly used in the continuous optimization community to evaluate their algorithms is the Black-Box Optimization Benchmark (BBOB) [9]. This benchmark suite is embedded into the platform Comparing Continuous Optimizers (COCO) [8], which provides data from various competitions on the BBOB test bed. The accompanying results of these competitions are uploaded to COCO and offer further opportunity for scrutiny and study.

To test and compare our feature-free methods w.r.t. automated algorithm selection, we use data of Kerschke and Trautmann [14], which in turn was collected from COCO. This data set consists of the performance data of twelve different optimization algorithms evaluated on BBOB. An enumeration of these algorithms can be found in Fig. 4 or in [14]. BBOB is structured as a set of different problem **functions** (FID) which can be initialized in different **dimensions** (D), and can also be subjected to slight modification such as shifts and rotations. The latter leads to different **instances** (IID) of a given problem function. In total, the set of benchmark problem consists of 480 unique problem instances i, constituted by the tuple $i := $ (FID, D, IID) with FID $\in \{1, 2, ..., 24\}$, $D \in \{2, 3, 5, 10\}$, and IID $\in \{1, 2, ..., 5\}$.

For a given problem instance i, an algorithm is considered to be successful if it reaches the objective value of the global optimum up to a given precision value. In [14], the authors deemed a precision value of 10^{-2} as reasonable. A smaller precision value would otherwise lead to many unsuccessful runs across all algorithms. For a given FID and D, the five different runs (over the instances) are aggregated using the expected running time (ERT) [7].

$$ERT = \frac{1}{s} \sum_k FE_k, \tag{1}$$

where k refers to the different BBOB instances, FE_k denotes the spent function evaluations on k, and s is the number of successful runs. Since we aggregated over the instances, the algorithmic performance data set is reduced from 480 to $480/5 = 96$ observations.

While an ERT is useful in its own right, it is hard to compare these absolute ERT values between functions of different dimensionality simply because algorithms are allowed a larger budget in higher dimensions. As a remedy, we employ a slight adaption which is called relative ERT (relERT). For any given problem $p := $ (FID, D), we divide the ERT values of the twelve algorithms by its respective lowest one. Thereby, we obtain a measure which indicates by which factor each algorithm is more expensive (in terms of function evaluations) compared to the best performing algorithm on that specific problem p. Furthermore, there are cases where none of the five instances of a problem was solved by an algorithm. This causes the value s (from Eq. 1) to be zero, resulting in an infinite ERT. For the purpose of our algorithm selector, we impute these values similar to Kerschke and Trautmann [14] by taking the largest observed (i.e., finite) relERT value overall and multiplying it by the factor 10 as penalty [14].

The single-best solver (SBS) across all problems is the HCMA with an relERT value of 30.37 [17]. This offers ample opportunity to close the gap between the SBS and theoretically achievable performance of 1. The latter is also referred to as the virtual-best solver (VBS). The VBS serves as an upper bound whereas the SBS is a baseline which needs to be surpassed.

3.2 Landscape Features

Here, the term 'landscape features' comprises each kind of input to our models, i.e., either a set of ELA features, a fitness map, or a fitness cloud. While these three input variants differ substantially, the underlying sample, upon which they are computed, largely remains the same. Meaning, these methods act as a surrogate of an individual problem instance i, which is constructed from an initial sample of size $50D$. The fitness cloud differs in this respect because the sample size is a constant of either 100 or 500 – the lower ($D = 2$) and upper bound ($D = 10$) of our sampling sizes – to estimate the lower and upper performance bounds. This is done for every problem instance i, i.e., we create 480 different samples.

We use latin hypercube sampling which is inherently stochastic and aim to avoid propagating this stochasticity to the training phase of our algorithm selectors as much as is reasonably possible by sampling independently ten times for each i. Thereby, we not only reduce the stochasticity but also artificially increase our training data by the factor ten. This is especially beneficial since DL generally requires a more extensive set of training data. In short, we increase the number of observations for a specific surrogate variant from 480 to 4 800 (= 480 · 10) observations. This is in stark contrast to the work of Kerschke and Trautmann [14]. There, the numerical landscape features where computed once for each problem instance i and aggregated using the median to match the performance data set with 96 observations. Here, we increase the performance data to match our landscape feature data set. To give an example, let us assume that for a given problem p (e.g., FID 1 in dimension 2 of BBOB), we have an algorithm a which performs best in terms of relERT. We surmise that this fact also holds true for the underlying five BBOB instances and we duplicate these relERT values onto these instances. With that, we increase the algorithm performance data set from 96 observations to 480(= 96 · 5) and duplicate these entries ten times, to match our landscape data set of size 4 800.

We compute the numerical landscape features, i.e., features described in Sect. 2.1, with the Python package `pflacco`[1]. Features which suffer from missing values are removed in their entirety. These belong mostly to the set of *ELA level* and sum up to 14 features. After removal, we are left with 48 distinct features. On the other hand, we create the fitness map as described in Sect. 2.2 with no further adjustment and an image resolution of 224×224 pixels. The fitness clouds, however, require the selection of a predetermined and static input size. For this, we ascertain two variants 100 and 500 as adequate based on [33].

3.3 Experimental Setup

We model our algorithm selection problem as a classification task with a typical data set of $\mathcal{S} = (X, y)$. Meaning, our algorithm selection models, which are described in more detail in the following, do not predict the performance for a given algorithm and problem instance, but rather predict which algorithm to choose for a given problem instance. Hence, our final data set does not consist of twelve algorithms for each problem instance, but only the best one on that specific problem. In this scenario, the algorithm operates as a label y, whereas the landscape features represent X within our machine learning model. In cases of misclassification, however, it is apparent that not each false prediction is equally costly. Therefore, we utilize a cost matrix C, where an individual value $c_{i,a}$

---

[1] https://github.com/Reiyan/pflacco_experiment.

defines the cost of a chosen algorithm a on a problem instance i. This granularity of costs, which is relegated to individual observations, is also known as example-specific cost-sensitivity [35] (see Eq. 2).

The performance of each used model is evaluated by using five-fold cross validation. Since our data set is comprised of five instances per BBOB function, each fold consists of the predetermined instance of each function.

Models Based on ELA Features. We designed two classification models based on ELA features. The first approach is similar to our previous work in [14], i.e., we use a gradient boosting classifier, a support vector machine (SVM) and a random forest (RF) in conjunction with feature selection. These experiments are conducted in Python where the machine learning models stem from `sklearn` [27]. Furthermore, we use a 'greedy forward-backward' feature selection strategy which is implemented in the package `mlxtend` [30]. This process deems most of the features unnecessary or redundant, reducing the set to 15 features. As feature selection is a computationally expensive task, we did not perform any hyper-parameter tuning (similar to [14]). A major distinction compared to the remaining models is that the aforementioned three classifiers are not cost-sensitive during their training. Through this, we intend to highlight the potential of cost-sensitivity on an individual observation-level.

The alternative strategy uses a multilayer-perceptron (MLP) [5] incorporating cost-sensitive training. Feature selection is omitted, as the training duration of an MLP is significantly larger than that other machine learning models, e.g., RF, and the MLP internally includes feature selection due to its working mechanism. In general, the models predict a probability distribution $\hat{p}_a \sim \hat{P}$ where $\hat{p}_a$ is the probability to choose the a-th solver out of all $\mathcal{A}$ solvers. To implement example-specific cost-sensitive training, we choose the loss function

$$\mathcal{L}_{\mathcal{A}}(\hat{P}, T, C; \Theta) = \sum_{a=1}^{\mathcal{A}} |t_{i,a} - \hat{p}_{i,a}| \cdot c_{i,a}. \tag{2}$$

Here, $t_{i,a} \in T_i$ are the true labels for the best performing algorithm on instance i, $c_{i,a} \in C$ are the costs for predicting algorithm a based on the current instance i, and Θ are the model's weights. The loss function has two properties: (1) it is continuously differentiable, and (2) $\mathcal{L}_K = 0$ for models that predict the best performing solver with a confidence of 1.

Next, we performed hyperparameter optimization using a coarse grid search. We tested for the learning rate $\in \{0.01, 0.001, 0.0001\}$, dropout $\in \{0.1, 0.3, 0.5\}$, the number of layers $\in [1, 6]$, hidden neurons $\in \{128, 256, 512, 1024\}$, and the learning rate decay $\in \{0.99, 0.97, 0.95\}$. The remaining setup (including the 5-fold cross validation) was chosen identically to the setup of the classical machine learning approach. We found that a model with 1 024 hidden neurons, 3 repetitions, a dropout of 0.1, trained with a learning rate of 0.01, and a learning rate decay of 0.99 performed best. In the following, we will refer to this model as ELA-MLP. All DL models were trained on a single Nvidia Quadro RTX 6000, and by using the `PyTorch` library in Python. In addition, the MLP models were trained for 500 epochs and the models with the lowest loss on the validation folds were selected as final.

Models Based on Fitness Maps. The training setup of the image-based DL models is similar as proposed in [33]. We used a *ShuffleNet V2* [19] (see Fig. 2 a) with a width-multiplicator of 1.5, and we used the same cost-sensitive loss function for these models (see Eq. 2). Further, we tested several different training approaches: learning rate $\in \{0.001, 0.0003, 0.0001, 0.00005, 0.00001\}$, an entropy regularization $\in \{0, 0.001, 0.01, 0.1, 1, 10\}$, and a scaled and un-scaled version of the loss function. The remaining training setup was identical to the ELA-MLP approach. We found, that a learning rate of 0.0003, no entropy regularization, and the un-scaled loss worked best. All image-based models were trained for 100 epochs, and the models with the lowest loss on the validation folds were selected as the final models.

Models Based on Fitness Clouds. The model topology and training setup for the PCTs was chosen identically to [33] (see Fig. 2 b). We tested for different kNNs, $k \in \{3, 5, 10\}$, and (as proposed by [1]) also considered $p = 0.5$ for the L_p distance function. However, we could not find any performance difference between the different PCTs, confirming the finding of [33] that the hyperparameters of the embedding layer have (at most) only a slight impact on performance for the BBOB data. The PCT models were trained for 50 epochs and the models with the lowest loss on the validation folds were chosen again as final models.

Next, due to a limitation of our implementation, we were not able to train the PCT models with $50D$ points but had to choose a consistent number of points for all dimensions. By using 100 and 500 points, we can estimate the upper and lower PCT model performance as $50D = 100$ for 2D and $50D = 500$ for 10D. As PCTs have lower or higher sampling costs compared to all other models, they may have an advantage or disadvantage.

4 Results and Discussion

Table 1 summarizes the results of the considered approaches for constructing AAS models. Model performances are provided in terms of mean relERT values, split by dimensionality and BBOB function groups as well as in total, i.e., considering all instances simultaneously. Next to the SBS (cf. 3.1), results of feature-based approaches are contrasted to feature-free concepts based on fitness maps and point clouds. Figure 3 visualizes the respective results on problem dimension level in terms of a parallel-coordinate-plot. Except from the SBS, the relERT values of the AAS models incorporate the costs of the initial sample.

It becomes obvious that the classical ELA feature based AAS model approach (ELA-RF) is largely outperformed. However, this is the only strategy not considering cost-sensitive learning in terms of integrating the loss in relERT induced

Table 1. Performance results of difference AAS models represented by their relERT. The first two columns represent information about the underlying problems, i.e., the different dimensions (D) and the BBOB function groups (FGrp) of similar characteristics. The best option based on the ELA features (ELA-RF and ELA-MLP) for a given row is highlighted in green, and the best model consisting of either fitness map (columns MC to PCA-Func) or fitness cloud (PCT-100 and PCT-500) is highlighted in blue. Values displayed in red indicate the overall best option for a given group of problems.

D	FGrp	SBS	ELA-RF	ELA-MLP	MC	Conv-rMC	rMC	PCA	PCA-Func	PCT-100	PCT-500
2	1	3.71	10.41	10.59	23.95	23.95	23.95	14.99	15.12	23.95	61.15
	2	5.80	8.51	3.72	3.54	3.54	3.54	6.97	6.40	3.54	11.04
	3	6.29	1473.16	4.72	4.02	4.02	4.02	4.98	5.64	4.02	16.11
	4	25.34	3.89	9.25	8.90	8.90	8.90	26.98	24.00	8.90	12.21
	5	44.95	148.39	3.32	4.33	4.33	4.33	30.01	30.97	4.33	6.18
	all	17.69	342.22	6.43	9.17	9.17	9.17	17.19	16.84	9.17	21.77
3	1	356.10	1480.68	11.87	16.58	16.56	191.76	365.54	344.65	13.32	39.24
	2	4.46	8.33	3.50	3.32	3.32	3.51	5.32	5.20	2.85	6.65
	3	4.98	7.07	3.82	3.61	3.58	4.11	5.21	5.51	2.72	9.58
	4	2.63	441.96	5.06	11.16	11.54	9.36	2.77	3.11	11.49	11.87
	5	66.81	1.22	2.54	2.53	2.53	2.53	50.46	50.20	2.46	3.04
	all	90.43	403.67	5.44	7.61	7.68	43.87	89.22	84.92	6.72	14.39
5	1	11.99	14.14	11.97	22.69	22.69	22.70	22.89	22.80	16.27	33.39
	2	3.90	369.26	2.62	4.74	4.78	4.60	4.64	4.70	4.25	5.66
	3	4.21	150.44	3.97	6.72	6.72	6.49	6.40	6.56	5.21	9.23
	4	4.29	1470.28	6.81	4.38	4.38	4.38	4.38	4.38	4.33	4.47
	5	7.67	1.13	1.83	4.22	7.80	3.38	4.08	4.90	7.72	7.93
	all	6.52	402.38	5.56	8.71	9.46	8.46	8.64	8.83	7.69	12.41
10	1	2.74	14.64	15.27	16.34	16.34	16.34	16.39	16.41	5.46	16.34
	2	2.16	1.62	1.76	2.71	2.71	2.71	2.69	2.67	2.27	2.71
	3	2.76	2.87	4.35	4.48	4.48	4.48	4.48	4.48	3.10	4.48
	4	2.02	442.01	1.96	2.09	2.09	2.09	2.07	2.09	2.03	2.09
	5	23.64	148.01	3.25	21.77	23.74	14.61	5.08	10.21	23.66	23.74
	all	6.85	126.84	5.46	9.76	10.17	8.27	6.29	7.36	7.51	10.17
all	1	93.63	379.97	12.43	19.89	19.88	63.68	104.96	99.74	14.75	37.53
	2	4.08	96.93	2.90	3.58	3.59	3.59	4.91	4.75	3.23	6.52
	3	4.56	408.38	4.21	4.71	4.70	4.78	5.27	5.55	3.76	9.85
	4	8.57	589.54	5.77	6.63	6.73	6.18	9.05	8.40	6.69	7.66
	5	35.77	74.69	2.74	8.22	9.60	6.21	22.41	24.07	9.54	10.22
	all	30.37	318.78	5.72	8.81	9.12	17.44	30.34	29.49	7.78	14.68

by wrong predictions of the best suited solver. We exemplary report on ELA-RF performance, as the best option among alternative standard classifiers such as SVM or gradient boosting, which qualitatively show comparable performance. However, integrating cost-sensitive learning into a multilayer-perceptron model based on ELA features (ELA-MLP) has a tremendously positive impact. ELA-

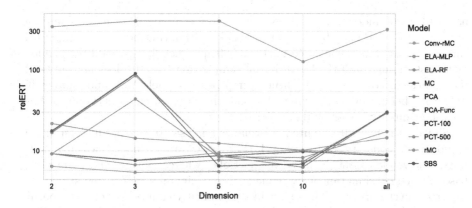

Fig. 3. Parallel coordinate plot visualizing AAS model results per problem dimension in terms of relERT based on Table 1.

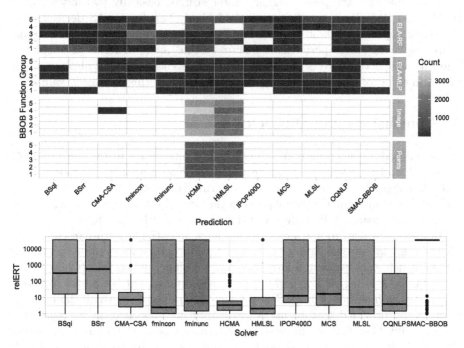

Fig. 4. Top Figure: Absolute frequencies of predicted solvers by AAS model type ('actual' solver portfolio) grouped by BBOB function groups. Bottom Figure: Individual solver performances (relERT) across all instances and dimensions.

MLP not only beats the SBS on almost all instances but rather largely reduces the SBS-VBS gap (across all instances and dimensionalities) by 78%. While this by itself is an outstanding result, even more notable is the performance of our feature-free alternatives. Without requiring feature information, the point-cloud based PCT-100 model is second-best with really comparable results, closing the respective SBS-VBS gap by 74%. Surprisingly, within this model class, this PCT variant outperforms the more costly PCT-500 approach, i.e., the additional information stemming from the larger initial sample (for $D < 10$) does not outweigh the additional costs. Also, the fitness map based models MC- and Conv-rMC do not lag behind by much, while rMC and the PCA-based models do not show consistent quality across all dimensions.

Shifting the focus from pure performance analysis to a deeper understanding of model decisions, we see from Fig. 4, that models based on ELA features make use of the whole extent of the solver portfolio, whereas image- and point-cloud based strategies simultaneously manage to meaningfully reduce the size of the *required* solver portfolio. The upper image explicitly counts how often a specific solver has been selected within the AAS models, split by model type. Note that all image based approaches are aggregated within the *image* row, while the same holds for the *point* category. Interestingly, basically two solvers, i.e., HCMA and HMLSL, are sufficient to yield highly beneficial AAS decisions. A third, less important candidate is CMA-CSA. This is supported by the bottom figure displaying solver performances across all instances in terms of boxplots. The three respective solvers are performing consistently well, but also are most robust in terms of substantially smaller variability in terms of relERT values.

5   Conclusions and Outlook

We introduced a conceptually novel feature-free alternative to the conventional numerical landscape features and, thereby, provide an innovative solution to the ASP. Although the feature-free models perform slightly worse than the ELA feature models, they significantly outperform the single-best solver of our algorithm portfolio and meaningfully reduce the size of the algorithm portfolio.

However, our feature-free methods offer huge potential for future research. As discussed in [33], a considerable challenge of our proposed image-based methods is the balancing act between information loss for higher dimensions (which holds for PCA, PCA-Func, MC, and rMC) and increasing sparsity of points in lower dimensions (which holds for MC and Conv-rMC). Thus, for small dimensions most of the provided feature maps are empty and the weights get partially trained, only. Therefore, we proposed an alternative point cloud-based approach (PCT) which can also be considered as highly competitive to the conventional approaches. Summarizing, the classical ELA-MLP approach shows a slightly better overall AAS model performance compared to PCT-100, but comes with the trade-off of requiring a larger solver portfolio.

However, there is also large improvement potential for PCT based methods. First, the embedding layer may play a crucial role for the model's performance, yet we know little about the impact of the different hyperparameters. Thus, we will closer investigate the respective impact and experiment with different distance metrics. Also, we will investigate the potential and usability of deep features for cross-domain knowledge transfer. Deep features can be obtained after the last hidden layer of a trained model applied to a new task. Afterwards, a small MLP or any other ML algorithm can be trained efficiently on the obtained deep features. This is especially useful if the quantity of data or the training resources for the new tasks are limited.

References

1. Aggarwal, C.C., Hinneburg, A., Keim, D.A.: On the surprising behavior of distance metrics in high dimensional space. In: Van den Bussche, J., Vianu, V. (eds.) ICDT 2001. LNCS, vol. 1973, pp. 420–434. Springer, Heidelberg (2001). https://doi.org/10.1007/3-540-44503-X_27
2. Alissa, M., Sim, K., Hart, E.: Algorithm selection using deep learning without feature extraction. In: Proceedings of the Genetic and Evolutionary Computation Conference, pp. 198–206 (2019)
3. Bischl, B., Mersmann, O., Trautmann, H., Preuß, M.: Algorithm selection based on exploratory landscape analysis and cost-sensitive learning. In: Proceedings of the 14th Annual Conference on Genetic and Evolutionary Computation, pp. 313–320 (2012)
4. Bossek, J., Doerr, C., Kerschke, P.: Initial design strategies and their effects on sequential model-based optimization: an exploratory case study based on BBOB. In: Proceedings of the 22nd Annual Conference on Genetic and Evolutionary Computation (GECCO), pp. 778–786 (2020)
5. Goodfellow, I., Bengio, Y., Courville, A.: Deep Learning. MIT Press (2016). http://www.deeplearningbook.org
6. Guo, M.H., Cai, J.X., Liu, Z.N., Mu, T.J., Martin, R.R., Hu, S.M.: PCT: point cloud transformer. Comput. Visual Media 7(2), 187–199 (2021)
7. Hansen, N., Auger, A., Finck, S., Ros, R.: Real-Parameter Black-Box Optimization Benchmarking 2010: Experimental Setup. Research Report RR-7215, INRIA (2010). https://hal.inria.fr/inria-00462481
8. Hansen, N., Auger, A., Ros, R., Mersmann, O., Tušar, T., Brockhoff, D.: COCO: a platform for comparing continuous optimizers in a black-box setting. Optimi. Meth. Software 36(1), 114–144 (2021)
9. Hansen, N., Finck, S., Ros, R., Auger, A.: Real-Parameter Black-Box Optimization Benchmarking 2009: Noiseless Functions Definitions. Technical report RR-6829, INRIA (2009). https://hal.inria.fr/inria-00362633/document
10. Ioffe, S., Szegedy, C.: Batch normalization: accelerating deep network training by reducing internal covariate shift. In: International Conference on Machine Learning, pp. 448–456. PMLR (2015)
11. Jones, T., Forrest, S.: Fitness distance correlation as a measure of problem difficulty for genetic algorithms. In: Proceedings of the 6th International Conference on Genetic Algorithms (ICGA), pp. 184–192. Morgan Kaufmann Publishers Inc. (1995)
12. Kerschke, P., Hoos, H.H., Neumann, F., Trautmann, H.: Automated algorithm selection: survey and perspectives. Evol. Comput. (ECJ) 27(1), 3–45 (2019)
13. Kerschke, P., Preuss, M., Wessing, S., Trautmann, H.: Detecting funnel structures by means of exploratory landscape analysis. In: Proceedings of the 17th Annual Conference on Genetic and Evolutionary Computation (GECCO), pp. 265–272. ACM, July 2015

14. Kerschke, P., Trautmann, H.: Automated algorithm selection on continuous black-box problems by combining exploratory landscape analysis and machine learning. Evol. Comput. (ECJ) **27**(1), 99–127 (2019)
15. Kerschke, P., Trautmann, H.: Comprehensive feature-based landscape analysis of continuous and constrained optimization problems using the r-package flacco. In: Bauer, N., Ickstadt, K., Lübke, K., Szepannek, G., Trautmann, H., Vichi, M. (eds.) Applications in Statistical Computing. SCDAKO, pp. 93–123. Springer, Cham (2019). https://doi.org/10.1007/978-3-030-25147-5_7
16. LeCun, Y., Bengio, Y., et al.: Convolutional networks for images, speech, and time series. Handbook Brain Theory Neural Networks **3361**(10), 1995 (1995)
17. Loshchilov, I., Schoenauer, M., Sèbag, M.: Bi-population CMA-ES algorithms with surrogate models and line searches. In: Proceedings of the 15th Annual Conference Companion on Genetic and Evolutionary Computation. GECCO 2013 Companion, pp. 1177–1184. ACM (2013)
18. Lunacek, M., Whitley, L.D.: The dispersion metric and the CMA evolution strategy. In: Proceedings of the 8th Annual Conference on Genetic and Evolutionary Computation (GECCO), pp. 477–484. ACM (2006)
19. Ma, N., Zhang, X., Zheng, H.-T., Sun, J.: ShuffleNet V2: practical guidelines for efficient CNN architecture design. In: Ferrari, V., Hebert, M., Sminchisescu, C., Weiss, Y. (eds.) Computer Vision – ECCV 2018. LNCS, vol. 11218, pp. 122–138. Springer, Cham (2018). https://doi.org/10.1007/978-3-030-01264-9_8
20. Malan, K.M., Engelbrecht, A.P.: A survey of techniques for characterising fitness landscapes and some possible ways forward. Inf. Sci. (JIS) **241**, 148–163 (2013)
21. Mersmann, O., Bischl, B., Trautmann, H., Preuss, M., Weihs, C., Rudolph, G.: Exploratory landscape analysis. In: Proceedings of the 13th Annual Conference on Genetic and Evolutionary Computation (GECCO), pp. 829–836. ACM (2011). Recipient of the 2021 ACM SigEVO Impact Award
22. Muñoz Acosta, M.A., Kirley, M., Halgamuge, S.K.: Exploratory landscape analysis of continuous space optimization problems using information content. IEEE Trans. Evol. Comput. (TEVC) **19**(1), 74–87 (2015)
23. Muñoz Acosta, M.A., Sun, Y., Kirley, M., Halgamuge, S.K.: Algorithm selection for black-box continuous optimization problems: a survey on methods and challenges. Inf. Sci. (JIS) **317**, 224–245 (2015)
24. Muñoz, M.A., Kirley, M.: Sampling effects on algorithm selection for continuous black-box optimization. Algorithms **14**(1), 19 (2021). https://doi.org/10.3390/a14010019
25. Nair, V., Hinton, G.E.: Rectified linear units improve restricted Boltzmann machines. In: ICML 2010, Madison, WI, USA, pp. 807–814. Omnipress (2010)
26. Pearson, K.: On lines and planes of closest fit to system of points in space. Philos. Mug 6th ser. **2**, 559–572 (1901)
27. Pedregosa, F., et al.: Scikit-learn: machine learning in python. J. Mach. Learn. Res. **12**, 2825–2830 (2011)
28. Prager, R.P., Seiler, M.V., Trautmann, H., Kerschke, P.: Towards feature-free automated algorithm selection for single-objective continuous black-box optimization. In: Proceedings of the IEEE Symposium Series on Computational Intelligence. Orlando, Florida, USA (2021)
29. Prager, R.P., Trautmann, H., Wang, H., Bäck, T.H.W., Kerschke, P.: Per-instance configuration of the modularized CMA-ES by means of classifier chains and exploratory landscape analysis. In: Proceedings of the IEEE Symposium Series on Computational Intelligence (SSCI), pp. 996–1003. IEEE (2020)

30. Raschka, S.: MLxtend: providing machine learning and data science utilities and extensions to python's scientific computing stack. J. Open Source Software (JOSS) **3**(24), 638 (2018)
31. Rice, J.R.: The algorithm selection problem. Adv. Comput. **15**(65–118), 5 (1976)
32. Seiler, M., Pohl, J., Bossek, J., Kerschke, P., Trautmann, H.: Deep learning as a competitive feature-free approach for automated algorithm selection on the traveling salesperson problem. In: Bäck, T., et al. (eds.) PPSN 2020. LNCS, vol. 12269, pp. 48–64. Springer, Cham (2020). https://doi.org/10.1007/978-3-030-58112-1_4
33. Seiler, M.V., Prager, R.P., Kerschke, P., Trautmann, H.: A collection of deep learning-based feature-free approaches for characterizing single-objective continuous fitness landscapes. arXiv preprint (2022)
34. Shaw, P., Uszkoreit, J., Vaswani, A.: Self-Attention with Relative Position Representations. arXiv preprint arXiv:1803.02155 (2018)
35. Turney, P.D.: Types of Cost in Inductive Concept Learning. arXiv preprint cs/0212034 (2002)
36. Vaswani, A., et al.: Attention is all you need. In: Advances in Neural Information Processing Systems, pp. 5998–6008 (2017)

Improving Nevergrad's Algorithm Selection Wizard NGOpt Through Automated Algorithm Configuration

Risto Trajanov[1], Ana Nikolikj[2,3], Gjorgjina Cenikj[2,3], Fabien Teytaud[4], Mathurin Videau[5], Olivier Teytaud[5], Tome Eftimov[2(✉)], Manuel López-Ibáñez[6], and Carola Doerr[7]

[1] Faculty of Computer Science and Engineering, Ss. Cyril and Methodius Skopje, Skopje, North Macedonia
[2] Jožef Stefan Institute, Ljubljana, Slovenia
`tome.eftimov@ijs.si`
[3] Jožef Stefan International Postgraduate School, Ljubljana, Slovenia
[4] Université Littoral Côte d'Opale, Dunkirk, France
[5] Facebook AI Research, Paris, France
[6] ITIS Software, Universidad de Málaga, Málaga, Spain
[7] Sorbonne Université, CNRS, LIP6, Paris, France

Abstract. Algorithm selection wizards are effective and versatile tools that automatically select an optimization algorithm given high-level information about the problem and available computational resources, such as number and type of decision variables, maximal number of evaluations, possibility to parallelize evaluations, etc. State-of-the-art algorithm selection wizards are complex and difficult to improve. We propose in this work the use of automated configuration methods for improving their performance by finding better configurations of the algorithms that compose them. In particular, we use elitist iterated racing (irace) to find CMA configurations for specific artificial benchmarks that replace the hand-crafted CMA configurations currently used in the NGOpt wizard provided by the Nevergrad platform. We discuss in detail the setup of irace for the purpose of generating configurations that work well over the diverse set of problem instances within each benchmark. Our approach improves the performance of the NGOpt wizard, even on benchmark suites that were not part of the tuning by irace.

Keywords: Algorithm configuration · Algorithm selection · Black-box optimization · Evolutionary computation

1 Introduction

In the context of black-box optimization, the use of a portfolio of optimization algorithms [23], from which an algorithm is selected depending on the features of the particular problem instance to be solved, is becoming increasingly popular. The algorithm that encapsulates the selection rules and the portfolio of algorithms is sometimes referred as a "wizard" [20]. Algorithm selection wizards provide versatile, robust and convenient tools, particularly for black-box optimization, where problems that arise in practice show a large variety of different requirements, both with respect to problem models, but also with respect to the (computational) resources that are available to solve them.

Building a competitive algorithm selection wizard is not an easy task, because it not only requires devising the rules for selecting an algorithm for a given problem instance, but also configuring the parameters of the algorithms to be selected, which is a difficult task by itself [3]. Although there are examples of wizards that were build automatically for SAT problems [29], many algorithm selection wizards are still hand-crafted. An example of a successful hand-crafted algorithm selection wizard is NGOpt [20], provided by the optimization platform Nevergrad [22].

NGOpt was designed by carefully studying the performance of tens of optimizers on a wide range of benchmark suites to design hand-crafted rules that aim to select the best optimizer for particular problem features. Through an iterative improvement process, new versions of NGOpt are designed by refining the hand-crated rules and replacing optimizers by others that lead to a better performance of the new NGOpt version. The result is a complex algorithm selection wizard that outperforms many well-known stand-alone optimizers on a wide range of benchmark suites.

In this work, we attempt to investigate the potential of improving NGOpt via automated algorithm design techniques. Rebuilding NGOpt from scratch using automated methods is a daunting task that would waste the knowledge already encoded in it and the human effort already invested in its creation. Instead, we examine how automatic algorithm configuration (AC) methods may be used in a judicious manner to improve NGOpt by focusing on improving one of its most critical underlying optimizers.

AC methods such as irace [16], aim to find a configuration, i.e., a setting of the parameters of an algorithm, that gives good expected performance over a large space of problem instances by evaluating configurations on a *training* subset from such space. AC methods, including irace, are typically designed to handle categorical and numerical parameter spaces and stochastic performance metrics, not only due to the inherent stochasticity of randomized optimization algorithms but also because the training instances represent a random sample of the problems of interest. Traditionally, AC methods have been used to tune the parameters of specific algorithms [3], however, more recently, they have been used to automatically design algorithms from flexible frameworks [1,12,18,19] and to build algorithm portfolios [28]. For more background on AC, we refer the interested reader to two recent surveys [8,25].

As mentioned above, we do not wish to re-build nor replace NGOpt but, instead, iteratively improve it. To do so, we first answer the question of whether the hand-crafted NGOpt may be further improved by replacing some of its components by algorithm configurations obtained from the application of an AC method, in particular, irace. When applied to a complex algorithm selection wizard such as NGOpt and a diverse black-box optimization scenario, such as the one tackled by NGOpt, the correct setup of an AC method is crucial. We discuss in detail our approach here, which we believe is applicable to other similar scenarios. Our results show that our proposed approach clearly improves the hand-crafted NGOpt, even on benchmark suites not considered in the tuning process by irace.

This paper is structured as follows. Section 2 gives an introduction to the concept of algorithm selection wizard and describes, in particular, the one used in our experiments, NGOpt. Section 3 presents the experimental setup and the benchmark suites used for tuning the target algorithm. In Sect. 4 we present the integration of the tuned algorithms into the new algorithm selection wizard. The evaluation results are presented in Sect. 5. Section 6 concludes the paper.

2   Preliminaries

Algorithm Selection Wizards. Modern algorithm selection wizards make use of a number of different criteria to chose which algorithm(s) are executed on a given problem instance, and for how long. Algorithm selection wizards take into account a priori information about the problem and the resources that are available to solve it. Using this information, the algorithm selection wizards recommend one or several algorithms to be run, along with a protocol to assign computational budget to each of these.

NGOpt is built atop of several dozens of state-of-the-art algorithms that have been added to Nevergrad over the last years. However, NGOpt does not only select which algorithms to execute on which problem instances, but it also combines algorithms in several ways, e.g., by enriching classic approaches with local surrogate models. Since all algorithms submitted to Nevergrad are periodically run on all fitting benchmark problems available within the environment, a very large amount of performance data is available. In light of this rich data set, and in light of the tremendous performance gains obtained through automated algorithm designs [2,12,21] and selection rules [11], it is therefore surprising that both the decision rules of NGOpt and the configuration of the algorithms available in Nevergrad are hand-picked.

CMA. Among the algorithms that compose NGOpt, a key component is CMA, an instance of the family of covariance matrix adaptation evolution strategies (CMA-ES [7]). In Nevergrad's CMA implementation, the following parameters are explicitly exposed, making them straightforward candidates for the configuration task. The `scale` parameter controls the scale of the search of the algorithm in the search domain: a small value leads to start close to the center. The `elitist` parameter is a Boolean variable that can have values 'True' or 'False' and it represents a switch for elitist mode, i.e., preserving the best individual even if it includes a parent. The `diagonal` parameter is another Boolean setting that controls the use of diagonal CMA [24]. Finally, the population size is another crucial parameter that is usually set according to the dimension of the problem. Instead of setting its value directly, we decided to create a higher-level integer parameter `popsize_factor` and let the population size be $\lfloor 4 + $ `popsize_factor` $\cdot \log d \rfloor$, where d is the problem dimension. The parameter search space of CMA is shown in Table 3.

Automated Algorithm Configuration. Given a description of the parameter space of a target algorithm, a class of problems of interest, and a computational budget (typically measured in target algorithm runs), the goal of an AC method is to find a parameter configuration of the target algorithm that is expected to perform well on the problem class of interest. Due to the stochasticity of randomized algorithms and the fact that only a limited sample of (*training*) problem instances can be evaluated in practice, specialized methods for AC have been developed in recent years [8,9,25] that try to avoid inherent human biases and limitations in manual parameter tuning. We have selected here the *elitist iterated racing* algorithm implemented by the irace package [16], since it has shown good performance for configuring black-box optimization algorithms in similar scenarios [13–15,17].

3 Experimental Setup

To show the influence of the parameter tuning of the CMA included in NGOpt [22], we have performed several parameter tuning experiments using irace. The setup and results of this process are described in this section.

3.1 Setup of Irace for Tuning CMA

A training *instance* is defined by a benchmark function, a dimension, a rotation and the budget available to CMA. We also define "blocks" of instances: all instances within a block are equal except for the benchmark function and there are as many instances within a block as benchmark functions. We setup irace so that, within each race, the first elimination test (*FirstTest*) happens after seeing 5 blocks and subsequent elimination tests (*EachTest*) happen after each block. Moreover, configurations are evaluated by irace on blocks in order of increasing budget first and increasing dimension second, such that we can quickly discard poor-performing configurations on small budgets and only good configurations are evaluated on large ones [26]. The performance criterion optimized by irace is the objective value of the point recommended by CMA after it has exhausted its budget. Since Nevergrad validates performance according to the mean loss (as explained later), the elimination test used by irace is set to t-test. Finally, we set a maximum of 10 000 individual runs of CMA as the termination criterion of each irace run. By parallelizing each irace run across 4 CPUs, the runtime of a single run of irace was around 8 h.

3.2 Benchmark Suites Used for Tuning

Nevergrad contains a very large number of benchmark *suites*. Each suite is a collection of benchmark problems that share some common properties such as similar domains or constraints, similar computational budgets, etc. A large number of different academic and real-world problems are available. We selected five

Table 1. Artificial benchmarks used as training instances by irace. The third column gives the name of the CMA variant obtained by irace on that particular benchmark (for the parameter values, see Table 4). The detailed code can be found in [22].

Benchmark name	Context	Optimized CMA
YABBOB	Dimension $\in [2, 50]$, budget $\in [50, 12800]$	CMAstd
YASMALLBBOB	Budget < 50	CMAsmall
YATUNINGBBOB	Budget < 50 and dim ≤ 15	CMAtuning
YAPARABBOB	Num-workers=100	CMApara
YABOUNDEDBBOB	Box-constrained, budget ≤ 300, dim ≤ 40	CMAbounded

suites (YABBOB, YASMALLBBOB, YATUNINGBBOB, YAPARABBOB, and YABOUND-EDBBOB) as training instances for irace, all of them derived from the BBOB benchmark suite [5] of the COCO benchmarking environment [6]. See Table 1 for their various characteristics. We ran irace once for each benchmark suite separately, by setting up their problem instances as described above. All runs of irace used the parameter space of CMA shown in Table 3. For each benchmark suite, we selected one CMA configuration from the ones returned by irace by doing 10 validation runs using the best configurations obtained by irace and the default CMA configuration. Each run consisted of running each CMA configuration on the whole instance space. For each instance we declared a winner configuration that yielded the smallest result. Using majority vote then we determined a winner from the 10 validation runs. The optimized configurations selected are named CMAstd for YABBOB, CMAsmall for YASMALLBBOB, CMAtuning for YATUN-INGBBOB, CMApara for YAPARABBOB, and CMAbounded for YABOUNDEDBBOB (Table 1). Their parameter values are shown in Table 4 (Table 2).

Table 2. Artificial and real-world benchmarks used for testing. The detailed code can be found in [22].

Type	Name	Context
Artificial	YABIGBBOB	Budget 40000 to 320000
	YABOXBBOB	Box constrained
	YAHDBBOB	Dimension 100 to 3000
	YAWIDEBBOB	Different settings (multi-objective, noisy, discrete...)
	Deceptive	Hard benchmark, far from YABBOB
Real-world	SeqMLTuning	Hyperparameter tuning for SciKit models
	SimpleTSP	Traveling Salesman Problem, black-box
	ComplexTSP	Traveling Salesman Problem, black-box, nonlinear terms
	UnitCommitment	Unit Commitment for power systems
	Photonics	Simulators of nano-scale structural engineering
	GP	Gym environments used in [27]
	007	Game simulator for the 007 game

Algorithm 1. Pseudocode for MetaCMA.

1: **procedure** METACMA(*budget, dimension, num_workers, fully_bounded*)
2:     **if** *fully_bounded* **then return** CMAbounded   ▷ All variables have bounded domain.
3:     **if** *budget* < 50 **then**
4:         **if** *dimension* ≤ 15 **then return** CMAtuning **else return** CMAsmall
5:     **if** *num_workers* > 20 **then return** CMApara **else return** CMAstd

---

Table 3. Search space for the tuning of CMA by irace.

Parameter	Default value	Domain
Scale	1	$(0.1, 10) \subset \mathbb{R}$
Popsize-factor	3	$[1, 9] \subset \mathbb{N}$
Elitist	False	{True, False}
Diagonal	False	{True, False}

4 MetaCMA and Its Integration in Nevergard

The CMA configurations found by irace were further combined into a new algorithm selection wizard. For this purpose, we propose MetaCMA (Algorithm 1), a deterministic ensemble for noiseless single-objective continuous optimization that, based on deterministic rules, switches between different tuned CMA configurations. For problems where all variables have bounded domain, MetaCMA selects CMAbounded. Otherwise, for low budget (function evaluations) available, the MetaCMA model selects either the CMAtuning configuration (for low dimension) or the CMAsmall configuration (otherwise). If number of workers is more than 20, then the MetaCMA uses the CMApara. If neither of the above-mentioned rules is met, the MetaCMA will switch to CMAstd, which was tuned on YABBOB.

Given that NGOpt considers many more cases than MetaCMA, we have to integrate it inside NGOpt so that we have both the improved performance of our MetaCMA and the generality of NGOpt. NGTuned corresponds to NGOpt with all instances of CMA in NGOpt replaced by MetaCMA.

5 Experimental Results

To evaluate the performance of the MetaCMA model and its integration in NGOpt (NGTuned), we compared it with the previous variant of NGOpt and other state-of-the-art algorithms in three different scenarios. The first scenario consists of the benchmark suites that were used by irace during tuning. All algorithms have been rerun independently of the tuning process. The second scenario involves artificial benchmark suites whose problem instances were not used in the tuning process. The third scenario involves real-world problems not

Table 4. Best parameters found by irace for each benchmark suite.

Parameter	CMAstd (Yabbob)	CMAsmall (Yasmallbbob)	CMAtuning (Yatuningbob)	CMApara (Yaparabbob)	CMAbounded (Yaboundedbbob)
Scale	0.3607	0.4151	0.4847	0.8905	1.5884
Popsize-factor	3	9	1	8	1
Elitist	False	False	True	True	True
Diagonal	False	False	False	True	True

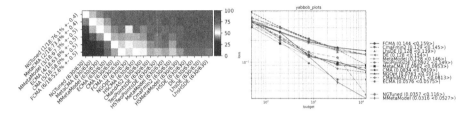

Fig. 1. Results on YABBOB: suite used in the tuning, moderate dimension and budget.

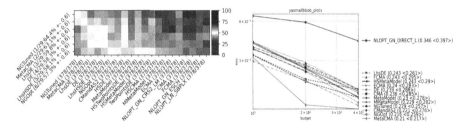

Fig. 2. Results on YASMALLBBOB: benchmark suite used in the tuning, low budget.

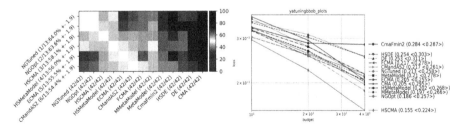

Fig. 3. Results on YATUNINGBBOB: benchmark suite used in the tuning, counterpart of YABBOB with low budget and low dimension.

presented in the tuning process of irace. Details about these benchmark suites are given in Table 1.

The selection of algorithms involved in the evaluation for each benchmark suite differ because each suite has its own set of state-of-the-art algorithms.

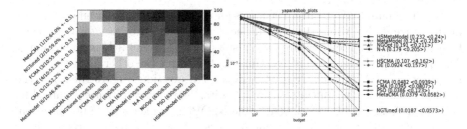

Fig. 4. Results on YAPARABBOB: benchmark suite used in the tuning, parallel case.

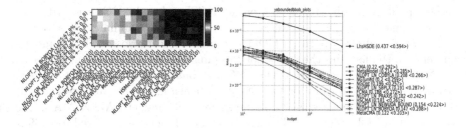

Fig. 5. YABOUNDEDBBOB: suite used in the tuning, bounded domain.

5.1 Evaluation on Artificial Benchmark Suites Presented in the Tuning Process of Irace

Figures 1, 2, 3, 4, and 5 present results on benchmark suites used for running irace. On the left side of each figure is a robustness heatmap, where the rows and columns represent different algorithms, and the values of the heatmap reflect the number of times the algorithm in each row outperforms the algorithm in each column. We use mean quantiles over multiple settings (compared to other optimization methods) in an anytime manner, meaning that all budgets are taken into account, not only the maximum one. This is robust to arbitrary scalings of objective functions. The algorithms' global score, defined as their mean scores (the number of times it performs better than the other algorithms), is used for ranking the rows and columns. The six best-performing algorithms are listed as rows and sorted by their mean score, while the columns contain all of the algorithms involved in the algorithm portfolio. The labels in the columns contain the mean score of each algorithm. For instance, in Fig. 1, the label for NGTuned is "1/18:76.1% +− 0.4", meaning that it was ranked first, out of 18 evaluated algorithms, and its mean score is 76.1%, which is the mean number of times it performed better than the rest of the algorithms, on all problems and all budgets in the benchmark suite. On the right side of each figure, we present the algorithms' performance with different budgets. The x-axis represents the budget, while the y-axis represents the rescaled losses (normalized objective value, averaged over all settings), for a given budget, and linearly normalized in the range $[0, 1]$. Apart from the algorithm name, the label of each algorithm in the legend contains two values in brackets. The first one denotes the mean loss achieved by the algorithm for the maximum budget (after the losses for each problem have been normalized in the range $[0, 1]$), while the second one denotes the mean loss for the second largest budget. Figure 1 features the results on the YABBOB benchmark suite. The heatmap (left) shows that NGTuned outperforms the other algorithms, followed by MetaCMA, while NGOpt is in the 7th position. Figure 2 features the results on the YASMALLBBOB benchmark suite. The best results are obtained by NGTuned, followed by MetaCMA, while NgOPt is ranked in the 6th position. In addition, with regard to different budgets, MetaCMA outperforms NGOpt by obtaining lower losses. In Fig. 3, the results from the evaluation on YATUNINGBBOB benchmark suite are presented. It follows the trend of NGTuned being the best performing algorithm. Comparing NGTuned and NGOpt according to the convergence it seems that till 2×10^1, both achieved similar loss, but later on for budgets above 2×10^1, NGOpt can provide better losses. The comparison between the left and the right plot (keeping the rescaling in mind) suggests that NGTuned has a better rank than NGOpt, but that the rescaled loss is better for NGOpt. The results for YAPARABBOB benchmark suite are presented in Fig. 4. In terms of mean scores obtained across all budgets, MetaCMA is the best algorithm, followed by NGTuned. In terms of the loss achieved for different budgets, NGTuned and MetaCMA provide similar results, with the biggest difference being in the largest budget, where NGTuned achieves a lower loss. Due to rescalings and the difference criteria (ranks on the left, normalized loss on the right) the visual comparison between algorithms can differ. Both outperform NGOpt. On the YABOUNDEDBBOB benchmark suite (Fig. 5), NGOpt achieves better performance than NGTuned. Here, several algorithms from [10] have been tested: the low budget makes them relevant. The best results are achieved by NLOPT_LN_BOBYQA [4], closely followed by MetaCMA, with a difference of 0.4% in the average quantiles referring to the number of times they outperform other methods. From the line plot on the right of the figure, we can see that MetaCMA consistently provides the lowest normalized loss, for different budgets.

5.2   Evaluation on Benchmark Suites Not Used for Tuning

Artificial Benchmarks. Figures 6, 7, 9, 8, and 10 present results on benchmark suites related to YABBOB though with different configurations. The problem instances used were not used in the tuning process.

Figure 6 shows the results on the YABIGBBOB benchmark suite, which consists of the same problem instances as the YABBOB benchmark suite with longer budgets. Here, NGOpt achieves the best performance, while NGTuned is second ranked. From the line plot on the right of Fig. 6, we can see that NGOpt and NGTuned are better than the other algorithms over all budgets. The results for the YABOXBBOB benchmark suite (box-constrained and low budget) are presented in Fig. 7. In this case, all of the proposed algorithms appear as top ranked, with MetaCMA being the best performing one. As for budget up to around 10^3, MetaCMA performs the best, and is later outperformed by NGTuned, which uses the MetaModel on top of MetaCMA. The YAHDBBOB benchmark suite is

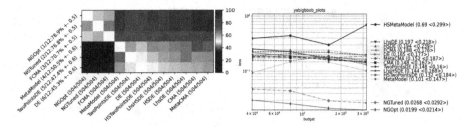

Fig. 6. Results on YABIGBBOB: longer budgets than for BBOB. Not used in the irace trainings, but same functions as YABBOB.

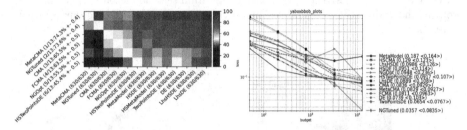

Fig. 7. Results on YABOXBBOB: box-constrained and low budget: not used in the irace training but has similarities.

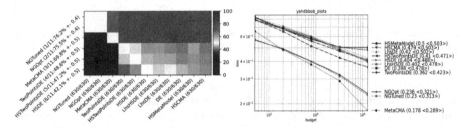

Fig. 8. YAHDBBOB: high-dim. counterpart of YABBOB, not used for training.

a high-dimensional counter part of YABBOB. On it, NGTuned is selected as the best algorithm in Fig. 8, followed by NGOpt with 0.7% difference referring to the % of times they outperform other algorithms. The right subplot shows that these algorithms together with MetaCMA achieved the best losses, across all budgets.

Figure 9 shows results on the YAWIDEBBOB benchmark suite, which is considered to be specially difficult and covers many different test problem instances (continuous, discrete, noisy, noise-free, parallel or not, multi-objective or not). On YAWIDEBBOB, NGTuned is selected as the best algorithm, followed by NGOpt, with a difference of 1.7% in the average score related to the number of times they outperform the other algorithms. MetaCMA is ranked as third. From the line plot we can see that these algorithms achieved lowest losses, however the behavior is not stable over different budgets. We need to point out

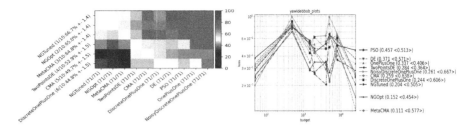

Fig. 9. Results on YAWIDEBBOB: this benchmark matters particularly as it covers many different test cases, continuous, discrete, noisy, noise-free, parallel or not, multi-objective or not. It is designed for validating algorithm selection wizards. The different problems have different budgets, so that it is not surprising that the curve is not decreasing.

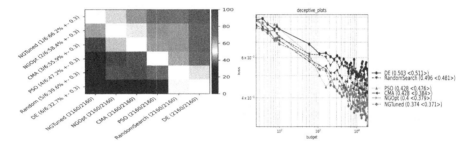

Fig. 10. Deceptive functions: this benchmark [22] is quite orthogonal to the YA*BBOB benchmarks, and very hard (infinitely many local minima arbitrarily close to the optimum/infinite condition number/infinitely tiny path to the optimum).

here that NGTuned is not using CMA for multi-objective problems, so that some parts of YAWIDEBBOB are not modified by our work. The results indicate that the automated configuration that is performed for such benchmark suites also helps to improve the performance of the NGOpt on benchmark suites that contain a mix of different problem instances. The results for the Deceptive functions benchmark suite are presented in Fig. 10, where NGTuned is the winning algorithm in terms of the number of times it performs better than the other algorithms and the mean loss achieved for different budgets over all settings. This benchmark suite is quite orthogonal to the YABBOB variants and very hard to optimize, since the deceptive functions contain infinitely many local minima arbitrarily close to the optimum or have an infinite condition number or have an infinite tiny path to the optimum.

Real-World Problems. We also tested our algorithms on some of the real-world benchmark suites presented in Nevergrad. Due to length constraints, the figures displaying these results are available only in the full version of this paper. On the SimpleTsp benchmark suite, NGTuned outperforms NGOpt in terms of the average frequency of winning over other algorithms and mean loss achieved for different budgets over all settings. On the SeqMlTuning benchmark NGTuned appears among the best-performing algorithms with rank six. It outperforms NGOpt and MetaCMA which do not appear in the set of best performing algorithms. On the ComplexTSP problem, NGTuned is found to be the best performing algorithm. It also surpasses NGOpt. With regard to mean performance per budget, they are very similar and consistently better than the other algorithms. On the Unit Commitment benchmark suite, NGTuned is the third ranked algorithm. It outperforms NGOpt but not enough for competing with the best variants of DE. According to the convergence analysis, it seems that NGTuned and NGOpt have similar behavior. On the Photonics problem, NGOpt and NGTuned appear between the top ranked algorithms showing very similar performance. On th benchmark suite (GP), which is a challenge in genetic programming, NGOpt and NGTuned show almost identical but very bad performance overall: it is mentioned in [22] that NGOpt has counterparts dedicated to neural reinforcement learning (MixDeterministic RL and ProgNoisyRL) that perform better: we confirm this. On the 007 benchmark suite, NGTuned is the best performing algorithm. Regarding the mean loss achieved for different budgets, NGTuned shows similar performance to NGOpt.

6 Conclusion

We have shown in this paper how automated algorithm configuration methods like irace may be used to improve an algorithm selection wizard like NGOpt. NGTuned was better than NGOpt in most benchmark suites, included those not used for tuning our MetaCMA. Our results improve the algorithm selection wizard not only for the benchmark suites used during the tuning process but also for a wider set of benchmark suites, including real-world test cases. Given the complexity of NGOpt, we have an impact on few cases, so that the overall impact on NGOpt across all Nevergrad suites remains moderate: in some cases, NGTuned is equivalent to NGOpt, and in some cases (in which CMA is crucial) it performs clearly better. In some cases, the differences between the two algorithms are small. If we consider only clear differences, we have gaps for YABBOB, 007, YASMALLBBOB, YAPARABBOB, YABOXBBOB, SimpleTSP, and Deceptive: in all these cases, NGTuned performs better than NGOpt.

Our tuning process still contains a few manual steps that require an expert, such as deciding which benchmark suites to use for tuning, as well as the integration of the tuned CMA configurations into NGOpt. Our ultimate goal is to automatize all steps of the process, and to also tune other base components as well as the decision rules that underlie the NGOpt algorithm selection process.

Acknowledgments. M. López-Ibáñez is a "Beatriz Galindo" Senior Distinguished Researcher (BEAGAL 18/00053) funded by the Spanish Ministry of Science and Innovation (MICINN). C. Doerr is supported by the Paris Ile-de-France Region (AlgoSelect) and by the INS2I institute of CNRS (RandSearch). T. Eftimov, A. Nikolikj, and G. Cenikj is supported by the Slovenian Research Agency: research core fundings No. P2-0098 and project No. N2-0239. G. Cenikj is also supported by the Ad Futura grant for postgraduate study.

References

1. Aziz-Alaoui, A., Doerr, C., Dréo, J.: Towards large scale automated algorithm design by integrating modular benchmarking frameworks. In: Chicano, F., Krawiec, K. (eds.) Proceedings of the Genetic and Evolutionary Computation Conference Companion, GECCO Companion 2021, New York, NY, pp. 1365–1374. ACM Press (2021). https://doi.org/10.1145/3449726.3463155
2. Bezerra, L.C.T., López-Ibáñez, M., Stützle, T.: Automatically designing state-of-the-art multi- and many-objective evolutionary algorithms. Evol. Comput. **28**(2), 195–226 (2020). https://doi.org/10.1162/evco_a_00263
3. Birattari, M.: Tuning Metaheuristics: A Machine Learning Perspective, Studies in Computational Intelligence, vol. 197. Springer, Heidelberg (2009). https://doi.org/10.1007/978-3-642-00483-4
4. Cartis, C., Fiala, J., Marteau, B., Roberts, L.: Improving the flexibility and robustness of model-based derivative-free optimization solvers (2018)
5. Hansen, N., Auger, A., Finck, S., Ros, R.: Real-parameter black-box optimization benchmarking 2009: Experimental setup. Technical report, RR-6828, INRIA, France (2009). https://hal.inria.fr/inria-00362633/document
6. Hansen, N., Auger, A., Ros, R., Mersmann, O., Tušar, T., Brockhoff, D.: COCO: a platform for comparing continuous optimizers in a black-box setting. Optim. Meth. Software **36**(1), 1–31 (2020). https://doi.org/10.1080/10556788.2020.1808977
7. Hansen, N., Ostermeier, A.: Completely derandomized self-adaptation in evolution strategies. Evol. Comput. **9**(2), 159–195 (2001). https://doi.org/10.1162/106365601750190398
8. Hoos, H.H.: Automated algorithm configuration and parameter tuning. In: Hamadi, Y., Monfroy, E., Saubion, F. (eds.) Autonomous Search, pp. 37–71. Springer, Heidelberg (2011). https://doi.org/10.1007/978-3-642-21434-9_3
9. Huang, C., Li, Y., Yao, X.: A survey of automatic parameter tuning methods for metaheuristics. IEEE Trans. Evol. Comput. **24**(2), 201–216 (2020). https://doi.org/10.1109/TEVC.2019.2921598
10. Johnson, S.G.: The nlopt nonlinear-optimization package (1994). http://github.com/stevengj/nlopt
11. Kerschke, P., Hoos, H.H., Neumann, F., Trautmann, H.: Automated algorithm selection: survey and perspectives. Evol. Comput. **27**(1), 3–45 (2019). https://doi.org/10.1162/evco_a_00242
12. KhudaBukhsh, A.R., Xu, L., Hoos, H.H., Leyton-Brown, K.: SATenstein: automatically building local search SAT solvers from components. Artif. Intell. **232**, 20–42 (2016). https://doi.org/10.1016/j.artint.2015.11.002
13. Liao, T., Molina, D., Stützle, T.: Performance evaluation of automatically tuned continuous optimizers on different benchmark sets. Appl. Soft Comput. **27**, 490–503 (2015)
14. Liao, T., Montes de Oca, M.A., Stützle, T.: Computational results for an automatically tuned CMA-ES with increasing population size on the CEC 2005 benchmark set. Soft Comput. **17**(6), 1031–1046 (2013). https://doi.org/10.1007/s00500-012-0946-x
15. Liao, T., Stützle, T., Montes de Oca, M.A., Dorigo, M.: A unified ant colony optimization algorithm for continuous optimization. Eur. J. Oper. Res. **234**(3), 597–609 (2014)
16. López-Ibáñez, M., Dubois-Lacoste, J., Pérez Cáceres, L., Stützle, T., Birattari, M.: The irace pacskage: iterated racing for automatic algorithm configuration. Oper. Res. Perspect. **3**, 43–58 (2016). https://doi.org/10.1016/j.orp.2016.09.002

17. López-Ibáñez, M., Liao, T., Stützle, T.: On the anytime behavior of IPOP-CMA-ES. In: Coello, C.A.C., Cutello, V., Deb, K., Forrest, S., Nicosia, G., Pavone, M. (eds.) PPSN 2012. LNCS, vol. 7491, pp. 357–366. Springer, Heidelberg (2012). https://doi.org/10.1007/978-3-642-32937-1_36

18. López-Ibáñez, M., Stützle, T.: The automatic design of multi-objective ant colony optimization algorithms. IEEE Trans. Evol. Comput. **16**(6), 861–875 (2012). https://doi.org/10.1109/TEVC.2011.2182651

19. Mascia, F., López-Ibáñez, M., Dubois-Lacoste, J., Stützle, T.: Grammar-based generation of stochastic local search heuristics through automatic algorithm configuration tools. Comput. Oper. Res. **51**, 190–199 (2014). https://doi.org/10.1016/j.cor.2014.05.020

20. Meunier, L., et al.: Black-box optimization revisited: improving algorithm selection wizards through massive benchmarking. IEEE Trans. Evol. Comput. **26**(3), 490–500 (2022). https://doi.org/10.1109/TEVC.2021.3108185

21. Pagnozzi, F., Stützle, T.: Automatic design of hybrid stochastic local search algorithms for permutation flowshop problems. Eur. J. Oper. Res. **276**, 409–421 (2019). https://doi.org/10.1016/j.ejor.2019.01.018

22. Rapin, J., Teytaud, O.: Nevergrad: a gradient-free optimization platform (2018). https://github.com/FacebookResearch/Nevergrad

23. Rice, J.R.: The algorithm selection problem. Adv. Comput. **15**, 65–118 (1976)

24. Ros, R., Hansen, N.: A simple modification in CMA-ES achieving linear time and space complexity. In: Rudolph, G., Jansen, T., Beume, N., Lucas, S., Poloni, C. (eds.) PPSN 2008. LNCS, vol. 5199, pp. 296–305. Springer, Heidelberg (2008). https://doi.org/10.1007/978-3-540-87700-4_30

25. Schede, E., et al.: A survey of methods for automated algorithm configuration (2022). https://doi.org/10.48550/ARXIV.2202.01651

26. Styles, J., Hoos, H.H.: Ordered racing protocols for automatically configuring algorithms for scaling performance. In: Blum, C., Alba, E. (eds.) Proceedings of the Genetic and Evolutionary Computation Conference, GECCO 2013, New York, NY, pp. 551–558. ACM Press (2013). ISBN 978-1-4503-1963-8, https://doi.org/10.1145/2463372.2463438

27. Videau, M., Leite, A., Teytaud, O., Schoenauer, M.: Multi-objective genetic programming for explainable reinforcement learning. In: Medvet, E., Pappa, G., Xue, B. (eds.) EuroGP 2022. LNCS, vol. 13223, pp. 256–281. Springer, Cham (2022). https://doi.org/10.1007/978-3-031-02056-8_18

28. Xu, L., Hoos, H.H., Leyton-Brown, K.: Hydra: automatically configuring algorithms for portfolio-based selection. In: Fox, M., Poole, D. (eds.) Proceedings of the AAAI Conference on Artificial Intelligence. AAAI Press (2010)

29. Xu, L., Hutter, F., Hoos, H.H., Leyton-Brown, K.: SATzilla: portfolio-based algorithm selection for SAT. J. Artif. Intell. Res. **32**, 565–606 (2008). https://doi.org/10.1613/jair.2490

Non-elitist Selection Can Improve the Performance of Irace

Furong Ye[1(✉)], Diederick Vermetten[1], Carola Doerr[2],
and Thomas Bäck[1]

[1] LIACS, Leiden University, Leiden, The Netherlands
{f.ye,d.l.vermetten,t.h.w.baeck}@liacs.leidenuniv.nl
[2] Sorbonne Université, CNRS, LIP6, Paris, France
carola.doerr@lip6.fr

Abstract. Modern optimization strategies such as evolutionary algorithms, ant colony algorithms, Bayesian optimization techniques, etc. come with several parameters that steer their behavior during the optimization process. To obtain high-performing algorithm instances, automated algorithm configuration techniques have been developed. One of the most popular tools is irace, which evaluates configurations in sequential races, making use of iterated statistical tests to discard poorly performing configurations. At the end of the race, a set of *elite* configurations are selected from those *survivor* configurations that were not discarded, using greedy truncation selection. We study two alternative selection methods: one keeps the best survivor and selects the remaining configurations uniformly at random from the set of survivors, while the other applies entropy to maximize the diversity of the elites. These methods are tested for tuning ant colony optimization algorithms for traveling salesperson problems and the quadratic assignment problem and tuning an exact tree search solver for satisfiability problems. The experimental results show improvement on the tested benchmarks compared to the default selection of irace. In addition, the obtained results indicate that non-elitist can obtain diverse algorithm configurations, which encourages us to explore a wider range of solutions to understand the behavior of algorithms.

Keywords: Parameter tuning · Algorithm configuration · Black-box optimization · Evolutionary computation

1 Introduction

Algorithm configuration (AC) addresses the issue of determining a well-performing parameter configuration for a given algorithm on a specific set of optimization problems. Many techniques such as local search, Bayesian optimization, and racing methods have been proposed and applied to solve the AC problem. The corresponding software packages, such as ParamILS [14], SMAC [13], SPOT [3], MIP-EGO [29], and irace [19] have been applied to problem domains

© The Author(s), under exclusive license to Springer Nature Switzerland AG 2022
G. Rudolph et al. (Eds.): PPSN 2022, LNCS 13398, pp. 32–45, 2022.
https://doi.org/10.1007/978-3-031-14714-2_3

such as combinatorial optimization [19], software engineering [4], and machine learning [15].

Irace, one of the most popular tools, has shown its ability to improve the performance of the algorithms for various optimization problems [2,7,19,27]. However, we can still intuitively expect to improve the performance of irace considering contemporary optimization techniques. Premature convergence is a common problem for optimization methods resulting in being trapped into local optima, which can also present irace from finding the optimal configurations. For example, irace fails to find the optimal configuration of a family of genetic algorithms (GAs) for ONEMAX in [30]. There exists more than one type of competitive configuration of the GA for ONEMAX, which is known due to the extensive body of theoretical work [11,28]. However, irace converges to a specific subset of configurations that share similar algorithm characteristics. In order to avoid issues like this, one could aim to increase the exploration capabilities of irace. However, this does not necessarily address the concern of finding well-performing configurations located in different parts of the space. Instead, we would want to allow irace to automatically explore search space around a diverse set of well-performing configurations to avoid converging on one specific type of configuration.

A "soft-restart" mechanism has been introduced for irace to avoid premature convergence in [19], which partially reinitializes the sampling distribution for the configurations that are almost identical to others. However, evaluations can be wasted on testing similar configurations before the restart, and the configuration may converge on the type of configurations that were found before restarting. Therefore, we investigate alternative selection mechanisms which take into account the diversity of the selected elite configurations. In addition, the observations from [30] inspire a discussion on searching for various competitive configurations with different patterns, which is addressed by our discussion that more knowledge can be obtained by searching diverse configurations.

1.1 Our Contributions

In this paper, we show that an alternative random selection of elites can result in performance benefits over the default selection mechanism in irace. Moreover, we propose a selection operator maximizing the entropy of the selected elites. These alternative selection operators are compared to default irace on the tested scenarios.

The alternative approaches are tested on three scenarios: tuning the Ant Colony Optimization (ACO) algorithm for the traveling salesperson problem (TSP) and the quadratic assignment problem (QAP) and minimizing the computational cost of the SPEAR tool (an exact tree search solver for the satisfiability (SAT) problem).

Experimental results show that (1) randomly selecting elites among configurations that survived the racing procedure performs better than the greedy truncation selection, and (2) the irace variant that uses the entropy metric obtains diverse configurations and outperforms the other approaches. Finally, the obtained configurations encourage us to (3) use such a diversity-enhancing approach to find better configurations and understand the relationship between parameter settings and algorithm behavior for future work.

Reproducibility: We provide the full set of logs from the experiments described in this paper in [31]. Additionally, our implementation of the modified irace versions described in this paper is available at https://github.com/FurongYe/irace-1.

2   Related Work

2.1   Algorithm Configuration

Traditionally, the AC problem, as defined below, aims at finding a *single* optimal configuration for solving a set of problem instances [9].

Definition 1 (Algorithm Configuration Problem). *Given a set of problem instances Π, a parametrized algorithm A with parameter configuration space Θ, and a cost metric $c : \Theta \times \Pi \to \mathbb{R}$ that is subject to minimization, the objective of the AC problem is to find a configuration $\theta^* \in \arg\min_{\theta \in \Theta} \sum_{\pi \in \Pi} c(\theta, \pi)$.*

The parameter space can be continuous, integer, categorical, or mixed-integer. In addition, some parameters can be conditional.

Many configurators have been proposed for the AC problem [3,13,14,17,19, 29], and they usually follow Definition 1 by searching for a *single* optimal solution, although the solvers may apply population-based methods. However, in some cases it can be desirable to find a set of diverse, well-performing solutions to the AC problem. For example, previous studies [21,30] found that algorithm configurators can obtain different results when tuning for different objectives (i.e., expected running time, best-found fitness, and anytime performance), which suggests that a bi- or multi-objective approach to algorithm configuration can be a promising research direction. For such multi-objective configuration tasks, having diverse populations of configurations is a necessity to understand the Pareto front.

2.2   Diversity Optimization

To address the objective of obtaining a set of diverse solutions, certain evolutionary algorithms have been designed specifically to converge to more than one solution in a single run. For example, the Niching Genetic Algorithms are applied for solving multimodular functions [6,12] and searching diverse solutions of association rules [24], chemical structures [23], etc. Diversity optimization also addresses the problem of searching for multiple solutions. Quality-diversity optimization [8] was introduced to aim for a collection of well-performing and diverse solutions. The method proposed in [8] measures the quality of solutions based on their performance (i.e., quality) and distance to other solutions (i.e., novelty) dynamically. The novelty score of solutions is measured by the average distance of the k-nearest neighbors [16]. Also, to better understand the algorithm's behavior and possible solutions, feature-based diversity optimization was introduced for problem instance classification [10]. A discrepancy-based diversity optimization was studied on evolving diverse sets of images and TSP instances [25]. The approaches in both studies measure the solutions regarding their features instead of performance. Unfortunately, the AC problem usually deals with a mixed-integer search space, which is often not considered in the methods described in this section.

3   Irace

In this section, we describe the outline of irace. Irace is an iterated racing method that has been applied for hyperparameter optimization problems in many domains. It samples configurations (i.e., hyperparameter values) from distributions that evolve along the configuration process. Iteratively, the generated configurations are tested across a set of instances and are selected based on a racing method. The racing is based on statistical tests on configurations' performance for each instance, and elite configurations are selected from the configurations surviving from the racing. The sampling distributions are updated after selection. The distributions from sampling hyperparameter values are independent unless specific conditions are defined. As a result, irace returns one or several elite configurations at the end of the configuration process.

Algorithm 1: Algorithm Outline of irace

---

1 **Input:** Problem instances $\Pi = \{\pi_1, \pi_2, \ldots\}$, parameter configuration space X, cost metric c, and tuning budget B;
2 Generate a set of Θ_1 sampling from X uniformly at random;
3 $\Theta^{\text{elite}} = \text{Race}(\Theta_1, B_1)$;
4 **while** *The budget B is not used out* **do**
5 $\quad$ $j = j + 1$;
6 $\quad$ $\Theta_j = \text{Sample}(X, \Theta^{\text{elite}})$;
7 $\quad$ $\Theta^{\text{elite}} = \text{Race}(\Theta_j \cup \Theta^{\text{elite}}, B_j)$;
8 **Output:** Θ^{elite}

Algorithm 1 presents the outline of irace [19]. Irace determines the number of racing iterations $N^{\text{iter}} = \lfloor 2 + \log_2(N^{\text{param}}) \rfloor$ before performing the race steps, where N^{param} is the number of parameters. For each $\text{Race}(\Theta_j, B_j)$ step, the budget of the number of configuration evaluations $B_j = (B - B_{\text{used}})/(N^{\text{iter}} - j + 1)$, where B_{used} is the used budget, and $j = \{1, \ldots, N^{\text{iter}}\}$. After sampling a set of new configurations in each iteration, $\text{Race}(\Theta, B)$ selects a set of elite configurations Θ^{elite} (elites). New configurations are sampled based on the parent selected from elites Θ^{elite} and the corresponding self-adaptive distributions of hyperparameters. Specific strategies have been designed for different types (numerical and categorical) of parameters.

Each race starts with a set of configurations Θ_j and performs with a limited computation budget B_j. Precisely, each candidate configuration of Θ_j is evaluated on a single instance π_i, and the configurations that perform statistically worse than at least another one will be discarded after being evaluated on a number of instances. Note that the irace package provides multiple statistical test options for eliminating worse configurations such as the F-test and the t-test. The race terminates when the remaining budget is not enough for evaluating the surviving configurations on a new problem instance, or when $N^{\min}$ or fewer configurations survived after the test. At the end of the race, N_j^{surv} configurations survive and are ranked based on their performance. Irace selects $\min\{N^{\min}, N_j^{\text{surv}}\}$ configurations with the best ranks to form Θ^{elite} for the next iteration. Note that irace applies here a *greedy elitist mechanism*, and this is the essential step where our irace variants alter in this paper.

To avoid confusion, we note that an "elitist iterated racing" is described in the paper introducing the irace package [19]. The "elitist" there indicates preserving the best configurations found so far. The idea is to prevent "elite" configurations from being eliminated due to poor performance on specific problem instances during racing, and the best surviving "elite" configurations are selected to form Θ^{elite}. We apply this "elitist racing" for our experiments in this paper, while the alternative methods select diverse surviving "elite" configurations instead of the best ones.

4   Random Survivor Selection

To investigate the efficacy of the greedy truncation selection mechanism used by default within irace, we compare the baseline version of irace to a version of irace that uses a random selection process. In particular, we adopt the selection of elites by *taking the best-performing configuration and randomly selecting the remaining* $N_j^{\text{surv}} - N^{\min} - 1$ distinct ones from the best $\sigma N^{\min}$ surviving configurations when $N^{\text{surv}} \geq \sigma N^{\min}$, for some $\sigma \geq 1$. The implementation of our variants is built on the default irace package [20].

4.1   Tuning Scenario: ACOTSP

ACOTSP [27] is a package implementing ACO for the symmetric TSP. We apply irace variants in this paper to configure 11 parameters (three categorical, four continuous, and four integer variables) of ACO for lower solution costs (fitness). The experimental results reported in the following are from 20 independent runs of each irace variant. Each run is assigned with a budget of $5,000$ runs of ACOTSP, and ACOTSP executes 20s of CPU-time per run following the suggestion in [19]. We set $\sigma N^{\min} = N^{\text{surv}}$ indicating that the irace variant, irace-rand, randomly selects survivor configurations to form elites. Other settings remain as default: the "elitist iterated racing" is applied, and $N^{min} = 5$. We apply the benchmark set of Euclidean TSP instances of size $2,000$ with 200 train and 200 test instances.

Figure 1 plots the deviations of the best configurations, which are obtained by each run, from the best-found (optimum) configuration obtained by 20 (60 in total) runs of the irace variants. The results are averaged across 200 TSP instances. We observe that the median and mean of irace-rand results are smaller than those of irace, but the performance variance among these 20 irace-rand runs is significantly larger.

Though irace is initially proposed for searching configurations that generally perform well across a whole set of problem instances, we are nevertheless interested in the performance of the obtained configurations on individual instances. Therefore, we plot in Fig. 2 the performance of all obtained configurations on nine instances. Still, we observe comparable performance between irace and irace-rand. It is not surprising that the performance of irace-rand presents larger variance because the configurations that do not perform the best get a chance to be selected. Moreover, we spot significant improvement on instances "2000-6" and "2000-9", on which the configurations obtained by irace-rand generally perform closer to the optima, compared to irace.

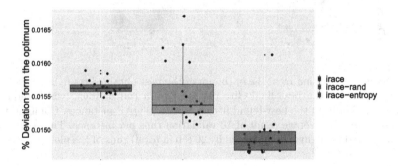

Fig. 1. Average deviation from the optimum of the best obtained configurations. Each dot corresponds to the best final elite obtained by a run of irace, which plots the average deviation from the best-found fitness across 200 TSP instances. Configurations are measured by the average result of 10 validation runs per instance. The "optimum" for each instance is the best-found configuration obtained by 20 (60 in total) runs of the plotted methods.

4.2 Tuning Scenario: ACOQAP

We apply the same ACO implementation in [19] for solving QAP [22]. ACOQAP executes 60s CPU-time per run following the default setting of the package, and we apply the benchmark set of 50 train and test instances, respectively. The other settings remain the same with the ACOTSP scenario.

Unfortunately, we do not observe similar improvement of using irace-rand for ACOQAP. Irace-rand present worse performance than irace, comparing the

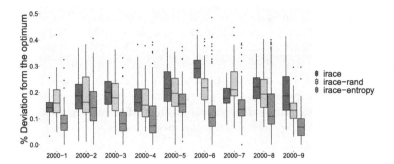

Fig. 2. Boxplots of the deviation from the optimum of the obtained configurations for TSP instances. Results are from the average fitness of 10 validation runs for each obtained configuration.

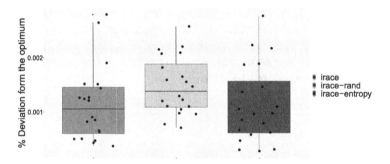

Fig. 3. Average deviation from the optimum of the best obtained configurations. Each dot corresponds to the best final elite obtained by a run of irace, which plots the average deviation from the best-found fitness across 50 QAP instances. Configurations are measured by the average result of 10 validation runs per instance. The "optimum" is the best-found configuration obtained by 20 (60 in total) runs of the plotted methods.

average results across 50 instances in Fig. 3. While looking at Fig. 4, which plots the results on ten randomly picked instances, we do not observe improvement using irace-rand on ACOQAP, either. These observations indicate that using this random selection to select elite configurations may deteriorate the performance of irace for ACOQAP, though it does not necessarily mean that diverse configurations are not helpful for the configuring process. We will discuss this topic in more detail in Sect. 5.

4.3   Tuning Scenario: SPEAR

SPEAR [2] is a custom-made SAT solver configurable with 26 categorical parameters, of which nine are conditional, i.e., their activation depends on the values of one or several of the other parameters. Our goal here is to minimize the runtime of SPEAR. We run each irace variant 20 independent times. Each run of irace

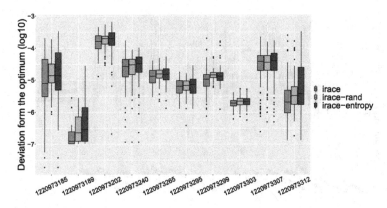

Fig. 4. Boxplots of the deviation from the optimum of the obtained configurations for ACOQAP instances. Results are from the average fitness of 10 validation runs for each obtained configuration.

is assigned with a budget of 10 000 runs of SPEAR, and the maximal runtime of SPEAR is 30 s CPU-time per run. Other irace settings remain default: the "elitist iterated racing" is applied, and $N^{\min} = 6$. The training and test set are 302 different SAT instances, respectively [1]. Note that the number of survivor configurations is large ($\sim$250) during racing, and experimental results show that randomly selecting with such a large population deteriorates the performance of irace. Therefore, for this scenario, we cap the size of survivor candidates by $2N^{\min}(\sigma = 2)$ to select from a relatively well-performing population.

Overall, we observe that the performance difference between the two methods is tiny for most instances, though irace-rand can not obtain better average results of runtime across all tested instances than irace. Note that the obtained configurations may use much runtime ($\sim$30 s) for a few instances, resulting in the comparison among the average runtime ($\sim$3 s) across all instances can be significantly affected by the results on those particular instances. Therefore, we plot only the runtime for the first two instances of each class of instances in Fig. 5. Compared to irace, though the performance of irace-rand deteriorates on "itox" instances, significant improvements using irace-rand can be observed on more instances such as "dspam_vc9400", "winedump" instances, and "xinetd_vc56633".

5   Selecting Diverse Elites

The optimistic results of ACOTSP and SPEAR scenarios introduced in Sect. 4 indicate that, while keeping the best configuration, randomly selecting from well-performing survivor configurations to form elites can have positive impacts on the performance of irace. An intuitive explanation is that irace-rand allows exploring search space around those *non-elitist* configurations to avoid premature convergence on specific types of configurations, which matches our expectation following the motivation introduced in Sect. 1. However, the failure to

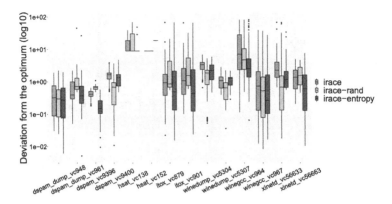

Fig. 5. Boxplots of the deviation from the optimum of the obtained configurations for SPEAR instances. Results are from the average fitness of 10 validation runs for each obtained configuration. Results of "gzip" class is omitted because runtime of all obtained configurations are identical.

achieve improvements for ACOQAP requires us to consider explicitly controlling the selected elite configurations' diversity. To this end, we study an alternative selection strategy based on entropy [5] as a diversity measure.

5.1 Maximizing Population Entropy

In information theory, entropy represents random variables' information and uncertainty level [26]. The larger the entropy, the more information the variables deliver, e.g., the more diverse the solutions are. Our irace-entropy configurator makes use of this idea, by using the Shannon entropy as criterion for selecting survivor configurations to form elites.

For a random variable X with distribution $P(X)$, the normalized entropy of X is defined as:

$$H(X) = \sum_{i=1}^{n} P(X_i) \log P(X_i) / \log(n),$$

In this paper, we estimate the entropy of integer and categorical variables from the probability of each value. For continuous variables, the values are discretized into bins, and entropy is estimated based on the counts of each bin. Precisely, the domain of a continuous variable is equally divided into n bins, where n is the number of observations (i.e., configurations). Finally, we calculate the diversity level $D(\Theta)$ of a set of configuration Θ using the mean entropy across p variables (i.e., parameters), which is defined as:

$$D(\Theta) = \frac{\sum_{j=1}^{p} H(\Theta^j)}{p}, \quad \Theta^j = \{\theta_1^j, \theta_2^j, \ldots, \theta_n^j\}$$

We introduce a variant of irace (irace-entropy) maximizing $D(\Theta^{\text{elite}})$ for each race step. Recall that N^{surv} configurations survive at the end of race, and the

$N^{\min}$ best-ranked configurations are selected to form Θ^{elite} in Algorithm 1 (line 7). Irace-entropy adapts this step by selecting a subset of configurations Θ with the maximal $D^*(\Theta)$, where $|\Theta| = N^{\min}$ and the best-ranked configuration $\theta^* \in \Theta$. In practice, we replace the greedy truncation selection in Algorithm 1 (line 7) with Algorithm 2. Note that we do not explicitly handle conditional parameters.

Algorithm 2: Entropy-maximization selection

---

1 **Input:** A set of ranked configurations Θ^{surv}, the maximal size N^{min} of Θ^{elite} ;

2 **if** $|\Theta^{surv}| \leq N^{min}$ **then**

3 $\quad |\quad \Theta^{\text{elite}} = \Theta^{surv}$

4 **else**

5 $\quad |\quad \Theta^{\text{elite}} = \{\theta^*\}, \Theta^{surv} = \Theta^{surv} \backslash \{\theta^*\}$, where $\theta^* \in \Theta^{surv}$ is the best-ranked;

6 $\quad |\quad \Theta^{\text{elite}} = \Theta^{\text{elite}} \cup S^*$, where $S^* = \underset{S \subset \Theta^{surv}, |S| = N^{min}-1}{\arg\max} D(\Theta^{\text{elite}} \cup S)$

7 **Output:** Θ^{elite}

5.2 Experimental Results

We present the results of irace-entropy in this section. All the settings remain the same as reported in Sect. 4 while applying the introduced alternative selection method.

For ACOTSP, we observe in Fig. 1 that irace-entropy performs better than irace and irace-rand, obtaining significantly smaller deviations from the optimum than those of irace for 19 out of 20 runs. Regarding the results on individual problem instances, irace-entropy also shows in Fig. 2 significant advantages against irace and irace-rand across all the plotted instances.

Recall that, for ACOQAP, the performance of irace-rand deteriorates compared to irace by randomly selecting survivor configurations to form elites. However, through using entropy as the metric to control the diversity explicitly, irace-entropy shows comparable results to irace in Fig. 3, obtaining a smaller median of deviations from the optimum for 20 runs. We also observe that the performance of irace-entropy is comparable to irace for individual instances in Fig. 4. In addition, irace-entropy can obtain the best-found configurations for some instances such as "1220973202" and "1220973265".

For SPEAR, we observe in Fig. 5 that irace-entropy outperforms irace. For 12 out of the 14 plotted instances, irace-entropy obtains better median results than irace. Moreover, irace-entropy achieves improvements compared to irace-rand for most instances. Especially for the "itox" instances, in which irace-rand does not perform as well as irace, irace-entropy obtains better results while also keeping the advantages over irace on other instances.

According to these results, we conclude that non-elitist selection can help improve the performance of irace. By using entropy as the metric to maximize the diversity of the selected elite configurations, irace-entropy achieves improvements compared to irace. However, irace-entropy does not obtain significant advantages against irace for ACOQAP and performs worse than irace-rand on some SPEAR instances, indicating potential improvements for non-elitist selection through further enhancements in regards to controlling the diversity of elites for specific problem instances.

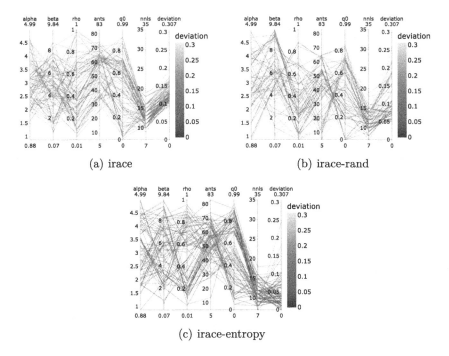

(a) irace

(b) irace-rand

(c) irace-entropy

Fig. 6. Parameter values and deviations from the optimum of the ACOTSP configurations obtained by the irace variants. Each configuration is represented as a polyline with vertices on the parallel axes. The point of the vertex on each x-axis corresponds to the parameter value. We process "Null" values of the conditional parameter as 0. The color of the lines indicates the deviation of their average solution costs across all tested instances from that of the best-found one. Darker lines indicate better configurations.

5.3 Benefits from Diverse Configurations

While the irace variants, i.e., irace-rand and irace-entropy, achieve improvements by using non-elitist selection, significant variances are noticeable in Figs. 2, 4, and 5 for results of the obtained configurations. Recall that AC techniques have been applied in [30] for exploring promising configurations of the GA on diverse problems. Apart from analyzing a single optimal configuration, such benchmarking studies can also benefit from diverse configurations to investigate algorithms' performance with specific parameter settings. Therefore, we illustrate in this section that non-elitist selection can not only improve the performance of irace but also help understand the behavior of algorithms.

Using ACOTSP as an example, we show the configurations obtained by each irace variant in Fig. 6. The color of the configuration lines are scaled by the deviation $\frac{f-f^*}{f^*}\%$ from the optimum, where f is the average solution cost across 200 problem instances.

We observe that irace-entropy obtains most of the competitive configurations while covering a wider range of performance with deviations from 0 to 0.25. However, the performance of the configurations obtained by the irace cluster in a range of deviations from 0.1 to 0.2. Moreover, regarding the parameters of the obtained configurations, the range of $beta$ and $q0$ are narrower for irace compared to the other methods. However, the configurations with $beta > 8$ and $q0 > 0.9$, outside the range obtained by irace, generally perform well.

We will not investigate how the parameter values practically affect the performance of ACOTSP since it is beyond the scope of this paper. Nonetheless, Fig. 6 provides evidence that irace-entropy can provide more knowledge concerning the distribution of obtained performance (i.e., fitness) and parameter values, which is helpful for understanding algorithms' behavior.

6   Conclusions and Discussions

In this paper, we have demonstrated that randomly selecting survivor configurations can improve the performance of irace, as illustrated on the cases of tuning ACO on TSP and tuning SPEAR to minimize the runtime of a SAT solver. Moreover, we have proposed an alternative selection method to form diverse elite configurations, using Shannon entropy as the diversity metric. Experimental results show significant advantages of maximizing entropy in this way.

While the irace-entropy presents improvement in the performance of irace via exploring diverse configurations in all the tested scenarios, irace-rand obtain better configurations for specific SPEAR instances. Therefore, there is still room for further study of incorporating diversity into the selection operators. More in-depth analysis on a wider set of algorithm configuration problems can help us better understand the benefits of considering diversity in selection. In addition, we did not modify the procedure of sampling new configurations. Nevertheless, we believe effectively generating diverse configurations can be beneficial and shall be studied for future work.

Apart from boosting the performance of irace via focusing more on diversity, we can find a diverse portfolio of well-performing algorithm configurations while keeping the benefits of the iterated racing approach, by changing the objective of the tuning from finding the best performing configuration to find a diverse portfolio of well-performing algorithm configurations. In the context of algorithm selection, such approaches are studied under the notion of *algorithm portfolio selection* [18].

References

1. Babić, D., Hu, A.J.: Structural abstraction of software verification conditions. In: Damm, W., Hermanns, H. (eds.) CAV 2007. LNCS, vol. 4590, pp. 366–378. Springer, Heidelberg (2007). https://doi.org/10.1007/978-3-540-73368-3_41
2. Babic, D., Hutter, F.: Spear theorem prover. Solver description, SAT competition 2007 (2007)
3. Bartz-Beielstein, T.: SPOT: an R package for automatic and interactive tuning of optimization algorithms by sequential parameter optimization. CoRR abs/1006.4645 (2010)
4. Basmer, M., Kehrer, T.: Encoding adaptability of software engineering tools as algorithm configuration problem: a case study. In: Proceedings of International Conference on Automated Software Engineering Workshop (ASEW 2019), pp. 86–89. IEEE (2019)
5. Bromiley, P., Thacker, N., Bouhova-Thacker, E.: Shannon entropy, Renyi entropy, and information. In: Statistics and Information Series, pp. 1–8 (2004)
6. Cavicchio, D.: Adaptive search using simulated evolution. Ph.D. thesis, University of Michigan (1970)
7. Cintrano, C., Ferrer, J., López-Ibáñez, M., Alba, E.: Hybridization of racing methods with evolutionary operators for simulation optimization of traffic lights programs. In: Zarges, C., Verel, S. (eds.) EvoCOP 2021. LNCS, vol. 12692, pp. 17–33. Springer, Cham (2021). https://doi.org/10.1007/978-3-030-72904-2_2
8. Cully, A., Demiris, Y.: Quality and diversity optimization: a unifying modular framework. IEEE Trans. Evol. Comput. **22**(2), 245–259 (2017)
9. Eggensperger, K., Lindauer, M., Hutter, F.: Pitfalls and best practices in algorithm configuration. J. Artif. Intell. Res. **64**, 861–893 (2019)
10. Gao, W., Nallaperuma, S., Neumann, F.: Feature-based diversity optimization for problem instance classification. In: Handl, J., Hart, E., Lewis, P.R., López-Ibáñez, M., Ochoa, G., Paechter, B. (eds.) PPSN 2016. LNCS, vol. 9921, pp. 869–879. Springer, Cham (2016). https://doi.org/10.1007/978-3-319-45823-6_81
11. Gießen, C., Witt, C.: The interplay of population size and mutation probability in the $(1 + \lambda)$ EA on OneMax. Algorithmica **78**(2), 587–609 (2017)
12. Goldberg, D.E., Richardson, J., et al.: Genetic algorithms with sharing for multimodal function optimization. In: Proceedings of International Conference on Genetic Algorithms (ICGA 1987), vol. 4149 (1987)
13. Hutter, F., Hoos, H.H., Leyton-Brown, K.: Sequential model-based optimization for general algorithm configuration. In: Coello, C.A.C. (ed.) LION 2011. LNCS, vol. 6683, pp. 507–523. Springer, Heidelberg (2011). https://doi.org/10.1007/978-3-642-25566-3_40
14. Hutter, F., Hoos, H.H., Leyton-Brown, K., Stützle, T.: ParamILS: an automatic algorithm configuration framework. J. Artif. Intell. Res. **36**, 267–306 (2009)
15. Kotthoff, L., Thornton, C., Hoos, H.H., Hutter, F., Leyton-Brown, K.: Auto-WEKA: automatic model selection and hyperparameter optimization in WEKA. In: Hutter, F., Kotthoff, L., Vanschoren, J. (eds.) Automated Machine Learning. TSSCML, pp. 81–95. Springer, Cham (2019). https://doi.org/10.1007/978-3-030-05318-5_4
16. Lehman, J., Stanley, K.O.: Abandoning objectives: evolution through the search for novelty alone. Evol. Comput. **19**(2), 189–223 (2011)
17. Li, L., Jamieson, K., DeSalvo, G., Rostamizadeh, A., Talwalkar, A.: Hyperband: a novel bandit-based approach to hyperparameter optimization. J. Mach. Learn. Res. **18**(1), 6765–6816 (2017)

18. Lindauer, M., Hoos, H., Hutter, F., Leyton-Brown, K.: Selection and Configuration of Parallel Portfolios. In: Handbook of Parallel Constraint Reasoning, pp. 583–615. Springer, Cham (2018). https://doi.org/10.1007/978-3-319-63516-3_15

19. López-Ibáñez, M., Dubois-Lacoste, J., Cáceres, L.P., Birattari, M., Stützle, T.: The Irace package: iterated racing for automatic algorithm configuration. Oper. Res. Perspect. **3**, 43–58 (2016)

20. López-Ibáñez, M., Dubois-Lacoste, J., Cáceres, L.P., Birattari, M., Stützle, T.: Irace: iterated racing for automatic algorithm configuration. github.com/cran/irace (2020). commit: bae6ae86f2ee0fab9e3270801343482600f095e7

21. López-Ibánez, M., Stützle, T.: Automatically improving the anytime behaviour of optimisation algorithms. Eur. J. Oper. Res. **235**(3), 569–582 (2014)

22. López-Ibáñez, M., Stützle, T., Dorigo, M.: Ant colony optimization: a component-wise overview. In: Marti, R., Panos, P., Resende, M. (eds.) Handbook of Heuristics, pp. 1–37. Springer, Cham (2016). https://doi.org/10.1007/978-3-319-07153-4_21-1

23. de Magalhães, C.S., Almeida, D.M., Barbosa, H.J.C., Dardenne, L.E.: A dynamic niching genetic algorithm strategy for docking highly flexible ligands. Inf. Sci. **289**, 206–224 (2014)

24. Martín, D., Alcalá-Fdez, J., Rosete, A., Herrera, F.: NICGAR: a niching genetic algorithm to mine a diverse set of interesting quantitative association rules. Inf. Sci. **355–356**, 208–228 (2016)

25. Neumann, A., Gao, W., Doerr, C., Neumann, F., Wagner, M.: Discrepancy-based evolutionary diversity optimization. In: Proceedings of the Genetic and Evolutionary Computation Conference (GECCO 2018), pp. 991–998. ACM (2018)

26. Shannon, C.E.: A mathematical theory of communication. ACM SIGMOBILE: Mob. Comput. Commun. Rev. **5**(1), 3–55 (2001)

27. Stützle, T.: ACOTSP: a software package of various ant colony optimization algorithms applied to the symmetric traveling salesman problem (2002). www.aco-metaheuristic.org/aco-code

28. Sudholt, D.: Crossover speeds up building-block assembly. In: Proceedings of Genetic and Evolutionary Computation Conference (GECCO 2012), pp. 689–702. ACM (2012)

29. Wang, H., van Stein, B., Emmerich, M., Bäck, T.: A new acquisition function for bayesian optimization based on the moment-generating function. In: Proceedings of International Conference on Systems, Man, and Cybernetics (SMC 2017), pp. 507–512. IEEE (2017)

30. Ye, F., Doerr, C., Wang, H., Bäck, T.: Automated configuration of genetic algorithms by tuning for anytime performance. IEEE Trans. Evol. Comput. (2022). https://doi.org/10.1109/TEVC.2022.3159087

31. Ye, F., Vermetten, D., Doerr, C., Bäck, T.: Data Sets for the study "Non-Elitist Selection Can Improve the Performance of Irace" (2022). https://doi.org/10.5281/zenodo.6457959

Per-run Algorithm Selection
with Warm-Starting Using
Trajectory-Based Features

Ana Kostovska[1,2], Anja Jankovic[3(✉)], Diederick Vermetten[4],
Jacob de Nobel[4], Hao Wang[4], Tome Eftimov[1], and Carola Doerr[3]

[1] Jožef Stefan Institute, Ljubljana, Slovenia
[2] Jožef Stefan International Postgraduate School, Ljubljana, Slovenia
[3] LIP6, Sorbonne Université, CNRS, Paris, France
anja.jankovic@lip6.fr
[4] LIACS, Leiden University, Leiden, The Netherlands

Abstract. Per-instance algorithm selection seeks to recommend, for a given problem instance and a given performance criterion, one or several suitable algorithms that are expected to perform well for the particular setting. The selection is classically done offline, using openly available information about the problem instance or features that are extracted from the instance during a dedicated feature extraction step. This ignores valuable information that the algorithms accumulate during the optimization process. In this work, we propose an alternative, online algorithm selection scheme which we coin as "per-run" algorithm selection. In our approach, we start the optimization with a default algorithm, and, after a certain number of iterations, extract instance features from the observed trajectory of this initial optimizer to determine whether to switch to another optimizer. We test this approach using the CMA-ES as the default solver, and a portfolio of six different optimizers as potential algorithms to switch to. In contrast to other recent work on online per-run algorithm selection, we warm-start the second optimizer using information accumulated during the first optimization phase. We show that our approach outperforms static per-instance algorithm selection. We also compare two different feature extraction principles, based on exploratory landscape analysis and time series analysis of the internal state variables of the CMA-ES, respectively. We show that a combination of both feature sets provides the most accurate recommendations for our test cases, taken from the BBOB function suite from the COCO platform and the YABBOB suite from the Nevergrad platform.

Keywords: Algorithm selection · Black-box optimization · Exploratory landscape analysis · Evolutionary computation

1 Introduction

It is widely known that optimization problems are present in many areas of science and technology. A particular subset of these problems are the black-box

© The Author(s), under exclusive license to Springer Nature Switzerland AG 2022
G. Rudolph et al. (Eds.): PPSN 2022, LNCS 13398, pp. 46–60, 2022.
https://doi.org/10.1007/978-3-031-14714-2_4

problems, for which a wide range of optimization algorithms has been developed. However, it is not always clear which algorithm is the most suitable one for a particular problem. Selecting which algorithm to use comes with its own cost and challenges, so the choice of an appropriate algorithm poses a meta-optimization problem that has itself become an increasingly important area of study.

Moreover, a user needs to be able to select different algorithms for different *instances* of the same problem, which is a scenario that very well reflects real-world conditions. This **per-instance algorithm selection** most often relies on being able to compute a set of features which capture the relevant properties of the problem instance at hand. A popular approach is the *landscape-aware* algorithm selection, where the problem features' definition stems from the field of *exploratory landscape analysis* (ELA) [25]. In this approach, an initial set of points is sampled and evaluated on the problem instance to identify its global properties. However, this induces a significant overhead cost to the algorithm selection procedure, since the initial sample of points used to extract knowledge from the problem instance is usually not directly used by the chosen algorithm in the subsequent optimization process.

Previous research into landscape-aware algorithm selection suggests that, as opposed to creating a separate set of samples to compute ELA features in a dedicated preprocessing step, one could use the samples observed by some initial optimization algorithm. This way, the algorithm selection changes from being a purely offline procedure into being one which considers whether or not to switch between different algorithms during the search procedure. This is an important step towards dynamic (online) algorithm selection, in which the selector is able to track and adapt the choice of the algorithm throughout the optimization process in an intelligent manner.

In this paper, we coin the term **per-run algorithm selection** to refer to the case where we make use of information gained by running an initial optimization algorithm ($A1$) during a single run to determine which algorithm should be selected for the remainder of the search. This second algorithm ($A2$) can then be warm-started, i.e., initialized appropriately using the knowledge of the first one. The pipeline of the approach is shown in Fig. 1.

Following promising results from [15], in this work we apply our *trajectory-based algorithm selection* approach to a broader set of algorithms and problems. To extract relevant information about the problem instances, we rely on ELA features computed using samples and evaluations observed by the initial algorithm's search trajectory, i.e., *local* landscape features. Intuitively, we consider the problem instance as perceived from the algorithm's viewpoint.

In addition, we make use of an alternative aspect that seems to capture critical information during the search procedure – the algorithm's internal state, quantified through a set of state variables at every iteration of the initial algorithm. To this end, we choose to track their evolution during the search by computing their corresponding *time-series* features.

Using the aforementioned values to characterize problem instances, we build algorithm selection models based on the prediction of the fixed-budget performance of the second solver on those instances, for different budgets of function evaluations. We train and test our algorithm selectors on the well-known BBOB problem collection of the COCO platform [11], and extend the testing on the YABBOB collection of the Nevergrad platform [28]. We show that our approach leads to promising results with respect to the selection accuracy and we also point out interesting observations about the particularities of the approach.

State of the Art. Given an optimization problem, a specific instance of that problem which needs to be solved, and a set of algorithms which can be used to solve it, the so-called per-instance algorithm selection problem arises. How does one determine which of those algorithms can be expected to perform best on that particular instance? In other words, one is not interested in having an algorithm recommendation for a whole problem class (such as TSP or SAT in the discrete domain), but for a specific instance of some problem. A large body of work exists in this line of research [2,5,13,21,23,36]. All of these deal predominantly with offline AS. An effort towards online AS has been recently proposed [24], where the switching rules between algorithms were defined based on non-convex ratio features extracted during the optimization process. However, this particular study is not based on using supervised machine learning techniques to define the switching rule, which is the key difference presented in our approach.

Paper Outline. In Sect. 2, we introduce the problem collections and the algorithm portfolio, and give details about the raw data generation for our experiments. The full experimental pipeline is more closely presented in Sect. 3. We discuss the main results on two benchmark collections in Sects. 4 and 5, respectively. Finally, Sect. 6 gives several possible avenues for future work.

Data and Code Availability. Our source code, raw data, intermediate artefacts and analysis scripts have been made available on our Zenodo repository [17]. In this paper, we highlight only selected results for reasons of space.

2 Data Collection

Problem Instance Portfolio. To implement and verify our proposed approach, we make use of a set of black-box, single-objective, noiseless problems. The data set is the BBOB suite from the COCO platform [11], which is a very common benchmark set within numerical optimization community. This suite consists of a total of 24 functions, and each of these functions can be changed by applying pre-defined transformations to both its domain and objective space, resulting in a set of different instances of each of these problems that share the same global characteristics [10].

Another considered benchmark set is the YABBOB suite from the Nevergrad platform [28], that contains 21 black-box functions, out of which we keep 17 for this paper. By definition, YABBOB problems do not allow for generating different instances.

Algorithm Portfolio. As our algorithm portfolio, we consider the one used in [18,32]. This gives us a set of 5 black-box optimization algorithms: MLSL [19, 31], BFGS [3,8,9,33], PSO [20], DE [34] and CMA-ES [12]. Since for the CMA-ES

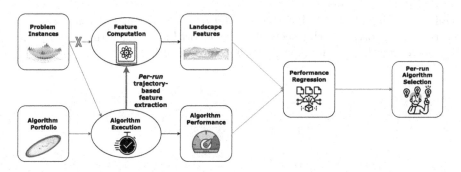

Fig. 1. Per-run algorithm selection pipeline. The overhead cost of computing ELA features per problem instance is circumvented via collecting information about the instance during the default optimization algorithm run.

we consider two versions from the modular CMA-ES framework [29] (elitist and non-elitist), this gives us a total portfolio of 6 algorithm variants. The implementation of the algorithms used can be found in more detail in our repository [17].

Warm-Starting. To ensure we can switch from our initial algorithm (A1) to any of the others (A2), we make use of a basic warm-starting approach specific to each algorithm. For the two versions of modular CMA-ES, we do not need to explicitly warm-start, since we can just continue the run with the same internal parameters and turn on elitist selection if required. The detailed warm-start mechanisms are discussed in [18], and the implementations are available in our repository [17].

Performance Data. For our experiments, we consider a number of data collection settings, based on the combinations of dimensionality of the problem, where we use both 5- and 10-dimensional versions of the benchmark functions, and budget for A1, where we use $30 \cdot D$ budget for the initial algorithm. This is then repeated for all functions of both the BBOB and the YABBOB suite. For BBOB, we collect 100 runs on each of the first 10 instances, resulting in 1 000 runs per function. For YABBOB (only used for testing), we collect 50 runs on each function (as there are no instances in Nevergrad).

We show the performance of the six algorithms in our portfolio in the 5-dimensional case in Fig. 2. Since the A1 budget is $30 \cdot D = 150$, the initial part of the search is the same for all algorithms until this point. In the figure, we can see that, for some functions, clear differences in performance between the algorithm appear very quickly, while for other functions the difference only becomes apparent after some more evaluations are used. This difference leads us to perform our experiments with three budgets for the A2 algorithm, namely $20 \cdot D$, $70 \cdot D$ and $170 \cdot D$.

To highlight the differences between the algorithms for each of these scenarios, we can show in what fraction of runs each algorithm performs best. This is visualized in Fig. 3. From this figure, we can see that while some algorithm are clearly more impactful than others, the differences between them are still

significant. This indicates that there would be a significant difference between
a virtual best solver which selects the best algorithm for each run and a single
best solver which uses only one algorithm for every run.

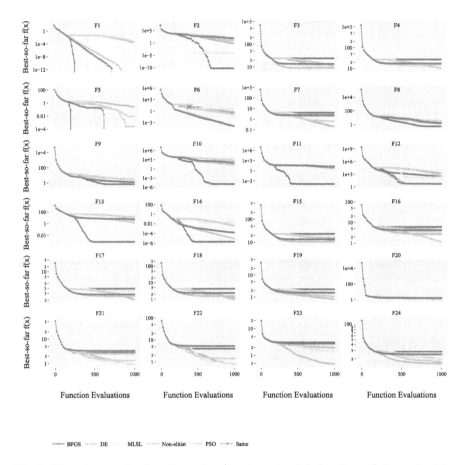

Fig. 2. Mean best-so-far function value (precision to global optimum) for each of the
six algorithms in the portfolio. For computational reasons, each line is calculated based
on a subset of 10 runs on each of the 10 instances used, for a total of 100 runs. Note
that the first 150 evaluations for each algorithm are identical, since this is the budget
used for A1. Figure generated using IOHanalyzer [35].

3   Experimental Setup

Adaptive Exploratory Landscape Analysis. As previously discussed, the
per-run trajectory-based algorithm selection method consists of extracting ELA
features from the search trajectory samples during a single run of the initial
solver. A vector of numerical ELA feature values is assigned to each run on the

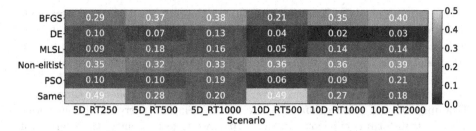

Fig. 3. Matrix showing for each scenario (with respect to the dimensionality and A2 budget) in what proportion of runs each algorithm reaches the best function value. Note that these value per scenario can add to more than 1 because of ties.

problem instance, and can be then used to train a predictive model that maps it to different algorithms' performances on the said run. To this end, we use the ELA computation library named FLACCO [22].

Among over 300 different features (grouped in feature sets) available in FLACCO, we only consider features that do not require additional function evaluations for their computation, also referred to as *cheap features* [1]. They are computed using the fixed initial sample, while *expensive features*, in contrast, need additional sampling during the run, an overhead that makes them more inaccessible for practical use. For the purpose of this work, as suggested in preliminary studies [15,18], we use 38 cheap features most commonly used in the literature, namely those from *y-Distribution, Levelset, Meta-Model, Dispersion, Information Content* and *Nearest-Better Clustering* feature sets.

We perform this per-run feature extraction using the initial $A1 = 30 \cdot D$ budget of samples and their evaluations per each run of each of the first 10 instances of each of the 24 BBOB problems, as well aas 17 YABBOB problems (that have no instances) in $5D$ and $10D$.

Time-Series Features. In addition to ELA features computed during the optimization process, we consider an alternative – *time-series* features of the internal states of the CMA-ES algorithm. Since the internal variables of an algorithm are adapted during the optimization, they could potentially contain useful information about the current state of the optimization. Specifically, we consider the following internal variables: the step size σ, the eigenvalues of covariance matrix $\vec{v}$, the evolution path $\vec{p}_c$ and its conjugate $\vec{p}_\sigma$, the Mahalanobis distances from each search point to the center of the sampling distribution $\vec{\gamma}$, and the log-likelihood of the sampling model $\mathcal{L}\left(\vec{m}, \sigma^2, \mathbf{C}\right)$. We consider these dynamic strategy parameters of the CMA-ES as a multivariate real-valued time series, for which at every iteration of the algorithm, we compute one data point of the time series as follows: $\forall t \in [L]$:

$$\vec{\psi}_t := \left(\sigma, \mathcal{L}(\vec{m}, \sigma^2, \mathbf{C}), ||\vec{v}||, ||\vec{p}_\sigma||, ||\vec{p}_c||, ||\vec{\gamma}||, \text{ave}(\vec{v}), \text{ave}(\vec{p}_\sigma), \text{ave}(\vec{p}_c), \text{ave}(\vec{\gamma})\right)^\top,$$

where L represents the number of iterations these data points were sampled, which equals the A1 budget divided by the population size of the CMA-ES. In order to store information invariant to the problem dimension, we compute the component-wise average ave($\cdot$) and norm $||\vec{x}|| = \sqrt{\vec{x}^{\top}\vec{v}}$ of each vector variable.

Given a set of m feature functions $\{\phi_i\}_{i=1}^{m}$ from TSFRESH (where $\phi_i \colon \mathbb{R}^L \to \mathbb{R}$), we apply each feature function over each variable in the collected time series. Examples of such feature functions are autocorrelation, energy and continuous wavelet transform coefficients. In this paper, we take this entire time series (of length L) as the feature window. We employ all 74 feature functions from the TSFRESH library [4], to compute a total of 9 444 time-series features per run. After the feature generation, we perform a feature selection method using a Random Forests classifier trained to predict the function ID, for computing the feature importance. We then select only the features whose importance is larger than 2×10^{-3}. This selection procedure yields 129 features, among which features computed on the Mahalanobis distance and the step-size σ are dominant. More details on this approach can be found in [26].

Regression Models. To predict the algorithm performance after the A2 budget, we use as performance metric the target precision reached by the algorithm in the fixed-budget context (i.e., after some fixed number of function evaluations). We create a mapping between the input feature data, which can be one of the following: (1) the trajectory-based representation with 38 ELA features per run (ELA-based AS), (2) the trajectory-based representation with 129 time-series (TS) features per run (TS-based AS), or (3) a combination of both (ELA+TS-based AS) and the target precision of different algorithm runs. We then train supervised machine learning (ML) regression models that are able to predict target precision for different algorithms on each of the trajectories involved in the training data. Following some strong insights from [14] and subsequent studies, we aim at predicting the logarithm ($log10$) of the target precision, in order to capture the order of magnitude of the distance from the optimum. In our case, since we are dealing with an algorithm portfolio, we have trained a separate single target regression (STR) model for each algorithm involved in our portfolio. As an ML regression method, we opt for using a random forest (RF), as previous studies have shown that they can provide promising results for automated algorithm performance prediction [16]. To this end, we use the RF implementation from the Python package SCIKIT-LEARN [27].

Evaluation Scenarios. To find the best RF hyperparameters and to evaluate the performance of the algorithm selectors, we have investigated two evaluation scenarios:

(1) Leave-instance out validation: in this scenario, 70% of the instances from each of the 24 BBOB problems are randomly selected for training and 30% are selected for testing. Put differently, all 100 runs for the selected instance will either appear in the train or the test set. We thus end up with 16 800 trajectories used for training and 7 200 trajectories for testing.

(2) Leave-run out validation: in this scenario, 70% of the runs from each BBOB problem instance are randomly selected for training and 30% are selected for testing. Again, we end up with 16 800 trajectories used for training and 7 200 trajectories for testing.

We repeat each evaluation scenario five independent times, in order to analyze the robustness of the results. Each time, the training data set was used to find the best RF hyperparameters, while the test set was used only for evaluation of the algorithm selector.

Table 1. RF hyperparameter names and their corresponding values considered in the grid search.

Hyperparameter	Search space
n_estimators	$[100, 300]$
max_features	$[\text{AUTO}, \text{SQRT}, \text{LOG2}]$
max_depth	$[3, 5, 15, \text{NONE}]$
min_samples_split	$[2, 5, 10]$

Hyperparameter Tuning for the Regression Models. The best hyperparameters are selected for each RF model via grid search for a combination of an algorithm and a fixed A2 budget. The training set for finding the best RF hyperparameters for each combination of algorithm and budget is the same. Four different RF hyperparameters are selected for tuning: (1) *n_estimators*: the number of trees in the random forest; (2) *max_features*: the number of features used for making the best split; (3) *max_depth*: the maximum depth of the trees, and (4) *min_samples_split*: the minimum number of samples required for splitting an internal node in the tree. The search spaces of the hyperparameters for each RF model utilized in our study are presented in Table 1.

Per-run Algorithm Selection. In real-world dynamic AS applications, we rely on the information obtained within the current run of the initial solver on a particular problem instance to make our decision to switch to a better suited algorithm. A randomized component of black-box algorithms comes into play here, as one algorithm's performance can vastly differ from one run to another on the very same problem instance.

We estimate the quality of our algorithm selectors by comparing them to standard baselines, the virtual best solver (VBS) and the single best solver (SBS). As we make a clear distinction between per-run and per-instance perspective, in order to compare we need to suitably aggregate the results. Our baseline is the *per-run VBS*, which is the selector that always chooses the real best algorithm for a particular run on a certain problem (i.e., function) instance. We then define VBS_{iid} and VBS_{fid} as virtual best solvers on instance and problem levels, i.e., selectors that always pick the real best algorithm for a certain instance (across all runs) or a certain problem (across all instances). Last, we define the SBS as the algorithm that is most often the best one across all runs.

For each of these methods, we can define their performance relative to the *per-run VBS* by considering their performance ratio, which is defined on each run as taking the function value achieved by the VBS and dividing it by the value reach by the considered selector. As such, the performance ratio for the *per-run VBS* is 1 by definition, and in $[0,1]$ for each other algorithm selector.

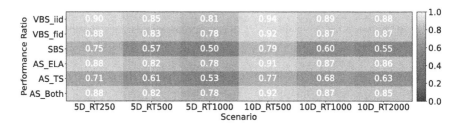

Fig. 4. Heatmap showing for each scenario the average performance ratio relative to the per-run virtual best solver of different versions of VBS, SBS and algorithm selectors (based on the per-instance folds). Scenario names show the problem dimensionality and the total used budget.

To measure the performance ratio for the algorithm selectors themselves, we calculate this performance ratio on every run in the test-set of each of the 5 folds, and average these values. We point out here that the performance of different AS models are not statistically compared, since the obtained performance values from the folds are not independent [6].

4   Evaluation Results: COCO

For our first set of experiments, we train our algorithm selectors on BBOB functions using the evaluation method described in Sect. 3. Since we consider 2 dimensionalities of problems and 3 different A2 budgets, we have a total of 6 scenarios for each of the 3 algorithm selectors (ELA-, TS-, and ELA+TS-based). In Fig. 4, we show the performance ratios of these selectors, as well as the performance ratios of the previously described VBS and SBS baselines. Note that for this figure, we make use of the *instance-run* folds, but results are almost identical for the *per-run* case.

Based on Fig. 4, we can see that the ELA-based algorithm selector performs almost as well as the per-function VBS, which itself shows only minor performance differences to the per-instance VBS. We also notice that as the total evaluation budget increases, the performance of every selector deteriorates. This seems to indicate that as the total budget becomes larger, there are more cases where runs on the same instance have different optimal switches.

To study the performance of the algorithm selectors in more detail, we can consider the performance ratios for each function separately, as is visualized in Fig. 5. From this figure, we can see that for the functions where there is a clearly optimal A2, all algorithm selectors are able to achieve near-optimal performance. However, for the cases where the optimal A2 is more variable, the discrepancy between the ELA and TS-based algorithm selectors increases.

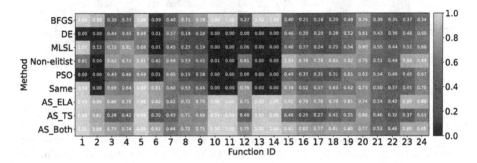

Fig. 5. Heatmap showing for each 5-dimensional BBOB function the mean performance ratio at 500 total evaluations relative to the per-run virtual best solver, as well as the average performance ratio of each of the 3 algorithm selectors.

5   Evaluation Results: Nevergrad

We now study how a model trained on BBOB problem trajectories can be used to predict the performances on trajectories not included in the training. We do so by considering the YABBOB suite from the Nevergrad platform. While there is some overlap between these two problem collections, introducing another sufficiently different validation/test suite allows us to verify the stability of our algorithm selection models. We recall that for the performance data of the same algorithm portfolio on YABBOB functions, we have target precisions for 850 runs, 50 runs per 17 problems, in all considered A2 budgets.

Training on COCO, Testing on Nevergrad. This experiment has resulted in somewhat poorer performance of the algorithm selection models on an inherently different batch of problems. The comparison of the similarity between BBOB and YABBOB problems presented below nicely shows how the YABBOB problems are structurally more similar to one another than to the BBOB ones.

To investigate performance flaws of our approach when testing on Nevergrad, we compare, for each YABBOB problem, how often a particular algorithm is selected by the algorithm selection model trained on the BBOB data with how often that algorithm was actually the best one. This comparison is exhibited in Fig. 6. We observe that MLSL in particular is not selected often enough in the case of a large A2 budget, as well as a somewhat strong preference of the selector towards BFGS. An explanation for these results may be the (dis)similarities

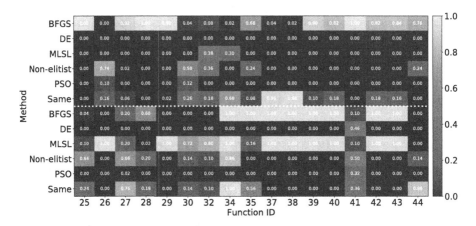

Fig. 6. Heatmap showing for each 5-dimensional YABBOB/Nevergrad function the fraction of times each algorithm was optimal to switch to when considering a total budget of 500 evaluations (bottom) and how often each of these algorithm was selected by the algorithm selector trained on BBOB/COCO (top). Note that the columns of the bottom part can sum to more than 1 in case of ties.

between the benchmarks. Only for some YABBOB functions in the second half of the set we might have similarities in the trajectories already seen from the second half of the BBOB data, but this is anecdotal as the overall tendency is that there are few parallels between BBOB and YABBOB.

Analyzing the Complementarity Between the COCO and Nevergrad Suites. We illustrate the intra-similarity among the YABBOB test trajectories from the Nevergrad suite, which are not part of the training data set. This is shown via correlation between the YABBOB trajectories (test data) and the BBOB trajectories (train data). For this purpose, we first find the subspace that is covered by the training trajectories, where we then project the testing trajectories. To find the subspace that is covered by training data, we apply singular value decomposition (SVD), following an approach presented in [7]. For the training and testing data, we summarize the trajectories on a problem level using the median values for each ELA feature by using all trajectory instances that belong to the same problem. Next, we map the BBOB trajectories to a linear vector space they cover (found by the SVD decomposition), where the trajectories are represented in different uncorrelated dimensions of the data. We then project each of the YABBOB trajectories to the linear subspace that is covered by the 24 BBOB problems, which allows us to find their correlation.

The Pearson correlation values between the trajectories obtained for $5D$ and $10D$ problem instances are showcased in Fig. 7. We opt for the Pearson correlation coefficient since the trajectories are projected in a linear subspace. The trajectories from 1 to 24 correspond to the BBOB suite, and the trajectories starting from 25 to 44 correspond to the YABBOB suite. It is important to recall here that the YABBOB problems F31, F33, F36 and F45 were omitted from further analysis due to missing values. This figure shows that the BBOB trajectories are not correlated (the white square portion of the lower left part of the heatmap), which confirms high diversity in the training trajectory portfolio. However, there are lower positive and negative correlations between BBOB and YABBOB trajectories, which indicate that the properties of the YABBOB trajectories are not captured in the training data. This might be a possible explanation for the poor performance for the algorithm selection models which is trained on the BBOB trajectories, but only tested on the YABBOB trajectories.

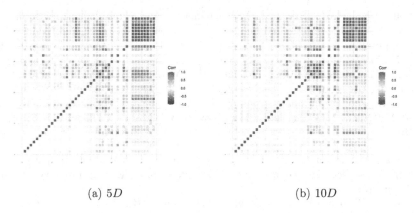

(a) 5D   (b) 10D

Fig. 7. Pearson correlation between BBOB (lower left portion) and YABBOB (bright red upper right corner) trajectories for 5D and 10D.

6 Conclusions and Future Work

We have shown the feasibility of building an algorithm selector based on a very limited amount of samples from an initial optimization algorithm. Results within the BBOB benchmark suite show performance comparable to the per-function virtual best solver when using a selector based on ELA features. While these results did not directly transfer to other benchmark suites, this seems largely caused by the relatively low similarity between the collections.

Since this work is based on warm-starting the algorithms using the information of the initial search trajectory, further improvement in warm-starting would be highly beneficial to the overall performance of this feature-based selection mechanism. In addition, identifying exactly what features contribute to the decisions being made can show us what properties might be important to the performance of the switching algorithm, which in turn can support the development of better warm-starting mechanisms.

While the time-series based approach did not perform as well as the one based on ELA, it still poses an interesting avenue for future research. In particular, it would be worthwhile to consider the combined model in more detail, and aim to identify the level of complementarity between landscape and algorithm state features, which would help gain insight into the complex interplay between problems and algorithms.

Acknowledgment. The authors acknowledge financial support by the Slovenian Research Agency (research core grants No. P2-0103 and P2-0098, project grant No. N2-0239, and young researcher grant No. PR-09773 to AK), by the EC (grant No. 952215 - TAILOR), by the Paris Ile-de-France region, and by the CNRS INS2I institute.

References

1. Belkhir, N., Dréo, J., Savéant, P., Schoenauer, M.: Per instance algorithm configuration of CMA-ES with limited budget. In: Proceedings of Genetic and Evolutionary Computation (GECCO 2017), pp. 681–688. ACM (2017). https://doi.org/10.1145/3071178.3071343

2. Bischl, B., Mersmann, O., Trautmann, H., Preuss, M.: Algorithm selection based on exploratory landscape analysis and cost-sensitive learning. In: Proceedings of Genetic and Evolutionary Computation Conference, GECCO'12. pp. 313–320. ACM (2012). https://doi.org/10.1145/2330163.2330209

3. Broyden, C.G.: The convergence of a class of double-rank minimization algorithms. J. Inst. Math. Appl. **6**, 76–90 (1970)

4. Christ, M., Braun, N., Neuffer, J., Kempa-Liehr, A.W.: tsfresh package for time-series feature engineering. http://tsfresh.readthedocs.io/en/latest/text/listoffeatures.html

5. Cosson, R., Derbel, B., Liefooghe, A., Aguirre, H., Tanaka, K., Zhang, Q.: Decomposition-based multi-objective landscape features and automated algorithm selection. In: Zarges, C., Verel, S. (eds.) EvoCOP 2021. LNCS, vol. 12692, pp. 34–50. Springer, Cham (2021). https://doi.org/10.1007/978-3-030-72904-2_3

6. Demšar, J.: Statistical comparisons of classifiers over multiple data sets. J. Mach. Learn. Res. **7**, 1–30 (2006)

7. Eftimov, T., Popovski, G., Renau, Q., Korošec, P., Doerr, C.: Linear matrix factorization embeddings for single-objective optimization landscapes. In: Proceedings of IEEE Symposium Series on Computational Intelligence (SSCI 2020), pp. 775–782. IEEE (2020). https://doi.org/10.1109/SSCI47803.2020.9308180

8. Fletcher, R.: A new approach to variable metric algorithms. Comput. J. **13**, 317–322 (1970)

9. Goldfarb, D.F.: A family of variable-metric methods derived by variational means. Math. Comput. **24**, 23–26 (1970)

10. Hansen, N., Auger, A., Finck, S., Ros, R.: Real-Parameter Black-Box Optimization Benchmarking: Experimental Setup. RR-7215, INRIA (2010)

11. Hansen, N., Auger, A., Ros, R., Mersmann, O., Tušar, T., Brockhoff, D.: COCO: a platform for comparing continuous optimizers in a black-box setting. Optim. Meth. Software **36**, 1–31 (2020). https://doi.org/10.1080/10556788.2020.1808977

12. Hansen, N., Ostermeier, A.: Completely derandomized self-adaptation in evolution strategies. Evol. Comput. **9**(2), 159–195 (2001)

13. Hutter, F., Kotthoff, L., Vanschoren, J. (eds.): Automated Machine Learning. TSSCML, Springer, Cham (2019). https://doi.org/10.1007/978-3-030-05318-5

14. Jankovic, A., Doerr, C.: Landscape-aware fixed-budget performance regression and algorithm selection for modular CMA-ES variants. In: Proceedings of Genetic and Evolutionary Computation Conference (GECCO 2020), pp. 841–849. ACM (2020). https://doi.org/10.1145/3377930.3390183

15. Jankovic, A., Eftimov, T., Doerr, C.: Towards feature-based performance regression using trajectory data. In: Castillo, P.A., Jiménez Laredo, J.L. (eds.) EvoApplications 2021. LNCS, vol. 12694, pp. 601–617. Springer, Cham (2021). https://doi.org/10.1007/978-3-030-72699-7_38
16. Jankovic, A., Popovski, G., Eftimov, T., Doerr, C.: The impact of hyper-parameter tuning for landscape-aware performance regression and algorithm selection. In: Proceedings of Genetic and Evolutionary Computation Conference (GECCO 2021), pp. 687–696. ACM (2021). https://doi.org/10.1145/3449639.3459406
17. Jankovic, A., et al.: Per-Run Algorithm Selection with Warm-starting using Trajectory-based Features - Data, April 2022. https://doi.org/10.5281/zenodo.6458266
18. Jankovic, A., Vermetten, D., Kostovska, A., de Nobel, J., Eftimov, T., Doerr, C.: Trajectory-based algorithm selection with warm-starting (2022). https://doi.org/10.48550/arxiv.2204.06397
19. Kan, A., Timmer, G.: Stochastic global optimization methods Part II: multi level methods. Math. Program. **39**, 57–78 (1987). https://doi.org/10.1007/BF02592071
20. Kennedy, J., Eberhart, R.: Particle swarm optimization. In: Proceedings of ICNN 1995 - International Conference on Neural Networks, vol. 4, pp. 1942–1948 (1995). https://doi.org/10.1109/ICNN.1995.488968
21. Kerschke, P., Hoos, H.H., Neumann, F., Trautmann, H.: Automated algorithm selection: survey and perspectives. Evol. Comput. **27**(1), 3–45 (2019)
22. Kerschke, P., Trautmann, H.: The R-package FLACCO for exploratory landscape analysis with applications to multi-objective optimization problems. In: CEC, pp. 5262–5269. IEEE (2016). https://doi.org/10.1109/CEC.2016.7748359
23. Lindauer, M., Hoos, H.H., Hutter, F., Schaub, T.: Autofolio: an automatically configured algorithm selector. J. Artif. Intell. Res. **53**, 745–778 (2015). https://doi.org/10.1613/jair.4726
24. Meidani, K., Mirjalili, S., Farimani, A.B.: Online metaheuristic algorithm selection. Exp. Syst. Appl. **201**, 117058 (2022)
25. Mersmann, O., Bischl, B., Trautmann, H., Preuss, M., Weihs, C., Rudolph, G.: Exploratory landscape analysis. In: Proceedings of Genetic and Evolutionary Computation Conference (GECCO 2021), pp. 829–836. ACM (2011). https://doi.org/10.1145/2001576.2001690
26. Nobel, J., Wang, H., Bäck, T.: Explorative data analysis of time series based algorithm features of CMA-ES variants, pp. 510–518, June 2021. https://doi.org/10.1145/3449639.3459399
27. Pedregosa, F., et al.: Scikit-learn: machine learning in Python. JMLR **12**, 2825–2830 (2011)
28. Rapin, J., Teytaud, O.: Nevergrad - A gradient-free optimization platform (2018). http://GitHub.com/FacebookResearch/Nevergrad
29. van Rijn, S.: Modular CMA-ES framework from [30], v0.3.0 (2018). http://github.com/sjvrijn/ModEA. Available also as PyPi package at http://pypi.org/project/ModEA/0.3.0/
30. van Rijn, S., Wang, H., van Leeuwen, M., Bäck, T.: Evolving the structure of evolution strategies. In: Proceedings of IEEE Symposium Series on Computational Intelligence (SSCI 2016), pp. 1–8. IEEE (2016). https://doi.org/10.1109/SSCI.2016.7850138
31. RinnooyKan, A.H.G., Timmer, G.T.: Stochastic global optimization methods. Part 1: clustering methods. Math. Program. **39**(1), 27–56 (1987)

32. Schröder, D., Vermetten, D., Wang, H., Doerr, C., Bäck, T.: Chaining of numerical black-box algorithms: Warm-starting and switching points (2022). https://doi.org/10.48550/arxiv.2204.06539

33. Shanno, D.: Conditioning of quasi-newton methods for function minimization. Math. Comput. **24**, 647–656 (1970)

34. Storn, R., Price, K.: Differential evolution - a simple and efficient heuristic for global optimization over continuous spaces. J. Global Optim. **11**(4), 341–359 (1997). https://doi.org/10.1023/A:1008202821328

35. Wang, H., Vermetten, D., Ye, F., Doerr, C., Bäck, T.: Iohanalyzer: Detailed performance analysis for iterative optimization heuristic. ACM Trans. Evol. Learn. Optim. (2022). https://doi.org/10.1145/3510426, to appear. IOHanalyzer is available at CRAN, on GitHub, and as web-based GUI, see http://iohprofiler.github.io/IOHanalyzer/ for links

36. Xu, L., Hutter, F., Hoos, H., Leyton-Brown, K.: Evaluating component solver contributions to portfolio-based algorithm selectors. In: Cimatti, A., Sebastiani, R. (eds.) SAT 2012. LNCS, vol. 7317, pp. 228–241. Springer, Heidelberg (2012). https://doi.org/10.1007/978-3-642-31612-8_18

Bayesian- and Surrogate-Assisted Optimization

A Systematic Approach to Analyze the Computational Cost of Robustness in Model-Assisted Robust Optimization

Sibghat Ullah[1]([✉]), Hao Wang[1], Stefan Menzel[2], Bernhard Sendhoff[2], and Thomas Bäck[1]

[1] Leiden Institute of Advanced Computer Science (LIACS), Leiden University, Leiden, The Netherlands
{s.ullah,h.wang,t.h.w.baeck}@liacs.leidenuniv.nl
[2] Honda Research Institute Europe GmbH (HRI-EU), Offenbach/Main, Germany
{stefan.menzel,bernhard.sendhoff}@honda-ri.de

Abstract. Real-world optimization scenarios under uncertainty and noise are typically handled with robust optimization techniques, which reformulate the original optimization problem into a robust counterpart, e.g., by taking an average of the function values over different perturbations to a specific input. Solving the robust counterpart instead of the original problem can significantly increase the associated computational cost, which is often overlooked in the literature to the best of our knowledge. Such an extra cost brought by robust optimization might depend on the problem landscape, the dimensionality, the severity of the uncertainty, and the formulation of the robust counterpart. This paper targets an empirical approach that evaluates and compares the computational cost brought by different robustness formulations in Kriging-based optimization on a wide combination (300 test cases) of problems, uncertainty levels, and dimensions. We mainly focus on the CPU time taken to find the robust solutions, and choose five commonly-applied robustness formulations: "mini-max robustness", "mini-max regret robustness", "expectation-based robustness", "dispersion-based robustness", and "composite robustness" respectively. We assess the empirical performance of these robustness formulations in terms of a fixed budget and a fixed target analysis, from which we find that "mini-max robustness" is the most practical formulation w.r.t. the associated computational cost.

Keywords: Optimization under uncertainty · Robust optimization · Surrogate-assisted optimization · Kriging

1 Introduction

Solving a real-world optimization problem entails dealing with uncertainties and noise within the system [2,9,16]. Due to various reasons, various types of uncer-

This research has received funding from the European Union's Horizon 2020 research and innovation programme under grant agreement number 766186.

tainties and noise can emerge in optimization problems, e.g., uncertainty and noise in the output of the system if the system is non-deterministic in nature. Hence, for practical scenarios, optimization methods are needed which can deal with these uncertainties, and solutions have to be found which take into account the impact of the unexpected drifts and changes in the optimization setup. The practice of optimization that accounts for uncertainties and noise is often referred to as Robust Optimization (RO) [2,13,14].

Despite its significance, achieving robustness in modern engineering applications is quite challenging due to several reasons [4,8]. One of the major reasons is the computational cost involved to find the robust solution. The computational cost mainly depends on the problem landscape, high dimensionality [15], the type and structure of the uncertainty [3], and the robustness formulation (RF) or criterion among others.

While the impact of high dimensionality and problem landscape in RO is understood to some extent [3,4,9,15], the impact of the chosen RF, e.g., "mini-max robustness", on the computational cost, has been overlooked in the literature. For expensive-to-evaluate black-box problems, the chosen robustness criterion can have a significant impact on the computational cost. This is due to the fact that obtaining a robust solution requires additional computational resources as opposed to a nominal one, since the optimizer has to take into account the impact of uncertainty and noise as well. This need for additional computational resources is based on the robustness criterion[1] chosen, e.g., RO based on the "mini-max" principle requires an internal optimization loop to compute the worst impact of the uncertainty, whereas RO based on the "expectation" of a noisy function requires computing an integral [4,16].

Since the Computational Cost of Robustness (CCoR) – the need for additional computational resources to find the robust instead of a nominal optimal solution – depends on the robustness criterion chosen, it is desirable to evaluate and compare commonly-employed robustness criteria with regards to computational cost, where the computational cost is based on the CPU time taken to find the robust solution. By evaluating and comparing different robustness criteria based on computational cost and quality of the solution, a novel perspective is provided to practitioners for choosing the suitable robustness criterion for practical scenarios. To the best of our knowledge, there are no systematic studies dealing with this issue so far.

Our contribution in this paper is as following. First, we generalize the Kriging-based RO proposed in [11] (for the "mini-max robustness") to other RFs. Second, we evaluate and compare the empirical performance of Kriging-based RO based on five of the most common RFs, namely "mini-max robustness", "mini-max regret robustness", "expectation-based robustness", "dispersion-

---

[1] An underlying assumption in this study is the non-existence of hard constraints on the choice of RF. In some practical scenarios of RO, this assumption is not valid. Note, however, that, there are plenty of situations where the assumption is valid, and the lack of information on the computational aspects of the RFs makes it harder for practitioners to choose a suitable robustness criterion.

based robustness", and "composite robustness" respectively. Note that the performance assessment is based on 300 test cases, owing to the combinations of ten well-known benchmark problems from the continuous optimization domain, two noise levels, three different settings of dimensionality, and five RFs reported. Additionally, the performance in this context is characterized by a fixed budget and a fixed target analysis, as well as the analysis on the average and maximum of CPU time. Based on the findings of our investigation, we provide a novel perspective on the computational aspects of these RFs, which is useful when employing these RFs in practical scenarios.

The remainder of this paper is organized as follows. Section 2 describes the basic working mechanism of Kriging-based optimization, and introduces the RFs mentioned above. Section 3 extends the nominal Kriging-based optimization to the robust scenario to account for parametric uncertainties in the search variables. Section 4 describes the experimental setup of our study. This is followed by our experimental results in Sect. 5. The discussion on these results is presented in Sect. 6. Finally, we discuss the conclusions of the paper along side the potential future research in Sect. 7.

2   Background

In this paper, we aim to minimize an unconstrained numerical black-box optimization problem, i.e., $f: \mathcal{S} \subseteq \mathbb{R}^D \to \mathbb{R}$, using Kriging-based optimization (KBO) [7,11]. KBO works on the principle of adaptive sampling, whereby the Kriging model is sequentially updated according to a sampling infill criterion, such as the "expected improvement" (EI) criterion. The sampling infill criterion tries to balance the search behavior - exploration and exploitation - to find a globally optimal solution on the model surface.

KBO starts by generating an initial data set $\mathcal{D} = (X, \mathbf{y})$ on the objective function f. The locations: $X = \{\mathbf{x}_1, \mathbf{x}_2, \ldots, \mathbf{x}_N\}$, can be determined by the Design of Experiment (DoE) methodologies, such as the Latin Hyper-cube Sampling (LHS) scheme [12]. After this, function responses: $\mathbf{y} = [f(\mathbf{x}_1), f(\mathbf{x}_2), \ldots, f(\mathbf{x}_N)]^\top$, are computed on these locations. The next step involves constructing the Kriging model $\mathcal{K}_f$ based on the available data set $\mathcal{D}$. Following this, the next query point $\mathbf{x}_{new}$ (to sample the function) is determined with the help of a sampling infill criterion, such as the EI criterion. The function response $f(\mathbf{x}_{new})$ is computed at this location, and the data set is extended. Finally, the Kriging model $\mathcal{K}_f$ is updated based on the extended data set. This process is repeated until either a satisfactory solution is obtained, or a predetermined computational budget or other termination criterion is reached.

When optimizing the function f in real-world applications, we note that it is surrounded by the parametric uncertainties in the decision variables [13,14]. These uncertainties, commonly denoted as $\Delta_\mathbf{x}$, are assumed to be structurally symmetric, additive in nature, and can be modelled in a deterministic or a probabilistic fashion [9]. For the objective function f, the notion of robustness refers to the quality of the solution with respect to these uncertainties. In the following, we define robustness with respect to the five RFs discussed in our paper.

2.1 Robustness Formulations

We start with the so-called "mini-max robustness" (MMR), which deals with deterministic uncertainty [2,9,16]. Given a real-parameter objective function: $f(\mathbf{x})$, and the additive uncertainty in the decision variables: $\Delta_{\mathbf{x}}$, the "mini-max" treatment considers minimizing the worst-case scenario of each search point $\mathbf{x}$, where the worst-case is defined as to taking account all possible perturbations to $\mathbf{x}$, which are restricted in a compact set $U \subseteq \mathbb{R}^D$ (containing a neighborhood of $\mathbf{x}$). Effectively, this is to minimize the following objective function

$$f_{\text{eff}}(\mathbf{x}) = \max_{\Delta_{\mathbf{x}} \in U} f(\mathbf{x} + \Delta_{\mathbf{x}}). \tag{1}$$

Note that the radius of the compact set U is based on the anticipated scale of the uncertainty, i.e., based on the maximum anticipated deviation of the decision variables from their nominal values. The worst-case scenario refers to the fact that we consider the maximal f-value under additive uncertainty at each search point, and try to minimize that [16].

As opposed to MMR, the "mini-max Regret Robustness" (MMRR) [8] focuses on minimizing the maximum regret under uncertainty. The regret can be defined as the difference between the best obtainable value of the function f^* for an uncertainty event $\Delta_{\mathbf{x}}$, and the actual function value under that uncertainty event $f(\mathbf{x} + \Delta_{\mathbf{x}})$. The best obtainable response f^* of the function under an uncertainty event $\Delta_{\mathbf{x}}$ can be defined as

$$f^*(\Delta_{\mathbf{x}}) = \min_{\mathbf{x} \in \mathcal{S}} f(\mathbf{x} + \Delta_{\mathbf{x}}), \tag{2}$$

and the effective (robust) objective function can be defined as

$$f_{\text{eff}}(\mathbf{x}) = \max_{\Delta_{\mathbf{x}} \in U} (f(\mathbf{x} + \Delta_{\mathbf{x}}) - f^*(\Delta_{\mathbf{x}})). \tag{3}$$

Minimizing Eq. (3) refers to the fact that firstly, the best achievable response value for each uncertainty event $\Delta_{\mathbf{x}} \in U$ is subtracted from the actual outcome $f(\mathbf{x} + \Delta_{\mathbf{x}})$. Then, the worst-case is determined similar to the MMR. As a conclusion, the optimal solution is identified as the one for which the worst-case has a minimal deviation from f^* as defined in Eq. (2). The benefit of employing MMRR is that even in the worst-case scenario, we do not compromise significantly in terms of optimality. The biggest challenge, however, is the prohibitively high computational demand. Note that solving Eq. (3) inside an iterative optimization framework, e.g., Kriging-based optimization, implies a quadrupled nested loop, which is computationally infeasible even for a modest setting of dimensionality.

Different from the first two RFs, the expected output of a noisy function can also serve as a robustness criterion [4,8,9]. The focus of this robustness criterion is the overall good performance rather than the minimal deviation of the optimal solution under uncertainty. Note, however, that, this RF requires the uncertainty to be defined in a probabilistic manner. The uncertainty can be modelled according to a continuous uniform probability distribution if no prior information is available. The effective counterpart of the original function based on the "expectation-based robustness" (EBR) is defined as

$$f_{\text{eff}}(\mathbf{x}) = \mathbb{E}_{\Delta_{\mathbf{x}} \sim \mathcal{U}(a,b)}[f(\mathbf{x} + \Delta_{\mathbf{x}})], \tag{4}$$

where the bounds a and b can be set according to the anticipated scale of the uncertainty.

As opposed to EBR, the "dispersion-based robustness" (DBR) emphasizes on minimizing the performance variance under variation of the uncertain search variables [8,9]). In this case, the original objective function $f(\mathbf{x})$ can be remodelled into a robust objective function $f_{\text{eff}}(\mathbf{x})$ by minimizing the variance as

$$f_{\text{eff}}(\mathbf{x}) = \sqrt{\text{Var}_{\Delta_{\mathbf{x}} \sim \mathcal{U}(\text{a,b})}[f(\mathbf{x} + \Delta_{\mathbf{x}})]}. \tag{5}$$

Note that this RF also requires the uncertainty to be defined in a probabilistic manner, similar to the previous case.

Different from the robustness criteria mentioned above, practitioners may also optimize the expected output of a noisy function while minimizing the dispersion simultaneously. We refer to this formulation as the "composite robustness" (CR), similar to [16]. CR requires the uncertainty to be specified in the form of a probability distribution. The expectation and dispersion of the noisy function are combined at each search point $\mathbf{x}$ in $\mathcal{S}$ to produce a robust scalar output. The optimization goal thus becomes to find a point $\mathbf{x}^*$ in $\mathcal{S}$, which minimizes this scalar

$$f_{\text{eff}}(\mathbf{x}) := \mathbb{E}_{\Delta_{\mathbf{x}} \sim \mathcal{U}(a,b)}[f(\mathbf{x} + \Delta_{\mathbf{x}})] + \sqrt{\text{Var}_{\Delta_{\mathbf{x}} \sim \mathcal{U}(a,b)}[f(\mathbf{x} + \Delta_{\mathbf{x}})]}. \tag{6}$$

3 Kriging-Based Robust Optimization

When aiming to find a solution based on the RFs discussed above, we note that the standard KBO approach as described in Sect. 2 cannot be employed. There are mainly two reasons for that. Firstly, the potential "improvement" which is defined in the nominal scenario renders inapplicable. This is due to the fact that this improvement is defined with respect to the "best-so-far" observed value of the function: $f_{\min}$, which has no clear meaning and usage when aiming for a robust solution. Rather, in the case of RO, the improvement must be defined with respect to the current best known "robust" value of the function: $\hat{f}^*(\mathbf{x})$, which by implication can only be estimated on the Kriging surface (as opposed to observed or fully known in the nominal case). Secondly, the Kriging surrogate $\mathcal{K}_f$ in the nominal scenario does not model the robust/effective response of the function[2], which is desirable when aiming for a robust solution. Therefore, the standard KBO approach must be extended to the robust scenario, which is henceforth referred to as Kriging-based Robust Optimization (KB-RO) in this paper.

---

[2] The robust or effective function response has already been defined in Sect. 2 for five of the most common RFs.

Following the approach in [11], the adaptation of the KBO algorithm to KB-RO is done in the following manner. Firstly, one must substitute the "best-so-far" observed value of the function: $f_{\min}$, with its robust Kriging counterpart: $\hat{f}^*(\mathbf{x})$, which is defined as: $\hat{f}^*(\mathbf{x}) = \min_{\mathbf{x} \in \mathcal{S}} \hat{f}_{\mathrm{eff}}(\mathbf{x})$. Note that $\hat{f}_{\mathrm{eff}}(\mathbf{x})$ is the approximation of the true robust response of the function: $f_{\mathrm{eff}}(\mathbf{x})$. In the context of deterministic uncertainty – MMR and MMRR, this estimation merely refers to the substitution of true function responses with their Kriging predictions in Eqs. (1)–(3). On the other hand, in the context of probabilistic uncertainty – EBR, DBR, and CR, it also encompasses the monte-carlo approximations for the corresponding statistical quantities of interests, e.g., in Eq. (4), $\hat{f}_{\mathrm{eff}}(\mathbf{x})$ is approximated with monte-carlo samples based on the Kriging prediction at each search point $\mathbf{x} + \Delta_{\mathbf{x}}$.

We model the robust Kriging response of the function using a normally distributed random variable: $Y_{\mathrm{eff}}(\mathbf{x})$, with mean $\hat{f}_{\mathrm{eff}}(\mathbf{x})$ and variance $s_{\mathrm{eff}}^2(\mathbf{x})$, i.e., $Y_{\mathrm{eff}}(\mathbf{x}) \sim \mathcal{N}(\hat{f}_{\mathrm{eff}}(\mathbf{x}), s_{\mathrm{eff}}^2(\mathbf{x}))$. Note that the assumption that $Y_{\mathrm{eff}}(\mathbf{x})$ is normally distributed is not entirely rigorous, but rather a practical compromise [11]. Ideally, we should have attempted to estimate the actual posterior distribution of the robust Kriging response of the function: $\hat{f}_{\mathrm{eff}}(\mathbf{x})$, which would require additional assumptions on the joint distribution of all search points. However, the computational costs of finding this generally non-Gaussian distribution several times on the original Kriging surface $\mathcal{K}_f$ are prohibitively high. Additionally, numerically computing the integral for the expectation of the improvement for this generally non-Gaussian distribution would also be computationally expensive. To add to that, we note that the Kriging surface $\mathcal{K}_f$ only ever provides an approximation, and hence the true distribution of the robust response of the function for each RF can never be described with certainty in KB-RO.

Modelling the true robust response of the function with a normally distributed random variable: $Y_{\mathrm{eff}}(\mathbf{x})$, we note that in the context of deterministic uncertainty, the value $s_{\mathrm{eff}}^2(\mathbf{x})$ merely refers to the Kriging mean squared error at point $\mathbf{x} + \Delta_{\mathbf{x}}^*$, where $\Delta_{\mathbf{x}}^*$ indicates the worst setting of the uncertainty – which maximizes Eq. (1) or (3) as the case may be. In the context of EBR, $s_{\mathrm{eff}}^2(\mathbf{x})$ has a closed form expression as: $s_{\mathrm{eff}}^2 = \frac{1}{J^2} \sum_{i,j}^{J} \mathcal{C}$, where $\mathcal{C}$ is a co-variance matrix with elements $C(\mathbf{x}_i', \mathbf{x}_j')$. The entries $C(\mathbf{x}_i', \mathbf{x}_j')$ in the matrix $\mathcal{C}$ are computed with the help of posterior Kernel (with optimized hyper-parameters), and the point $\mathbf{x}_j'$ is defined as: $\mathbf{x}_j' = \mathbf{x} + \Delta_{\mathbf{x}}^j$, where $\Delta_{\mathbf{x}}^j$ indicates the j-th sample for $\Delta_{\mathbf{x}}$. In the context of DBR and CR, $s_{\mathrm{eff}}^2(\mathbf{x})$ does not have a closed form expression, and should be computed numerically.

After substituting the "best-so-far" observed value of the function: $f_{\min}$, with its robust Kriging counterpart: $\hat{f}^*(\mathbf{x})$, and modelling the true robust response of the function with a normally distributed random variable: $Y_{\mathrm{eff}}(\mathbf{x}) \sim \mathcal{N}(\hat{f}_{\mathrm{eff}}(\mathbf{x}), s_{\mathrm{eff}}^2(\mathbf{x}))$, we can define the improvement and its expectation in the robust scenario as

$$\mathcal{I}_{\mathrm{eff}}(\mathbf{x}) = \max\{0, \hat{f}^*(\mathbf{x}) - Y_{\mathrm{eff}}(\mathbf{x})\}, \tag{7}$$

Algorithm 1: Kriging-based Robust Optimization

1: **procedure** $(f, \mathcal{S}, \mathscr{A}_{\text{eff}}, \Delta_{\mathbf{x}})$ ▷ f: objective function, $\mathcal{S}$: search space, $\mathscr{A}_{\text{eff}}$: robust acquisition function, $\Delta_{\mathbf{x}}$: uncertainty in the search variables

2:     Generate the initial data set $\mathcal{D} = \{X, \mathbf{y}\}$ on the objective function.

3:     Construct the Kriging model $\mathcal{K}_f$ on $\mathcal{D} = \{X, \mathbf{y}\}$.

4:     **while** the stop criteria are not fulfilled **do**

5:         Find robust optimum on the Kriging surface $\mathcal{K}_f$
$$\hat{f}^*(\mathbf{x}) = \min_{\mathbf{x} \in \mathcal{S}} \ \hat{f}_{\text{eff}}(\mathbf{x}).$$

6:         Choose a new sample $\mathbf{x}_{\text{new}}$ by maximizing the robust (effective) acquisition function
$$\mathbf{x}_{\text{new}} \leftarrow \operatorname{argmax}_{\mathbf{x} \in \mathcal{S}} \ \mathscr{A}_{\text{eff}}(\mathbf{x}).$$

7:         Compute function response $f(\mathbf{x}_{\text{new}})$.

8:         Extend the data set $\mathcal{D}$ by appending the pair $(\mathbf{x}_{\text{new}}, f(\mathbf{x}_{\text{new}}))$ to $\mathcal{D} = \{X, \mathbf{y}\}$.

9:         Reconstruct the Kriging model $\mathcal{K}_f$ on $\mathcal{D} = \{X, \mathbf{y}\}$.

10:     **end while**

11: **end procedure**

---

and

$$\mathbb{E}[\mathcal{I}_{\text{eff}}(\mathbf{x})] := (\hat{f}^*(\mathbf{x}) - \hat{f}_{\text{eff}}(\mathbf{x}))\Phi\left(\frac{\hat{f}^*(\mathbf{x}) - \hat{f}_{\text{eff}}(\mathbf{x})}{s_{\text{eff}}(\mathbf{x})}\right) + s_{\text{eff}}(\mathbf{x})\phi\left(\frac{\hat{f}^*(\mathbf{x}) - \hat{f}_{\text{eff}}(\mathbf{x})}{s_{\text{eff}}(\mathbf{x})}\right), \quad (8)$$

where $\Phi(\cdot)$ and $\phi(\cdot)$ in Eq. (8) represent the cumulative distribution function and probability density function of the standard normal random variable respectively. An important thing to note here is that in the context of MMR and MMRR, the point $\mathbf{x} + \Delta_{\mathbf{x}}$ can become infeasible with respect to the original search space $\mathcal{S}$ if $\mathbf{x}$ is already close to the boundary of $\mathcal{S}$. To address this issue, we compute the robust optimum within a restricted search space $\mathcal{S}'$ of the original domain to avoid extrapolation [17]. The working mechanism of KB-RO is summarized in Algorithm 1, where the only significant difference to the nominal KBO is the evaluation of steps 5 and 6, which emphasize on robustness.

4   Experimental Setup

Our aim in this paper is to understand the impact of RF in KB-RO with regards to computational efficiency. Intuitively, RF can have a significant impact on the performance of KB-RO since steps 5 and 6 in Algorithm 1 require much more computational resources as opposed to the nominal KBO [7]. This need for additional computational resources is based on the chosen RF. Through our experimental setup[3], we aim to better understand this impact for each of the five RFs discussed in the paper. To make our setup comprehensive, we take into account the variability in external factors such as problem landscape, dimensionality, and the scale/severity of the uncertainty.

---

[3] The source code to reproduce the experimental setup and results is available at: https://github.com/SibghatUllah13/UllahPPSN2022.

For our study, we select ten unconstrained, noiseless, single-objective optimization problems from the continuous benchmark function test-bed known as "Black-Box-Optimization-Benchmarking" (BBOB) [6]. Note that BBOB provides a total of twenty four such functions divided in five different categories, namely "Separable Functions", "Functions with low or moderate conditioning", "Functions with high conditioning and unimodal", "Multi-modal functions with adequate global structure", and "Multi-modal functions with weak global structure" respectively. We select two functions from each of these categories to cover a broad spectrum of test cases. The set of selected test functions is given as: $\mathscr{F} = \{f_2, f_3, f_7, f_9, f_{10}, f_{13}, f_{15}, f_{16}, f_{20}, f_{24}\}$. An important thing to note is that each of the test functions in $\mathscr{F}$ is subject to minimization, and is evaluated on three different settings of dimensionality as: $\mathscr{D} = \{2, 5, 10\}$. Apart from the test functions and dimensionality, we also vary the uncertainty level based on two distinct settings as: $\mathscr{L} = \{0.05, 0.1\}$, which indicate the maximum % deviation in the nominal values of the search variables.

For the deterministic setting of the uncertainty, i.e., MMR and MMRR, the compact set U is defined as: $U = [-(L \times R), (L \times R)]$, where $L \in \mathscr{L}$ denotes the choice of the uncertainty level, and R serves as the absolute range of the search variables. For the test functions in $\mathscr{F}$, the absolute range of the search variables is 10, since all test functions are defined from -5 to 5. For the probabilistic setting of the uncertainty, i.e., EBR, DBR and CR, the uncertainty is modelled according to a continuous uniform probability distribution: $\Delta_{\mathbf{x}} \sim \mathcal{U}(a, b)$, where the boundaries a and b are defined similar to the boundaries of the set U in the deterministic case. In our study, the size of the initial training data is set to be $2 \times D$, where $D \in \mathscr{D}$ denotes the corresponding setting of the dimensionality. Likewise, the maximum number of iterations for KB-RO is set to be $50 \times D$. The computational budget for each of the nested (internal) loop is set to be $10 \times D$, whereas the number of samples for the probabilistic setting of the uncertainty is set to be $100 \times D$. Note that our Kriging surrogate is based on the popular Matérn $3/2$ kernel [5], and we standardize the function responses: $\mathbf{y} = [f(\mathbf{x}_1), f(\mathbf{x}_2), \ldots, f(\mathbf{x}_N)]^\top$, before constructing the Kriging surrogate $\mathcal{K}_f$. In addition, we utilize the robust EI in Eq. (8) as the sampling infill criterion for our experiments.

For the parallel execution of KB-RO for each of the 300 test cases considered, we utilize the Distributed ASCI Supercomputer 5 (DAS-5) [1], where each standard node has a dual 8-core 2.4 GHz (Intel Haswell E5-2630-v3) CPU configuration and 64 GB memory. We implement our experiments in python 3.7.0 with the help of "scikit-learn" module [10]. The performance assessment of the solutions in our experiments is based on 15 independent runs $\mathscr{R}$ of the KB-RO for each of the 300 test cases considered. Note that for each trial, i.e., the unique combination of the independent run and the test case, we ensure the same configuration of hardware and software to account for fairness. Furthermore, in each trial, we measure the CPU time for all iterations of KB-RO. After the successful parallel execution of all trials, we assess the quality of the optimal solutions based on Relative Mean Absolute Error (RMAE) as: $\text{RMAE} = \left(\frac{|f' - \hat{f}'|}{|f'|} \right)$, where f' denotes the true robust optimal function value obtained from solving the original function under uncertainty with the help of a benchmark numerical optimization algorithm (without the use of a surrogate model), also referred to as the ground truth for the particular choice of the test case, and $\hat{f}'$ serves as the robust optimal function value obtained from KB-RO (step 5 in Algorithm 1). As noted in [16], the benefit of utilizing RMAE is that the quality of the optimal solution is always determined relative to the corresponding ground truth, and the performance of KB-RO across different RFs can be easily compared.

Having obtained the quality of the optimal solution and CPU time for all iterations of the KB-RO for each trial, we perform a fixed budget and a fixed target analysis. For the fixed budget analysis, we consider two possibilities. Firstly, we perform the analysis with respect to the running CPU time by fixing 50 different settings of the CPU time. For each such setting, we report the best quality solution (measured in terms of RMAE) obtained from KB-RO. The performance in this context is averaged over all 50 settings of the CPU time. Secondly, we perform the fixed budget analysis also with respect to the number of iterations. In this context, we identify 30 different settings of the number of iterations (checkpoints) to analyze the performance similar to the previous case.

Contrary to this, in fixed target analysis, we identify 10 distinct target values for the RMAE – a set of thresholds describing the minimum desirable quality of the solution. As soon as a particular target is achieved, we report the accumulated CPU time taken by KB-RO to reach that target. If such a target is never achieved, we report the penalized CPU time which is set to be $D \times T_{\max}$, where $D \in \mathcal{D}$ is the corresponding setting of the dimensionality, and $T_{\max}$ is the accumulated CPU time at the last iteration of that trial. In addition to the fixed budget and fixed target analysis, we also report the average CPU time per iteration (ACTPI), and $T_{\max}$ for each trial.

5 Results

We share the results originating from our experiments in Figs. 1 and 2. In particular, Fig. 1 focuses on four distinct analyses, which include fixed CPU time analysis, fixed iterations analysis, fixed target analysis, and the analysis on the ACTPI. On the other hand, Fig. 2 reports the accumulated CPU time at the last iteration of KB-RO: $T_{\max}$, which is averaged over 15 independent runs $\mathcal{R}$, and grouped by the RFs. Note that the results in Fig. 1 are presented in the form of empirical cumulative distribution function (ECDF) for each RF and for each type of analysis. The first row of plots in Fig. 1 illustrates the results on fixed CPU time and fixed iteration analyses respectively (from left to right). In a similar fashion, the second row of plots illustrates the performance with respect to the fixed target analysis and the analysis on the ACTPI. Note that each curve in these analyses is based on 900 data points (trials) due to 15 independent runs $\mathcal{R}$ of KB-RO, 10 test functions in $\mathcal{F}$, 3 settings of dimensionality in $\mathcal{D}$, and 2 noise levels in $\mathcal{L}$. The results in Fig. 2 are presented in the form of box plots, where each box inside a subplot presents $T_{\max}$ values for the test cases corresponding to the particular setting of the dimensionality and RF. The reported $T_{\max}$ in this context is averaged over 15 independent runs $\mathcal{R}$ of KB-RO.

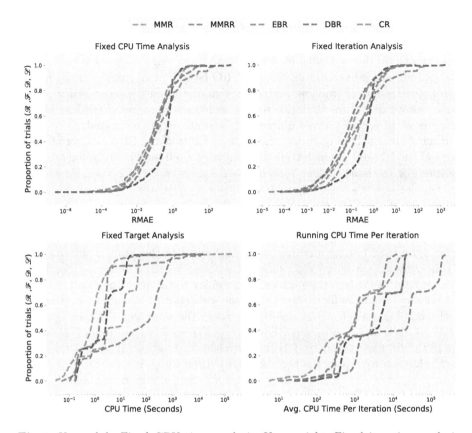

Fig. 1. Upper left: Fixed CPU time analysis, Upper right: Fixed iteration analysis, Lower left: Fixed target analysis, Lower right: Average running CPU time per iteration. For each analysis, the empirical cumulative distribution function (ECDF) for all five RFs is presented. Each ECDF curve is based on 900 data points (trials) owing to 15 independent runs $\mathscr{R}$, 10 test functions in $\mathscr{F}$, 3 settings of dimensionality in $\mathscr{D}$, and 2 noise levels in $\mathscr{L}$.

In terms of performance comparison with respect to the fixed CPU time analysis, we note the promising nature of all RFs except DBR, which performs poorly compared to its competitors in most trials. Furthermore, we also note the highest variation in quality (RMAE) for DBR. Although no RF is deemed a clear winner for this analysis, we note that MMR, MMRR and CR have high empirical success rates. Likewise, we note the highest variation in quality for DBR also in the context of fixed iteration analysis. In this case, MMRR and CR perform superior to the other RFs as we observe a high empirical success rate for both. For the performance measure with respect to the fixed target analysis, we observe that MMR outperforms the competitors, albeit the variation in the running CPU time for MMR is also deemed higher. Here, we note a clear distinction in the empirical success rate between MMRR and other RFs. For instance, if we cut-off

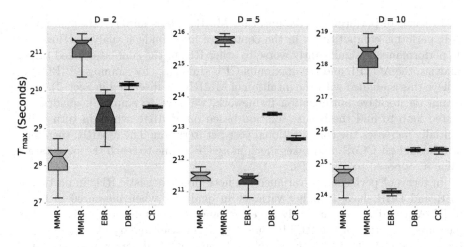

Fig. 2. Left: $2D$ test cases, Middle: $5D$ test cases, Right: $10D$ test cases. Each box plot shares the maximum CPU time spent to run KB-RO: $T_{\max}$, averaged over 15 independent runs in $\mathscr{R}$ and grouped by the RFs.

the running CPU time at 100 s, we observe that MMRR has an empirical success rate of 45 %, whereas MMR, DBR, and CR achieve a success rate of more than 99%. We also note that MMRR has the highest variance of ACTPI, and the lowest empirical success rate of all RFs. In this case, none of the MMR and EBR can be deemed a clear winner, although both perform superior to other RFs in most trials. When comparing the performance of RFs in the context of maximum CPU time spent: $T_{\max}$, we note that MMR and EBR in general, perform superior to other RFs, whereas MMRR performs the worst for each setting of dimensionality. Furthermore, we deem that $T_{\max}$ increases rapidly with respect to dimensionality in the context of deterministic uncertainty, i.e., MMR and MMRR, when compared with the probabilistic uncertainty, i.e., EBR, DBR, and CR. Lastly, we note that in general, the variance in $T_{\max}$ for the probabilistic setting is also significantly lower when compared to the deterministic case.

6 Discussion

Based on the observations from the fixed budget analyses, we deem MMR, MMRR, EBR and CR to be suitable RFs regarding the computational cost involved in finding the robust solution. This validates their applicability in practical scenarios where the computational resources are limited, and the designer cannot spend more than a fixed amount of computational budget (whether measured in terms of CPU time or the number of iterations). Note that MMR appears to be the most promising RF also in the scenarios where the designer aims for a fixed quality solution – where the designer cannot compromise on the quality below a certain threshold. In those situations, MMR can yield the desired quality robust solution with considerably less CPU time.

In terms of performance, we find that MMRR poses an interesting situation as it performs competitively in the context of fixed budget analyses. However, its performance is significantly worse to other RFs in the context of fixed target analysis, the ACTPI, and the maximum CPU time T_{max} for running KB-RO. We believe this is aligned with the intuition of MMRR (as discussed in Sect. 2), since within an iterative optimization framework, we have to employ a quadrupled nested loop to find the robust solution based on MMRR, which in turn exponentially increases the computational cost per iteration. The MMRR, therefore, has the highest CCoR, and takes much more CPU time to reach the same target value as opposed to other RFs.

In terms of performance variance, we note that stochastic RFs, in particular EBR and DBR, have a higher variance in quality – when measured in terms of RMAE, and a comparatively lower variance in computational cost – when measured in terms of the ACTPI and T_{max}. This can mainly be attributed to their intrinsic stochastic nature as they rely on numerical approximations. Since the sample size of the numerical approximations is fixed with respect to the corresponding setting of the dimensionality, we observe relatively lower variance in the CPU time. However, since we only ever approximate the robust response of the function, the quality of the solution may be deteriorated.

7   Conclusion and Outlook

This paper analyzes the computational cost of robustness in Kriging-based robust optimization for five of the most commonly employed robustness criteria. In a first approach of such kind, we attempt to evaluate and compare the robustness formulations with regards to the associated computational cost, where the computational cost is based on the CPU time taken to find the optimal solution under uncertainty. Our experimental setup constitutes 300 test cases, which are evaluated for 15 independent runs of Kriging-based robust optimization. A fixed budget analysis of our experimental results suggests the applicability of "minimax robustness", "mini-max regret robustness", "expectation-based robustness", and "composite robustness" in practical scenarios where the designer cannot afford the computational budget beyond a certain threshold. On the other hand, a fixed target analysis deems the 'mini-max robustness' as the most efficient robustness criterion in the scenario where the designer cannot compromise the quality of the optimal solution below a certain threshold. The analysis of the ACTPI and T_{max} also determines "mini-max robustness" as one of the most efficient robustness criteria. Overall, "mini-max robustness" is deemed as the most suitable robustness criterion with regards to the associated computational cost.

A limitation of our study is that we only emphasize on Kriging-based robust optimization. Therefore, the findings may not be valid for other meta-heuristics based approaches for numerical optimization. Furthermore, we fix the internal computational budget for each robustness formulation in Kriging-based robust optimization. Visualizing the impact of variability in the internal computational budget, e.g., the internal optimization loop in the context of "mini-max robustness", is the focus of our future research. Lastly, we note that each robustness formulation is intrinsically associated with another cost, namely the cost of compromising on optimality to ensure robustness or stability. Focusing on this cost of robustness will advance the state-of-the-art in this area, and help practitioners choose the most suitable formulation with regards to optimality.

References

1. Bal, H., et al.: A medium-scale distributed system for computer science research: infrastructure for the long term. Computer **49**(5), 54–63 (2016)
2. Ben-Tal, A., Ghaoui, L.E., Nemirovski, A.: Robust Optimization, Princeton Series in Applied Mathematics, vol. 28. Princeton University Press (2009)
3. Beyer, H.G.: Evolutionary algorithms in noisy environments: Theoretical issues and guidelines for practice. Comput. Methods Appl. Mech. Eng. **186**(2–4), 239–267 (2000)
4. Beyer, H.G., Sendhoff, B.: Robust optimization-a comprehensive survey. Comput. Methods Appl. Mech. Eng. **196**(33–34), 3190–3218 (2007)
5. Genton, M.G.: Classes of kernels for machine learning: a statistics perspective. J. Mach. Learn. Res. **2**, 299–312 (2001)
6. Hansen, N., Auger, A., Mersmann, O., Tusar, T., Brockhoff, D.: COCO: a platform for comparing continuous optimizers in a black-box setting. CoRR abs/1603.08785 (2016)
7. Jones, D.R., Schonlau, M., Welch, W.J.: Efficient global optimization of expensive black-box functions. J. Glob. Optim. **13**(4), 455–492 (1998)
8. Jurecka, F.: Robust design optimization based on metamodeling techniques. Ph.D. thesis, Technische Universität München (2007)
9. Kruisselbrink, J.W.: Evolution strategies for robust optimization. Leiden Institute of Advanced Computer Science (LIACS), Faculty of Science (2012)
10. Pedregosa, F., et al.: Scikit-learn: machine learning in Python. J. Mach. Learn. Res. **12**, 2825–2830 (2011)
11. ur Rehman, S., Langelaar, M., van Keulen, F.: Efficient kriging-based robust optimization of unconstrained problems. J. Comput. Sci. 5(6), 872–881 (2014)
12. Stein, M.: Large sample properties of simulations using Latin hypercube sampling. Technometrics **29**(2), 143–151 (1987)
13. Taguchi, G., Konishi, S.: Taguchi Methods: Orthogonal Arrays and Linear Graphs. Tools for Quality Engineering. American Supplier Institute Dearborn, MI (1987)
14. Taguchi, G., Phadke, M.S.: Quality engineering through design optimization. In: Dehnad, K. (eds.) Quality Control, Robust Design, and the Taguchi Method, pp. 77–96. Springer, Boston (1989). https://doi.org/10.1007/978-1-4684-1472-1_5
15. Ullah, S., Nguyen, D.A., Wang, H., Menzel, S., Sendhoff, B., Bäck, T.: Exploring dimensionality reduction techniques for efficient surrogate-assisted optimization. In: 2020 IEEE Symposium Series on Computational Intelligence (SSCI), pp. 2965–2974. IEEE (2020)
16. Ullah, S., Wang, H., Menzel, S., Sendhoff, B., Back, T.: An empirical comparison of meta-modeling techniques for robust design optimization. In: 2019 IEEE Symposium Series on Computational Intelligence (SSCI), pp. 819–828. IEEE (2019)
17. Ullah, S., Wang, H., Menzel, S., Sendhoff, B., Bäck, T.: A new acquisition function for robust bayesian optimization of unconstrained problems. In: Proceedings of the Genetic and Evolutionary Computation Conference Companion, pp. 1344–1345 (2021)

Adaptive Function Value Warping for Surrogate Model Assisted Evolutionary Optimization

Amir Abbasnejad$^{(\boxtimes)}$ ⓘ and Dirk V. Arnold ⓘ

Faculty of Computer Science, Dalhousie University, Halifax, NS B3H 4R2, Canada
{a.abbasnejad,dirk}@dal.ca

Abstract. Surrogate modelling techniques have the potential to reduce the number of objective function evaluations needed to solve black-box optimization problems. Most surrogate modelling techniques in use with evolutionary algorithms today do not preserve the desirable invariance to order-preserving transformations of objective function values of the underlying algorithms. We propose adaptive function value warping as a tool aiming to reduce the sensitivity of algorithm behaviour to such transformations.

1 Introduction

Many evolutionary algorithms are designed to be invariant to order-preserving transformations of objective function values: the behaviour of the algorithm applied to a problem with objective function $f : \mathbb{R}^n \to \mathbb{R}$ is identical to its behaviour when applied to objective function $g \circ f$, where $\circ$ denotes function composition and $g : \mathbb{R} \to \mathbb{R}$ is an arbitrary strictly monotonically increasing function. This invariance is commonly achieved by using objective function values in a way that is comparison-based rather than value-based: objective function values of candidate solutions are compared in order to determine which solution is better, but at no point are objective function values used in arithmetic operations. As a consequence, the algorithms are subject to the fundamental constraints on their convergence rates that have been shown by Teytaud and Gelly [18]. At the same time, they are relatively insensitive to small disturbances in objective function values, and they are potentially useful for solving problems that are not locally well approximated by low-order Taylor polynomials.

Cost in black-box optimization is commonly measured in terms of the number of objective function evaluations expended to solve a problem. Surrogate modelling techniques are often used to reduce the number of objective function evaluations. The majority of candidate solutions sampled by evolutionary algorithms are relatively poor, and models of the objective function that are built based on information gained from objective function evaluations made in prior iterations can often be used to either discard poor candidate solutions without evaluating them using the objective function, or to suggest potentially good candidate solutions to evaluate. Commonly used types of surrogate models include

G. Rudolph et al. (Eds.): PPSN 2022, LNCS 13398, pp. 76–89, 2022.
https://doi.org/10.1007/978-3-031-14714-2_6

low-order polynomials, Gaussian processes, and radial basis function (RBF) networks. Regression using any of those involves arithmetic operations applied to function values, and when used in evolutionary algorithms that are otherwise invariant to order-preserving transformations of the objective function, their use breaks that invariance. In recognition of this, Loshchilov et al. [11] advocate the use of purely comparison-based ranking support vector machines (SVMs) as surrogate models. Their algorithm preserves the invariance properties of the underlying evolution strategy. However, presumably at least in part due to their simplicity, Gaussian processes, RBF networks, and polynomial models continue to be commonly used as surrogate models.

Snelson et al. [17] have proposed function value warping as a tool for achieving a better fit of Gaussian process models applied to regression tasks. The central idea is to employ a warping function to transform function values before fitting the model to the data points. If the transformation is such that the warped data are a better fit for the type of model used, then function value warping may result in superior models. In this paper we propose to employ function value warping in connection with surrogate model assisted evolutionary optimization. If warping function $g^{-1} : \mathbb{R} \to \mathbb{R}$ is used, then the behaviour of the algorithm employing that function applied to a problem with objective function $g \circ f : \mathbb{R}^n \to \mathbb{R}$ is identical to its behaviour when applied to objective function f without warping. In order to derive a suitable warp, we select a parameterized family of warping functions and adapt the parameters by optimizing the goodness of fit of the warped surrogate model. The proposed approach is evaluated experimentally by incorporating it in a surrogate model assisted covariance matrix adaptation evolution strategy (CMA-ES).

2   Related Work

Surrogate model assisted CMA-ES variants include the local meta model assisted CMA-ES (lmm-CMA-ES) proposed by Kern et al. [9] and the linear/quadratic CMA-ES (lq-CMA-ES) due to Hansen [5]. Both employ polynomial models. Bajer et al. [3] and Toal and Arnold [19] use Gaussian processes as surrogate models instead; the latter authors refer to their algorithm as GP-CMA-ES. None of those algorithms is invariant to order-preserving transformations of objective function values. In contrast, Loshchilov et al. [12,13] propose ^{s*}ACM-ES, which employ ranking support vector machines and are invariant to such transformations. Toal and Arnold [19] conduct a comparison of several of the above algorithms using three classes of objective functions: convex, spherically symmetric functions where a parameter controls how much the function deviates from being quadratic, convex-quadratic functions with variable degrees of conditioning, and a variant of the generalized Rosenbrock function. They find that lq-CMA-ES excel predominantly on the quadratic test problems, where they are able to locate near optimal solutions with very few objective function evaluations once accurate models of the objective have been built. For the spherically symmetric problems that are not quadratic, the invariance to order-preserving transformations of objective function values of ^{s*}ACM-ES is seen to be a valuable asset. GP-CMA-ES, which do not employ quadratic models, are less efficient than lq-CMA-ES when applied to quadratic problems, and they do not share the robustness to function value transformations of ^{s*}ACM-ES. However, they exhibit at least the second best performance of all of the surrogate-model assisted CMA-ES variants across all problems considered.

The use of function value warping in connection with Gaussian processes was proposed by Snelson et al. [17]. They consider regression tasks and apply warps consisting of sums of parameterized hyperbolic tangent functions to the data points. The warp parameters are set using maximum likelihood estimation. The tasks considered by Snelson et al. are static in the sense that models are fit to static, unchanging data sets. There is thus no need to continually adapt the warp parameters. In the context of surrogate model assisted optimization, a more rudimentary version of function value warping is implemented in COBRA, a surrogate model assisted algorithm for constrained optimization proposed by Regis [14]. That algorithm potentially subjects the objective or constraint functions either to division by a constant or to a logarithmic transformation. The goal is to "make the function values less extreme and avoid problems with fitting RBF surrogates". SACOBRA, a self-adaptive variant of COBRA proposed by Bagheri et al. [2], makes that transformation adaptive in the sense that the algorithm itself determines whether or not it is applied. However, the decision is binary, and no attempt is made to parameterize the transformation and evolve appropriate settings for the parameters.

A somewhat orthogonal approach that attempts to enable surrogate functions to better model given data is to optimize the choice of kernel function used in the models. Kronberger and Kommenda [10] employ genetic programming to evolve Gaussian process kernel functions for regression tasks. Roman et al. [16] propose a similar approach for time series extrapolation. A preliminary exploration of the use of different kernel functions in a surrogate model assisted CMA-ES has been presented by Repický et al. [15].

3   Algorithm

In order to explore the use of adaptive function value warping for surrogate model assisted evolutionary optimization, we incorporate it in the algorithm of Toal and Arnold [19]. That algorithm integrates covariance matrix adaptation as proposed by Hansen et al. [6,7] into algorithms by Kayhani and Arnold [8] and Yang and Arnold [20]. While both of the latter papers refer to the algorithms as Gaussian process model assisted, only the mean values of the stochastic processes are used and the models are more simply described using RBF terminology. A single iteration of the algorithm, which we refer to as warped surrogate CMA-ES (ws-CMA-ES) is shown in Fig. 1. The pseudocode is almost identical to that presented by Toal and Arnold [19], with differences only in Lines 4, 6, and 7, where the warping function is used, and in Line 11, where its parameters are updated. The following description is adapted from that reference.

Required: candidate solution $\mathbf{x} \in \mathbb{R}^n$, step size parameter $\sigma \in \mathbb{R}_{>0}$, covariance matrix $\mathbf{C} \in \mathbb{R}^{n \times n}$, search path $\mathbf{s} \in \mathbb{R}^n$, archive $\mathcal{A} = \{(\mathbf{x}_k, f(\mathbf{x}_k)) \,|\, k = 1, 2 \ldots, m\}$, warping function Ω_ω, warp parameters $\boldsymbol{\omega}$

1: Build a surrogate model from archive $\mathcal{A}$.
2: Compute $\mathbf{A} = \mathbf{C}^{1/2}$.
3: Generate trial step vectors $\mathbf{z}_i \sim \mathcal{N}(\mathbf{0}, \mathbf{I}_{n \times n})$, $i = 1, 2, \ldots, \lambda$.
4: Evaluate $\mathbf{y}_i = \mathbf{x} + \sigma \mathbf{A} \mathbf{z}_i$ using the warped surrogate model, yielding $\tilde{f}(\mathbf{y}_i)$, $i = 1, 2, \ldots, \lambda$.
5: Let $\mathbf{z} = \sum_{j=1}^{\lambda} w_j \mathbf{z}_{j;\lambda}$, where $j; \lambda$ is the index of the jth smallest of the $\tilde{f}(\mathbf{y}_i)$.
6: Evaluate $\mathbf{y} = \mathbf{x} + \sigma \mathbf{A} \mathbf{z}$ using the warped surrogate model, yielding $\tilde{f}(\mathbf{y})$.
7: **if** $\tilde{f}(\mathbf{y}) > \Omega_\omega(f(\mathbf{x}))$ **then**
8:     Let $\sigma \leftarrow \sigma \, e^{-d_1/D}$.
9: **else**
10:     Evaluate $\mathbf{y}$ using the objective function, yielding $f(\mathbf{y})$.
11:     Add $(\mathbf{y}, f(\mathbf{y}))$ to $\mathcal{A}$ and update the warp parameters $\boldsymbol{\omega}$.
12:     **if** $f(\mathbf{y}) > f(\mathbf{x})$ **then**
13:         Let $\sigma \leftarrow \sigma \, e^{-d_2/D}$.
14:     **else**
15:         Let $\mathbf{x} \leftarrow \mathbf{y}$ and $\sigma \leftarrow \sigma \, e^{d_3/D}$.
16:         Update $\mathbf{s}$ and $\mathbf{C}$.
17:     **end if**
18: **end if**

Fig. 1. Single iteration of the ws-CMA-ES.

The state of the algorithm consists of candidate solution $\mathbf{x} \in \mathbb{R}^n$, step size parameter $\sigma \in \mathbb{R}_{>0}$, positive definite $n \times n$ matrix $\mathbf{C}$ that is referred to as the covariance matrix, vector $\mathbf{s} \in \mathbb{R}^n$ that is referred to as the search path, an archive of m candidate solutions that have been evaluated in prior iterations along with their objective function values, and warp parameters $\boldsymbol{\omega}$. In Line 1, a warped surrogate model is constructed from the archive, resulting in a function $\tilde{f} : \mathbb{R}^n \to \mathbb{R}$ that approximates the objective function in the vicinity of previously evaluated points, but is assumed to be much cheaper to evaluate. Details of this step are described below. The algorithm then proceeds to compute positive definite matrix $\mathbf{A}$ as the principal square root of $\mathbf{C}$. Line 3 generates $\lambda \geq 1$ trial step vectors that are independently drawn from a multivariate Gaussian distribution with zero mean and unit covariance matrix. Line 4 uses the warped surrogate model to evaluate the corresponding trial points $\mathbf{y}_i = \mathbf{x} + \sigma \mathbf{A} \mathbf{z}_i$, and Line 5 computes $\mathbf{z} \in \mathbb{R}^n$ as a weighted sum of the trial vectors, using rank based weights. Line 6 generates candidate solution $\mathbf{y} = \mathbf{x} + \sigma \mathbf{A} \mathbf{z}$ that is then evaluated using the warped surrogate model. If the warped surrogate model suggests that $\mathbf{y}$ is not superior to parental candidate solution $\mathbf{x}$ (the function value of which is transformed using the warping function for the comparison), then the step size is reduced and the iteration is complete. Otherwise, $\mathbf{y}$ is evaluated using the true objective function and added to the archive. The warp parameters are updated if needed as described below. If **y** is inferior to the parental candidate solution, then the step size is reduced and the iteration is complete. Otherwise, the offspring candidate solution replaces the parental one, the step size is increased, and the search path **s** and covariance matrix **C** are updated as proposed by Hansen [4] and described by Toal and Arnold [19]. Notice that at most one evaluation of the objective function is performed in each iteration of the algorithm.

Parameter Settings Parameter λ determines the degree of surrogate model exploitation as discussed by Yang and Arnold [20]. Toal and Arnold [19] found its setting to be uncritical for the test problems they considered, with larger values affording a moderate speed-up on some problems. Throughout this paper, we use $\lambda = 10$ and set the rank based weights used in Line 5 as proposed by Hansen [4]. That is, weights w_1 through $w_{\lfloor \lambda/2 \rfloor}$ form a strictly monotonically decreasing sequence of positive values that sum to one; the remaining weights are zero. The parameters that determine the relative rates of change of the step size parameter in Lines 8, 13, and 15 are set to $d_1 = 0.2$, $d_2 = 1.0$, and $d_3 = 1.0$ according to the recommendation by Yang and Arnold [20]. In accordance with prior work [8, 20], parameter D, which scales the rates of the step size parameter changes, is set to $\sqrt{1 + n}$.

Warped Surrogate Model. Warping functions are strictly monotonically increasing functions $\Omega_\omega : \mathbb{R} \to \mathbb{R}$ that are parameterized with warp parameters ω. To ensure flexibility, warping function families should contain functions with both positive and negative curvature. As the warp needs to be optimized throughout a run of the algorithm, it is desirable to have a small number of warp parameters. In what follows we employ warping functions

$$\Omega_{\langle p,q \rangle}(y) = (y - q)^p$$

that are parameterized with warp parameters $\omega = \langle p, q \rangle$. We refer to q and p as the shift and the power of the warp, respectively. The shift needs to be set such that the warping function is never applied to function values less than q. Setting the parameters according to the procedure described below satisfies that requirement.

As surrogate models for the objective function we employ RBF networks. In Line 1 of the algorithm in Fig. 1, in order to build a warped surrogate model from the points in the archive we generate $m \times m$ matrix **K** with entries $k_{ij} = k(\mathbf{x}_i, \mathbf{x}_j)$, where kernel function k is defined as

$$k(\mathbf{x}, \mathbf{y}) = \exp\left(-\frac{(\mathbf{x} - \mathbf{y})^{\mathrm{T}} \mathbf{C}^{-1} (\mathbf{x} - \mathbf{y})}{2h^2 \sigma^2} \right)$$

with the length scale parameter set to $h = 8n$. Notice that this is the commonly employed squared exponential kernel with the Mahalanobis distance using matrix $\sigma^2 \mathbf{C}$ replacing the Euclidean distance. Toal and Arnold [19] point to the similar use of the Mahalanobis distance for quadratic regression by Kern et al. [9]

and, considering spherically symmetric test functions, find that choice of kernel function results in surrogate models that are a better fit for quadratic functions than for those where function values do not scale quadratically. The setting of the length scale parameter is adopted from Toal and Arnold, who have experimented with maximum likelihood estimation without having been able to consistently improve on the performance of the constant setting.

To evaluate the warped surrogate model at a point $\mathbf{y} \in \mathbb{R}^n$, in Lines 4 and 6 of the algorithm in Fig. 1 we compute $m \times 1$ vector $\mathbf{k}$ with entries $k_i = k(\mathbf{x}_i, \mathbf{y})$ and define

$$\tilde{f}(\mathbf{y}) = \Omega_\omega \left(f(\mathbf{x}) \right) + \mathbf{k}^T \mathbf{K}^{-1} \tilde{\mathbf{f}},$$

where $m \times 1$ vector $\tilde{\mathbf{f}}$ has entries $\tilde{f}_i = \Omega_\omega (f(\mathbf{x}_i)) - \Omega_\omega (f(\mathbf{x}))$. That is, the warping function is applied to the function values before using them in the RBF model, and the function value of the best solution evaluated so far is used as an offset.

In order to avoid increasing computational costs from the need to invert matrix $\mathbf{K}$ as the size of the archive grows, we use at most the most recent $m = 8n$ points from the archive in order to construct the surrogate model. For $n \geq 3$ that number is smaller than the value of $m = (n + 2)^2$ used by Toal and Arnold [19]. The primary reason for our choice is the computational cost of optimizing the warp parameters, which grows with increasing archive size.

Whenever a point is added to the archive, it may become necessary to adapt the warp parameters in Line 11 of the algorithm in Fig. 1. We employ leave-one-out cross validation in order to judge the quality of the current warp parameter settings. That is, for each point in the archive we build a warped surrogate model from the remaining $m - 1$ points and use it to determine the warped surrogate model value of the point in question. The Kendall rank correlation coefficient τ between those m warped surrogate model values and the points' true objective function values (which have been computed and stored in prior iterations) is determined. If $\tau \geq 0.9$, then the current warp parameter settings are considered adequate and the warp parameters remain unchanged. If $\tau < 0.9$, then the algorithm attempts to find better parameter settings using a simple coordinate search: first a new setting for the shift parameter is obtained by computing leave-one-out rank correlation coefficients for 31 uniformly spaced shift parameter values and selecting the shift that results in the highest correlation. If the resulting value of τ exceeds 0.9, then that setting is adopted and the power parameter is left unchanged. Otherwise, the shift parameter setting is adopted and a new setting for the power parameter is obtained analogously. In case the new setting does not result in a rank correlation coefficient in excess of 0.9, the strategy defaults to no warp and sets $p = 1$ and $q = 0$.

Figure 2 shows data from three typical runs of ws-CMA-ES on objective functions $f(\mathbf{x}) = (\mathbf{x}^T \mathbf{x})^{\alpha/2}$ with $n = 16$ and $\alpha \in \{1, 2, 4\}$. Once the archive of evaluated points has reached a sufficient length, linear convergence is observed. All runs succeed in locating warp parameter settings that result in leave-one-out rank correlations of at least 0.9 throughout. The values of the power parameter that are generated are approximately equal to $2/\alpha$, effectively resulting in the surrogate model "seeing" function values close to those obtained for $\alpha = 2$.

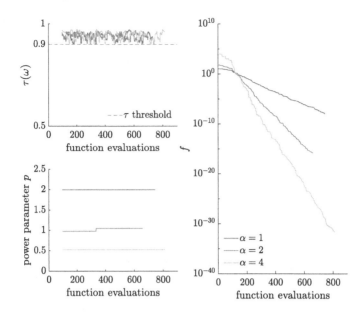

Fig. 2. Convergence plots of ws-CMA-ES on sphere functions with $\alpha \in \{1, 2, 4\}$ and $n = 16$. The subfigure on the right shows the evolution of objective function values against the number of iterations. The subfigures on the left display rank correlation values τ encountered in those runs and the values of power parameter p.

Initialization and Start-Up. The search path **s** is initialized to the zero vector, covariance matrix **C** to the identity matrix. The initialization of **x** and σ usually is problem specific. We avoid constructing models based on insufficient data by not using surrogate models in the first $2n$ iterations of the algorithm. For the duration of this start-up phase, the algorithm is thus a model-free $(1 + 1)$-CMA-ES and we use parameter settings $d_2 = 0.25$ and $d_3 = 1.0$ for step size adaptation. The warp parameters are initialized once the archive contains sufficiently many data points. An illustration of the dependence of the value of the leave-one-out rank correlation coefficient on the warp parameter settings is shown in Fig. 3 for a sphere function with $\alpha = 4$ and $n = 8$. While more efficient approaches to selecting warp parameter settings with near optimal values of τ are conceivable, we simply enumerate 31×31 settings and select the setting that results in the highest leave-one-out rank correlation coefficient.

4   Evaluation

This section experimentally evaluates the performance of ws-CMA-ES relative to that of several other algorithms. Section 4.1 briefly outlines the comparator algorithms. Section 4.2 describes the testing environment, and Sect. 4.3 presents experimental data along with a discussion of the findings.

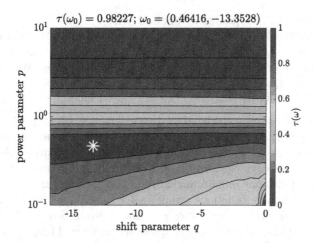

Fig. 3. Selection of warp parameters p and q on a sphere function with $\alpha = 4$ and $n = 8$. Rank correlation values obtained from leave-one-out cross validation are colour-coded. The selected warp parameter setting is marked by the asterisk.

4.1 Comparator Algorithms

We compare ws-CMA-ES with four other algorithms:

- CMA-ES without surrogate model assistance; we use Version 3.61.beta of the MATLAB implementation provided by N. Hansen at `cma-es.github.io`.
- ^{s*}ACM-ES; we use Version 2 of the MATLAB implementation provided by I. Loshchilov at `loshchilov.com`.
- lq-CMA-ES; we use Version 3.0.3 of the Python implementation provided by N. Hansen on `GitHub`.
- GP-CMA-ES as described by Toal and Arnold [19].

All algorithm-specific parameters are set to their default values, with one exception: the archive size of GP-CMA-ES is set to $m = 8n$ rather than $m = (n+2)^2$. The change was made in order to better isolate the effect of warping in ws-CMA-ES, which use the smaller archive size in order to reduce the cost of computing effective warp parameters. As used here, GP-CMA-ES are identical to ws-CMA-ES with the warp power fixed at $p = 1$. All evolution strategies but lq-CMA-ES, which employ quadratic models with diagonal scaling before switching to fully quadratic models once sufficiently many data points are available, are invariant to rotations of the coordinate system. Both CMA-ES and ^{s*}ACM-ES are invariant to order-preserving transformations of objective function values; GP-CMA-ES and lq-CMA-ES are not. ws-CMA-ES are not fully invariant, but it will be seen that their sensitivity to transformations of objective function values is reduced significantly compared to GP-CMA-ES.

4.2 Test Environments

We consider the following three basic test problems:

– Spherically symmetric quadratic function $f(\mathbf{x}) = \mathbf{x}^T\mathbf{x}$.
– Convex quadratic function $f(\mathbf{x}) = \mathbf{x}^T\mathbf{B}\mathbf{x}$ where symmetric $n \times n$ matrix $\mathbf{B}$ has eigenvalues $b_{ii} = \beta^{(i-1)/(n-1)}$, $i = 1,\ldots,n$, with condition number $\beta = 10^6$. The eigenvectors of $\mathbf{B}$ coincide with the coordinate axes. We refer to this function as the Ostermeier ellipsoid.
– Generalized Rosenbrock function $f(\mathbf{x}) = \sum_{i=1}^{n-1}\left[100(x_{i+1} - x_i^2)^2 + (1 - x_i)^2\right]$.

All of those functions are considered in dimensions $n \in \{2, 4, 8\}$. Function value transformations $g(y) = y^{\alpha/2}$ with $\alpha \in [1, 4]$ are applied, creating three families of test problems. For the former two functions varying α controls the degree to which the problems deviate from being quadratic. In all three cases the contour lines of the functions remain unchanged. The optimal function value for all problems is zero. Sphere functions are perfectly conditioned and do not require the learning of axis scales, but for $\alpha \neq 2$ may pose difficulties for algorithms that internally build quadratic or near-quadratic models. The primary difficulty inherent in ellipsoid functions is their conditioning and thus the need to learn appropriate axis scales. The generalized Rosenbrock function is only moderately ill-conditioned, but in contrast to the other function families requires a constant relearning of axis scales. For $n \geq 4$ it possesses a local minimizer different from the global one and, depending on initialization, a minority of the runs of all of the algorithms converge to that merely local optimizer. As global optimization ability is outside of the focus of this paper, we discard such runs and repeat them until convergence to the global optimizer is observed.

All runs of all algorithms are initialized by sampling starting points uniformly at random in $[-4, 4]^n$. The step size parameter is initially set to $\sigma = 2$ for all strategies. Runs are terminated when a candidate solution with an objective function value no larger than $10^{-8\alpha}$ is generated and evaluated. As all optimal function values are zero, that stopping criterion is well within reach with double precision floating-point accuracy. An advantage of requiring such high accuracy is that the coefficient of variation of the observed running times is relatively low. With the stopping criterion as chosen, running times (as measured in objective function evaluations) are independent of α for those algorithms that are invariant to order-preserving transformations of the objective. Except for those runs on the (value-transformed) Rosenbrock function that were discarded for converging to the merely local minimizer, all runs of all algorithms located the globally optimal solutions with no restarts required. Some further experimental results are presented by Abbasnejad [1].

4.3   Results

We have performed fifteen runs of each of the algorithms for each test problem instance considered. Figures 4, 5, and 6 plot the numbers of objective function evaluations required to locate the optimal solutions to within the required accuracy divided by the dimension of the problem against parameter α. Lines connect the median values and error bars illustrate the ranges observed (i.e., they span the range from the smallest to the largest values observed across the fifteen runs). Results are discussed in detail in the following paragraphs.

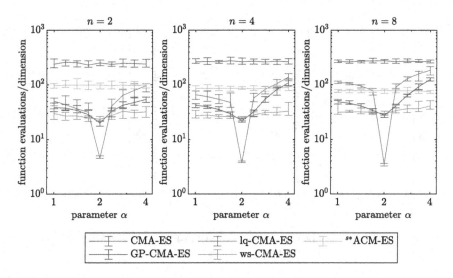

Fig. 4. Number of objective function evaluations per dimension required to optimize sphere functions with parameters $\alpha \in [1,4]$. The lines connect median values; the error bars reflect the full range of values observed for the respective algorithms.

Sphere Functions: Figure 4 shows that all surrogate model assisted strategies improve on the performance of basic CMA-ES across the entire range of parameter values considered. ^{s*}ACM-ES achieve a constant speed-up factor of about four, reflecting the invariance of the algorithm to strictly increasing transformations of function values. As observed by Toal and Arnold [19], lq-CMA-ES excel for $\alpha = 2$, where the quadratic models that they build perfectly match the objective function. Their relative performance deteriorates the more α deviates from that value. GP-CMA-ES exhibit a similar but less pronounced pattern. To judge the performance of ws-CMA-ES, recall that they differ from GP-CMA-ES solely in their use of warping. Their performance nearly matches that of GP-CMA-ES for $\alpha = 2$, but it deteriorates to a much lesser degree as values of α deviate from 2. While not invariant to the choice of α, the sensitivity to the value of that parameter is significantly reduced, showing that the algorithm succeeds in locating beneficial warps. The speed-up factor of ws-CMA-ES over CMA-ES without surrogate model assistance ranges between six and ten throughout.

Ostermeier Ellipsoids: Running times on the Ostermeier ellipsoids are represented Fig. 5 and show essentially the same patterns. All of the surrogate model assisted algorithms improve on the basic CMA-ES. ^{s*}ACM-ES provide a speed-up factor of about four. lq-CMA-ES excel for $\alpha = 2$, but their performance is matched or exceeded by that of ^{s*}ACM-ES for other values of that parameter. Applying a rotation to the coordinate system would further tilt the scale in favour of ^{s*}ACM-ES. GP-CMA-ES for $\alpha = 2$ perform second to lq-CMA-ES. For values of α significantly larger than 2 they are eventually outperformed by

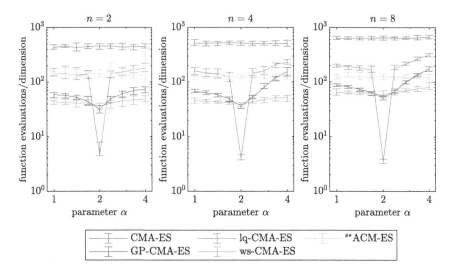

Fig. 5. Number of objective function evaluations per dimension required to optimize ellipsoid functions with parameters $\alpha \in [1,4]$. The lines connect median values; the error bars reflect the full range of values observed for the respective algorithms.

^{s*}ACM-ES, which are entirely unaffected by the function value transformation. ws-CMA-ES again nearly match the performance of GP-CMA-ES for $\alpha = 2$, where no warp is needed. Due to the beneficial warping functions that are generated, they significantly outperform that algorithm for values of α different from 2.

Rosenbrock Functions: Figure 6 shows results on function value-transformed generalized Rosenbrock functions. With few outliers, all of the surrogate model assisted algorithms require reduced numbers of function evaluations compared to the algorithm that does not employ surrogate models. As the objective is not quadratic for any value of α, lq-CMA-ES do not dominate the other algorithms as they do for $\alpha = 2$ in Figs. 4 and 5. Their performance appears broadly comparable with that of ^{s*}ACM-ES. GP-CMA-ES again exhibit their best performance for values of α near 2, where the Gaussian process models that they use appear to be particularly effective. ws-CMA-ES succeed in reducing the slowdown resulting from function value transformations particularly for larger values of α. Notably, their advantage over the other algorithms appears to decrease with increasing dimension. We hypothesize that this is at least partly due to the limited archive size employed in the experiments.

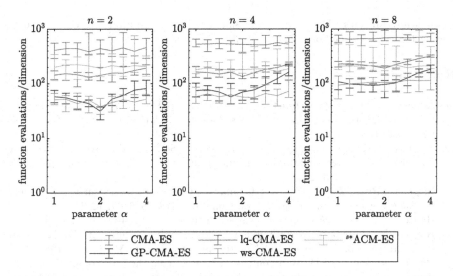

Fig. 6. Number of objective function evaluations per dimension required to optimize Rosenbrock functions with parameters $\alpha \in [1, 4]$. The lines connect median values; the error bars reflect the full range of values observed for the respective algorithms.

5 Conclusions

To conclude, we have proposed adaptive function value warping as an approach to potentially improving the performance of surrogate model assisted evolutionary algorithms. Many types of surrogate models are better suited to modelling some classes of functions than others. Clearly, quadratic models best fit quadratic functions. RBF models with squared exponential kernels have also been observed to better fit quadratic functions than function value-transformed versions thereof. By subjecting function values to a nonlinear warp before applying the model, better fitting models can be generated provided that the warping function is well chosen. ws-CMA-ES generate suitable warps by using leave-one-out cross validation to quantify the quality of a model. Warp parameters are adapted throughout the runs of the algorithm. Our experimental results suggest that ws-CMA-ES successfully generate beneficial warps for the three classes of function value-transformed test problems considered.

The computational cost of leave-one-out cross validation as implemented effectively restricts the use of the algorithm to relatively low-dimensional problems. Considering problems with $n > 10$ is feasible, but may limit the length of the archive used when fitting surrogate models. Developing either faster implementations of leave-one-out cross validation or alternative approaches to evaluating the quality of a surrogate model would enable the approach to be applied to higher-dimensional problems without having to compromise on the archive size. A further subject of our ongoing research is the development of other techniques for the adaptation of warp parameters.

Acknowledgements. This research was supported by the Natural Sciences and Engineering Research Council of Canada (NSERC).

References

1. Abbasnejad, A.: Adaptive function value warping for surrogate model assisted evolutionary optimization. Master's thesis, Faculty of Computer Science, Dalhousie University (2021)
2. Bagheri, S., Konen, W., Emmerich, M., Bäck, T.: Self-adjusting parameter control for surrogate-assisted constrained optimization under limited budgets. Appl. Soft Comput. **61**, 377–393 (2017)
3. Bajer, L., Pitra, Z., Repický, J., Holeňa, M.: Gaussian process surrogate models for the CMA evolution strategy. Evol. Comput. **27**(4), 665–697 (2019)
4. Hansen, N.: The CMA evolution strategy: a tutorial. arxiv:1604.00772 (2016)
5. Hansen, N.: A global surrogate assisted CMA-ES. In: Genetic and Evolutionary Computation Conference – GECCO 2019, pp. 664–672. ACM Press (2019)
6. Hansen, N., Müller, S.D., Koumoutsakos, P.: Reducing the time complexity of the derandomized evolution strategy with covariance matrix adaptation (CMA-ES). Evol. Comput. **11**(1), 1–18 (2003)
7. Hansen, N., Ostermeier, A.: Completely derandomized self-adaptation in evolution strategies. Evol. Comput. **9**(2), 159–195 (2001)
8. Kayhani, A., Arnold, D.V.: Design of a surrogate model assisted (1 + 1)-ES. In: Auger, A., Fonseca, C.M., Lourenço, N., Machado, P., Paquete, L., Whitley, D. (eds.) PPSN 2018. LNCS, vol. 11101, pp. 16–28. Springer, Cham (2018). https://doi.org/10.1007/978-3-319-99253-2_2
9. Kern, S., Hansen, N., Koumoutsakos, P.: Local meta-models for optimization using evolution strategies. In: Runarsson, T.P., Beyer, H.-G., Burke, E., Merelo-Guervós, J.J., Whitley, L.D., Yao, X. (eds.) PPSN 2006. LNCS, vol. 4193, pp. 939–948. Springer, Heidelberg (2006). https://doi.org/10.1007/11844297_95
10. Kronberger, G., Kommenda, M.: Evolution of covariance functions for gaussian process regression using genetic programming. In: Moreno-Díaz, R., Pichler, F., Quesada-Arencibia, A. (eds.) EUROCAST 2013. LNCS, vol. 8111, pp. 308–315. Springer, Heidelberg (2013). https://doi.org/10.1007/978-3-642-53856-8_39
11. Loshchilov, I., Schoenauer, M., Sebag, M.: Comparison-based optimizers need comparison-based surrogates. In: Schaefer, R., Cotta, C., Kołodziej, J., Rudolph, G. (eds.) PPSN 2010. LNCS, vol. 6238, pp. 364–373. Springer, Heidelberg (2010). https://doi.org/10.1007/978-3-642-15844-5_37
12. Loshchilov, I., Schoenauer, M., Sebag, M.: Self-adaptive surrogate-assisted covariance matrix adaptation evolution strategy. In: Genetic and Evolutionary Computation Conference – GECCO 2012, pp. 321–328. ACM Press (2012)
13. Loshchilov, I., Schoenauer, M., Sebag, M.: Intensive surrogate model exploitation in self-adaptive surrogate-assisted CMA-ES. In: Genetic and Evolutionary Computation Conference – GECCO 2013, pp. 439–446. ACM Press (2013)
14. Regis, R.G.: Constrained optimization by radial basis function interpolation for high-dimensional expensive black-box problems with infeasible initial points. Eng. Optim. **46**(2), 218–243 (2014)
15. Repický, J., Holeňa, M., Pitra, Z.: Automated selection of covariance function for Gaussian process surrogate models. In: Krajci, S. (ed.) Information Technologies: Applications and Theory – ITAT 2018, pp. 64–71. CEUR Workshop Proceedings (2018)

16. Roman, I., Santana, R., Mendiburu, A., Lozano, J.A.: Evolving Gaussian process kernels from elementary mathematical expressions for time series extrapolation. Neurocomputing **462**, 426–439 (2021)
17. Snelson, E., Rasmussen, C.E., Ghahramani, Z.: Warped Gaussian processes. In: Thrun, S., et al. (eds.) Conference on Neural Information Processing Systems – NeurIPS. pp. 337–344. MIT Press (2003)
18. Teytaud, O., Gelly, S.: General lower bounds for evolutionary algorithms. In: Runarsson, T.P., Beyer, H.-G., Burke, E., Merelo-Guervós, J.J., Whitley, L.D., Yao, X. (eds.) PPSN 2006. LNCS, vol. 4193, pp. 21–31. Springer, Heidelberg (2006). https://doi.org/10.1007/11844297_3
19. Toal, L., Arnold, D.V.: Simple surrogate model assisted optimization with covariance matrix adaptation. In: Bäck, T., et al. (eds.) PPSN 2020. LNCS, vol. 12269, pp. 184–197. Springer, Cham (2020). https://doi.org/10.1007/978-3-030-58112-1_13
20. Yang, J., Arnold, D.V.: A surrogate model assisted (1 + 1)-ES with increased exploitation of the model. In: Genetic and Evolutionary Computation Conference – GECCO 2019, pp. 727–735. ACM Press (2019)

Efficient Approximation of Expected Hypervolume Improvement Using Gauss-Hermite Quadrature

Alma Rahat[1]([✉])[ID], Tinkle Chugh[2][ID], Jonathan Fieldsend[2][ID],
Richard Allmendinger[3][ID], and Kaisa Miettinen[4][ID]

[1] Swansea University, Swansea SA1 8EN, UK
a.a.m.rahat@swansea.ac.uk
[2] University of Exeter, Exeter EX4 4QD, UK
{T.Chugh,J.E.Fieldsend}@exeter.ac.uk
[3] The University of Manchester, Manchester M15 6PB, UK
richard.allmendinger@manchester.ac.uk
[4] Faculty of Information Technology, University of Jyvaskyla, P.O. Box 35 (Agora),
40014 Jyväskylä, Finland
kaisa.miettinen@jyu.fi

Abstract. Many methods for performing multi-objective optimisation of computationally expensive problems have been proposed recently. Typically, a probabilistic surrogate for each objective is constructed from an initial dataset. The surrogates can then be used to produce predictive densities in the objective space for any solution. Using the predictive densities, we can compute the expected hypervolume improvement (EHVI) due to a solution. Maximising the EHVI, we can locate the most promising solution that may be expensively evaluated next. There are closed-form expressions for computing the EHVI, integrating over the multivariate predictive densities. However, they require partitioning of the objective space, which can be prohibitively expensive for more than three objectives. Furthermore, there are no closed-form expressions for a problem where the predictive densities are dependent, capturing the correlations between objectives. Monte Carlo approximation is used instead in such cases, which is not cheap. Hence, the need to develop new accurate but cheaper approximation methods remains. Here we investigate an alternative approach toward approximating the EHVI using Gauss-Hermite quadrature. We show that it can be an accurate alternative to Monte Carlo for both independent and correlated predictive densities with statistically significant rank correlations for a range of popular test problems.

Keywords: Gauss-Hermite · Expected hypervolume improvement · Bayesian optimisation · Multi-objective optimisation · Correlated objectives

© The Author(s), under exclusive license to Springer Nature Switzerland AG 2022
G. Rudolph et al. (Eds.): PPSN 2022, LNCS 13398, pp. 90–103, 2022.
https://doi.org/10.1007/978-3-031-14714-2_7

1 Introduction

Many real-world optimisation problems have multiple conflicting objectives [10,23,28]. In many cases, these objective functions can take a substantial amount of time for one evaluation. For instance, problems involving computational fluid dynamic simulations can take minutes to days for evaluating a single design (or decision vector/candidate solution) [2,7]. Such problems do not have analytical or closed-form expressions for the objective functions and are termed as *black-box* problems. To alleviate the computation time and obtain solutions with few expensive function evaluations, surrogate-assisted optimisation methods [3,6], e.g. Bayesian optimisation (BO) [27], have been widely used. In such methods, a surrogate model (also known as a metamodel) is built on given data (which is either available or can be generated with some design of experiments technique [24]). If one builds independent models for each objective function [15,31], the correlation between the objective functions is not directly considered. Multi-task surrogates [5,26] have been used recently to consider the correlation.

In BO, the surrogate model is usually a Gaussian process (GP) because GPs provide uncertainty information in the approximation in addition to the point approximation. These models are then used in optimising an acquisition function (or infill criterion) to find the next best decision vector to evaluate expensively. The acquisition function usually balances the convergence and diversity. Many acquisition functions have been proposed in the literature. Here, we focus on using expected hypervolume improvement (EHVI) [13], which has become a popular and well-studied acquisition function for expensive multi-objective optimisation largely due to its reliance on the hypervolume [20,32] (the only strictly Pareto compliant indicator known so far). The EHVI relies on a predictive distribution of solutions (with either independent [13] or correlated objective functions [26]). An optimiser is used to maximise the EHVI to find a decision vector with maximum expected improvement in hypervolume. The EHVI can be computed analytically for any number of objectives assuming the objective functions $f_1, \ldots, f_m$ are drawn from independent GPs [15]. However, this computation is expensive for more than three objectives. Monte Carlo (MC) approximation of EHVI is often used instead in such cases but this is not cheap. Consequently, there is a need for accurate but cheaper approximation methods for EHVI. We propose and investigate a novel way of approximating the EHVI using Gauss-Hermite (GH) quadrature [19,22]. In essence, GH approximates the integral of a function using a weighted sum resulting in fewer samples to approximate the EHVI.

The rest of the article is structured as follows. In Sect. 2, we briefly describe multivariate predictive densities and EHVI, and then introduce the GH method in Sect. 3. In Sect. 4, we show the potential of the proposed idea of using GH by comparing it with analytical and MC approximations (for 2–3 objectives). Finally, conclusions are drawn in Sect. 5.

2 Background

For multi-objective optimisation problems with m objective functions to be minimised, given two vectors $\mathbf{z}$ and $\mathbf{y}$ in the objective space, we say that $\mathbf{z}$ dominates $\mathbf{y}$ if $z_i \leq y_i$ for all $i = 1, \ldots, m$ and $z_j < y_j$ for at least one index j. A solution is Pareto optimal if no feasible solution dominates it. The set of Pareto optimal solutions in the objective space is called the Pareto front.

In multi-objective BO, the predictive distribution due to a solution with independent models is defined as:

$$\mathbf{y} \sim \mathcal{N}(\boldsymbol{\mu}, \mathrm{diag}(\sigma_1^2, \ldots, \sigma_m^2)),$$

where m is the number of objectives and $\boldsymbol{\mu} = (\mu_1, \ldots, \mu_m)^\top$ is the mean vector, with μ_i and σ_i being the mean and standard deviations of the predictive density for the ith objective. To quantify the correlation between objectives, a multi-task surrogate model can be used. The distribution of a solution with a single multi-task model is defined with a multi-variate Gaussian distribution:

$$\mathbf{y} \sim \mathcal{N}(\boldsymbol{\mu}, \Sigma),$$

where $\boldsymbol{\mu}$ is the vector of means and Σ is the covariance matrix that quantifies the correlation between different objectives. It should be noted that considering only the diagonal elements of Σ would ignore any correlations between objectives, and result in an independent multivariate predictive density.

The hypervolume measure [20,32] is a popular indicator to assess the quality of a set of solutions to a multi-objective optimisation problem. Thus it is often used to compare multi-objective optimisation algorithms or for driving the search of indicator-based multi-objective optimisation algorithms. The interested reader is referred to [4] for an investigation of the complexity and running time of computing the hypervolume indicator for different Pareto front shapes, number of non-dominated solutions, and number of objectives m. The EHVI answers the question of what the expected improvement of the hypervolume is if some new candidate solution $\mathbf{x}$ would be added to an existing set of solutions. Consequently, the solution with the highest EHVI may be the one worth an expensive function evaluation. To avoid ambiguity, in the following, we provide formal definitions of the concepts discussed here, before discussing methods to calculate the EHVI.

Definition 1 (Hypervolume indicator). *Given a finite set of k points (candidate solutions) $P = \{\mathbf{p}_1, \ldots, \mathbf{p}_k\} \subset \mathbf{R}^m$ where $\mathbf{p}_i = (f_1(\mathbf{x}_i), \ldots, f_m(\mathbf{x}_i))^\top$ for an optimisation problem with m objectives, the hypervolume indicator (HI) of P is defined as the Lebesgue measure of the subspace (in the objective space) dominated by P and a user-defined reference point $\mathbf{r}$ [31]:*

$$HI(P) = \Lambda(\cup_{\mathbf{p} \in P}[\mathbf{p}, \mathbf{r}]),$$

where Λ is the Lebesgue measure on $\mathbf{R}^m$, and $\mathbf{r}$ chosen such that it is dominated by all points in P, and ideally also by all points of the Pareto front.

Definition 2 (Hypervolume contribution). *Given a point* $\mathbf{p} \in P$, *its hypervolume contribution with respect to* P *is* $\Delta HI(P, \mathbf{p}) = HI(P) - HI(P\backslash\{\mathbf{p}\})$.

Definition 3 (Hypervolume improvement). *Given a point* $\mathbf{p} \notin P$, *its hypervolume improvement with respect to* P *is* $I(\mathbf{p}, P) = HI(P \cup \{\mathbf{p}\}) - HI(P)$.

Definition 4 (Expected hypervolume improvement). *Given a point* $\mathbf{p} \notin P$, *its expected hypervolume improvement (EHVI) with respect to* P *is*

$$\int_{\mathbf{p}\in\mathbb{R}^m} HI(P, \mathbf{p}) \cdot PDF(\mathbf{p})d\mathbf{p},$$

where $PDF(\mathbf{p})$ *is the predictive distribution function of* $\mathbf{p}$ *over points in the objective space.*

The EHVI can be computed analytically for any number of objectives assuming they are uncorrelated, but this requires partitioning the objective space, which can be prohibitively expensive for $m > 3$ objectives. Consequently, there is considerable interest in finding more efficient ways to compute EHVI, see e.g. [8,9,14,15,18,30]. MC integration is an alternative to an exact computation of EHVI. It is easy to use in practice and has been the method of choice for problems with $m > 3$ objectives. Given a multivariate Gaussian distribution from which samples are drawn, or $\mathbf{p}_i \sim \mathcal{N}(\boldsymbol{\mu}, \Sigma)$, and a set of points P (e.g. an approximation of the Pareto front), then the MC approximation of EHVI across c samples is

$$\frac{1}{c}\sum_{i=1}^{c} I(\mathbf{p}_i, P), \tag{1}$$

where $I(\mathbf{p}_i, P)$ is the hypervolume improvement (see Definition 3). The approximation error is given by $e = \sigma_M/\sqrt{c}$, where σ_M is the sample standard deviation [21]. Clearly, as the sample size c increases, the approximation error reduces, namely in proportion to $1/\sqrt{c}$. In other words, a hundred times more samples will result in improving the accuracy by ten times.

Typically, evaluating the improvement due to a single sample can be rapid. Even if we consider a large c, it is often not that time-consuming to compute the EHVI for a single predictive density. However, when we are optimising EHVI to locate the distribution that is the most promising in improving the current approximation of the front, an MC approach may become prohibitively expensive with a large enough c for an acceptable error level. Therefore, alternative approximation methods that are less computationally intensive are of interest. In the next section, we discuss such an approach based on GH quadrature.

3 Gauss-Hermite Approximation

The idea of GH approximation is based on the concept of Gaussian quadratures, which implies that if a function f can be approximated well by a polynomial of order $2n - 1$ or less, then a quadrature with n nodes suffices for a good approximation of a (potentially intractable) integral [19, 22], i.e.

$$\int_D f(\mathbf{x})\psi(\mathbf{x})\, dx \approx \sum_{i=1}^{n} w_i f(\mathbf{x}_i),$$

where D is the domain over which $f(\mathbf{x})$ is defined, and ψ a known weighting kernel (or probability density function). The domain D and weighing kernel ψ define a set of n weighted nodes $\mathcal{S} = \{\mathbf{x}_i, w_i\}, i = 1, \ldots, n$, where $\mathbf{x}_i$ is the ith deterministically chosen node and w_i its associated quadrature weight. We refer to this concept as *Gauss-Hermite* if D is infinite, i.e., $D \in (-\infty, \infty)$, and the weighting kernel ψ is given by the density of a standard Gaussian distribution.

The location of the nodes $\mathbf{x}_i$ are determined using the roots of the polynomial of order n, while the weights w_i are computed from a linear system upon computing the roots [19]; the interested reader is referred to [25] for technical details of this calculation. Intuitively, one can think of the selected nodes as representatives of the Gaussian distribution with the weights ensuring convergence as n increases and a low approximation error for a given n [12].

Extending the one-dimensional GH quadrature calculations to multivariate integrals is achieved by expanding the one-dimensional grid of nodes to form an m-dimensional grid, which is then pruned, rotated, scaled, and, finally, the nodes are translated. Figure 1 illustrates the key steps of this process for a two-dimensional ($m = 2$) integral. The weights of the m-variate quadrature points are the product of the corresponding m one-dimensional weights; for $m = 2$, this leads to the following two-dimensional Gaussian quadrature approximation:

$$\sum_{i=1, j=1}^{n,n} w_i w_j f(\mathbf{x}_i, \mathbf{x}_j).$$

The pruning step eliminates nodes that are associated with a low weight (i.e., points on the diagonal as they are further away from the origin); such nodes do not contribute significantly to the total integral value, hence eliminating them improves computational efficiency. Rotating, scaling and translating nodes account for correlations across dimensions, which is often present in practice. The rotation and scaling are conducted using a rotation matrix constructed from the dot product of the eigenvector and the eigenvalues of the covariance matrix, and the translation is performed by adding the mean vector.

Note that this approach may result in a combinatorial explosion of nodes in approximating a high-dimensional multivariate Gaussian distribution. Given n nodes per dimension for an m-dimensional space, the total number of nodes generated is $K = \lfloor n^m(1 - r) \rfloor$, where $r \in [0, 1]$ is the pruning rate (the greater r, the more nodes are discarded with $r = 0$ corresponding to not discarding any nodes). For instance, using $n = 5$, $m = 10$ and $r = 0.2$, we have $K = 7812500$ nodes. Clearly, this would be computationally more expensive than MC approximation. Therefore, for high-dimensional integration the default GH approach may not be suitable. As a rule of thumb, we propose that if the number

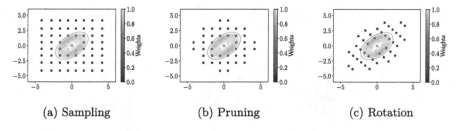

(a) Sampling   (b) Pruning   (c) Rotation

Fig. 1. An illustration of the process of generating the nodes and the associated weights using the GH quadrature for a two-dimensional ($m = 2$) Gaussian density with the mean vector $\boldsymbol{\mu} = (0,0)^{\top}$ and the covariance matrix $\Sigma = \begin{pmatrix} 1 & 0.5 \\ 0.5 & 1 \end{pmatrix}$. The parameters used are: $n = 8$ points per dimension and a pruning rate of $r = 0.2$ (i.e., 20%). The dots represent the nodes, and the colours represent the respective weights. The contours show the Gaussian density with the outermost contour corresponding to two-standard deviation.

of nodes from GH goes beyond the number of samples required for a good MC approximation, then one should use the latter, instead.

It should be noted that there is some work on high-dimensional GH approximations, e.g. [16], but we do not investigate these in this paper.

3.1 Gauss-Hermite for Approximating EHVI

To approximate the EHVI, we use K samples (nodes) and associated weights from GH quadrature as follows:

$$\sum_{i=1}^{K} \omega_i I(\mathbf{p}_i, P), \tag{2}$$

where P is the approximated Pareto front, $\mathbf{p}_i$ is the ith sample, and $\omega_i = \prod_{j=1}^{m} w_j(\mathbf{x}_i)$ is the weight in an m-dimensional objective space corresponding to the sample $\mathbf{x}_i$. This is effectively a weighted sum of the contributions, where the weights vary according to the probability density. This is also illustrated in the *right panel* of Fig. 2: The dots show the GH samples (nodes). The grid of points covers an area that is consistent with the underlying Gaussian distribution. Since we know how the probability density varies, we can generate proportional weights, which in turn permits us to derive a good approximation with only a few points in the grid.

On the other hand, with MC in (1), every sample (dots in Fig. 2, *left panel*) contributes equally to the average EHVI. Hence, a sample is somewhat unrelated to the intensity of the underlying probability density at that location. As such, with few samples, we may not derive a good approximation. It should be noted that the gray diamonds (in both panels) are dominated by the approximation of the Pareto front, and therefore there is no improvement (see Definition 3) due to these solutions. Hence, these gray diamonds do not contribute to the EHVI for either of the methods.

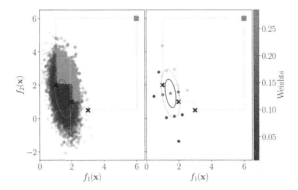

Fig. 2. A visual comparison of MC (*left*) and GH (*right*) samples in the objective space (assuming a minimisation problem). The approximation of the Pareto front is depicted with red crosses, and the reference vector **r** for hypervolume computation is shown with a magenta square. The blue dashed line outlines the dominated area. We used a two-variate Gaussian distribution to generate samples with the mean vector $\boldsymbol{\mu} = (1.5, 1.5)^\top$ and the covariance matrix $\Sigma = \begin{pmatrix} 0.16 & -0.15 \\ -0.15 & 1.01 \end{pmatrix}$; the pick contours represent this density. The gray diamonds are dominated by the approximated front. (Color figure online)

4   Experimental Study

In this section, we focus on comparing the accuracy of GH and MC approximations with respect to the analytical calculation of EHVI (A) introduced in [8,15]. As the analytical method is only suitable for independent multivariate Gaussian densities, we firstly investigate the efficacy of the approximation methods for uncorrelated densities for $m = 2$ and 3, and then expand our exploration to correlated multivariate densities. We use popular test problems: DTLZ 1–4, 7 [11], and WFG 1–9 [17]. They were chosen as they allow us to validate the efficacy of the approximation methods for Pareto fronts with diverse features; e.g., DTLZ2 and WFG4 have concave, DTLZ7 and WFG2 have disconnected, and DTLZ1 and WFG3 have linear Pareto fronts.

Our strategy was to first generate a random multivariate distribution, and then, for an approximation of the known Pareto front, compute the EHVI due to this random distribution analytically and with the two approximation methods (GH and MC). Using this approach, we aimed to collect data on a range of randomly generated multivariate distributions and inspect the agreement between analytical measurements and approximations. To quantify this, we used Kendall's τ rank correlation test [1], which varies between $[-1, 1]$ with 1 showing perfectly (positively) correlated ordering of the data by a pair of competing methods. The test also permits the estimation of a p-value, which, if below a predefined level α indicates that results are significant. In this paper, we set $\alpha = 0.05$, however, in all cases, we found the p-value to be practically zero, hence indicating significance in the results.

To implement the GH approximation, we converted existing R code[1] into Python; our code is available to download at github.com/AlmaRahat/ EHVI_Gauss-Hermite. If not stated otherwise, MC uses 10,000 samples, and GH uses a pruning rate of $r = 0.2$. For GH, we investigate different numbers of nodes (points) n per dimension, and use the notation GH_n to indicate this number. Any results reported are results obtained across 100 randomly generated multivariate Gaussian distributions to generate as many EHVIs.

4.1   Uncorrelated Multivariate Gaussian Distribution

To generate a random multivariate distribution, we first take a reference front P. We then calculate the maximum p^i_{max} and minimum p^i_{min} values along each objective function f_i. The span along the ith objective is thus $s = p^i_{max} - p^i_{min}$. Using this, we construct a hyper-rectangle H which has lower and upper bounds at vectors $\mathbf{l} = (l_1, \ldots, l_m)$ and $\mathbf{u} = (u_1, \ldots, u_m)$, respectively, with $l_i = p^i_{min} - 0.3s$ and $u_i = p^i_{max} + 0.3s$. We take a sample from H uniformly at random to generate a mean vector $\boldsymbol{\mu}$. The covariance matrix must be a diagonal matrix with positive elements for an independent multivariate distribution. Hence, we generate the ith diagonal element by sampling uniformly at random in the range $[0, u_i - l_i]$.

Figure 3 shows an example comparison between the analytical (A), MC and GH computations of EHVI for DTLZ2. The comparisons clearly show that the performances of MC and GH_{15} are reliable with respect to A with a Kendall's τ coefficient of over 0.97 and associated p-value of (almost) zero. To investigate if there is an increase in accuracy with the number of points per dimension, we repeated the experiment by varying the number of points per dimension between 3 and 15 (see Fig. 4 for results on the DTLZ2 problem with $m = 2$). Interestingly, there is a difference between having an odd or an even number of points per dimension: there is often a dip in performance when we go from even to odd. In Fig. 4, we see that there is a slight decrease in the rank coefficient between 4 and 5 points per dimension. We attribute this decrease to how the points are distributed for odd and even numbers of points per dimension. When we have an odd number of points for GH, it produces a node at the mean of the distribution. If there is an even number of points per dimensions, there is no node at the mean (see Fig. 5). Because of this, the approximation may vary between odd and even number of points. Nonetheless, the monotonicity in accuracy improvement is preserved when the number of points is increased by two.

---

[1] https://biostatmatt.com/archives/2754.

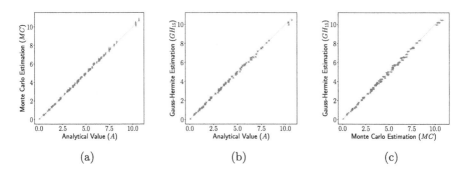

Fig. 3. Efficacy of MC and GH (with 15 points per dimension, GH_{15}) approximations in comparison to analytical measurements of EHVI for the DTLZ2 problem with 2 objectives and 100 randomly generated multivariate Gaussian distributions. The dotted red-line depicts the performance of the perfect approximations. MC approximations used 10,000 samples. In all cases, we observe strong rank correlations with Kendall's τ coefficient over 0.97 with practically zero p-values.

We took the same approach to investigate the efficacy of GH and MC in all the test problems for $m = 2$ and 3. The results of the comparison are summarised in Fig. 6. We observed the same trends that with the increase in the number of points per dimension, we increase the accuracy. Even with a small number of points per dimension we are able to derive coefficients of over 0.85 for all the problems. Interestingly, in some instances, e.g., WFG3 ($m = 3$) and WFG4 ($m = 2$), we clearly get better approximations from GH in comparison to MC.

4.2   Correlated Multivariate Gaussian Distribution

The key issue with the analytical formula for EHVI is that it does not cater for correlated multivariate predictive distributions. However, both MC and GH, even though they are computationally relatively intensive, do not suffer from this issue. To investigate the efficacy of different methods, again, we take the same approach as before. We generate random distributions and compute the EHVI values with A, MC and GH, and then evaluate the rank correlations using Kendall's τ coefficient. Importantly, the most reliable method in this case is MC.

In this instance, the process to generate a random mean vector remains the same. However, for a valid covariance matrix, we must ensure that the randomly generated matrix remains positive definite. We, therefore, use Wishart distribution [29] to generate a positive definite matrix that is scaled by diag($u_1 - l_1, \ldots, u_m - l_m$). To demonstrate that the analytical version for uncorrelated distributions generates a poor approximation for the EHVI due to a correlated distribution, we use the diagonal of the covariance matrix and ignore the off-diagonal elements, and compute the EHVI. This allows us to quantitatively show that GH may be a better alternative to MC from an accuracy perspective.

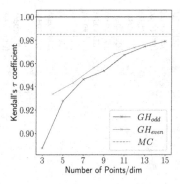

Fig. 4. Increase in accuracy with the increase in the number of points per dimension between 3 to 15 for the GH approximation with respect to the analytical result for DTLZ2 ($m = 2$) and 100 random multivariate distributions. The black horizontal line shows the theoretical upper bound for Kendall's τ coefficient. The red dashed horizontal line shows the coefficient for the EHVI using MC. The blue and red lines depict the increase in coefficient as we increase the number of points per dimensions for odd and even numbers, respectively, for GH.

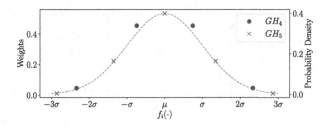

Fig. 5. An example of the distribution of GH nodes for 4 (GH_4) and 5 (GH_5) points per dimension (before pruning) for the standard Gaussian density with the mean $\mu = 0$ and the standard deviation $\sigma = 1$ (shown in dashed black line).

In Figs. 7a–7c, we show the comparison between different methods for computing EHVI for the DTLZ2 problem with $m = 2$. Here, A somewhat agrees with MC and GH_{15} with a correlation coefficient of approximately 0.84 in each case. However, MC and GH_{15} are essentially producing the same ranking of solutions with a coefficient of just over 0.97. Therefore, GH_{15} with 180 nodes is an excellent alternative to MC with 10,000 samples. The results on DTLZ2 do not appear too bad for A. To ensure that this is the case for all test problems under scrutiny, we repeated the experiments, but this time generating 100 random multivariate *correlated* Gaussian distributions in each instance. Here, we assumed that MC is the most reliable measure, and computed the Kendall's τ coefficient with respect to MC. The correlation coefficient distributions for A and GH_ns are given in Fig. 7d. Clearly, there is a wide variance in the performance of A, with the minimum being 0.39 for WFG7 ($m = 3$) and maximum being 0.9 for DTLZ1 ($m = 2$). On the other hand, GH_3 produced the worst performance

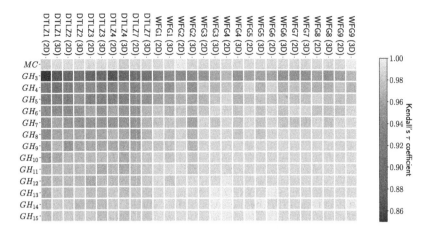

Fig. 6. Performance comparison of MC and GH (GH_n, where n is the number of points per dimension) with respect to the analytical EHVI for a range of test problems with 100 randomly generated multivariate Gaussian distributions in each instance. Lighter colours correspond to better coefficient values. (Color figure online)

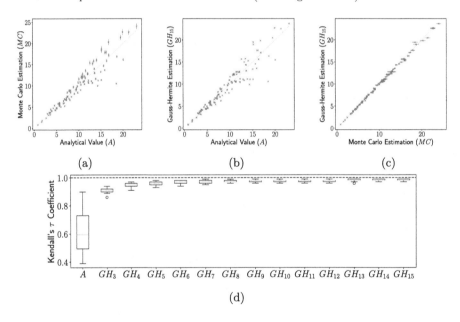

Fig. 7. Illustration of the efficacy of GH for correlated multivariate Gaussian distributions as Fig. 3 in 7a–7c for DTLZ2 ($m = 2$). In 7d, we show the summary of efficacies for different approximation methods when compared to MC across all DTLZ and WFG problems (for $m = 2$ and $m = 3$). Analytical approximations were generated using the diagonal of the covariance matrix.

for GH across the board, but that was at 0.86 for DTLZ ($m = 2$), which shows a strong rank correlation. This shows that just considering the diagonal of the covariance matrix and computing the analytical EHVI is not a reliable approximation method under a *correlated* multivariate predictive density. Instead, GH can produce a solid approximation with very few points.

5   Conclusions

EHVI is a popular acquisition function for expensive multi-objective optimisation. Computing it analytically is possible for independent objectives (predictive densities). However, this can be prohibitively expensive for more than 3 objectives. Monte Carlo approximation can be used instead, but this is not cheap. We proposed an approach using GH quadrature as an alternative to approximating EHVI. Our experimental study showed that GH can be an accurate alternative to MC for both independent and correlated predictive densities with statistically significant rank correlations for a range of popular test problems. Future work can look at improving the computational efficiency of GH for high-dimensional problems, and validating GH within BO using EHVI as the acquisition function.

Acknowledgements. This work is a part of the thematic research area Decision Analytics Utilizing Causal Models and Multiobjective Optimization (DEMO, jyu.fi/demo) at the University of Jyvaskyla. Dr. Rahat was supported by the Engineering and Physical Research Council [grant number EP/W01226X/1].

References

1. Abdi, H.: The Kendall rank correlation coefficient. In: Salkind, N.J. (ed.) Encyclopedia of Measurement and Statistics, pp. 508–510. Sage Publications, Thousand Oaks (2007)
2. Allmendinger, R.: Tuning evolutionary search for closed-loop optimization. Ph.D. thesis, The University of Manchester (2012)
3. Allmendinger, R., Emmerich, M.T., Hakanen, J., Jin, Y., Rigoni, E.: Surrogate-assisted multicriteria optimization: complexities, prospective solutions, and business case. J. Multi-Criteria Decis. Anal. **24**(1–2), 5–24 (2017)
4. Allmendinger, R., Jaszkiewicz, A., Liefooghe, A., Tammer, C.: What if we increase the number of objectives? Theoretical and empirical implications for many-objective combinatorial optimization. Comput. Oper. Res. **145**, 105857 (2022)
5. Bonilla, E.V., Chai, K., Williams, C.: Multi-task Gaussian process prediction. In: Proceedings of the 20th International Conference on Neural Information Processing Systems, pp. 153–160 (2007)
6. Chugh, T., Sindhya, K., Hakanen, J., Miettinen, K.: A survey on handling computationally expensive multiobjective optimization problems with evolutionary algorithms. Soft. Comput. **23**(9), 3137–3166 (2017). https://doi.org/10.1007/s00500-017-2965-0
7. Chugh, T., Sindhya, K., Miettinen, K., Jin, Y., Kratky, T., Makkonen, P.: Surrogate-assisted evolutionary multiobjective shape optimization of an air intake ventilation system. In: Proceedings of the 2017 IEEE Congress on Evolutionary Computation (CEC), pp. 1541–1548. IEEE (2017)

8. Couckuyt, I., Deschrijver, D., Dhaene, T.: Fast calculation of multiobjective probability of improvement and expected improvement criteria for pareto optimization. J. Global Optim. **60**(3), 575–594 (2014)

9. Daulton, S., Balandat, M., Bakshy, E.: Differentiable expected hypervolume improvement for parallel multi-objective Bayesian optimization. Adv. Neural. Inf. Process. Syst. **33**, 9851–9864 (2020)

10. Deb, K.: Multi-objective optimization. In: Burke, E.K., Kendall, G. (eds.) Search Methodologies, pp. 403–449. Springer, Boston (2014). https://doi.org/10.1007/978-1-4614-6940-7_15

11. Deb, K., Thiele, L., Laumanns, M., Zitzler, E.: Scalable test problems for evolutionary multiobjective optimization. In: Abraham, A., Jain, L., Goldberg, R. (eds.) Evolutionary multiobjective optimization, pp. 105–145. Springer, London (2005). https://doi.org/10.1007/1-84628-137-7_6

12. Elvira, V., Closas, P., Martino, L.: Gauss-Hermite quadrature for non-Gaussian inference via an importance sampling interpretation. In: Proceedings of the 2019 27th European Signal Processing Conference (EUSIPCO), pp. 1–5. IEEE (2019)

13. Emmerich, M.: Single- and multi-objective evolutionary design optimization assisted by Gaussian random field metamodels. Ph.D. thesis, TU Dortmund (2005)

14. Emmerich, M., Yang, K., Deutz, A., Wang, H., Fonseca, C.M.: A multicriteria generalization of Bayesian global optimization. In: Pardalos, P.M., Zhigljavsky, A., Žilinskas, J. (eds.) Advances in Stochastic and Deterministic Global Optimization. SOIA, vol. 107, pp. 229–242. Springer, Cham (2016). https://doi.org/10.1007/978-3-319-29975-4_12

15. Emmerich, M.T., Deutz, A.H., Klinkenberg, J.W.: Hypervolume-based expected improvement: Monotonicity properties and exact computation. In: 2011 IEEE Congress of Evolutionary Computation (CEC), pp. 2147–2154. IEEE (2011)

16. Holtz, M.: Sparse grid quadrature in high dimensions with applications in finance and insurance. Ph.D. thesis, Institut für Numerische Simulation, Universität Bonn (2008)

17. Huband, S., Hingston, P., Barone, L., While, L.: A review of multiobjective test problems and a scalable test problem toolkit. IEEE Trans. Evol. Comput. **10**(5), 477–506 (2006)

18. Hupkens, I., Deutz, A., Yang, K., Emmerich, M.: Faster exact algorithms for computing expected hypervolume improvement. In: Gaspar-Cunha, A., Henggeler Antunes, C., Coello, C.C. (eds.) EMO 2015. LNCS, vol. 9019, pp. 65–79. Springer, Cham (2015). https://doi.org/10.1007/978-3-319-15892-1_5

19. Jäckel, P.: A note on multivariate Gauss-Hermite quadrature. ABN-Amro. Re, London (2005)

20. Knowles, J.D.: Local-search and hybrid evolutionary algorithms for Pareto optimization. Ph.D. thesis, University of Reading Reading (2002)

21. Koehler, E., Brown, E., Haneuse, S.J.P.: On the assessment of Monte Carlo error in simulation-based statistical analyses. Am. Stat. **63**(2), 155–162 (2009)

22. Liu, Q., Pierce, D.A.: A note on Gauss-Hermite quadrature. Biometrika **81**(3), 624–629 (1994)

23. Miettinen, K.: Nonlinear Multiobjective Optimization. Kluwer Academic Publishers, New York (1999)

24. Montgomery, D.C.: Design and Analysis of Experiments. Wiley, Hoboken (2017)

25. Press, W.H., Teukolsky, S.A., Vetterling, W.T., Flannery, B.P.: Numerical Recipes in C. Oxford University Press, Oxford (1992)

26. Shah, A., Ghahramani, Z.: Pareto frontier learning with expensive correlated objectives. In: Proceedings of the 33rd International Conference on Machine Learning, pp. 1919–1927 (2016)
27. Shahriari, B., Swersky, K., Wang, Z., Adams, R.P., de Freitas, N.: Taking the human out of the loop: a review of Bayesian optimization. Proc. IEEE **104**(1), 148–175 (2016)
28. Stewart, T., et al.: Real-world applications of multiobjective optimization. In: Branke, J., Deb, K., Miettinen, K., Słowiński, R. (eds.) Multiobjective Optimization. LNCS, vol. 5252, pp. 285–327. Springer, Heidelberg (2008). https://doi.org/10.1007/978-3-540-88908-3_11
29. Wishart, J.: The generalised product moment distribution in samples from a normal multivariate population. Biometrika **20A**(1–2), 32–52 (1928)
30. Yang, K., Emmerich, M., Deutz, A., Fonseca, C.M.: Computing 3-D expected hypervolume improvement and related integrals in asymptotically optimal time. In: Trautmann, H., et al. (eds.) EMO 2017. LNCS, vol. 10173, pp. 685–700. Springer, Cham (2017). https://doi.org/10.1007/978-3-319-54157-0_46
31. Yang, K., Gaida, D., Bäck, T., Emmerich, M.: Expected hypervolume improvement algorithm for PID controller tuning and the multiobjective dynamical control of a biogas plant. In: Proceedings of the 2015 IEEE Congress on Evolutionary Computation (CEC), pp. 1934–1942. IEEE (2015)
32. Zitzler, E., Thiele, L., Laumanns, M., Fonseca, C.M., Da Fonseca, V.G.: Performance assessment of multiobjective optimizers: an analysis and review. IEEE Trans. Evol. Comput. **7**(2), 117–132 (2003)

Finding Knees in Bayesian
Multi-objective Optimization

Arash Heidari[1]([✉]), Jixiang Qing[1], Sebastian Rojas Gonzalez[1,3],
Jürgen Branke[2], Tom Dhaene[1], and Ivo Couckuyt[1]

[1] Faculty of Engineering and Architecture, Ghent University - imec, Ghent, Belgium
`arash.heidari@ugent.be`
[2] Warwick Business School, University of Warwick, Coventry, UK
[3] Data Science Institute, Hasselt University, Hasselt, Belgium

Abstract. Multi-objective optimization requires many evaluations to
identify a sufficiently dense approximation of the Pareto front. Especially
for a higher number of objectives, extracting the Pareto front might not
be easy nor cheap. On the other hand, the *Decision-Maker* is not always
interested in the entire Pareto front, and might prefer a solution where
there is a desirable trade-off between different objectives. An example
of an attractive solution is the knee point of the Pareto front, although
the current literature differs on the definition of a knee. In this work, we
propose to detect knee solutions in a data-efficient manner (i.e., with a
limited number of time-consuming evaluations), according to two definitions of knees. In particular, we propose several novel acquisition functions in the Bayesian Optimization framework for detecting these knees,
which allows for scaling to many objectives. The suggested acquisition
functions are evaluated on various benchmarks with promising results.

Keywords: Multi-objective optimization · Knee finding · Bayesian
optimization · Surrogate modeling

1 Introduction

Optimization is an important topic in many domains, from engineering design
to economics and even biology. Real-world problems often involve multiple conflicting objectives. For example, in engineering, minimization of cost and maximization of efficiency are looked for simultaneously. As a result, there will be
a set of solutions, each better in one or more objectives and worse in at least
one objective. In the other words, they do not dominate each other. Hence,
these solutions are referred to as *non-dominated* or *Pareto-optimal* and form
the so-called Pareto set and Pareto front in the decision and objective spaces,

This work has been supported by the Flemish Government under the 'Onderzoekspro-
gramma Artificiële Intelligentie (AI) Vlaanderen' and the 'Fonds Wetenschappelijk
Onderzoek (FWO)' programmes.

G. Rudolph et al. (Eds.): PPSN 2022, LNCS 13398, pp. 104–117, 2022.
https://doi.org/10.1007/978-3-031-14714-2_8

respectively. A multi-objective optimization, without the loss of generality, can be defined as:

$$minimize \ f_1(\mathbf{x}), f_2(\mathbf{x}), ..., f_m(\mathbf{x}) \quad \mathbf{x} \in \Omega \subseteq \mathbb{R}^n \qquad (1)$$

Finding a set of non-dominated solutions is challenging, or even infeasible, especially with an increasing number of objectives, as the number of solutions to cover the entire Pareto front usually grows exponentially with the number of objectives [5]. In practice, the *Decision-Maker (DM)* is not interested in the whole front of the solutions and might *prefer* a solution where there is a desirable trade-off between different objectives. One approach to tackle this problem is to transform the multi-objective setting into a single-objective problem [9,14], for example, by using a (non)linear utility function, but identifying the appropriate weights with no prior information is not an easy task.

One set of attractive solutions are the *knees* of the Pareto front (see Fig. 1), first defined in [11]. However, definitions of what *a knee* is differ in the literature; depending on the definition, a knee might hold different properties. For example, the ratio of gain and loss in each objective might be the same at a knee point.

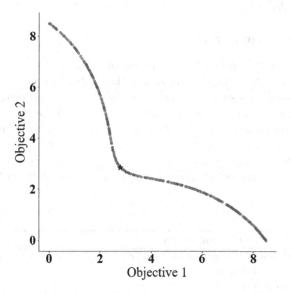

Fig. 1. Pareto front approximation of a bi-objective minimization problem. Intuitively, the knee (red star) is an attractive solution as it strikes a good trade-off between objectives. (Color figure online)

Most of the current literature on knee-oriented optimization focuses on Evolutionary Algorithms (EAs) to estimate the location of the knee. While EAs are a good solution for high-dimensional and intractable problems, they are data-hungry methods, as they evaluate the objective functions many times during optimization. However, EAs are still preferred in some situations, e.g., when the objective exhibits complex non-linear behavior.

Evaluating objective functions are often computationally expensive, severely limiting the number of function evaluations during optimization. In engineering design, for instance, high-fidelity models can take hours to days for one simulation. Thus, it is of interest to solve the problem in a data-efficient manner, i.e., finding the most interesting Pareto-optimal solutions with minimal computational budget.

In this paper, we investigate two definitions of a knee in multi-objective optimization, and propose **three novel acquisition functions** for the Bayesian Optimization (BO) framework to detect them in a **data-efficient** way. BO is an optimization technique that utilizes a surrogate model to reduce the number of time-consuming evaluations.

This paper is structured as follows. In Sect. 2, we briefly review the related work. Proposed algorithms are covered in detail in Sect. 3. Section 4 summarizes the experimental setup, while the results are discussed in Sect. 5. Finally, in the last section, we conclude with a discussion on further improvements.

2   Related Work

2.1   Bayesian Optimization

A powerful option for finding the optimum of a black-box and expensive-to-evaluate function is Bayesian Optimization (BO) [22]. BO employs an acquisition function based on a surrogate model to quantify how interesting a solution is. The point that maximizes the acquisition function will be chosen as the next candidate for evaluation (Algorithm 1). Popular choices for the acquisition function are Expected Improvement (EI) [13,18] and Probability of Improvement (PoI) [12,17]. BO can also be used to find the complete Pareto front of the solutions, using e.g., the Expected Hyper-Volume Improvement (EHVI) acquisition function [6,8].

Algorithm 1. Bayesian Optimization

Input Evaluated design of experiment using, e.g., Halton sampling
Input An acquisition function
 1: **while** *Budget* left **do**
 2:     Train a surrogate model
 3:     Prediction of the surrogate model in the decision space
 4:     $K \leftarrow$ The point that maximizes the acquisition function
 5:     Evaluate K using time consuming function
 6:     Reduce *Budget*
 7: **end while**

2.2 Gaussian Process

The Gaussian Process (GP) surrogate model [20] is a common choice in Bayesian Optimization. The GP provides the prediction for a new input as well as uncertainty. The GP is fully specified by a mean function and kernel matrix $k(\mathbf{x}_i, \mathbf{x}_j)$. Assuming a zero mean function, the posterior of the GP is given as:

$$\mu(\mathbf{x}_*) = k_* K_{xx}^{-1} y \tag{2}$$

$$\sigma^2(\mathbf{x}_*) = k_{**} - k_* K_{xx}^{-1} k_*^T \tag{3}$$

where $\mathbf{x}_*$ is a new input, $K_{xx} = k(\mathbf{x}_i, \mathbf{x}_j)$, $k_* = k(\mathbf{x}_*, \mathbf{x}_i)$, and $k_{**} = k(\mathbf{x}_*, \mathbf{x}_*)$. Different kernel functions, such as Matérn kernels [16] or an RBF kernel, can be used. Which one to choose is problem-dependent. For example, the Matérn 5/2 kernel has less strong assumptions on the smoothness of the target function and found to be more suitable for real-life problems [19].

2.3 Knee Finding Using Evolutionary Algorithms

Multi-objective Evolutionary Algorithms (MOEAs) are popular to find the Pareto front of a problem. Yu et al. [25] classify the MOEAs into four different categories, i.e., dominance relations based, decomposition based, indicator based, and secondary-criterion based methods. Interested readers can refer to [10,15,23,25] for more details.

There is no unique definition of what a knee point is. A knee point is an attractive solution of the Pareto front that will often be chosen by the DM in the absence of prior knowledge of the problem [4]. In [21], the methods to quantitatively measure a knee are classified into two different groups: (1) based on the geometric characteristics of the Pareto front, and (2) based on the trade-off information. In [25], knee-oriented MOEAs are classified into five categories, i.e., utility-based, angle-based, dominance-based, niching-based, and visualization-based approaches. Each of the algorithms has its own definition of the knee, making it difficult to compare them. For example, Branke et al. [4] defines the knee as a point in the Pareto front that has the largest angle to its neighbours, while other works take a utility-based approach for specifying a knee point [2,26]. In this work we focus on the definition of knee as described in [11] to develop the proposed acquisition functions. We also propose another definition of the knee and construct an acquisition function based on that. These are described in detail in the next section.

3 Proposed Algorithms

We investigate two definitions of a knee point: (1) based on the Hyper-Volume (i.e., the volume of objective space dominated by a given set of solutions [27]) with respect to a reference point, and (2) based on the distance to a reference line.

3.1   Hyper-Volume-Based Knee (HV-Knee)

The Hyper-Volume can be used to define knees in the Pareto front. A trade-off between various objectives can be observed by calculating the Hyper-Volume between different solutions and a reference point. Solutions with a high Hyper-Volume are intuitively more interesting for the DM. Accordingly, the knee point is the point on the Pareto front that has the maximum Hyper-Volume with respect to a fixed reference point. The corresponding regret function is calculated as follows:

$$Regret_{HV} = HV(\mathbf{y}^*{}_{HV}, N^*) - HV(\mathbf{y}_{best}, N^*) \tag{4}$$

where N^* is the true Nadir point, $\mathbf{y}^*{}_{HV}$ is the point in the Pareto front that has the maximum hyper-volume with respect to N^* (ground truth), and $\mathbf{y}_{best}$ is the point that the algorithm found and has the maximum hyper-volume with respect to N^*.

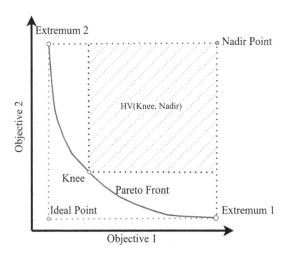

Fig. 2. Illustration of a knee point. Based on the HV-Knee definition, the point on the Pareto front that has the maximum Hyper-Volume with respect to the reference point is the best knee (yellow point). The striped region represents the Hyper-Volume (HV) between the Knee and the Nadir Point. (Color figure online)

The identified knee depends on the reference point. Due to this sensitivity, selecting the reference point is a critical part of the proposed algorithm. It can be defined upfront by the DM (informed), which might be unrealistic for many problems. Hence, we set the reference point the same as the nadir point as a sensible default (see Fig. 2) which in turn depends on an accurate estimation of the extrema. However, locating the extrema is not an easy task. We propose an interleaved approach to find the extrema. Algorithm 2 shows how two steps iterate.

Algorithm 2. The HV-Knee Algorithm

Input Evaluated design of experiment using, e.g., Halton sampling
1: **while** *Budget* left **do**
2:      Train a surrogate model for each objective
3:      $E \leftarrow ExtremaSampler()$
4:      Evaluate E.
5:      Update surrogate models
6:      $K \leftarrow HVKneeSampler()$
7:      Evaluate K.
8:      Reduce *Budget*
9: **end while**

Both *ExtremaSampler* and *HVKneeSampler* are acquisition functions. To find the extrema, first, the ideal point, which is the minimum of each objective (see Fig. 2) is extracted from the current dataset. A large reference point will be chosen to focus more on the extrema. Finally, a derivation of the standard EHVI [6] is used. EHVI tries to evaluate the point that contributes the most to the expected Hyper-volume of the Pareto front given a fixed reference point and the extracted Pareto front so far. We modify EHVI with the ideal point as the only point in the Pareto front and a sufficiently large vector as the reference point. Reference point should have large values in a way that is dominated by all of the extremum points. For example, a vector such as $(1e6, \ldots, 1e6)$ can be used as the reference point. Algorithm 3 shows the implementation of *ExtremaSampler*.

Algorithm 3. *ExtremaSampler*: Optimizing the Extrema acquisition function

Input R : A sufficiently large vector
1: $I \leftarrow$ Ideal Point extracted from the current Pareto front
2: $E \leftarrow$ Maximize $EHVI$ with R and I as the reference point and the Pareto front, respectively.
3: Return E as the next candidate point

HVKneeSampler modifies the standard EHVI to estimate the location of the knee as well. EHVI is evaluated with the nadir point as both the reference point and the only point in the Pareto front. Algorithm 4 shows the implementation of *HVKneeSampler*.

Algorithm 4. *HVKneeSampler*: Optimizing the HVKnee acquisition function

1: $N \leftarrow$ Nadir Point extracted from the current Pareto front
2: $K \leftarrow$ Maximize $EHVI$ with N as the reference point and the Pareto front.
3: Return K as the next candidate point

3.2 Distance to Line-Based Knee

Another intuitive definition of a knee is based on the distance to an imaginary line connecting the extrema of the front in a bi-objective setting (see Fig. 3) first proposed by [11]. It is in the interest of DM to maximize the gap between a solution and the reference line. The regret function is calculated as follows:

$$Regret_{DL} = Distance(\mathbf{y}^*{}_{DL}, L^*) - Distance(\mathbf{y}_{best}, L^*) \tag{5}$$

where L^* is the true reference line, $\mathbf{y}^*{}_{DL}$ is the point in the Pareto front that has the maximum distance to the L^* (ground truth), and $\mathbf{y}_{best}$ is the current best solution.

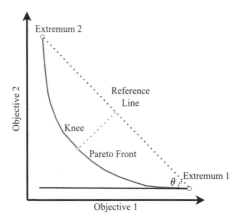

Fig. 3. Illustration of a knee based on the distance to the reference line between the two extrema. The point on the Pareto front that has the maximum distance to the line constructed by connecting the two extrema is considered the knee (yellow point). (Color figure online)

Similarly to the HV-Knee approach, the location of the extrema is unknown beforehand, and estimating them is a vital part. The two-step approach from the previous section is reused, replacing $HVKneeSampler$ with $D2LKneeSampler$.

Algorithm 5 shows the implementation of $D2LKneeSampler$. First, the current Pareto front and extrema are extracted, and the reference line will be constructed. The point in the current Pareto front that has the largest distance to the reference line is designated as the current best knee. To calculate the probability of improving over the current best knee, a naive approach is to solve a double integration requiring Monte Carlo integration. Instead, we propose to transform to a new coordinate system based on the reference line. In particular, we consider a line parallel to the reference line that passes through the current knee. The system is rotated so reference line is aligned with the horizontal axis. As a result, it is much easier to analytically integrate the (transformed) multivariate Gaussian distribution of the GPs. Now, the equation can be simplified to a single variable probability of improvement (or expected improvement) acquisition function (Fig. 4). Keep in mind that, the point should be below the reference line (assuming minimization). If the point is above the line, we use the negative of the distance.

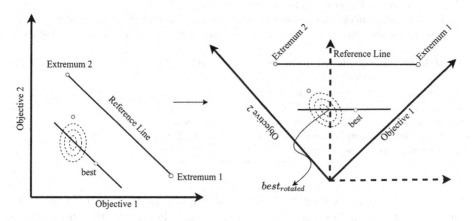

Fig. 4. The probability of improvement (or expected improvement) is calculated by transforming the problem to a new coordinate system, in such a way that the reference line becomes horizontal. A one-dimensional integration similar to the standard probability of improvement and expected improvement can be applied to the rotated Gaussian distribution.

If μ_1, σ_1^2, μ_2, and σ_2^2 are the predicted mean and variance of a candidate point using the GPs for objective 1 and objective 2, respectively, then Eqs. 6–11 can be used to calculate lines 4–9 of Algorithm 5.

$$Cov = \begin{bmatrix} \sigma_1^2 & 0 \\ 0 & \sigma_2^2 \end{bmatrix} \tag{6}$$

$$Means = \begin{bmatrix} \mu_1 & \mu_2 \end{bmatrix} \tag{7}$$

$$mean_{rotated} = \mu_2 \times cos(\theta) + \mu_1 \times sin(\theta) \tag{8}$$

$$Cov_{rotated} = \begin{bmatrix} cov_{11} & cov_{12} \\ cov_{21} & cov_{22} \end{bmatrix} = \begin{bmatrix} cos(\theta) & -sin(\theta) \\ sin(\theta) & cos(\theta) \end{bmatrix} \times Cov \times \begin{bmatrix} cos(\theta) & sin(\theta) \\ -sin(\theta) & cos(\theta) \end{bmatrix} \tag{9}$$

$$\sigma_{rotated}^2 = cov_{22} \tag{10}$$

If the coordinates of the current knee before the rotation is $best_1$ and $best_2$, then:

$$best_{rotated} = (best_2 - tan(\theta) \times best_1) \times cos(\theta) \tag{11}$$

Algorithm 5. *D2LKneeSampler*: Optimizing the D2LKnee acquisition function

---
1: *alpha* ← Acquisition function (probability of improvement or expected improvement).
2: E_1, E_2 ← Extract extrema from the current Pareto front.
3: L ← Reference line constructed using E_1 and E_2.
4: θ ← The angle between the constructed reference line and the horizontal axis.
5: Cov ← The diagonal covariance matrix composed of the predicted variances of the GPs.
6: $Means$ ← The predicted mean vector of the GPs.
7: $\mu_{rotated}$ ← Rotated mean vector by θ degrees
8: $Cov_{rotated}$ ← Rotated Cov by θ degrees.
9: $\sigma^2_{rotated}$ ← cov_{22} element of $Cov_{rotated}$
10: $best_{rotated}$ ← Vertical coordinate of the rotated best knee.
11: K ← Maximize *alpha* with $best_{rotated}$, $\mu_{rotated}$ and $\sigma^2_{rotated}$.
12: Return K as the next candidate point.

It is possible to use either Probability of Improvement (PoI) or Expected Improvement (EI) in the 11th line of Algorithm 5, leading to the Probability of Improving with respect to the Distance to Line (PID2L), and the Expected Improvement with respect to the Distance to Line (EID2L).

4 Experimental Setup

Experiments have been conducted with three various benchmark functions, namely DO2DK, DEB2DK, and DEB3DK [4]. We configure the functions to have input dimensions 9, 5, and 7, respectively. DO2DK has an additional parameter, s, that skews the front, which is set to 1 in the experiments.

The Pareto fronts of various benchmark functions have been approximated using the NSGA-II algorithm, and the knee(s) are calculated based on the Hyper-Volume and Distance to Line knee definitions as shown in Table 1. The extracted knee(s) using NSGA-II are designated as ground truth knee(s) and used for regret calculation. The nadir point N^*, and the reference line, L^*, are constructed using extracted extrema from NSGA-II results as well. For each benchmark function, these two definitions might end up choosing the same point as the knee, but generally this is not true, however, they are often remarkably close to each other.

For extracting the Pareto front, we configure NSGA-II with population size 200 and 200 generations (DO2DK, DEB2DK), and population size 500 and 1000 generations (DEB3DK).

We compare the proposed acquisition functions for knee detection against the standard Expected Hyper-Volume Improvement (EHVI), which extracts the whole Pareto front. We use the RBF kernel for the GPs, and the number of initialization points is ten times the input dimension. To optimize the acquisition function, a Monte Carlo approach with one thousand times input dimension samples is used and L-BFGS-B optimizer is utilized to fine-tune the best point.

Table 1. Summary of the benchmark functions to validate the proposed approaches and their truth ground knee extracted using the NSGA-II algorithm.

Name	Input dimension	Output dimension	Ground truth
DEB2DK [4]	9	2	$(2.83, 2.83)^*$
DO2DK [4]	5	2	$(1.07, 1.02)^*$
DEB3DK [4]	7	3	$(2.85, 2.83, 3.52)^*$

*Rounded to two decimal places

Due to ambiguous definitions of a knee, as well as the data-hungry nature of EAs, no other knee-oriented methods could be included in the comparison (evaluating the initial population would exceed the computation budget). $Regret_{HV}$ and $Regret_{DL}$ are used to measure the performance during the optimization process. Each experiment was repeated 15 times for DEB2DK and DO2DK, and 10 times for DEB3DK, and the 50th, 20th, and 80th percentiles were calculated.

The Pymoo python package [3], and Trieste framework [1] have been used for NSGA-II and the BO methods, respectively.

5   Results

The results are shown in Fig. 5. For DO2DK, a small value, 1.7×10^{-4} and 2×10^{-4}, is added to all Hyper-Volume and Distance Regrets, respectively, since the regret was negative for HV-knee and PID2L. This means that both PID2L and HV-Knee were successful in finding a point that performs better than the ground truth knee found by NSGA-II.

For DEB2DK the EID2L acquisition function shows a quick improvement in the early stages, but the HV-Knee and PID2L show a continuous improvement leading to better results near the end of the optimization process. All the acquisition functions exhibit the same behavior for DO2DK as well.

The last benchmark function, DEB3DK, has three objectives, and, hence, can only be used with the HV-Knee method. Note that the best regret is also negative for this case. The shaded area at the 170th iteration is between -21 and -25, which means the HV-Knee acquisition function was able to find a point that performs much better than the knee point found by NSGA-II.

Figure 6 shows the extracted Pareto front for DTLZ2 [7] benchmark function. Pareto front of the DTLZ2 function is concave. In this case, the DM might prefer one of the extrema, known as the edge knee [24]. As all of the proposed acquisition functions are able to estimate the location of the extrema, and since the shape of the extracted Pareto front using the proposed acquisition functions is concave, the DM can choose one of the extrema as the final solution. PID2L (and also EID2L) return one of the extrema as the best point, but HV-Knee acquisition function prefers a point that is almost in the middle of the front.

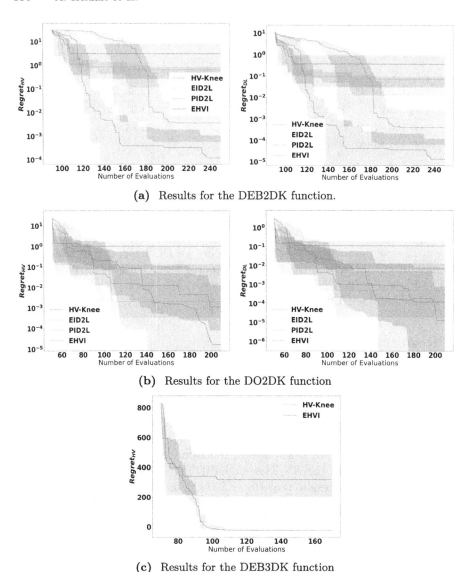

(a) Results for the DEB2DK function.

(b) Results for the DO2DK function

(c) Results for the DEB3DK function

Fig. 5. Results for the various benchmark function. Each experiment is repeated 15 times for the DO2DK and the DEB2DK functions, and 10 times for the DEB3DK function. The medians are denoted by the solid lines, while the shaded area represents the area between 20th and 80th percentile.

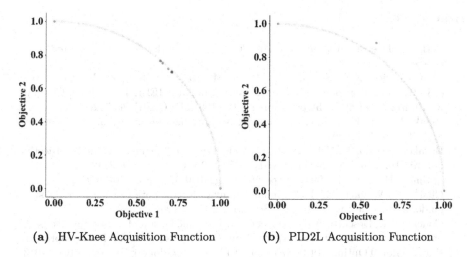

(a) HV-Knee Acquisition Function   **(b)** PID2L Acquisition Function

Fig. 6. Extracted Pareto front of the DTLZ2 benchmark function using HV-Knee and PID2L acquisition function (in blue) and NSGA-II algorithm (in red). In both cases, it is clear that the Pareto front is concave, and since both acquisition functions are able to estimate the location of the extrema, one of the extrema can be chosen as the final solution. (Color figure online)

6   Conclusion and Next Steps

In this work we have proposed three acquisition functions for Bayesian Optimization to find attractive solutions (knees) on the Pareto front. These acquisition functions were able to identify the correct knee in a data-efficient manner, using about 200 evaluations (or even less) for a satisfying solution. Identifying a single solution is more efficient than the complete Pareto front, allowing Bayesian Optimization to scale up to more inputs and objectives. The proposed acquisition functions outperformed the ground truth obtained using an expensive NSGA-II approach. However, in some cases EAs are still preferred, for example, when at least one of the objective functions is hard to model with a GP (intractable function, high-input dimension), or when the evaluation of the objective functions is cheap and fast.

The developed acquisition functions alternated between two steps which is more time-consuming than it needs to be. More rigorous approaches will be developed to achieve an automatic balance between finding the extrema and identifying the knee. This will reduce the number of required evaluations further. Moreover, the proposed methods only focused on the global knee. If there is more than one knee in the Pareto front, they often remain unexplored. Current approaches will be extended, so other knees are also explored.

References

1. Berkeley, J., et al.: Trieste, February 2022. https://github.com/secondmind-labs/trieste
2. Bhattacharjee, K.S., Singh, H.K., Ryan, M., Ray, T.: Bridging the gap: many-objective optimization and informed decision-making. IEEE Trans. Evol. Comput. **21**(5), 813–820 (2017). https://doi.org/10.1109/TEVC.2017.2687320
3. Blank, J., Deb, K.: pymoo: multi-objective optimization in python. IEEE Access **8**, 89497–89509 (2020)
4. Branke, J., Deb, K., Dierolf, H., Osswald, M.: Finding knees in multi-objective optimization. In: Yao, X., et al. (eds.) PPSN 2004. LNCS, vol. 3242, pp. 722–731. Springer, Heidelberg (2004). https://doi.org/10.1007/978-3-540-30217-9_73
5. Chand, S., Wagner, M.: Evolutionary many-objective optimization: a quick-start guide. Surv. Oper. Res. Manage. Sci. **20**(2), 35–42 (2015)
6. Couckuyt, I., Deschrijver, D., Dhaene, T.: Fast calculation of multiobjective probability of improvement and expected improvement criteria for Pareto optimization. J. Global Optim. **60**(3), 575–594 (2013). https://doi.org/10.1007/s10898-013-0118-2
7. Deb, K., Thiele, L., Laumanns, M., Zitzler, E.: Scalable multi-objective optimization test problems. In: Proceedings of the 2002 Congress on Evolutionary Computation. CEC 2002 (Cat. No.02TH8600). vol. 1, pp. 825–830 (2002). https://doi.org/10.1109/CEC.2002.1007032
8. Emmerich, M., Giannakoglou, K., Naujoks, B.: Single- and multiobjective evolutionary optimization assisted by gaussian random field metamodels. IEEE Trans. Evol. Comput. **10**(4), 421–439 (2006). https://doi.org/10.1109/TEVC.2005.859463
9. Hakanen, J., Knowles, J.D.: On using decision maker preferences with ParEGO. In: Trautmann, H., et al. (eds.) EMO 2017. LNCS, vol. 10173, pp. 282–297. Springer, Cham (2017). https://doi.org/10.1007/978-3-319-54157-0_20
10. Hua, Y., Liu, Q., Hao, K., Jin, Y.: A survey of evolutionary algorithms for multi-objective optimization problems with irregular pareto fronts. IEEE/CAA J. Automatica Sinica **8**(2), 303–318 (2021). https://doi.org/10.1109/JAS.2021.1003817
11. Indraneel, D.: On characterizing the "knee" of the Pareto curve based on normal-boundary intersection. Struct. Optim. **18**(2), 107–115 (1999). https://doi.org/10.1007/BF01195985
12. Jones, D.: A taxonomy of global optimization methods based on response surfaces. J. Global Optim. **21**, 345–383 (2001). https://doi.org/10.1023/A:1012771025575
13. Jones, D.R., Schonlau, M., Welch, W.J.: Efficient global optimization of expensive black-box functions. J. Global Optim. **13**, 455–492 (1998)
14. Knowles, J.: ParEGO: a hybrid algorithm with on-line landscape approximation for expensive multiobjective optimization problems. IEEE Trans. Evol. Comput. **10**(1), 50–66 (2006)
15. Ma, X., Yu, Y., Li, X., Qi, Y., Zhu, Z.: A survey of weight vector adjustment methods for decomposition-based multiobjective evolutionary algorithms. IEEE Trans. Evol. Comput. **24**(4), 634–649 (2020). https://doi.org/10.1109/TEVC.2020.2978158
16. Minasny, B., McBratney, A.B.: The matérn function as a general model for soil variograms. Geoderma **128**(3), 192–207 (2005). https://doi.org/10.1016/j.geoderma.2005.04.003, pedometrics 2003
17. Mockus, J.: Bayesian Approach to global Optimization: Theory and Applications, vol. 37. Springer (1989). https://doi.org/10.1007/978-94-009-0909-0

18. Mockus, J., Tiesis, V., Zilinskas, A.: The application of Bayesian methods for seeking the extremum. Towards Global Optim. **2**(117–129), 2 (1978)
19. Picheny, V., Wagner, T., Ginsbourger, D.: A benchmark of kriging-based infill criteria for noisy optimization. Struct. Multidiscip. Optim. **48**, 607–626 (2013). https://doi.org/10.1007/s00158-013-0919-4
20. Rasmussen, C.E., Williams, C.K.I.: Gaussian Processes for Machine Learning, 1st edn. The MIT Press, Cambridge (2005)
21. Ray, T., Singh, H.K., Rahi, K.H., Rodemann, T., Olhofer, M.: Towards identification of solutions of interest for multi-objective problems considering both objective and variable space information. Appl. Soft Comput. **119**, 108505 (2022). https://doi.org/10.1016/j.asoc.2022.108505
22. Rojas-Gonzalez, S., Van Nieuwenhuyse, I.: A survey on kriging-based infill algorithms for multiobjective simulation optimization. Comput. Oper. Res. **116**, 104869 (2020)
23. Trivedi, A., Srinivasan, D., Sanyal, K., Ghosh, A.: A survey of multiobjective evolutionary algorithms based on decomposition. IEEE Trans. Evol. Comput. **21**(3), 440–462 (2017). https://doi.org/10.1109/TEVC.2016.2608507
24. Yu, G., Jin, Y., Olhofer, M.: Benchmark problems and performance indicators for search of knee points in multiobjective optimization. IEEE Trans. Cybern. **50**(8), 3531–3544 (2020). https://doi.org/10.1109/TCYB.2019.2894664
25. Yu, G., Ma, L., Jin, Y., Du, W., Liu, Q., Zhang, H.: A survey on knee-oriented multi-objective evolutionary optimization. IEEE Trans. Evol. Comput. 1 (2022). https://doi.org/10.1109/TEVC.2022.3144880
26. Zhang, K., Yen, G.G., He, Z.: Evolutionary algorithm for knee-based multiple criteria decision making. IEEE Trans. Cybern. **51**(2), 722–735 (2021). https://doi.org/10.1109/TCYB.2019.2955573
27. Zitzler, E., Thiele, L., Laumanns, M., Fonseca, C., da Fonseca, V.: Performance assessment of multiobjective optimizers: an analysis and review. IEEE Trans. Evol. Comput. **7**(2), 117–132 (2003). https://doi.org/10.1109/TEVC.2003.810758

High Dimensional Bayesian Optimization with Kernel Principal Component Analysis

Kirill Antonov[1,3]([✉]) [iD], Elena Raponi[2,4] [iD], Hao Wang[3] [iD], and Carola Doerr[4] [iD]

[1] ITMO University, Saint Petersburg, Russia
k.antonov@liacs.leidenuniv.nl
[2] TUM School of Engineering and Design, Technical University of Munich, Munich, Germany
[3] LIACS Department, Leiden University, Leiden, Netherlands
[4] Sorbonne Université, CNRS, LIP6, Paris, France

Abstract. Bayesian Optimization (BO) is a surrogate-based global optimization strategy that relies on a Gaussian Process regression (GPR) model to approximate the objective function and an acquisition function to suggest candidate points. It is well-known that BO does not scale well for high-dimensional problems because the GPR model requires substantially more data points to achieve sufficient accuracy and acquisition optimization becomes computationally expensive in high dimensions. Several recent works aim at addressing these issues, e.g., methods that implement online variable selection or conduct the search on a lower-dimensional sub-manifold of the original search space. Advancing our previous work of PCA-BO that learns a linear sub-manifold, this paper proposes a novel kernel PCA-assisted BO (KPCA-BO) algorithm, which embeds a non-linear sub-manifold in the search space and performs BO on this sub-manifold. Intuitively, constructing the GPR model on a lower-dimensional sub-manifold helps improve the modeling accuracy without requiring much more data from the objective function. Also, our approach defines the acquisition function on the lower-dimensional sub-manifold, making the acquisition optimization more manageable.

We compare the performance of KPCA-BO to a vanilla BO and to PCA-BO on the multi-modal problems of the COCO/BBOB benchmark suite. Empirical results show that KPCA-BO outperforms BO in terms of convergence speed on most test problems, and this benefit becomes more significant when the dimensionality increases. For the 60D functions, KPCA-BO achieves better results than PCA-BO for many test cases. Compared to the vanilla BO, it efficiently reduces the CPU time required to train the GPR model and to optimize the acquisition function compared to the vanilla BO.

Keywords: Bayesian optimization · Black-box optimization · Kernel principal component analysis · Dimensionality reduction

---

K. Antonov—Work done while visiting Sorbonne Université in Paris.

© The Author(s), under exclusive license to Springer Nature Switzerland AG 2022
G. Rudolph et al. (Eds.): PPSN 2022, LNCS 13398, pp. 118–131, 2022.
https://doi.org/10.1007/978-3-031-14714-2_9

1   Introduction

Numerical black-box optimization problems are challenging to solve when the dimension of the problem's domain becomes high [1]. The well-known *curse of dimensionality* implies that exponential growth of the data points is required to maintain a reasonable coverage of the search space. This is difficult to accommodate in numerical black-box optimization, which aims to seek a well-performing solution with a limited budget of function evaluations. Bayesian optimization (BO) [15,18] suffers from high dimensionality more seriously compared to other search methods, e.g., evolutionary algorithms, since it employs a surrogate model of the objective function internally, which scales poorly with respect to the dimensionality (see Sect. 2 below). Also, BO proposes new candidate solutions by maximizing a so-called acquisition function (see Sect. 2), which assesses the potential of each search point for making progresses. The maximization task of the acquisition function is also hampered by high dimensionality. As such, BO is often taken only for small-scale problems (typically less than 20 search variables), and it remains an open challenge to scale it up for high-dimensional problems [3].

Recently, various methods have been proposed for enabling high-dimensional BO, which can be categorized into three classes: (1) variable selection methods that only execute BO on a subset of search variables [27], (2) methods that leverage the surrogate model to high dimensional spaces, e.g., via additive models [6,7], and (3) conducting BO on a sub-manifold embedded in the original search space [14,29]. Notably, in [9], a kernel-based approach is developed for parametric shape optimization in computer-aided design systems, which is not a generic approach since the kernel function is based on the representation of the parametric shape and is strongly tied to applications in mechanical design. In [21] we proposed PCA-BO, in which we conduct BO on a linear sub-manifold of the search space that is learned from the linear principal components analysis (PCA) procedure.

This paper advances the PCA-BO algorithm by considering the kernel PCA procedure [25], which is able to construct a non-linear sub-manifold of the original search space. The proposed algorithm - *Kernel PCA-assisted BO* (KPCA-BO) adaptively learns a nonlinear forward map from the original space to the lower-dimensional sub-manifold for reducing dimensionality and constructs a backward map that converts a candidate point found on the sub-manifold to the original search space for the function evaluation. We evaluate the empirical performance of KPCA-BO on the well-known BBOB problem set [12], focusing on the multi-modal problems.

This paper is organized as follows. In Sect. 2, we will briefly recap Bayesian optimization and some recent works on alleviating the issue of high dimension for BO. In Sect. 3, we describe the key components of KPCA-BO in detail. The experimental setting, results, and discussions are presented in Sect. 4, followed by the conclusion and future works in Sect. 5.

2   Related Work

Bayesian Optimization (BO) [15,26] is a sequential model-based optimization algorithm which was originally proposed to solve single-objective black-box optimization problems that are expensive to evaluate. BO starts with sampling a small initial design of experiment (DoE, obtained with e.g., Latin Hypercube Sampling [24] or low-discrepancy sequences [20]) $X \subseteq S$. After evaluating $f(x)$ for all $x \in X$, it proceeds to construct a probabilistic model $\mathbb{P}(f \mid X, Y)$ (e.g., Gaussian process regression, please see the next paragraph). BO balances exploration and exploitation of the search by considering, for a decision point $\mathbf{x}$, two quantities: the predicted function value $\hat{f}(\mathbf{x})$ and the uncertainty of this prediction (e.g., the mean squared error $\mathrm{E}(f(\mathbf{x}) - \hat{f}(\mathbf{x}))^2$). Both of them are taken to form the acquisition function $\alpha \colon S \to \mathbb{R}$ used in this work, i.e., the expected improvement [15], which quantifies the potential of each point for making progresses. BO chooses the next point to evaluate by maximizing the acquisition function. After evaluating $\mathbf{x}^*$, we augment the data set with $(\mathbf{x}^*, f(\mathbf{x}^*))$ and proceed with the next iteration.

Gaussian Progress Regression (GPR) [22] models the objective function f as the realization of a Gaussian process $f \sim \mathrm{gp}(0, c(\cdot, \cdot))$, where $c \colon S \times S \to \mathbb{R}$ is the covariance function, also known as kernel. That is, $\forall \mathbf{x}, \mathbf{x}' \in S$, it holds that $\mathrm{Cov}\{f(\mathbf{x}), f(\mathbf{x}')\} = c(\mathbf{x}, \mathbf{x}')$. Given a set X of evaluated points and the corresponding function values Y, GPR learns a posterior Gaussian process to predict the function value at each point, i.e., $\forall \mathbf{x} \in S, f(\mathbf{x}) \mid X, Y, \mathbf{x} \sim \mathcal{N}(\hat{f}(\mathbf{x}), \hat{s}^2(\mathbf{x}))$, where $\hat{f}$ and $\hat{s}^2$ are the posterior mean and variance functions, respectively. When equipped with a GPR and the expected improvement, BO has a convergence rate of $O(n^{-1/d})$ [4], which decreases quickly when the dimension increases.

High-Dimensional Bayesian optimization. High dimensionality negatively affects the performance of BO in two aspects, the quality of the GPR model and the efficiency of acquisition optimization. For the quality of the GPR model, it is well-known[4] that many more data points are needed to maintain the modeling accuracy in higher dimensions. Moreover, acquisition optimization is a high-dimensional task requiring more surrogate evaluations to obtain a reasonable optimum. Notably, each surrogate evaluation takes $O(d)$ time to compute, making the acquisition optimization more time-consuming.

Depending on the expected structure of the high-dimensional problem, various strategies for dealing with this curse of dimensionality have been proposed in the literature, often falling into one of the following classes:

1. Variable selection or screening. It may be the case that a subset of parameters does not have any significant impact on solutions' quality, and it is convenient to identify and keep only the most influential ones. Different approaches may be considered: discarding variables uniformly [17], assigning weights to the variables based on the dependencies between them [27], identifying the most descriptive variables based on their length-scale value in the model [2], etc.

2. Additive models. They keep all the variables but limit their interaction as they are based on the idea of decomposing the problem into blocks. For example, the model kernels can be seen as the sum of univariate ones [6,7], the high-dimensional function can decompose as a sum of lower-dimensional functions on subsets of variables [23], or the additive model can be based on an ANOVA decomposition [10,19].
3. Linear/nonlinear embeddings. They are based on the hypothesis that a large percentage of the variation of a high-dimensional function can be captured in a low-dimensional embedding of the original search space. The embedding can be either linear [21,29] or nonlinear [9,11].

We point the reader to [3] for a comprehensive overview of the state-of-the-art in high-dimensional BO.

3   Kernel-PCA Assisted by Bayesian Optimization

In this paper, we deal with numerical black-box optimization problems $f\colon S \subseteq \mathbb{R}^d \to \mathbb{R}$, where the search domain is a hyperbox, i.e., $S = [l_1, u_1] \times [l_2, u_2] \times \cdots \times [l_d, u_d]$. We reduce the dimensionality of the optimization problem *on-the-fly*, using a *Kernel Principal Component Analysis* (KPCA) [25] for learning, from the evaluated search points, a non-linear sub-manifold $\mathcal{M}$ on which we optimize the objective function. Ideally, such a sub-manifold $\mathcal{M}$ should capture important information of f for optimization. In other words, $\mathcal{M}$ should "traverse" several basins of attractions of f. Loosely speaking, in contrast to a linear sub-manifold (e.g., our previous work [21]), the non-linear one would solve the issue that the correlation among search variables is non-linear (e.g., on multimodal functions), where it is challenging to identify a linear sub-manifold that passes through several local optima simultaneously. KPCA tackles this issue by first casting the search points to a high-dimensional Hilbert space $\mathcal{H}$ (typically infinite-dimensional), where we learn a linear sub-manifold thereof. We consider a positive definite function $k\colon S \times S \to \mathbb{R}$, which induces a reproducing kernel Hilbert space (RKHS) $\mathcal{H}$ constructed as the completion of span$\{k(\mathbf{x}, \cdot)\colon \mathbf{x} \in S\}$. The function $\phi(\mathbf{x}) := k(\mathbf{x}, \cdot)$ maps a point $\mathbf{x}$ from the search space to $\mathcal{H}$, which we will refer as *the feature map*. An inner product on $\mathcal{H}$ is defined with k, i.e., $\forall \mathbf{x}, \mathbf{x}' \in S$, $\langle \phi(\mathbf{x}), \phi(\mathbf{x}') \rangle_{\mathcal{H}} = k(\mathbf{x}, \mathbf{x}')$, known as the *kernel trick*.

The KPCA-BO Algorithm. Figure 1 provides an overview of the proposed KPCA-BO algorithm. We also present the pseudo-code of KPCA-BO in Algorithm 1. Key differences to our previous work that employs the linear PCA method [21] are highlighted. Various building blocks of the algorithm will be described in the following paragraphs. Notably, the sub-routine KPCA indicates performing the standard kernel PCA algorithm, which returns a set of selected principal components, whereas, GPR represents the training of a Gaussian process regression model. In the following discussion, we shall denote by $X = \{\mathbf{x}_i\}_{i=1}^n$ and $Y = \{f(\mathbf{x}_i)\}_{i=1}^n$ the set of the evaluated points and their function values, respectively.

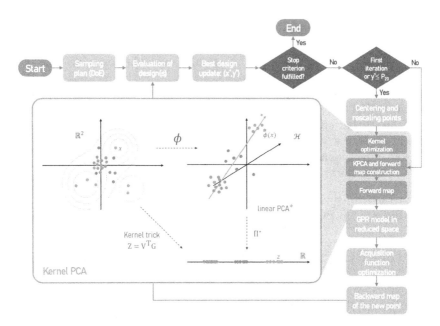

Fig. 1. Flowchart of the KPCA-BO optimization algorithm with detailed graphical representation of the KPCA subroutine.

Rescaling Data Points. As an unsupervised learning method, KPCA disregards the distribution of the function values if applied directly, which contradicts our aim of capturing the objective function in the dimensionality reduction. To mitigate this issue, we apply the weighting scheme proposed in [21], which scales the data points in S with respect to their objective values. In detail, we compute the rank-based weights for all points: w_i is proportional to $\ln n - \ln R_i, i = 1, \ldots, n$, where $R_1, R_2, \ldots, R_n$ are the rankings of points with respect to Y in increasing order (minimization is assumed). Then we rescale each point with its weight, i.e., $\mathbf{x}_i = w_i(\mathbf{x}_i - n^{-1} \sum_{k=1}^{n} \mathbf{x}_k), i = 1, \ldots, n$. It is necessary to show that the feature map ϕ respects the rescaling operation performed in S. For any stationary and monotonic kernel (i.e., $k(\mathbf{x}, \mathbf{y}) = k(D_S(\mathbf{x}, \mathbf{y}))$ (D_S is a metric in S) and k decreases whenever $D_S(\mathbf{x}, \mathbf{y})$ increases), for all $\mathbf{x}, \mathbf{y}, \mathbf{c} \in X$ it holds that $D_S(\mathbf{x}, \mathbf{c}) \leq D_S(\mathbf{y}, \mathbf{c})$ implies that $D_{\mathcal{H}}(\phi(\mathbf{x}), \phi(\mathbf{c})) \leq D_{\mathcal{H}}(\phi(\mathbf{y}), \phi(\mathbf{c}))$. Consequently, the point pushed away from the center of the data in X will still have a large distance to the center of the data in $\mathcal{H}$ after the feature map. The rescaling (or alternatives that incorporate the objective values into the distribution of data points in the domain) is an essential step in applying PCA to BO.

On the one hand, DoE aims to span the search domain as evenly as possible and thereby the initial random sample from it has almost the same variability in all directions, which provides no information for PCA to learn. On the other hand, new candidates are obtained by the global optimization of the acquisition function in each iteration, which is likely to produce multiple clusters and/or

Algorithm 1 KPCA-assisted Bayesian Optimization. Highlighted are those lines in which KPCA-BO differs from the linear PCA method [21]

1: **procedure** KPCA-BO(f, S)   ▷ f: objective function, $S \subseteq \mathbb{R}^d$: search space
2:     Create X = $\{\mathbf{x}_1, \mathbf{x}_2, \ldots, \mathbf{x}_{n_0}\} \subset S$ with Latin hypercube sampling
3:     Y $\leftarrow \{f(\mathbf{x}_1), \ldots, f(\mathbf{x}_{n_0})\}$, $n \leftarrow n_0$
4:     **while** the stop criteria are not fulfilled **do**
5:         **if** $n = n_0$ or $y^* \leq 20\%$-percentile of Y **then**
6:             $R_1, R_2, \ldots, R_n$ are the rankings of points in X w.r.t. Y (increasing order)
7:             $\mathbf{x}'_i \leftarrow \mathbf{x}_i - n^{-1} \sum_{k=1}^n \mathbf{x}_k, i = 1, \ldots, n$ ▷ centering
8:             $\mathbf{x}'_i \leftarrow w_i \mathbf{x}'_i, w_i \propto \ln n - \ln R_i, i = 1, \ldots, n$ ▷ rescaling
9:             $\gamma^* \leftarrow$ OPTIMIZE-RBF-KERNEL($\{\mathbf{x}'_i\}_{i=1}^n$)   ▷ Eq. (2)
10:        **end if**
11:        $v_1, \ldots, v_r \leftarrow$ KPCA($\{\mathbf{x}'_i\}_{i=1}^n, \gamma^*$)   ▷ $r < d \ll n$
12:        construct the forward map $\mathcal{F}$ from span$\{v_1, \ldots, v_r\}$.   ▷ Eq. (1)
13:        $\mathbf{z}_i \leftarrow \mathcal{F}(\mathbf{x}_i), i = 1, \ldots, n$ ▷ map the data to $T := \text{span}\{v_1, \ldots, v_r\}$
14:        $\hat{f}, \hat{s}^2 \leftarrow$ GPR($\{\mathbf{z}_i\}_{i=1}^n$, Y)   ▷ Gaussian process regression
15:        $\mathbf{z}^* \leftarrow \arg\max_{\mathbf{z} \in T} \text{EI}(\mathbf{z}; \hat{f}, \hat{s}^2)$
16:        $y^* \leftarrow f(\mathbf{x}^*), \mathbf{x}^* \leftarrow \mathcal{B}(\mathbf{z}^*)$ ▷ the backward map; Eq. (4)
17:        X $\leftarrow$ X $\cup \{\mathbf{x}^*\}$, Y $\leftarrow$ Y $\cup \{y^*\}$, $n \leftarrow n + 1$
18:     **end while**
19: **end procedure**

isolated points that are not meaningful to the PCA procedure. This is in contrast to the direct application of PCA to evolutionary algorithms [16], where we apply PCA to the current population. Since the population is usually generated from a unimodal mutation distribution, it is well-suited for applying the PCA procedure.

Dimensionality reduction in Hilbert spaces. After the rescaling operation in X, we map the points to the feature space $\mathcal{H}$: $\phi(\text{X}) = \{\phi(\mathbf{x}_i)\}_i, 1 = 1, \ldots, n$. After centering the feature points in $\mathcal{H}$, i.e., $\tilde{\phi}(\mathbf{x}_i) = \phi(\mathbf{x}_i) - n^{-1} \sum_{i=1}^n \phi(\mathbf{x}_i)$, we express the sample covariance[1] of the feature points: $C = n^{-1} \sum_{i=1}^n \tilde{\phi}(\mathbf{x}_i)\tilde{\phi}(\mathbf{x}_i)^\top$. KPCA essentially computes the eigenvalues and eigenfunctions of C, namely $\forall i \in [1..n]$, $Cv_i = \lambda_i v_i, v_i \in \mathcal{H}, ||v_i||_\mathcal{H} = 1$, and $\langle v_i, v_j \rangle_\mathcal{H} = 0$, if $i \neq j$. Note that (1) C is positive semi-definite; (2) since rank(C) $\leq \sum_{i=1}^n \text{rank}(\tilde{\phi}(\mathbf{x}_i)\tilde{\phi}(\mathbf{x}_i)^\top) = n$, there are maximally n nonzero eigenvalues and eigenfunctions; (3) the eigenfunction takes the following form $v_i = \sum_{j=1}^n a_j^{(i)} \tilde{\phi}(\mathbf{x}_j), a_j^{(i)} \in \mathbb{R}$ and thereby all eigenfunctions can be represented by a matrix $\mathbf{V} = (a_j^{(i)})_{ij}$. Assume the eigenvalues are ordered in the decreasing manner (i.e., $\lambda_1 \geq \lambda_2 \geq \cdots \geq \lambda_n \geq 0$. Eigenfunctions are sorted accordingly). It is not hard to show that the variance of $\phi(\text{X})$ along v_i is exactly λ_i: $n^{-1} \sum_{k=1}^n \langle \tilde{\phi}(\mathbf{x}_k), v_i \rangle_\mathcal{H}^2 = \langle v_i, n^{-1} \sum_{k=1}^n [\tilde{\phi}(\mathbf{x}_k)\tilde{\phi}(\mathbf{x}_k)^\top] v_i \rangle_\mathcal{H} = \lambda_i$. Therefore, the eigenvalues can be used to select a linear subspace of $\mathcal{H}$ which keeps the majority of the variability of $\phi(\text{X})$. Specifically, we choose a subspace

---

[1] The outer product is a linear operator defined as $\forall h \in \mathcal{H}, [\phi(\mathbf{x})\phi(\mathbf{x})^\top](h) \colon h \mapsto \langle \phi(\mathbf{x}), h \rangle_\mathcal{H} \phi(\mathbf{x})$. Hence, the sample covariance is also a linear operator $C \colon \mathcal{H} \to \mathcal{H}$.

$T := \text{span}\{v_1, \ldots, v_r\} \subset \mathcal{H}$ as the reduced search space of BO, where r is chosen as the smallest integer such that the first-r eigenvalues explain at least η percent of the total variability (we use $\eta = 90\%$ in our experiments). For a point $\mathbf{x} \in S$, we can formulate a *forward map* that projects $\phi(\mathbf{x})$ onto the reduced space T: $\mathcal{F}: \mathbf{x} \mapsto \sum_{i=1}^{r} \langle \tilde{\phi}(\mathbf{x}), v_i \rangle_{\mathcal{H}} v_i$. Let $g_i(\mathbf{x}) = \langle \tilde{\phi}(\mathbf{x}), \tilde{\phi}(\mathbf{x}_i) \rangle_{\mathcal{H}} = k(\mathbf{x}, \mathbf{x}_i) - n^{-1} \sum_{j=1}^{n} k(\mathbf{x}, \mathbf{x}_j) - n^{-1} \sum_{j=1}^{n} k(\mathbf{x}_i, \mathbf{x}_j) + n^{-2} \sum_{i=1}^{n} \sum_{j=1}^{n} k(\mathbf{x}_i, \mathbf{x}_j)$, the forward map can be re-expressed as

$$\mathcal{F}: \mathbf{x} \mapsto \mathbf{V}\mathbf{g}(\mathbf{x}), \quad \mathbf{g}(\mathbf{x}) = (g_1(\mathbf{x}), \ldots, g_n(\mathbf{x}))^{\top}. \tag{1}$$

Computationally, the eigenfunction representation $\mathbf{V}$ can be calculated via eigendecomposition of the Gram matrix ($\mathbf{G}_{ij} = \langle \tilde{\phi}(\mathbf{x}_i), \tilde{\phi}(\mathbf{x}_j) \rangle_{\mathcal{H}}$, i.e., $\mathbf{G} = \mathbf{V}^{\top}\mathbf{D}\mathbf{V}$, $\mathbf{D}$ is a $n \times n$ diagonal matrix with the eigenvalues of C on its nonzero entries.

Learning the forward map. We use the *radial basis function* (RBF, a.k.a. Gaussian kernel) for KPCA in this paper. The RBF kernel $k(\mathbf{x}, \mathbf{x}') = \exp(-\gamma \|\mathbf{x} - \mathbf{x}'\|_2^2)$, contains a single length-scale hyperparameter $\gamma \in \mathbb{R}_{>0}$. To determine this length-scale, we minimize the number of eigenvalues/functions chosen to keep at least η percent of the variance, which effectively distributes more information of $\phi(X)$ on the first few eigenfunctions and hence allows for constructing a lower-dimensional space T. Also, we reward γ values which choose the same number of eigenfunctions and also yield a higher ratio of explained variance. In all, the cost function for tuning γ is:

$$\gamma^* = \underset{\gamma \in (0, \infty)}{\arg\min} \, r - \frac{\sum_{i=1}^{r} \lambda_i}{\sum_{i=1}^{n} \lambda_i}, \quad r = \inf\left\{ k \in [1..n] : \sum_{i=1}^{k} \lambda_i \geq \eta \sum_{i=1}^{n} \lambda_i \right\}. \tag{2}$$

This equation is solved numerically, using a quasi-Newton method (the L-BFGS-B algorithm [5]) with $\gamma \in [10^{-4}, 2]$ and maximally $200d$ iterations. It is worth noting that we do not consider anisotropic kernels (e.g., individual length-scales for each search variable) since such a kernel increases the number of hyperparameters to learn.

Also, note that the choice of the kernel can affect the smoothness of the manifold in the feature space $\mathcal{H}$, e.g., the Matérn $5/2$ kernel induces a C^2 atlas for the manifold $\phi(S)$. We argue that the smoothness of $\phi(S)$ is less important to the dimensionality reduction task, comparing to the convexity and connectedness thereof. In this work, we do not aim to investigate the impact of the kernel on the topological properties of $\phi(S)$. Therefore, we use the RBF kernel for the construction of the forward map for its simplicity.

Learning the Backward Map. When performing the Bayesian optimization in the reduced space T, we need to determine a "pre-image" of a candidate point $\mathbf{z} \in T$ for the function evaluation. To implement such a backward map $\mathcal{B}: T \to S$, we base our construction on the approach proposed in [8], in which the pre-image of a point $\mathbf{z} \in T$ is a conical combination of some points in S: $\sum_{i=1}^{d} w_i \mathbf{p}_i, w_i \in \mathbb{R}_{>0}$. In this paper, the points $\{\mathbf{p}_i\}_{i=1}^{d}$ are taken as a random subset of the data points $\{\mathbf{x}_i\}_{i=1}^{n}$. The conical weights are determined by minimizing the distance between $\mathbf{z}$ and the image of the conical combination under the forward map:

$$w_1^*, \ldots, w_d^* = \underset{\{w_i\}_{i=1}^d \subset \mathbb{R}_{>0}^d}{\arg\min} \left\| \mathbf{z} - \mathcal{F}\left(\sum_{i=1}^d w_i \mathbf{p}_i\right) \right\|_2^2 + Q\left(\sum_{i=1}^d w_i \mathbf{p}_i\right), \qquad (3)$$

$$Q(\mathbf{x}) = \exp\left(\sum_{i=1}^d \max(0, l_i - x_i) + \max(0, x_i - u_i)\right).$$

where the function Q penalizes the case that the pre-image is out of S. As with Eq. (2), the weights are optimized with the L-BFGS-B algorithm (starting from zero with $200d$ maximal iterations). Taking the optimal weights, we proceed to define the *backward map*: $\forall \mathbf{z} \in T$,

$$\mathcal{B} \colon \mathbf{z} \mapsto \mathrm{CLIP}\left(\sum_{i=1}^d w_i^* \mathbf{p}_i\right), \qquad (4)$$

where the function $\mathrm{CLIP}(\mathbf{x})$ cuts off each component of $\mathbf{x}$ at the lower and upper bounds of S, which ensures the pre-image is always feasible.

Remark. The event that $\{\mathbf{p}_i\}_{i=1}^d$ contains a co-linear relation is of measure zero and the conical form can procedure pre-images outside S, allowing for a complete coverage thereof. There exist multiple solutions to Eq. (3) (and hence multiple pre-images) since the forward map $\mathcal{F}$ contains an orthogonal projection step, which is not injective. Those multiple pre-images can be obtained by randomly restarting the quasi-Newton method used to solve Eq. (3). However, since those pre-images do not distinguish from each other for our purpose, we simply take a random one in this work.

Bayesian Optimization in the Reduced Space. Given the forward and backward maps, we are ready to perform the optimization task in the space T. Essentially, we first map the data set $\mathrm{X} \subset S$ to T using the forward map (Eq. (1)): $\mathcal{F}(\mathrm{X}) = \{\mathcal{F}(\mathbf{x}_i)\}_{i=1}^n$, which implicitly defines the counterpart $f' := f \circ \mathcal{B}$ of the objective function in T. Afterwards, we train a Gaussian process model with the data set $(\mathcal{F}(\mathrm{X}), \mathrm{Y})$ to model f', i.e., $\forall \mathbf{z} \in T, f'(\mathbf{z}) \mid \mathcal{F}(\mathrm{X}), \mathrm{Y}, \mathbf{z} \sim \mathcal{N}(\hat{f}(\mathbf{z}), \hat{s}^2(\mathbf{z}))$. The search domain in the reduced space T is determined as follows. Since the RBF kernel monotonically decreases w.r.t. the distance between its two input points, we can bound the set $\phi(\mathrm{X})$ by first identifying the point $\mathbf{x}_{\max}$ with the largest distance to the center of data points $\mathbf{c}$ and secondly computing the distance r between $\phi(\mathbf{x}_{\max})$ and $\phi(\mathbf{c})$ in the feature space $\mathcal{H}$. Since S is a hyperbox in $\mathbb{R}^d$, we simply take an arbitrary vertex of the hyperbox for $\mathbf{x}_{\max}$. Note that, as the orthogonal projection (from $\mathcal{H}$ to T) does not increase the distance, the open ball $B := \{\mathbf{z} \in T \colon \|\mathbf{z}\|_2 < r\}$ always covers $\mathcal{F}(\mathrm{X})$. For the sake of optimization in T, we take the smallest hyperbox covering B as the search domain in T.

After the GPR model is created on the data set $(\mathcal{F}(\mathrm{X}), \mathrm{Y})$, we maximize the expected improvement function $\mathrm{EI}(\mathbf{z}; \hat{f}, \hat{s}) = \hat{s}(\mathbf{z}) u\, \mathrm{CDF}(u) + \hat{s}(\mathbf{z})\, \mathrm{PDF}(u), u = (\min \mathrm{Y} - \hat{f}(\mathbf{z}))/\hat{s}(\mathbf{x})$ to pick a new candidate point $\mathbf{z}^*$, where CDF and PDF stand for the cumulative distribution and probability distribution functions of a standard normal random variable, respectively. Due to our construction of the search domain in T, it is possible that the global optimum $\mathbf{z}^*$ of EI is associated with an infeasible pre-image in S. To mitigate this issue, we propose a multi-restart optimization strategy for maximizing EI (with different starting points in each restart), in which we only take the best outcome (w.r.t. its EI value) whose pre-image belongs to S. In our experiments, we used 10 random restarts of the optimization. Also, it is unnecessary to optimize kernel's hyperparameter γ in each iteration of BO since the new point proposed by EI would not make a significant impact on learning the feature map, if its quality is poor relative to the observed ones in Y (and hence assigned with a small weight). Therefore, it suffices to only re-optimize γ whenever we find a new point whose function value is at least as good as the 20% percentile of Y. Also, the 20% threshold is manually chosen to balance the convergence and computation time of KPCA-BO, after experimenting several different values on BBOB test problems.

4   Experiments

Experimental Setup. We evaluate the performance of KPCA-BO on ten multi-modal functions from the BBOB problem set [12] (F15 - F24), which should be sufficiently representative of the objective functions handled in real-world applications. We compare the experimental result of KPCA-BO to standard BO and the PCA-BO in our previous work [21] on three problem dimensions $d \in \{20, 40, 60\}$ with the evaluation budget in $\{100, 200, 300\}$, respectively. We choose a relatively large DoE size of $3d$, to ensure enough information for learning the first sub-manifold. On each function, we consider five problem instances (instance ID from 0 to 4) and conduct 10 independent runs of each algorithm. We select the Matérn 5/2 kernel for the GPR model. The L-BFGS-B algorithm [5] is employed to maximize the likelihood of GPR as well as the EI acquisition function at each iteration. We add to our comparison results for CMA-ES [13], obtained by executing the pycma package (https://github.com/CMA-ES/pycma) with 16 independent runs on each problem. The implementation of BO, PCA-BO, and KPCA-BO can be accessed at https://github.com/wangronin/Bayesian-Optimization/tree/KPCA-BO.

Results. All our data sets are available for interactive analysis and visualization in the IOHanalyzer [28] repository, under the *bbob-largescale* data sets of the *IOH* repository. In Fig. 2, we compare the convergence behavior of all four algorithms, where we show the evolution of the best-so-far target gap $(f_{\text{best}} - f^*)$ with respect to the iteration for each function-dimension pair. In all dimensions, it is clear that both KPCA-BO and PCA-BO outperform BO substantially across functions F17 - F20, while both KPCA-BO and PCA-BO exhibit about the same convergence with BO on F23, and are surpassed by BO significantly on F16. The poor performance of PCA-BO and KPCA-BO on the Weierstrass function (F16) can be attributed to the nature of its landscape, which is highly rugged and

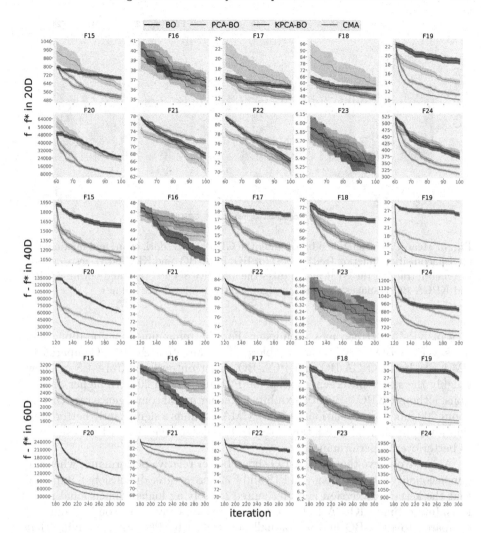

Fig. 2. The best-so-far target gap $(f_{best} - f^*)$ against the iteration for CMA-ES (purple), BO (red), PCA-BO (blue), and KPCA-BO (green) is averaged over 50 independent runs on each test problem. We compare the algorithms in three dimensions, 20D (top), 40D (middle), and 50D (bottom). The shaded area indicates the standard error of the mean target gap. CMA-ES data is obtained from running the `pycma` package with the same evaluation budgets as BO. (Color figure online)

moderately periodic with multiple global optima. Therefore, a basin of attraction is not clearly defined, which confuses both PCA variants. On functions F17, F19, and F24, we observe that KPCA-BO is outperformed by PCA-BO in 20D, while as the dimensionality increases, KPCA-BO starts to surpass the convergence rate of PCA-BO. For functions F21 and F22, KPCA-BO's performance is indistinguishable from PCA-BO in 20D, and in higher dimensions, the advan-

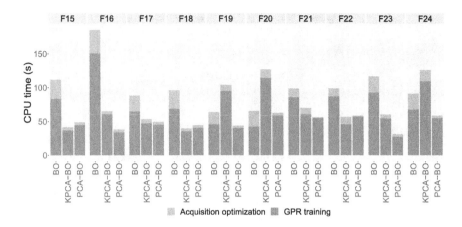

Fig. 3. Mean CPU time taken by training the GPR model (dark cyan) and maximizing the EI acquisition function (red) in 60D for BO, PCA-BO, and KPCA-BO, respectively. In general, training the GPR model takes the majority of CPU time and both PCA-BO and KPCA-BO manages to reduce it significantly. (Color figure online)

tage of KPCA-BO becomes prominent. For the remaining function-dimension pairs, KPCA-BO shows a comparable performance to PCA-BO. Compared to CMA-ES, both KPCA-BO and PCA-BO either outperform CMA-ES or show roughly the same convergence except on F21 and F22 in 20D. In higher dimensions, although KPCA-BO still exhibits a steeper initial convergence rate (before about 200 function evaluations in 60D), CMA-ES finds a significantly better solution after the first $3d$ evaluations (the DoE phase of the BO variants), leading to better overall performance.

Also, we observe that KPCA-BO shows better relative performance when the dimensionality increases (e.g., on F18 and F22 across three dimensions), implying that the kernelized version is better suited for solving higher-dimensional problems. Interestingly, KPCA-BO shows an early faster convergence on F21 and F22 compared to PCA-BO and is gradually overtaken by PCA-BO, implying that KPCA-BO stagnates earlier than PCA-BO. We conjecture that the kernel function of KPCA-BO (and consequently the sub-manifold) stabilizes much faster than the linear subspace employed in PCA-BO, which might attribute to such a stagnation behavior. Therefore, KPCA-BO is more favorable than PCA-BO in higher dimensions (i.e., $d \geq 60$), while in lower dimensions ($d \approx 20$), it is competitive to PCA-BO when the budget is small for most test cases (it only loses a little to PCA-BO on F24).

In Fig. 3, we depict the CPU time (in seconds) taken to train the GPR model as well as maximize EI in 60D. As expected, the majority of the CPU time is taken by the training of the GPR model. PCA-BO and KPCA-BO achieve substantially smaller CPU time of GPR training than the vanilla BO with exception on F19, F20, and F24, where KPCA-BO actually takes more time. In all cases, the CPU time for maximizing EI is smaller for KPCA-BO and PCA-BO than for BO.

5 Conclusions

In this paper, we proposed a novel KPCA-assisted Bayesian optimization algorithm, the KPCA-BO. Our algorithm enables BO in high-dimensional numerical optimization tasks by learning a nonlinear sub-manifold of the original search space from the evaluated data points and performing BO directly in this sub-manifold. Specifically, to capture the information about the objective function when performing the kernel PCA procedure, we rescale the data points in the original space using a weighting scheme based on the corresponding objective values of the data point. With the help of the KPCA procedure, the training of the Gaussian process regression model and the acquisition optimization – the most costly steps in BO – are performed in the lower-dimensional space. We also implement a backward map to convert the candidate point found in the lower-dimensional manifold to the original space.

We empirically evaluated KPCA-BO on the ten multimodal functions (F15-F24) from the BBOB benchmark suite. We also compare the performance of KPCA-BO with the vanilla BO, the PCA-BO algorithm from our previous work, and CMA-ES, a state-of-the-art evolutionary numerical optimizer. The results show that KPCA-BO performs better than PCA-BO in capturing the contour lines of the objective functions when the variables are not linearly correlated. The higher the dimensionality, the more significant this better capture becomes for the optimization of functions F20, F21, F22, and F24. Also, the mean CPU time measured in the experiments shows that after reducing the dimensionality, the CPU time needed to train the GPR model and maximize the acquisition is greatly reduced in most cases.

The learning of the lower-dimensional manifold is the crux of the proposed KPCA-BO algorithm. However, we observe that this manifold stabilizes too quickly for some functions, leading to unfavorable stagnation behavior. In further work, we plan to investigate the cause of this premature convergence and hope to identify mitigation methods. Since the manifold is learned from the data points evaluated so far, a viable approach might be to use a random subset of data points to learn the manifold, rather than taking the entire data set.

Another future direction is to improve the backward map proposed in this paper. Since an orthogonal projection is involved when mapping the data to the manifold, the point on the manifold has infinitely many pre-images. In our current approach, we do not favor one direction or another, whereas it might be preferable to bias the map using the information of the previously sampled points. For example, the term to minimize (to exploit) or maximize (to explore) the distance between the candidate pre-image point and the lower-dimensional manifold can be added to the cost function used in the backward map. These two approaches can be combined or switched in the process of optimization.

Acknowledgments. Our work is supported by the Paris Ile-de-France region (via the DIM RFSI project AlgoSelect), by the CNRS INS2I institute (via the Rand-Search project), by the PRIME programme of the German Academic Exchange Service (DAAD) with funds from the German Federal Ministry of Education and Research (BMBF), and by RFBR and CNRS, project number 20-51-15009.

References

1. Bellman, R.: Dynamic programming. Science **153**(3731), 34–37 (1966). https://doi.org/10.1126/science.153.3731.34
2. Ben Salem, M., Bachoc, F., Roustant, O., Gamboa, F., Tomaso, L.: Sequential dimension reduction for learning features of expensive black-box functions (2019). https://hal.archives-ouvertes.fr/hal-01688329, preprint
3. Binois, M., Wycoff, N.: A survey on high-dimensional Gaussian process modeling with application to Bayesian optimization. arXiv:2111.05040 [math], November 2021
4. Bull, A.D.: Convergence rates of efficient global optimization algorithms. J. Mach. Learn. Res. **12**, 2879–2904 (2011). http://dl.acm.org/citation.cfm?id=2078198
5. Byrd, R.H., Lu, P., Nocedal, J., Zhu, C.: A limited memory algorithm for bound constrained optimization. SIAM J. Sci. Comput. **16**(5), 1190–1208 (1995). https://doi.org/10.1137/0916069
6. Delbridge, I., Bindel, D., Wilson, A.G.: Randomly Projected Additive Gaussian Processes for Regression. In: Proc. of the 37th International Conference on Machine Learning (ICML), pp. 2453–2463. PMLR, November 2020
7. Duvenaud, D.K., Nickisch, H., Rasmussen, C.: Additive Gaussian Processes. In: Advances in Neural Information Processing Systems, vol. 24. Curran Associates, Inc. (2011)
8. García-González, A., Huerta, A., Zlotnik, S., Díez, P.: A kernel principal component analysis (kpca) digest with a new backward mapping (pre-image reconstruction) strategy. CoRR abs/2001.01958 (2020)
9. Gaudrie, D., Le Riche, R., Picheny, V., Enaux, B., Herbert, V.: Modeling and optimization with Gaussian processes in reduced eigenbases. Struct. Multidiscip. Optim. **61**(6), 2343–2361 (2020). https://doi.org/10.1007/s00158-019-02458-6
10. Ginsbourger, D., Roustant, O., Schuhmacher, D., Durrande, N., Lenz, N.: On ANOVA decompositions of kernels and Gaussian random field paths. arXiv:1409.6008 [math, stat], October 2014
11. Guhaniyogi, R., Dunson, D.B.: Compressed gaussian process for manifold regression. J. Mach. Learn. Res. **17**(69), 1–26 (2016). http://jmlr.org/papers/v17/14-230.html
12. Hansen, N., Auger, A., Ros, R., Mersmann, O., Tušar, T., Brockhoff, D.: COCO: a platform for comparing continuous optimizers in a black-box setting. Optimization Methods and Software, pp. 1–31 (2020)
13. Hansen, N., Ostermeier, A.: Completely derandomized self-adaptation in evolution strategies. Evol. Comput. **9**(2), 159–195 (2001). https://doi.org/10.1162/106365601750190398
14. Huang, W., Zhao, D., Sun, F., Liu, H., Chang, E.: Scalable Gaussian process regression using deep neural networks. In: Proceedings of the 24th International Conference on Artificial Intelligence (IJCAI), pp. 3576–3582. AAAI Press (2015)

15. Jones, D.R., Schonlau, M., Welch, W.J.: Efficient global optimization of expensive black-box functions. J. Global Optim. **13**(4), 455–492 (1998). https://doi.org/10.1023/A:1008306431147

16. Kapsoulis, D., Tsiakas, K., Asouti, V., Giannakoglou, K.C.: The use of kernel PCA in evolutionary optimization for computationally demanding engineering applications. In: 2016 IEEE Symposium Series on Computational Intelligence, SSCI 2016, Athens, Greece, December 6–9, 2016, pp. 1–8. IEEE (2016). https://doi.org/10.1109/SSCI.2016.7850203

17. Li, C., Gupta, S., Rana, S., Nguyen, V., Venkatesh, S., Shilton, A.: High dimensional bayesian optimization using dropout. In: Proceedings of the 26th International Joint Conference on Artificial Intelligence (IJCAI), pp. 2096–2102. AAAI Press (2017)

18. Močkus, J.: On bayesian methods for seeking the extremum. In: Marchuk, G.I. (ed.) Optimization Techniques 1974. LNCS, vol. 27, pp. 400–404. Springer, Heidelberg (1975). https://doi.org/10.1007/3-540-07165-2_55

19. Muehlenstaedt, T., Roustant, O., Carraro, L., Kuhnt, S.: Data-driven Kriging models based on FANOVA-decomposition. Stat. Comput. **22**(3), 723–738 (2012). https://doi.org/10.1007/s11222-011-9259-7

20. Niederreiter, H.: Low-discrepancy and low-dispersion sequences. J. Number Theory **30**(1), 51–70 (1988)

21. Raponi, E., Wang, H., Bujny, M., Boria, S., Doerr, C.: High dimensional bayesian optimization assisted by principal component analysis. In: Bäck, T., Preuss, M., Deutz, A., Wang, H., Doerr, C., Emmerich, M., Trautmann, H. (eds.) PPSN 2020. LNCS, vol. 12269, pp. 169–183. Springer, Cham (2020). https://doi.org/10.1007/978-3-030-58112-1_12

22. Rasmussen, C.E., Williams, C.K.I.: Gaussian processes for machine learning. Adaptive computation and machine learning, MIT Press (2006), https://www.worldcat.org/oclc/61285753

23. Rolland, P., Scarlett, J., Bogunovic, I., Cevher, V.: High-dimensional bayesian optimization via additive models with overlapping groups. In: Proceedings of the Twenty-First International Conference on Artificial Intelligence and Statistics, pp. 298–307. PMLR, March 2018

24. Santner, T.J., Williams, B.J., Notz, W.I.: The Design and Analysis of Computer Experiments. Springer (2003). https://doi.org/10.1007/978-1-4757-3799-8

25. Schölkopf, B., Smola, A., Müller, K.R.: Nonlinear component analysis as a kernel eigenvalue problem. Neural Comput. **10**(5), 1299–1319 (1998). https://doi.org/10.1162/089976698300017467

26. Shahriari, B., Swersky, K., Wang, Z., Adams, R.P., de Freitas, N.: Taking the Human Out of the Loop: A Review of Bayesian Optimization. Proc. IEEE **104**(1), 148–175 (2016). https://doi.org/10.1109/JPROC.2015.2494218

27. Ulmasov, D., Baroukh, C., Chachuat, B., Deisenroth, M., Misener, R.: Bayesian optimization with dimension scheduling: application to biological systems. Comput. Aided Chem. Eng. **38**, November 2015. https://doi.org/10.1016/B978-0-444-63428-3.50180-6

28. Wang, H., Vermetten, D., Ye, F., Doerr, C., Bäck, T.: IOHanalyzer: performance analysis for iterative optimization heuristic. ACM Trans. Evol. Learn. Optim. (2022). https://doi.org/10.1145/3510426

29. Wang, Z., Hutter, F., Zoghi, M., Matheson, D., De Freitas, N.: Bayesian optimization in a billion dimensions via random embeddings (2016)

Single Interaction Multi-Objective Bayesian Optimization

Juan Ungredda[1(✉)], Juergen Branke[1], Mariapia Marchi[2], and Teresa Montrone[2]

[1] University of Warwick, Coventry, UK
J.Ungredda@warwick.ac.uk, juergen.branke@wbs.ac.uk
[2] ESTECO SpA, Trieste, Italy
{marchi,montrone}@esteco.com

Abstract. When the decision maker (DM) has unknown preferences, the standard approach to a multi-objective problem is to generate an approximation of the Pareto front and let the DM choose from the non-dominated designs. However, if the evaluation budget is very limited, the true best solution according to the DM's preferences is unlikely to be among the small set of non-dominated solutions found. We address this issue with a multi-objective Bayesian optimization algorithm and allowing the DM to select solutions from a *predicted* Pareto front, instead of the final population. This allows the algorithm to understand the DM's preferences and make a final attempt to identify a more preferred solution that will then be returned without further interaction. We show empirically that significantly better solutions can be found in terms of true DM's utility than if the DM would pick a solution at the end.

Keywords: Preference elicitation · Simulation optimization · Gaussian processes · Bayesian optimization

1 Introduction

Many real-world optimization problems have multiple, conflicting objectives. A popular way to tackle such problems is to search for a set of Pareto-optimal solutions with different trade-offs, and allow the decision maker (DM) to pick their most preferred solution from this set. This has the advantage that the DM doesn't have to specify their preferences explicitly before the optimization, which is generally considered very difficult.

In case of expensive multi-objective optimization problems, where the number of solutions that can be evaluated during optimization is small, the Pareto front, which may consist of thousands of Pareto-optimal solutions or even be continuous, can only be approximated by a small set of solutions. It is thus unlikely that the solution most preferred by the DM would be among the small set of solutions found by the optimization algorithm, even if these are truly Pareto-optimal solutions.

© The Author(s), under exclusive license to Springer Nature Switzerland AG 2022
G. Rudolph et al. (Eds.): PPSN 2022, LNCS 13398, pp. 132–145, 2022.
https://doi.org/10.1007/978-3-031-14714-2_10

We suggest tackling this issue by using Bayesian Optimization (BO), a surrogate-based global optimization technique, and letting the DM choose a solution from a *predicted* Pareto front rather than from the identified non-dominated solutions at the end of the run. BO is not only known to be very suitable for expensive optimization as it carefully selects points to evaluate through an acquisition function that explicitly balances exploration and exploitation. It also generates a surrogate model of each objective function. These surrogate models can be optimized by a multi-objective evolutionary algorithm to generate an approximated Pareto front, and as evaluation of the surrogate model is cheap relative to a fitness evaluation, we can generate a fine-granular representation of the approximated Pareto front, consisting of very many solution candidates. This approximated Pareto front with many hypothetical solutions can then be shown to a DM to select from. While we cannot guarantee that the picked solution is actually achievable, the location of the picked solution should still give us a very good idea about the DM's preferences. Essentially, it provides a reference point which we expect to be quite close to what should be achievable. We then continue to run BO for a few more steps, aiming to generate the desired solution or something better.

We believe that the cognitive burden for the DM is not much higher than in standard multi-objective optimization: rather than having to identify the most preferred solution from a discrete approximation of the Pareto front at the end of the run, they now pick the most preferred out of the predicted (larger) set of solutions, but the size of the set presented to the DM should not make a big difference in terms of cognitive effort if the problem has only 2 (perhaps 3) objectives, where the interesting region can be identified easily by inspecting the Pareto front visually. After the final optimization step, the algorithm has to return a single recommended solution based on the elicited preference information. Of course it would be possible to ask the DM again to choose from all non-dominated solutions found, and this would probably further enhance the quality of the identified solution. However, we deliberately limit the preference elicitation in this paper to a *single* interaction.

Compared to the existing literature, we offer the following contributions:

- We are the first to question the common practice of returning the non-dominated solutions *after* optimization.
- Instead, we demonstrate that it is beneficial to let the DM choose from an approximated Pareto front *before* the end of optimization. While we cannot guarantee to be able to find this solution, the found solution still has a better utility than the best solution in the Pareto front obtained in the usual way.
- We examine the influence of the point in time when the DM is asked to pick a solution and show that asking too early may be detrimental, while asking too late may forfeit some of the benefits of the proposed approach.
- We explore the impact of a model mismatch between assumed and true utility function and show that the benefit of our approach, while reduced, remains significant despite model mismatch.

The paper is structured as follows. After a literature review, we formally define the problem considered in Sect. 3. The proposed algorithm is described in Sect. 4, followed by empirical results in Sect. 5. The paper concludes with a summary.

2 Literature Review

Depending on the involvement of the DM in the optimization process, multi-objective optimization can be classified into a priori approaches, a posteriori approaches, and interactive approaches [8,21]. The field is very large, so we can only mention some of the most relevant papers here. A priori approaches ask the DM to specify their preferences ahead of optimization. This allows to turn the multi-objective optimization problem into a single objective optimization problem, but it is usually very difficult for a DM to specify their preferences before having seen the alternatives. Most multi-objective EAs are a-posteriori approaches, attempting to identify a good approximation of the Pareto frontier, and the DM can then pick the most preferred solution from this set. This is much easier for a DM, but identifying the entire Pareto front may be computationally expensive. Interactive approaches attempt to learn the DM's preferences during optimization and then focus the search on the most preferred region of the Pareto front. While this may yield solutions closer to the DM's true preferences, it also requires additional cognitive effort from the DM.

Our proposed algorithm lies in between a priori and interactive approaches: It generates an initial approximation of the Pareto front, and only requires the DM to pick a solution from this front. It then makes a final attempt to find a more preferred solution based on what the DM has picked, and returns this single final solution to the DM rather than an entire frontier.

BO is a global optimization technique that builds a Gaussian process (GP) surrogate model of the fitness landscape, and then uses the estimated mean and variance at each location to decide which solution to evaluate next. It uses an acquisition function to explicitly make a trade-off between exploitation and exploration (e.g., [17]). A frequently used acquisition function is the expected improvement (EI) [9] which selects the point with the largest expected improvement over the current best known solution as the next solution to evaluate. Recently, BO has been adapted to the multi-objective case, for a survey see [15]. One of the earliest approaches, ParEGO [11] simply uses the Tchebychev scalarization function to turn the multi-objective problem into a single objective problem, but it uses a different scalarization vector in every iteration where the next solution is decided according to EI. Other multi-objective algorithms fit separate models for each individual objective. [5] trains a GP model for each objective, then chooses the next solution to be evaluated according to a hypervolume-based acquisition criterion. Other multi-objective BO approaches include [2,10,12].

Recently, a few interactive multi-objective BO approaches have also been proposed [1,7,19]. Gaudrie et al. [6] allow the DM to specify a reference point a priori, and use this to subsequently focus the BO search.

3 Problem Definition

The standard multi-objective optimization problem with respect to a particular DM is defined as follows.We assume a D-dimensional real-valued space of possible *solutions*, i.e., $\mathbf{x} \in X \subset \mathbb{R}^D$. The objective function is an arbitrary black box $\mathbf{f} : X \to \mathbb{R}^K$ which returns a deterministic vector *output* $\mathbf{y} \in \mathbb{R}^K$. The (unknown) DM preference over the outputs can be characterized by a utility function $U : \mathbb{R}^K \to \mathbb{R}$. Thus, of all solutions in X, the DM's most preferred solution is $\mathbf{x}^* = \arg\max_{\mathbf{x} \in X} U(f(\mathbf{x}))$. There is a budget of B objective function evaluations, and we denote the n-th evaluated design by $\mathbf{x}^n$ and the n-th output by $\mathbf{y}^n = \mathbf{f}(\mathbf{x}^n)$ where, for convenience, we define the sampled solutions and outputs as $\tilde{X}^n = \{\mathbf{x}^1, \ldots, \mathbf{x}^n\}$ and $\tilde{Y}^n = \{\mathbf{y}^1, \ldots, \mathbf{y}^n\}$, respectively.

In a standard EMO procedure, after consuming the budget, the algorithm returns the set of evaluated non-dominated solutions $\Gamma \subset \tilde{X}^n$ and the DM chooses a preferred solution $\mathbf{x}_p$ according to $\mathbf{x}_p = \arg\max_{x \in \Gamma} U(\mathbf{f}(x))$. Figure 1 shows that a solution set may not contain any solution close to the DM's true preferred Pareto-optimal solution, and thus the DM will choose a sub-optimal solution.

Interactive multi-objective optimization algorithms attempt to learn the DM's preferences and then focus the search effort onto the most preferred region of the Pareto front, which allows them to provide a more relevant set of solutions. However, the multiple interactions mean additional cognitive effort for the DM. In this paper, we restrict the interaction to a *single* selection of a most preferred solution from a non-dominated front, as in the standard, non-interactive case. However, we allow to ask the DM for this information *before* the end of optimization, after $B - p$ evaluations. This allows the algorithm to identify potentially more relevant solutions in the final p evaluations. At the end, the algorithm has to return a *single* recommended solution $\mathbf{x}_r$ (rather than asking the DM to choose again), so that the cognitive effort is equivalent to the non-interactive case. The aim is then to minimize the Opportunity Cost (OC) of the chosen sample,

$$OC = U(\mathbf{f}(\mathbf{x}^*)) - U(\mathbf{f}(\mathbf{x}_r)).$$

4 Proposed Approach

This section describes details of the proposed algorithm. Section 4.1 and Sect. 4.2 show the statistical models used for the objectives and the utility, respectively. Then, Sect. 4.3 provides background on EI-UU and Sect. 4.4 considers the case when the DM utility model is different from the model assumed by EI-UU. Finally, Sect. 4.5 provides a summary of the algorithm.

4.1 Statistical Model of the Objectives

Let us denote the set of evaluated points and their objective function values up to iteration n as $\mathscr{F}^n = \{(\mathbf{x}, \mathbf{y})^1, \ldots, (\mathbf{x}, \mathbf{y})^n\}$. To model each objective function $y_j = f_j(\mathbf{x}), \forall j = 1, \ldots, K$, we use an independent GP defined by a mean function $\mu_j^0(\mathbf{x}) : X \to \mathbb{R}$ and a covariance function $k_j^0(\mathbf{x}, \mathbf{x}') : X \times X \to \mathbb{R}$. Given $\mathscr{F}^n$, predictions at new locations $\mathbf{x}$ for output y_j are given by the posterior GP mean $\mu_j^n(\mathbf{x})$ and covariance $k_j^n(\mathbf{x}, \mathbf{x}')$. We use the popular squared exponential kernel that assumes $f_j(\mathbf{x})$ is a smooth function, and we estimate the hyper-parameters from $\mathscr{F}^n$ by maximum marginal likelihood. Details can be found in [13].

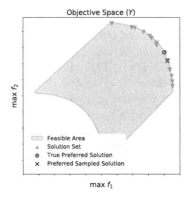

Fig. 1. HOLE test problem (b = 0). (Orange triangles) Solution set Γ shown to the DM. (Red cross) Solution picked by the DM x_p. (Red dot) True most preferred solution of the DM. (Color figure online)

4.2 Statistical Model over the Utility

After $B-p$ evaluations, the DM selects a point μ^* from an estimated Pareto front $\mathscr{P}$. Therefore, an interaction I is defined by the generated front and the selected solution, $I = (\mathscr{P}, \mu^*)$. The estimated front is generated using an evolutionary algorithm (NSGA-II [3]) on the posterior mean response surfaces of the Gaussian processes.

Let us assume that the DM's utility can be described by a parametric utility function $U(\mathbf{x}, \theta)$ with parameters $\theta \in \Theta$. Similar to [1] and [16], we adopt a Bayesian approach to obtain a distribution over parameters $\theta \in \Theta$. Commonly used likelihood functions include probit and logit [20]. However, for simplicity we assume fully accurate preference responses. Then, if we consider a utility model $U(\theta)$, we would only accept a candidate parameter θ if the best solution from the generated front $\mathscr{P}$ according to $\mu(\theta) = \arg\max_{\mu \in \mathscr{P}} U(\mu, \theta)$ is the solution selected by the DM. The likelihood for θ is then

$$\mathscr{L}(\theta) = \mathbb{I}_{\mu(\theta)=\mu^*}.$$

Figure 2b shows the above process by drawing three randomly generated linear utility functions.

Depending on the utility function model, the set of "compatible" ($\mathscr{L}(\theta) = 1$) parameterizations can be determined easily and in a deterministic way. More generally, if we assume a flat Dirichlet prior on the parameter space Θ, then a posterior distribution over θ is given by Bayes rule as

$$\mathbb{P}[\theta|I] \propto \mathscr{L}(\theta)\mathbb{P}[\theta].$$

We may obtain samples from the posterior distribution $\mathbb{P}[\theta|I]$ by simply generating Dirichlet samples from $\mathbb{P}[\theta]$ and accepting only those that are compatible with the DM interaction ($\mathscr{L}(\theta) = 1$). This approach may also be immediately

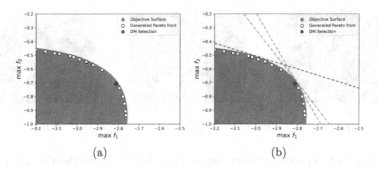

(a)   (b)

Fig. 2. (a) Estimated Pareto front shown to the DM (white dots) and a single point selected by the DM (red dot). (b) Three scalarizations are generated using a Linear utility function. Some are compatible with the information elicited from the decision maker (dashed green), and thus accepted. Scalarizations that are not compatible would be rejected (dashed red). (Color figure online)

extended to multiple and independent interactions where $\mathscr{L}$ may be expressed as the product of all different interactions. Thus, only those parameters compatible with all interactions would be accepted.

4.3   EI-UU with Preference Information

EI-UU [1] is a recently proposed multi-objective BO algorithm that is able to include uncertain preference information. Similar to ParEGO [11] it translates the multi-objective problem into a single-objective problem using an achievement scalarization function. However, ParEGO uses Tchebycheff scalarizations and randomly picks a different scalarization in every iteration to ensure coverage of the entire Pareto front, EI-UU uses linear scalarizations and integrates the expected improvement over all possible scalarizations, so it takes into account different scalarizations simultaneously rather than sequentially over iterations.

Given a parameterization θ, it is possible to determine the utility of the most preferred solution out of the solutions sampled so far, $\mathscr{F}^n$, as $u^*(\theta) = \max_{i=1,\dots,n} U(\mathbf{x}_i, \theta)$. Then, EI-UU simply computes the expected improvement over all possible realizations for θ and outputs $\mathbf{y} = \mathbf{f}(\mathbf{x})$, i.e.,

$$\text{EI-UU}(\mathbf{x}) = \mathbb{E}_{\theta,\mathbf{y}}[\max\{U(\mathbf{x},\theta) - u^*(\theta), 0\}].$$

Figure 3 provides an example in two-dimensional solution space, showing two EI landscapes for different θ (part a and b), as well as the corresponding overall improvement over several such realizations (c).

If the utility is linear (as suggested in [1]), the computations are essentially reduced to standard expected improvement, where we integrate over θ using a Monte-Carlo (MC) approximation. Otherwise, the whole expectation must be computed using Monte-Carlo with realisations of each objective at $\mathbf{x}$ generated

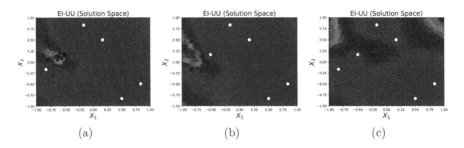

(a)   (b)   (c)

Fig. 3. Expected Improvement over the solution space, with brighter colors indicating higher expected improvement. Circles indicate sampled solutions. (a-b) Expected Improvement according to two different realizations of θ. Most preferred solution according to the specific θ is highlighted in red. (c) shows the MC average over several realizations of θ to obtain EI-UU. The recommended solution by the algorithm is selected according to the maximum value of EI-UU. (Color figure online)

as $f_j(\mathbf{x}) = \mu_j^n(\mathbf{x}) + k_j^n(\mathbf{x}, \mathbf{x})Z$, where $Z \sim N(0,1)$. Then, the overall expectation over θ and Z is computed using a MC average,

$$\text{EI-UU}(\mathbf{x}) \approx \frac{1}{N_{\Theta} N_Z} \sum_{w=1, t=1} \max \{U(\mathbf{x}, \theta_w, Z_t) - u^*(\theta_w)\}.$$

It is straightforward to accommodate preference information in this approach simply by adapting the distribution of the different scalarizations considered.

4.4 Utility Mismatch

All proposed approaches require a parameterized model of the utility function. In reality, this assumed model may not be able to accurately represent the true utility function of the DM. To mitigate the risk of the DM being misrepresented, we can use more flexible models (e.g., Cobb-Douglas utility function, Choquet Integral, or artificial neural network) or simply allow any of a set of simple models, which is the approach we take below. In general, there is a trade-off: The more restrictive the utility model, the more informative the elicited DM preference information, and the more focused the search in the final p steps may be. On the other hand, if the assumed utility model was wrong, the focus would be put on the wrong area, and the approach could fail. Thus, a more flexible model will provide less benefit, but also smaller risk of getting it wrong.

In this paper, let us consider a given set of L utility model candidates, as $\{U_1(\theta_1), \ldots, U_l, \ldots, U_L(\theta_L)\}$. In the Bayesian framework, one complete utility model U_l is formed by a likelihood function $\mathscr{L}_{U_l}(\theta)$ and a prior probability density function $\mathbb{P}(\theta|U_l)$, where now we explicitly emphasize the dependence of a model U_l on the likelihood and prior distribution. Therefore, to obtain a posterior distribution over models given the interaction I, i.e., $\mathbb{P}(U_l|I)$, we must compute the marginal likelihood, or Bayesian evidence, as

$$\mathbb{P}[I|U_l] = \int_{\theta_l \in \Theta_l} \mathscr{L}_{U_l}(\theta_l)\mathbb{P}(\theta_l|U_l)d\theta_l.$$

This quantity represents the probability of the data collected, I, given a utility model assumption U_l. However, we rely on approximating each term $\mathbb{P}[I|U_l]$ through MC integration. Hence, a posterior distribution over the different candidate utilities may be computed as

$$\mathbb{P}[U_l|I] = \frac{\mathbb{P}[I|U_l]\mathbb{P}[U_l]}{\sum_{i=1}^{L}\mathbb{P}[I|U_i]\mathbb{P}[U_i]},$$

where $\mathbb{P}[U_l]$ represents a prior distribution for the candidate utility models. If we also consider a uniform distribution over the candidate models as $\mathbb{P}[U_l] = 1/L$, then the posterior distribution only depends on the likelihood distribution terms. This shows that models with higher evidence values tend to have higher weight than other competing candidate models. Finally, EI-UU may be adapted by simply taking the expectation over the posterior $\mathbb{P}[U_l|I]$,

$$\text{EI-UU}(\mathbf{x}) = \sum_{i=1}^{L}\mathbb{E}_{\theta_i,\mathbf{y}}[\max\{U_i(\mathbf{x},\theta_i) - u_i^*(\theta_i), 0\}]\mathbb{P}[U_i|I]$$

4.5 Algorithm

The proposed algorithm follows the standard EI-UU algorithm with a single exception. Instead of letting the DM choose their most preferred solution at the end of optimization (after the budget of B samples has been depleted), we let the DM choose their most preferred solution from an approximated frontier already after $B - p$ samples. This approximated frontier is generated by NSGA-II on the posterior mean function obtained from the GPs for each objective. From the interaction, we derive user preferences as explained above, and evaluate an additional p solutions using this preference information. Finally, at the end of the optimization run, the algorithm returns to the DM the single solution $\mathbf{x}_r$ that it thinks has the best expected performance.

5 Results and Discussion

To assess the performance of the proposed approach, we investigate the Opportunity Cost (OC) dependence on the time when the DM is asked to pick a solution.

5.1 Experimental Setup

In all experiments, EI-UU is seeded with an initial stage of evaluations using $2(D+1)$ points allocated by Latin hypercube sampling over X. These evaluations are in addition to the budget B. NSGA-II is run for 300 generations with a population size of 100 to produce a Pareto front approximation.

Algorithm 1: Overall Algorithm.

Input: black-box function, size of Monte-Carlo N_Θ and N_Z for EI-UU, and the number of function evaluations $B - p$ before asking the DM.

0. Collect initial simulation data, $\mathscr{F}^n$, and fit an independent Gaussian process for each black-box function output.

1. **While** $b < $ B **do:**

2.   **If** $b = B - p$ **do:**

3.   Generate approximated Pareto frontier.

4.   DM selects preferred solutions.

5.   Compute posterior distribution $\mathbb{P}[\theta|I]$.

7.   Compute $\mathbf{x}^{n+1} = \arg\max_{\mathbf{x} \in X} \text{EI-UU}(\mathbf{x}; N_\Theta, N_Z)$.

8.   Update $\mathscr{F}^n$, with sample $\{(\mathbf{x}, \mathbf{y})^{n+1}\}$

9.   Update each Gaussian process with $\mathscr{F}^n$

10.   Update budget consumed, $b \leftarrow b + 1$

11. **Return:** Recommend solution, $\mathbf{x}_r = \arg\max_{\mathbf{x} \in X} \sum_{i=1}^{L} \mathbb{E}_{\theta_l}[U_i(\mathbf{x}, \theta)]\mathbb{P}[U_i|I]$

We use four different test functions:

1. The HOLE function [14] is defined over $X = [-1, 1]^2$ and has $K = 2$ objectives. We use $b > 0$, which produces two unconnected Pareto fronts, and the following function parameters: $q = 0.2$, $p = 2$, $d_0 = 0.02$, and $h = 0.2$.
2. An instance of the DTLZ2 function [4] with $K = 3$ objectives and defined over $X = [0, 1]^3$.
3. The ZDT1 function [4] with $k = 2$ objectives and defined over $X = [0, 1]^3$.
4. The rocket injector design problem [18]. This problem consists of minimizing the maximum temperature of the injector face, the distance from the inlet, and the maximum temperature on the post tip over $X = [0, 1]^4$.

All results are averaged over 20 independent replications and the figures below show the mean and 95% confidence intervals for the OC.

5.2   Preference Elicitation Without Model Mismatch

In this subsection, we look at the benefit that can be gained from letting the DM choose from the approximated Pareto front. We assume the DM has a Tchebychev utility with true underlying (but unknown to the algorithm) parameters θ. The true underlying parameters are generated randomly for every replication of a run using a different random seed. We include an optimistic setting ("Perfect Information"), such that once we interact with the DM, we receive the true underlying parameter θ and use this information in the following optimization steps. This represents the best performance we can hope for.

Figure 4 (**first column**) shows results depending on when the DM selects a point from the Pareto front. The benchmark is the case where the DM picks from the final set of non-dominated solutions after $B - p = 100$ iterations. The figure clearly indicates a trade-off about when the preference information should be elicited. The earlier the DM is involved, the longer it is possible for EI-UU to exploit the preference information gained. On the other hand, the earlier the DM is shown an approximated Pareto front, the less accurate is this Pareto front, and thus the learned preference information may be wrong. For the HOLE function, asking the DM too early (small $B - p$ values) even leads to substantially worse results than letting the DM pick a solution after the end of the optimization ($B - p = 100$). However, for all four functions considered, there is a broad range of settings that yield a significantly better result than when asking the DM after optimization. The intuitive explanation of the above trade-off is confirmed when comparing results with the case when perfect preference information is obtained. In this case, it is best to interact as early as possible, as there is no risk of learning something wrong or less informative due to a poor approximation of the Pareto front.

Figure 4 (**second column**) looks at the dependence of the observed benefit of letting the DM pick a solution early on the budget B, comparing $B = 20, 100, 200$. As expected, the OC decreases with increasing number of function evaluations. If the budget is very small ($B = 20$), it seems not possible to gain much by asking the DM early. Following the intuition above this is not really surprising, as after a very short optimization run the Pareto front is not well approximated and any information gained from the DM may be misleading. For a large budget ($B = 200$), there is still a small benefit to be gained for some of the test functions. However, intuitively, as the available budget tends to infinity, we can expect that the algorithm will return a very dense and accurate Pareto frontier, so the DM will be able to choose their true most preferred solution (or some solution very close to that), and so it is not possible to improve over that. The figure also allows us to appreciate the magnitude of the benefit of letting the DM choose earlier. For example, using a medium budget of $B = 100$ for the HOLE function and letting the DM choose after $80\% = 80$ samples yields an OC almost as small as increasing the budget to $B = 200$ samples and letting the DM choose at the end of the run.

5.3   Preference Elicitation with Model Mismatch

If the DM's true utility model is different from the parameterized model used by the algorithm there is a mismatch between the DM's true utility model (unknown to the algorithm) and the learned utility model in the algorithm. The consequences of model mismatch are explored next.

So far, we assumed that the true DM utility model is *Tchebychev*, and this model is used when the DM picks a solution from the approximated Pareto front, and again to evaluate the OC at the end. Now, in each replication of the algorithm, a random true *linear* utility model is generated for the DM. However, all approaches still assume a Tchebychev utility function and use this to focus the search over the last p iterations and also to recommend a final solution.

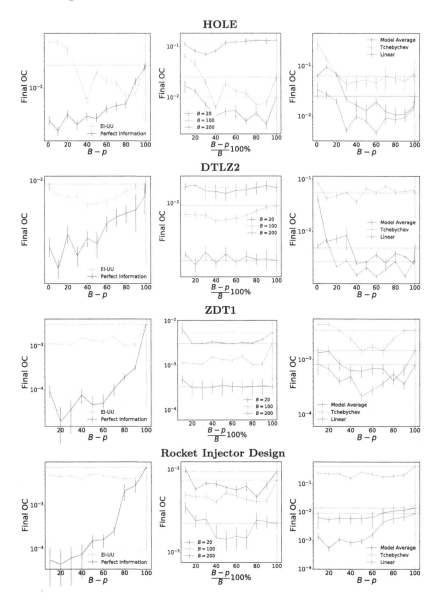

Fig. 4. First column: Final true utility of the generated solution set after $B = 100$ iterations, depending on how many iterations from the start the DM was presented with an approximation of the front $(B - p)$. True and assumed DM utility was *Tchebychev*. **Second column**: Final true utility of the generated solution set after different budgets, depending on the *percentage* of consumed budget before presenting an approximation of the front to the DM. True and assumed DM utility was *Tchebychev*. **Third column**: Final true utility of the generated solution set after $B = 100$ iterations, depending on how many iterations from the start the DM was presented with an approximation of the front $(B-p)$. The DM has a *Linear* utility and we show results for three different utility model assumptions (Linear, Tchebychev, and a model average). The dashed horizontal lines show the OC when the DM selects the best solution after optimization.

We consider two candidate utility functions to average EI-UU, Tchebychev and Linear. As Fig. 4 (**third column**) shows, model mismatch (orange) leads to substantially worse solutions, even when the DM only picks a solution *after* optimization. This seems counter-intuitive at first, but actually, EI-UU implicitly assumes a distribution of utility functions for scalarization, even if the distribution of parameter values θ is uniform. It appears that even with this relatively weak assumption, a model mismatch leads to significantly worse solutions. It is also clear that if there is a severe model mismatch (the algorithm assumes Tchebychev but the true utility is linear), letting the DM choose a solution before the end of the optimization is not always helpful (for the HOLE and DTLZ2 function, no setting of $B - p$ leads to significantly better results than $B - p = 100$). On the other hand, except for letting the DM choose very early, it is also not significantly worse. For the correct and the more flexible model encompassing Tchebychev and Linear, better results can be obtained for all four test problems by letting the DM choose before the end of optimization. But as expected, the benefit is smaller with the more flexible model, as its flexibility means it can focus less narrowly than if a less flexible but correct model is assumed.

6 Conclusion

For the case of expensive multi-objective optimization, we show how the surrogate models generated by Bayesian optimization can be used not only to speed up optimization, but also to show the DM a *predicted* Pareto front *before* the end of optimization, rather than the sampled non-dominated solutions at the end of the optimization run. Then, the information on the most preferred solution can be used to focus the final iterations of the algorithm to try and find this predicted most preferred solution, or even a better solution, which is then returned to the DM as recommended solution (no more interaction from the DM required - as in the standard case for EMO, the DM only once picks a most preferred solution from a frontier). We demonstrate empirically on four test problems that for various scenarios, the benefit in terms of true utility to the DM is significant.

Future directions of research may include the evaluation on a wider range of test problems, making the time to ask the DM self-adaptive, and to turning the approach into a fully interactive approach with multiple interactions with the DM during the optimization.

Acknowledgements. The first author would like to acknowledge funding from ESTECO SpA and EPSRC through grant EP/L015374/1.

References

1. Astudillo, R., Frazier, P.: Multi-attribute bayesian optimization with interactive preference learning. In: International Conference on Artificial Intelligence and Statistics, pp. 4496–4507. PMLR (2020)

2. Daulton, S., Balandat, M., Bakshy, E.: Parallel bayesian optimization of multiple noisy objectives with expected hypervolume improvement. In: Neural Information Processing Systems (2021)
3. Deb, K., Pratap, A., Agarwal, S., Meyarivan, T.: A fast and elitist multiobjective genetic algorithm: Nsga-ii. IEEE Trans. Evol. Comput. **6**(2), 182–197 (2002). https://doi.org/10.1109/4235.996017
4. Deb, K., Thiele, L., Laumanns, M., Zitzler, E.: Scalable test problems for evolutionary multiobjective optimization. In: Abraham, A., Jain, L., Goldberg, R. (eds.) Evolutionary Multiobjective Optimization. Advanced Information and Knowledge Processing, pp. 105–145. Springer, London (2005). https://doi.org/10.1007/1-84628-137-7_6
5. Emmerich, M., Deutz, A., Klinkenberg, J.: Hypervolume-based expected improvement: monotonicity properties and exact computation. In: IEEE Congress of Evolutionary Computation (CEC), pp. 2147–2154 (2011)
6. Gaudrie, D., Riche, R.L., Picheny, V., Enaux, B., Herbert, V.: Targeting solutions in bayesian multi-objective optimization: sequential and batch versions. Ann. Math. Artif. Intell. **88**(1), 187–212 (2020)
7. Hakanen, J., Knowles, J.D.: On using decision maker preferences with ParEGO. In: Trautmann, H., Rudolph, G., Klamroth, K., Schütze, O., Wiecek, M., Jin, Y., Grimme, C. (eds.) EMO 2017. LNCS, vol. 10173, pp. 282–297. Springer, Cham (2017). https://doi.org/10.1007/978-3-319-54157-0_20
8. Branke, J.: MCDA and multiobjective evolutionary algorithms. In: Greco, S., Ehrgott, M., Figueira, J.R. (eds.) Multiple Criteria Decision Analysis. ISORMS, vol. 233, pp. 977–1008. Springer, New York (2016). https://doi.org/10.1007/978-1-4939-3094-4_23
9. Jones, D.R., Schonlau, M., Welch, W.J.: Efficient global optimization of expensive black-box functions. J. Global Optim. **13**(4), 455–492 (1998). https://doi.org/10.1023/A:1008306431147
10. Keane, A.J.: Statistical improvement criteria for use in multiobjective design optimization. AIAA J. **44**(4), 879–891 (2006)
11. Knowles, J.: Parego: a hybrid algorithm with on-line landscape approximation for expensive multiobjective optimization problems. IEEE Trans. Evol. Comput. **10**(1), 50–66 (2006)
12. Picheny, V.: Multiobjective optimization using Gaussian process emulators via stepwise uncertainty reduction. Stat. Comput. **25**(6), 1265–1280 (2015)
13. Rasmussen, C.E., Williams, C.K.I.: Gaussian Processes for Machine Learning. MIT Press, Cambridge (2006)
14. Rigoni, E., Poles, S.: NBI and MOGA-II, two complementary algorithms for multi-objective optimizations. In: Dagstuhl seminar proceedings. Schloss Dagstuhl-Leibniz-Zentrum für Informatik (2005)
15. Rojas-Gonzalez, S., Van Nieuwenhuyse, I.: A survey on kriging-based infill algorithms for multiobjective simulation optimization. Comput. Oper. Res. **116**, 104869 (2020). https://doi.org/10.1016/j.cor.2019.104869
16. Schoenauer, M., Akrour, R., Sebag, M., Souplet, J.C.: Programming by feedback. In: Xing, E.P., Jebara, T. (eds.) Proceedings of the 31st International Conference on Machine Learning. Proceedings of Machine Learning Research, 22–24 June 2014, vol. 32, pp. 1503–1511. PMLR, Bejing (2014), https://proceedings.mlr.press/v32/schoenauer14.html
17. Shahriari, B., Swersky, K., Wang, Z., Adams, R.P., de Freitas, N.: Taking the human out of the loop: a review of bayesian optimization. Proc. IEEE **104**(1), 148–175 (2016). https://doi.org/10.1109/JPROC.2015.2494218

18. Tanabe, R., Ishibuchi, H.: An easy-to-use real-world multi-objective optimization problem suite. Appl. Soft Comput. **89**, 106078 (2020). https://doi.org/10.1016/j.asoc.2020.106078,   https://www.sciencedirect.com/science/article/pii/S1568494620300181
19. Taylor, K., Ha, H., Li, M., Chan, J., Li, X.: Bayesian preference learning for interactive multi-objective optimisation. In: Proceedings of the Genetic and Evolutionary Computation Conference, pp. 466–475 (2021)
20. Wirth, C., Akrour, R., Neumann, G., Fürnkranz, J.: A survey of preference-based reinforcement learning methods. J. Mach. Learn. Res. **18**(1), 4945–4990 (2017)
21. Xin, B., Chen, L., Chen, J., Ishibuchi, H., Hirota, K., Liu, B.: Interactive multiobjective optimization: a review of the state-of-the-art. IEEE Access **6**, 41256–41279 (2018)

Surrogate-Assisted LSHADE Algorithm Utilizing Recursive Least Squares Filter

Mateusz Zaborski$^{(\boxtimes)}$ and Jacek Mańdziuk

Faculty of Mathematics and Information Science, Warsaw University of Technology,
Warsaw, Poland
{M.Zaborski,mandziuk}@mini.pw.edu.pl

Abstract. Surrogate-assisted (meta-model based) algorithms are dedicated to expensive optimization, i.e., optimization in which a single Fitness Function Evaluation (FFE) is considerably time-consuming. Meta-models allow to approximate the FFE value without its exact calculation. However, their effective incorporation into Evolutionary Algorithms remains challenging, due to a trade-off between accuracy and time complexity. In this paper we present the way of recursive meta-model incorporation into LSHADE (rmmLSHADE) using a Recursive Least Squares (RLS) filter. The RLS filter updates meta-model coefficients on a sample-by-sample basis, with no use of an archive of samples. The performance of rmmLSHADE is measured using the popular CEC2021 benchmark in expensive scenario, i.e. with the optimization budget of $10^3 \cdot D$, where D is the problem dimensionality. rmmLSHADE is compared with the baseline LSHADE and with psLSHADE – a novel algorithm designed specifically for expensive optimization. Experimental evaluation shows that rmmLSHADE distinctly outperforms both algorithms. In addition, the impact of the forgetting factor (RLS filter parameter) on algorithm performance is examined and the runtime analysis of rmmLSHADE is presented.

Keywords: Surrogate model · LSHADE · Recursive Least Squares

1 Introduction

Continuous optimization is a well-known but still intensively explored field. The variety of problem structures implies that universal general purpose approach does not exist. In this work, we consider single-objective continuous global optimization in a black-box manner, i.e., no information about function gradient is provided, and the only interaction with the algorithm is through the Fitness Function Evaluation (FFE) calculation. Furthermore, the solution space is constrained by given search boundaries. Formally, the goal is to find a solution $\boldsymbol{x}^*$ that minimizes the function $f : \mathbb{R}^D \to \mathbb{R}$, where D is the problem dimensionality.

Evolutionary algorithms are metaheuristics commonly employed for solving optimization problems [26]. Various relatively uncomplicated algorithms such as

© The Author(s), under exclusive license to Springer Nature Switzerland AG 2022
G. Rudolph et al. (Eds.): PPSN 2022, LNCS 13398, pp. 146–159, 2022.
https://doi.org/10.1007/978-3-031-14714-2_11

Differential Evolution [23] (DE) or Particle Swarm Optimization [17] (PSO) are not only versatile but also provide a base for numerous extensions. For instance, parameter adaptation is a mechanism that resulted in the formation of Adaptive Differential Evolution with Optional External Archive [32] (JADE) and its successor Success-History Based Parameter Adaptation for Differential Evolution [24] (SHADE). Covariance Matrix Adaptation Evolution Strategy [13] (CMA-ES) is, by its principle, based on multiple parameters adaptation. Another commonly used mechanism is a dynamic reduction or increasing of a population size. In this context, SHADE evolved into SHADE with Linear Population Size Reduction, known as LSHADE [25]. Variations of the two above-mentioned methods (SHADE and CMA-ES) have resulted in even more sophisticated algorithms (e.g. LSHADE_cnEpSin [4], MadDE [7], IMODE [21], IPOP-CMA-ES [2], KL-BIPOP-CMA-ES [28], or PSA-CMA-ES [20]). The heuristics listed above are computationally efficient and their complexity does not depend on the number of FFEs, i.e., the time and memory required for an algorithm iteration are approximately constant during the whole optimization run. Therefore, the well-known benchmarks assume, by default, a relatively high FFEs budget. For example, employed in this work, CEC2021 Special Session and Competition on Single Objective Bound Constrained Numerical Optimization benchmark suite [19] (CEC2021) considers a budget of $2 \cdot 10^5$ for $D = 10$ dimensions.

A significantly different type of optimization is an expensive optimization, where a single FFE is costly, so high optimization budgets are inapplicable. CEC2015 benchmark for Computationally Expensive Numerical Optimization [10] assumes $5 \cdot 10^2$ FFEs for $D = 10$. Consequently, an entirely different class of algorithms are utilized in such optimization (e.g. Kriging [11] or Efficient Global Optimization [16] (EGO)). They usually operate on complex surrogate models of the function being evaluated. The purpose of the surrogate model is to estimate the fitness function value without its costly evaluation. Increasing the number of FFEs over the rational threshold makes the model computationally inefficient. Nevertheless, the idea of surrogate-assisted optimization was incorporated into evolutionary algorithms [15]. The CMA-ES extensions: S-CMA-ES [5] and DTS-CMA-ES [6], which rely heavily on Kriging, are designed for regular expensive optimization. Likewise, lmm-CMA-ES [18], which utilizes Locally Weighted Regression [1]. In contrast, LS-CMA-ES [3] uses a more efficient quadratic approximation of fitness function, which results in the ability to perform considerably more FFEs.

The concept of efficient local meta-models hybridized with PSO and DE was presented in [31]. The lq-CMA-ES [12] employs a global quadratic model with interactions, but its structure relies on simplicity and computational efficiency. However, when analyzing its experimental results, the benefits of incorporating the meta-model are visible up to $10^3 \cdot D$ FFEs. Meta-models can also be incorporated into the LSHADE algorithm. An idea of pre-screening solutions supported by a quadratic meta-model with interactions and inverse transformations was incorporated into the psLSHADE algorithm [29,30]. The psLSHADE outperformed both LSHADE and MadDE using $10^3 \cdot D$ FFEs budget. The psLSHADE is the reference point for the algorithm presented in this work.

Moreover, the above works introducing psLSHADE suggest that the gains from meta-model utilization can vanish with an optimization budget greater than $10^3 \cdot D$ FFEs. In contrast, complex surrogate-assisted algorithms, such as, mentioned above, Kriging or EGO, are overly ineffective in budgets greater than $10^2 \cdot D$ FFEs. To sum up, we perceive and try to explore a niche in research on expensive optimization in budgets greater than $10^2 \cdot D$ FFEs but still less than $10^5 \cdot D$ - $10^6 \cdot D$ FFEs.

Most of the surrogate-assisted algorithms are based on some form of linear regression. Therefore, we decided to investigate how to employ meta-model estimation without maintaining an archive of already evaluated samples. The utilization of the Recursive Least Squares (RLS) filter [22] appeared to be an engaging idea. The RLS filter has been successfully applied to mixed-variable optimization [8] or combinatorial problems with integer constraints [9].

We investigate the usefulness of the RLS filter in continuous optimization by integrating it with the LSHADE algorithm. In effect, we propose the rmmL-SHADE algorithm: recursive meta-model LSHADE utilizing the Recursive Least Squares filter. The rmmLSHADE is an extension of LSHADE and similarly to psLSHADE employs the pre-screening mechanism, albeit in a different, more efficient manner. In particular, the meta-model coefficients are re-estimated with the help of RLS, instead of the Ordinary Least Squares applied in psLSHADE. Furthermore, unlike psLSHADE, rmmLSHADE does not use an archive of already evaluated samples. Accordingly, it re-estimates meta-model coefficients once per evaluation, not once per iteration as is the case of psLSHADE.

2   Related Work

In this section, LSHADE and psLSHADE are briefly described. The LSHADE algorithm is the basis of the proposed rmmLSHADE. The psLSHADE extends LSHADE by adding a pre-screening mechanism. For a detailed description of LSHADE and psLSHADE, please see the original works ([25] and [29,30], resp.).

2.1   LSHADE

LSHADE is an iterative population-based meta-heuristic. In each iteration g, population $P^g = [\boldsymbol{x}_1^g, \ldots, \boldsymbol{x}_{N^g}^g]$, where each individual i represents a solution $\boldsymbol{x}_i^g = [x_{i,1}^g, \ldots, x_{i,D}^g]$, is subjected to three successive phases: mutation, crossover, and selection. In the mutation phase, each parent vector $\boldsymbol{x}_i^g$ is randomly mutated into vector $\boldsymbol{v}_i^g$. In the crossover phase, each mutated vector $\boldsymbol{v}_i^g$ is crossed with its parent vector $\boldsymbol{x}_i^g$, resulting in trial vector $\boldsymbol{u}_i^g$. Finally, in the selection phase, vector $\boldsymbol{u}_i^g$ is evaluated using fitness function and replaces its parent $\boldsymbol{x}_i^g$ if its value is better than the parent's value ($f(\boldsymbol{u}_i^g) < f(\boldsymbol{x}_i^g)$). Both mutation and selection are based on parameters (F_i^g and CR_i^g, respectively) obtained independently for each individual i. In a nutshell, memory, included in LSHADE, stores successful F_i^g and CR_i^g parameters values that have resulted in fitness function value improvement in the selection phase. Parameter adaptation shifts the mean value of F_i^g and CR_i^g during generation towards successful values. Additionally, during the optimization run the population size decreases from N_{init} to N_{min}.

2.2 psLSHADE

The psLSHADE extends the baseline LSHADE by means of the pre-screening mechanism. In the mutation phase, each individual i generates independently N_s mutated vectors $v_i^{g,j}$, $j = 1, \ldots, N_s$. Next, they are subjected to the same (per individual) rules in the crossover phase. Right before the selection phase, the meta-model estimates the surrogate value f^{surr} of each trial vector $u_i^{g,j}$ and designates the best one $u_i^{g,best}$, independently for each i. Finally, all N^g best trial vectors $u_i^{g,best}$, $i = 1, \ldots, N^g$ are evaluated, and their parameters $F_i^{g,best}, CR_i^{g,best}$ are stored if they have improved the fitness function value compared to the parent's one. The whole pre-screening process is transparent for parameter adaptation and memory features.

The meta-model utilizes an archive of limited size $N_a = 2 \cdot df_{mm}$, where df_{mm} is the number of degrees of freedom of the meta-model. The archive contains N_a best already evaluated solutions and corresponding fitness function values. The meta-model is estimated using Ordinary Least Squares [27], once per iteration, before pre-screening, i.e., the same meta-model is utilized in all N^g pre-screening procedures. The meta-model consists of the following transformations: linear (x_d), quadratic (x_d^2), interactions $(x_d \cdot x_{d'})$, inverse linear (x_d^{-1}) and inverse quadratic (x_d^{-2}), resulting in $df_{mm} = 0.5 \cdot (D^2 + 7D) + 1$.

3 Proposed rmmLSHADE

The underlying idea of rmmLSHADE is discussed below, along with a description of the meta-model integration and the pre-screening mechanism. A pseudocode of the method is presented in Algorithm 1. The source code of rmmLSHADE is available at https://bitbucket.org/mateuszzaborski/rmmlshade/.

In rmmLSHADE, the basic population P^g consists of N^g individuals $x_i^g = [x_{i,1}^g, \ldots, x_{i,D}^g]$, $i = 1, \ldots, N^g$. Before the mutation phase population P^g is multiplicated N_m times resulting in the extended population:

$$P_{ext}^g = [P^g, \ldots, P^g] = [x_1^g, \ldots, x_{N^g}^g, \ldots, x_1^g, \ldots, x_{N^g}^g] \qquad (1)$$

where $|P_{ext}^g| = N_m \cdot N^g$. The same extension rules apply to a vector containing fitness values. Next, the mutation is applied to each individual $k = 1, \ldots, N_m \cdot N^g$ as follows:

$$v_k^g = x_k^g + F_k^g(x_{pbest_k}^g - x_k^g) + F_k^g(x_{r1_k}^g - x_{r2_k}^g) \qquad (2)$$

The $pbest_k$ index denotes a randomly selected individual from the set of $p \cdot N^g$ best individuals from the original population P^g, where $p \in [0, 1]$ is a parameter.

Algorithm 1. rmmLSHADE high-level pseudocode

1: Set all parameters $N_{init}, N_{min}, M_F, M_{CR}, p, a, H, N_a, N_m, \lambda$
2: Initialize $M^0_{F,m}$ and $M^0_{CR,m}$ memory entries with default values of M_F and M_{CR}
3: $P^0 = [\boldsymbol{x}^0_1, \ldots, \boldsymbol{x}^0_{N^0}]$ ▷ Population initialization using Latin Hypercube
   Sampling [14]
4: Estimate meta-model's coefficients $\boldsymbol{w}_0$ from P^0 and fitness function values of P^0
   using Ordinary Least Squares [27]
5: g=0
6: **while** evaluation budget left **do**
7:     Extend P^g to $P^g_{ext} = [P^g, P^g, \ldots, P^g]$, where $|P^g_{ext}| = N_m \cdot N^g$ using eq. (1)
8:     Generate $N_m \cdot N^g$ mutated vectors $\boldsymbol{v}^g_k$ using eq. (2)
9:     Generate $N_m \cdot N^g$ trial vectors $\boldsymbol{u}^g_k$ using eq. (3)
10:    **for** i = 1 to N^g **do**
11:        Calculate $N_m \cdot N^g$ surrogate values $f^{surr}(\boldsymbol{u}^g_k)$
12:        Designate the best (not already chosen) trial vector $\boldsymbol{u}^{g,best}_i$
13:        Evaluate trial vector $\boldsymbol{u}^{g,best}_i$ using true fitness function
14:        Update coefficients $\boldsymbol{w}_g$ using RLS eq. (9)
15:    **end for**
16:    Do selection of all N^g chosen trial vectors $\boldsymbol{u}^{g,best}_i$ using eq. (4)
17:    Update memory with $M^g_{F,m}$ and $M^g_{CR,m}$ using eq. (6)
18:    Set new population size N^{g+1} using eq. (5)
19:    $g = g + 1$
20: **end while**

Index $r1_k \in \{1, \ldots, N^g\}$ indicates the individual from the original population P^g. Index $r2_k \in \{1, \ldots, N^g + |A|\}$, denotes an individual from the union of the original population P^g and an external archive A. The principles of external archive A are discussed in Sect. 3.2. The following inequalities: $k \neq r1_k$, $k \neq r2_k$, $r2_k \neq r1_k$, $k \not\equiv r1_k \pmod{N^g}$ and $k \not\equiv r2_k \pmod{N^g}$ must be fulfilled. The way of setting the scaling factor F^g_k in (2) is discussed in Sect. 3.1.

In the crossover phase $N_m \cdot N^g$ trial vectors $\boldsymbol{u}^g_k = [u^g_{k,1}, \ldots, u^g_{k,D}]$ are generated as follows:

$$u^g_{k,d} = \begin{cases} v^g_{k,d}, & \text{if } rand(0,1) \leq CR^g_k \text{ or } d = d^{rand}_k \\ x^g_{k,d}, & \text{otherwise} \end{cases} \tag{3}$$

where $rand(0,1)$ is a uniformly sampled number from $(0,1)$, $d^{rand}_k \in \{1, \ldots, D\}$ is a randomly selected index and CR^g_k is a crossover rate described in Sect. 3.1. All listed above parameters are generated independently per individual k.

A four-step loop occurs after the crossover phase (cf. Algorithm 1). The loop is executed N^g times in order to evaluate N^g subsequent best trial vectors $\boldsymbol{u}^{g,best}_i$, $i = 1, \ldots, N^g$. In the first step, the meta-model estimates surrogate values f^{surr} of all trial vectors $\boldsymbol{u}^g_k$. Then, the best, not already chosen, trial vector $\boldsymbol{u}^{g,best}_i$ is picked for evaluation. Finally, after fitness function evaluation, the meta-model is updated using the rules presented in Sect. 3.3.

The selection phase (4) results in restoring the population size to the original value N^g.

$$x_i^{g+1} = \begin{cases} u_i^{g,best}, & \text{if } f(u_i^{g,best}) < f(x_i^g) \\ x_i^g, & \text{otherwise} \end{cases} \tag{4}$$

Finally, Linear Population Size Reduction (LPSR), derived from LSHADE, changes the population size P^g after the selection phase. A target population size is defined by the following equation:

$$N^{g+1} = round\left(\left(\frac{N_{min} - N_{init}}{MAX\_NFE}\right) \cdot NFE + N_{init}\right) \tag{5}$$

where MAX_NFE is the optimization budget and NFE is the number of FFEs made so-far. N_{init} and N_{min} are algorithm parameters. The worst individuals are removed from the population to achieve the desired population size N^{g+1}.

3.1 Parameter Adaptation

Both the scaling factor F_k^g in (2) and crossover rate CR_k^g in (3) are adaptive parameters. Analogously to LSHADE, the adaptation is provided by a memory mechanism. The memory stores those historical values of $F_i^{g,best}$ and $CR_i^{g,best}$ that succeeded, i.e., their corresponding trial vector $u_i^{g,best}$ improved the fitness function value $f(u_i^{g,best})$ compared to its parent $f(x_i^g)$.

The memory consists of H (H is a parameter) entry pairs $(M_{F,m}^g, M_{CR,m}^g)$, where $m = 1, \ldots, H$. In each iteration, after the selection phase, all successful values $F_i^{g,best}$ and $CR_i^{g,best}$ are stored in the sets S_F and S_{CR}, resp. Both sets are transformed using weighted Lehmer means to obtain two scalar values: $mean_{W_L}(S_F)$ and $mean_{W_L}(S_{CR})$:

$$mean_{W_L}(S) = \frac{\sum_{s=1}^{|S|} w_s S_s^2}{\sum_{s=1}^{|S|} w_s S_s}, \qquad w_s = \frac{\Delta f_s}{\sum_{r=1}^{|S|} \Delta f_r} \tag{6}$$

where $\Delta f_p = f(x_p^g) - f(u_i^{g,best})$, $p \in \{s, r\}$.

The memory entry pairs $(M_{F,m}^g, M_{CR,m}^g)$ are updated cyclically with pairs $(mean_{W_L}(S_F), mean_{W_L}(S_{CR}))$, using $m = 1, 2, \ldots, H, 1, 2, \ldots$ entry indexing scheme. Moreover, if all $CR_i^{g,best}$ values in set S_{CR} are equal to 0, the entry $M_{CR,m}^g$ is permanently set to terminal value $\perp$ (the notation follows LSHADE's description [25]) instead of $mean_{W_L}(S_{CR})$.

The values F_k^g and CR_k^g are generated randomly using Cauchy distribution and Normal distribution, respectively, using M_{F,r_k}^g and M_{CR,r_k}^g as distribution parameters. Both M_{F,r_k}^g and M_{CR,r_k}^g are taken from the memory. A random index $r_k \in 1, \ldots H$ denotes the memory index and is designated independently for each individual k of population P_{ext}^g. Concluding, F_k^g and CR_k^g are generated using the following formulas:

$$F_k^g = rand_{Cauchy}(M_{F,r_k}^g, 0.1) \tag{7}$$

$$CR_k^g = \begin{cases} 0 & \text{if } M_{CR,r_k}^g = \bot \\ rand_{Normal}(M_{CR,r_k}^g, 0.1) & \text{otherwise} \end{cases} \tag{8}$$

Both values F_k^g and CR_k^g are limited. Scaling factor F_k^g is truncated from top to 1. If scaling factor $F_k^g \leq 0$, a random generation is repeated. Crossover rate CR_k^g is truncated to $[0, 1]$.

3.2 External Archive

An external archive A is incorporated into rmmLSHADE. Its principle is inspired by the one employed in baseline LSHADE. However, the proposed pre-screening mechanism enforces some adjustments. Archive A extends the current population P_{ext}^g with parent vectors replaced by a better trial vector in the selection phase. The parent vector is an $\boldsymbol{x}_k^g$ vector utilized in the successful $\boldsymbol{u}_i^{g,best}$ generation (2). The size of the archive is $a \cdot N^g$, where a is a parameter. Randomly selected elements are removed when the archive is full in order to insert new element. A gradual decrease of the population size leads to shrinking the archive size. Similarly, randomly selected elements are when necessary.

3.3 Recursive Meta-model Description

The rmmLSHADE utilizes a global linear meta-model that pre-screens trial vectors to choose the best ones according to its surrogate value. Although the model is technically linear, it contains additional nonlinear transformations of variables: constant, quadratic, and interactions. Compared to the meta-model incorporated into psLSHADE, rmmLSHADE meta-model does not include inverse linear and inverse quadratic transformations. The absence of these inverse transformations is due to the higher risk of numerical instability in Recursive Linear Squares (RLS) compared to Ordinary Least Squares. We observed this phenomenon in preliminary trials while including extra inverse transformations. The final form of the meta-model utilized in rmmLSHADE is presented in Table 1.

Table 1. A description of transformations and the final form of the recursive meta-model (rmm.) For the sake of readability, the estimated coefficients applied to each variable are omitted.

Name	Form	DoF
Constant	$\boldsymbol{z}_c = [1]$	$df_c = 1$
Linear	$\boldsymbol{z}_l = [x_1, \ldots, x_D]$	$df_l = D$
Quadratic	$\boldsymbol{z}_q = [x_1^2, \ldots, x_D^2]$	$df_q = D$
Interactions	$\boldsymbol{z}_i = [x_1 x_2, \ldots, x_{D-1} x_D]$	$df_i = \frac{D(D-1)}{2}$
Final rmm	$\boldsymbol{z}_{mm} = [\boldsymbol{z}_c + \boldsymbol{z}_l + \boldsymbol{z}_q + \boldsymbol{z}_i]$	$df_{mm} = \frac{D^2 + 3D}{2} + 1$

The RLS filter [22, Eqs. (21.36)–(21.39)] is an adaptive algorithm that recursively solves the least squares problem. It provides a procedure that computes coefficients $\boldsymbol{w}_t$ from $\boldsymbol{w}_{t-1}$ considering input signal $\boldsymbol{a}_t$, output signal d_t, and output signal estimation $\hat{d}_t = \boldsymbol{a}_i^T \boldsymbol{w}_{t-1}$. $e_t = d_t - \boldsymbol{a}_t^T \boldsymbol{w}_{t-1}$ is an output error. The sum of error squares is a cost function being minimized.

The rmmLSHADE employs RLS filter to recursively estimate meta-model coef-
ficients. Hence: $\boldsymbol{a}_t = \boldsymbol{z}_{mm}$, $d_t = f(\boldsymbol{u}_i^{g,best})$, and $\hat{d}_t = f^{surr}(\boldsymbol{u}_i^{g,best}) = \boldsymbol{z}_{mm}^\top \boldsymbol{w}_{t-1}$, where $\boldsymbol{z}_{mm}$ denotes the $\boldsymbol{u}_i^{g,best}$ vector transformation (cf. Table 1). The following system of equations expresses the recursive estimation of the meta-model coefficients $\boldsymbol{w}_t$:

$$e_t = f(\boldsymbol{u}_i^{g,best}) - \boldsymbol{z}_{mm}^\top \boldsymbol{w}_{t-1}, \qquad \boldsymbol{g}_t = \frac{Q_{t-1}\boldsymbol{z}_{mm}}{\lambda + \boldsymbol{z}_{mm}^\top Q_{t-1}\boldsymbol{z}_{mm}}$$
$$Q_t = \frac{1}{\lambda}(Q_{t-1} - \boldsymbol{g}_t \boldsymbol{z}_{mm}^\top Q_{t-1}), \qquad \boldsymbol{w}_t = \boldsymbol{w}_{t-1} + \boldsymbol{g}_t e_t \tag{9}$$

where $\lambda \in (0, \ldots 1]$ is a forgetting factor and Q_t is a matrix of size $|\boldsymbol{z}_{mm}| \times |\boldsymbol{z}_{mm}|$.

The RLS filter requires two initial components: Q_0 and $\boldsymbol{w}_0$. The initial matrix Q_0 is an identity matrix and $\boldsymbol{w}_0$ is estimated using OLS regression [27] performed for the initial population $P^0 = [\boldsymbol{x}_1^0, \ldots, \boldsymbol{x}_{N^0}^0]$ on the corresponding $N^0 = N_{init}$ fitness function values $f(\boldsymbol{x}_i^0)$. P^0 is generated using Latin Hyper-cube Sampling [14] to provide better search space coverage. Therefore, the initial population size must at least equal the number of the meta-model's degrees of freedom $df_{mm} = 0.5 \cdot (D^2 + 3D) + 1$.

The forgetting factor λ directly affects the meta-model performance. As the population converges, the global meta-model is evolving into the local one. In other words, since we are dealing with a time-varying fitness landscape, $\lambda = 1$ may decline the adaptation ability of the meta-model. In contrast, undersized $\lambda < 1$ may cause instability due to the overfitting, i.e., insufficient noise signal filtering. Section 4 includes experimental evaluation of $\lambda = 0.98, 0.99, 1.00$.

4 Experimental Evaluation

In the experimental evaluation the *CEC2021 Special Session and Competition on Single Objective Bound Constrained Numerical Optimization benchmark suite* was used, described in the technical report [19], hereafter referred to as CEC2021.

CEC2021 consists of 10 functions defined for $D = 10$ and $D = 20$. The functions are split into 4 categories: unimodal (F_1), basic ($F_2 - F_4$), hybrid ($F_5 - F_7$), and composition functions ($F_8 - F_{10}$). All of them are bounded to $[-100, 100]^D$. Furthermore, three function transformations are applied: bias (B), shift (S), and rotation (R) to increase the complexity of the benchmark problems. Besides the variant without transformations, four additional combinations are considered in the tests: S, B+S, S+R, and B+S+R. Each experiment was repeated 30 times.

Although CEC2021 proposes, by default, an optimization budget of $2 \cdot 10^5$ for $D = 10$ and 10^6 for $D = 20$, we follow the experimental procedure used for the psLSHADE evaluation [29,30] and assume an optimization budget of $10^3 \cdot D$.

Moreover, to ensure a fair comparison, all rmmLSHADE parameters shared with psLSHADE were not changed with respect to psLSHADE parameterization [29,30]. Also, the baseline LSHADE included in the comparison follows the same parametrization: initial population size $N_{init} = 18 \cdot D$, final population size $N_{min} = 4$, initial value of $M_F = 0.5$, initial value of $M_{CR} = 0.5$, best-from-population rate $p = 0.11$, archive rate $a = 1.4$, memory size $H = 5$. Since psLSHADE utilizes $N_s = 5$ number of trial vectors per individual, despite principal operational differences in rmmLSHADE, we set $N_m = 5$, as well.

We followed the performance metrics described in detail in the technical report [19]. The final $Score = Score1 + Score2$, where $Score \in (0, 100]$, utilizes four partial measures: SNE, SR, $Score1$, and $Score2$. SNE represents the sum of the smallest errors obtained by a given algorithm for each problem. SR indicates the sum of the algorithm's ranks among all problems. Both SNE and SR are relative, referring to the performance of all evaluated algorithms. $Score1$ transforms SNE by comparing it to the best SNE obtained among other algorithms. $Score2$ transforms SR in the same manner.

The comparison results of rmmLSHADE with LSHADE and psLSHADE are presented in Table 2. Three rmmLSHADE variants, for $\lambda = 0.98$, 0.99, and 1.00 are included. In addition, a variant denoted $\lambda = \varnothing$, without the RLS filter adaptation is presented. In this case coefficients $\boldsymbol{w}_0$, once-estimated using OLS, remain unchanged $(\boldsymbol{w} = \boldsymbol{w}_0)$ during the whole optimization run.

Table 2. Scores of LSHADE, psLSHADE and rmmLSHADE $(\lambda = 0.98, 0.99, 1.0)$ for $10^3 \cdot D$ optimization budget. $\lambda = \varnothing$ is a variant without RLS adaptation.

Score	Algo					
	LSHADE	psLSHADE	$\lambda = \varnothing$	$\lambda = 0.98$	$\lambda = 0.99$	$\lambda = 1.0$
SNE	30.06	17.86	36.72	15.75	14.81	21.62
SR	217.00	154.25	245.50	138.50	122.75	172.00
Score 1	24.64	41.47	20.16	47.02	50.00	34.25
Score 2	28.28	39.79	25.00	44.31	50.00	35.68
Score	**52.92**	**81.26**	**45.16**	**91.33**	**100.00**	**69.93**

Not surprisingly, the variant with $\lambda = 0.99$ turned out to be the best performing. In 57 out of 100 test cases (functions × transformations × dimensions) the difference between this variant and psLSHADE was significant (Mann-Whitney test, $p\text{-}value=0.05$). A smaller or larger value of λ causes a clear performance degradation. The variant with no $\boldsymbol{w}$ adaptation is distinctly weakest, which supports the assumption from Sect. 3.3 about the significantly time-varying fitness landscape. The proposed rmmLSHADE outperformed the reference psLSHADE. The advantage over LSHADE is, unsurprisingly, more prominent.

Figure 1 shows four convergence plots for functions: F_1, F_3, F_7, and F_9 $(D = 20)$ one per each of the four categories. The comparison includes rmmLSHADE

($\lambda = 0.99$), psLSHADE, and LSHADE. The plots present the error obtained after every 100 FFEs. The error is averaged over 5 transformations and 30 independent runs, i.e., each point is the average of 150 values.

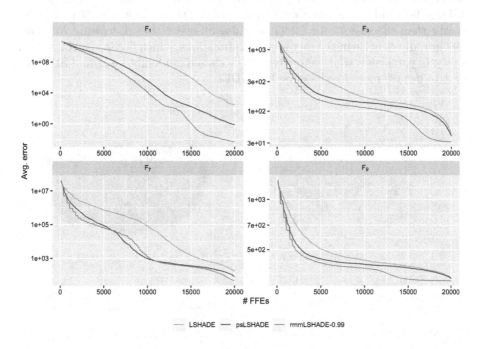

Fig. 1. The averaged convergence of rmmLSHADE, psLSHADE and LSHADE for F_1, F_3, F_7, and F_9 ($D = 20$) with $10^3 \cdot D$ optimization budget. The x-axis represents the number of FFEs, and the y-axis the error (a difference from the optimum) averaged across 150 runs (5 transformation selections $\times$ 30 runs).

Function F_1 is unimodal, so meta-model utilization is particularly noticeable in this case. For the remaining functions, better convergence of rmmLSHADE over LSHADE and psLSHADE is clearly observable with the greatest advantage is occurring in the first stage of the optimization run. Further, the distinction for the last $2 \cdot 10^3$ - $5 \cdot 10^3$ FFEs (depending on the function) tends to decrease. It is worth emphasizing that for F_3 and F_9, psLSHADE starts to converge approximately in the same time as rmmLSHADE but later loses its momentum and approaches the baseline LSHADE. The above observation suggests better adaptability of the meta-model utilized in rmmLSHADE than the one of psLSHADE.

Figure 2 illustrates the convergence of LSHADE and the four rmmLSHADE variants from Table 2 for F_2 and F_4, with $D = 20$ (the $\lambda = \emptyset$ variant is denoted rmmLSHADE-no-adapt). Additionally, $3D$ maps of $2D$ versions of F_2 and F_4 are shown. Function F_2 was chosen due to the absence of a significant global structure. In turn, function F_4 contains a noticeable flat area. Both cases make meta-model adaptation pertinent. The lack of model adaptation (rmmLSHADE-no-adapt) results in dramatic performance degradation for both functions. A tendency to convergence to local optimum could be observed when the meta-model was not fitted well during experiments. Similarly, a slightly smaller but still significant phenomenon occurs for $\lambda = 1.00$. The $\lambda = 0.98$ variant is slightly weaker than the case of $\lambda = 0.99$. We also observed that decreasing λ below 0.99 introduces noise in the meta-model estimations, which declines meta-model utility. In particular, trial vectors tend to be chosen randomly, i.e., rmmLSHADE start to behave more like baseline LSHADE.

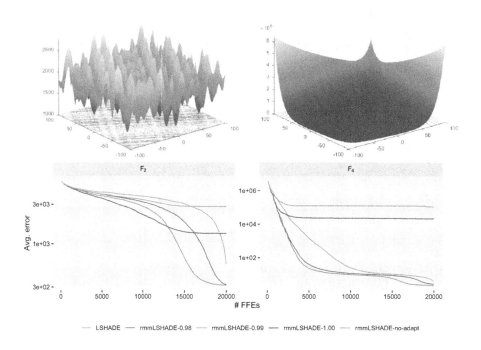

Fig. 2. The averaged convergence of rmmSHADE with $\lambda = 0.98, 0.99, 1.00$, and without w adaptation (denoted as rmmLSHADE-no-adapt) for F_2 and F_4 ($D = 20$) with $10^3 \cdot D$ optimization budget. The x-axis represents the number of FFEs, and the y-axis the error (a difference from the optimum) averaged across 150 runs (5 transformation selections × 30 runs). Additionally, $3D$ maps from [19] of $2D$ versions of F_2 and F_4 are presented.

4.1   The Algorithm's Runtime

We examined the runtime of the methods in accordance with detailed instructions presented in the technical report [19]. The results are presented in Table 3. The core measure is T_3 which represents the final runtime assessment (named *algorithm complexity* in [19]). The final runtime is understood as the time required by a pure algorithm computations (excluding the time for FFEs calculation) in relation to test program runtime T_0 (see [19] for the details). In addition, we present extra measure T_3' which indicates the relation of algorithm's final runtime to LSHADE's final runtime. psLSHADE is 18.7 times slower than LSHADE for $D = 10$ and 35.1 for $D = 20$. The proposed rmmLSHADE is characterized by 36.2 and 169.9 ratios. It follows that rmmLSHADE is approximately two times slower than psLSHADE for $D = 10$ and 5 times slower for $D = 20$.

Please note that rmmLSHADE estimates the meta-model after each evaluation, not each iteration, which results in 46 and 79 times more meta-model estimations for $D = 10$ and $D = 20$, resp. compared to psLSHADE (column n^{mm}). Hence, we consider the estimated runtime highly satisfactory.

Table 3. Runtime of rmmLSHADE, psLSHADE and LSHADE calculated according to the recommendation proposed in [19]. T_0 is time of a test program run. T_1 - time of pure $2 \cdot 10^5$ FFEs of F_2 function. T_2 - average running time of the algorithm for F_2 with $2 \cdot 10^5$ evaluation budget. $T_3 = (T_2 - T_1)/T_0$ is the final runtime. T_3' is T_3 related to LSHADE's final runtime. n^{mm} denotes the average number of meta-model estimates during an optimization run.

D	$Algorithm$	$T_0[s]$	$T_1[s]$	$T_2[s]$	T_3	T_3'	n^{mm}
10	rmmLSHADE	0.002323	10.4592	83.3448	31364.19	**36.2**	**199821**
	psLSHADE	0.002323	10.4592	48.0409	16172.15	**18.7**	**4329**
	LSHADE	0.002323	10.4592	12.4722	866.23	1	0
20	rmmLSHADE	0.002323	11.8945	279.9270	115339.96	**169.9**	**199641**
	psLSHADE	0.002323	11.8945	67.2213	23808.24	**35.1**	**2528**
	LSHADE	0.002323	11.8945	13.4723	678.94	1	0

5   Conclusion

In this paper the rmmLSHADE algorithm enhancing the popular LSHADE with a recursive meta-model based on the RLS filter is introduced. The rmmLSHADE is archive-free and its meta-model is estimated sample-by-sample. The algorithm is evaluated using the popular CEC2021 benchmark, in expensive scenario, i.e., $10^3 \cdot D$ FFEs optimization budget. rmmLSHADE is compared to baseline LSHADE and to psLSHADE – which proven to be efficient in expensive scenarios. The proposed algorithm outperforms both LSHADE and psLSHADE when utilizing exactly the same experiment settings and common parameterization. We also demonstrate how the forgetting factor λ affects the performance.

Future work includes using less accurate but faster recursive algorithms, e.g. Least Mean Squared filter.

Acknowledgments. Studies were funded by BIOTECHMED-1 project granted by Warsaw University of Technology under the program Excellence Initiative: Research University (ID-UB).

References

1. Atkeson, C.G., Moore, A.W., Schaal, S.: Locally weighted learning. In: Aha, D.W. (ed.) Lazy Learning, pp. 11–73. Springer, Dordrecht (1997). https://doi.org/10.1007/978-94-017-2053-3_2
2. Auger, A., Hansen, N.: A restart CMA evolution strategy with increasing population size. In: 2005 IEEE Congress on Evolutionary Computation, vol. 2, pp. 1769–1776. IEEE (2005)
3. Auger, A., Schoenauer, M., Vanhaecke, N.: LS-CMA-ES: a second-order algorithm for covariance matrix adaptation. In: Yao, X., et al. (eds.) PPSN 2004. LNCS, vol. 3242, pp. 182–191. Springer, Heidelberg (2004). https://doi.org/10.1007/978-3-540-30217-9_19
4. Awad, N.H., Ali, M.Z., Suganthan, P.N.: Ensemble sinusoidal differential covariance matrix adaptation with Euclidean neighborhood for solving CEC2017 benchmark problems. In: 2017 IEEE Congress on Evolutionary Computation (CEC), pp. 372–379. IEEE (2017)
5. Bajer, L., Pitra, Z., Holeňa, M.: Benchmarking gaussian processes and random forests surrogate models on the bbob noiseless testbed. In: Proceedings of the Companion Publication of the 2015 Annual Conference on Genetic and Evolutionary Computation, pp. 1143–1150 (2015)
6. Bajer, L., Pitra, Z., Repický, J., Holeňa, M.: Gaussian process surrogate models for the CMA evolution strategy. Evol. Comput. **27**(4), 665–697 (2019)
7. Biswas, S., Saha, D., De, S., Cobb, A.D., Das, S., Jalaian, B.A.: Improving differential evolution through Bayesian hyperparameter optimization. In: 2021 IEEE Congress on Evolutionary Computation (CEC), pp. 832–840. IEEE (2021)
8. Bliek, L., Guijt, A., Verwer, S., De Weerdt, M.: Black-box mixed-variable optimisation using a surrogate model that satisfies integer constraints. In: Proceedings of the Genetic and Evolutionary Computation Conference Companion, pp. 1851–1859 (2021)
9. Bliek, L., Verwer, S., de Weerdt, M.: Black-box combinatorial optimization using models with integer-valued minima. Ann. Math. Artif. Intell. **89**(7), 639–653 (2020). https://doi.org/10.1007/s10472-020-09712-4
10. Chen, Q., Liu, B., Zhang, Q., Liang, J., Suganthan, P., Qu, B.: Problem definitions and evaluation criteria for CEC 2015 special session on bound constrained single-objective computationally expensive numerical optimization. Technical report, Computational Intelligence Laboratory, Zhengzhou University, Zhengzhou, China and Technical report, Nanyang Technological University (2014)
11. Cressie, N.: The origins of kriging. Math. Geol. **22**(3), 239–252 (1990)
12. Hansen, N.: A global surrogate assisted CMA-ES. In: Proceedings of the Genetic and Evolutionary Computation Conference, pp. 664–672 (2019)
13. Hansen, N., Müller, S.D., Koumoutsakos, P.: Reducing the time complexity of the derandomized evolution strategy with covariance matrix adaptation (CMA-ES). Evol. Comput. **11**(1), 1–18 (2003)
14. Helton, J.C., Davis, F.J.: Latin hypercube sampling and the propagation of uncertainty in analyses of complex systems. Reliabil. Eng. Syst. Saf. **81**(1), 23–69 (2003)
15. Jin, Y.: Surrogate-assisted evolutionary computation: recent advances and future challenges. Swarm Evol. Comput. **1**(2), 61–70 (2011)
16. Jones, D.R., Schonlau, M., Welch, W.J.: Efficient global optimization of expensive black-box functions. J. Global Optim. **13**(4), 455–492 (1998)

17. Kennedy, J., Eberhart, R.: Particle swarm optimization. In: Proceedings of ICNN'95-International Conference on Neural Networks, vol. 4, pp. 1942–1948. IEEE (1995)
18. Kern, S., Hansen, N., Koumoutsakos, P.: Local meta-models for optimization using evolution strategies. In: Runarsson, T.P., Beyer, H.-G., Burke, E., Merelo-Guervós, J.J., Whitley, L.D., Yao, X. (eds.) PPSN 2006. LNCS, vol. 4193, pp. 939–948. Springer, Heidelberg (2006). https://doi.org/10.1007/11844297_95
19. Mohamed, A.W., Hadi, A.A., Mohamed, A.K., Agrawal, P., Kumar, A., Suganthan, P.: Problem definitions and evaluation criteria for the CEC 2021 special session and competition on single objective bound constrained numerical optimization. https://github.com/P-N-Suganthan/2021-SO-BCO/blob/main/CEC2021%20TR_final%20(1).pdf
20. Nishida, K., Akimoto, Y.: Benchmarking the PSA-CMA-ES on the BBOB noiseless testbed. In: Proceedings of the Genetic and Evolutionary Computation Conference Companion, pp. 1529–1536 (2018)
21. Sallam, K.M., Elsayed, S.M., Chakrabortty, R.K., Ryan, M.J.: Improved multi-operator differential evolution algorithm for solving unconstrained problems. In: 2020 IEEE Congress on Evolutionary Computation (CEC), pp. 1–8. IEEE (2020)
22. Sayed, A.H., Kailath, T.: Recursive least-squares adaptive filters. Digit. Sig. Process. Handb. **21**(1) (1998)
23. Storn, R., Price, K.: Differential evolution - a simple and efficient heuristic for global optimization over continuous spaces. J. Global Optim. **11**(4), 341–359 (1997)
24. Tanabe, R., Fukunaga, A.: Success-history based parameter adaptation for differential evolution. In: 2013 IEEE Congress on Evolutionary Computation, pp. 71–78. IEEE (2013)
25. Tanabe, R., Fukunaga, A.S.: Improving the search performance of SHADE using linear population size reduction. In: 2014 IEEE Congress on Evolutionary Computation (CEC), pp. 1658–1665. IEEE (2014)
26. Vikhar, P.A.: Evolutionary algorithms: a critical review and its future prospects. In: 2016 International Conference on Global Trends in Signal Processing, Information Computing and Communication (ICGTSPICC), pp. 261–265. IEEE (2016)
27. Weisberg, S.: Applied Linear Regression. Wiley, Hoboken (2013)
28. Yamaguchi, T., Akimoto, Y.: Benchmarking the novel CMA-ES restart strategy using the search history on the BBOB noiseless testbed. In: Proceedings of the Genetic and Evolutionary Computation Conference Companion, pp. 1780–1787 (2017)
29. Zaborski, M., Mańdziuk, J.: Improving LSHADE by means of a pre-screening mechanism. In: Proceedings of the Genetic and Evolutionary Computation Conference, GECCO 2022, Boston, Massachusetts, pp. 884–892. Association for Computing Machinery (2022). https://doi.org/10.1145/3512290.3528805
30. Zaborski, M., Mańdziuk, J.: LQ-R-SHADE: R-SHADE with quadratic surrogate model. In: Proceedings of the 21st International Conference on Artificial Intelligence and Soft Computing (ICAISC 2022), Zakopane, Poland (2022)
31. Zaborski, M., Okulewicz, M., Mańdziuk, J.: Analysis of statistical model-based optimization enhancements in generalized self-adapting particle swarm optimization framework. Found. Comput. Decis. Sci. **45**(3), 233–254 (2020)
32. Zhang, J., Sanderson, A.C.: JADE: adaptive differential evolution with optional external archive. IEEE Trans. Evol. Comput. **13**(5), 945–958 (2009)

Towards Efficient Multiobjective Hyperparameter Optimization: A Multiobjective Multi-fidelity Bayesian Optimization and Hyperband Algorithm

Zefeng Chen[1], Yuren Zhou[2(✉)], Zhengxin Huang[3], and Xiaoyun Xia[4]

[1] School of Artificial Intelligence, Sun Yat-sen University, Zhuhai 519082, China
chenzef5@mail.sysu.edu.cn
[2] School of Computer Science and Engineering, Sun Yat-sen University,
Guangzhou 510006, China
zhouyuren@mail.sysu.edu.cn
[3] Department of Computer Science,
Youjiang Medical University for Nationalities, Baise 533000, China
huangzhx26@mail2.sysu.edu.cn
[4] College of Information Science and Engineering,
Jiaxing University, Jiaxing 314001, China
xiaxiaoyun@zjxu.edu.cn

Abstract. Developing an efficient solver for hyperparameter optimization (HPO) can help to support the environmental sustainability of modern AI. One popular solver for HPO problems is called BOHB, which attempts to combine the benefits of Bayesian optimization (BO) and Hyperband. It conducts the sampling of configurations with the aid of a BO surrogate model. However, only the few high-fidelity measurements are utilized in the building of BO surrogate model, leading to the fact that the built BO surrogate cannot well model the objective function in HPO. Especially, in the scenario of multiobjective optimization (which is more complicated than single-objective optimization), the resultant BO surrogates for modelling all conflicting objective functions would be more likely to mislead the configuration search. To tackle this low-efficiency issue, in this paper, we propose an efficient algorithm, referred as Multiobjective Multi-Fidelity Bayesian Optimization and Hyperband, for solving multiobjective HPO problems. The key idea is to fully consider the contributions of computationally cheap low-fidelity surrogates and expensive high-fidelity surrogates, and enable effective utilization of the integrated information of multi-fidelity ensemble model in an online manner. The weightages for distinct fidelities are adaptively determined based on the approximation performance of their corresponding surrogates. A range of experiments on diversified real-world multiobjective HPO problems (including the HPO of multi-label/multi-task learning models and the HPO of models with several performance metrics) are carried out to investigate the performance of our proposed algorithm.

G. Rudolph et al. (Eds.): PPSN 2022, LNCS 13398, pp. 160–174, 2022.
https://doi.org/10.1007/978-3-031-14714-2_12

Experimental results showcase that the proposed algorithm outperforms more than 10 state-of-the-art peers, while demonstrating the ability of our proposed algorithm to efficiently solve real-world multiobjective HPO problems at scale.

Keywords: Hyperparameter optimization · Multiobjective optimization · Bayesian optimization · Hyperband · Surrogate

1   Introduction

In the field of AutoML, there are two fundamental problems, namely, hyperparameter optimization (HPO) and neural architecture search (NAS) [9]. The former aims to automate the search for well-behaved hyperparameter configurations based on the data at hand, and hence it plays a crucial role in enabling the underlying machine learning (ML) model perform effectively. However, in today's era, the HPO problem generally suffers from heavy evaluation costs for training the underlying model due to the *"large-instance"* character of datasets and/or the *"deep"* depth of models. Resultantly, a substantial amount of energy is usually required in the procedure of HPO, and the attention has been increasingly drawn to the carbon footprint of deep learning [1,25]. Developing an efficient solver could help to alleviate the computational bottleneck in HPO, thus supporting the *environmental sustainability* of modern AI [5].

In the literature, there is a large body of work on single-objective HPO. However, there is relatively less work on the more challenging HPO problem considering multiple objective functions, which is called multiobjective HPO (MOHPO). In essence, MOHPO is a multiobjective optimization problem (MOOP), which is a more general, challenging and realistic scenario. For example, when tuning the hyperparameters of a neural network model, one may be interested in maximizing the prediction accuracy while minimizing the prediction time. In this scenario, it is unlikely to be possible to optimize all of the objectives simultaneously, since they may be conflicting. Thus, for an MOHPO problem, we aim to find a set of best trade-off configurations rather than a single best configuration. This would provide a decision maker an even greater degree of flexibility. Furthermore, the multiobjective optimization technique has attracted more and more attention within the ML area, since [24] has demonstrated the feasibility of modeling multi-task learning problem as multiobjective optimization. Along this line, we could model the HPO of a multi-task/multi-label learning model as an MOHPO problem. Therefore, in terms of the research on MOHPO, there is a lot more practical potential that has yet to be explored.

Among existing solvers for HPO problems, one popular method called BOHB [7] attempts to combine the benefits of Bayesian optimization (BO) and Hyperband (HB) [17]. And the MO-BOHB [12] for solving MOHPO problems is a simple extension of BOHB to the multiobjective case. Both them conduct the sampling of configurations with the aid of a BO surrogate model. However, only the few high-fidelity measurements are utilized in the building of BO surrogate model, leading to the fact that the built BO surrogate cannot well model the objective function in HPO. This issue is more serious when simultaneously optimizing multiple objectives, as the resultant BO surrogates for modelling all conflicting objective functions would be of higher approximation error and more likely to mislead the configuration search.

Of particular interest in this paper is: *how to well tackle the low-efficiency issue of Hyperband-style HPO solver due to the scarcity of high-fidelity measurements, in the scenario of multiobjective optimization?* With this in mind, we draw inspirations from some state-of-the-art multi-fidelity single-objective HPO solvers [11,14,18], and propose a Multiobjective Multi-Fidelity Bayesian Optimization and Hyperband (called *MoFiBay* for short) algorithm to efficiently solve MOHPO problems. **This is the first work that attempts to combine the benefits of both HB and multi-fidelity multiobjective BO.** The key idea of *MoFiBay* is to fully consider the contributions of computationally cheap low-fidelity surrogates and expensive high-fidelity surrogates, and enable effective utilization of the integrated information of *multi-fidelity ensemble model* in an online manner. To differentiate the contributions of the surrogates with distinct fidelities, the weightages for all fidelities are adaptively determined based on the approximation performance of their corresponding surrogates. To validate the performance of the proposed *MoFiBay* algorithm, we conducted a series of experiments on real-world MOHPO problems (including the HPO of multi-label/multi-task learning models and the HPO of models with several performance metrics), covering various sizes of real-world datasets and distinct models. Moreover, we have also compared *MoFiBay* to more than 10 state-of-the-art MOHPO solvers, to showcase the efficacy of the proposed algorithm.

2   Related Work

In essence, MOHPO is generally considered as a type of expensive black-box optimization problem. Without loss of generality, it can be expressed as a multiobjective optimization problem (MOOP), shown as follows:

$$\min_{\mathbf{x} \in \Omega} \mathbf{F}(\mathbf{x}) = \left(f_1(\mathbf{x}), \ldots, f_M(\mathbf{x}) \right) \tag{1}$$

where $\mathbf{x}$ is an n-dimensional decision vector and $f_i(\mathbf{x})$ denotes the i-th objective function (whose computation requires some data at hand). $\Omega \subseteq R^n$ represents the decision space (also known as search space), and the image set $S = \{\mathbf{F}(\mathbf{x}) | \mathbf{x} \in \Omega\}$ is called the objective space.

In the literature, there are a wide variety of surrogate model based multiobjective BO algorithms that can be used to solve the expensive MOHPO problems. One classical solver called ParEGO [15] employs random scalarization technique to transform the original MOOP to single-objective optimization, and utilizes the expected improvement (EI) as the acquisition function to select the next input for evaluation. In addition to scalarization technique, many methods attempt to optimize the Pareto hypervolume (PHV) metric [6] that has the ability to capture the quality of a candidate Pareto set. To achieve this, one may extend the standard acquisition functions to the scenario of multiobjective optimization, for example, expected improvement in PHV (EHI) [6] and probability of improvement in PHV (SUR) [22]. SMSego [23] attempts to find a limited set of points by optimizing the posterior means of the Gaussian processes, and computes the gain in hypervolume over those set of points.

On the other hand, some principled algorithms (such as PAL [28], PESMO [10], USeMO [4] and MESMO [2]) attempt to reduce the uncertainty by virtue of information theory. PAL iteratively selects the candidate input for evaluation towards the goal of minimizing the size of uncertain set [28]. Although PAL is theoretically guaranteed, it is only applicable for the input space with finite set of discrete points. PESMO depends on input space entropy search, and selects the input maximizing the information gain about the Pareto set in each iteration [10]. USeMO is a general framework that iteratively generates a cheap Pareto front using the surrogate models and then selects the point with highest uncertainty as the next query [4]. MESMO improves over PESMO by extending max-entropy search (MES) to the multi-objective setting. Till now, there exists only one multi-fidelity multiobjective BO algorithm within ML literature. This algorithm, called MF-OSEMO [3], employs an output space entropy based acquisition function to select the sequence of candidate input and fidelity-vector pairs for evaluation.

Although BOHB has been a popular solver for single-objective HPO problems, there exists only one Hyperband-style solver (namely, MO-BOHB [12]) for MOHPO problems. MO-BOHB is a simple extension of the popular BOHB [7] to the multiobjective case. However, this solver did not attempt to tackle the low-efficiency issue of BOHB due to the scarcity of high-fidelity measurements. Thus, **the low-efficiency issue of Hyperband-style solver for MOHPO problems still remains to be settled.**

3   Proposed *MoFiBay* Algorithm

This section first provides an overview of the proposed *MoFiBay* algorithm for solving MOHPO problems, and then introduces its important components.

3.1   Framework of *MoFiBay*

First of all, we assume that an MOHPO problem is modeled as an MOOP as shown in Eq. (1). Then, we can accordingly specify the type of training resource (such as the size of training subset or the number of iterations) and also the unit of resources.

The pseudocode of the proposed *MoFiBay* algorithm for solving this type of MOHPO problem are shown in Algorithm 1. On the whole, the *MoFiBay* follows the basic flows of BOHB. The pseudocode shown in Lines 4–10 constitute a whole call of a multiobjective version of BOHB algorithm, which can be called several times until the termination criterion (in terms of the total budget for solving

---

Algorithm 1. Pseudocode of Proposed *MoFiBay*

Input: F: objective functions; Ω: hyperparameter space; R_{max}: maximum amount of resources for a hyperparameter configuration; η: discarding faction;

Output: Non-dominated solutions (best trade-off configurations);

1: Initialization: $s_{max} = \lfloor \log_\eta(R_{max}) \rfloor$, $B = (s_{max} + 1)R_{max}$, $K = \lfloor \log_\eta(R_{max}) \rfloor + 1$;

2: Initialize the multi-fidelity measurements $D_i = \emptyset$ $(i = 1, ..., K)$ and the multi-fidelity ensemble model $\mathcal{M}_{ens} = None$;

3: **while** *termination criterion is not fulfilled* **do**

4:    **for** $s \in \{s_{max}, s_{max} - 1, ..., 0\}$ **do**

5:       $n_1 = \lceil \frac{B}{R_{max}} \cdot \frac{\eta^s}{s+1} \rceil$, $r_1 = R_{max} \cdot \eta^{-s}$;

6:       Sample n_1 configurations from Ω by using **Algorithm 2**;

7:       Execute the SuccessiveHalving procedure (the inner loop) with the sampled n_1 configurations and r_1 as input, and collect the new multi-fidelity measurements $D'_{1:K}$;

8:       $D_i = D_i \bigcup D'_i$ $(i = 1, ..., K)$;

9:       Update the multi-fidelity ensemble model $\mathcal{M}_{ens}$ with $D_{1:K}$;

10:    **end for**

11: **end while**

12: Output the non-dominated solutions (w.r.t. **F**) in D_K.

Algorithm 2. Pseudocode of Model-based Sampling

Input: F: objective functions; Ω: hyperparameter space; $\mathcal{M}_{ens}$: multi-fidelity ensemble model; N_s: number of random configurations to optimize EHVI; $D_{1:K}$: multi-fidelity measurements; ρ: fraction of random configuration;

Output: Next configuration to evaluate;

1: **if** $rand() < \rho$ or $\mathcal{M}_{ens} = None$ **then**

2:    Output a random configuration;

3: **else**

4:    Draw N_s configurations randomly, and then compute their values of EHVI metric (where $\mathcal{M}_{ens}$ is adopted as the surrogate model);

5:    Output the configuration with the largest EHVI value.

6: **end if**

---

an MOHPO problem) is fulfilled. Next, we will illustrate how the multiobjective version of BOHB (Lines 4–10 in Algorithm 1) is conducted.

In the outer loop, a grid search over feasible values of n_1 (the number of hyperparameter configurations to evaluate in the inner loop) and r_1 (the amount of resources) is conducted, which is the same as in Hyperband [17]. For each specific pair (n_1, r_1), Algorithm 2 is invoked to sample n_1 configurations from Ω with the aid of a multi-fidelity ensemble model M_{ens}. Then, the SuccessiveHalving procedure (i.e., the inner loop) is executed with the sampled n_1 configurations and r_1 as input. Each of the new multi-fidelity quality measurements D'_i appeared in the SuccessiveHalving procedure is collected, and then merged with the old D_i so as to form an augmented D_i. With the augmented measurements $D_{1:K}$, the multi-fidelity ensemble model M_{ens} is updated. Finally, the non-dominated solutions (best trade-off configurations w.r.t. **F**) in D_K are output as final results.

The important components of *MoFiBay* (including the SuccessiveHalving procedure, the building and update of multi-fidelity ensemble model, and the sampling of configurations) will be elaborated in the following subsections.

3.2 Successive Halving Based on Random Scalarization

Each time when entering the inner loop, a search direction vector λ is drawn uniformly at random from the set of evenly distributed vectors defined by

$$\Lambda = \left\{ \lambda = (\lambda_1, ..., \lambda_M) \left| \sum_{j=1}^{M} \lambda_j = 1 \wedge \forall j, \lambda_j = \frac{l}{H}, l \in \{0, ..., H\} \right. \right\} \quad (2)$$

with $|\Lambda| = C_{H+M-1}^{M-1}$.

Recall that a specific pair (n_1, r_1) serves as the input of the SuccessiveHalving procedure, and n_1 hyperparameter configurations have been sampled before the SuccessiveHalving procedure. The sampled n_1 hyperparameter configurations are firstly evaluated with the initial r_1 units of resources, and the performance score of each configuration $\mathbf{x}$ along the search direction λ is computed with the augmented Tchebycheff function:

$$f_\lambda(\mathbf{x}) = \max_{j=1}^{M} (\lambda_j \cdot f_j(\mathbf{x})) + 0.05 \sum_{j=1}^{M} \lambda_j \cdot f_j(\mathbf{x}). \quad (3)$$

The n_1 configurations are ranked based on the performance score. Then, only the top η^{-1} configurations remain (while the evaluations of the other configurations are stopped in advance), and their evaluations are continued with η times larger resources, that is, $n_2 = n_1 * \eta - 1$ and $r_2 = r_1 * \eta$. The n_2 configurations are ranked again by the performance score. This process repeats until the maximum amount of training resources R_{max} is reached, that is, $r_i = R_{max}$. In this way, the badly-behaved configurations which have lower performance score along the specific search direction vector are gradually discarded. And the superior configurations have more chance to be evaluated with higher amount of training resources. Therefore, this SuccessiveHalving procedure could help to accelerate the configuration evaluations.

3.3 Construction and Update of Multi-fidelity Ensemble Model

Multi-fidelity Measurements: Note that after the SuccessiveHalving procedure, different hyperparameter configurations have finally gone through objective function evaluations with different amounts of training resources. This actually classifies all configurations into several distinct configuration sets with different fidelities. Denote these sets as $D_1, ..., D_K$, where the value of K can be determined based on the successive halving rule, that is, $K = \lfloor \log_\eta(R_{max}) \rfloor + 1$. Specifically, the measurement $(\mathbf{x}, \mathbf{y})$ in each set D_i ($i = 1, ..., K$) is acquired through evaluating $\mathbf{x}$ with $r_i = \eta^{i-1}$ units of resources. The former $K - 1$ sets (i.e., $D_{1:K-1}$) represent the low-fidelity measurements obtained from the early-stopped evaluations, while the last one D_K represents the high-fidelity measurements obtained from the evaluations with the maximum amount of training resources (i.e., $r_K = R_{max}$). Thus, the low-fidelity measurements in $D_{1:K-1}$ are actually the biased measurements about the true objective functions $\mathbf{F}$, whereas all the high-fidelity measurements in D_K are the unbiased measurements.

With the above in mind, if we use D_i ($i < K$) to fit a BO surrogate model (denoted as $\mathcal{M}_i$), then $\mathcal{M}_i$ should be to model the objective functions $\mathbf{F}^i$ with r_i units of training resources, rather than to model the true objective functions $\mathbf{F}^K = \mathbf{F}$ with the maximum amount of training resources. When i increases, the measurements in D_i are obtained from the evaluations with a larger amount of training resources, and then the corresponding model $\mathcal{M}_i$ has the ability to provide a more accurate approximation to the true objective functions $\mathbf{F}$. Thus, it may be expected that the low-fidelity measurement sets $D_{1:K-1}$, which are computationally cheaper than D_K, can provide a certain degree of instrumental information for surrogate modeling. In this regard, we can consider to integrate the biased yet informative low-fidelity measurements with high-fidelity measurements, when building a BO surrogate to model the true objective functions $\mathbf{F}$.

Construction of Multi-fidelity Ensemble Model: With the aim of integrating all the instrumental information from low- and high-fidelity measurements, we construct a multi-fidelity ensemble model called $\mathcal{M}_{ens}$. To be specific, we let each BO surrogate $\mathcal{M}_i$ (fitted on D_i, $i = 1, ..., K$) act as a base surrogate, and assign each base surrogate with a weightage $w_i \in [0, 1]$. Note that all weightages need to satisfy $\sum_{i=1}^{K} w_i = 1$. It's expected that the weightages can reflect the online contributions of different base surrogates with different fidelities. Specifically, for the base surrogate $\mathcal{M}_i$ that can provide a more accurate approximation to the true objective functions $\mathbf{F}$, a larger w_i should be assigned to it. Otherwise, the corresponding w_i should be smaller.

Here, we employ the MOTPE [21] to serve as the BO surrogate $\mathcal{M}_i$ fitted on D_i. Therefore, similar to [18], we also can use the generalized product of experts framework to combine the predictions from the base surrogates $\mathcal{M}_{1:K}$ (MOTPEs) with weightages. The predictive mean and variance of the multi-fidelity ensemble model $\mathcal{M}_{ens}$ at $\mathbf{x}$ are computed by the following equations:

$$\mu_{ens}(\mathbf{x}) = \left(\sum_{i=1}^{K} \mu_i(\mathbf{x}) w_i \sigma_i^{-2}(\mathbf{x}) \right) \sigma_{ens}^2(\mathbf{x}),$$

$$\sigma_{ens}^2(\mathbf{x}) = \left(\sum_{i=1}^{K} w_i \sigma_i^{-2}(\mathbf{x}) \right)^{-1}. \tag{4}$$

Weightage Computation Method Based on Generalized Mean Square Cross-validation Error: As illustrated above, the value of the weightage w_i should be proportional to the performance of $\mathcal{M}_i$ when approximating $\mathbf{F}$. Moreover, all the weightages should be adaptively determined by means of an online computation method. An intelligent weightage computation method is important for building a superior ensemble of surrogates.

It is claimed in [8] that the weightages should not only reflect the confidence in the surrogates, but also filter out the adverse effects of surrogates which perform poorly in sampling sparse regions. A weightage computation method based on generalized mean square cross-validation error (GMSE) is proposed to address these two issues. Here, we modify this method to make it suitable for the multiobjective case, which can be formulated as follows:

$$w_i = w_i^* \Big/ \sum_{j=1}^{K} w_j^*, w_i^* = \left(E_i + \alpha \bar{E} \right)^{\beta},$$

$$\bar{E} = \frac{1}{K} \sum_{i=1}^{K} E_i, E_i = \sqrt{\frac{1}{|D_K|} \sum_{l=1}^{|D_K|} \|\mathbf{y}_l - \hat{\mathbf{y}}_{il}\|^2}, \tag{5}$$

where $(\mathbf{x}_l, \mathbf{y}_l)$ is the l-th measurement in D_K, and $\hat{\mathbf{y}}_{il}$ is the i-th surrogate model (i.e., $\mathcal{M}_i$)'s corresponding prediction value using cross-validation. E_i is the given error measure of $\mathcal{M}_i$ in the context of multiobjective optimization, and $\bar{E}$ indicates the average value of all base surrogates' error measures. Two parameters α and β control the importance of averaging and individual surrogates, respectively. And we adopt the setting of $\alpha = 0.05$ and $\beta = -1$ according to the suggestion in [8].

Update of Multi-fidelity Ensemble Model: With the augmented measurements $D_{1:K}$, the multi-fidelity ensemble model $\mathcal{M}_{ens}$ is updated as follows: (1) refitting each base surrogate $\mathcal{M}_i$ with all measurements in the augmented D_i; (2) computing the weightage for each base surrogate; (3) combining all base surrogates to form the new multi-fidelity ensemble model $\mathcal{M}_{ens}$.

4 Experimental Studies

4.1 Comparative Algorithms

Our proposed *MoFiBay* is compared with 10 peers: (1) MO-BOHB [12]; (2) ParEGO [15]; (3) SMSego [23]; (4) EHI [6]; (5) SUR [22]; (6) PESMO [10]; (7) MESMO [2]; (8) PFES [26]; (9) MF-OSEMO-TG; (10) MF-OSEMO-NI [3]. In addition, the baselines (i.e., random search, SH-EMOA, MS-EHVI, MO-BANANAS, and BULK & CUT) proposed in [12] for solving multiobjective joint NAS+HPO problems are also compared. The parameter settings for the above algorithms follow the suggestions in their original literature.

As the true Pareto fronts of the MOHPO problems constructed in the experiments are unknown, we use the hypervolume as the indicator to compare the performances of various algorithms. To be specific, we report the average hypervolume values obtain by each algorithm over 10 independent runs on each MOHPO problem.

4.2 Experiment on the HPO of Models with Several Performance Metrics

Construction of MOHPO Problems: In this experiment, we aim to solve the type of MOHPO problem where several performance metrics are to be simultaneously optimized. To be specific, we consider the *MNIST* dataset [16]. The MOHPO problem at hand is to find a neural network model with low prediction error and low prediction time. These two performance metrics are conflicting, since reducing the prediction error will involve larger networks which will inevitably take longer time to predict. The underlying neural network model for MOHPO is set as a feed-forward network (FFN) with ReLus in the hidden layers and a soft-max output layer, and Adam [13] (with a mini-batch size of 4,000 instances during 150 epochs) is selected as the optimizer. The hyperparameters being optimized are as follows: the number of layers (between 1 and 3), the number of hidden units per layer (between 50 and 300), the learning rate, the amount of dropout, and the level of l_1 and l_2 regularization. The prediction error is measured on a set of 10,000 instances extracted from the dataset. The rest of the dataset, i.e., 50,000 instances, is used for training. We consider a logit transformation of the prediction error, since the error rates are very small. The prediction time is measured as the average time required for doing 10,000 predictions. We compute the logarithm of the ratio between the prediction time of the network and the prediction time of the fastest network, (i.e., a single hidden layer and 50 units). When measuring the prediction time, we do not train the network and consider random weights (the time objective is also set to ignore irrelevant parameters). Thus, the problem is suited for a decoupled evaluation because both objectives can be evaluated separately.

In this experiment, the resource type is the number of iterations; the maximum amount of training resource R_{max} is 81; one unit of resource corresponds to 0.5 epoch. The termination criterion for each solver is set to 4 h.

Results: The average hypervolume values obtained by our proposed *MoFiBay* algorithm and 10 peers on MOHPO of FFN model with minimizing prediction error and minimizing prediction time are shown in Fig. 1. As can be seen, after 2 h, the hypervolume obtained by the three solvers involving multi-fidelity evaluations (i.e., MF-OSEMO-TG, MF-OSEMO-NI and our proposed *MoFiBay*) are much larger than the results obtained by other solvers, meaning that these three solvers converge faster than other solvers. After 4 h, the hypervolume difference among most solvers are small. This phenomenon indicates that most solvers have converged. Among all solvers, our proposed *MoFiBay* algorithm has the best performance in terms of efficiency, as it can obtain faster convergence and larger hypervolume values.

4.3 Experiment on the HPO of Multi-label/Multi-task Learning Models

Construction of MOHPO Problems: We employ a convolution neural network (CNN) to serve as the underlying ML model, and use it to conduct multi-

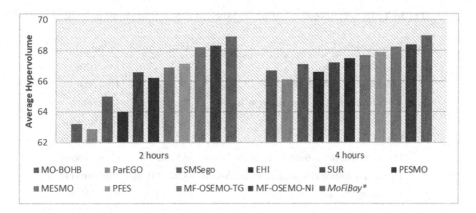

Fig. 1. Performance comparisons among different algorithms on MOHPO of FFN model (minimizing prediction error & time).

label/multi-task classification. Multi-task learning, as the name suggests, needs to simultaneously deal with multiple learning tasks. In essence, multi-label learning is a special form of multi-task learning where each task represents a different label. The feasibility of modeling multi-task learning problem as multi-objective optimization has been demonstrated in [24]. Thus, we can use multi-label/multi-task learning datasets to conduct MOHPO experiments.

For the sake of simplicity, we limit the number of convolution layers in the used CNN to 2. For the training of CNN, we choose the cross entropy loss with dropout regularization as the loss function and select Adam (with 20 epochs) as a state-of-the-art CNN optimizer. The hyperparameters being optimized and their ranges are listed in Table 1. All hyperparameters are encoded in the range $[0, 1]$, and they can be transformed into their corresponding ranges through the transformation method used in [19].

To construct an MOHPO problem, we let the classification error for each label/task act as each objective function to be minimized. Two types of real-world datasets are adopted:

1. Four multi-label learning datasets (including *scene*, *yeast*, *delicious* and *tmc2007*) downloaded from the Mulan website[1]. For simplicity, we restrict each of the original datasets to three labels by following the steps below: selecting the top three labels (in terms of the number of instances in each label) from all labels and then deleting the data instances without any label. In this way, the MOHPO on each dataset would be modeled as a 3-objective optimization problem. In addition, each dataset is split into a training set and a validation set with a splitting ratio of 80% and 20%, respectively.
2. One multi-task learning dataset (i.e., the *MultiMNIST* dataset). Concretely, we adopt the construction method introduced in [24] to build the *MultiMNIST* dataset (where the training set and validation set contain 60,000

---

[1] http://mulan.sourceforge.net/datasets-mlc.html.

Table 1. Hyperparameters of multi-label/multi-task learning models.

No.	Description	Range	Type
1	Mini-batch size	$[2^4, 2^8]$	Discrete
2	Size of convolution window	$\{1, 3, 5\}$	Discrete
3	# of filters in the convolution layer	$[2^3, 2^6]$	Discrete
4	Dropout rate	$[0, 0.5]$	Continuous
5	Learning rate	$[10^{-4}, 10^{-1}]$	Continuous
6	Decay parameter $beta\_1$ used in Adam	$[0.8, 0.999]$	Continuous
7	Decay parameter $beta\_2$ used in Adam	$[0.99, 0.9999]$	Continuous
8	Parameter ϵ used in Adam	$[10^{-9}, 10^{-3}]$	Continuous

Table 2. Average hypervolume values obtained by different algorithms on HPO of multi-label/multi-task learning models.

	Scene	Yeast	Delicious	tmc2007	MultiMNIST
MO-BOHB	0.644	0.523	0.266	0.520	0.917
ParEGO	0.640	0.521	0.260	0.478	0.908
SMSego	0.650	0.529	0.269	0.510	0.922
EHI	0.645	0.525	0.265	0.522	0.915
SUR	0.651	0.529	0.275	0.548	0.923
PESMO	0.654	0.532	0.274	0.556	0.928
MESMO	0.656	0.533	0.277	0.568	0.930
PFES	0.658	0.535	0.278	0.577	0.933
MF-OSEMO-TG	**0.664**	0.538	0.282	0.586	0.938
MF-OSEMO-NI	0.663	0.539	0.282	0.588	0.940
*MoFiBay**	0.663	**0.543**	**0.285**	**0.593**	**0.945**

and 10,000 instances, respectively), which is a two-task learning version of *MNIST* dataset. Hence, the MOHPO on *MultiMNIST* is a 2-objective optimization problem.

In this experiment, the resource type is data subset; the maximum amount of training resource R_{max} is 27; one unit of resource corresponds to $1/27$ times the number of samples. The termination criterion for each solver is set to 4 h.

Results: Table 2 displays the average hypervolume values obtained by different solvers on HPO of five multi-label/multi-task learning models. From this table, we can observe that our proposed *MoFiBay* algorithm performs best on all considered datasets except *scene*. On the *scene* dataset, *MoFiBay* (Hyperband-style solver) and MF-OSEMO-NI (not belonging to Hyperband-style solver) share the second place, while MF-OSEMO-TG ranks first.

Table 3. Architectural and non-architectural hyperparameters being optimized.

No.	Description	Range	Log scale
1	# of convolution layers	$\{1, 2, 3\}$	No
2	# of filters in the convolution layer	$[2^4, 2^{10}]$	Yes
3	Kernel size	$\{3, 5, 7\}$	No
4	Batch normalization	$\{true, false\}$	No
5	Global average pooling	$\{true, false\}$	No
6	# of fully connected layers	$\{1, 2, 3\}$	No
7	# of neurons in the fully connected layer	$[2^1, 2^9]$	Yes
8	Learning rate	$[10^{-5}, 10^0]$	Yes
9	Mini-batch size	$[2^0, 2^9]$	Yes

Table 4. Average hypervolume values obtained by different algorithms on mutiobjective joint NAS + HPO.

	Flowers	Fashion-MNIST
Random search	283.10	296.40
SH-EMOA	304.73	308.25
MO-BOHB	301.02	332.52
MS-EHVI	306.45	360.57
MO-BANANAS	302.09	301.55
BULK & CUT	311.96	350.08
*MoFiBay**	**316.19**	**363.48**

4.4 Experiment on Mutiobjective Joint NAS+HPO

Construction of MOHPO Problems: In this experiment, we further test our algorithm on the mutiobjective joint NAS+HPO benchmark proposed in [12]. We select the network size (measured by the number of parameters in the network) and the classification accuracy to act as the objectives of our multi-objective optimization. And we define a maximum budget of 25 epochs for training a single configuration. We used the *Flowers* dataset [20] and *Fashion-MNIST* dataset [27]. We split these two datasets according to [12]. The hyperparameters being optimized and their ranges are listed in Table 3. As for more details about the search space, readers can refer to [12].

In this experiment, the resource type is data subset; the maximum amount of training resource R_{max} is 27; one unit of resource corresponds to 1/27 times the number of samples. The termination criterion for each solver is set to 10 h.

Results: Table 4 lists the average hypervolume values obtained by our proposed *MoFiBay* algorithm and 6 baselines on the mutiobjective joint NAS+HPO problem. We can see that our proposed *MoFiBay* algorithm has the best performance on both datasets. This phenomenon demonstrates that the proposed *MoFiBay* also can well optimize the architectural hyperparameters.

5   Conclusions

This paper has proposed an efficient MOHPO solver named *MoFiBay*, and it is the first work attempting to combine the benefits of both Hyperband and multi-fidelity multiobjective BO. We have conducted three main types of experiments on real-world MOHPO problems, covering various sizes of real-world datasets and distinct models, to validate the performance of the proposed algorithm. For future work, we would like to conduct research on how to design a more efficient acquisition function for multiobjective BO. In addition, we also would like to further verify the performance and efficiency of our proposed algorithm on datasets with more than one hundred thousand data instances and on a much richer variety of multi-objective AutoML problems under large-instance data.

Acknowledgements. This work is supported by the Fundamental Research Funds for the Central Universities, Sun Yat-sen University (22qntd1101). It is also supported by the National Natural Science Foundation of China (62162063, 61703183), Science and Technology Planning Project of Guangxi (2021AC19308), and Zhejiang Province Public Welfare Technology Application Research Project of China (LGG19F030010).

References

1. Anthony, L.F.W., Kanding, B., Selvan, R.: Carbontracker: tracking and predicting the carbon footprint of training deep learning models. arXiv preprint arXiv:2007.03051 (2020)
2. Belakaria, S., Deshwal, A.: Max-value entropy search for multi-objective Bayesian optimization. In: International Conference on Neural Information Processing Systems (NeurIPS) (2019)
3. Belakaria, S., Deshwal, A., Doppa, J.R.: Multi-fidelity multi-objective Bayesian optimization: an output space entropy search approach. In: Proceedings of the AAAI Conference on Artificial Intelligence, vol. 34, pp. 10035–10043 (2020)
4. Belakaria, S., Deshwal, A., Jayakodi, N.K., Doppa, J.R.: Uncertainty-aware search framework for multi-objective bayesian optimization. In: Proceedings of the AAAI Conference on Artificial Intelligence. vol. 34, pp. 10044–10052 (2020)
5. Cai, H., Gan, C., Wang, T., Zhang, Z., Han, S.: Once-for-all: train one network and specialize it for efficient deployment. arXiv preprint arXiv:1908.09791 (2019)
6. Emmerich, M., Klinkenberg, J.W.: The computation of the expected improvement in dominated hypervolume of pareto front approximations. Technical report, Leiden University, p. 34 (2008)
7. Falkner, S., Klein, A., Hutter, F.: BOHB: robust and efficient hyperparameter optimization at scale. In: International Conference on Machine Learning, pp. 1437–1446. PMLR (2018)

8. Goel, T., Haftka, R.T., Shyy, W., Queipo, N.V.: Ensemble of surrogates. Struct. Multidiscip. Optim. **33**(3), 199–216 (2007)
9. He, X., Zhao, K., Chu, X.: AutoML: a survey of the state-of-the-art. Knowl.-Based Syst. **212**, 106622 (2021)
10. Hernández-Lobato, D., Hernandez-Lobato, J., Shah, A., Adams, R.: Predictive entropy search for multi-objective Bayesian optimization. In: International Conference on Machine Learning, pp. 1492–1501. PMLR (2016)
11. Hu, Y.Q., Yu, Y., Tu, W.W., Yang, Q., Chen, Y., Dai, W.: Multi-fidelity automatic hyper-parameter tuning via transfer series expansion. In: Proceedings of the AAAI Conference on Artificial Intelligence, vol. 33, pp. 3846–3853 (2019)
12. Izquierdo, S., et al.: Bag of baselines for multi-objective joint neural architecture search and hyperparameter optimization. In: 8th ICML Workshop on Automated Machine Learning (AutoML) (2021)
13. Kingma, D.P., Ba, J.: Adam: a method for stochastic optimization. arXiv preprint arXiv:1412.6980 (2014)
14. Klein, A., Falkner, S., Bartels, S., Hennig, P., Hutter, F.: Fast Bayesian optimization of machine learning hyperparameters on large datasets. In: Artificial Intelligence and Statistics, pp. 528–536. PMLR (2017)
15. Knowles, J.: Parego: a hybrid algorithm with on-line landscape approximation for expensive multiobjective optimization problems. IEEE Trans. Evol. Comput. **10**(1), 50–66 (2006)
16. LeCun, Y., Bottou, L., Bengio, Y., Haffner, P.: Gradient-based learning applied to document recognition. Proc. IEEE **86**(11), 2278–2324 (1998)
17. Li, L., Jamieson, K., DeSalvo, G., Rostamizadeh, A., Talwalkar, A.: Hyperband: a novel bandit-based approach to hyperparameter optimization. J. Mach. Learn. Res. **18**(1), 6765–6816 (2017)
18. Li, Y., Shen, Y., Jiang, J., Gao, J., Zhang, C., Cui, B.: MFES-HB: efficient hyperband with multi-fidelity quality measurements. In: Proceedings of the AAAI Conference on Artificial Intelligence, vol. 35, pp. 8491–8500 (2021)
19. Loshchilov, I., Hutter, F.: CMA-ES for hyperparameter optimization of deep neural networks. arXiv preprint arXiv:1604.07269 (2016)
20. Nilsback, M.E., Zisserman, A.: A visual vocabulary for flower classification. In: 2006 IEEE Computer Society Conference on Computer Vision and Pattern Recognition (CVPR 2006), vol. 2, pp. 1447–1454. IEEE (2006)
21. Ozaki, Y., Tanigaki, Y., Watanabe, S., Onishi, M.: Multiobjective tree-structured Parzen estimator for computationally expensive optimization problems. In: Proceedings of the 2020 Genetic and Evolutionary Computation Conference, pp. 533–541 (2020)
22. Picheny, V.: Multiobjective optimization using gaussian process emulators via stepwise uncertainty reduction. Stat. Comput. **25**(6), 1265–1280 (2015)
23. Ponweiser, W., Wagner, T., Biermann, D., Vincze, M.: Multiobjective optimization on a limited budget of evaluations using model-assisted S-metric selection. In: Rudolph, G., Jansen, T., Beume, N., Lucas, S., Poloni, C. (eds.) PPSN 2008. LNCS, vol. 5199, pp. 784–794. Springer, Heidelberg (2008). https://doi.org/10.1007/978-3-540-87700-4_78
24. Sener, O., Koltun, V.: Multi-task learning as multi-objective optimization. In: Proceedings of the 32nd International Conference on Neural Information Processing Systems, pp. 525–536 (2018)
25. Strubell, E., Ganesh, A., McCallum, A.: Energy and policy considerations for deep learning in NLP. arXiv preprint arXiv:1906.02243 (2019)

26. Suzuki, S., Takeno, S., Tamura, T., Shitara, K., Karasuyama, M.: Multi-objective Bayesian optimization using pareto-frontier entropy. In: International Conference on Machine Learning, pp. 9279–9288. PMLR (2020)
27. Xiao, H., Rasul, K., Vollgraf, R.: Fashion-MNIST: a novel image dataset for benchmarking machine learning algorithms. arXiv preprint arXiv:1708.07747 (2017)
28. Zuluaga, M., Sergent, G., Krause, A., Püschel, M.: Active learning for multi-objective optimization. In: International Conference on Machine Learning, pp. 462–470. PMLR (2013)

Benchmarking and Performance Measures

A Continuous Optimisation Benchmark Suite from Neural Network Regression

Katherine M. Malan[1]([⊠])[iD] and Christopher W. Cleghorn[2][iD]

[1] Department of Decision Sciences, University of South Africa,
Pretoria, South Africa
malankm@unisa.ac.za
[2] School of Computer Science and Applied Mathematics,
University of the Witwatersrand, Johannesburg, South Africa
christopher.cleghorn@wits.ac.za

Abstract. Designing optimisation algorithms that perform well in general requires experimentation on a range of diverse problems. Training neural networks is an optimisation task that has gained prominence with the recent successes of deep learning. Although evolutionary algorithms have been used for training neural networks, gradient descent variants are by far the most common choice with their trusted good performance on large-scale machine learning tasks. With this paper we contribute CORNN (Continuous Optimisation of Regression tasks using Neural Networks), a large suite for benchmarking the performance of any continuous black-box algorithm on neural network training problems. Using a range of regression problems and neural network architectures, problem instances with different dimensions and levels of difficulty can be created. We demonstrate the use of the CORNN Suite by comparing the performance of three evolutionary and swarm-based algorithms on over 300 problem instances, showing evidence of performance complementarity between the algorithms. As a baseline, the performance of the best population-based algorithm is benchmarked against a gradient-based approach. The CORNN suite is shared as a public web repository to facilitate easy integration with existing benchmarking platforms.

Keywords: Benchmark suite · Unconstrained continuous optimisation · Neural network regression

1 Introduction

The importance of a good set of benchmark problem instances is a critical component of a meaningful benchmarking study in optimisation [4]. As a consequence of the No Free Lunch Theorems for optimisation [49], if an algorithm is tuned to improve its performance on one class of problems it will most likely perform worse on other problems [22]. Therefore, to develop algorithms that perform well in general or are able to adapt to new scenarios, a wide range of different problem instances are needed for experimental algorithm design.

© The Author(s), under exclusive license to Springer Nature Switzerland AG 2022
G. Rudolph et al. (Eds.): PPSN 2022, LNCS 13398, pp. 177–191, 2022.
https://doi.org/10.1007/978-3-031-14714-2_13

Most of the benchmark problems available for continuous optimisation are artificial, so performance achieved through tuning algorithms on these problems cannot be assumed to transfer to real-world problems. On the other hand, testing algorithms on real-world problems is not always feasible and has the disadvantage of not covering the wide range of problem characteristics needed for different problem scenarios. To bridge this gap, we introduce a benchmark suite from the real-world domain of neural network (NN) training that includes some of the advantages of artificial benchmark problems.

Metaheuristics frequently suffer from the *curse of dimensionality* with performance degrading as the number of decision variables increases [30,31,37,46]. This is in part due to the increased complexity of the problem, but also due to the exponential growth in the size of the search space [30]. The training of NNs presents an ideal context for high-dimensional optimisation as even a medium-sized network will have hundreds of weights to optimise. Most NN training studies use a limited number of problem instances (classification datasets or regression problems), which brings into question the generalisability of the results. For example, in six studies using population-based algorithms for NN training [6,15,33–35,43], the number of problem instances used for testing ranged from a single real-world instance [15] to eight classification or regression problems [33,35]. To facilitate the generalisability of NN training studies, we provide a suite of hundreds of problem instances that can easily be re-used for benchmarking algorithm performance.

Stochastic gradient descent [29] is the default approach to training NNs with its trusted good performance on large-scale learning problems [10]. Population-based algorithms have been proposed for training NNs [6,15,33–35,43], but they are seldom used in practice. One of the challenges is that the search space of NN weights is unbounded and algorithms such as particle swarm optimisation may fail due to high weight magnitudes leading to hidden unit saturation [39,40]. The one domain where population-based metaheuristics have shown competitive results compared to gradient-based methods is in deep reinforcement learning tasks [45]. More benchmarking of gradient-free methods against gradient-based methods is needed to highlight the possible benefits of different approaches.

This paper proposes CORNN, Continuous Optimisation of Regression tasks using Neural Networks, a software repository of problem instances, including benchmarking of population-based algorithms against a gradient-based algorithm (Adam). Over 300 regression tasks are formed from 54 regression functions with different network architectures generating problem instances with dimensions ranging from 41 to 481. The source code and datasets associated with the benchmark suite are publicly available at github.com/CWCleghornAI/CORNN.

2   Continuous Optimisation Benchmark Suites

When benchmarking algorithms, it is usually not practical to use real-world problems, due to the limited range of problem instances available and the domain knowledge required in constructing the problems. Individual real-world problem instances also do not effectively test the limits of an algorithm because they will not usually cover all the problem characteristics of interest [41]. Artificial benchmark suites and problem instance generators have therefore become popular alternatives for testing optimisation algorithms. In the continuous optimisation domain, the most commonly used artificial benchmark suites include the ACM Genetic and Evolutionary Computation Conference (GECCO) BBOB suites [13] and IEEE Congress on Evolutionary Computation (CEC) suites [14]. These artificial suites have been criticised for having no direct link to real-world settings [17], resulting in a disconnect between the performance of algorithms on benchmarks and real-world problems [47].

To address the limitations of artificial benchmarks, suites that are based on real-world problems or involve tasks that are closer to real-world problems have been proposed, such as from the domains of electroencephalography (EEG) [21], clustering [20], crane boom design [18], and games [48]. The CORNN Suite proposed in this paper extends these sets to include the class of problems for solving NN regression tasks. These tasks are unique in that the decision variables are unbounded and the scenario makes it possible to benchmark black-box algorithms against gradient-based approaches.

3 Neural Network Training Landscapes

Training NNs involves adjusting weights on the connections between neurons to minimise the error of the network on some machine learning task. Since weight values are real numbers, the search space is continuous with dimension equal to the number of adjustable weights in the network. Training NNs is known to be NP-complete even for very small networks [9] and the properties of error landscapes are still poorly understood [12] with conflicting theoretical claims on the presence and number of local minima [2,3,23,32]. Some studies have suggested that these landscapes have large flat areas with valleys that radiate outwards [11,19,28,36] and a prevalence of saddle points rather than local minima [12,16]. Saddle points present a challenge for search, because they are generally surrounded by high error plateaus [16] and, being stationary points, can create the illusion of being local optima. It has, however, also been found that failure of gradient-based deep learning is not necessarily related to an abundance of saddle points, but rather to aspects such as the level of informativeness of gradients, signal-to-noise ratios and flatness in activation functions [42].

To better understand the nature of NN error landscapes, investigations are needed into the behaviour of different algorithms on a wide range of problems. The CORNN Suite can be used to complement existing suites or as a starting point for this kind of analysis.

4 The CORNN Benchmark Suite

With sufficient neurons, NNs are able to model an arbitrary mathematical function [7,25], so are a suitable model for solving complex regression problems. The optimisation task of the CORNN Suite involves fitting a fully-connected feed-forward NN to a real-valued function, $f(\mathbf{x})$. Each network has an n-dimensional real-valued input,$\mathbf{x}$, and a single real-valued output, which is the prediction of the target value, $f(\mathbf{x})$. The CORNN Suite uses 54 two-dimensional functions as the basis for regression fitting tasks. These functions are specified on the CORNN Suite repository on github[1]. The functions cover a range of characteristics with respect to modality, separability, differentiability, ruggedness, and so on. Note, however, that although these characteristics will no doubt have some effect on the difficulty of the regression task, we cannot assume that the features of the functions relate to the characteristics of the higher dimensional search space of NN weights for fitting the functions.

4.1 Training and Test Sets

Datasets were generated for each of the 54 functions as follows: 5000 (x_1, x_2) pairs were sampled from a uniform random distribution of the function's domain. A value of 5000 was used to be large enough to represent the actual function, while still allowing for reasonable computational time for simulation runs. The true output for each (x_1, x_2) pair was calculated using the mathematical function and stored as the target variable. Each dataset was split randomly into training (75% of samples) and testing (25% of samples) sets. Two forms of preprocessing were performed on the CORNN Suite datasets: (1) Input values were normalised to the range $[-1, 1]$ using the domain of each function; (2) To compare results of problem instances with different output ranges, output values were normalised using simple min-max scaling based on the training data to the range $(0, 1)$.

In addition, the CORNN Suite's implementation allows for the use of custom datasets; either generated from analytic functions or existing datasets.

4.2 Neural Network Models

The architecture used in the CORNN Suite is a fully connected feed-forward network with 2 inputs; 1-, 3-, or 5-hidden layers, each with 10 neurons plus a bias unit; and 1 output neuron. This results in 41, 261, and 481 weights to be optimised for the 1-, 3-, and 5-layer networks respectively. Each architecture uses one of two hidden layer activation functions: the conventional hyperbolic tangent (Tanh) and the rectified linear unit (ReLU). ReLU is currently the most commonly used activation function in deep learning [29], but Tanh has been recommended above ReLU for reinforcement learning tasks [1]. The output layer uses a linear activation function in all cases. These six topologies are referred to as Tanh1, Tanh3, Tanh5, ReLU1, ReLU3, and ReLU5, specifying the activation function and number of hidden layers. The CORNN Suite therefore consists of $54 \times 6 = 324$ problem instances, since each function has six NN models for fitting the function. Note that the CORNN Suite's implementation allows for complete customisation of architectures to create any desired topology for further analysis.

---

[1] https://github.com/CWCleghornAI/CORNN/blob/main/CORNN_RegressionFunc tions.pdf.

4.3 Performance Evaluation

Performance of an algorithm is measured using mean squared error (MSE) of the trained model on the test set given a set budget of function evaluations. Note that in the analysis presented in this paper, no evidence of overfitting was observed. If overfitting becomes a consideration as more specialised optimisers are developed/considered, it may become necessary to hold out a portion of the training set to employ techniques such as early stopping etc. When using hold-out training instances, the total number of function evaluations should be seen as the maximum number of times any one training instance has been used. If no hold-out instances are used this measurement is equivalent to the number of full passes of the training data. A similar consideration should be made if an optimiser requires hyper-parameter tuning; in such cases hold-out instances from the training set should be used for tuning and not the test set instances.

4.4 Implementation Details

The CORNN Suite was developed in Python 3 using PyTorch. The user selects a regression task and a model architecture, after which the library constructs a problem instance object with a callable function to which the user can pass a candidate solution for evaluation on either the training set during optimisation, or on the test set after optimisation. A user of CORNN therefore does not have to concern themselves with any data processing or NN computation. The complexity of the problem instances are abstracted away to the point where a user of CORNN can just work with an objective function after setup.

The GitHub repository provides installation instructions with a detailed example of how to construct and use a CORNN problem instance. The suite is easily extended beyond the 324 problem instances presented in this paper to include other regression tasks and/or NN architectures through reflection. The datasets for all 54 regression tasks are also provided in CSV format, but when using the CORNN Suite it is not necessary to directly interact with these files.

5 Experimentation and Results

To demonstrate CORNN, we provide results of metaheuristics and contrast these with a gradient-based method. The aim is not to compare algorithms, but to provide a use-case of the suite. No tuning of algorithm parameters was done, so the results are not representative of the best performance of the algorithms.

5.1 Experimental Setup

The algorithms used in this study were: particle swarm optimisation (PSO) [26], differential evolution (DE) [44], covariance matrix adaptation evolution strategy (CMA-ES) [24], Adam [27], and random search. For the population-based algorithms, standard versions and hyperparameters defined in the Nevergrad [5] library were used to facilitate reproducibility[2]. Adam was used for the gradient-based approach (PyTorch [38] implementation with default parameters). Each algorithm had a function evaluation (FE) budget of 5000 per problem instance, where an FE is defined as one complete pass through the training dataset. We used full batch learning, but the suite is not limited to this approach. Optimisation runs were repeated 30 times for each algorithm/problem instance pair.

5.2   Analysis of Population-Based Algorithms

The first set of results contrasts the performance of the three population-based algorithms: CMA-ES, DE and PSO against random search. We only present the performance on the testing datasets, because we found no evidence of overfitting by any of the algorithms on the problem set. Each algorithm is given a performance score at each evaluation using the following scoring mechanism per problem instance against each competing algorithm:

- 1 point is awarded for a draw (when there is no statistically significant difference based on a two-tailed Mann-Whitney U test with 95% confidence).
- 3 points are awarded for a win and 0 for a loss. In the absence of a draw, we determine whether a win or loss occurred using one-tailed Mann-Whitney U tests (with 95% confidence).

This results in a maximum score for an algorithm on a single instance of 9, or $3(n-1)$ in general, where n is the number of algorithms. The scores per instance were normalised to the range $[0, 1]$.

Figure 1 plots the normalised mean performance score over all 54 problem instances for the full budget of 5000 evaluations for the six NN models. Solid lines denote the mean performance score with shaded bands depicting the standard deviation around the mean. Two general observations from Fig. 1 are that the three metaheuristics all performed significantly better than random search and that no single algorithm performed the best on all NN models. On the Tanh models (plots in the left column of Fig. 1), CMA-ES performed the best on the 1-layer network, while DE performed the best on the 3- and 5-layer networks after the full budget of evaluations. On the ReLU models (plots in the right column of Fig. 1), PSO ultimately performed the best on all three models. CMA-ES was the quickest to find relatively good solutions, but converged to solutions that were inferior to those ultimately found by PSO.

Figure 2 visualises the performance of the algorithms on individual problem instances (the 3-layer models are omitted to save space). Each dot represents a single problem instance (54 of them in each column) and the vertical position corresponds to the average normalised MSE on the testing set from 30 runs of the algorithm after 5000 evaluations. Algorithms are plotted in a different colour for clarity. The final column (purple dots) shows the results of the best performing

---

[2] Nevergrad version 0.4.2, PSO with *optimizers.PSO*, DE with *optimizers.TwoPoints DE*, CMA-ES with *optimizers.CMA*, and random search with *optimizers.Random Search*.

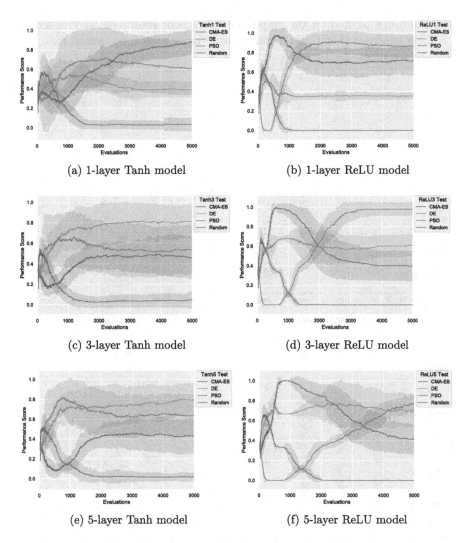

Fig. 1. Summarised relative performance on the test set of a 1-, 3- and 5-layer Tanh and ReLU models over 54 function datasets. Lines denote mean performance with shaded bands denoting the standard deviation around the mean.

algorithm for each problem instance. In these plots, better performing algorithms have a concentration of dots closer to the horizontal axis (lower MSE values).

For all models, random search (red dots) performs worse than other algorithms. Contrasting Fig. 2a and 2c, shows that for Tanh, the performance across all algorithms deteriorates (fewer dots lower down) as the number of layers increase from 1 to 5, indicating an increase in problem difficulty. The same can be observed for ReLU architectures in Figs. 2b to 2d. Note that in Fig. 2d, only a single dot is shown for random search, because the other MSE values are above the range plotted on the graph. The very poor results of random search on ReLU5 indicates that these instances are more challenging overall than the other architectures.

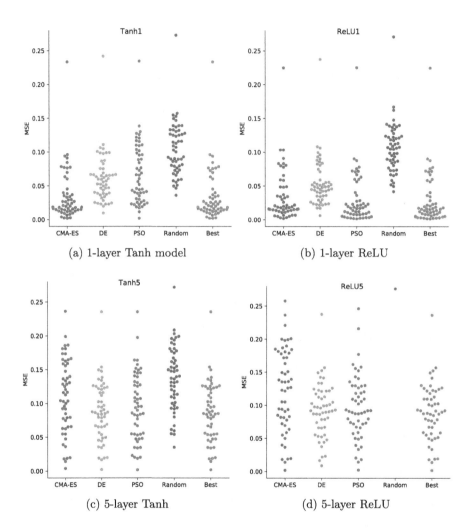

(a) 1-layer Tanh model   (b) 1-layer ReLU

(c) 5-layer Tanh   (d) 5-layer ReLU

Fig. 2. Performance per function for each optimiser on two Tanh and ReLU models (Color figure online)

Figure 2 shows interesting outliers in the individual high dots at approximately 0.25 MSE. This corresponds to function 34, Periodic, on which most algorithms perform markedly worse than on the other problem instances. In contrast, the lowest dots in Fig. 2c correspond to function 20, Easom, on which all algorithms (except random search) achieved close to 0.00 MSE. Figure 3 plots these two functions, clearly illustrating why it was easier for the search algorithms to fit models to Easom than to Periodic.

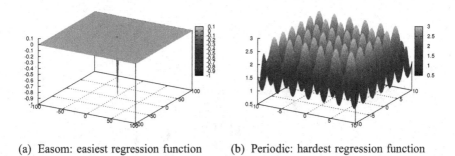

(a) Easom: easiest regression function   (b) Periodic: hardest regression function

Fig. 3. Two regression tasks on which all algorithms performed the best and the worst.

5.3 Analysis of Individual Problem Instances

The next set of results highlights the range of difficulty of problems in the CORNN Suite for population-based algorithms in relation to a gradient-based approach. As a baseline, we provide the results of Adam, a form of gradient-descent with adaptive learning-rate that is popular for training deep NNs.

Figures 4, 5, 6 and 7 show violin plots of the distribution of MSE values from 30 runs on each of the 54 problem instances for Tanh1, Tanh5, Relu1 and Relu5, respectively. The blue violins represent the performance of the best population-based algorithm (of the three discussed in Sect. 5.2) on the problem instance, while the red violins represent the performance of Adam. Note that most of the violins for Adam are very small due to the small variance in the performance over the 30 runs – except for the random initial weights, Adam is a deterministic algorithm. The median MSE values appear as tiny white dots in the centre of each violin and the maximal extent of the violins are cropped to reflect the actual range of the data.

The functions are sorted from left to right by the difference between the median MSE of the two approaches. For example, in Fig. 4, the first function on the left is function 26 (Himmelblau), where the median MSE of the best population-based algorithm was slightly lower than the median MSE of Adam. From about the eighth function onwards, it can however be seen that Adam out-performed the best population-based algorithm. The superior performance of Adam is even more marked for the 5-layer Tanh model (Fig. 5), with function 43 (Schwefel 2.22) resulting in the worst relative performance of the population-based algorithms. The most difficult function to fit for both Adam and population-based algorithms is evident by the high MSE values on function 34 (Periodic).

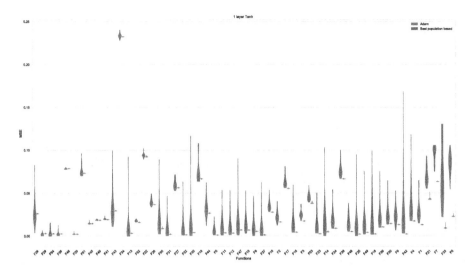

Fig. 4. Best performing population based optimiser compared to Adam on the 1-layer Tanh model (Color figure online)

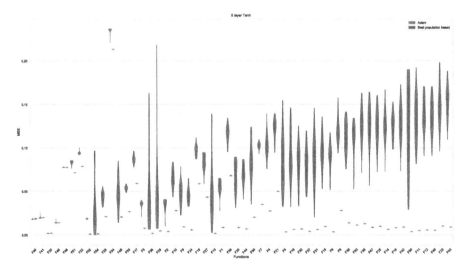

Fig. 5. Best performing population based optimiser compared to Adam on the 5-layer Tanh model (Color figure online)

Figures 6 to 7 show slightly better relative performance of the population based algorithms on the ReLU architectures compared to the Tanh architectures. On the left of Fig. 7 we can see that the median MSE of the best population based algorithm is lower than Adam on about the first nine functions (37, 8, 47, 54, 25, 43, 46, and 52).

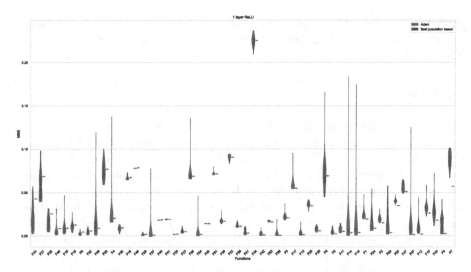

Fig. 6. Best performing population based optimiser compared to Adam on the 1-layer ReLU model (Color figure online)

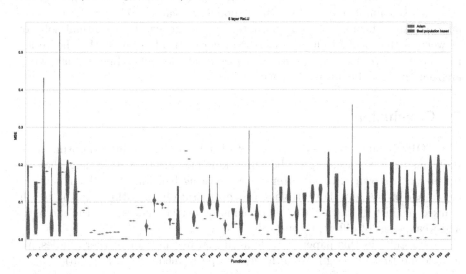

Fig. 7. Best performing population based optimiser compared to Adam on the 5-layer ReLU model (Color figure online)

In this way, these results provide a ranking of CORNN problem instances from the easiest to the hardest in terms of relative performance against gradient-based techniques. Future studies can focus on black-box algorithm development to reduce the gap in performance compared to gradient-based approaches. The CORNN Suite can be used in different types of analysis and not just in the way illustrated in this paper. For example, to simulate cases where the analytical gradient is not available for using gradient-based techniques, black-box optimisers can be benchmarked against one another to investigate the effectiveness on NN training tasks.

6   Discussion

The CORNN Suite complements existing benchmark sets with NN training tasks that can be used to benchmark the performance of any continuous black-box algorithm. Ideally, a benchmarking suite should be [4]: (1) diverse, (2) representative, (3) scalable and tunable, and (4) should have known solutions/best performance. The problem instances of the CORNN Suite are *diverse* as demonstrated by the wide range of performances by different algorithms. In addition, the problems are partly-*representative* of real-world problems in being continuous (which is more common in real-world settings than combinatorial problems [8]), with computationally expensive evaluation and involving the real-world task of NN training. Problems are *scalable* and *tunable* through the selection of different NN models coupled with different regression tasks. For each problem instance the theoretical *optimal solution is known* (where MSE = 0 on the test set). However, given a fixed number of neurons in a model, we cannot rely on the universal approximator theorem [25] to guarantee the existence of an optimal solution of weights that will result in an error of zero. In addition to the theoretical minimum, we provide the performance of Adam as a baseline against which alternative algorithms can be benchmarked.

7   Conclusion

The CORNN Suite is an easy-to-use set of unbounded continuous optimisation problems from NN training for benchmarking optimisation algorithms that can be used on its own or as an extension to existing benchmark problem sets. An advantage of the suite is that black-box optimisation algorithms can be benchmarked against gradient-based algorithms to better understand the strengths and weaknesses of different approaches.

The results in this paper provide an initial baseline for further studies. We have found that although Adam in general performed better, population-based algorithms did out-perform Adam on a limited set of problem instances. Further studies could analyse the characteristics of these instances using landscape analysis to better understand which NN training tasks are better suited to population-based approaches than gradient-based approaches. The CORNN Suite can also be used to try to improve population-based algorithms on NN training tasks. It would be interesting to analyse whether parameter configurations from tuning on the CORNN Suite can be transferred to other contexts to improve black-box metaheuristic algorithm performance on NN training tasks in general.

Acknowledgments. This work was supported by the National Research Foundation, South Africa, under Grant 120837. The authors acknowledge the use of the High

Performance Computing Cluster of the University of South Africa. The authors also acknowledge the contribution of Tobias Bester for his initial implementation of the underlying regression functions.

References

1. Andrychowicz, M., et al.: What matters in on-policy reinforcement learning? A large-scale empirical study. CoRR abs/2006.05990 (2020). https://arxiv.org/abs/2006.05990
2. Auer, P., Herbster, M., Warmuth, M.K.: Exponentially many local minima for single neurons. In: Advances in Neural Information Processing Systems (NIPS 1996), vol. 9 (1996). http://papers.nips.cc/paper/1028-exponentially-many-local-minima-for-single-neurons.pdf
3. Baldi, P., Hornik, K.: Neural networks and principal component analysis: learning from examples without local minima. Neural Netw. **2**(1), 53–58 (1989). https://doi.org/10.1016/0893-6080(89)90014-2
4. Bartz-Beielstein, T., et al.: Benchmarking in optimization: best practice and open issues. arXiv 2007.03488v2 (2020)
5. Bennet, P., Doerr, C., Moreau, A., Rapin, J., Teytaud, F., Teytaudt, O.: Nevergrad: black-box optimization platform. ACM SIGEVOlution **14**(1), 8–15 (2021). https://doi.org/10.1145/3460310.3460312
6. den Bergh, F.V., Engelbrecht, A.: Cooperative learning in neural networks using particle swarm optimizers. South Afr. Comput. J. **2000**(26), 84–90 (2000)
7. Bishop, C.M.: Neural Networks for Pattern Recognition. Oxford University Press, Oxford (1995)
8. van der Blom, K., et al.: Towards realistic optimization benchmarks. In: Proceedings of the 2020 Genetic and Evolutionary Computation Conference Companion, pp. 293–294. ACM, July 2020. https://doi.org/10.1145/3377929.3389974
9. Blum, A.L., Rivest, R.L.: Training a 3-node neural network is NP-complete. Neural Netw. **5**(1), 117–127 (1992). https://doi.org/10.1016/s0893-6080(05)80010-3
10. Bottou, L., Bousquet, O.: The tradeoffs of large scale learning. In: Optimization for Machine Learning (chap. 13), pp. 351–368. The MIT Press (2012)
11. Chaudhari, P., et al.: Entropy-SGD: biasing gradient descent into wide valleys. J. Stat. Mech: Theory Exp. **2019**(12), 124018 (2019). https://doi.org/10.1088/1742-5468/ab39d9
12. Choromanska, A., Henaff, M., Mathieu, M., Arous, G.B., LeCun, Y.: The loss surfaces of multilayer networks. In: Proceedings of the 18th International Conference on Artificial Intelligence and Statistics, pp. 192–204 (2015)
13. COCO: Black-box optimisation benchmarking (BBOB) (2021). https://coco.gforge.inria.fr
14. Yue, C.T., et al.: IEEE CEC Bound Constrained benchmark suite (2020). https://github.com/P-N-Suganthan/2020-Bound-Constrained-Opt-Benchmark
15. Das, G., Pattnaik, P.K., Padhy, S.K.: Artificial neural network trained by particle swarm optimization for non-linear channel equalization. Expert Syst. Appl. **41**(7), 3491–3496 (2014). https://doi.org/10.1016/j.eswa.2013.10.053
16. Dauphin, Y.N., Pascanu, R., Gulcehre, C., Cho, K., Ganguli, S., Bengio, Y.: Identifying and attacking the saddle point problem in high-dimensional non-convex optimization. In: Ghahramani, Z., Welling, M., Cortes, C., Lawrence, N.D., Weinberger, K.Q. (eds.) Advances in Neural Information Processing Systems, vol. 27, pp. 2933–2941. Curran Associates, Inc. (2014)

17. Fischbach, A., Bartz-Beielstein, T.: Improving the reliability of test functions generators. Appl. Soft Comput. **92**, 106315 (2020). https://doi.org/10.1016/j.asoc.2020.106315

18. Fleck, P., et al.: Box-type boom design using surrogate modeling: introducing an industrial optimization benchmark. In: Andrés-Pérez, E., González, L.M., Periaux, J., Gauger, N., Quagliarella, D., Giannakoglou, K. (eds.) Evolutionary and Deterministic Methods for Design Optimization and Control With Applications to Industrial and Societal Problems. CMAS, vol. 49, pp. 355–370. Springer, Cham (2019). https://doi.org/10.1007/978-3-319-89890-2_23

19. Gallagher, M.R.: Multi-layer perceptron error surfaces: visualization, structure and modelling. Ph.D. thesis, University of Queensland, Australia (2000)

20. Gallagher, M.: Towards improved benchmarking of black-box optimization algorithms using clustering problems. Soft Comput. **20**(10), 3835–3849 (2016). https://doi.org/10.1007/s00500-016-2094-1,https://doi.org/10.1007/s00500-016-2094-1

21. Goh, S.K., Tan, K.C., Al-Mamun, A., Abbass, H.A.: Evolutionary big optimization (BigOpt) of signals. In: 2015 IEEE Congress on Evolutionary Computation (CEC). IEEE, May 2015. https://doi.org/10.1109/cec.2015.7257307

22. Haftka, R.T.: Requirements for papers focusing on new or improved global optimization algorithms. Struct. Multidiscip. Optim. **54**(1), 1–1 (2016). https://doi.org/10.1007/s00158-016-1491-5

23. Hamey, L.G.: XOR has no local minima: a case study in neural network error surface analysis. Neural Netw. **11**(4), 669–681 (1998). https://doi.org/10.1016/s0893-6080(97)00134-2

24. Hansen, N., Ostermeier, A.: Adapting arbitrary normal mutation distributions in evolution strategies: the covariance matrix adaptation. In: Proceedings of the IEEE Congress on Evolutionary Computation, pp. 312–317. IEEE Press, Piscataway (1996)

25. Huang, G.B., Chen, L., Siew, C.K.: Universal approximation using incremental constructive feedforward networks with random hidden nodes. IEEE Trans. Neural Netw. **17**(4), 879–892 (2006). https://doi.org/10.1109/tnn.2006.875977

26. Kennedy, J., Eberhart, R.: Particle swarm optimization. In: Proceedings of the IEEE International Joint Conference on Neural Networks, pp. 1942–1948. IEEE Press, Piscataway (1995)

27. Kingma, D., Ba, J.: Adam: a method for stochastic optimization. arXiv abs/1412.6980 (2014)

28. Kordos, M., Duch, W.: A survey of factors influencing MLP error surface. Control. Cybern. **33**, 611–631 (2004)

29. LeCun, Y., Bengio, Y., Hinton, G.: Deep learning. Nature **521**(7553), 436–444 (2015). https://doi.org/10.1038/nature14539

30. Lozano, M., Molina, D., Herrera, F.: Editorial scalability of evolutionary algorithms and other metaheuristics for large-scale continuous optimization problems. Soft. Comput. **15**(11), 2085–2087 (2010). https://doi.org/10.1007/s00500-010-0639-2

31. Mahdavi, S., Shiri, M.E., Rahnamayan, S.: Metaheuristics in large-scale global continues optimization: A survey. Inf. Sci. **295**, 407–428 (2015). https://doi.org/10.1016/j.ins.2014.10.042

32. Mehta, D., Zhao, X., Bernal, E.A., Wales, D.J.: Loss surface of XOR artificial neural networks. Phys. Rev. E **97**(5) (2018). https://doi.org/10.1103/physreve.97.052307

33. Mirjalili, S.: How effective is the Grey Wolf optimizer in training multi-layer perceptrons. Appl. Intell. **43**(1), 150–161 (2015). https://doi.org/10.1007/s10489-014-0645-7

34. Mirjalili, S., Hashim, S.Z.M., Sardroudi, H.M.: Training feedforward neural networks using hybrid particle swarm optimization and gravitational search algorithm. Appl. Math. Comput. **218**(22), 11125–11137 (2012). https://doi.org/10. 1016/j.amc.2012.04.069

35. Mousavirad, S.J., Schaefer, G., Jalali, S.M.J., Korovin, I.: A benchmark of recent population-based metaheuristic algorithms for multi-layer neural network training. In: Proceedings of the 2020 Genetic and Evolutionary Computation Conference Companion. ACM, July 2020. https://doi.org/10.1145/3377929.3398144

36. Keskar, N.S., Mudigere, D., Nocedal, J., Smelyanskiy, M., Tang, P.T.P.: On large-batch training for deep learning: generalization gap and sharp minima. In: Proceedings of the International Conference for Learning Representations (2017)

37. Oldewage, E.T.: The perils of particle swarm optimization in high dimensional problem spaces. Master's thesis, University of Pretoria, South Africa (2017). https://hdl.handle.net/2263/66233

38. Paszke, A., et al.: Pytorch: an imperative style, high-performance deep learning library. In: Advances in Neural Information Processing Systems, vol. 32, pp. 8024–8035. Curran Associates, Inc. (2019)

39. Rakitianskaia, A., Engelbrecht, A.: Training high-dimensional neural networks with cooperative particle swarm optimiser. In: 2014 International Joint Conference on Neural Networks (IJCNN). IEEE, July 2014. https://doi.org/10.1109/ijcnn.2014. 6889933

40. Rakitianskaia, A., Engelbrecht, A.: Saturation in PSO neural network training: good or evil? In: 2015 IEEE Congress on Evolutionary Computation (CEC). IEEE, May 2015. https://doi.org/10.1109/cec.2015.7256883

41. Rardin, R.L., Uzsoy, R.: Experimental evaluation of heuristic optimization algorithms: a tutorial. J. Heurist. **7**(3), 261–304 (2001). https://doi.org/10.1023/a: 1011319115230

42. Shalev-Shwartz, S., Shamir, O., Shammah, S.: Failures of gradient-based deep learning. In: Proceedings of the 34th International Conference on Machine Learning, pp. 3067–3075. PMLR, 06–11 August 2017)

43. Socha, K., Blum, C.: An ant colony optimization algorithm for continuous optimization: application to feed-forward neural network training. Neural Comput. Appl. **16**(3), 235–247 (2007). https://doi.org/10.1007/s00521-007-0084-z

44. Storn, R., Price, K.: Differential evolution: a simple evolution strategy for fast optimization. J. Glob. Optim. **11**, 341–359 (1997)

45. Such, F., Madhavan, V., Conti, E., Lehman, J., Stanley, K.O., Clune, J.: Deep neuroevolution: genetic algorithms are a competitive alternative for training deep neural networks for reinforcement learning. arXiv abs/1712.06567 (2018)

46. Tang, K., Li, X., Suganthan, P.N., Yang, Z., Weise, T.: Benchmark functions for the CEC 2010 special session and competition on large-scale global optimization. Technical report, Nature Inspired Computation and Applications Laboratory (2009). https://titan.csit.rmit.edu.au/~e46507/publications/lsgo-cec10.pdf

47. Tangherloni, A., etal.: Biochemical parameter estimation vs. benchmark functions: a comparative study of optimization performance and representation design. Appl. Soft Comput. **81**, 105494 (2019). https://doi.org/10.1016/j.asoc.2019.105494

48. Volz, V., Naujoks, B., Kerschke, P., Tušar, T.: Single- and multi-objective game-benchmark for evolutionary algorithms. In: Proceedings of the Genetic and Evolutionary Computation Conference, pp. 647–655. ACM (2019)

49. Wolpert, D.H., Macready, W.G.: No free lunch theorems for optimization. IEEE Trans. Evol. Comput. **1**(1), 67–82 (1997)

BBE: Basin-Based Evaluation of Multimodal Multi-objective Optimization Problems

Jonathan Heins[1], Jeroen Rook[2]($\boxtimes$), Lennart Schäpermeier[1],
Pascal Kerschke[1], Jakob Bossek[3], and Heike Trautmann[2,4]

[1] Big Data Analytics in Transportation, TU Dresden, Dresden, Germany
{jonathan.heins,lennart.schaepermeier,
pascal.kerschke}@tu-dresden.de
[2] Data Management and Biometrics, University of Twente,
Enschede, The Netherlands
j.g.rook@utwente.nl
[3] AI Methodology, RWTH Aachen University, Aachen, Germany
bossek@aim.rwth-aachen.de
[4] Data Science: Statistics and Optimization, University of Münster,
Münster, Germany
trautmann@wi.uni-muenster.de

Abstract. In multimodal multi-objective optimization (MMMOO), the focus is not solely on convergence in objective space, but rather also on explicitly ensuring diversity in decision space. We illustrate why commonly used diversity measures are not entirely appropriate for this task and propose a sophisticated basin-based evaluation (BBE) method. Also, BBE variants are developed, capturing the anytime behavior of algorithms. The set of BBE measures is tested by means of an algorithm configuration study. We show that these new measures also transfer properties of the well-established hypervolume (HV) indicator to the domain of MMMOO, thus also accounting for objective space convergence. Moreover, we advance MMMOO research by providing insights into the multimodal performance of the considered algorithms. Specifically, algorithms exploiting local structures are shown to outperform classical evolutionary multi-objective optimizers regarding the BBE variants and respective trade-off with HV.

Keywords: Multi-objective optimization · Multimodality · Performance metric · Benchmarking · Continuous optimization · Anytime behavior

---

J Heins and J. Rook—Equal contributions.

Supplementary Information The online version contains supplementary material available at https://doi.org/10.1007/978-3-031-14714-2_14.

G. Rudolph et al. (Eds.): PPSN 2022, LNCS 13398, pp. 192–206, 2022.
https://doi.org/10.1007/978-3-031-14714-2_14

1   Introduction

Multi-objective optimization (MOO), i.e., the simultaneous optimization of multiple (often competing) objectives, is challenging for both research and industrial applications [19]. Despite the practical relevance of MOO and decades of research in this area, multi-objective optimization problems (MOPs) have long been treated as black boxes – probably due to their numerous dimensions in the decision and objective space. This view made it very difficult to study a MOP's properties or the algorithmic behavior on it. As a result, MOPs were often visualized only by their Pareto fronts (i.e., a representation of the non-dominated solutions of the MOP in the objective space), algorithms were designed to converge to this Pareto front as fast as possible, and visualization of this search behavior was often based on point clouds evolving towards the Pareto front.

In related domains, such as single-objective continuous optimization, knowledge of a problem's characteristics has proven to be critical for a better problem understanding and for designing appropriate algorithms. For example, in single-objective optimization (SOO), it is widely accepted that multimodality can pose difficult obstacles [21]. Despite the insights gained in SOO, research in MOO has only recently begun to focus on multimodality [9]. Nevertheless, several visualization methods capable of revealing multimodal structures of MOPs [13,24,25,30], definitions that provide a theoretical description of a MOP's structural characteristics such as locally efficient sets [15,16], a couple of benchmark suites consisting mainly of multimodal MOPs [8,12,17,34], and optimization algorithms with a particular focus on (finding or at least exploiting local structures of) multimodal MOPs [10,18,26,29] have been proposed in recent years.

All these advances ultimately help to gain a better understanding of MOPs and to develop more efficient algorithms. For example, combining visualizations and theoretical definitions helps categorize MOPs into four categories of multimodality: (1) *Unimodal* MOPs consist of a single locally efficient set (i.e., the multi-objective counterpart of a local optimum in single-objective optimization) that naturally maps to the (single) Pareto front of the MOP; (2) *Multiglobal* MOPs contain multiple efficient sets that are all mapped to the same Pareto front; (3) *Multilocal* MOPs that contain multiple locally efficient sets that map to different fronts in the objective space; (4) (Truly) *Multimodal* MOPs, where some efficient sets map to the same (Pareto) front and others map to different fronts. A schematic representation of multiglobal and multilocal MOPs is shown in Fig. 1. Note that due to space limitations, we refrain from showing unimodal and multimodal MOPs as those are special variants of the shown ones.

Multimodal solutions may, e.g., be interesting to consider if the decision space values of the optimal points are not feasible, but the values of only slightly worse non-optimal solutions are. The problem with assessing algorithm performance concerning multimodality is that classical evaluation methods like hypervolume (HV) [36] cannot account for dominated solutions. Therefore, measures that consider decision space diversity like the Solow-Polasky (SP) indicator [28] are used to express to how distributed the solutions are over the efficient sets. A recent study on the two aforementioned indicators showed that these indicators

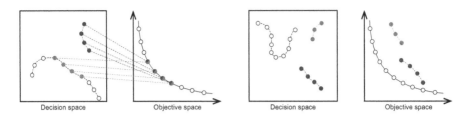

Fig. 1. Schematic differentiation of a multiglobal (left) and a multilocal MOP (right). There are two further specializations: MOPs containing only a single efficient set (and front) are called unimodal, whereas MOPs, which are both multilocal and multiglobal, are called multimodal. This figure is inspired by Figure 1 in [9].

alone are not achieving the desired performance assessment in multimodal problems by MOO algorithms [23]. However, diversity measures, also those specifically considering multimodality [22], can neither capture the properties of the HV nor problem-specific aspects. Thus, there is a need for a new indicator developed in this paper.

Also in the light of obtaining better problem understanding there is a shortage of multi-objective landscape features. These features, however, are needed for a variety of tasks, such as automated algorithm selection [14].

The remainder of this paper is organized as follows. Section 2 describes the considered algorithms and performance indicators from the literature. Subsequently, Sect. 3 presents our proposed measure. Finally, our experimental study is described in Sect. 4 before Sect. 5 concludes our work.

2 Background

2.1 Algorithms

Until recently, the main focus of algorithmic developments was on the approximation of globally optimal solutions of MOPs. Well-known and commonly used respective evolutionary MOO algorithms (EMOAs) are NSGA-II [4] and SMS-EMOA [7]. NSGA-II uses non-dominated sorting in a first step and crowding distance in a second step to focus on global convergence and diversity in objective space. SMS-EMOA implements a $(\mu + 1)$ steady-state approach where NSGA-II's second step is replaced by a procedure which drops the individual with the least contribution to the dominated hypervolume. Again, this algorithm does not consider diversity in decision space. Omni-Optimizer [6] was developed with the idea in mind to be very generic in the sense that it allows for optimization of both single- and multi-objective problems. It operates very much like NGSA-II, but also adopts diversity preservation in decision space. NSGA-II, SMS-EMOA and Omni-Optimizer are all evolutionary optimization algorithms. Recently, Schäpermeier et al. [26] proposed MOLE as a member of the family of gradient-based multi-objective optimizers, which refines upon earlier work on the MOGSA concept [10]. MOLE takes a very different approach to multimodality by actively modeling locally efficient sets and exploiting interactions between their attraction basins, leading to a sequential exploration of the MO optimization landscape. Note that due to this sequential nature, MOLE does not maintain a "population" and thus points need to be sampled from its return values to enable a fair comparison to the other algorithms, which have fixed-size populations.

2.2 Indicators

Performance assessment of multi-objective optimizers is a non-trivial task. Research came up with a plethora of indicators, i.e., functions which map an approximation set to the real-valued numbers. Usually, an indicator measures either cardinality, convergence, spread/diversity, or a combination of these. Prominent examples are the Inverted Generational Distance [36] or the dominated hypervolume in the objective space or the Solow-Polasky measure in decision space. In this study we focus on the latter two for which we provide more details:

The *hypervolume (HV)* [36] is arguably one of the most often used performance indicators im MOO. The dominated hypervolume can be interpreted as the (hyper-)space enclosed by the approximation set and the reference point. HV rewards both convergence to the Pareto front and diversity and brings along many desirable properties, e.g., Pareto compliance [37].

In 1994, Solow and Polasky introduced their eponymous *Solow-Polasky* (SP) indicator to measure the amount of diversity between species in biology [28]. Its first application in evolutionary computation dates back to work by Ulrich and Thiele [33] in the context of *Evolutionary Diversity Optimization* to guide a single-objective EA towards diversity in (continuous) decision space subject to a minimum quality threshold. Given a set of points $X = \{x_1, \ldots, x_\mu\}$ and pairwise (Euclidean) distances $d(x_i, x_j)$, $1 \leq, i, j \leq \mu$ let M be a $(\mu \times \mu)$ matrix with $M_{ij} = \exp(-\theta \cdot d(x_i, x_j))$. Here, θ serves for normalization of the relation between d and the number of species; its choice is not critical [28]. Now the Solow-Polasky diversity is defined as the sum of all elements of the Moore-Penrose generalized inverse M^{-1} of the matrix M. The measure can be interpreted "as the number of different species in the population" [28]. Note, however, that the measure calculates a real-valued diversity in $[1, \mu]$ and no integer value. As pointed out in [32], SP is maximized if points are aligned in a grid.

3   The BBE Measure(s)

For classical MOO, the HV serves as an excellent measure capturing the coverage of the Pareto front. However, in MMMOO, the local efficient sets that, per definition, cannot contribute to the HV are of interest as well. One approach to achieve this is measuring decision space diversity with SP. SP, however, does not focus on the coverage of the local efficient sets but rather on the coverage on the whole decision space. Therefore, we introduce a *basin-based evaluation (BBE)* method in this paper, which focuses on the coverage of the Pareto front as well

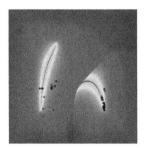

Fig. 2. If a set of solution points (the black points in the left image) is to be evaluated with BBE, first (middle image) only the points in the first basin (on the left hand side) are evaluated and then (right image) only the points in the next basin (on the right hand side). This continues until all basins of interest are evaluated.

as local efficient sets simultaneously. The main idea is to compute the HV per basin and not only globally.

The division of the decision space into basins is done based on the technique for decision space visualization by Schäpermeier et al. [24]. In order to visualize the optimization landscape, they divide the decision space into equal-sized regions arranged in a grid. Every region is represented by the contained point with the lowest value in all decision variables. This enables the computation of multi-objective gradients for all parts of the grid. Then, based on the hull spanned by the gradients, regions which likely contain parts of the efficient sets can be identified. With the gradients and the approximation of the efficient sets, the path from a region to an efficient set can be traced. With this, the accumulated gradient length along the path can be calculated as a measure of distance to the attracting set. Based on this measure of distance, the visualization of the regions is determined. For more detailed information on this procedure we refer the interested reader to [24]. As a by-product, the affiliation of a region to a basin is determined as well. The latter information is used to evaluate a set of returned solution points separately per basin for the proposed measure. For every point of the solution set, the region, and thus, the corresponding basin they are encapsulated in, is identified. This allows to filter out all points that are not contained in a basin and calculate a specific metric only for the points of interest. In the default case, this metric is the HV. See Fig. 2 for a visualization.

As not all basins can be included in the evaluation, only the first k are evaluated. The order of the basins is determined by non-dominated sorting [4] of the regions that approximate the efficient sets. The basin which contributes the most points to the approximated Pareto front thus will be the first in the constructed order. In case of an equal number of regions, the number of regions attributed to the next domination layer is decisive. Basins that are not part of the Pareto front have an equal number of regions (i.e., zero) constituting the Pareto front. In case of many basins, they can also be joined based on the contained

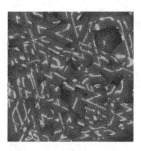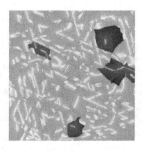

Fig. 3. The shown gradient field heatmap belongs to the same highly multimodal problem instance as the PLOT visualization in Fig. 8: the bi-objective BBOB function with FID 55 and IID 11. If the basins are considered individually the area covered by the first basin is relatively small (image in the middle). To circumvent too many individual basins the efficient sets can be joined, which leads to a potentially distributed area covered by the first basin (right image).

regions attributed to the domination layer with the lowest number. The joint basins are regarded as a single one during evaluation, see Fig. 3.

To aggregate the attained HV from the k basins of interest, the arithmetic mean is taken. Note that this includes a natural weighting between the basins, as a higher HV is attainable in the basins closer to the Pareto front in case the reference point is fixed. To capture an algorithm's anytime behavior, this mean is recorded in every interval of a specific number of function calls needed by the algorithm. Here, one can decide if the accumulation of all the points evaluated by the algorithm up until this point (the solution archive) or only the ones evaluated in the interval should be considered. A visualization of the latter case can be seen in Fig. 4. In order to aggregate those intermediate results, the area under the curve is computed. This value captures the anytime behavior of an algorithm regarding the convergence to k local efficient sets of interest when we focus on multimodal multi-objective optimization.

4   Analysis

To test our basin-based indicators we conduct a benchmark study with four MO optimizers on a set of multimodal problem instances. First, we provide the experimental setup, followed by our experimental results and we end with a discussion and interpretation of these results.

4.1   Experimental Setup

Hardware and Software. All experiments were conducted on PALMA, the high-performance compute cluster of the WWU Münster. Each optimizer run had access to 1 CPU core and 4 GB of memory. In total, all experiments required 20 000 CPU hours. All experimental code for reproducibility is available online[1].

---

[1] Available at: github.com/jeroenrook/BBE-experiments.

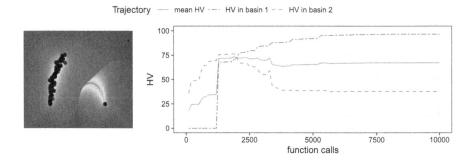

Fig. 4. Shown is a run of NSGA-II on the gradient field heatmap of the Aspar Function [9] together with the corresponding BBE scores. Note that only the most recent points are considered, and therefore, the achieved HV of the points in the second basin eventually declines. (See supplementary material for an animated version of the figure.)

Resources. To run our analysis we use the R implementations of the optimizers SMS-EMOA [1], NSGA-II [1], Omni-Optimizer [3], and MOLE [27]. Furthermore, we compiled a set of 35 well-established, mainly multimodal MOP instances. We selected all instances from ZDT [35], DLTZ [5], MMF [34], with exception of MMF13 and ZDT5, which are provided by smoof [2]. Furthermore, we selected 5 problem instances (FID $\in \{46, 47, 50, 10, 55\}$, IID $= 1$) from BiOb-jBBOB [31], which are provided by moPLOT [24]. All problem instances have a 2D decision and objective space. For each instance, we approximated the Pareto front and chose the reference point such that it covered the whole reachable objective space with moPLOT. These reference points are needed to compute the HV and the BBE indicators. The approximated Pareto front is used to compute the maximum obtainable HV. In turn, the maximum HV is used to normalize the BBE and HV indicators to make them comparable.

Indicators. For computing the BBE measures we use our own R package[2]. We used 3 variants of BBE; 1) the mean HV of the basins with the population returned by the optimizers (BBE(HV)), 2) the mean HV of the basins with the complete archive of function calls the optimizer made (BBE^{cum}(HV)), and 3) the area under the curve of the convergence of the mean HV across all basins during search (BBE^{cum,auc}(HV)). For each variant we considered the 5 most important basins (automatically derived by the BBE package with the landscape exploration of moPLOT) and we did not merge basins of joined fronts.

Configuration. To maximize optimizer performance on the problem instances w.r.t. the indicators we make use of the automated algorithm configurator SMAC [11] (through Sparkle[3]). Here, each configuration scenario aims to maximize

---

[2] BBE available at: github.com/jonathan-h1/BBE.
[3] Accessible through ada.liacs.nl/projects/sparkle.

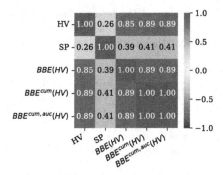

Fig. 5. Spearman correlation between performance metrics. Points taken from 25 runs with all algorithms in all configuration scenarios and on all instances.

performance for one indicator in 10 separate configuration runs. Each configuration run had a budget of 250 algorithm calls and the configuration run with the highest performance score on the whole training set is used for validation. We used leave-one-out validation, i.e., we configured on 34 instances to derive the parameters and then validated the performance on the left out instance. Separate configuration experiments were conducted, each aiming to maximize one of the 5 indicator scores: HV, SP, BBE(HV), BBE^{cum}(HV), and $BBE^{cum,auc}$(HV).

Validation. We validated each optimizer configuration on each instance 25 times with fixed random seeds per run. The median score over these runs was used to represent the configuration's performance. Furthermore, each run was given a budget of 25 000 function evaluations. In total, we validated on all 840 pairs that can be made out of the 6 configurations (including the default configuration), 4 algorithms, and 35 problem instances.

4.2 Indicator Similarity

We start our analysis by focusing on the similarity of the measures. We specifically look at the Spearman correlations between the indicators over all conducted optimizer runs, which are visualized in Fig. 5. A high correlation score indicates that the two indicators yield a similar ordering within the underlying optimizer runs.

The correlation matrix shows that the global convergence in objective space measured by hypervolume (HV) is highly correlated with the variants of our basin-based approach. These high correlations indicate that, despite the reduction in focus on obtaining global convergence, the basin-based approaches are still able to measure this property. Another observation is a clear trade-off between

Table 1. The mean indicator score of the best ranked optimizer under different configuration scenarios. Before the mean was taken over the runs, all indicators, except SP, were first normalized against the maximum approximated HV for each instance.

Indicator	Default	Configuration target				
		HV	SP	BBE(HV)	BBE^{cum}(HV)	$BBE^{cum,auc}$(HV)
HV	1.031	1.071	1.067	1.048	**1.072**	1.053
SP	2.226	5.195	**5.568**	3.598	5.197	3.572
BBE(HV)	0.417	0.515	0.497	**0.531**	0.517	0.488
BBE^{cum}(HV)	0.613	**0.651**	0.650	0.647	0.650	**0.651**
$BBE^{cum,auc}$(HV)	15 074	16 092	16 069	15 891	16 064	**16 103**

HV and the diversity in decision space (SP). Interestingly, the correlation scores between the basin-based indicators and SP are higher than between HV and SP. This suggests that the basin-based indicators are more considerate of the global decision space diversity than HV is.

We bolster the latter observations by looking at the mean indicator values of the best-ranked optimizers after automatically configuring for a particular indicator, depicted in Table 1. Here, we see that the best indicator score (or within proximity to the best) is obtained when explicitly configuring for that indicator. Additionally, the untargeted indicators tend to improve as well by configuration. Disappointingly, when we configure for the mean basin-based HV, both HV and SP are behind compared to the improvements we see in the other configuration scenarios. Speculatively, this is because the points of the last populations of the optimizer are more distributed over the different basins. Further, we see that the configurations of the cumulative BBE variants have excellent performance across all indicators compared to the BBE variant that focuses only on the points in the last population. This could potentially be explained by the fact that these variants aim to visit at least all the basins during search and not on maintaining a population across the different basins. Thus, by keeping an archive of all visited points, one can easily obtain a good coverage across all basins retrospectively.

4.3 Rankings

We now shift our focus to the rankings between optimizers under varying circumstances. Specifically, we compute the average rankings between the 4 optimizers (Sect. 2.1) for each indicator-configuration target pair, resulting in a total of 30 rankings. Figure 6 plots these rankings where each column indicates the indicator by which the ranking was generated, and the rows refer to the configuration target indicator which was used to tune the optimizers' parameters. These rankings reveal that NSGA-II and SMS-EMOA rank best for default parameters. After configuration, both optimizers are almost always exceeded by MOLE or Omni-Optimizer for all indicators, except for HV. There, SMS-EMOA also remains the best-ranked optimizer after configuration. MOLE and Omni-Optimizer have a larger parameter space compared to SMS-EMOA and NSGA-II, likely making them more configurable. However, these two optimizers are also conceptually dif-

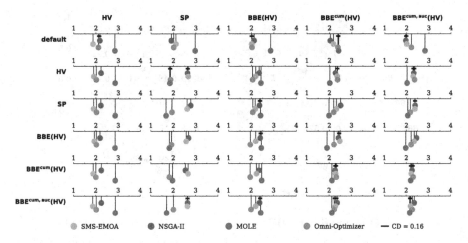

Fig. 6. Average rankings of the optimizers for each measure (columns) and configuration scenario (rows). The confidence distance (CD) to be significantly differently ranked, as determined by a Nemenyi test [20] with $\alpha < 0.1$, is 0.16.

ferent because they exploit structural knowledge of the problem instance during search. Especially for Omni-Optimizer this causes larger ranking improvements when it is configured for the basin-based indicators.

4.4 Ranking Changes

The visualized average rankings in Fig. 6 revealed significant changes in the ranking concerning a measure if the algorithms are configured based on different desired properties. However, the relationship of the ranking shifts, and therefore the mutual configuration impact on two measures cannot be assessed with this figure. Thus, we consider the correlation between the ranking shifts with regard to the measures aggregated over all runs per algorithm and problem instance. Here, the ranking shift is the difference between the average ranking if the algorithms are run with default settings and the average ranking if the algorithms are configured for HV or SP. Therefore, a high correlation between two measures means that the shifts in algorithm rankings per problem instance are similar. The corresponding correlation heatmaps can be seen in Fig. 7. Independent of whether the aim is convergence to the Pareto front (configuring for HV) or maximizing diversity (configuring for SP), the BBE measure variants yield a higher correlation of ranking changes with the diversity indicator SP than with HV. The correlation with HV is relatively small in general, which may be caused by the small changes in the ranking w.r.t. HV independent of the configuration (see Fig. 6). Nevertheless, the correlation of the BBE measures with SP indicates that the former are more similar to a diversity measure than HV and can result in more similar ranking changes in relation to diversity.

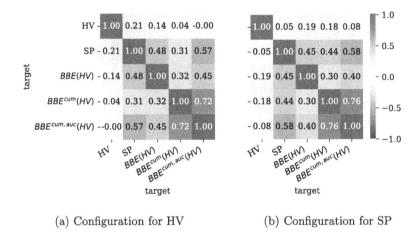

(a) Configuration for HV   (b) Configuration for SP

Fig. 7. The Spearman correlations of the differences between the resulting rankings after configuration for a certain target and the rankings with default parameters. In the left plot the configuration target is the HV and in the right one the target is the SP.

4.5 Discussion

From the presented experimental results, we derive the following insights:

First, the proposed basin-based evaluation captures the main properties of the classical HV evaluation. This is supported by the observed high correlation between HV and the introduced BBE variants. Theoretically, this is explained by the fact that HV is measured per basin. In case of only one basin containing the efficient set which makes up the Pareto front, the found solution points in other basins do not matter for the achieved HV score. In general, the proposed measure variants are generalizations of HV, where HV corresponds to the special case in which the whole decision space is regarded as one basin.

Second, the proposed measure additionally captures diversity aspects that HV cannot capture. The HV, even though a reliable measure for global convergence, does not cover decision space diversity, as shown by the low correlation with the SP. Further, the low correlation of rank shifts, if the algorithms are configured for SP, demonstrates that the HV lacks the ability to score an algorithm run based on the diversity in decision space of the proposed solution set as all points that are dominated by others cannot contribute to the HV. The BBE variants alleviate this issue, as can be seen by the higher correlation with the SP regarding the general scores and the rank shifts.

Third, the actual aim of MMMOO is not to find an algorithm that is performing well w.r.t. diversity over the complete decision space but *conditional diversity* as explained in the following. So far, in MMMOO, a form of general diversity is used to measure how points are distributed in the complete decision space and not just areas in decision space that have objective values along the Pareto front (i.e., multiglobal basins). However, this general diversity can only

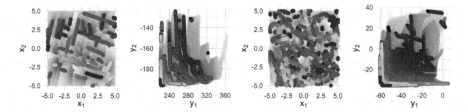

Fig. 8. PLOT visualizations of two functions (left: FID 10, IID 5; right: FID 55, IID 11) from the bi-objective BBOB in decision (x) and objective space (y). The colored points are locally efficient, with the blue points being globally efficient. The gray-scale background illustrates the attraction basins of each locally efficient set, with darker colors indicating more optimal points within a basin. While the left one contains the good-performing locally efficient sets mostly in the lower right corner, the globally efficient regions are much further spread in the right function. (Color figure online)

serve as a proxy of the actual goal to find points other than the global efficient sets. Evenly distributed points may allow choosing the favored combination of decision variables, but the corresponding objectives may be arbitrarily bad. For some problem instances, the interesting space may be a fraction of the complete decision space. As, e.g., shown in Fig. 8, the interesting basins with efficient sets close to the Pareto front are distributed in the decision space but cover only a fraction. Thus, the actual desired diversity is a conditional one. The proposed measure enforces this conditional diversity by only considering the user-defined basins of interest and neglecting the other parts of the decision space.

5 Conclusions

This paper provides different perspectives on multimodality in multi-objective optimization and explicitly contributes to problem characterization and algorithm performance evaluation. We not only introduce a specific method for acquiring comprehensive information on decision-space basins, but also propose variants of a performance indicator BBE which addresses both convergence in objective space as well as decision space diversity. In this regard, not overall diversity but rather conditional diversity adhering to local efficient sets and basin coverage is accounted for. Classical EMOAs and algorithms exploiting local problem structures are experimentally compared and automatically configured. We experimentally show that especially the latter algorithms extremely profit from being configured w.r.t. BBE. By this means, global HV as well as conditional decision space diversity are optimized and the improvements for classical EMOAs lag behind. Moreover, it has to be noted that BBE explicitly needs underlying basin information and is thus not an indicator to be incorporated into an indicator-based EMOA or that can be computed on-the-fly. However, it can be used for optimally configuring EMOAs and for getting an increased understanding on problem hardness w.r.t. multimodality. Also, it can perspectively contribute to deriving multi-objective landscape features. For future work

204 J. Heins et al.

it would be interesting to see how BBE behaves for different classes of multimodality (e.g., problems with only multiglobal basins) and for problems with higher dimensionality in both decision or objective space. Furthermore, instead of using HV, also other measures can potentially be used for computing the basin performance (e.g., BBE(SP)).

References

1. Bossek, J.: ECR 2.0: a modular framework for evolutionary computation in R. In: Proceedings of the Genetic and Evolutionary Computation Conference (GECCO) Companion, pp. 1187–1193. ACM (2017). https://doi.org/10.1145/3067695.3082470
2. Bossek, J.: smoof: Single- and multi-objective optimization test functions. R J. **9**(1), 103–113 (2017). https://doi.org/10.32614/RJ-2017-004
3. Bossek, J., Deb, K.: omnioptr: Omni-Optimizer (2021). , R package version 1.0.0: https://github.com/jakobbossek/omnioptr
4. Deb, K., Pratap, A., Agarwal, S., Meyarivan, T.: A fast and elitist multiobjective genetic algorithm: NSGA-II. IEEE Trans. Evol. Comput. (TEVC) **6**(2), 182–197 (2002). https://doi.org/10.1109/4235.996017
5. Deb, K., Thiele, L., Laumanns, M., Zitzler, E.: Scalable test problems for evolutionary multiobjective optimization. In: Abraham, A., Jain, L., Goldberg, R. (eds.) Evolutionary Multiobjective Optimization. Advanced Information and Knowledge Processing. Springer, London (2005). https://doi.org/10.1007/1-84628-137-7_6
6. Deb, K., Tiwari, S.: Omni-optimizer: a generic evolutionary algorithm for single and multi-objective optimization. Eur. J. Oper. Res. (EJOR) **185**, 1062–1087 (2008). https://doi.org/10.1016/j.ejor.2006.06.042
7. Emmerich, M., Beume, N., Naujoks, B.: An EMO algorithm using the hypervolume measure as selection criterion. In: Coello Coello, C.A., Hernández Aguirre, A., Zitzler, E. (eds.) EMO 2005. LNCS, vol. 3410, pp. 62–76. Springer, Heidelberg (2005). https://doi.org/10.1007/978-3-540-31880-4_5
8. Fieldsend, J.E., Chugh, T., Allmendinger, R., Miettinen, K.: A feature rich distance-based many-objective visualisable test problem generator. In: Proceedings of the Genetic and Evolutionary Computation Conference (GECCO), pp. 541–549. ACM (2019). https://doi.org/10.1145/3321707.3321727
9. Grimme, C., et a.: Peeking beyond peaks: challenges and research potentials of continuous multimodal multi-objective optimization. Comput. Oper. Res. (COR) **136**, 105489 (2021). https://doi.org/10.1016/j.cor.2021.105489
10. Grimme, C., Kerschke, P., Trautmann, H.: Multimodality in multi-objective optimization – more boon than bane? In: Deb, K., et al. (eds.) EMO 2019. LNCS, vol. 11411, pp. 126–138. Springer, Cham (2019). https://doi.org/10.1007/978-3-030-12598-1_11
11. Hutter, F., Hoos, H.H., Leyton-Brown, K.: Sequential model-based optimization for general algorithm configuration. In: Coello, C.A.C. (ed.) LION 2011. LNCS, vol. 6683, pp. 507–523. Springer, Heidelberg (2011). https://doi.org/10.1007/978-3-642-25566-3_40
12. Ishibuchi, H., Peng, Y., Shang, K.: A scalable multimodal multiobjective test problem. In: Proceedings of the IEEE Congress on Evolutionary Computation (CEC), pp. 310–317. IEEE (2019). https://doi.org/10.1109/CEC.2019.8789971

13. Kerschke, P., Grimme, C.: An expedition to multimodal multi-objective optimization landscapes. In: Trautmann, H., et al. (eds.) EMO 2017. LNCS, vol. 10173, pp. 329–343. Springer, Cham (2017). https://doi.org/10.1007/978-3-319-54157-0_23
14. Kerschke, P., Hoos, H.H., Neumann, F., Trautmann, H.: Automated algorithm selection: survey and perspectives. Evol. Comput. (ECJ) **27**, 3–45 (2019). https://doi.org/10.1162/evco_a_00242
15. Kerschke, P., et al.: Towards analyzing multimodality of continuous multiobjective landscapes. In: Handl, J., Hart, E., Lewis, P.R., López-Ibáñez, M., Ochoa, G., Paechter, B. (eds.) PPSN 2016. LNCS, vol. 9921, pp. 962–972. Springer, Cham (2016). https://doi.org/10.1007/978-3-319-45823-6_90
16. Kerschke, P., et al.: Search dynamics on multimodal multi-objective problems. Evol. Comput. (ECJ) **27**, 577–609 (2019). https://doi.org/10.1162/evco_a_00234
17. Li, X., Engelbrecht, A.P., Epitropakis, M.G.: Benchmark functions for cec'2013 special session and competition on niching methods for multimodal function optimization. Technical report, Evolutionary Computation and Machine Learning Group, RMIT University, Australia (2013). http://goanna.cs.rmit.edu.au/~xiaodong/cec13-niching/competition/
18. Maree, S.C., Alderliesten, T., Bosman, P.A.N.: Real-valued evolutionary multimodal multi-objective optimization by Hill-Valley clustering. In: Proceedings of the Genetic and Evolutionary Computation Conference (GECCO), pp. 568–576. ACM (2019). https://doi.org/10.1145/3321707.3321759
19. Miettinen, K.: Nonlinear Multiobjective Optimization. International Series in Operation Research and Management Science, vol. 12. Springer, Heidelberg (1998). https://doi.org/10.1007/978-1-4615-5563-6
20. Nemenyi, P.B.: Distribution-free multiple comparisons. Ph.D. thesis, Princeton University (1963)
21. Preuss, M.: Multimodal Optimization by Means of Evolutionary Algorithms. Natural Computing Series (NCS). Springer, Cham (2015). https://doi.org/10.1007/978-3-319-07407-8
22. Preuss, M., Wessing, S.: Measuring multimodal optimization solution sets with a view to multiobjective techniques. In: Emmerich, M. et al. (eds.) EVOLVE - A Bridge Between Probability, Set Oriented Numerics, and Evolutionary Computation IV, pp. 123–137. Springer, Heidelberg (2013). https://doi.org/10.1007/978-3-319-01128-8_9
23. Rook, J., Trautmann, H., Bossek, J., Grimme, C.: On the potential of automated algorithm configuration on multi-modal multi-objective optimization problems. In: Proceedings of the Genetic and Evolutionary Computation Conference (GECCO) Companion. p. tbd. ACM (2022). https://doi.org/10.1145/3520304.3528998, accepted
24. Schäpermeier, L., Grimme, C., Kerschke, P.: One PLOT to show them all: visualization of efficient sets in multi-objective landscapes. In: Bäck, T., et al. (eds.) PPSN 2020. LNCS, vol. 12270, pp. 154–167. Springer, Cham (2020). https://doi.org/10.1007/978-3-030-58115-2_11
25. Schäpermeier, L., Grimme, C., Kerschke, P.: To boldly show what no one has seen before: a dashboard for visualizing multi-objective landscapes. In: Proceedings of the International Conference on Evolutionary Multi-criterion Optimization (EMO), pp. 632–644 (2021). https://doi.org/10.1007/978-3-030-72062-9_50
26. Schäpermeier, L., Grimme, C., Kerschke, P.: MOLE: digging tunnels through multimodal multi-objective landscapes. In: Proceedings of the Genetic and Evolutionary Computation Conference (GECCO). p. tbd. ACM (2022). https://doi.org/10.1145/3512290.3528793, accepted

27. Schäpermeier, L.: An R package implementing the multi-objective landscape explorer (MOLE), February 2022. https://github.com/schaepermeier/moleopt
28. Solow, A.R., Polasky, S.: Measuring biological diversity. Environ. Ecol. Stat. **1**, 95–103 (1994). https://doi.org/10.1007/BF02426650
29. Tanabe, R., Ishibuchi, H.: A Niching indicator-based multi-modal many-objective optimizer. Swarm Evol. Comput. (SWEVO) **49**, 134–146 (2019). https://doi.org/10.1016/j.swevo.2019.06.001
30. Tušar, T., Filipič, B.: Visualization of pareto front approximations in evolutionary multiobjective optimization: a critical review and the prosection method. IEEE Trans. Evol. Comput. (TEVC) **19**(2), 225–245 (2015). https://doi.org/10.1109/TEVC.2014.2313407
31. Tušar, T., Brockhoff, D., Hansen, N., Auger, A.: COCO: the bi-objective black box optimization benchmarking (BBOB-BIOBJ) test suite. arXiv preprint abs/1604.00359 (2016). https://doi.org/10.48550/arXiv.1604.00359
32. Ulrich, T., Bader, J., Thiele, L.: Defining and optimizing indicator-based diversity measures in multiobjective search. In: Schaefer, R., Cotta, C., Kołodziej, J., Rudolph, G. (eds.) PPSN 2010. LNCS, vol. 6238, pp. 707–717. Springer, Heidelberg (2010). https://doi.org/10.1007/978-3-642-15844-5_71
33. Ulrich, T., Thiele, L.: Maximizing population diversity in single-objective optimization. In: Proceedings of the Genetic and Evolutionary Computation Conference (GECCO), pp. 641–648. ACM (2011). https://doi.org/10.1145/2001576.2001665
34. Yue, C., Qu, B., Yu, K., Liang, J., Li, X.: A novel scalable test problem suite for multimodal multiobjective optimization. Swarm Evol. Comput. **48**, 62–71 (2019). https://doi.org/10.1016/j.swevo.2019.03.011
35. Zitzler, E., Deb, K., Thiele, L.: Comparison of multiobjective evolutionary algorithms: empirical results. Evol. Comput. (ECJ) **8**(2), 173–195 (2000). https://doi.org/10.1162/106365600568202
36. Zitzler, E., Thiele, L.: Multiobjective optimization using evolutionary algorithms - a comparative case study. In: Eiben, A.E., Bäck, T., Schoenauer, M., Schwefel, H.P. (eds.) PPSN 1998. LLNCS, vol. 1498. pp. 292–301. Springer, Heidelberg (1998). https://doi.org/10.1007/bfb0056872
37. Zitzler, E., Thiele, L., Laumanns, M., Fonseca, C.M., da Fonseca, V.G.: Performance assessment of multiobjective optimizers: an analysis and review. IEEE Trans. Evol. Comput. (TEVC) **7**(2), 117–132 (2003). https://doi.org/10.1109/TEVC.2003.810758

Evolutionary Approaches to Improving the Layouts of Instance-Spaces

Kevin Sim[iD] and Emma Hart[(⊠)][iD]

Edinburgh Napier University, Edinburgh, UK
{k.sim,e.hart}@napier.ac.uk

Abstract. We propose two new methods for evolving the layout of an instance-space. Specifically we design three different fitness metrics that seek to: (i) reward layouts which place instances won by the same solver close in the space; (ii) reward layouts that place instances won by the same solver *and* where the solver has similar performance close together; (iii) simultaneously reward proximity in both class and distance by combining these into a single metric. Two optimisation algorithms that utilise these metrics to evolve a model which outputs the coordinates of instances in a 2d space are proposed: (1) a multi-tree version of GP (2) a neural network with the weights evolved using an evolution strategy. Experiments in the TSP domain show that both new methods are capable of generating layouts in which subsequent application of a classifier provides considerably improved accuracy when compared to existing projection techniques from the literature, with improvements of over 10% in some cases. Visualisation of the the evolved layouts demonstrates that they can capture some aspects of the performance gradients across the space and highlight regions of strong performance.

Keywords: Instance-space · Dimensionality-reduction · Algorithm-selection

1 Introduction

Instance Space Analysis is a methodology first proposed by Smith-Miles *et al.* in a series of papers [14,15,18] with the purpose of (1) providing a means of visualising the location of benchmark instances in a 2d space; (2) illustrating the 'footprint' of an algorithm (i.e. the regions of the space in which it performs well) and (3) calculating objective (unbiased) metrics of algorithmic power via analysis of the aforementioned footprints. Once an instance-space has been created, it can be used in various ways: for example, to identify regions in the space that are lacking representative data (followed by generation of instances targeted at filling these gaps) or developing automated algorithm selection tools to determine the best algorithm for solving a new instance.

A critical step of the methodology is clearly the projection of an instance described by a high-dimensional feature-vector into a 2d instance-space, with

G. Rudolph et al. (Eds.): PPSN 2022, LNCS 13398, pp. 207–219, 2022.
https://doi.org/10.1007/978-3-031-14714-2_15

the quality of this projection having significant influence of the utility of the resulting space [14]. Three factors are important. Firstly, the projection method should be model-based, i.e. it should learn a model that can be used to project future unseen instances into the space once it has been created—this rules out *embedding* approaches such as the popular t-sne [10] technique. Secondly, if instances are labelled according to the solver which produces the 'best' performance, instances with the same label should be co-located in the space (potentially in multiple distinct regions). Finally, within a subset of instances with the same label, we propose that the projection should locate instances where the solver provides similar performance closer together than those where the performance differs widely, i.e. performance should vary smoothly across a cluster of instances with the same label, thereby indicating regions in which the winning solver is particularly strong and vice-versa.

The vast majority of work in the area of instance-space creation has used Principal Component Analysis (PCA) [12] as the means of projecting to a 2d-space—despite the fact that this method is unsupervised and therefore does not take into account either instance-labels or relative performance of solvers. Alternative methods for dimensionality-reduction such as manifold-learning methods (e.g. UMAP [3]) which can be used in a supervised manner seek to place instances that are close in the high-dimensional space close in low-dimensional instance space: that is, an instance should retain the same nearest neighbours in the embedded space as in the input space. However, mapping from an high-dimensional *feature-space* to a low-dimensional feature-space that also smoothly captures variation in the *performance-space* for the purpose of instance-space creation poses a problem for most manifold-learning methods: neighbours in the performance-space are not necessarily neighbours in the feature-space. This is clearly illustrated in Fig. 1 which plots the distance in a feature-space against distance in the performance-space for all pairs of instances taken from a large set of 950 TSP instances, and shows there is very little correlation.

Therefore, in order to produce instance-spaces which attempt to satisfy all three criteria outlined above, we propose two evolutionary approaches to learn a model which maps from a high-dimensional feature-space to a low-dimensional instance-space. The first uses a multi-tree genetic programming (GP) algorithm to output the coordinates in a 2d space, while the second evolves the weights of a neural-network which outputs the 2d coordinates, using an evolution strategy (ES) to train the network. We propose three novel fitness functions that can be combined with either optimiser to address this. To evaluate the quality of the evolved layouts, *post-evolution* we apply multiple classifiers to the space to determine whether it facilitates algorithm-selection, and visualise the space to gain a qualitative understanding of whether the space smoothly reflect the performance gradient of the solvers. Both approaches are demonstrated to evolve layouts that improve the accuracy of classifiers trained on the 2d space compared to using layouts created using PCA and UMAP, with some progress towards improving layouts with respect to the performance gradients within a class.

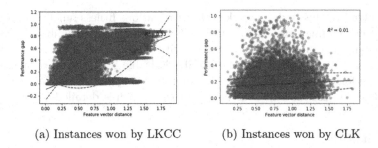

(a) Instances won by LKCC   (b) Instances won by CLK

Fig. 1. Scatter plots showing distance in feature-space (x-axis) vs performance-gap for pairwise comparisons of TSP instances, illustrating lack of correlation between these quantities

2 Related Work

A long line of work by Smith-Miles and her collaborators [14,15,18] has gradually refined the approach to defining an instance-space into a rigorous methodology, culminating in the freely available MATILDA toolbox [2]. The method has been applied in multiple domains within combinatorial optimisation (e.g. TSP [16], timetabling [17], knapsack [2], graph-colouring [18]) and more recently to machine-learning datasets [11]. In the vast majority of the work described in combinatorial optimisation, instance-spaces are created using PCA to project into low-dimensions. As this is an unsupervised method, in order to find projections that place instances won by the same solver in similar regions, an evolutionary algorithm is used to select a subset of features that - when projected via PCA - maximise classification accuracy using an SVM classifier. However, this potentially biases the projection towards SVM. A more recent approach in which a single instance-space is developed to represent multiple *machine-learning* datasets uses a new projection method that tries to minimise the approximation error $|F - f| + |P - p|$ where F, P correspond to the feature/performance vector in a high-d space and f, p to the same vectors in the low-d space, but cannot be directly translated to laying out an instance-space to reflect the performance gradients of *multiple* solvers.

More recently, Lensen *et al.* [7] proposed a GP approach to manifold learning of machine-learning datasets [19]. A multi-tree GP method is used to learn a 2d projection using a fitness function that attempts to maintain the same ordering between neighbours of an instance in both spaces. The quality of a learned embedding is estimated via a proxy measure calculated post-evolution—applying a classifier to the newly projected data and measuring classification accuracy. In more recent work, the same authors propose further extensions that (1) optimise the embedding learned by GP to match a pre-computed UMAP embedding, and (2) optimise UMAP's own cost-function directly [13]. Most recently they adapt their approach to consider how local structure within an embedding can be better reflected, proposing a modified fitness functions that seeks to measure how well local topology is preserved by the evolved mapping [8]. However, a common trait in this body of work is that the various fitness functions proposed optimise embeddings such that the input and embedded neighbourhood of a given instance contain the same instances with the same ordering. As previously discussed in Sect. 1, this is not necessarily desirable if the instance-space is to capture performance gradients as well as maintain neighbourhoods in the feature-space. These issues motivate our new approach which is detailed below.

3 Methods

Our goal is to evolve an instance-space layout that places instances won by the same solver in the same regions of a low-d feature-space while also attempting to maintain an ordering in the performance-space, i.e. *instances won by the same algorithm and eliciting similar performance from that algorithm should placed close together.* We investigate two approaches for creating the layout: the first uses an evolution-strategy to evolve the weights of a neural network that outputs the 2d coordinates of each instances. The second, inspired by the work of Lensen *et al.*, uses a multi-tree genetic programming approach to achieve the same goal but with new fitness function(s). Assuming a set of instances, each of which is described by a high-dimensional feature-vector and is labelled with the solver that 'wins' the instance, we propose the following novel fitness functions:

1. Label-based ($\mathcal{L}$): maximise the proportion of the k nearest neighbours of an instance i in the embedded space that have the same label as i (averaged over all instances)
2. Distance-based ($\mathcal{D}$) :minimise the normalised average distance between the performance of an instance and that of its k nearest neighbours *that have the same label.* A penalty of 1 is added to this quantity for every neighbour that is wrongly labelled, in order to prevent the fitness function favouring a small number of correctly labelled neighbours with very small distance.
3. Combined ($\mathcal{L} + \mathcal{D}$): maximise $\mathcal{L} + (1 - \mathcal{D})$, i.e. a linear combination of the previous fitness functions

We evolve mappings using each combination of fitness-function and optimisation algorithm (ES/GP), i.e. 6 combinations in total. Post-optimisation, we follow the approach of Lensen *et al.* and estimate the quality of an evolved instance-space using a proxy measure: we apply three off-the-shelf classifiers to the evolved layout to predict the best solver on an *unseen* set of instances, hence determining whether the layout facilitates algorithm-selection. Secondly, we provide a qualitative view of the extent to which instances that have similar performance are mapped to similar regions of the space using visualisation. Results are compared to PCA and UMAP. PCA is chosen as it is the method of choice to produce an instance-space in MATILDA. UMAP is selected as it can be used in a supervised fashion and therefore offers a comparison to our proposed supervised techniques.

3.1 Mapping Using an Evolution Strategy

An ES is used to learn the hyper-parameters of a neural-network that given a set of features describing an instance, outputs the coordinates of the instance in a 2-d space. A population encodes a set of individuals that specify the real-valued weights of a fixed-size neural network. The neural network is a feed-forward neural network with f inputs corresponding to the number of features describing an instance, and $o = 2$ outputs specifying the new coordinates in a 2d space. The network has one hidden layer with $(f + o)/2$ neurons. The hidden neurons use a *relu* activation function, while the two output neurons have *sigmoid* activation.

To evolve the population to optimise the chosen fitness function(s) (as defined above), we apply the standard evolution strategy CMA-ES [4] due its prevalence in the literature in the context of neuro-evolution [5,6,9]. We use the default implementation of CMA-ES provided in the DEAP library [1]. This requires three parameters to be set: the centroid (set to 0.0), the value of *sigma* (set to 1.0) and the number of offspring *lambda* which is set to 50 in all experiments. The algorithm was run for 50 generations.

3.2 Mapping Using a Multi-tree Genetic Programming

We use a multi-tree GP representation in which each individual contains 2 GP trees, each representing a single dimension in the embedding, following the general approach of Lensen *et al.* [7]. The terminal set contains the n features describing an instance, with a mix of linear/non-linear functions (Table 1). The algorithm is implemented using DEAP. All parameter settings are given in Table 1. To enable direct comparison with the ES, the number of generations was fixed at 25, resulting in the same number of individual evaluations for both methods.

Table 1. GP parameters and settings

GP parameters	
Population size	100
Initialisation	GenHalfandHalf (5,10)
Selection	Tournament (size = 7)
Crossover	1 pt
Mutation	GenFull (3, 5)
Max tree depth	17

GP function set	
ADD	COSINE
SUBTRACT	SIN
MULTIPLY	TANH
ProtectedDIVIDE	RELU
NEG	Eph. Const.(0,1)

3.3 Instance Data

We use the instances from the TSP domain provided via MATILDA [2]. This includes the meta-data associated with an instance that defines the values of features identified as relevant for the domain and the performance data from two solvers (Chained Lin-Kernighan (CLK) an Lin-Kernighan with Cluster Compensation (LKCC) [18]). 950 instances are provided which are synthetically generated to be either easy or hard for each solver (see [18] for a detailed description of the generation method and solver), labelled by the best-solver determined according to the time taken to solve.

We conduct experiments using the 17 features given in the MATILDA metadata (denoted 'full features'). In addition we repeat experiments using the subset of 6 features that were selected by MATILDA to produce the instance-space projection using PCA, as described in the Sect. 2, referred from here on as the 'reduced set'. Recall that this reduced set was specifically chosen to optimise a PCA projection but is used in all experiments to enable direct comparison. The reader is referred to the MATILDA toolbox [2] for a detailed description of each feature and the features designated as the 'reduced' feature-set.

4   Experiments

Experiments are conducted for each optimiser (ES/GP) combined with each of the three proposed fitness functions. For each combination we evolve layouts using both the full and reduced feature set. All experiments are repeated 10 times. The feature-data is normalised (specifically as this provides input in a suitable range to the neural network). For PCA and UMAP, standardised scaling is applied to the data to remove the mean and scale to unit variance as this is widely accepted as best-practice for these methods. The 950 instances are split into a 'training' set using 60% of the data (selected using stratified sampling) to preserve class distribution in each split. The same training dataset is used in all experiments to evolve the instance-space layout. In all experiments reported, the k nearest neighbour parameter required to calculate fitness was set to 15. (A series of preliminary experiments that varied k between 15 and 45 did not provide any statistical evidence that the setting influenced results).

Following evolution, we calculate a proxy measure of quality as described in Sect. 2 to quantify the effectiveness of the evolved layout: three off-the-shelf classifiers are trained to predict the best solver (a binary classification problem) using the evolved 2d projection as input to the classifier. The classifier is trained using the same training data used to evolve the layout, then results are reported on the *held-out test set*. Three classifiers are chosen: Random Forest (generally cited as providing strong performance), support vector machines (as used in the MATILDA methodology), and finally a k-nearest neighbour classifier (which would be expected to perform well in the spaces evolved by the label-based method which also relies on neighbourhoods). For each evolved layout, we record the accuracy and F1 score (which combines the precision and recall of a classifier into a single metric by taking their harmonic mean) of each classifier. This is repeated using layouts created using PCA and UMAP on the same data as comparison. The standard scikit-learn implementation of PCA is used which does not require any parameter setting. UMAP requires a parameter *nearest_neighbor* which controls the balance between local versus global structure in the data (where low values emphasise local structure) which was set to

5 again following a brief empirical investigation. All other UMAP settings were left in their default setting as provided in the Python implementation [3]. As previously noted, UMAP is used in its supervised form.

5 Results

This section reports quantitative and qualitative results that (1) compare the quality of embedding (using classification accuracy/F1-score as a proxy) using each combination of optimiser/fitness-function to off-the-shelf methods (PCA, UMAP) and (2) provide a qualitative evaluation of the evolved instance-spaces with respect to the extent to which they appear to separate the two classes, and illustrate gradients in the performance data.

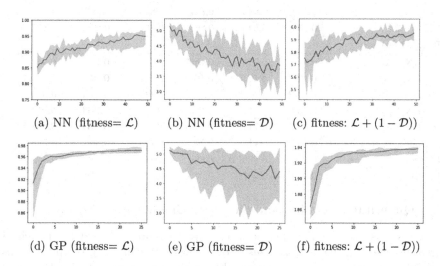

(a) NN (fitness= $\mathcal{L}$)   (b) NN (fitness= $\mathcal{D}$)   (c) fitness: $\mathcal{L} + (1 - \mathcal{D})$)

(d) GP (fitness= $\mathcal{L}$)   (e) GP (fitness= $\mathcal{D}$)   (f) fitness: $\mathcal{L} + (1 - \mathcal{D})$)

Fig. 2. Convergence curves for combinations of (optimiser, fitness function), obtained using the reduced feature-set as input. Top row shows results for the NN method, the bottom row for GP. Red line shows median value over 10 runs (Color figure online)

5.1 Insights into Evolutionary Progress

Figure 2 plots convergence curves for each combination of optimiser/fitness function applied to evolving a layout in the reduced feature-space[1]. The GP method combined with label-based fitness measure $\mathcal{L}$ or the combined fitness measure exhibits less variance than the neural network approach, although both methods show wide variance using the distance based function which only implicitly accounts for class-labels. Furthermore, the GP approaches converge more quickly than the neural-network equivalents.

---

[1] Similar trends are observed in the plots obtained in the full feature space but not shown due to space limitations.

Table 2. Median accuracy and F1 score per classifier using projection obtained by each combination of optimiser/fitness function, with comparison to PCA and UMAP. Top row - full feature set; bottom row - reduced feature set. **Bold** indicates that *accuracy* is better than both PCA/UMAP, *italics* that F1 is better than both. All results reported on held-out test set

	A(SVM)	F1(SVM)	A(KNN)	F1(KNN)	A(RndF)	F1(RndF)
Full features						
GP-KNN ($\mathcal{L}$)	**0.920**	*0.911*	**0.972**	*0.972*	**0.959**	*0.958*
GP-DIST ($\mathcal{D}$)	**0.953**	*0.951*	**0.966**	*0.966*	**0.962**	*0.962*
GP-DIST-ADD ($\mathcal{L}+\mathcal{D}$)	**0.942**	*0.939*	**0.966**	*0.966*	**0.961**	*0.960*
NN-KNN ($\mathcal{L}$)	**0.912**	*0.912*	**0.920**	*0.919*	**0.915**	*0.913*
NN-DIST ($\mathcal{D}$)	0.833	0.782	**0.933**	*0.933*	**0.917**	*0.917*
NN-DIST-ADD ($\mathcal{L}+\mathcal{D}$)	0.800	0.711	0.778	0.710	0.800	0.711
PCA	0.850	0.843	0.847	0.845	0.858	0.844
UMAP	0.855	0.845	0.861	0.852	0.855	0.845
Reduced features						
GP-KNN ($\mathcal{L}$)	0.891	0.897	**0.965**	*0.964*	**0.958**	*0.957*
GP-DIST ($\mathcal{D}$)	**0.950**	*0.949*	**0.970**	*0.969*	**0.961**	*0.960*
GP-DIST-ADD ($\mathcal{L}+\mathcal{D}$)	**0.952**	*0.952*	**0.968**	*0.968*	**0.961**	*0.960*
NN-KNN ($\mathcal{L}$)	0.800	0.711	**0.955**	*0.955*	**0.951**	*0.951*
NN-DIST ($\mathcal{D}$)	0.800	0.711	**0.958**	*0.957*	**0.951**	*0.951*
NN-DIST-ADD ($\mathcal{L}+\mathcal{D}$)	0.800	0.711	0.772	0.708	0.800	0.711
PCA	0.874	0.875	0.889	0.888	0.868	0.846
UMAP	0.909	0.907	0.903	0.902	0.913	0.911

5.2 Quantitative Evaluation via Proxy Classification Metrics

Table 2 shows the classification accuracy and F1-score obtained from each of three classifiers on the *unseen dataset* for each experiment using the projections evolved in the prior step, and compared to projections obtained from PCA and UMAP. With the exception of two combinations that use the neural network method ($\mathcal{L}$, $\mathcal{L}+\mathcal{D}$), it is clear that the evolved layouts enable all three classifiers tested to produce significantly better results that PCA and UMAP, with performance gains of over 10% in several cases. This demonstrates that the evolved layouts provided a good general basis for classification, in eliciting good performance from mutiple different types of classifier. As expected, the neighbour-based $\mathcal{L}$ fitness function creates layouts that favour the KNN classifier, which also relies on a neighbourhood method, but it is clear that the other classifiers (particularly Random Forest) are also competitive in a space evolved to favour similar neighbours. The GP approach generally outperforms the neural-network approaches. The SVM classifier generally provides weaker results than the other two classifiers, although still markedly better than PCA/UMAP in 4 out of 6 experiments.

Figure 3 shows boxplots of results obtained on the test set from the 10 runs for each combination, plotted per classifier. Pairwise significance tests were conduct on each set of results per classifiers (Mann-Whitney test using Bonferroni correction). Similar plots of p-values were obtained for all three classifiers, however only one is shown due to space restrictions. The boxplots demonstrate that the SVM classifier has more variable performance in the evolved spaces while the other two classifiers appear robust to the projection; similarly, the neural approach tends to result in layouts in which classification performance is more variable than the spaces evolved using GP; this result is consistent across all three classifiers.

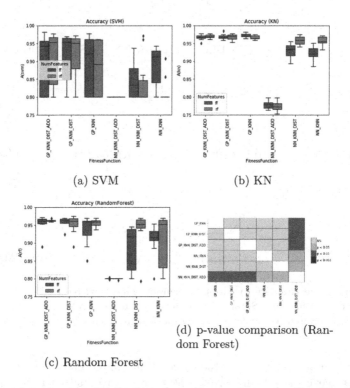

(a) SVM   (b) KN

(c) Random Forest

(d) p-value comparison (Random Forest)

Fig. 3. Boxplots showing distribution of results per proxy classifier for each layout combination. (d) Shows a typical plot of p-values obtained from comparing pairs of methods: similar plots were obtained for all 3 classifiers.

5.3   Qualitative Evaluation: Visualisation of Layouts (by Label)

Figure 4 shows examples of layouts obtained by the single run from the 10 runs of each combination that resulted in the best fitness for each combination of optimiser/fitness function. The plots are shaded by class label. A wide variety of layouts are observed. Note that UMAP (as expected) produces plots that favour local structure within the data. The fitness function $\mathcal{L} + \mathcal{D}$ that favours both positioning instance with the same label *and* similar performance tends to spread

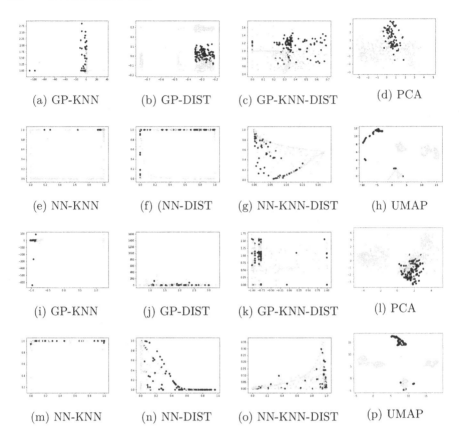

Fig. 4. Embeddings from run with best accuracy. First row: reduced feature-set, GP plus PCA reduced features. Second row: reduced feature-set, NN, plus UMAP reduced features. Third/fourth row: as above using full feature set.

the instances more widely across the space. While all but one of the evolutionary methods (the exception being $NN(\mathcal{L} + \mathcal{D})$ produce layouts that result in considerably higher classification accuracy than PCA/UMAP, the layouts are perhaps less easy to interpret to the human eye. In several cases, multiple instances are mapped to the same coordinates which improves classification but does not easily enable 'similar' instances (from an algorithm-selection perspective) to be easily identified. A trade-off thus exists: ultimately the choice of method depends on the priorities of the user, i.e. whether the goal is simply to select the best algorithm or from a more scientific perspective, to gain insights into algorithm performance relative to algorithm features.

Furthermore, while the boxplots show low variance in classification accuracy across multiple runs, there is considerably variation in the layouts themselves (see Fig. 5 as an example). The different biases of the two fitness functions is clearly observed: note the tighter clustering of instances per class in the top

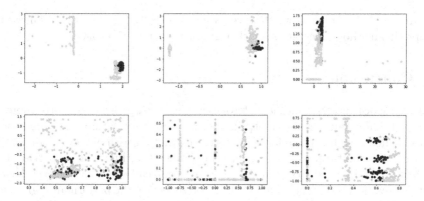

Fig. 5. Layouts from repeated runs of the same optimiser/fitness function. Top row - GP with nearest-neighbour fitness; bottom - GP with distance based fitness

row from rewarding near neighbours of the same class, while the distance based fitness function results in a wider spread of instances. This variability further emphasises the trade-offs to be considered as mentioned above.

5.4   Visualisation of Layouts (by Performance)

Finally we evaluate the extent to which the distance based fitness function results in layouts that place instances that elicit similar performance from the same algorithm close together. Figure 6 shows three examples obtained from separate runs of the GP optimiser with the distance-based fitness function. Instances 'won' by each class are shown on separate plots for clarity with the shading representing the relative performance of the algorithm (normalised between 0 and 1). Some clear clusters of similar performance are visible (e.g. particularly regarding the dark colours representing very strong performance), while there is some evidence of instances with weak performance (lightest colours) appearing

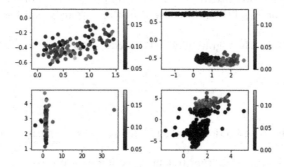

Fig. 6. Layouts from 2 runs of GP optimiser (fitness=Distance), one run per row. Instances shaded by relative performance and separated into two plots according to the class label of the instance

towards the edge of clusters. However, there is clearly further work required to adapt this method to create smoother gradients across the space.

6 Conclusion

Instance-space analysis methods have been attracting increasing attention as a way of understanding the 'footprints' of an algorithm within a feature-space, enabling new insights into relative performance and algorithmic power [14], while additionally facilitating algorithm-selection. We first outlined the properties that we believe an instance-space should embody, specifically that it should co-locate instances with the same class label, while also reflecting the performance-gradient across a cluster, i.e. placing instances that elicit the similar performance from an algorithm close together. These factors are peculiar to the goal of instance-space analysis in optimisation. Specifically, they differ from the goals of standard dimensionality reduction methods which usually try to creating a mapping in which the ordering of neighbours of a point in a high-dimensional space is reflected in the low-dimensional space. As described in Sect. 1 however, an ordering in the feature-space of a set of instances can differ extensively from the ordering within the performance-space, hence manifold-learning techniques might not be appropriate for instance-space analysis.

We demonstrate that both proposed optimisation methods are capable of generating layouts that provide considerable improvement in classification accuracy to UMAP and PCA, of over 10% in some cases. We also provide a more qualitative analysis of the visualisations in terms of their ability to reflect performance gradients. Results suggest that the approach shows promise with the respect to the latter goal, although there is scope for refinement. The calculation of the k nearest-neighbours is time-consuming for a large dataset but this can be significantly improved using an efficient implementation (e.g. k-d trees) and is therefore not limiting. We deliberately chose not to conduct a wide exploration of possible neural architectures or to spend a large amount of effort in tuning the parameters of the proposed algorithms. This will be further investigated in future work. Finally, we intend to repeat the experiments in other combinatorial domains, using the MATILDA generated instance-spaces as a baseline.

Acknowledgments. Hart gratefully acknowledges the support EPSRC EP/V02 6534/1.

References

1. Deap: Distributed evolutionary algorithms in Python. https://deap.readthedocs. io/en/master/
2. Matilda: Melbourne algorithm test instance library with data analytics. https:// matilda.unimelb.edu.au/matilda/
3. Umap: Uniform manifold approximation and projection for dimension reduction. https://umap-learn.readthedocs.io/en/latest/index.html

4. Hansen, N., Müller, S.D., Koumoutsakos, P.: Reducing the time complexity of the derandomized evolution strategy with covariance matrix adaptation (CMA-ES). Evol. Comput. **11**(1), 1–18 (2003)
5. Hasselmann, K., Ligot, A., Ruddick, J., Birattari, M.: Empirical assessment and comparison of neuro-evolutionary methods for the automatic off-line design of robot swarms. Nat. Commun. **12**(1), 1–11 (2021)
6. Le Goff, L.K., et al.: Sample and time efficient policy learning with CMA-ES and Bayesian optimisation. In: Artificial Life Conference Proceedings, pp. 432–440. MIT Press (2020)
7. Lensen, A., Xue, B., Zhang, M.: Can genetic programming do manifold learning too? In: Sekanina, L., Hu, T., Lourenço, N., Richter, H., García-Sánchez, P. (eds.) EuroGP 2019. LNCS, vol. 11451, pp. 114–130. Springer, Cham (2019). https://doi.org/10.1007/978-3-030-16670-0_8
8. Lensen, A., Xue, B., Zhang, M.: Genetic programming for manifold learning: preserving local topology. IEEE Trans. Evol. Comput. (2021)
9. Loshchilov, I., Hutter, F.: CMA-ES for hyperparameter optimization of deep neural networks. arXiv preprint arXiv:1604.07269 (2016)
10. Van der Maaten, L., Hinton, G.: Visualizing data using t-SNE. J. Mach. Learn. Res. **9**(11) (2008)
11. Muñoz, M.A., Villanova, L., Baatar, D., Smith-Miles, K.: Instance spaces for machine learning classification. Mach. Learn. **107**(1), 109–147 (2017). https://doi.org/10.1007/s10994-017-5629-5
12. Partridge, M., Calvo, R.A.: Fast dimensionality reduction and simple PCA. Intell. Data Anal. **2**(3), 203–214 (1998)
13. Schofield, F., Lensen, A.: Using genetic programming to find functional mappings for UMAP embeddings. In: 2021 IEEE Congress on Evolutionary Computation (CEC), pp. 704–711. IEEE (2021)
14. Smith-Miles, K., Baatar, D., Wreford, B., Lewis, R.: Towards objective measures of algorithm performance across instance space. Comput. Oper. Res. **45**, 12–24 (2014)
15. Smith-Miles, K., Bowly, S.: Generating new test instances by evolving in instance space. Comput. Oper. Res. **63**, 102–113 (2015)
16. Smith-Miles, K., van Hemert, J., Lim, X.Y.: Understanding TSP difficulty by learning from evolved instances. In: Blum, C., Battiti, R. (eds.) LION 2010. LNCS, vol. 6073, pp. 266–280. Springer, Heidelberg (2010). https://doi.org/10.1007/978-3-642-13800-3_29
17. Smith-Miles, K., Lopes, L.: Generalising algorithm performance in instance space: a timetabling case study. In: Coello, C.A.C. (ed.) LION 2011. LNCS, vol. 6683, pp. 524–538. Springer, Heidelberg (2011). https://doi.org/10.1007/978-3-642-25566-3_41
18. Smith-Miles, K., Lopes, L.: Measuring instance difficulty for combinatorial optimization problems. Comput. Oper. Res. **39**(5), 875–889 (2012)
19. Wang, Y., Huang, H., Rudin, C., Shaposhnik, Y.: Understanding how dimension reduction tools work: an empirical approach to deciphering t-SNE, UMAP, TriMap, and PaCMAP for data visualization. J. Mach. Learn. Res. **22**(201), 1–73 (2021)

Combinatorial Optimization

A Novelty-Search Approach to Filling an Instance-Space with Diverse and Discriminatory Instances for the Knapsack Problem

Alejandro Marrero[1(✉)], Eduardo Segredo[1], Coromoto León[1], and Emma Hart[2]

[1] Departamento de Ingeniería Informática y de Sistemas, Universidad de La Laguna, San Cristóbal de La Laguna, Tenerife, Spain
{amarrerd,esegredo,cleon}@ull.edu.es
[2] School of Computing, Edinburgh Napier University, Edinburgh, UK
e.hart@napier.ac.uk

Abstract. We propose a new approach to generating synthetic instances in the knapsack domain in order to fill an instance-space. The method uses a novelty-search algorithm to search for instances that are diverse with respect to a feature-space but also elicit discriminatory performance from a set of target solvers. We demonstrate that a single run of the algorithm per target solver provides discriminatory instances and broad coverage of the feature-space. Furthermore, the instances also show diversity within the performance-space, despite the fact this is not explicitly evolved for, i.e. for a given 'winning solver', the magnitude of the performance-gap between it and other solvers varies across a wide-range. The method therefore provides a rich instance-space which can be used to analyse algorithm strengths/weaknesses, conduct algorithm-selection or construct a portfolio solver.

Keywords: Instance generation · Novelty search · Evolutionary algorithm · Knapsack problem · Optimisation

1 Introduction

The term *instance-space*—first coined by Smith-Miles *et al.* [16]—refers to a high-dimensional space that summarises a set of instances according to a vector containing a list of measures features derived from the instance-data. Projecting the feature-vector into a lower-dimensional space (ideally 2D) enables the instance-space to be visualised. Solver-performance can be superimposed on the visualisation to reveal regions of the instance-space in which a potential solver outperforms other candidate solvers. The low-dimensional visualisation of the instance-space can then be used in multiple ways: to understand areas of the space in which solvers are strong/weak, to perform algorithm-selection or to assemble a portfolio of solvers.

© The Author(s), under exclusive license to Springer Nature Switzerland AG 2022
G. Rudolph et al. (Eds.): PPSN 2022, LNCS 13398, pp. 223–236, 2022.
https://doi.org/10.1007/978-3-031-14714-2_16

The ability to generate a useful instance-space however depends on the availability of a large set of instances. These instance sets should ideally (1) cover a high proportion of the 2D-space, i.e. instances are diverse with respect to the *features* defining the 2D-space; (2) contain instances on which the portfolio of solvers of interest exhibit discriminatory performance; (3) contain instances that highlight diversity in the *performance-space* (i.e. highlight a range of values for the performance-gap between the winning solver and the next best solver), in order to gain further insight into the relative performance of different solvers.

On the one hand, previous research has focused on evolving new instances that are maximally discriminative with respect to solvers [1,3,13] (i.e. maximise the performance-gap between a target and other solvers), but tend not to have explicit mechanisms for creating instances that are diverse w.r.t feature-space. On the other hand, space-filling approaches [14] directly attempt to fill gaps in the feature-space, but tend not to account for discriminatory behaviour. The main contribution of our work is therefore in proposing an approach based on Novelty Search (NS) [9] that is simultaneously capable of generating a set of instances which are diverse with respect to a feature space *and* exhibit discriminatory but diverse performance with respect to a portfolio of solvers (where diversity in this case refers to variation in the magnitude of the performance gap). The latter results from forcing the search to explore areas of the feature-space in which one solver outperforms others only by a small amount, which would be overlooked by methods that attempt to optimise this. Furthermore, only one run of the method is required to generate the instance set targeted to each particular solver considered, where each solution of the NS corresponds to one KP instance.

We evaluate the approach in the Knapsack Problem (KP) domain, using a portfolio of stochastic solvers (Evolutionary Algorithms - EAs), extending previous work on instance generation which has tended to use deterministic portfolios. Finally, we explain in the concluding section why we believe the method can easily be generalised both to other domains and other solvers.

2   Related Work

The use of EAs to target generation of a set of instances where one solver outperforms others in a portfolio is relatively common. For instance, in the bin-packing domain, Alissa *et al.* [1] evolve instances that elicit discriminatory performance from a set of four heuristic solvers. Plata *et al.* [13] synthesise discriminatory instances in the knapsack domain, while there are multiple examples of this approach to generate instances for Travelling Salesman Problem (TSP) [3,15]. All of these methods follow a similar approach: the EA attempts to maximise the performance gap between a target solver and others in the portfolio. Hence, while the methods are successful in discovering instances that *optimise* this gap, depending on the search-landscape (i.e. number and size of basins of local optima), multiple runs can converge to very similar solutions. Furthermore, these methods focus only on discrimination and therefore there is no pressure to explore the 'feature-space' of the domain. An implicit attempt to address this in TSP is described in [6], in which the selection method of the EA is altered to favour offspring that maintain diversity with respect to a chosen feature, as long as the offspring have a performance gap over a given threshold. Again working in TSP, Bossek *et al.* [3] tackle this issue by proposing novel mutation operators that are designed to provide better exploration of the feature-space however while still optimising for performance-gap. In contrast to the above, Smith-Miles *et al.* [14] describe a method for evolving new instances to directly fill gaps in an *instance-space* which is defined on a 2D plane, with each axis representing a feature derived from the instance data. While this targets filling the instance-space, it does not pay attention to whether the generated instances show discriminatory behaviour on a chosen portfolio.

As noted in the previous section, our proposed approach addresses the above issue using a novelty-search algorithm to generate diverse instances that demonstrate statistically superior performance for a specified target algorithm compared to the other solvers in the portfolio.

3   Novelty Search for Instance Generation: Motivation

NS was first introduced by Lehman *et al.* [9] as an attempt to mitigate the problem of finding optimal solution in deceptive landscapes, with a focus on the control problems. The core idea replaces the objective function in a standard evolutionary search process with a function that rewards novelty rather than a performance-based fitness value to force exploration of the search-space. A 'pure' novelty-search algorithm rewards only novelty: in the case of knapsack instances, this can be defined w.r.t a set of user-defined features describing the instance. However, as we wish to generate instances that are both diverse but also illuminate the region in which a single solver outperforms others in a portfolio, we use a modified form of NS in which the objective function reflects a weighted balance between diversity and performance, where the latter term quantifies the performance difference between a target algorithm and the others in the portfolio.

Given a descriptor x, i.e., typically a multi-dimensional vector capturing features of a solution, the most common approach to quantify novelty of an individual is via the *sparseness* metric which measures the average distance between the individual's descriptor and its k-nearest neighbours. The main motivation behind the usage of descriptors is to obtain a deeper representation of solutions via their features. These features are problem dependent.

The k nearest-neighbours are determined by comparing an individual's descriptor to the descriptors of all other members of the current population and to those stored in an external *archive* of past individuals whose descriptors were highly novel when they originated. Sparseness s is then defined as:

$$s(x) = \frac{1}{k} \sum_{i=0}^{k} dist(x, \mu_i) \tag{1}$$

where μ_i is the ith-nearest neighbour of x with respect to a user-defined distance metric *dist*.

The archive is supplemented at each generation in two ways. Firstly, a sample of individuals from the current population is randomly added to the archive with a probability of 1% following common practice in the literature [17]. Secondly, any individual from the current generation with sparseness greater than a predefined threshold t_a is also added to the archive.

In addition to the archive described above which is used to calculate the sparseness metric that drives evolution, a separate list of individuals (denoted as the solution set) is incrementally built as the algorithm runs: this constitutes the final set of instances returned when the algorithm terminates, again following the method of [17]. At the end of each generation, each member of the current population is scored against the solution set by finding the distance to the nearest neighbour ($k = 1$) in the solution set. Those individuals that score above a particular threshold t_{ss} are added to the solution set. The solution set forms the output of the algorithm.

It is important to note that the solution set does not influence the sparseness metric driving the evolutionary process: instead, this approach ensures that each solution returned has a descriptor that differs by at least the given threshold t_{ss} from the others in the final collection. Finally, both the archive and the solution set grow randomly on each generation depending on the diversity discovered without any limit in their final size.

4   Methods

We apply the approach to generating instances for the KP, a commonly studied combinatorial optimisation problem with many practical applications. The KP requires the selection of a subset of items from a larger set of N items, each with profit p and weight w in such a way that the total profit is maximised while respecting a constraint that the weight remains under the knapsack capacity C. The main motivation behind choosing the KP over other optimisation problems is the lack of literature about discriminatory instance generation for this problem in contrast to other well-known NP-hard problems such as the TSP.

4.1   Instance Representation and Novelty Descriptor

A knapsack instance is described by an array of integer numbers of size $N \times 2$, where N is the dimension (number of items) of the instance of the KP we want to create, with the weights and profits of the items stored at the even and odd positions of the array, respectively. The capacity C of the knapsack is determined for each new individual generated as 80% of the total sum of weights, as using a fixed capacity would tend to create insolvable instances. From each instance, we extract a set of features to form a vector that defines the descriptor used in the sparseness calculation shown in Eq. 1. The features chosen are shown below, i.e. the descriptor is a 8-dimensional vector taken from [13] containing: *capacity of the knapsack; minimum weight/profit; maximum weight/profit; average item efficiency; mean distribution of values between profits and weights ($N \times 2$ integer values representing the instance); standard deviation of values between profits*

Table 1. Parameter settings for EA_{solver}. The crossover rate is the distinguishing feature for each configuration.

Parameter	Value
Population size	32
Max. evaluations	1e5
Mutation rate	1/N
Crossover rate	0.7, 0.8, 0.9, 1.0
Crossover	Uniform crossover
Mutation	Uniform one mutation
Selection	Binary tournament selection

and weights. We evolve fixed size instances containing $N = 50$ items, hence each individual describing an instance contains 100 values describing pairs of (profit, weight). In addition, upper and lower bounds were set to delimit the maximum and minimum values of both profits and weights.[1] All algorithms were written in C++.[2]

4.2 Algorithm Portfolio

While in principle the portfolio can contain any number or type of solvers, we restrict experiments to a portfolio containing four differently configured versions of an EA. Parameter tuning can significantly impact EA performance on an instance [12]. That is the reason why we are interested in addressing the generation of instances for specific EA configurations rather than different heuristics or algorithmic schemes. As a result, it is expected that different configurations of the same approach cover different regions of the instance space. Each EA (EA_{solver}) is a standard generational with elitism GA [10] with parameters defined in Table 1. The four solvers differ only in the setting of the crossover rate, i.e. $\in$ 0.7, 0.8, 0.9, 1.0, which are common values used in the literature.

4.3 Novelty Search Algorithm

The NS approach ($EA_{instance}$), described by Algorithm 1, evolves a population of instances: one run of the algorithm evolves a diverse set of instances that are tailored to a chosen target algorithm. All parameters are given in Table 2. We note that, since $EA_{instance}$ is time-consuming, its population size, as well as its number of evaluations, were set by trying to get a suitable trade-off between the results obtained and the time invested for attaining them.

---

[1] The description of an instance follows the general method of [13], except that they converted the real-valued profits/weights to a binary representation.

[2] The source code, instances generated and results obtained are available in a GitHub repository: https://github.com/PAL-ULL/ns_kp_generation.

Table 2. Parameter settings for $EA_{instance}$ which evolves the diverse population of instances. This approach was executed 10 times for each targeted algorithm for statistical purposes.

Parameter	Value
Knapsack items (N)	50
Weight and profit upper bound	1,000
Weight and profit lower bound	1
Population size	10
Crossover rate	0.8
Mutation rate	$1/(N \times 2)$
Evaluations	2,500, 5,000, 10,000, 15,000
Repetitions (R)	10
Distance metric	Euclidean distance
Neighbourhood size (k)	3
Thresholds (t_a, t_{ss})	3.0

To calculate the fitness of an *instance* in the population (Algorithm 2), two quantities are required: (1) the novelty score measuring the sparseness of the instance and (2) the performance score measuring difference in average performance over R repetitions between the target algorithm and the best of the remaining algorithms. The *novelty* score s (sparseness) for an instance is calculated according to Eq. 1 using the descriptor x detailed in Sect. 4.1, and the Euclidean distance between the vectors as the *dist* function. The *performance* score ps is calculated according to Eq. 2, i.e., the difference between the mean profit achieved in R repetitions by the target algorithm and the maximum of the mean profits achieved in R repetitions by the remaining approaches of the portfolio (where profit is the sum of the profits of items included in a knapsack). The reader should consider that, in order to generate discriminatory instances for different algorithms, the target algorithm must vary from one execution to another. In other words, our approach does not generate biased instances for different algorithms in one single execution.

$$ps = \text{target\_mean\_profit} - max(\text{other\_mean\_profit}) \qquad (2)$$

Finally, the fitness f used to drive the evolutionary process is calculated as a linearly weighted combination of the novelty score s and the performance score ps of an instance, where ϕ is the performance/novelty balance weighting factor.

$$f = \phi * ps + (1 - \phi) * s \qquad (3)$$

5   Experiments and Results

Experiments address the following questions:

Algorithm 1: Novelty Search

Input: N, k, $MaxEvals$, $portfolio$
1 initialise(*population*, N);
2 evaluate(*population*, *portfolio*);
3 archive = $\emptyset$;
4 feature_list = $\emptyset$;
5 **for** $i = 0$ *to MaxEvals* **do**
6   parents = select(*population*);
7   offspring = reproduce(*parents*);
8   offspring = evaluate(*offspring*, *portfolio*, *archive*, k) (Algorithm 2);
9   population = update(*population*, *offspring*);
10   archive = update_archive(*population*, *archive*);
11   solution_set = update_ss(*population*, *solution_set*);
12 **end**
13 **return** *solution_set*

Algorithm 2: Evaluation method

Input: *offspring*, *portfolio*, *archive*, k
1 **for** *instance in offspring* **do**
2   **for** *algorithm in portfolio* **do**
3   apply *algorithm* to solve *instance* R times;
4   calculate mean profit of *algorithm*
5   **end**
6   calculate the novelty score(*offspring*, archive, k) (Equation 1);
7   calculate the performance score(*offspring*) (Equation 2);
8   calculate fitness(*offspring*) (Equation 3);
9 **end**
10 **return** *offspring*

---

1. What influence does the number of generations have on the distribution of evolved instances?
2. To what extent do the evolved instances provide diverse coverage of the instance space?
3. What effect does the parameter ϕ that governs the balance between novelty and performance have on the diversity of the evolved instances?
4. How diverse are the instances evolved for each target with respect to the performance difference between the target algorithm and the best of the other algorithms, i.e. according to Eq. 2?
5. Given a set of instances evolved to be tailored to a specific target algorithm, to what extent is the performance of the target on the set statistically significant compared to the other algorithms in the portfolio?

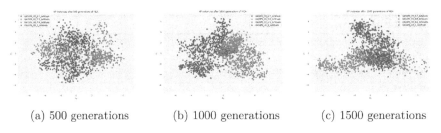

| (a) 500 generations | (b) 1000 generations | (c) 1500 generations |

Fig. 1. Instance representation in a 2D space after applying PCA. Colours reflect the 'winning' algorithm for an instance: red (crossover rate 1.0); green (0.9), orange (0.8), blue (0.7). For more detail about this and following figures please refer to the GitHub repository previously mentioned. (Color figure online)

5.1 Influence of Generation Parameter

$EA_{instance}$ was run for 250, 500, 1,000 and 1,500 generations (2,500, 5,000, 10,000 and 15,000 evaluations, respectively). $EA_{instance}$ was run 10 times for each of the four target algorithms and the instances for each run per target were combined. The parameter ϕ describing the performance/novelty balance was set to 0.6. Principal Component Analysis (PCA) was then applied to the feature-descriptor detailed in Sect. 4.1 to reduce each instance to two dimensions. The results are shown in Fig. 1.

From a qualitative perspective, the most separated clusters are seen for the cases of 500 and 1,000 generations. More overlap is observed when running for too few or too many generations. A plausible explanation often noted in the novelty search literature, e.g. [4], is that if the novelty procedure is run for too long, it eventually becomes difficult to locate novel solutions as the search reaches the boundaries of the feasible space. As a result, the algorithm tends to fill in gaps in the space already explored. On the other hand, considering 250 generations does not allow sufficient time for the algorithm to discover solutions that are both novel and high performing. That is the reason why those results are not shown in Fig. 1. In the remaining experiments, we fix the generations at 1,000.

In order to quantitatively evaluate the extent to which the evolved instances cover the instance space, we calculate the exploration uniformity (U) metric, previously proposed in [7,8]. This enables a comparison of the distribution of solutions in the space with a hypothetical Uniform Distribution (UD). First, the environment is divided into a grid of 25 x 25 cells, after which the number of solutions in each cell is counted. Next, the Jensen-Shannon divergence (JSD) [5] is used to compare the distance of the distribution of solutions with the ideal UD. The U metric is then calculated according to Eq. 4. The higher the U metric score, the more uniformly distributed the instances and better covered the instance space in a given region. Obtaining a score of 1 proves a perfect uniformity distributed set of solutions.

$$U(\delta) = 1 - JSD(P_\delta, UD) \tag{4}$$

Table 3. Average number of instances generated after running $EA_{instance}$ 10 times for each target algorithm, average coverage metric (U) per run, and total number of unique instances obtained from combining the instances over multiple runs with its corresponding coverage metric (U). No duplicated instances were found when comparing the individual's descriptors.

Target	Avg. instances	Avg. U	Tot. instances	Tot. U
GA_0.7	33.4	0.404	334	0.673
GA_0.8	37.6	0.416	376	0.678
GA_0.9	33.9	0.405	339	0.647
GA_1.0	38.3	0.410	383	0.647

In Eq. 4, δ denotes a *descriptor* associated with a solution. Following common practice in the literature and to simplify the computations, this descriptor is defined as the two principal components of each solution extracted after applying PCA to the feature-based descriptor described in Sect. 4.1.

Table 3 summarises the average number of instances generated per run and the average coverage metric U per each target algorithm, as well as the total number of unique instances generated and its corresponding coverage metric U per each target approach. Since the portfolio approaches only differ in the crossover rate, we use the term GA_cr with $cr \in \{0.7, 0.8, 0.9, 1.0\}$ to refer to each target algorithm. Considering the instance space, it can be observed that the method is robust in terms of the number of instances generated, as well as in terms of the corresponding U metric values. Values are similar regardless of the particular target approach for which instances were generated.

5.2 Instance Space Coverage

The left-hand side of Fig. 2 shows results from running $EA_{instance}$ with the performance/novelty weighting factor ϕ set to 1, i.e., the EA only attempts to

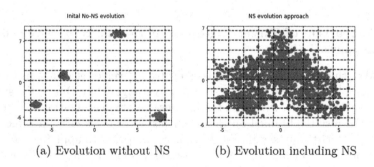

(a) Evolution without NS   (b) Evolution including NS

Fig. 2. Instance representation in a 2D search space after applying PCA comparing two methods of instances generation.

maximise the performance gap and ignores novelty, i.e. an equivalent experiment to that described in [13]. The target algorithm in this case has crossover rate = 0.7. Small groups of clustered instances are observed, with a U value of 0.3772. In contrast, the right-hand side (again with target crossover=0.7) demonstrates that running $EA_{instance}$ with a performance/novelty weighting factor ϕ set to 0.6 clearly results in a large coverage of the space with a corresponding U value of 0.7880.

5.3 Influence of the Balance Between Novelty and Performance

Recall that the evolutionary process in $EA_{instance}$ is guided by Eq. 3, where parameter ϕ balances the contribution of performance and novelty to calculate an individual's fitness. Low values favour feature-diversity, while high values large performance-gaps. We test eight different weighting settings, showing three examples with $\phi \in \{1, 0.7, 0.3\}$ in Fig. 3. The target approach was GA_0.7. Results show that a reasonable compromise is obtained with a performance/novelty balance weighting factor ϕ equal to 0.7: instances are clustered according to the target algorithm while maintaining diversity. As ϕ reduces to favour novelty, as expected, coverage increases at the expense of clustered instances.

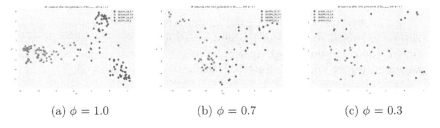

(a) $\phi = 1.0$ (b) $\phi = 0.7$ (c) $\phi = 0.3$

Fig. 3. Instance representation in a 2D search space after applying PCA for three examples of performance/novelty balance weighting factors used to calculate fitness.

5.4 Comparison of Target Algorithm Performance on Evolved Instances

The goal of the approach presented is to evolve a diverse set of instances whose performance is tailored to favour a specific target algorithm. Due to the stochastic nature of the solvers, we conduct a rigorous statistical evaluation to determine whether the results obtained on the set of instances evolved for a target algorithm show statistically significant differences compared to applying each of the other algorithms to the same set of instances (Table 4). First, a *Shapiro-Wilk test* was performed to check whether the values of the results followed a normal (Gaussian) distribution. If so, the *Levene test* checked for the homogeneity of the variances. If the samples had equal variances, an ANOVA *test* was done; if not, a

Table 4. Statistical analysis. A **win** (↑) indicates significance difference between two configurations and that the mean performance value of the target was higher. A **draw** (↔) indicates no significance difference between both configurations. The number of instances generated for each target approach was 90 (GA_0.7), 101 (GA_0.8), 110 (GA_0.9) and 80 (GA_1.0).

	GA_0.7	GA_0.8	GA_0.9	GA_1.0
GA_0.7		↑ 87 ↔ 3	↑ 69 ↔ 21	↑ 25 ↔ 65
GA_0.8	↑ 100 ↔ 1		↑ 77 ↔ 24	↑ 21 ↔ 80
GA_0.9	↑ 107 ↔ 3	↑ 87 ↔ 23		↑ 18 ↔ 92
GA_1.0	↑ 21 ↔ 59	↑ 61 ↔ 19	↑ 76 ↔ 4	

Welch test was performed. For non-Gaussian distributions, the non-parametric *Kruskal-Wallis* test was used [11]. For every test, a significance level $\alpha = 0.05$ was considered. The comparison was carried out considering the mean profits achieved by each approach at the end of 10 independent executions for each instance generated.

For each target approach A in the first column, the number of 'wins' (↑) and 'draws' (↔) of each target algorithm with respect to other approach B is shown. A 'win' means that approach A provides statistically better performance in comparison to approach B, according to the procedure described above, when solving a particular instance. A 'draw' indicates no significant difference. For example, GA_0.7 provides statistically better performance than GA_0.8 in 87 out of the 90 instances generated for the former. Note that in no case did the target algorithm lose on an instance to another algorithm.

For the three algorithms with crossover rates {0.7, 0.8, 0.9} then for the vast majority of instances, the target algorithm outperforms the other algorithms. However for these three algorithms, it appears harder to find diverse instances where the respective algorithm outperforms the algorithm with configuration 1.0. Thus the results provide some insights into the relative strengths and weaknesses of each algorithm in terms of the size of their footprint within the space (approximated by the number of generated instances).

5.5 Performance Diversity

Finally, we provide further insight into the diversity of the evolved instances with respect to the *performance* space (see Figs. 4 and 5). That is, we consider instances that are 'won' by a target algorithm and consider the spread in the magnitude of the performance gap as defined in Eq. 2. We note that the approach is able to generate diverse instances in terms of this metric: while a significant number of instances have a relatively small gap (as seen, for instance, by the left skew to the distribution in Fig. 4), we also find instances spread across the range (see Fig. 5). The instances therefore exhibit performance diversity as well as diversity in terms of coverage of the instance space.

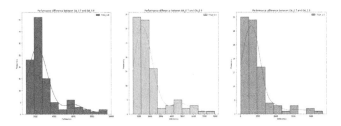

Fig. 4. Histogram showing the distribution of performance gap between the approach GA_0.7 and the remaining approaches by considering the instances generated for the former.

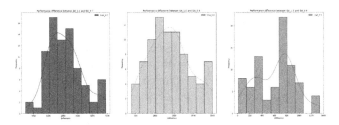

Fig. 5. Histogram showing the distribution of performance gap between the approach GA_1.0 and the remaining approaches by considering the instances generated for the former.

6   Conclusions and Further Research

The paper proposed an NS-based algorithm to generate sets of instances tailored to work well with a specific solver that are diverse with respect to a feature-space and also diverse with respect to the magnitude of the performance-gap between the target solvers and others in a portfolio.

The results demonstrate that the NS-based method provides larges sets of instances that are considerably more diverse in a feature-space in comparison to those generated by an evolutionary method that purely focuses on maximising the performance gap (i.e. following the method of [13]). It also provides instances that demonstrate diversity in the performance space (Figs. 4 and 5). A major advantage of the proposed method is that a single run returns a *set* of diverse instances per target algorithm, in contrast to previous literature for instance generation [1,13,15]) that requires repeated runs due to EA convergence, with no guarantee that repeated runs will deliver unique solutions.

The results also shed new insights into the strengths and weaknesses of the four algorithms used, in terms of the size of their footprint in the instance-space, while also emphasising the benefits of algorithm-configuration. Despite only changing one parameter (crossover rate) per configuration, we are able to generate a large set of instances per configuration that are specifically tailored to that configuration, demonstrating that even small changes in parameter values can lead to different performance.

Although our results are restricted to evolving knapsack instances in conjunction with a portfolio of EA-based approaches for generating solutions, we suggest that the method is generalisable. The underlying core of the approach is an EA to evolve new instances: this has already been demonstrated to be feasible in multiple other domains, e.g. binpacking and TSP [1,3]. Secondly, it requires the definition of a feature-vector: again the literature describes numerous potential features relevant to a range of combinatorial domains[3] At the same time, a basic version of NS was recently used by Alissa *et al.* [2] to evolve instances that are diverse in the *performance-space* for 1D bin-packing, suggesting that other descriptors and other domains are plausible.

Finally, regarding the KP domain, it would be interesting to add the capacity C of the knapsack as a feature of the instances being evolved.

Acknowledgement. Funding from Universidad de La Laguna and the Spanish Ministerio de Ciencia, Innovación y Universidades is acknowledged [2022_0000580]. Alejandro Marrero was funded by the Canary Islands Government "Agencia Canaria de Investigación Innovación y Sociedad de la Información - ACIISI" [TESIS2020010005] and by HPC-EUROPA3 (INFRAIA-2016-1-730897), with the support of the EC Research Innovation Action under the H2020 Programme. This work used the ARCHER2 UK National Supercomputing Service (https://www.archer2.ac.uk).

References

1. Alissa, M., Sim, K., Hart, E.: Algorithm selection using deep learning without feature extraction. In Genetic and Evolutionary Computation Conference (GECCO 2019), 13–17 July 2019, Prague, Czech Republic. ACM, New York (2019). https://doi.org/10.1145/3321707
2. Alissa, M., Sim, K., Hart, E.: Automated algorithm selection: from feature-based to feature-free approaches (2022). https://doi.org/10.48550/ARXIV.2203.13392
3. Bossek, J., Kerschke, P., Neumann, A., Wagner, M., Neumann, F., Trautmann, H.: Evolving diverse TSP instances by means of novel and creative mutation operators. In: Proceedings of the 15th ACM/SIGEVO Conference on Foundations of Genetic Algorithms, pp. 58–71 (2019)
4. Doncieux, S., Paolo, G., Laflaquière, A., Coninx, A.: Novelty search makes evolvability inevitable. In: Proceedings of the 2020 Genetic and Evolutionary Computation Conference. GECCO 2020, pp. 85–93. Association for Computing Machinery, New York (2020). https://doi.org/10.1145/3377930.3389840
5. Fuglede, B., Topsoe, F.: Jensen-Shannon divergence and Hilbert space embedding. In: International Symposium on Information Theory. ISIT 2004. Proceedings, p. 31 (2004). https://doi.org/10.1109/ISIT.2004.1365067
6. Gao, W., Nallaperuma, S., Neumann, F.: Feature-based diversity optimization for problem instance classification. In: Handl, J., Hart, E., Lewis, P.R., López-Ibáñez, M., Ochoa, G., Paechter, B. (eds.) PPSN 2016. LNCS, vol. 9921, pp. 869–879. Springer, Cham (2016). https://doi.org/10.1007/978-3-319-45823-6_81

---

[3] Features for creating instance-spaces in a broad range described in detail from MATILDA https://matilda.unimelb.edu.au/matilda/.

7. Gomes, J., Mariano, P., Christensen, A.L.: Devising effective novelty search algorithms: a comprehensive empirical study. In: GECCO 2015 - Proceedings of the 2015 Genetic and Evolutionary Computation Conference, pp. 943–950 (2015). https://doi.org/10.1145/2739480.2754736

8. Le Goff, L.K., Hart, E., Coninx, A., Doncieux, S.: On Pros and Cons of evolving topologies with novelty search. In: The 2020 Conference on Artificial Life, pp. 423–431 (2020)

9. Lehman, J., Stanley, K.O.: Abandoning objectives: evolution through the search for novelty alone. Evol. Comput. **19**(2), 189–222 (2011)

10. Marrero, A., Segredo, E., Leon, C.: A parallel genetic algorithm to speed up the resolution of the algorithm selection problem. In: Proceedings of the Genetic and Evolutionary Computation Conference Companion. GECCO 2021, pp. 1978–1981. Association for Computing Machinery, New York (2021). https://doi.org/10.1145/3449726.3463160

11. Marrero, A., Segredo, E., León, C., Segura, C.: A memetic decomposition-based multi-objective evolutionary algorithm applied to a constrained menu planning problem. Mathematics **8**(11) (2020). https://doi.org/10.3390/math8111960

12. Nannen, V., Smit, S.K., Eiben, A.E.: Costs and benefits of tuning parameters of evolutionary algorithms. In: Rudolph, G., Jansen, T., Beume, N., Lucas, S., Poloni, C. (eds.) PPSN 2008. LNCS, vol. 5199, pp. 528–538. Springer, Heidelberg (2008). https://doi.org/10.1007/978-3-540-87700-4_53

13. Plata-González, L.F., Amaya, I., Ortiz-Bayliss, J.C., Conant-Pablos, S.E., Terashima-Marín, H., Coello Coello, C.A.: Evolutionary-based tailoring of synthetic instances for the Knapsack problem. Soft. Comput. **23**(23), 12711–12728 (2019). https://doi.org/10.1007/s00500-019-03822-w

14. Smith-Miles, K., Bowly, S.: Generating new test instances by evolving in instance space. Comput. Oper. Res. **63**, 102–113 (2015). https://doi.org/10.1016/j.cor.2015.04.022https://www.sciencedirect.com/science/article/pii/S0305054815001136

15. Smith-Miles, K., van Hemert, J., Lim, X.Y.: Understanding TSP difficulty by learning from evolved instances. In: Blum, C., Battiti, R. (eds.) LION 2010. LNCS, vol. 6073, pp. 266–280. Springer, Heidelberg (2010). https://doi.org/10.1007/978-3-642-13800-3_29

16. Smith-Miles, K.A.: Cross-disciplinary perspectives on meta-learning for algorithm selection. ACM Comput. Surv. **41**(1), 1–25 (2009). https://doi.org/10.1145/1456650.1456656, http://doi.acm.org/10.1145/1456650.1456656

17. Szerlip, P.A., Morse, G., Pugh, J.K., Stanley, K.O.: Unsupervised feature learning through divergent discriminative feature accumulation. In: AAAI 2015: Proceedings of the Twenty-Ninth AAAI Conference on Artificial Intelligence, June 2014. http://arxiv.org/abs/1406.1833

Co-evolutionary Diversity Optimisation
for the Traveling Thief Problem

Adel Nikfarjam[1]([✉]), Aneta Neumann[1], Jakob Bossek[2], and Frank Neumann[1]

[1] Optimisation and Logistics, School of Computer Science,
The University of Adelaide, Adelaide, Australia
`adel.nikfarjam@adelaide.edu.au`
[2] AI Methodology, Department of Computer Science,
RWTH Aachen University, Aachen, Germany

Abstract. Recently different evolutionary computation approaches have been developed that generate sets of high quality diverse solutions for a given optimisation problem. Many studies have considered diversity 1) as a mean to explore niches in behavioural space (quality diversity) or 2) to increase the structural differences of solutions (evolutionary diversity optimisation). In this study, we introduce a co-evolutionary algorithm to simultaneously explore the two spaces for the multi-component traveling thief problem. The results show the capability of the co-evolutionary algorithm to achieve significantly higher diversity compared to the baseline evolutionary diversity algorithms from the literature.

Keywords: Quality diversity · Co-evolutionary algorithms ·
Evolutionary diversity optimisation · Traveling thief problem

1 Introduction

Diversity has gained increasing attention in the evolutionary computation community in recent years. In classical optimisation problems, researchers seek a single solution that results in an optimal value for an objective function, generally subject to a set of constraints. The importance of having a diverse set of solutions has been highlighted in several studies [15,17]. Having such a set of solutions provides researchers with 1) invaluable information about the solution space, 2) robustness against imperfect modelling and minor changes in problems and 3) different alternatives to involve (personal) interests in decision-making. Traditionally, diversity is seen as exploring niches in the fitness space. However, two paradigms, namely quality diversity (QD) and evolutionary diversity optimisation (EDO), have been formed in recent years.

QD achieves diversity in exploring niches in behavioural space. QD maximises the quality of a set of solutions that differ in a few predefined features. Such a set of solutions can aid in the grasp of the high-quality solutions' behaviour in the feature space. QD has a root in novelty search, where researchers seek solutions with new behaviour without considering their quality [10]. For the first time,

© The Author(s), under exclusive license to Springer Nature Switzerland AG 2022
G. Rudolph et al. (Eds.): PPSN 2022, LNCS 13398, pp. 237–249, 2022.
https://doi.org/10.1007/978-3-031-14714-2_17

a mechanism is introduced in [5] to keep best-performing solutions whereby, searching for unique behaviours. At the same time, the MAP-Elites framework was introduced in [4] to plot the distribution of high-performing solutions over a behavioural space. It has been shown that MAP-Elites is efficient in evolving behavioural repertoires. Later, the problem of computing a set of best-performing solutions differing in terms of some behavioural features is formulated and named QD in [22,23].

In contrast to QD, the goal of EDO is to explicitly maximise the structural diversity of a set of solutions that all have a desirable minimum quality. This approach was first introduced in [25] in the context of continuous optimisation. Later, EDO was adopted to generate images and benchmark instances for the traveling salesperson problem (TSP) [3,9]. Star-discrepancy and performance indicators from multi-objective evolutionary optimisation were adopted to achieve the same goals in [14,15]. In recent years EDO was studied in the context of well-known combinatorial optimisation problems, such as the quadratic assignment problem [7], the minimum spanning tree problem [2], the knapsack problem [1], and the optimisation of monotone sub-modular functions [13]. Distance-based diversity measures and entropy have been incorporated into EDO to evolve diverse sets of high-quality solutions for the TSP [6,18]. Nikfarjam et al. [17] introduced an EAX-based crossover focusing on structural diversification of TSP solutions. Most recently, Neumann et al. [12] introduced a co-evolutionary algorithm to find Pareto-front for bi-objective optimisation problem and simultaneously evolve another population to maximise structural diversity.

In this paper, we introduce a co-evolutionary algorithm (Co-EA) to compute two sets of solutions simultaneously; one employs the QD concept and the other evolves towards EDO. We consider the traveling thief problem (TTP) as a well-studied multi-component optimisation problem. QD and EDO have separately been studied in the context of TTP in [19] and [20], respectively. However, the Co-EA has several advantages:

- QD provides researchers with invaluable information about the distribution of best-performing solutions in behavioural space and enables decision-makers to select the best solution having their desirable behaviour. On the other hand, EDO provides us with robustness against imperfect modelling and minor changes in problems. We can benefit from both paradigms by using the Co-EA.
- Optimal or close-to optimal solutions are required in most EDO studies for initialization. The Co-EA eliminates this restriction.
- We expect the Co-EA brings about better results, especially in terms of structural diversity since the previous frameworks are built upon a single solution (the optimal solution). The Co-EA eliminates this drawback.
- The Co-EA benefits from a self-adaptation method to tune and adjust some hyper-parameters during the search improving the results meaningfully.

The remainder of the paper is structured as follows. We formally define the TTP and diversity in Sect. 2. The Co-EA is introduced in Sect. 3. We conduct comprehensive experimental investigation to evaluate Co-EA in Sect. 4. Finally, we finish with concluding remarks.

2   Preliminaries

In this section, we introduce the traveling thief problem and outline different diversity optimisation approaches established for this problem.

2.1   The Traveling Thief Problem

The traveling thief problem (TTP) is a multi-component combinatorial optimization problem. I. e., it is a combination of the classic traveling salesperson problem (TSP) and the knapsack problem (KP). The TSP is defined on a graph $G = (V, E)$ with a node set V of size n and a set of pairwise edges E between the nodes, respectively. Each edge, $e = (u, v) \in E$ is associated with a non-negative distance $d(e)$. In the TSP, the objective is to compute a tour/permutation $x : V \rightarrow V$ which minimizes the objective function

$$f(x) = d(x(n), x(1)) + \sum_{i=1}^{n-1} d(x(i), x(i+1)).$$

The KP is defined on a set of items I with $m := |I|$. Each item $i \in I$ has a profit p_i and a weight w_i. The goal is to determine a selection of items, in the following encoded as a binary vector $y = (y_1, \ldots, y_m) \in \{0, 1\}$, that maximises the profit, while the selected items' total weight does not exceed the capacity $W > 0$ of the knapsack:

$$g(y) = \sum_{j=1}^{m} p_j y_j \text{ s. t. } \sum_{j=1}^{m} w_j y_j \leq W.$$

Here, $y_j = 1$ if the jth item is included in the selection and $y_j = 0$ otherwise.

The TTP is defined on a graph G and a set of items I. Each node i except the first one includes a set of items $M_i \subseteq I$. In TTP, a thief visits each city exactly once and picks some items into the knapsack. A rent R is to be paid for the knapsack per time unit, and the speed of thief non-linearly depends on the weight W_{x_i} of selected items so far. Here, the objective is to find a solution $p = (x, y)$ including a tour x and a packing list (the selection of items) y that maximises the following function subject to the knapsack capacity:

$$z(p) = g(y) - R \left(\frac{d(x(n), x(1))}{\nu_{\max} - \nu W_{x_n}} + \sum_{i=1}^{n-1} \frac{d(x(i), x(i+1))}{\nu_{\max} - \nu W_{x_i}} \right) \text{ s. t. } \sum_{j=1}^{m} w_j y_j \leq W.$$

where $\nu_{\max}$ and $\nu_{\min}$ are the maximal and minimal traveling speed, and $\nu = \frac{\nu_{\max} - \nu_{\min}}{W}$.

2.2   Diversity Optimisation

This study simultaneously investigates QD and EDO in the context of the TTP. For this purpose, two populations P_1 and P_2 co-evolve. P_1 explores niches in the behavioural space and the P_2 maximises its structural diversity subject to a quality constraint. In QD, a behavioural descriptor (BD) is defined to determine to which part of the behavioural space a solution belongs. In line with [19], we consider the length of tours $f(x)$, and the profit of selected items $g(y)$, to serve as the BD. To explore niches in the behavioural space, we propose a MAP-Elites-based approach in the next section.

For maximising structural diversity, we first require a measure to determine the diversity. For this purpose, we employ the entropy-based diversity measure in [20]. Let $E(P_2)$ and $I(P_2)$ denote the set of edges and items included in population P_2. The structural entropy of P_2 defines on two segments, the frequency of edges and items included in $E(P_2)$ and $I(P_2)$, respectively. let name these two segments edge and item entropy and denote them by H_e and H_i. H_e and H_i are calculated as

$$H_e(P_2) = \sum_{e \in E(P_2)} h(e) \text{ with } h(e) = - \left(\frac{f(e)}{\sum_{e \in I} f(e)} \right) \cdot \ln \left(\frac{f(e)}{\sum_{e \in I} f(e)} \right)$$

and

$$H_i(P_2) = \sum_{i \in I(P_2)} h(i) \text{ with } h(i) = - \left(\frac{f(i)}{\sum_{i \in I} f(i)} \right) \cdot \ln \left(\frac{f(i)}{\sum_{i \in I} f(i)} \right)$$

where $h(e)$ and $h(i)$ denote the contribution of edge e and item i to the entropy of P_2, respectively. Also, the terms $f(e)$ and $f(i)$ encode the number of solutions in P_2 that include e and i. It has been shown that $\sum_{e \in E(P_2)} f(e) = 2n\mu$ in [18], where $\mu = |P_2|$, while the number of selected items in P_2 can fluctuate. The overall entropy of P_2 is calculated by summation

$$H(P_2) = H_e(P_2) + H_i(P_2).$$

P_2 evolves towards maximisation of $H(P_2)$ subject to $z(p) \geq z_{\min}$ for all $p \in P_2$. Overall, we maximise the solutions' quality and their diversity in the feature-space through P_1, while we utilise P_2 to maximises the structural diversity.

3   Co-evolutionary Algorithm

This section presents a co-evolutionary algorithm – outlined in Algorithm 1 – to simultaneously tackle QD and EDO problems in the context of TTP. The algorithm involves two populations P_1 and P_2, employing MAP-Elite-based and EDO-based selection procedures.

3.1   Parent Selection and Operators

A bi-level optimisation procedure is employed to generate offspring. A new tour is generated by crossover at the first level; then, $(1 + 1)$ EA is run to optimise the packing list for the tour. The crossover is the only bridge between P_1 and

P_2. For the first parent we first select P_1 or P_2 uniformly at random. Then, one individual, $p_1(x_1, y_1)$ is selected again uniformly at random from the chosen population; the same procedure is repeated for the selection of the second parent $p_2(x_2, y_2)$. To generate a new solution $p'(x', y')$ from $p_1(x_1, y_1)$ and $p_2(x_2, y_2)$, a new tour $x' \leftarrow crossover(x_1, x_2)$ is first generated by EAX-1AB crossover. Edge-assembly crossover (EAX) is a high-performing operator and yields strong results in solving TSP. Nikfarjam et al. [19] showed that the crossover performs decently for the TTP as well.

EAX-1AB includes three steps: It starts with generating a so-called AB-Cycle of edges by alternatively selecting the edges from parent one and parent two. Next, an intermediate solution is formed. Having the first parent's edges copied to the offspring, we delete parent one's edges included in the AB-cycle and add the rest of edges in the AB-cycle. In this stage, we can have either a complete tour or a number of sub-tours. In latter case, we connect all the sub-tours one by one stating from the sub-tour with minimum number of edges. For connecting two sub-tours, we discard one edge from each sub-tour and add two new edges, a 4-tuple of edges. The 4-tuple is selected by following local search by choosing

$$(e_1, e_2, e_3, e_4) = \arg \min\{-d(e_1) - d(e_2) + d(e_3) + d(e_4)\}.$$

Note that if $E(t)$ and $E(r)$ respectively show the set of edges of the intermediate solution t and the sub-tour r, $e_1 \in E(r)$ $e_2 \in E(t) \setminus E(r)$. We refer interested readers to [11] for details on the implementation of the crossover.

Then, an internal $(1 + 1)$ EA is started to optimise a packing list y' for the new tour x' and form a complete TTP solution $p'(x', y')$. The new solution first inherits the first parent's packing list, $y' \leftarrow y_1$. Next, a new packing list is generated by standard bit-flip mutation ($y'' \leftarrow mutation(y')$). If $z(x', y'') > z(x', y')$, the new packing list is replaced with old one, $y' \leftarrow y''$. These steps repeats until an internal termination criterion for the $(1 + 1)$ EA is met. The process of generating a new solution $p'(x', y')$ is complete here, and we can ascend to survival selection.

3.2 Survival Selection Procedures

In MAP-elites, solutions with similar BD compete, and usually, the best solution survives to the next generation. To formally define the similarity and tolerance of acceptable differences in BD, the behavioural space is split into a discrete grid, where each solution belongs to only one cell. Only the solution with the highest objective value is kept in a cell in survival selection. The map not only contributes to the grasp of the high-quality solutions' behaviour but also does maintain the diversity of the population and aids to avoid premature convergence.

In this study, we discretize the behavioural space in the same way [19] did. They claimed that it is beneficial for the computational costs if we focus on a promising portion of behavioural space. In TTP, solely solving either TSP or KP is insufficient to compute a high-quality TTP solution. However, a solution $p(x, y)$ should score fairly good in both $f(x)$ and $g(y)$ in order to result in a high

TTP value $z(p)$. Thus, we limit the behavioural space to the neighbourhood close to optimal/near-optimal values of the TSP and the KP sub-problems. In other words, a solution $p(x, y)$ should result in $f(x) \in [f^*, (1 + \alpha_1) \cdot f^*]$ and $g(y) \in [(1 - \alpha_2) \cdot g^*, g^*]$. Note that f^* and g^* are optimal/near-optimal values of the TSP and the KP sub-problems, and α_1 and α_2 are acceptable thresholds to f^* and g^*, respectively. We obtain f^* and g^* by EAX [11] and dynamic programming [24]. Next, We discretize the space into a grid of size $\delta_1 \times \delta_2$. Cell (i, j), $1 \leq i \leq \delta_1$, $1 \leq j \leq \delta_2$ contains the best solution, with

$$f(x) \in \left[f^* + (i - 1) \cdot \left(\frac{\alpha_1 f^*}{\delta_1} \right), f^* + i \cdot \left(\frac{\alpha_1 f^*}{\delta_1} \right) \right]$$

and

$$g(y) \in \left[(1 - \alpha_2) \cdot g^* + (j - 1) \cdot \left(\frac{\alpha_2 g^*}{\delta_2} \right), (1 - \alpha_2) \cdot g^* + j \cdot \left(\frac{\alpha_2 g^*}{\delta_2} \right) \right].$$

After generating a new solution $p'(x', y')$, we find the cell corresponding with its BD $(f(x'), g(y'))$; if the cell is empty, p' is added to the cell. Otherwise, the solution with highest TTP value is kept in the cell.

Having defined the survival selection of P_1, we now look at P_2's survival selection based on EDO. We add $p'(x', y')$ to P_2 if the quality criterion is met, i. e., $z(p') \geq z_{\min}$. If $|P_2| = \mu + 1$, a solution with the least contribution to $H(P)$ will be discarded.

3.3   Initialisation

Population P_1 only accepts solutions with fairly high BDs $(f(x), g(y))$, and there is a quality constraint for P_2. Random solutions are unlikely to have these characteristics. As mentioned, we use the GA in [11] to obtain f^*; since the GA is a population-based algorithm, we can derive the tours in the final population resulting in a fairly good TSP score. Afterwards, we run the $(1+1)$ EA described above to compute a high-quality packing list for the tours. These packing lists also bring about a high KP score that allows us to populate P_1. Depending on the quality constraint $z_{\min}$, the initial solutions may not meet the quality constraint. Thus, it is likely that we have to initialize the algorithm with only P_1 until the solutions comply with the quality constraint; then, we can start to populate P_2. Note that both parents are selected from P_1 while P_2 is still empty. We stress that in most previous EDO-studies an optimal (or near-optimal solution) was required to be known a-priori and for initialization. In the proposed Co-EA, this strong requirement is no longer necessary.

3.4   Self Adaptation

Generating offspring includes the internal $(1+1)$ EA to compute a high-quality packing list for the generated tour. In [19], the $(1+1)$ EA is terminated after a fixed number of $t = 2m$ fitness evaluations. However, improving the quality of

---

Algorithm 1. The Co-Evolutionary Diversity Algorithm

1: Find the optimal/near-optimal values of the TSP and the KP by algorithms in [11, 24], respectively.

2: Generate an empty map and populate it with the initialising procedure.

3: **while** termination criterion is not met **do**

4:    Select two individuals based on the parent selection procedure and generate offspring by EAX and $(1 + 1)$ EA.

5:    **if** The offspring's TSP and the KP scores are within α_1, and α_2 thresholds to the optimal values of BD. **then**

6:        Find the corresponding cell to the TSP and the KP scores in the QD map.

7:        **if** The cell is empty **then**

8:            Store the offspring in the cell.

9:        **else**

10:            Compare the offspring and the individual occupying the cell and store the best individual in terms of TTP score in the cell.

11:    **if** The offspring complies with the quality criterion **then**

12:        Add the offspring to the EDO population.

13:        **if** The size of EDO population is equal to $\mu + 1$ **then**

14:            Remove one individual from the EDO population with the least contribution to diversity.

---

solutions is easier in the beginning and gets more difficult as the search goes on. Thus, we adopt a similar self-adaptation method proposed in [8,16] to adjust t during the search. Let $Z = \arg\max_{p \in P_1}\{z(p)\}$. Success defines an increase in Z. We discretize the search to intervals of u fitness evaluations. An interval is successful if Z increases; otherwise it is a failure. We reset t after each interval; t decreases if Z increases during the last interval. Otherwise, t increases to give the internal $(1 + 1)$ EA more budget in the hope of finding better packing lists and better TTP solutions. Here, we set $t = \gamma m$ where γ can take any value in $[\gamma_{\min}, \gamma_{\max}]$. We set

$$\gamma := \max\{\gamma \cdot F_1, \gamma_{\min}\} \text{ and } \gamma := \min\{\gamma \cdot F_2, \gamma_{\max}\}$$

in case of success and failure respectively. In our experiments, we use $F_1 = 0.5$, $F_2 = 1.2$, $\gamma_{\min} = 1$, $\gamma_{\max} = 10$, and $u = 2000m$ based on preliminary experiments. We refer to this method as $Gamma_1$.

Moreover, we propose an alternative terminating criterion for the internal $(1 + 1)$ EA, and denote it $Gamma_2$. Instead of running the $(1 + 1)$ EA for $t = \gamma m$, we terminate $(1 + 1)$ EA when it fails in improving the packing list in $t' = \gamma' m$ consecutive fitness evaluations. γ' is updated in the same way as γ. Based on the preliminary experiments, we set $\gamma'_{\min}$ and $\gamma'_{\max}$ to 0.1 and 1, respectively.

4   Experimental Investigation

We empirically study the Co-EA in this section. We run the Co-EA on eighteen TTP instances from [21], the same instances are used in [19]. We first illustrate

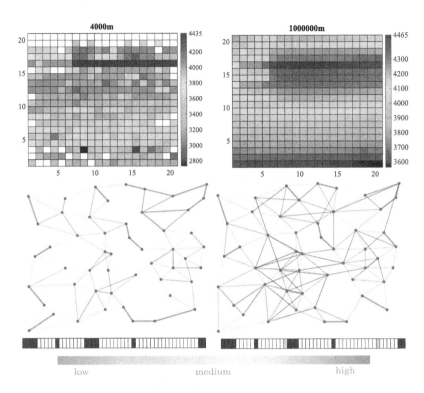

Fig. 1. Evolution of P_1 and P_2 over $4000m$ and $1000000m$ fitness evaluations on instance 1 with $\alpha = 2\%$. The first row depicts the distribution of high-quality solutions in the behavioural space (P_1). The second and the third rows show the overlay of all edges and items used in exemplary P_2, respectively. Edges and items are coloured by their frequency. (Color figure online)

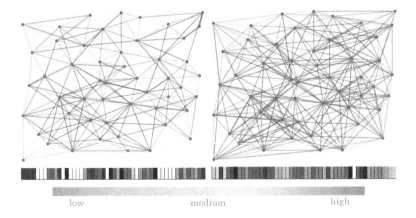

Fig. 2. Overlay of all edges and items used in an exemplary final population P_2 on instance 1 with $\alpha = 10\%$ (left) and $\alpha = 50\%$ (right). Edges and items are coloured by their frequency. (Color figure online)

the distribution of solutions in P_1, and the structural diversity of solutions in P_2. Then, we compare the self-adaptation methods with the fixed parameter setting. Afterwards, we conduct a comprehensive comparison between P_1 and P_2 and the populations obtained by [19] and [20]. Here, the termination criterion and α are set on $1000000m$ fitness evaluations and 10%, respectively.

MAP-Elite selection can be beneficial to illustrate the distribution of high-quality solutions in the behaviour space. On the other hand, EDO selection aims to understand which elements in high-quality solutions is easy/difficult to be replaced. Figure 1 depicts exemplary populations P_1 and P_2 after $4000m$ and $1000000m$ fitness evaluations of Co-EA on instance 1, where $\alpha = 2\%$. The first row illustrates the distribution P_1's high-performing solutions over the behavioural space of $f(x)$ and $g(y)$. The second and the third rows represent the overlay of edges and items in P_2, respectively. The figure shows the solutions with highest quality are located on top-right of the map on this test instance where the gaps of $f(x)$ and $g(y)$ to f^* and g^* are in $[0.0150.035]$ and $[0.150.18]$, respectively. In the second row of the figure, we can observe that Co-EA successfully incorporates new edges into P_2 and reduces the edges' frequency within the population. However, it is unsuccessful in incorporating new items in P_2. The reason can be that there is a strong correlation between items in this particular test instance, and the difference in the weight and profit of items is significant. It means that there is not many other good items to be replaced with the current selection. Thus, we cannot change the items easily when the quality criterion is fairly tight ($\alpha = 2\%$). As shown on the third row of the figure, the algorithm can change i_8 with i_{43} in some packing lists.

Figure 2 reveals that, as α increases, so does the room to involve more items and edges in P_2. In other words, there can be found more edges and items to be included in P_2. Figure 2 shows the overlays on the same instances, where α is set to 10% (left) and 50% (right). Not only more edges and items are included in P_2 with the increase of α, but also Co-EA reduces the frequency of the edges and items in P_2 to such a degree that we can barely see any high-frequent edges or items in the figures associated with $\alpha = 50\%$. Moreover, the algorithm can successfully include almost all items in P_2 except item i_{39}. Checking the item's weight, we notice that it is impossible to incorporate the item into any solution. This is because, $w_{i_{39}} = 4\,400$, while the capacity of the knapsack is set to $4\,029$. In other words, $w_{i_{39}} > W$.

4.1 Analysis of Self-Adaptation

In this sub-section, we compare the two proposed termination criteria and self-adaptation methods $Gamma_1$ and $Gamma_2$ with the fixed method employed in [19]. We incorporate these methods into the Co-EA and run it for ten independent runs. Table 1 summarises the mean of P_2's entropy obtained from the competitors. The table indicates that both $Gamma_1$ and $Gamma_2$ outperform the fixed method on all test instances. Kruskal-Wallis statistical tests at significance level 5% and Bonferroni correction also confirm a meaningful difference in median of results for all instances except instance 15 where there is no significant

Table 1. Comparison of $Gamma_1$ (1) and $Gamma_2$ (2), and the fixed method (3). The instances are numbered as Table 1 in [19]. In columns Stat the notation X^+ means the median of the measure is better than the one for variant X, X^- means it is worse, and X^* indicates no significant difference. Stat shows the results of Kruskal-Wallis statistical test at a significance level of 5% and Bonferroni correction. Also, $p^* = \max_{p \in P_1} \{z(p)\}$.

	$H(P_2)$						$z(p^*)$					
Inst.	$Gamma_1$ (1)		$Gamma_2$ (2)		fixed (3)		$Gamma_1$ (1)		$Gamma_2$ (2)		fixed (3)	
	mean	Stat	mean	Stat	mean	Stat	mean	Stat	mean	Stat	mean	Stat
1	8.7	2^-3^+	8.8	1^+3^+	8.2	1^-2^-	4452.4	2^-3^*	4465	1^+3^*	4461.1	1^*2^*
2	9.3	2^*3^+	9.3	1^*3^+	9.1	1^-2^-	8270.4	2^*3^*	8232.2	1^*3^*	8225.2	1^*2^*
3	9.9	2^+3^+	9.8	1^-3^+	9.6	1^-2^-	13545.4	2^*3^*	13607.5	1^*3^*	13609	1^*2^*
4	7.7	2^-3^+	7.7	1^+3^+	7.4	1^-2^-	1607.1	2^*3^*	1607.5	1^*3^*	1607.5	1^*2^*
5	9	2^*3^+	9	1^*3^+	8.8	1^-2^-	4814.7	2^*3^*	4805.3	1^*3^*	4811	1^*2^*
6	9.4	2^*3^+	9.4	1^*3^+	9.2	1^-2^-	6834.5	2^*3^*	6850	1^*3^*	6850	1^*2^*
7	8	2^*3^+	8.1	1^*3^+	7.6	1^-2^-	3200.8	2^*3^*	3218.4	1^*3^*	3165	1^*2^*
8	9	2^*3^+	9	1^+3^+	8.8	1^-2^-	7854.2	2^*3^*	7854.2	1^*3^*	7850.9	1^*2^*
9	9.5	2^*3^+	9.5	1^*3^+	9.3	1^-2^-	13644.8	2^+3^+	13644.8	1^-3^+	13644.8	1^-2^-
10	10.5	2^-3^+	10.5	1^+3^+	10.1	1^-2^-	11113.6	2^*3^*	11145.7	1^*3^-	11148	1^*2^+
11	11.2	2^*3^+	11.2	1^*3^+	11	1^-2^-	25384.6	2^+3^+	25416.6	1^-3^-	25401.3	1^*2^+
12	9.3	2^*3^+	9.4	1^*3^+	9.2	1^-2^-	3538.2	2^*3^*	3564.4	1^*3^*	3489.4	1^*2^*
13	10.7	2^*3^+	10.7	1^*3^+	10.5	1^-2^-	13369.3	2^+3^+	13310.4	1^-3^*	13338.4	1^*2^*
14	7.7	2^-3^*	9.7	1^+3^+	8.5	1^*2^*	5261.9	2^*3^*	5410.3	1^*3^*	5367.1	1^*2^*
15	10.9	2^*3^+	10.9	1^*3^+	10.7	1^-2^-	20506.8	2^*3^*	20506.8	1^*3^*	20385.3	1^*2^*
16	11.6	2^-3^*	11.7	1^+3^+	11.4	1^*2^*	18622.2	2^*3^*	18609.6	1^*3^*	18641.4	1^*2^*
17	11.2	2^*3^+	11.2	1^*3^+	11.1	1^-2^-	9403.8	2^*3^*	9448.3	1^*3^*	9428.1	1^*2^*
18	11.4	2^*3^+	11.4	1^*3^+	11.1	1^-2^-	19855.3	2^-3^*	19943.8	1^+3^*	19879.3	1^*2^*

difference in the mean of $Gamma_1$ and the fixed method. In comparison between $Gamma_1$ and $Gamma_2$, the latter outperforms the first in 4 test instances, while it is surpassed in only one case. In conclusion, Table 1 indicates that $Gamma_2$ works the best with respect to the entropy of P_2.

Moreover, Table 1 also shows the mean TTP score of the best solution in P_1 obtained from the three competitors. Although Table 1 indicates that the statistical test cannot confirm a significant difference in the mean of the best TTP solutions, $Gamma_2$'s results are slightly better in 7 cases, while $Gamma_1$ and $fixed$ have better results in 3 cases. Overall, all three competitors perform almost equally in terms of the best TTP score. Since $Gamma_2$ outperforms other methods in entropy, we employ it for the Co-EA in the rest of the study.

4.2   Analysis of Co-EA

This section compares P_1 and P_2 with the QD-based EA in [19] and the standard EDO algorithm, respectively. Table 2 summarises this series of experiments. The results indicate that the Co-EA outperforms the standard EDO in 14 instances, while the EDO algorithm has a higher entropy average in only two cases. In the two other test instances, both algorithms performed equally. Moreover, the Co-EA yields competitive results in terms of the quality of the best solution

Table 2. Comparison of the Co-EA and QD from [19] in terms of $z(p^*)$, and EDO algorithm from [20] in $H(P_2)$. Stat shows the results of Mann-Whitney U-test at significance level 5%. The notations are in line with Table 1.

Inst.	Co-EA (1)		QD (2)		Co-EA (1)		EDO (2)	
	Q	Stat	Q	Stat	H	Stat	H	Stat
1	4465	2*	4463.5	1*	8.8	2*	8.6	1*
2	8232.2	2*	8225.7	1*	9.3	2⁻	9.4	1⁺
3	13607.5	2*	13544.9	1*	9.8	2⁻	9.8	1⁺
4	1607.5	2⁺	1607.5	1⁻	7.7	2⁺	7.7	1⁻
5	4805.3	2*	4813.2	1*	9	2*	9	1*
6	6850	2*	6806.8	1*	9.4	2⁺	9.3	1⁻
7	3218.4	2*	3191.9	1*	8.1	2⁺	8	1⁻
8	7854.2	2*	7850.9	1*	9	2*	9	1*
9	13644.8	2⁺	13644.8	1⁻	9.5	2*	9.5	1*
10	11145.7	2⁻	11149.2	1⁺	10.5	2⁺	10.2	1⁻
11	25416.6	2⁻	25555.2	1⁺	11.2	2⁺	11	1⁻
12	3564.4	2*	3514	1*	9.4	2⁺	8.8	1⁻
13	13310.4	2*	13338.6	1*	10.7	2⁺	10.2	1⁻
14	5410.3	2*	5364.6	1*	9.7	2⁺	9.5	1⁻
15	20506.8	2*	20499.2	1*	10.9	2⁺	10.7	1⁻
16	18609.6	2⁻	18666.4	1⁺	11.7	2⁺	11.1	1⁻
17	9448.3	2*	9407.7	1*	11.2	2⁺	10.4	1⁻
18	19943.8	2⁺	19861.8	1⁻	11.4	2⁺	11.1	1⁻

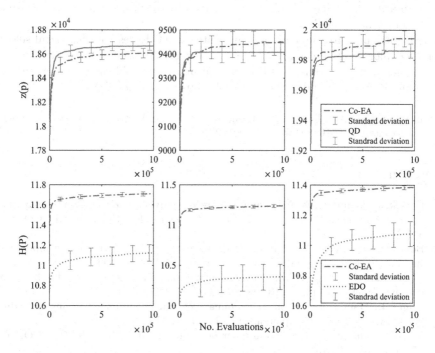

Fig. 3. Representative trajectories of Co-EA and standard EDO EA on instances 16, 17, 18. The top row shows $H(p_2)$ while the second row shows the best solution in P_1.

compared to the QD-based EA; in fact, the Co-EA results in a higher mean of TTP scores on 12 test instances. For example, the best solutions found by Co-EA score 19943.8 on average, whereby the figure stands at 19861.8 for the QD-based algorithms.

Figure 3 depicts the trajectories of Co-EA and the standard EDO algorithm in entropy of the population (the first row), and that of Co-EA and QD-based EA in quality of the best solution (the second row). Note that in the first row the x-axis shows fitness evaluations from $4000m$ to $1000000m$. This is because P_2 is empty in the early stages of running Co-EA and we cannot calculate the entropy of P_2 until $|P_2| = \mu$ for the sake of fair comparison. The figure shows that Co-EA converges faster and to a higher entropy than the standard EDO algorithm. Moreover, it also depicts results obtained by Co-EA has much less standard deviation. Regarding the quality of the best solution, both Co-EA and QD-based EA follow a similar trend.

5   Conclusion

We introduced a co-evolutionary algorithm to simultaneously evolve two populations for the traveling thief problem. The first population explore niches in a behavioural space and the other maximises structural diversity. The results showed superiority of the algorithm to the standard framework in the literature in maximising diversity. The co-evolutionary algorithm also yields competitive results in terms of quality.

It is intriguing to adopt more complicated MAP-Elites-based survival selection for exploring the behavioural space. Moreover, this study can be a transition from benchmark problems to real-world optimisation problems where imperfect modelling is common and diversity in solutions can be beneficial.

Acknowledgements. This work was supported by the Australian Research Council through grants DP190103894 and FT200100536.

References

1. Bossek, J., Neumann, A., Neumann, F.: Breeding diverse packings for the knapsack problem by means of diversity-tailored evolutionary algorithms. In: GECCO, pp. 556–564. ACM (2021)
2. Bossek, J., Neumann, F.: Evolutionary diversity optimization and the minimum spanning tree problem. In: GECCO, pp. 198–206. ACM (2021)
3. Chagas, J.B.C., Wagner, M.: A weighted-sum method for solving the bi-objective traveling thief problem. CoRR abs/2011.05081 (2020)
4. Clune, J., Mouret, J., Lipson, H.: Summary of "the evolutionary origins of modularity". In: GECCO (Companion), pp. 23–24. ACM (2013)
5. Cully, A., Mouret, J.: Behavioral repertoire learning in robotics. In: GECCO, pp. 175–182. ACM (2013)
6. Do, A.V., Bossek, J., Neumann, A., Neumann, F.: Evolving diverse sets of tours for the travelling salesperson problem. In: GECCO, pp. 681–689. ACM (2020)

7. Do, A.V., Guo, M., Neumann, A., Neumann, F.: Analysis of evolutionary diversity optimisation for permutation problems. In: GECCO, pp. 574–582. ACM (2021)
8. Doerr, B., Doerr, C.: Optimal parameter choices through self-adjustment: applying the 1/5-th rule in discrete settings. In: GECCO Companion, pp. 1335–1342 (2015)
9. Gao, W., Nallaperuma, S., Neumann, F.: Feature-based diversity optimization for problem instance classification. Evol. Comput. 29(1), 107–128 (2021)
10. Lehman, J., Stanley, K.O.: Abandoning objectives: evolution through the search for novelty alone. Evol. Comput. 19(2), 189–223 (2011)
11. Nagata, Y., Kobayashi, S.: A powerful genetic algorithm using edge assembly crossover for the traveling salesman problem. INFORMS J. Comput. 25(2), 346–363 (2013)
12. Neumann, A., Antipov, D., Neumann, F.: Coevolutionary Pareto diversity optimization. CoRR arXiv:2204.05457 (2022), accepted as full paper at GECCO 2022
13. Neumann, A., Bossek, J., Neumann, F.: Diversifying greedy sampling and evolutionary diversity optimisation for constrained monotone submodular functions. In: GECCO, pp. 261–269. ACM (2021)
14. Neumann, A., Gao, W., Doerr, C., Neumann, F., Wagner, M.: Discrepancy-based evolutionary diversity optimization. In: GECCO, pp. 991–998. ACM (2018)
15. Neumann, A., Gao, W., Wagner, M., Neumann, F.: Evolutionary diversity optimization using multi-objective indicators. In: GECCO, pp. 837–845. ACM (2019)
16. Neumann, A., Szpak, Z.L., Chojnacki, W., Neumann, F.: Evolutionary image composition using feature covariance matrices. In: GECCO, pp. 817–824 (2017)
17. Nikfarjam, A., Bossek, J., Neumann, A., Neumann, F.: Computing diverse sets of high quality TSP tours by EAX-based evolutionary diversity optimisation. In: FOGA, pp. 9:1–9:11. ACM (2021)
18. Nikfarjam, A., Bossek, J., Neumann, A., Neumann, F.: Entropy-based evolutionary diversity optimisation for the traveling salesperson problem. In: GECCO, pp. 600–608. ACM (2021)
19. Nikfarjam, A., Neumann, A., Neumann, F.: On the use of quality diversity algorithms for the traveling thief problem. CoRR abs/2112.08627 (2021)
20. Nikfarjam, A., Neumann, A., Neumann, F.: Evolutionary diversity optimisation for the traveling thief problem. CoRR abs/2204.02709 (2022)
21. Polyakovskiy, S., Bonyadi, M.R., Wagner, M., Michalewicz, Z., Neumann, F.: A comprehensive benchmark set and heuristics for the traveling thief problem. In: GECCO, pp. 477–484. ACM (2014)
22. Pugh, J.K., Soros, L.B., Stanley, K.O.: Quality diversity: a new frontier for evolutionary computation. Front. Robot. AI 3, 40 (2016)
23. Pugh, J.K., Soros, L.B., Szerlip, P.A., Stanley, K.O.: Confronting the challenge of quality diversity. In: GECCO, pp. 967–974. ACM (2015)
24. Toth, P.: Dynamic programming algorithms for the zero-one knapsack problem. Computing 25(1), 29–45 (1980)
25. Ulrich, T., Thiele, L.: Maximizing population diversity in single-objective optimization. In: GECCO, pp. 641–648. ACM (2011)

Computing High-Quality Solutions for the Patient Admission Scheduling Problem Using Evolutionary Diversity Optimisation

Adel Nikfarjam[1]([✉]), Amirhossein Moosavi[2], Aneta Neumann[1], and Frank Neumann[1]

[1] Optimisation and Logistics, School of Computer Science,
The University of Adelaide, Adelaide, Australia
adel.nikfarjam@adelaide.edu.au
[2] Telfer School of Management, University of Ottawa,
55 Laurier Avenue E, Ottawa, ON K1N 6N5, Canada

Abstract. Diversification in a set of solutions has become a hot research topic in the evolutionary computation community. It has been proven beneficial for optimisation problems in several ways, such as computing a diverse set of high-quality solutions and obtaining robustness against imperfect modeling. For the first time in the literature, we adapt the evolutionary diversity optimisation for a real-world combinatorial problem, namely patient admission scheduling. We introduce an evolutionary algorithm to achieve structural diversity in a set of solutions subjected to the quality of each solution, for which we design and evaluate three mutation operators. Finally, we demonstrate the importance of diversity for the aforementioned problem through a simulation.

Keywords: Evolutionary diversity optimisation · Combinatorial optimisation · Real-world problem · Admission scheduling

1 Introduction

Traditionally, researchers seek a single (near) optimal solution for a given optimisation problem. Computing a diverse set of high-quality solutions is gaining increasing attention in the evolutionary computation community. Most studies consider diversity as finding niches in either fitness landscape or a predefined-feature space. In contrast, a recently introduced paradigm, *evolutionary diversity optimisation* (EDO), explicitly maximises the structural diversity of a set of solutions subjected to constraints on their quality. EDO has shown to be beneficial in several aspects, such as creating robustness against imperfect modeling and minor changes in problems' features [20]. This paradigm was first defined by Ulrich and Thiele [26]. Afterwards, the use of EDO in generating images and *traveling salesperson problem* (TSP) benchmark instances are investigated

© The Author(s), under exclusive license to Springer Nature Switzerland AG 2022
G. Rudolph et al. (Eds.): PPSN 2022, LNCS 13398, pp. 250–264, 2022.
https://doi.org/10.1007/978-3-031-14714-2_18

in [1,13], respectively. The previous studies are extended by incorporation of the concepts of star discrepancy and indicators from the frameworks of multi-objective *evolutionary algorithms* (EAs) into EDO in [17,18]. More recently, diverse TSP solutions are evolved, using distance-based and entropy measure in [8,21]. Nikfarjam et al. [20] introduced A special EAX-based crossover that focuses explicitly on the diversity of TSP solutions. Several studies examine EDO in combinatorial problems, such as the quadratic assignment problem [9], the minimum spanning tree problem [4], the knapsack problem [3], the optimisation of monotone sub-modular functions [16], and the traveling thief problem [23]. Neumann et al. [15] introduced a co-evolutionary algorithm to compute Pareto-front for bi-objective optimisation problems and concurrently evolve another set of solutions to maximise structural diversity. For the first time, we incorporate EDO into a real-world combinatorial problem, namely the *patient admission scheduling* (PAS). PAS is a complex multi-component optimisation scheduling problem in healthcare, involving more features compared to the problems already studied in the literature of EDO.

A recent report on the global health spending of 190 countries shows that healthcare expenditure has continually increased and reached around US$ 10 trillion (or 10% of global GDP) [27]. Due to the ever-increasing demand and healthcare expenditures, there is a great deal of pressure on healthcare providers to increase their service quality and accessibility. Among the several obstacles involved in healthcare resource planning, the PAS problem is of particular significance, impacting organisational decisions at all decision levels [2]. PAS has been studied under different settings, but it generally investigates the allocation of patients to beds such that both treatment effectiveness and patients' comfort are maximised. A benchmark PAS problem is defined in [7], and an online database including 13 benchmark test instances, their best solutions and a solution validator are maintained in [6]. Various optimization algorithms have been proposed for the benchmark PAS problem, such as simulated annealing [5], tabu search [7], mixed-integer programming [2], model-based heuristic [25], and column generation [24]. The simulated annealing in [5] has demonstrated the best overall performance amongst the EAs for the PAS problem. The previous studies aimed for a single high-quality solution for the PAS problem. In contrast, our goal is to diversify a set of PAS solutions, all with desirable quality but some different structural properties.

Having a diverse set of solutions can be beneficial for PAS from different perspectives. PAS is a multi-stakeholder problem and requires an intelligent trade-off between their interests, a challenging issue because (i) stakeholders have conflicting interests, (ii) health departments have diverse admission policies and/or requirements, and (iii) decision-makers (or even health systems) have different values so they could have different decision strategies. A diverse set of high-quality solutions provides stakeholders with different options to reach an agreement. Moreover, most hospitals use manual scheduling methods [11]. Several subtle features are involved in such problems that are not general enough to be considered in the modelling but occasionally affect the feasibility of the optimal solutions (imperfect modelling). Once again, computing a diverse set of solutions can be highly beneficial in providing decision-makers with robustness against imperfect modelling and small changes in problems. Thus, the use of EDO can be a step forward in seeing applications of optimisation methods, particularly EAs, in the practice.

In this study, we first define an entropy-based measure to quantify the structural diversity of PAS solutions. Having proposed an EA maximising diversity, we introduce three variants of a random neighbourhood search operator for the EA. The first variant is a change mutation with fixed hyper-parameters, the second variant uses a self-adaptive method to adjust its hyper-parameters during the search process, and the last variant is a biased mutation to boost its efficiency in maximising diversity of PAS solutions. Finally, we conduct an experimental investigation to (i) examine the performance of the EA and its operators, and (ii) illustrate the effects of EDO in creating robustness against imperfect modelling.

We structure the remainder of the paper as follows. The PAS problem and diversity in this problem are formally defined in Sect. 2. The EA is introduced in Sect. 3. Section 4 presents experimental investigations, and finally, Sect. 5 provides conclusions and some remarks.

2   Patient Admission Scheduling

This section defines the benchmark PAS problem [7] to make our paper self-contained.

Problem Definition

The PAS problem assumes that each patient has a known gender, age, *Length of Stay* (LoS), specialty and room feature/capacity requirements/preferences. Similarly, the set of rooms is known in advance, each associated with a department, gender and age policies, and medical equipment (or features). The problem assumptions for the PAS can be further explained as follows:

- Room and department: Room is the resource that patients require during their treatment ($r \in \mathcal{R}$). Each room belongs to a unique department.
- Planning horizon: The problem includes a number of days where patients must be allocated to a room during their course of treatment ($t \in \mathcal{T}$).
- LoS: Each patient ($p \in \mathcal{P}$) has fixed admission and discharge dates, by which we can specify her LoS ($t \in \mathcal{T}_p$).
- Room capacity (A1): Each room has a fixed and known capacity (a set of beds) (CP_r).
- Gender policy (A2): The gender of each patient is known (either female or male). Each room has a gender policy. There are four different policies for rooms $\{D, F, M, N\}$: (i) rooms with policy F (resp. M) can only accommodate female (resp. male) patients, (ii) rooms with policy D can accommodate both genders, but they should include only one gender on a single day, and

(iii) rooms with policy N can accommodate both genders at the same time. There will be a penalty if cases (i) and (ii) are violated. While CG_{pr}^1 measures the cost of assigning patient p to room r regarding the gender policy for room types F and M, CG^2 specifies the cost of violating the gender policy for room type D.

- Age policy (A3): Each department is specialised in patients with a specific age group. Thus, patients should be allocated to the rooms of a department that respect their age policy. Otherwise, there will be a penalty. CA_{pr} determines the cost of assigning patient p to room r regarding the age policy.
- Department specialty (A4): Patients should be allocated to the rooms of a department with an appropriate specialty level. A penalty occurs if this assumption is violated. CD_{pr} specifies the cost of assigning patient p to room r regarding the department speciality level.
- Room specialty (A5): Like departments, patients should be allocated to rooms with an appropriate specialty level. CB_{pr} defines the penalty of assigning patient p to room r regarding the room speciality level.
- Room features (A6): Patients may need and/or desire some room features. A penalty occurs if the room features are not respected for a patient. Note that the penalty is greater for the required features compared to the desired ones. CF_{pr} shows the combined cost of assigning patient p to room r regarding the room features.
- Room capacity preference (A7): Patients should be allocated to rooms with either preferred or smaller capacity. CR_{pr} specifies the cost of assigning patient p to room r regarding the room capacity preference.
- Patient transfer: There exists a fixed penalty for each time that a patient is transferred during their LoS, which is shown by CT.

Assumption A1 must always be adhered (a hard constraint), while the rest of the assumptions (A2–A7) can be violated with a penalty (soft constraints). Demeester [6] provides complementary information on the calculation of these penalties. The objective function of the PAS problem minimizes the costs of violating assumptions A2–A7 and patient transfers between rooms within their LoS.

To simplify the cost coefficients of the PAS problem, we merge the costs associated with assumptions A2–A7 (except the penalty of violating the gender policy for room type D) into a single matrix CV_{pr} ($CV_{pr} = CA_{pr} + CD_{pr} + CB_{pr} + CF_{pr} + CR_{pr} + CG_{pr}^1$). Then, we formulate the objective function of this problem as below:

$$\text{Min}_{s \in \Xi}\, O(s) = O^1(s, CV_{pr}) + O^2(s, CG^2) + O^3(s, CT)$$

where Ξ denotes the feasible solution space. This objective function minimises: (i) costs of violations from assumptions A2 (for rooms with policies F and M) and A3-A7, (ii) costs of violations from assumption A2 (for rooms with policy D), and (iii) costs of patient transfers.

We use a two-dimensional integer solution representation in this paper. Figure 1 illustrate an example of the solution representation with four patients, two rooms (each with one bed), and a five-day planning horizon. Values within the figure represent the room numbers allocated to patients. With this solution representation, we can compute the above objective function and ensure the feasibility of solutions with respect to the room capacity (the number of times a room appears in each column - each day - must be less than or equal to its capacity). For a mixed-integer linear programming formulation of this problem, interested readers are referred to [7].

	t_1	t_2	t_3	t_4	t_5
p_1	1	1	1	0	0
p_2	0	2	2	2	2
p_3	2	0	0	0	0
p_4	0	0	0	1	1

Fig. 1. An illustrative example of the solution representation

This study aims to maximise diversity in a set of feasible PAS solutions subjected to a quality constraint. Let (i) S refers to a set of PAS solutions, (ii) $H(S)$ denotes diversity of population S, and (iii) c_{max} refers to the maximum acceptable cost of solution s. Then, we define the diversity-based optimization problem as follows:

$$\max H(S)$$

subjected to:

$$O(s) \leq c_{max} \qquad \forall s \in S$$

Align with most studies in the literature of EDO, such as [8,21], we assume that optimal/near-optimal solutions for given PAS instances are prior knowledge. Therefore, we set $c_{max} = (1+\alpha)O^*$, where O^* is the optimal/near-optimal value of $O(s)$, and α is an acceptable threshold for O^*.

2.1 Diversity in Patient Admission Scheduling

In this sub-section, we define a metric for the diversity of PAS solutions. We adopt a measure based on the concept of entropy. Here, entropy is defined on the number of solutions that patient p is assigned to the room r on the day t, n_{prt}. The entropy of population S can be calculated from:

$$H(S) = \sum_{p \in \mathcal{P}} \sum_{r \in \mathcal{R}} \sum_{t \in \mathcal{T}_p} h(n_{prt}) \text{ with } h(x) = -\left(\frac{x}{\mu}\right) \ln\left(\frac{x}{\mu}\right).$$

where, $h(n_{prt})$ is the contribution of n_{prt} to entropy of S, and $\mu = |S|$. Note that $h(n_{prt})$ is equal to zero if $n_{prt} = 0$. We, now, calculate the maximum achievable entropy. If $\mu \leq |\mathcal{R}|$, the maximum entropy (H_{max}) occurs when there are not two solutions in S where a patient p is assigned to the same room r at a day t. In other words, n_{prt} is at most equal to 1 if $H(S)$ is maximum and $\mu \leq |\mathcal{R}|$. Generally, $H \leftarrow H_{max} \iff max(n_{prt}) = \lceil \mu/|\mathcal{R}| \rceil$. Based on the pigeon holds principle, the maximum entropy can be calculated from:

$$H_{max} = W(\mu \mod |\mathcal{R}|)h(\lceil \mu/|\mathcal{R}| \rceil) + W|\mathcal{R}|h(\lfloor \mu/|\mathcal{R}| \rfloor)$$

where W is the total patient-day to be scheduled $(W = \sum_{p \in \mathcal{P}} |\mathcal{T}_p|)$.

3 Evolutionary Algorithm

We employ an EA to maximise the entropy of a set of PAS solutions (outlined in Algorithm 1). The algorithm starts with μ copies of an optimal/near-optimal solution. Having selected a parent uniformly at random, we generate an offspring by a mutation operator. If the quality of the offspring is acceptable, and it contributes to the entropy of the population S more than the parent does, we replace the parent with the offspring; otherwise, it will be discarded. These steps are continued until a termination criterion is met. Since the PAS problem is heavily constrained, we use a low mutation rate in order to maintain the quality of solutions. This results in parents and offspring similar to each other, and there is no point to keep two similar solutions in the population while maximising diversity. Thus, the parent could be only replaced with its offspring.

Algorithm 1. Diversity-Maximising-EA

Require: Initial population S, maximum quality threshold c_{max}
1: **while** The termination criterion is not met **do**
2:   Choose $s \in S$ uniformly at random and generate one offspring s' by the mutation operator
3:   **if** $O(s') \leq c_{max}$ **and** $H(\{S \setminus s\} \cup s') > H(S)$ **then**
4:     Replace s with s'

3.1 Mutation

In this sub-section, we introduce three mutation operators for the EA.

Fixed Change Mutation: This mutation first selects x patients uniformly at random and remove them from the solution. Let $\mathcal{P}_s$ denotes the set of selected patients. For each patient $p \in \mathcal{P}_s$, we identify those rooms that have enough capacity to accommodate them $(\mathcal{R}_p)$. If no room has the capacity for patient p, we terminate the operator and return the parent. Otherwise, we calculate the cost of allocating patient p to room $r \in \mathcal{R}_p$ (C_{pr}). We determine $q(r, p)$ for room r and patient p:

$$q(r, p) = 1 - \frac{C_{pr}}{\sum_{r' \in \mathcal{R}_p} C_{pr'}} \qquad \forall p \in \mathcal{P}_s; r \in \mathcal{R}_p$$

Finally, we use the normalized $q(r, p)$, $\tilde{p}r(r, p)$, and randomly allocate patient p to one of the eligible rooms. The above steps are repeated for all patients in $\mathcal{P}_s$. The fixed change mutation is outlined in Algorithm 2.

Hyper-parameter x should be passed to the operator, which will be tuned later for our problem. It is worth noting that we limit the search to the y best rooms identified based on matrix CV_{pr} for each patient to save computational cost. The standard swap, which has been used frequently in the PAS literature (e.g., see [5]), can be used as an alternative operator. In the standard swap, two patients are selected uniformly at random. Then, their rooms are swept for their whole LoS. While the standard swap might create infeasible offsprings, there is no such an issue for the fixed change mutation.

Algorithm 2. Fixed change mutation

1: Randomly select x patients ($\mathcal{P}_s$)
2: Remove patients in $\mathcal{P}_s$ from allocated rooms
3: **for** p in $\mathcal{P}_s$ **do**
4:     Make a list of candidate rooms R_p that they have the capacity for the LoS of patient p
5:     If R_p is empty, terminate the operator and return the parent
6:     **for** r in R_p **do**
7:         Evaluate the cost of allocating patient p to room r (C_r)
8:         Calculate $q(r, p) = 1 - \frac{C_{pr}}{\sum_{r' \in \mathcal{R}_p} C_{pr'}}$
9:         Normalize $q(r, p)$ ($\tilde{p}r(r, p)$)
10:     Randomly select room r for patient p using probability $\tilde{p}r(r, p)$
11:     Allocate patient p to room r

Self-adaptive Change Mutation: Adjusting hyper-parameters during the search procedure can boost efficiency of operators [10]. Based on initial experiments, we learned that the number of patients to be reallocated (i.e., hyper-parameter x) is the of great importance in the performance of the change mutation. Here, we employ a self-adaptive method to adjust parameter x (similar to [10,19]). As parameter x increases, so does the difference between the parent and its offspring. However, a very large value for x may result in poor quality offspring that do not meet the quality criterion. We assess $H(S)$ every u fitness evaluations. If the algorithm has successfully increased $H(S)$, we raise x to extend the changes in the offspring; otherwise, it usually indicates that offspring are low-quality and cannot contribute to the population. However, we limit x to take values between $[x_{\min}, x_{\max}]$. In successful intervals, we reset $x = \min\{x \cdot F, x_{\max}\}$.

And for unsuccessful intervals, $x = \max\{x_{\min}, x \cdot F^{-\frac{1}{k}}\}$. F and k here determine the adaptation scheme. While $x_{\min} = 4$ and $F = 2$, the rest of these hyperparameters will be tuned by the iRace framework [14].

Biased Change Mutation: Here, we make the fixed change mutation biased towards the entropy metric utilized for diversity maximisation. For a given room r, this new mutation operator selects patient $p \in \mathcal{P}_s$ according to probability $pr'(r,p)$.

$$pr'(r,p) = \frac{\sum_{t \in \mathcal{T}_p} n_{prt}}{\sum_{p' \in \mathcal{P}_s} \sum_{t \in \mathcal{T}_{p'}} n_{p'rt}} \qquad \forall p \in \mathcal{P}_s; r \in \mathcal{R}_p$$

Obviously, the higher $pr'(r,p)$, the higher chance patient p to be selected in the biased mutation. Thus, the most frequent assignments tend to occur less, which results in increasing the diversity metric $H(S)$.

4 Numerical Analysis

In this section, we tend to examine the performance of the EA using the existing benchmark instances. As mentioned earlier, there exist 13 test instances in the literature for the benchmark PAS problem. First, [7] introduced around half of test instances (instances 1–6), then, [6] introduced the other half (instances 7–13). Since test instances 1–6 have been further investigated in the literature, our study focuses on them. The specifications of these instances are reported in Table 1.

Table 1. Specifications of test instances: B, beds; R, rooms; P, patients; TP, total presence; PRC, average patient/room cost; BO, average percentage bed occupancy; SL, average LoS. [5]

In	B	R	D	P	TP	PRC	BO	SL
1	286	98	14	652	2390	32.16	59.69	3.66
2	465	151	14	755	3950	36.74	59.98	5.17
3	395	131	14	708	3156	35.96	57.07	4.46
4	471	155	14	746	3576	38.39	54.23	4.79
5	325	102	14	587	2244	31.23	49.32	3.82
6	313	104	14	685	2821	29.53	64.38	4.12

The diversity-based EA includes parameters whose values must be fully specified before use. The list of all parameters and their potential values for the fixed, adaptive and biased mutations are provided in Table 2. The choice of these parameters are significant for two main reasons. First, such algorithms are not parameter robust and might be inefficient with inappropriate parameter choices [12,22], and second, random choices of parameters would lead to an unfair comparison of the two algorithms [28]. For this problem, we apply the iterated racing package proposed by [14] for automatic algorithm configuration (available on R as the iRace package). After specifying the termination criterion of the iRace package equal to 300 iterations, the best configurations of the EA are found and reported in Table 2. Note that we set the termination criterion for each run of the EA equal to 100, 000 fitness evaluations during the hyper-parameter tuning experiments due to limited computational budget. However, we set this parameter equal to 1, 000, 000 fitness evaluations for the main numerical experiments (one fitness evaluation is equivalent to the computation of $H(S)$ for once). Since F and k are two dependent parameter and together determine the adaptation scheme in the self-adaptive mutation, we set F equal to two based on experiments to make parameter tuning easier. Also, the population size is set to 50 in the EA, and $\alpha \in \{0.02, 0.04, 0.16\}$. We consider 10 independent runs on each instances.

Table 2. Parameter tuning for the EA

Parameter	Type	Range	Elite configuration		
			Fixed	Adaptive	Biased
x	Integer	$[10, 30]$	14	-	-
x_{max}	Integer	$[10, 30]$	-	15	14
k	Categorical	$\{0.25, 0.5, 1, 2, 4, 8, 16\}$	-	8	1
u	Categorical	$\{10, 50, 100, 200, 1000\}$	-	200	200

The underlying aim of increasing $H(S)$ is to assign the patients to as many different rooms as possible. This increases the robustness of the solution population and provides decision-makers with more alternatives to choose. Figure 2 illustrates the distribution of patients over rooms in the final populations obtained by the EA on test instance 1 (first row) and 2 (second row) where $\alpha = 0.02$ (first column) and 0.16 (second column). Each cell represents a patients, and it is coloured based on the number of rooms allocated to them. Red cells are related to those patients with less changeable room assignments (given an acceptable cost threshold). On the other hand, blue cells associate with patients that the EA successfully assigned them to many different rooms. If $H(S) = 0$, all cells are coloured in red (like the initial population). Also, if $H(S) = H_{\max}$, the heat map only includes blue cells. Figure 2 illustrates that the EA successfully diversifies the PAS solutions (several cells are not red), from which stakeholders have access to different alternatives with desirable quality. Also, these solutions can aid with imperfect modeling. For example, the cost of violating assumption $A2$ (i.e., the gender policy) is considered to be equal for all patients in the benchmark test instances. However, cost of such violations

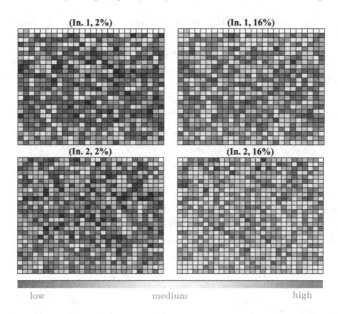

Fig. 2. The distribution of the patients over rooms in the final populations obtained by the introduced EA on test instances 1 and 2 where $\alpha \in \{0.02, 0.16\}$. (Color figure online)

depend on several factors, including - but not limited to - patient's health condition, stakeholder preference and hospital policy. Having said that, it may be impossible to set an appropriate penalty for every single patient. The diverse set of solutions enables decision-makers to take health condition of patients into account.

We now compare the standard swap and three variants of the change mutation. Table 3 summarises the results of the EA obtained using the swap (1), and fixed, self-adaptive and biased change mutations (2–4). In this table, H_{max} represents the maximum entropy where no quality constraint is imposed. It is highly unlikely to achieve H_{max} when the quality constraint is considered, but we use it as an upper-bound for $H(S)$. The table shows the superiority of all variants of the change mutation over the standard swap across all six instances. In fact, the change mutations result in 31% higher entropy on average compared to the standard swap. The self-adaptive mutation provides the average highest $H(S)$ over instances 1, 3, 4 and 6. This is while the fixed mutation and the biased have the best results for instances 2 and 5, respectively. Note that we can observe the same results for all $\alpha \in \{0.02, 0.04, 0.16\}$. All aforementioned observations are confirmed by the Kruskal-Wallis statistical test at a 5% significance level and with Bonferroni correction.

Table 3. Comparison of the standard swap and three variants of the change mutation. Stat shows the results of Kruskal-Wallis statistical test at a 5% significance level with Bonferroni correction. In row Stat, the notation X^+ means the median of the measure is better than the one for variant X, X^- means it is worse, and X^* indicates no significant difference.

		Swap		Fixed	Change	Self-adaptive	Change	Biased	Change	
α	In.	H(S)	Stat (1)	H(S)	Stat (2)	H(S)	Stat (3)	H(S)	Stat (4)	H_{max}
2	1	4865.8	$2^-3^-4^*$	6746.3	$1^+3^*4^*$	6752.4	$1^+2^*4^+$	6679.1	$1^*2^*3^-$	
4	1	4966.3	$2^-3^-4^*$	6895.9	$1^+3^*4^*$	6920.9	$1^+2^*4^+$	6818.9	$1^*2^*3^-$	13488.8
16	1	5426.7	$2^-3^-4^*$	7660	$1^+3^*4^*$	7691.4	$1^+2^*4^+$	7500.1	$1^*2^*3^-$	
2	2	7734.2	$2^-3^-4^*$	11563.3	$1^+3^*4^+$	11559.5	$1^+2^*4^+$	11374.1	$1^*2^-3^-$	
4	2	7889.4	$2^-3^-4^*$	11971.9	$1^+3^*4^*$	11957.3	$1^+2^*4^+$	11718.1	$1^*2^*3^-$	22039.2
16	2	8633.4	$2^-3^-4^*$	13597.5	$1^+3^*4^*$	13633.2	$1^+2^*4^+$	13448.1	$1^*2^*3^-$	
2	3	6000.6	$2^-3^-4^*$	9043.6	$1^+3^*4^*$	9061.5	$1^+2^*4^+$	9011.3	$1^*2^*3^-$	
4	3	6128	$2^-3^-4^*$	9332.7	$1^+3^*4^*$	9353.6	$1^+2^*4^+$	9242	$1^*2^*3^-$	17812
16	3	6706.5	$2^-3^-4^*$	10499.6	$1^+3^*4^+$	10503.3	$1^+2^*4^+$	10381.8	$1^*2^-3^-$	
2	4	6849.9	$2^-3^-4^*$	10273.3	$1^+3^*4^*$	10278.9	$1^+2^*4^+$	10225.2	$1^*2^*3^-$	
4	4	6986.5	$2^-3^-4^*$	10590.5	$1^+3^*4^*$	10620.2	$1^+2^*4^+$	10507.5	$1^*2^*3^-$	20812.4
16	4	6864.3	$2^-3^-4^*$	11989	$1^+3^*4^*$	12016.9	$1^+2^*4^+$	11881.1	$1^*2^*3^-$	
2	5	6446.5	$2^-3^-4^-$	7772	$1^+3^*4^-$	7774.9	$1^+2^*4^-$	7835.6	$1^+2^+3^+$	
4	5	6542.3	$2^*3^-4^-$	8081.4	$1^*3^*4^-$	8090.9	$1^+2^*4^*$	8142.7	$1^+2^+3^*$	12664.8
16	5	7012.7	$2^*3^-4^-$	7775.3	$1^*3^*4^-$	8965.8	$1^+2^*4^*$	9069.3	$1^+2^+3^*$	
2	6	6163.6	$2^-3^-4^*$	8368.2	$1^+3^*4^*$	8381.9	$1^+2^*4^+$	8328.5	$1^*2^*3^-$	
4	6	6293.1	$2^-3^-4^*$	8671.5	$1^+3^*4^*$	8680.9	$1^+2^*4^+$	8574.5	$1^*2^*3^-$	15921.3
16	6	6905.7	$2^-3^-4^*$	9857.8	$1^+3^*4^*$	9865.1	$1^+2^*4^+$	9772.9	$1^*2^*3^-$	

Figure 3 illustrates the trajectories of the average entropy of the population obtained by the EA using different operators over 10 runs. This figure indicates that all variants of the change mutation converge faster and to a higher value of $H(S)$ than the standard swap across all test instances. It also shows that the majority of improvements for the change mutations have occurred in the first $100,000$ fitness evaluations while the gradient is slower for the swap. Among the change mutations, the biased one has the highest $H(S)$ in the first $100,000$ fitness evaluations, but it gets overtaken by the self-adaptive mutation during the search over all instances except instance 5. The same observation (with smaller gap) can be seen between the fixed and self-adaptive mutations, where the fixed mutation is overtaken by the self-adaptive one during the last $200,000$ fitness evaluations except instance 2. Since instance 2 is the largest instance, the self-adaptive mutation may cross the fixed one if the EA runs for a larger number of fitness evaluations. Also, the figure depicts the variance of the entropy of populations obtained by the standard swap is considerably higher than those obtained by the fixed, adaptive, and biased change mutations.

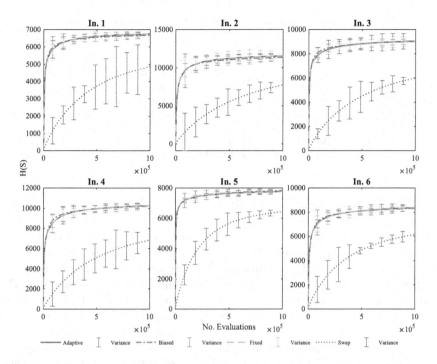

Fig. 3. Representation of trajectories of the EA employing different operators ($\alpha = 2\%$).

Earlier, we mentioned how the EA can provide decision-makers with robustness against imperfect modeling. To evaluate this robustness for population S, for instance, we investigate a scenario that pairs of patients should not share the same room. For this purpose, we randomly choose $b \in \{1, 4, 7\}$ pairs of patients that accommodate the same rooms in the initial population/solution. Repeating this selection for 100 times (e.g., the number of selected pairs is equal to 4×100 if $b = 4$), we then assess: (i) the ratio of times that all patient pairs are allocated to different rooms in at least one of the solutions in the population (Ratio), and (ii) the average number of solutions that all patient pairs occupy different rooms across 100 selections (Alt). Table 4 presents the results for this analysis. As shown in the table, the ratio is always higher than 95%. It also indicates that S includes more than 20 alternatives when $\alpha = 2\%$ and $b = 7$ (the worst case). All this means that the proposed EA can successfully diversify S and increase solution robustness against imperfect modeling.

Table 4. Simulation results for the solution robustness of the EA

b	1						4						7					
α	0.02		0.04		0.16		0.02		0.04		0.16		0.02		0.04		0.16	
In	Ratio	Alt	Ratio	Alt	Ratio	Alt	Ratio	Alt	Ratio	Alt	Ratio	Alt	Ratio	Alt	Ratio	Alt	Ratio	Alt
1	99	45.4	100	43.9	100	45.8	99	32.8	100	31.4	100	38.2	96	21.1	96	21.4	100	28.2
2	100	44.7	100	45.9	100	46.6	100	32	100	31.4	100	36.7	99	21.8	99	24	100	30.1
3	99	43.9	100	45	100	46.1	97	29.9	99	31.6	100	36.3	98	22.5	99	23	100	29.3
4	100	44	100	42.7	100	45.5	100	30.6	100	30.8	100	36.7	99	22.1	99	22.4	100	27.8
5	100	46.1	100	47	100	47.8	100	33.7	100	36.5	100	42.1	99	25.6	100	28.9	100	38
6	99	44.7	99	44.9	100	46.9	98	34.6	99	35.5	100	38.2	96	24	96	27.1	100	33.9

5   Conclusions and Remarks

In this study, we introduced a methodology to compute a highly diverse set of solutions for the PAS problem. We first defined an entropy-based measure to quantify the diversity of PAS solutions. Then, we proposed an EA to maximise the diversity of solutions and introduced three mutations for the problem. The iRace package was used to tune the hyper-parameters of the EA. Through a comprehensive numerical analysis, we demonstrated the efficiency of the proposed mutations in comparison to the standard swap. Our analyses revealed that the EA is capable of computing high-quality and diverse sets of solutions for the PAS problem. Finally, we showed the solution robustness of the proposed EA against imperfect modeling by performing a scenario analysis.

For future studies, it could be interesting to investigate the case that optimal/near-optimal solutions are not known a-priori for the PAS problem. It might be also valuable to apply diversity-based EAs to other real-world optimisation problems in healthcare, such as the operating room planning and scheduling, home-health care routing and scheduling, and nurse scheduling. These problems are usually multi-stakeholder (with conflicting interests) and require high solution robustness.

Acknowledgements. This work was supported by the Australian Research Council through grants DP190103894 and FT200100536.

References

1. Alexander, B., Kortman, J., Neumann, A.: Evolution of artistic image variants through feature based diversity optimisation. In: GECCO, pp. 171–178. ACM (2017)
2. Bastos, L.S., Marchesi, J.F., Hamacher, S., Fleck, J.L.: A mixed integer programming approach to the patient admission scheduling problem. Eur. J. Oper. Res. **273**(3), 831–840 (2019)
3. Bossek, J., Neumann, A., Neumann, F.: Breeding diverse packings for the knapsack problem by means of diversity-tailored evolutionary algorithms. In: GECCO, pp. 556–564. ACM (2021)

4. Bossek, J., Neumann, F.: Evolutionary diversity optimization and the minimum spanning tree problem. In: GECCO, pp. 198–206. ACM (2021)
5. Ceschia, S., Schaerf, A.: Local search and lower bounds for the patient admission scheduling problem. Comput. Oper. Res. **38**(10), 1452–1463 (2011)
6. Demeester, P.: Patient admission scheduling (2021). https://people.cs.kuleuven. be/~tony.wauters/wim.vancroonenburg/pas/
7. Demeester, P., Souffriau, W., De Causmaecker, P., Berghe, G.V.: A hybrid Tabu search algorithm for automatically assigning patients to beds. Artif. Intell. Med. **48**(1), 61–70 (2010)
8. Do, A.V., Bossek, J., Neumann, A., Neumann, F.: Evolving diverse sets of tours for the travelling salesperson problem. In: GECCO, pp. 681–689. ACM (2020)
9. Do, A.V., Guo, M., Neumann, A., Neumann, F.: Analysis of evolutionary diversity optimisation for permutation problems. In: GECCO, pp. 574–582. ACM (2021)
10. Doerr, B., Doerr, C.: Optimal parameter choices through self-adjustment: applying the 1/5-th rule in discrete settings. In: GECCO Companion, pp. 1335–1342 (2015)
11. Durán, G., Rey, P.A., Wolff, P.: Solving the operating room scheduling problem with prioritized lists of patients. Ann. Oper. Res. **258**(2), 395–414 (2016). https:// doi.org/10.1007/s10479-016-2172-x
12. Erfani, B., Ebrahimnejad, S., Moosavi, A.: An integrated dynamic facility layout and job shop scheduling problem: a hybrid NSGA-II and local search algorithm. J. Ind. Manag. Optim. **16**(4), 1801 (2020)
13. Gao, W., Nallaperuma, S., Neumann, F.: Feature-based diversity optimization for problem instance classification. Evol. Comput. **29**(1), 107–128 (2021)
14. López-Ibáñez, M., Dubois-Lacoste, J., Cáceres, L.P., Birattari, M., Stützle, T.: The irace package: iterated racing for automatic algorithm configuration. Oper. Res. Perspect. **3**, 43–58 (2016)
15. Neumann, A., Antipov, D., Neumann, F.: Coevolutionary Pareto diversity optimization. CoRR arXiv:2204.05457 (2022), accepted as full paper at GECCO 2022
16. Neumann, A., Bossek, J., Neumann, F.: Diversifying greedy sampling and evolutionary diversity optimisation for constrained monotone submodular functions. In: GECCO, pp. 261–269. ACM (2021)
17. Neumann, A., Gao, W., Doerr, C., Neumann, F., Wagner, M.: Discrepancy-based evolutionary diversity optimization. In: GECCO, pp. 991–998. ACM (2018)
18. Neumann, A., Gao, W., Wagner, M., Neumann, F.: Evolutionary diversity optimization using multi-objective indicators. In: GECCO, pp. 837–845. ACM (2019)
19. Neumann, A., Szpak, Z.L., Chojnacki, W., Neumann, F.: Evolutionary image composition using feature covariance matrices. In: GECCO, pp. 817–824 (2017)
20. Nikfarjam, A., Bossek, J., Neumann, A., Neumann, F.: Computing diverse sets of high quality TSP tours by EAX-based evolutionary diversity optimisation. In: FOGA, pp. 9:1–9:11. ACM (2021)
21. Nikfarjam, A., Bossek, J., Neumann, A., Neumann, F.: Entropy-based evolutionary diversity optimisation for the traveling salesperson problem. In: GECCO, pp. 600–608. ACM (2021)
22. Nikfarjam, A., Moosavi, A.: An integrated (1, t) inventory policy and vehicle routing problem under uncertainty: an accelerated benders decomposition algorithm. Transp. Lett. 1–22 (2020)
23. Nikfarjam, A., Neumann, A., Neumann, F.: Evolutionary diversity optimisation for the traveling thief problem. CoRR abs/2204.02709 (2022), accepted as full paper at GECCO 2022

24. Range, T.M., Lusby, R.M., Larsen, J.: A column generation approach for solving the patient admission scheduling problem. Eur. J. Oper. Res. **235**(1), 252–264 (2014)

25. Turhan, A.M., Bilgen, B.: Mixed integer programming based heuristics for the patient admission scheduling problem. Comput. Oper. Res. **80**, 38–49 (2017)

26. Ulrich, T., Thiele, L.: Maximizing population diversity in single-objective optimization. In: GECCO, pp. 641–648. ACM (2011)

27. World Health Organization: Global spending on health: weathering the storm (2022). https://www.who.int/publications/i/item/9789240017788

28. Yuan, B., Gallagher, M.: A hybrid approach to parameter tuning in genetic algorithms. In: 2005 IEEE Congress on Evolutionary Computation, vol. 2, pp. 1096–1103. IEEE (2005)

Cooperative Multi-agent Search on Endogenously-Changing Fitness Landscapes

Chin Woei Lim[1], Richard Allmendinger[1]($\boxtimes$) , Joshua Knowles[2] ,
Ayesha Alhosani[1], and Mercedes Bleda[1]

[1] Alliance Manchester Business School, The University of Manchester,
Manchester M15 6PB, UK
{richard.allmendinger,ayesha.alhosani,mercedes.bleda}@manchester.ac.uk
[2] Schlumberger Cambridge Research, Cambridge CB3 0EL, UK

Abstract. We use a multi-agent system to model how agents (representing firms) may collaborate and adapt in a business 'landscape' where some, more influential, firms are given the power to shape the landscape of other firms. The landscapes we study are based on the well-known NK model of Kauffman, with the addition of 'shapers', firms that can change the landscape's features for themselves and all other players. Our work investigates how firms that are additionally endowed with cognitive and experiential search, and the ability to form collaborations with other firms, can use these capabilities to adapt more quickly and adeptly. We find that, in a collaborative group, firms must still have a mind of their own and resist direct mimicry of stronger partners to attain better heights collectively. Larger groups and groups with more influential members generally do better, so targeted intelligent cooperation is beneficial. These conclusions are tentative, and our results show a sensitivity to landscape ruggedness and "malleability" (i.e. the capacity of the landscape to be changed by the shaper firms). Overall, our work demonstrates the potential of computer science, evolution, and machine learning to contribute to business strategy in these complex environments.

Keywords: Cooperative learning · NK models ·
Endogenously-changing landscape · Shaping · Searching · Adaptation

1 Introduction

Most non-trivial social systems are inherently challenging to gauge due to the potential complexity arising from interactions at both individual and collective levels [8]. Especially in the business context, the mechanics of interaction between competing firms (agents) are often based on rather coarse simplifications and an incomplete understanding of the business landscape. The sophistry embedded within the interplays of businesses, difficult to appreciate from the outside, produces counterintuitive resultant behaviours [7].

© The Author(s), under exclusive license to Springer Nature Switzerland AG 2022
G. Rudolph et al. (Eds.): PPSN 2022, LNCS 13398, pp. 265–278, 2022.
https://doi.org/10.1007/978-3-031-14714-2_19

Firms compete by developing new strategies, technologies or business models all of which involve solving complex problems and making a high number of interdependent choices. To solve these problems, managers need to search their firms' business landscapes and find a combination of these choices that allows them to outperform their competitors. Bounded rational managers cannot easily identify the optimal combination, and tend to engage in sequential search processes [13] and, via trial and error, learn and find what combinations are possible and perform well. Effective search for well-performing solutions in a business landscape is thus a source of competitive advantage for companies.

Conceptually, a business landscape dictates the effectiveness of a firm's search strategy by assigning them a fitness, which typically represents the level of return or performance. The active revision of a firm's choices is crucial in maintaining its competitive advantage, growth and profitability when competing against other firms on a business landscape. Such revisions are normally in the form of research and development of any aspect of a firm in order to find better choices (strategies, methods, and/or products), leading it towards a better path, and to higher local peaks on the landscape. Generalising, firms improve their performance by adapting to the business landscape within which they operate. However, actual business landscapes are dynamic, and they tend to change not only exogenously as a result of external factors (changes in government policies and regulations, in demographic and social trends, etc.) but also due to the behaviour and strategies of the firms competing within them. Firms simply do not limit themselves to only adapting and accepting the state of their current environment as it is. Capable firms might be able to *shape* the business landscape to their advantage (in addition to *search* the landscape) [6,17]. A quintessential example of this phenomenon was when Apple introduced the iPhone and swiftly shook the environment in its favour, demolishing Nokia, which was the incumbent cell-phone market leader at that time.

Management research has used the NK model [9,11] introduced in the management literature by [12] to build simulation models to represent business landscapes, and study different factors that influence the effectiveness of companies' search processes (see [2] for a review). Despite the usefulness of these models, most of them consider that the business landscape within which companies compete does not change or it changes in an exogenous manner, i.e. due to factors external to the companies. They thus do not account for the influence that endogenous changes rooted in firms' behaviour have in the business landscape, and in turn in the performance of firms within it. The first simulation model that has analysed companies search effectiveness when business landscapes change endogenously was proposed by Gavetti et al. in 2017 [10]. The authors extend the NK model to consider two types of firms (agents): agents that search a landscape only (referred to as *searchers*) and agents that can both search and shape the landscape (*shapers*). Consequently, searching firms need to adapt (search) on a landscape that is being shaped (changed) by the shaping firms. In other words, shapers have the power to change a landscape endogenously, while searchers perceive these changes as exogenous. Since all agents (shapers and searchers) search the same landscape, a change in the landscape (caused by a shaper) affects all agents. The study of Gavetti et al. [10] focused on studying how the impact of different levels of landscape ruggedness and complexity, and the proportion of shapers vs searchers operating in it, affect the performance of both types of firms. In real contexts, firms do not only compete within a business landscape but in many cases competing firms try to improve their performance by cooperating, i.e. via coopetition. Coopetition is the act of cooperation between competing companies. Businesses that engage in both competition and cooperation are said to be in coopetition. This paper extends Gavetti et al. [10] by allowing firms to cooperate and analyses how cooperation influence firms performance in endogenously changing business landscapes. This is achieved by incorporating cognitive and experiential search into the adaptation process.

The next section details the traditional NK model and the adapted version of that model by Gavetti et al. [10] (also referred to NKZE model), and explains the search rules. Section 3 proposes a cooperative approach with learning, and Sect. 4 provides details about the experimental study and then analyzes the proposed approach for different configurations of the simulated changing (and competitive) business environment. Finally, Sect. 5 concludes the paper, and discusses limitations of the work and areas of future research.

2   Preliminaries

2.1   Kauffman's NK(C) Model

The NK model of Kauffman [11] is a mathematical model of a tunably rugged fitness landscape. The ruggedness is encapsulated by the size of the landscape and the number of local optima, which are controlled by the parameters, N and K, respectively. Formally, in an NK model, the fitness $f(x)$ of an agent (firm) at location $g = (g_1, \ldots, g_N)$, $g_i \in \{0,1\}$, on the landscape can be defined as

$$f(g) = \frac{1}{N} \sum_{i=1}^{N} f_i(g_i, g_{i_1}, \ldots, g_{i_K}), \tag{1}$$

where g_i is the ith (binary) decision variable, and the fitness contribution f_i of the ith variable depends on its own value, g_i, and K other variable values, $g_{i_1}, \ldots, g_{i_K}$. The parameter K has a range of $0 \le K \le N-1$ that determines how many other K different g_i's will be affecting each g_i when computing fitness. The relationships between g_i's are determined randomly and recorded in an *interaction matrix* that shall be left unchanged. The function $f_i : \{0,1\}^{K+1} \to \mathbf{R}$ assigns a value drawn from the uniform distribution in the range $[0,1]$ to each of its 2^{K+1} inputs. The values $i_1, \ldots, i_K$ are chosen randomly (without replacement) from $\{1, \ldots, N\}$. Increasing the parameter K results in more variables interacting with each other, and hence a more rugged (epistatic) landscape. The two extreme cases, $K = 0$ and $K = N-1$, refer to the scenarios where the fitness contributions f_i depend only on g_i (i.e. each f_i can be optimized independently) and all variables, $g_1, \ldots, g_N$, respectively (maximum ruggedness).

Taking an arbitrary firm with a search policy string of $g = (011101)$, we can calculate the fitness contribution (f_i) of g_1 by forming a temporary string with the g_i's that are related to itself by referring to the interaction matrix. Let us assume that, in this example, g_1 was initially and randomly determined to be related to g_2, g_4 and g_5. Since $g_1 = 0$, $g_2 = 1$, $g_4 = 1$ and $g_5 = 0$, the string formed shall be (0110). The fitness contribution can then be extracted from the fitness matrix by taking the value from 6th row (0110 in decimal), and the ith column (1st column in this case). Understandably, the fitness contributions of subsequent g_i's are calculated similarly.

Kauffmann later extended the NK model to introduce coupled landscapes (a.k.a NKC model) [11], which allows multiple species to exist on different landscapes, and interact through a phenomena of niche construction.

2.2   Gavetti et al.'s NKZE Model

The conventional NK model allows firms to continually adapt on a fixed landscape until they reach some local or global optima. Additionally, the action of any firm has no consequence on other competing firms. However, in a realistic and dynamic business environment, the introduction of disruptive technologies and concepts can often drastically restructure the business landscape, thereby needing competing firms to re-strategize towards a new goal or face obsoleteness.

Gavetti et al. [10] introduced the concept of shapers, which have the ability to modify the business context to their own advantage on top of the standard agents (hereinafter known as searchers) in the baseline NK model. They then studied the effects of different levels of shaping on the performance of shapers themselves and on searchers. Unsurprisingly, shaper firms and the level of shaping have great effects on the performance of their competitors as landscape restructuring always tend to undermine competitor performance. However, a high level of shaping coupled with a great number of shapers were found to be highly non-beneficial for shapers and searchers alike, as constant landscape restructuring changes the objective too fast and too much, thus rendering local search obsolete and causing massive performance instabilities.

The key feature of shapers is their ability to influence the business context (hopefully) to its own advantage, and as a side-product, alter the fitness of their competitors. To achieve this, Gavetti et al. [10] extends the NK model with an additional Z (binary) decision variables, $e = (e_1, \ldots, e_Z), e_i \in \{0, 1\}$. Here, e is referred to as the shape policy string and is globally shared by all firms, differently to the search policy string, g, which is controlled by each firm (agent) independently. Similarly to the parameter K in the NK model, Gavetti et al. [10] use a parameter E to interlink the shape and search policy strings: each g_i is related to E randomly sampled e_i's or $e_{i_1} \ldots, e_{i_E}$. Such relationships are also recorded in an interaction matrix that shall be kept constant throughout a run. Notably, E has a range of $0 \leq E \leq Z$.

Accordingly (for the added dimensions), Gavetti et al. [10] update the fitness assignment function to $f_i : \{0, 1\}^{K+1+E} \rightarrow \mathbf{R}$, which assigns a value drawn from the uniform distribution in the range [0,1] to 2^{K+1+E} inputs. In practice, the fitness contributions are stored in a matrix of size $2^{K+1+E} \times N$, and the interactions are stored in a matrix of size $N \times K + 1 + E$. Now, the evaluation of a firm's fitness depends on E too. Taking the search policy string for an arbitrary firm to be $g = (011011)$, global shape policy string to be $e = (101000)$, $K = 3$ and $E = 3$, an interaction matrix was generated randomly, resulting in g_1 being related to g_2, g_4, g_5, e_2, e_3 and e_6. The fitness of contribution of g_1 (f_1) will be determined by firstly forming a temporary string in the order of $(g_1 g_2 g_4 g_5 e_2 e_3 e_6)$, making (0101010). Similarly, f_1 will be taken as the 42nd $(0101010$ in decimal) row and the 1st column (ith column) of the fitness contribution matrix.

Gavetti et al. [10] use the following approach to tackle the NKZE model: At the beginning of the simulation, a predetermined amount of firms will be turned into shapers based on a shaper proportion (β) parameter. Firms are reordered randomly at the beginning of each iteration. More specifically, all firms are allowed to make an action in accordance with the randomly determined order in each iteration. Thus, the number of actions within an iteration would be equal to that of the firm (agent) population. In terms of action, each (shaping and searching) firm is allowed to make one adaption move. A searching firm flips one randomly selected search policy bit (keeping the shaping policy unaltered) and if the resulting policy has a better fitness than the current policy, then the firm retains the new policy; otherwise, the firm will stick with the old policy. However, when it is the turn of a shaper, it has the choice of either adapt as a searcher would without altering the shape policy string, or randomly mutating a single bit of the shape policy string and evaluate fitness with its original unmutated search policy string. A shaper will then pick either choice that is better, or end its turn without adopting any mutation if both the choices were found to be unfit.

Intuitively, E also corresponds to the level of shaping and the malleability of the fitness landscape. A higher E means that the globally shared shape policy string has more influence on fitness contributions, and transitively, the fitness landscape itself. Under this condition, the extent of fitness landscape restructuring when a shaper acts on the shape policy string is high. Thus, the fitness landscape is said to be highly malleable at high E.

The NKZE model is a variation of Kaufmann's NKC model. While both models allow agents to dynamically change the environment, there are critical differences: in the NKC model (i) each species operate on a separate landscape, (ii) all species have the ability to change the landscape, and (iii) each species is represented by one agent only. Consequently, the two models are designed to simulate different (simplified) business environments.

3   Stealthy and Cooperative Learning

The transient environment of the NKZE model caused traditional myopic "hill-climbing" adaptation to underperform. Additionally, such myopic practices hardly capture the rationality of realistic firms. Motivated by this weakness, we extend the methodology to tackle the NKZE model to provide firms with the ability to learn from one another (stealthily or cooperatively), potentially allowing firms to adapt more quickly and adeptly to changing environments.

To implement such ideas, a strategy of exploiting multi-agent search in NKZE with population-based optimisation techniques, specifically particle swarm opti- misation (PSO) [5] and explicit direct memory genetic algorithm [20], was imple- mented. This was done by (i) allowing firms to quickly adapt to the environment by looking towards excellent firms during the exploration phase following con- cepts inspired by PSO (similar to neighborhood search [18]) and (ii) preserving good solutions in a memory and exploiting them at the end of exploration. We will refer to this strategy as stealthy global learning (StealthL). StealthL oper- ates in an idealistic environment where intel regarding the strategies (and success level) of competing firms is always freely and readily available without limitation (i.e., globally). However, such limitation does exist and is inherent to the nature of competition. Additionally, the NKZE and StealthL model do not share simi- lar dynamics, as the former had a single-mutation restriction whereas the latter allowed for very rapid adaptation by mutating multiple elements within policies of firms. The dynamics of NKZE is more realistic as a change in a firm's policy takes time and is limited by resources. A complete or near-complete revamp of policies continuously is not affordable.

As a result, the StealthL model was modified to allow firms to form col- laboration groups. Swarm intelligence and memory scheme were now restricted within the boundary of these groups, thus limiting the amount of information a firm gets. This new model will be referred to as the structured cooperation (StructC) model. Both StealthL and StructC were compared against the stan- dard adaptation used in the NKZE (hereinafter known as the standard model) in the next section. First, we will describe StealthL and StructC in the next two subsections.

3.1   Stealthy Global Learning

Our model of how stealthy learning occurs between firms is based on a simple information-sharing scheme used in the swarm intelligence method, PSO. This is augmented with the use of a memory of past policies, a technique reminiscent of poly-ploid organisms' storage of defunct (inactive) genetic material (chromo- somes) that can be resurrected quickly under environmental stress.

Swarm Intelligence. To implement swarm intelligence, the search policy string of an arbitrary firm was mutated based on a guiding vector that is unique to each firm [21]. Descriptively, the guiding vector of an arbitrary firm is $P = (p_1, \ldots, p_N)$ and has a length of N, matching that of the search policy string $g = (g_1, \ldots, g_N)$. Each element p_i in the guiding vector represents the probability of its corresponding element (g_i) in the search policy string to mutate to 1. Naturally, the probability of which g_i mutates to 0 is given by $1 - p_i$. All guiding vectors were randomly initialised with a uniform distribution with [0,1] range at the beginning. At its turn, the firm first learns towards the search policy string of the current global best performing firm $(g_{maxf,t})$ at time t using its guiding vector with a learning rate of α where $0 \leq \alpha \leq 1$:

$$\boldsymbol{P}_{t+1} = (1 - \alpha) \times \boldsymbol{P}_t + \alpha \times \boldsymbol{g}_{maxf,t} \, .$$

Subsequently, a string of random variates with length N is generated as $\boldsymbol{R} = (r_1, \ldots, r_N)$ using a uniform distribution with [0,1] range. Finally, the new search policy string is determined as follows:

$$g_i = \left\{ \begin{array}{l} 1, r_i < p_i \\ 0, r_i \geq p_i \, . \end{array} \right.$$

Note that p_i has a range of $0.05 \leq p_i \leq 0.95$ to allow for 5% random mutation after convergence. This new adaptation was designed to facilitate fast landscape adaptation via guided multiple mutations. The single random mutation of the shape policy string was kept without alteration to preserve the nature of the landscape-shaping dynamics. Finally, the firm chooses whether to adopt the mutated policy strings or to remain unchanged as in the standard NKZE model.

Learning from Experience, a.k.a. Polyploidy. In addition to swarm intelligence, the model was also extended to memorise the aggregated policy (search + shape) of the best performing firms in each iteration [20]. We limit the size of the database in which these memories are stored to Θ agents. To ensure environmental diversity within the database, newly memorized candidate memory should have a unique shape policy string. If the shape policy string of the candidate is already present in a memory, the fitter one will be adopted. At full capacity, replacement can only happen if an environmentally unique candidate was better than the worse performing memory. To prevent premature memory exploitation, a parameter ε representing the probability of not exploring the database was initialised to 1 at the beginning of the model. A decay parameter $\gamma < 1$ was then set to reduce ε at the end of each iteration using $\varepsilon_{t+1} = \varepsilon_t \times \gamma$.

At the turn of an arbitrary firm, the firm shall only exploit the memory if a random number generated is greater than ε without undergoing any exploration (searching and/or shaping). The firm then, without hesitation, adopts the best policies of the best performing memory. Note that a searcher can only adopt the search policy string of the best performing memory, whilst a shaper adopts both search and shape policy strings simultaneously.

Relevance to Practice. Coopetition emphasizes the mixed-motive nature of relationships in which two or more parties can create value by complementing each other's activity [3]. In the stealth model, organizations in the landscape are all competing to reach the best fitness, but also cooperating by sharing knowledge with each other. This resembles the scenario of organizations helping each other to reach a new common goal, such as tech giants Microsoft and SpaceX working together to explore space technology. In their collaboration, the organizations work together to provide satellite connectivity between field-deployed assets and cloud resources across the globe to both the public and private sector via SpaceX's Starlink satellite network [16]. At the same time, they are competing to dominate niche segments. If this collaboration succeeds, Microsoft and SpaceX will be dominating the space technology market.

3.2   Structured Cooperation

By forming random collaboration groups, firms are now equipped to exchange landscape knowledge amongst their partners, but have no information outside of the group. Thus, (i) a firm can only refer to the best performing policy within its group, and (ii) each group has a separate memory. Mutation-wise, only the element of the search policy string that has the highest probability of mutating can mutate. If multiple were present, one will be selected at random to mutate.

Relevance to Practice. Collaboration between companies can occur in numerous ways, one of which is sharing knowledge and expertise. In today's age, expertise and information are considered valuable strategic assets for organizations [15]. The StructC model mimics sharing knowledge among a closed pre-determined group of companies that falls under the same management/ownership (a.k.a conglomerate). Examples of this type of corporations are, Alphabet LLC who owns Google, DeepMind; Amazon who owns Audible, Amazon Fresh, Ring to list a few. Despite sharing their knowledge and expertise, collaborating companies work towards one goal while maintaining their independence and decision making [14]. Also, knowledge sharing becomes the natural required action for the company to reduce costs and save time, and improve efficiency [22].

4   Experimental Study

This section outlines the model and algorithm parameter settings followed by an analysis of results. Table 1 provides an overview of the default parameter settings for the the three models to be investigated, Standard Model, StealthL and StructC. Parameter N, Z and β were chosen in accordance to [10] for comparison purposes. We simulate a business environment with a population of $M = 10$

Table 1. Default algorithm parameter settings.

Parameter	Default values
N	12
Z	12
M	10
β	50%
α	0.2
$p_{i,ceil}$	0.95
$p_{i,floor}$	0.05
Θ	50
$\varepsilon_{t=0}$	1
γ	0.999
ω_{max}	4

Table 2. StructC group combinations.

$\omega = 1$	$\omega = 2$	$\omega = 3$	$\omega = 4$
1 searcher ($\beta = 0$)	2 searchers ($\beta = 0$)	3 searchers ($\beta =0$)	4 searchers ($\beta =0$)
1 shaper ($\beta = 1$)	1 searcher 1 shaper ($\beta = 0.5$)	2 searchers 1 shaper ($\beta = 0.33$)	3 searchers 1 shaper ($\beta = 0.25$)
-	2 shapers ($\beta = 1$)	1 searcher 2 shapers ($\beta = 0.67$)	2 searchers 2 shapers ($\beta = 0.5$)
-	-	3 shapers ($\beta = 1$)	1 searcher 3 shapers ($\beta = 0.75$)
-	-	-	4 shapers ($\beta = 1$)

agents (firms) and a maximum group size of $\omega_{max} = 4$. Table 2 lists the possible group compositions. The default parameters for the other parameters were set based on preliminary experimentation such that robust results are obtained.

4.1 Experimental Results

We investigate the performance of various models as a function of the problem complexity. We achieve this by visualizing and analysing the model behaviors during the search process. All models were validated using 50 runs with each run using a randomly generated fitness landscapes (same set of landscapes were used for each model) and lasting for 100 iterations. For StructC, group compositions were randomly sampled to preserve generality. As a result, it was ensured that each unique group composition will have appeared 50 times throughout the whole experiment. Each run takes around 2 minutes depending on problem complexity using an Intel i7 (8th gen.) CPU, 8 GB DDR3L RAM.[1]

Standard vs. Stealthy Global Learning. Figures 1, 2 and 3 compare the performance of searchers and shapers for the Standard Model and StealthL as well as for different learning rates (α). Following observations can be made:

Shapers and searchers in StealthL outperform their counterparts in the Standard Model significantly for rugged landscape ($K > 0$) and regardless of the learning rate α. However, for $K = 0$, the Standard Model achieved a better performance because StealthL is suffering from premature convergence caused by its weakened random perturbation, a trade-off of guided learning. Stronger mutation is necessary when peaks on the fitness landscape are rare (or a single peak is present as is the case for $K = 0$) and sufficiently far from one another since StealthL becomes complacent to the point in which exploration is inhibited when its corresponding agents all have roughly good and near solutions.

For StealthL, the performance of searchers and shapers is almost identical regardless of the level of ruggedness K. For the Standard Model, the performance gap between searchers and shapers depends on K with the most significant performance gap being observed for an intermediate level of landscape ruggedness.

The learning rate α has a significant impact on search performance. A decrease in α leads to a slower convergence but an improved final performance (if given sufficient optimization time). A high learning rate of $\alpha > 0.8$ leads to premature convergence. Generally, slightly higher learning rates perform better as the level of landscape ruggedness and malleability increases. Searchers and shapers are affected in a similar fashion by a changing learning rate, while the performance gap between searchers and shapers for a specific learning rate is minimal.

4.2 StructC Model at Low Landscape Ruggedness and Malleability

Figure 4 visualizes the impact of different group sizes on structured cooperation for $K = 0$, $E = 0$. Evidently, being in a bigger group with shapers helped with

---

[1] The code to replicate these experiment can be downloaded at https://github.com/BrandonWoei/NK-Landscape-Extensions.

a searcher's performance. Lone searchers suffered from severe premature convergence with no sign of any improvement. The advantage of having a shaper was especially prominent in a twin group. However, the relationship between shaper proportion and searcher performance was highly inconspicuous and non-linear. A group of size 3 gave the best searcher performance when it only had 1 shaper. The further increase in shapers decreased searcher performance but was still better than the complete searcher group. The results for size 4 groups were even more astonishing, with searchers only groups outperforming groups with 1 or 2 shapers in terms of searcher performance. However, a group with 3 shapers enabled its only searcher to outperform complete searcher groups by a noticeable degree. Whilst shapers are useful to searchers, the benefits were found to be highly dependent on group size and the number of shapers.

Albeit outperforming a lone searcher, a lone shaper suffers a similar premature convergence problem. For twin groups, a complete shaper group gave better shaper performance than a mixed group. For size 3 groups, a complete shaper

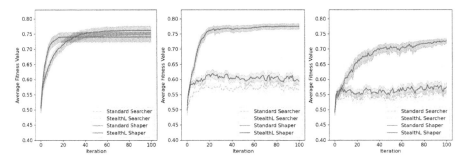

Fig. 1. Average best fitness (and 95% confidence interval in shading calculated using t statistic) obtained by the Standard Model and StealthL at (left plot) minimal ($K = 0$, $E = 0$), (middle plot) intermediate ($K = 5$, $E = 6$), and (right plot) high ($K = 11$, $E = 12$) landscape ruggedness and malleability. StealthL Searcher and Stealth Shaper have almost identical performance.

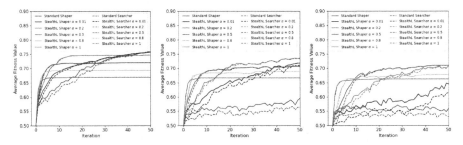

Fig. 2. Varying StealthL learning rate α at extreme malleability ($E = 12$) with (left plot) minimal ($K = 0$) (middle plot) intermediate ($K = 5$) (right plot) high ($K = 12$) landscape ruggedness. The confidence interval of the average fitness value has been omitted in this and following plots for the sake of readability.

gave the best shaper performance. In terms of the mixed groups, groups with 1 searcher were found give better shaper performance than groups with 2 searchers. For size 4 groups, shaper dominated groups gave better shaper performance than the balanced and searcher dominated groups. Unexpectedly, a shaper dominated group with 1 searcher gives slightly better shaper performance than a complete shaper group. Generally, a large group size dominated by shapers ($>75\%$) was

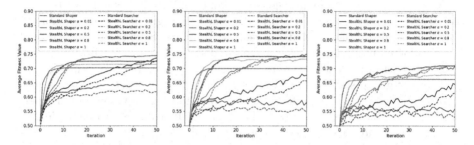

Fig. 3. Varying StealthL learning rate α at extreme ruggedness ($K = 11$) with (left plot) minimal ($E = 0$) (middle plot) intermediate ($E = 6$) (right plot) high ($E = 12$) landscape malleability.

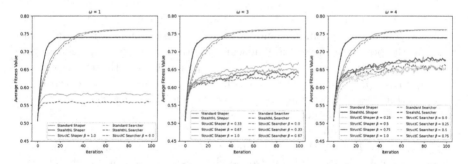

Fig. 4. Average fitness values achieved by StructC for different group sizes at low ruggedness and malleability ($K = 0$, $E = 0$).

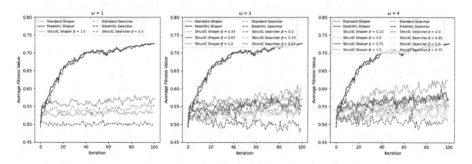

Fig. 5. Average fitness values achieved by StructC for different group sizes at high ruggedness and malleability ($K = 11$, $E = 12$).

found to be group configuration that produced excellent searchers and shapers. Predictably, all groups were not able to surpass the standard and StealthL model performance under all conditions.

4.3   StructC Model at High Landscape Ruggedness and Malleability

Figure 5 visualizes the impact of different group sizes on structured cooperation for $K = 11$, $E = 12$. At high ruggedness and malleability, lone searchers had no improvement from its initial fitness. Mixed twin groups were only able to marginally improve their initial searcher performance. Searcher performance of searcher dominated twin groups was seen to deteriorate slightly with time. Such temporal deterioration effects were also seen in all complete searcher groups regardless of group size. Universally, the increase in group size and shaper proportion improved searcher performance. At maximum group size, searcher performance was able to exceed that of the standard model by a noticeable degree. On the other hand, lone shapers found it difficult to improve themselves. Similarly, a larger group size with more shapers led to excellent shaper performance.

5   Conclusion

We have used a multi-agent system to model how agents (firms) may collaborate and adapt in a business 'landscape' where some, more influential, firms are given the power to *shape* the landscape of other firms. We have found that shapers outperform searchers under all landscape conditions. However, excessive landscape reshaping can lead to poor collective performance due to the instability it introduces. Additionally, both searchers and shapers perform best, even under dynamic business landscapes, when they can keep their organisational complexities at a minimum by reducing relationships between elements of their policies (encoded here by the two parameters K and E). Complex organisations can find it hard to cope with landscape changes especially when the changes are frequent and substantial. However, we found that this can be overcome to some extent by allowing for collaboration via experience-sharing between firms. Whilst mutual learning is beneficial, direct mimicry of best practices can lead to a reduction in collective knowledge as firm' shared inertia hinders exploration, thereby weakening the synergistic effects of collaboration. Lastly, the positive effects of collaboration were also found to be at their best when collaboration groups are dominated by shapers.

Despite having extended NK models, a limitation of our work is that reality is more complex than the abstraction considered here, particularly in terms of the kinds of strategy different firms might employ. Moreover, the model could be made more realistic by, for example, accounting for additional objectives and constraints, observing delays between evaluating an objective/constraint function and having its value available [1,4,19], and simulating more than two types of agents as firms may vary widely e.g. in terms of capabilities, goals, and how they engage with other firms.

References

1. Allmendinger, R., Handl, J., Knowles, J.: Multiobjective optimization: when objectives exhibit non-uniform latencies. Eur. J. Oper. Res. **243**(2), 497–513 (2015)
2. Baumann, O., Schmidt, J., Stieglitz, N.: Effective search in rugged performance landscapes: a review and outlook. J. Manag. **45**(1), 285–318 (2019)
3. Bonel, E., Rocco, E.: Coopeting to survive; surviving coopetition. Int. Stud. Manag. Org. **37**(2), 70–96 (1997)
4. Chugh, T., Allmendinger, R., Ojalehto, V., Miettinen, K.: Surrogate-assisted evolutionary biobjective optimization for objectives with non-uniform latencies. In: Proceedings of the 2018 Annual Conference on Genetic and Evolutionary Computation, pp. 609–616 (2018)
5. Engelbrecht, A.: Particle swarm optimization. In: Proceedings of the Companion Publication of the 2015 Annual Conference on Genetic and Evolutionary Computation. GECCO Companion 2015, pp. 65–91. Association for Computing Machinery, New York (2015). https://doi.org/10.1145/2739482.2756564
6. Felin, T., Kauffman, S., Koppl, R., Longo, G.: Economic opportunity and evolution: beyond landscapes and bounded rationality. Strateg. Entrep. J. **8**(4), 269–282 (2014)
7. Forrester, J.: Urban dynamics. IMR; Ind. Manag. Rev. (pre-1986) **11**(3), 67 (1970)
8. Forrester, J.: Counterintuitive behavior of social systems. Theor. Decis. **2**(2), 109–140 (1971)
9. Ganco, M., Agarwal, R.: Performance differentials between diversifying entrants and entrepreneurial start-ups: a complexity approach. Acad. Manag. Rev. **34**(2), 228–252 (2009)
10. Gavetti, G., Helfat, C.E., Marengo, L.: Searching, shaping, and the quest for superior performance. Strateg. Sci. **2**(3), 194–209 (2017)
11. Kauffman, S.A.: The Origins of Order: Self-organization and Selection in Evolution. Oxford University Press, Oxford (1993)
12. Levinthal, D.: Adaptation on rugged landscapes. Manage. Sci. **43**(7), 934–950 (1997)
13. March, J., Simon, H.: Organizations. Wiley, New York (1958)
14. Rakowski, N., Patz, M.: An Overview and Analysis of Strategic Alliances on the Example of the Car Manufacturer Renault. GRIN Verlag (2009)
15. Sen, Y.: Knowledge as a valuable asset of organizations: taxonomy, management and implications. Manag. Sci. Found. Innov. 29–48 (2019)
16. Sokolowsky, J.: Azure space partners bring deep expertise to new venture (2020). https://news.microsoft.com/transform/azure-space-partners-bring-deep-expertise-to-new-venture/#:~:text=Microsoft%20is%20working%20with%20SpaceX,via%20SpaceX's%20Starlink%20satellite%20network
17. Uzunca, B., Rigtering, J.C., Ozcan, P.: Sharing and shaping: a cross-country comparison of how sharing economy firms shape their institutional environment to gain legitimacy. Acad. Manag. Discov. **4**(3), 248–272 (2018)
18. Wang, H., Sun, H., Li, C., Rahnamayan, S., Pan, J.S.: Diversity enhanced particle swarm optimization with neighborhood search. Inf. Sci. **223**, 119–135 (2013)
19. Wang, X., Jin, Y., Schmitt, S., Olhofer, M., Allmendinger, R.: Transfer learning based surrogate assisted evolutionary bi-objective optimization for objectives with different evaluation times. Knowl.-Based Syst. **227**, 107190 (2021)
20. Yang, S.: Explicit memory schemes for evolutionary algorithms in dynamic environments. In: Yang, S., Ong, Y.S., Jin, Y. (eds.) Evolutionary Computation in

Dynamic and Uncertain Environments. SCI, vol. 51, pp. 3–28. Springer, Heidelberg (2007). https://doi.org/10.1007/978-3-540-49774-5_1

21. Yang, S.: Learning the dominance in diploid genetic algorithms for changing optimization problems. In: Proceedings of the 2nd International Symposium on Intelligence Computation and Applications (2007)

22. Ye, J.: The relations between knowledge sharing among technology centers within a conglomerate, its R&D synergy and performance: a case study of Dongfeng motor corporate. In: International Conference on Global Economy and Business Management (GEBM 2019), pp. 43–53 (2019)

Evolutionary Algorithm for Vehicle Routing with Diversity Oscillation Mechanism

Piotr Cybula[1]([✉])(iD), Andrzej Jaszkiewicz[2](iD), Przemysław Pełka[3](iD),
Marek Rogalski[1](iD), and Piotr Sielski[1](iD)

[1] Faculty of Mathematics and Computer Science, University of Lodz,
Banacha 22, 90-238 Lodz, Poland
`piotr.cybula@wmii.uni.lodz.pl`
[2] Faculty of Computing, Poznan University of Technology,
Piotrowo 3, 60-965 Poznan, Poland
[3] Faculty of Electrical, Electronic, Computer and Control Engineering,
Lodz University of Technology, Stefanowskiego 18/22, 90-924 Lodz, Poland

Abstract. We propose an evolutionary algorithm with a novel diversity oscillation mechanism for the Capacitated Vehicle Routing Problem with Time Windows (CVRPTW). Evolutionary algorithms are among state-of-the-art methods for vehicle routing problems and the diversity management is the key component of many of these algorithms. In our algorithm the diversity level slowly oscillates between its minimum and maximum value, however, whenever a new best solution is found the algorithm switches to decreasing the diversity level in order to intensify the search in the vicinity of the new best solution. We use also an additional population of high quality diverse solutions, which may be used to re-fill the main population when the diversification level is increased. The results of the computational experiment indicate that the proposed mechanism significantly improves the performance of our hybrid evolutionary algorithm on typical CVRPTW benchmarks and that the proposed algorithm is competitive to the state-of-the-art results presented in the literature.

Keywords: Diversity management · Vehicle routing · Hybrid evolutionary algorithms · Multi-population algorithms

1 Introduction

The family of vehicle routing problems (VRP) is an important class of NP-hard combinatorial optimization problems with a high practical importance in transportation and logistics, intensively studied by the research community [4, 17, 33]. In this paper we consider the Capacitated Vehicle Routing Problem with Time Windows (CVRPTW) which is defined by a depot node and a set of customer nodes with defined demands. For each node, a non-negative service time and

G. Rudolph et al. (Eds.): PPSN 2022, LNCS 13398, pp. 279–293, 2022.
https://doi.org/10.1007/978-3-031-14714-2_20

a time window is defined. For each pair of nodes (*edge*), a travel time and a travel distance are defined. A fleet of homogeneous vehicles with limited capacity is available. The goal is to find a set of routes that start and end at the depot, so that each customer node is visited exactly once, the sum of demands for each route does not exceed the vehicle capacity, and the time windows are respected. CVRPTW is usually formulated as a bi-objective problem with two lexicographically ordered objectives. The first objective is to minimize the number of routes and the second one is to minimize the total travel distance. Evolutionary algorithms (EAs) are among state-of-the-art methods to efficiently solve vehicle routing problems [17,20,22,35].

Premature convergence is well-known to be one of the major obstacles for EAs caused by the loss of population diversity that leads to getting trapped in suboptimal solutions [7,9]. To avoid this obstacle additional techniques promoting population diversity are often used [31,32]. For example, Squillero and Tonda [31] describe 25 diversity management techniques, e.g. clone extermination, crowding, fitness sharing, clearing, and island model. In particular in clearing the best individuals within a clearing radius are declared 'winners' and other individuals have their fitness cleared [24]. The risk of premature convergence naturally affects also EAs for VRP. Thus, diversity management techniques are often used in successful algorithms for VRP (see Sect. 2).

In this paper, we propose a novel diversity management scheme based on clearing with adaptive clearing radius. The clearing radius is an additional parameter of an EA. Larger values of the radius promote diversity (exploration), while lower values promote exploitation (intensification). Instead of trying to define a single value of this parameter we interlace the phases of exploration increasing and exploration decreasing. In the exploration increasing phase the clearing radius is gradually increased while in the exploration decreasing phase the radius is gradually decreased.

The main contributions of this work are summarized as follows.

- A novel diversity management mechanism based on the idea of oscillation of the diversity level with a dual-population model is proposed.
- The proposed diversity management mechanism is used in a hybrid evolutionary algorithm for the CVRPTW.
- Through a computational experiment it is shown that the proposed diversity management mechanism significantly improves results of the evolutionary algorithm.
- The computational experiment shows also that the results of our algorithm are competitive to the state-of-the-art results for the CVRPTW reported in the literature.

The paper is organized in the following way. In the next section we review the use of diversity management mechanisms in evolutionary algorithms for VRP. In Sect. 3, we present the proposed diversity management mechanism and the hybrid evolutionary algorithm. Then, in Sect. 4, we describe the order-based recombination used in the algorithm. Local search used in the algorithm is presented in Sect. 5. In Sect. 6, we describe and present results of a computational experiment. In the last section, we present conclusions and directions for further research.

2   Diversity Management in Vehicle Routing Problems

As previously stated, diversity management techniques are often used in successful EAs for VRP. Vidal et al. [35,36] proposed a hybrid genetic algorithm with adaptive diversity management for a large class of vehicle routing problems. Each individual in the population is characterized by both solution cost and its diversity contribution defined as the average distance to its closest neighbours. The evaluation biased fitness involves both the rank based on the solution cost and the rank based on the diversity contribution. In addition, a number of elite solutions are preserved, chosen randomly from among 25% of the best solutions. The authors conclude that the trade-off between diversity and elitism is critical for a thorough and efficient search. This mechanism became highly popular in algorithms for VRP and has also been applied in [2,5,10,13].

Prins [25] does not allow clones in the population. However, to speed-up the clone detection process clones are defined not in the decision space but in the objective space.

Segura et al. [30] use a multi-objective replacement strategy. First, they combine the population from the previous generation and the offspring in a temporary set. The best individual is selected to form part of the new population. Then, until the new population is filled with the required number of individuals, the following steps are executed. First, the distance to the closest neighbor is calculated. The calculation considers the currently selected individuals as the reference, i.e. for each pending individual the distance to the nearest individual previously selected is taken into account. Then, considering the individuals that have not been selected, the non-dominated front is calculated. The two objectives considered are the cost and the distance to the closest neighbor. This front is computed as a set with no repetitions. Finally, a non-dominated individual is randomly selected to survive.

Xu et al. [37] use a measure to evaluate population diversity. Once population diversity is below a given level, the evolutionary algorithm switches to the simulated annealing, which could avoid the drawback of premature convergence. The experimental data show the effectiveness of the algorithm and authenticate the search efficiency and solution quality of this approach.

Zhu [38] manages diversity by varying crossover and mutation rates. The authors observe, that increasing crossover and mutation rates promotes diversity and delays the convergence of the algorithm.

Sabar et al. [29] vary fitness function that takes into account both the fitness and diversity of each solution in the population. Furthermore, the authors propose to allow the EA to focus on exploration in the early stages and then gradually switch to exploitation.

Sabar et al. [28], in order to maintain solution diversity, utilize a memory mechanism that contains a population of both high-quality and diverse solutions that is updated during the problem-solving process.

Qu and Bard [26] assure high diversity of the initial population by generating a higher number of randomized solutions. Then a number of best solutions is selected based on the objective function and the remaining solutions are selected by solving the maximum diversity problem.

Repoussis et al. [27] initially place all solutions from the parent population in the new population. Then, each offspring is compared to the corresponding parent and if the offspring performs better, then it replaces the parent. Next, the best and the worst individuals are identified and the median similarity in the new population is determined. Subsequently, the remaining offsprings (if any) are competing against individuals currently placed into the new population with respect to the level of similarity. Precisely, if the offspring performs better than an existing solution and its level of similarity is lower than the current median, then the offspring replaces this solution. In addition, if the offspring performs better than any selected individual and its level of similarity is lower then the offspring also replaces this selected solution.

Liu et al. [18] divide the population equally into two subpopulations of winners and losers of a binary tournament. Then they use two different mutation strategies in the two subpopulations. The mutation strategy applied to winners is focused on exploitation, while the mutation strategy applied to losers is focused on exploration.

Boudia et al. [3] accept a new solution to replace a solution in the current population only if the minimum distance of the new solution to a solution in the current population exceeds a given threshold Δ. Furthermore, a decreasing policy is used to vary Δ. Starting from an initial value Δ_{max}, Δ is linearly decreased to reach 1 (the minimum value to avoid duplicate solutions) at the end of each cycle. In addition, after completion of each cycle, Boudia et al. use a renewal procedure which keeps the best solution and replaces the other solutions by new random ones. The acceptance threshold is then set to the initial, maximum value. This approach is perhaps the most similar to ours, however, we do not increase the diversity level immediately after achieving its minimum level, but rather we start to increase it slowly. Furthermore, we use a different population update mechanism based on clearing and we do not use a similar population renewal procedure but we rather try to preserve the current population.

3   The Diversity Management and the Evolutionary Algorithm

As it is clear from the previous section, diversity management mechanisms are typically used in EAs for VRP. The diversification mechanisms usually involve some parameters promoting either diversity (exploration) or elitism (exploitation, intensification). It is well acknowledged that the trade-off between exploration and exploitation has a crucial influence on the performance of EAs and other metaheuristics [15]. Too strong focus on exploitation may result in a hill-climber type algorithm which quickly becomes trapped in a local optimum. Too much exploration may result in a random search. Thus, the parameters of the diversity mechanisms may have a crucial influence on the performance of EAs for VRP. Some authors, instead of trying to set single values of such parameters, vary them during the run such that in the initial iterations the algorithm is more focused on exploration and gradually switches to exploitation [3,29]. This strategy is well-known from simulated annealing [16] and it has been also recently applied in a successful large neighborhood search for VRP [6]. However, the strategy exploration → exploitation may still lead to a convergence to suboptimal solutions. If such premature convergence happens it may be beneficial to increase the diversity of the current population in order to explore other regions of the search space. On the other hand, if the best solution has been improved, there is no need to increase the diversity level or it could be further reduced in order to intensify the search in the new promising region.

Thus, we propose to interlace the phases of increasing and decreasing exploration in our algorithm. The mechanism is based on the idea of clearing with a parameter clearing radius. In the exploration increasing phase, the clearing radius is gradually increased, while in the exploration decreasing phase the clearing radius is gradually decreased. The clearing radius is not changed if the algorithm improves the best solution in a given number of iterations, because we interpret this situation as an indicator that the algorithm is capable of finding new best solutions at the current diversification level. Otherwise, if the best solution has not been improved in a given number of iterations, the clearing radius is either increased or decreased, depending on the current phase of exploration decreasing/increasing. The phases are changed when the clearing radius achieves the minimum or maximum value. In addition, whenever a new best solution is found during the exploration increasing phase, we also switch to the exploration decreasing phase in order to intensify the search in the new promising region just found. In other words, the diversity level slowly oscillates between its minimum and maximum value, however, whenever a new best solution is found we switch to decreasing the diversity level.

As the distance measure we use the number of different edges in two solutions normalized to the range between 0 (equal solutions) to 1 (no common edges).

Figure 1 shows an exemplary evolution of the clearing radius and the best objective value in 1700 main iterations of the evolutionary algorithm (each main iteration involves k_{max} recombinations) for R1_10_8 instance (see Sect. 6). One can see that the clearing radius rarely achieves the maximum allowed value (0.15), that there are periods of search at relatively low diversity levels (periods of a high intensification) when the best solution is often improved, and that the diversity level increases when the search stagnates i.e. when the best solution is not improved for a longer time. Note that the best objective function value shown in this figure involves only the distance (since the number of vehicles was quickly reduced to the best known value) and constraint penalties with the current weights. Thus, the best solution may improve both due to the improvement of the distance and due to the reduction of constraint violations.

The proposed oscillation mechanism partially resembles the ideas of iterated local search (ILS) [19] and very large-scale neighborhood search (VLNS) [1]. Both algorithms alternate the phases of high intensification (local search or

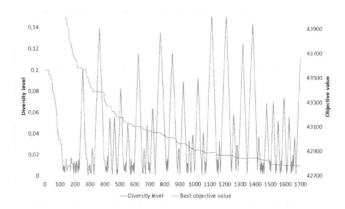

Fig. 1. Exemplary evolution of the clearing radius in 1700 iterations of the evolutionary algorithm.

greedy heuristics) with diversity increasing mechanisms (perturbation or destroy operators). However, these algorithms work with a single solution and the phases of intensification/exploration are much more fine-grained, while in our case oscillations are applied to the whole population and each of them involves multiple iterations of recombination and local search.

Motivated by Prins [25] we initialize current population with random solutions and apply local search only to several of them. Furthermore, similarly to [28] we maintain an additional small population (called reserve) of few best solutions for the maximum diversification level (the maximum clearing radius). After increasing the diversification level we try to add solutions from the reserve to the main population. Note, that our version of clearing differs slightly from the original version [24] because we immediately remove all losers from the current population, while in the original version their fitness is set to the minimum value (which leads to their further removal with a high probability). Thus, after increasing the diversification level, some solutions may be removed from the current population and the population will not be completely filled. The solutions from the reserve may either be added to the current population without removing any existing solutions or they may substitute some existing solutions. Note that a similar concept of an additional population serving as a reservoir of diversity was also proposed by Park and Ryu outside the scope of VRP [23]. They use, however the additional population for crossbreeding with solutions from the main population.

In addition, since CVRPTW is a constrained problem we adapt a simplified version of the penalty weights management mechanism proposed in [14]. Note, however, that we update the penalty weights only in the exploration increasing phase, while we do not change them in the exploration decreasing phase. The idea is, that in the latter case, we should intensify the search with the current level of penalty weights.

The algorithm is summarized in Algorithm 1. To decrease and increase the clearing radius it is multiplied by $\beta_1 < 1$ and $\beta_2 > 1$, respectively. The weights update mechanism is summarized in Algorithm 2 and the population update is summarized in Algorithm 3.

Solutions are represented by sets of routes, each defined by a list of nodes.

Algorithm 1. Main algorithm

$clearing\_radius \leftarrow$ maximum value
$current\_phase \leftarrow$ exploration decreasing
$penalty\_weights \leftarrow$ initial weights
Initialize the current population with random solutions and apply local search to some of them.
while stopping conditions are not met **do**
    for k_{max} iterations **do**
        $offspring \leftarrow$ recombine two randomly chosen solutions from the current population
        Apply local search to $offspring$.
        Try to insert $offspring$ to reserve population
        Try to insert $offspring$ to the current population.
    if the best solution has been improved **then**
        $current\_phase \leftarrow$ exploration decreasing
    else
        if $current\_phase =$ exploration decreasing **then**
            if $clearing\_radius$ has not achieved the minimum value **then**
                decrease $clearing\_radius$
            else
                $current\_phase \leftarrow$ exploration increasing
        if $current\_phase =$ exploration increasing **then**
            if $clearing\_radius$ has not achieved the maximum value **then**
                increase $clearing\_radius$
            else
                $current\_phase \leftarrow$ exploration decreasing
        Update $penalty\_weights$.
        Remove all solutions from the current population and try to reinsert them using the current $clearing\_radius$ and $penalty\_weights$.
        Try to insert solutions from the reserve population to the current population.
        while the current population is not full **do**
            Try to insert a random solution to the current population.
    if in the past 10 iterations of the while loop the best feasible solutions has not been improved **then**
    Increase k_{max} if its maximum value has not been achieved.

---

4 Order-Based Recombination

As the recombination operator we use Order-based recombination which is a new operator briefly introduced in [8]. It is motivated by the observation made

Algorithm 2. Update *penalty_weights*

---

for each constraint and corresponding *penalty_weight* **do**
    if the best solution in the current population is feasible on the current constraint
    then
        penalty_weight ← *penalty_weight* * α_1, where $\alpha_1 < 1$
    else
        With probability p, *penalty_weight* ← *penalty_weight*$*\alpha_2$, where $\alpha_2 > 1$

Algorithm 3. Update population (try to insert a new solution)

Parameters ↓: population (current or reserve), *clearing_radius*, new candidate solution x
if the set Y of solutions for the population within *clearing_radius* from x is not empty **then**
    if x is not better than the best solution in Y **then**
        Reject x.
    else
        Remove all solutions from Y from the population and add x to the population.

---

in preliminary experiments that pairs of local optima often have many common pairs of nodes (pairs of nodes assigned in both solutions to a single route) that appear in different orders in the two solutions, i.e. in one solution node a precedes (directly or indirectly) node b, while in the other solution node b precedes node a. Results for exemplary instances (see Sect. 6) are presented in Table 1. This table presents average numbers of all common pairs of nodes and common pairs with the same order observed between all pairs of 2000 local optima obtained by starting local search from random solutions. While majority of pairs have the same order in both solutions there are still quite many pairs with different orders. It suggests that new promising solutions may be obtained by inverting the order of pairs of nodes that have different order in the parents. The general idea of this operator is to find such pairs of nodes and to apply in parent p_A the order from parent p_B to all or some of these pairs. Thus, the proposed operator can also be interpreted as a kind of guided perturbation applied to parent p_A.

The Order-based operator is summarized in Algorithm 4. The sorting in this algorithm is performed via MergeSort algorithm, which simultaneously counts the number of inversions. We limit the number of modified routes for efficiency reasons concerning both the scope of the following local search procedure and the lack of notable improvements observed with $k > 3$.

The last **if** statement is only needed in longer runs when the space of inverted routes combinations may be exhausted, or in the rare instances when a single route is sufficient (otherwise the algorithm would produce solution p_B from p_A, which is blocked by the Tabu lists).

Table 1. Average numbers of all common pairs of nodes and common pairs with the same order between local optima

Instance	Number of all common pairs of nodes	Number of common pairs of nodes with the same order
RC2_4_1	1866.2	1819.2
RC2_4_2	2033.9	1849.9
RC2_6_1	2411.6	2358.9
RC2_6_2	2356.2	2125.7
RC2_8_1	3253.1	3187.2
RC1_4_10	651.0	465.4
RC1_8_10	1186.1	848.6
C1_10_10	3161.1	3045.1
RC2_10_1	3915.3	3836.3
RC2_10_2	3539.2	3227.7

5   Local Search

After each recombination a local search procedure is applied to the offspring. The local search is based on the results presented in [14,21,22,34]. We use a node based neighbourhood proposed in [21] (the strategy called LCRD): we maintain a list of changed nodes and apply local search operators until the list is empty. The best move that improves the objective function is chosen. If no such move is found then the considered node is removed from the list. The algorithm of [21] uses only one node exchanges and the well known [34] operators: cross, 2-opt*, relocate, Or-opt and intra route cross. Our version uses wider neighbourhoods and more operators:

1. Generalized cross - each of the exchanged fragments may be reversed. This effectively defines four operators corresponding to four combinations of reversal/non-reversal of two crossed fragments. The maximal number of exchanged nodes is limited to 3.
2. Generalized relocate - the relocated fragment may be reversed. This effectively defines two operators. The maximal number of relocated nodes is set to 5.
3. Generalized Or-opt. As in the previous operators the relocated fragment may be reversed. The ideas of [14] make it possible to evaluate this type of move in constant time. Moreover, [14] limited the moved fragment to 3 nodes and the distance to move the fragment forward or backward to 15 nodes. We observed that it may be beneficial to relocate longer fragments by a smaller number of nodes, however some single nodes with wide time windows may be moved freely inside the route. We therefore define an ascending sequence of distances: 16, 32, 64, 128 and descending sequence of fragment lengths to move: 8, 4, 2, 1. The time complexity depends on the product of these numbers (and is maintained constant and equal to 128).

Algorithm 4. Order-based recombination

Parameters ↓: Two parents p_A and p_B, global Tabu list T_G of forbidden groups of routes

Parameter ↑: Offspring solution O

For each node in p_A find its relative position in its corresponding route from p_B.

Find in p_A all the routes containing at least two pairs of nodes in the order inverted with respect to p_B and randomly choose up to ten such inverted routes.

Sort each of the inverted routes in the ascending order of the relative positions of its nodes in p_B (ties in the relative position numbers are resolved by adding very small random values).

for each k-combination of the set of inverted routes, where $k \in \{1, 2, 3\}$ **do**

    if the combination is not present in T_G **then**

        Exchange the original routes with the ones from current combination, obtaining O.

        Add the combination of the modified routes to T_G.

        Quit the loop.

if all combinations were already present in T_G **then**

    Consider the number I of all inversions in p_A with respect to p_B.

    Select at random a number $r \in \mathbb{N}$ from $[\frac{1}{4}I, \frac{3}{4}I]$.

    Perform at random order r neighbour transpositions.

return O

---

4. 2-opt* operator.
5. Intra route one node cross operator (as in [21,22]).
6. A new time feasibility operator. We use this operator only if the considered node belongs to an infeasible route and it is impossible to decrease the infeasibility by removing this node from the route. It was observed in [22] that if for some node the time infeasibility introduced by some route is equal to the sum of infeasibilities introduced by the fragments before and after the node, then removing this node from the route does not lead to improvement in time feasibility. It was noted in [22] that only the intra route cross move may improve feasibility in this situation. We propose another strategy to tackle this problem: move the node to some other route and introduce another node from the same route to this place. This allows us to reduce infeasibility caused by the starting or ending fragment by other type of operator than intra route cross. The reason for introducing this operator is that intra route cross is very rarely successful at improving feasibility of a route.

Such a strong family of operators would be impractical without a move filtering strategy. For generalized cross, generalized relocate, and 2-opt* operators we use neighbour lists as in [14]. The move evaluation is performed in constant time as in [14]. We also do not allow a move to increase time infeasibility of a solution. This allows to prune moves that increase time infeasibility without the full evaluation [22]. The neighbour lists and time based filtration allow evaluating larger neighbourhood without degrading the time of evaluation.

6   Computational Experiment

To evaluate the proposed algorithm we utilize commonly used Gehring and Homberger instances categorized into six groups with 200, 400, 600, 800, and 1000 nodes [11]. This benchmark includes instances of several types. C1 and C2 instances have geographically clustered nodes. In R1 and R2, nodes have random positions. In RC1 and RC2, a mix of random and clustered nodes is included. R1, C1 and RC1 have a short scheduling horizon, narrow time windows, and low vehicle capacities, while R2, C2 and RC2 have a long scheduling horizon, wider time windows, and greater vehicle capacities. As a result solutions for instances with suffix 2 have relatively low numbers of routes, while the solutions for the instances with suffix 1 have larger numbers of routes.

The goal of the experiment is twofold. First, the goal is to verify that the proposed diversity management mechanism improves the performance of the algorithm. Another goal is to verify if the results of our algorithm are competitive to the state-of-the-art results reported in the literature, namely the results of the algorithms of Vidal et al. (VCGP) [35], and Christiaens and Vanden Berghe (CVB) [6].

In Table 2 we present the comparison of the best results of 5 runs for each instance as stated in [6] - cumulative number of vehicles (CNV) and cumulative distance (CTD) for instances of each size. The best results are highlighted. Let us remind that the objectives are lexicographically ordered with minimization of the number of vehicles being the main objective. In online Appendix A[1] we present the details of the comparison: the parameters (Table 1), the detailed comparison of the average results (Table 2) and the best results (Table 3) obtained for each instance. Unfortunately, since the referenced papers do not provide results for particular instances, we were not able to test statistical significance of differences with respect to these methods. Nevertheless, we conclude that the results of our method are comparable to the state-of-the-art literature results. In fact, in both comparisons based on the average and the best values, our method generated the best cumulative results for 4 out of 5 sets of instances, with the exception of the smallest ones with 200 nodes.

Table 2. Comparison of the best results

Size	VCGP		CVB		Our method		Our method with constant diversity level							
							$R = 0.15$		$R = 0.1$		$R = 0.05$		$R = 0$	
	CNV	CTD	CNV	CTD	CNV	CTD	CNV	CTD	CNV	CTD	CNV	CTD	CNV	CTD
200	**694**	**168092**	694	168447	694	168290	694	168220	694	168264	694	168522	694	168982
400	1381	388013	1382	389861	**1381**	**388010**	1381	388375	1382	388058	1383	388376	1384	391438
600	2068	786373	2067	790784	**2067**	**785951**	2067	788231	2067	788136	2069	786939	2069	795718
800	2739	1334963	2743	1337802	**2738**	**1330989**	2738	1338609	2738	1337490	2740	1336973	2747	1347095
1000	3420	2036700	3421	2036358	**3420**	**2027373**	3421	2038359	3420	2036954	3422	2037095	3436	2053621

---

[1] https://math.uni.lodz.pl/~cybula/ppsn2022/ppsn2022supplement.pdf.

Furthermore, to verify the hypothesis that the proposed diversity management mechanism improves the performance of the evolutionary algorithm we run the proposed algorithm with constant normalized clearing radius R using values $R = 0.15$, $R = 0.1$, $R = 0.05$, and $R = 0$ (full elitism). We used Friedman post-hoc test with Bergmann and Hommel's correction at the significance level $\alpha = 0.05$. The results are presented in online Appendix A (Table 4). Except for the smallest instances, our method is significantly better than all methods with constant radius. In addition, the relative performance of the methods with constant clearing radius depends on the size of the instances. For example, $R = 0.15$ is the best setting for the smallest instances, but is the worst (except of $R = 0$) for the largest instances. This shows, that setting a good constant value of the clearing radius may be a difficult task even with the use of a normalized distance. In addition, the method with full elitism ($R = 0$) was in all cases outperformed by all other methods. This confirms that diversity management is crucial in evolutionary algorithms for CVRPTW.

7   Conclusions and Directions for Further Research

We have presented a hybrid evolutionary algorithm with a novel diversity management mechanism based on the idea of oscillation of the diversity level with a dual-population model for the Capacitated Vehicle Routing Problem with Time Windows. In the computational experiment the algorithm were competitive to the state-of-art algorithms for this problem. We have also shown that the proposed novel diversity management mechanism significantly improves results of the algorithm. The presented results confirm also the well-known fact that maintaining diversity of the population is necessary in evolutionary algorithms for CVRPTW.

The reported algorithm was used in a VRP solver that generated 124 best-known solutions for CVRPTW in multiple additional runs (which are not reported in this paper) that were verified and published by SINTEF[2].

The presented research was motivated by a commercial project with the aim to develop the aforementioned VRP solver. This is why we focus on CVRPTW. The proposed diversity management mechanism is, however, fairly general and could be potentially applied in evolutionary algorithms for other problems. Thus, such applications constitute one of the directions for further research. Furthermore, in this paper we use clearing as the underlying diversity management mechanism, however, the idea of oscillating the diversity level is quite general and could be potentially used with other diversity management mechanisms that include some parameters influencing the diversity levels.

Another interesting direction for further research would be to compare and combine our mechanism with some novel diversity management techniques which to our knowledge have not been applied in the context of VRP, e.g. Quality and Diversity Optimization [7, 12].

---

[2] https://www.sintef.no/projectweb/top/vrptw/.

Funding. Research reported in this article was supported by The National Centre for Research and Development (POIR.01.01.01-00-0222/16, POIR.01.01.01-00-0012/19). The computational experiments were made possible by the computing grant 358 awarded by the Poznan Supercomputing and Networking Center.

References

1. Ahuja, R.K., Özlem Ergun, Orlin, J.B., Punnen, A.P.: A survey of very large-scale neighborhood search techniques. Discrete Appl. Math. **123**(1), 75–102 (2002). https://doi.org/10.1016/S0166-218X(01)00338-9, https://www.sciencedirect.com/science/article/pii/S0166218X01003389

2. Borthen, T., Loennechen, H., Wang, X., Fagerholt, K., Vidal, T.: A genetic search-based heuristic for a fleet size and periodic routing problem with application to offshore supply planning. EURO J. Transp. Logist. **7**(2), 121–150 (2017). https://doi.org/10.1007/s13676-017-0111-x

3. Boudia, M., Prins, C., Reghioui, M.: An effective memetic algorithm with population management for the split delivery vehicle routing problem. In: Bartz-Beielstein, T., et al. (eds.) HM 2007. LNCS, vol. 4771, pp. 16–30. Springer, Heidelberg (2007). https://doi.org/10.1007/978-3-540-75514-2_2

4. Braekers, K., Ramaekers, K., Nieuwenhuyse, I.V.: The vehicle routing problem: state of the art classification and review. Comput. Ind. Engi. **99**, 300 – 313 (2016). https://doi.org/10.1016/j.cie.2015.12.007, http://www.sciencedirect.com/science/article/pii/S0360835215004775

5. Cherkesly, M., Desaulniers, G., Laporte, G.: A population-based metaheuristic for the pickup and delivery problem with time windows and lifo loading. Comput. Oper. Res. **62**, 23–35 (2015). https://doi.org/10.1016/j.cor.2015.04.002, https://www.sciencedirect.com/science/article/pii/S0305054815000829

6. Christiaens, J., Vanden Berghe, G.: Slack induction by string removals for vehicle routing problems. Transp. Sci. **54**(2), 417–433 (2020). https://doi.org/10.1287/trsc.2019.0914

7. Cully, A., Demiris, Y.: Quality and diversity optimization: a unifying modular framework. IEEE Trans. Evol. Comput. **22**(2), 245–259 (2018). https://doi.org/10.1109/TEVC.2017.2704781

8. Cybula, P., Rogalski, M., Sielski, P., Jaszkiewicz, A., Pełka, P.: Effective recombination operators for the family of vehicle routing problems. In: Proceedings of the Genetic and Evolutionary Computation Conference Companion. GECCO 2021, pp. 121–122. Association for Computing Machinery, New York (2021). https://doi.org/10.1145/3449726.3459574

9. Eiben, A.E., Smith, J.E.: Introduction to Evolutionary Computing, 2nd edn. Springer, Heidelberg (2015). https://doi.org/10.1007/978-3-662-44874-8

10. Folkestad, C.A., Hansen, N., Fagerholt, K., Andersson, H., Pantuso, G.: Optimal charging and repositioning of electric vehicles in a free-floating carsharing system. Comput. Oper. Res. **113**, 104771 (2020). https://doi.org/10.1016/j.cor.2019.104771, https://www.sciencedirect.com/science/article/pii/S0305054819302138

11. Gehring, H., Homberger, J.: A parallel hybrid evolutionary metaheuristic for the vehicle routing problem with time windows. In: University of Jyvaskyla, pp. 57–64 (1999)

12. Gravina, D., Liapis, A., Yannakakis, G.N.: Quality diversity through surprise. IEEE Trans. Evol. Comput. **23**(4), 603–616 (2019). https://doi.org/10.1109/TEVC.2018.2877215

13. Ha, Q.M., Deville, Y., Pham, Q.D., Hà, M.H.: A hybrid genetic algorithm for the traveling salesman problem with drone. J. Heurist. **26**(2), 219–247 (2019). https://doi.org/10.1007/s10732-019-09431-y

14. Hashimoto, H., Yagiura, M.: A path relinking approach with an adaptive mechanism to control parameters for the vehicle routing problem with time windows. In: van Hemert, J., Cotta, C. (eds.) EvoCOP 2008. LNCS, vol. 4972, pp. 254–265. Springer, Heidelberg (2008). https://doi.org/10.1007/978-3-540-78604-7_22

15. Hussain, K., Salleh, M.N.M., Cheng, S., Shi, Y.: On the exploration and exploitation in popular swarm-based metaheuristic algorithms. Neural Comput. Appl. **31**(11), 7665–7683 (2018). https://doi.org/10.1007/s00521-018-3592-0

16. Kirkpatrick, S., Gelatt, C., Vecchi, M.: Optimization by Simulated Annealing. Science **220**(4598), 671–680 (1983). https://doi.org/10.1126/science.220.4598.671

17. Konstantakopoulos, G.D., Gayialis, S.P., Kechagias, E.P.: Vehicle routing problem and related algorithms for logistics distribution: a literature review and classification. Oper. Res. (2020). https://doi.org/10.1007/s12351-020-00600-7

18. Liu, N., Pan, J.S., Chu, S.C.: A competitive learning QUasi affine TRansformation evolutionary for global optimization and its application in CVRP. J. Internet Technol. **21**(7), 1863–1883 (2020). https://doi.org/10.3966/160792642020122107002

19. Lourenço, H.R., Martin, O.C., Stützle, T.: Iterated local search: framework and applications. In: Gendreau, M., Potvin, J.Y. (eds.) Handbook of Metaheuristics. ISORMS, vol. 272, pp. 363–397. Springer, Cham (2010). https://doi.org/10.1007/978-1-4419-1665-5_12

20. Nagata, Y., Kobayashi, S.: A memetic algorithm for the pickup and delivery problem with time windows using selective route exchange crossover. In: Schaefer, R., Cotta, C., Kołodziej, J., Rudolph, G. (eds.) PPSN 2010. LNCS, vol. 6238, pp. 536–545. Springer, Heidelberg (2010). https://doi.org/10.1007/978-3-642-15844-5_54

21. Nagata, Y., Bräysy, O.: Efficient local search limitation strategies for vehicle routing problems. In: van Hemert, J., Cotta, C. (eds.) EvoCOP 2008. LNCS, vol. 4972, pp. 48–60. Springer, Heidelberg (2008). https://doi.org/10.1007/978-3-540-78604-7_5

22. Nagata, Y., Braysy, O., Dullaert, W.: A penalty-based edge assembly memetic algorithm for the vehicle routing problem with time windows. Comput. Oper. Res. **37**(4), 724–737 (2010)

23. Park, T., Ryu, K.R.: A dual-population genetic algorithm for adaptive diversity control. IEEE Trans. Evol. Comput. **14**(6), 865–884 (2010). https://doi.org/10.1109/TEVC.2010.2043362

24. Petrowski, A.: A clearing procedure as a niching method for genetic algorithms. In: Proceedings of IEEE International Conference on Evolutionary Computation, pp. 798–803 (1996). https://doi.org/10.1109/ICEC.1996.542703

25. Prins, C.: A simple and effective evolutionary algorithm for the vehicle routing problem. Comput. Oper. Res. **31**(12), 1985–2002 (2004). https://doi.org/10.1016/S0305-0548(03)00158-8, http://www.sciencedirect.com/science/article/pii/S0305054803001588

26. Qu, Y., Bard, J.F.: The heterogeneous pickup and delivery problem with configurable vehicle capacity. Transp. Res. Part C - Emerg. Technol. **32**(SI), 1–20 (2013). https://doi.org/10.1016/j.trc.2013.03.007

27. Repoussis, P.P., Tarantilis, C.D., Braysy, O., Ioannou, G.: A hybrid evolution strategy for the open vehicle routing problem. Comput. Oper. Res. **37**(3, SI), 443–455 (2010). https://doi.org/10.1016/j.cor.2008.11.003

28. Sabar, N.R., Ayob, M., Kendall, G., Qu, R.: Automatic design of a hyper-heuristic framework with gene expression programming for combinatorial optimization problems. IEEE Trans. Evol. Comput. **19**(3), 309–325 (2015). https://doi.org/10.1109/TEVC.2014.2319051

29. Sabar, N.R., Bhaskar, A., Chung, E., Turky, A., Song, A.: An adaptive memetic approach for heterogeneous vehicle routing problems with two-dimensional loading constraints. Swarm Evol. Comput. **58**, 100730 (2020). https://doi.org/10.1016/j.swevo.2020.100730, https://www.sciencedirect.com/science/article/pii/S2210650220303837

30. Segura, C., Botello Rionda, S., Hernandez Aguirre, A., Ivvan Valdez Pena, S.: A novel diversity-based evolutionary algorithm for the traveling salesman problem. In: Silva, S (ed.) GECCO 2015: Proceedings of the 2015 Genetic and Evolutionary Computation Conference, pp. 489–496. Assoc Comp Machinery SIGEVO (2015). https://doi.org/10.1145/2739480.2754802, 17th Genetic and Evolutionary Computation Conference (GECCO), Madrid, San Marino, 11–15 July 2015

31. Squillero, G., Tonda, A.: Divergence of character and premature convergence: a survey of methodologies for promoting diversity in evolutionary optimization. Inf. Sci. **329**, 782–799 (2016). https://doi.org/10.1016/j.ins.2015.09.056, https://www.sciencedirect.com/science/article/pii/S002002551500729X, special issue on Discovery Science

32. Sudholt, D., Squillero, G.: Theory and practice of population diversity in evolutionary computation. In: Proceedings of the 2020 Genetic and Evolutionary Computation Conference Companion. GECCO 2020, pp. 975–992. Association for Computing Machinery, New York (2020). https://doi.org/10.1145/3377929.3389892

33. Toth, P., Vigo, D.: Exact solution of the vehicle routing problem. In: Crainic, T.G., Laporte, G. (eds.) Fleet Management and Logistics. CRT, pp. 1–31. Springer, Boston (1998). https://doi.org/10.1007/978-1-4615-5755-5_1

34. Toth, P., Vigo, D. (eds.): The Vehicle Routing Problem. Society for Industrial and Applied Mathematics (2001)

35. Vidal, T., Crainic, T.G., Gendreau, M., Prins, C.: A hybrid genetic algorithm with adaptive diversity management for a large class of vehicle routing problems with time-windows. Comput. Oper. Res. **40**(1), 475–489 (2013)

36. Vidal, T., Crainic, T.G., Gendreau, M., Prins, C.: A unified solution framework for multi-attribute vehicle routing problems. Eur. J. Oper. Res. **234**(3), 658–673 (2014). https://doi.org/10.1016/j.ejor.2013.09.045, https://www.sciencedirect.com/science/article/pii/S037722171300800X

37. Xu, Z., Li, H., Wang, Y.: An improved genetic algorithm for vehicle routing problem. In: 2011 International Conference on Computational and Information Sciences, pp. 1132–1135 (2011). https://doi.org/10.1109/ICCIS.2011.78

38. Zhu, K.: A diversity-controlling adaptive genetic algorithm for the vehicle routing problem with time windows. In: Werner, B (ed.) 15TH IEEE International Conference on Tools with Artificial Intelligence, Proceedings. Proceedings-International Conference on Tools With Artificial Intelligence, pp. 176–183. IEEE Comp Sci (2003). https://doi.org/10.1109/TAI.2003.1250187, 15th IEEE International Conference on Tools with Artificial Intelligence (ICTAI 2003), Sacramento, CA, 03–05 November 2003

Evolutionary Algorithms for Limiting the Effect of Uncertainty for the Knapsack Problem with Stochastic Profits

Aneta Neumann[✉], Yue Xie, and Frank Neumann

Optimisation and Logistics, School of Computer Science,
The University of Adelaide, Adelaide, Australia
aneta.neumann@adelaide.edu.au

Abstract. Evolutionary algorithms have been widely used for a range of stochastic optimization problems in order to address complex real-world optimization problems. We consider the knapsack problem where the profits involve uncertainties. Such a stochastic setting reflects important real-world scenarios where the profit that can be realized is uncertain. We introduce different ways of dealing with stochastic profits based on tail inequalities such as Chebyshev's inequality and Hoeffding bounds that allow to limit the impact of uncertainties. We examine simple evolutionary algorithms and the use of heavy tail mutation and a problem-specific crossover operator for optimizing uncertain profits. Our experimental investigations on different benchmarks instances show the results of different approaches based on tail inequalities as well as improvements achievable through heavy tail mutation and the problem specific crossover operator.

Keywords: Stochastic knapsack problem · Chance-constrained optimization · Evolutionary algorithms

1 Introduction

Evolutionary algorithms [10] have been successfully applied to a wide range of complex optimization problems [5,17,20]. Stochastic problems play a crucial role in the area of optimization and evolutionary algorithms have frequently been applied to noisy environments [22].

Given a stochastic function to be optimized under a given set of constraints, the goal is often to maximize the expected value of a solution with respect to f. This however does not consider the deviation from the expected value. Guaranteeing that a function value with a good probability does not drop below a certain value is often more beneficial in real-world scenarios. For example, in the area of mine planning [3,15], profits achieved within different years should be maximized. However, it is crucial to not drop below certain profit values because

G. Rudolph et al. (Eds.): PPSN 2022, LNCS 13398, pp. 294–307, 2022.
https://doi.org/10.1007/978-3-031-14714-2_21

then the whole mine operation would not be viable and the company might go bankrupt.

We consider a stochastic version of the knapsack problem which fits the characteristics of the mine planning problem outline above. Here the profits are stochastic and the weights are deterministic. Motivated by the area of chance constrained optimization [4] where constraints can only be violated with a small probability, we consider the scenario where we maximize the function value P for which we can guarantee that the best solution x as a profit has less than P with probability at most α_p, i.e. $Prob(p(x) < P) \leq \alpha_p$. Note that determining whether $Prob(p(x) < P) \leq \alpha_p$ holds for a given solution x and values P and α_p is already hard for very simple stochastic settings where profits are independent and each profit can only take on two different values. Furthermore, finding a solution with a maximal P for which the condition holds poses in general a non-linear objective function that needs to take the probability distribution of $p(x)$ into account. Constraints of the beforehand mentioned type are known as chance constraints [4]. Chance constraints on stochastic components of a problem can only be violated with a small probability, in our case specified by the parameter α_p.

1.1 Related Work

Up to recently, only a few problems with chance constraints have been studied in the evolutionary computation literature [11,13,14]. They are based on simulations and sampling techniques for evaluating chance constraints. Such approaches require a relatively large computation time for evaluating the chance constraints. In contrast to this, tail inequalities can be used if certain characteristics such as the expected value and variance of a distribution are known. Such approaches have recently been examined for the chance constrained knapsack problem in the context of evolutionary computation [25,26]. The standard version of the chance-constrained knapsack problem considers the case where the profits are deterministic and the weights are stochastic. Here the constraint bound B can only be violated with a small probability. Different single and multi-objective approaches have recently been investigated [25,26].

Furthermore, chance constrained monotone submodular functions have been studied in [8,18]. In [8], greedy algorithms that use tail inequalities such as Chebyshev's inequality and Chernoff bounds have been analyzed. It has been shown that they almost achieve the same approximation guarantees in the chance constrained setting as in the deterministic setting. In [18], the use of evolutionary multi-objective algorithms for monotone submodular functions has been investigated and it has been shown that they achieve the same approximation guarantees as the greedy algorithms but perform much better in practice. In addition, different studies on problems with dynamically changing constraints have been carried out in the chance constrained [2] and deterministic constrained setting [23,24].

Finally, chance constrained problems have been further investigated through runtime analysis for special instances of the knapsack problem [19,27]. This includes a study on very specific instances showing when local optima arise [19] and a study on groups of items whose stochastic uniform weights are correlated with each other [27].

All previously mentioned studies concentrated on stochastic weights and how algorithms can deal with the chance constraints with respect to the weight bound of the knapsack problem. In [16], a version of the knapsack problem stochastic profits and deterministic weights has been considered where the goal is to maximize the probability that the profit meets a certain threshold value. In contrast to this, we will maximize the profit under the condition that it is achieved with high probability. We will provide the first study on evolutionary algorithms for giving guarantees when maximizing stochastic profits, a topic that is well motivated by the beforehand mentioned mine planning application but to our knowledge not studied in the literature before.

The paper is structured as follows. In Sect. 2, we introduce the problem formulation and tail bounds that will be used to construct fitness functions for dealing with stochastic profit. In Sect. 3, we derive fitness functions that are able to maximize the profit for which we can give guarantees. Section 4 introduces evolutionary algorithms for the problem and we report on our experimental investigations in Sect. 5. We finally finish with some conclusions.

2   Problem Definition

In this section, we formally introduce the problem and tail inequalities for dealing with stochastic profits that will later be used to design fitness functions. We consider a stochastic version of the classical NP-hard knapsack problem. In the classical problem, there are given n items $1, \ldots, n$ where each item has a profit p_i and a weight w_i, the goal is to maximize the profit $p(x) = \sum_{i=1}^{n} p_i x_i$ under the condition that $w(x) = \sum_{i=1}^{n} w_i x_i \leq B$ for a given weight bound B holds. The classical knapsack problem has been well studied in the literature. We consider the following stochastic version, where the profits p_i are stochastic and the weights are still deterministic. Our goal is to maximize the profit P for which we can guarantee that there is only a small probability α_p of dropping below P. Formally, we tackle the following problem:

$$\max P \tag{1}$$
$$s.t. \ Pr(p(x) < P) \leq \alpha_p \tag{2}$$
$$w(x) \leq B \tag{3}$$
$$x \in \{0, 1\}^n \tag{4}$$

Equation 2 is a chance constraint on the profit and the main goal of this paper is to find a solution x that maximize the value of P such that the probability of getting a profit lower than P is at most α_p. We denote by $\mu(x)$ the expected profit and by $v(x)$ the variance of the profit throughout this paper.

2.1 Concentration Bounds

In order to establish guarantees for the stochastic knapsack problem we make use of well-known tail inequalities that limit the deviation from the expected profit of a solution.

For a solution X with expected value $E[X]$ and variance $Var[X]$ we can use the lower tail of the following Chebyshev-Cantelli inequality.

Theorem 1 (One-sided Chebyshev's/Cantelli's inequality). *Let X be a random variable with expected value $E[X]$ and variance $\text{Var}[X] > 0$. Then, for all $\lambda > 0$,*

$$\Pr[X \geq E[X] - \lambda] \geq 1 - \frac{\text{Var}[X]}{\text{Var}[X] + \lambda^2} \tag{5}$$

We will refer to this inequality as Chebyshev's inequality in the following. Chebyshev's inequality only requires the expected value and variance of a solution, but no additional requirements such as the independence of the random variables.

We use the additive Hoeffding bound given in Theorem 1.10.9 of [7] for the case where the weights are independently chosen within given intervals.

Theorem 2 (Hoeffding bound). *Let $X_1, \ldots, X_n$ be independent random variables. Assume that each X_i takes values in a real interval $[a_i, b_i]$ of length $c_i := b_i - a_i$. Let $X = \sum_{i=1}^{n} X_i$. Then for all $\lambda > 0$,*

$$Pr(X \geq E[X] + \lambda) \leq e^{-2\lambda^2 / (\sum_{i=1}^{n} c_i^2)} \tag{6}$$

$$Pr(X \leq E[X] - \lambda) \leq e^{-2\lambda^2 / (\sum_{i=1}^{n} c_i^2)} \tag{7}$$

3 Fitness Functions for Profit Guarantees

The main task when dealing with the setting of chance constraint profits is to come up with fitness functions that take the uncertainty into account.

In this section, we introduce different fitness functions that can be used in an evolutionary algorithm to compute solutions that maximize the profit under the uncertainty constraint. We consider the search space $\{0,1\}^n$ and for a given search point $x \in \{0,1\}^n$, item i chosen iff $x_i = 1$ holds.

The fitness of a search point $x \in \{0,1\}^n$ is given by

$$f(x) = (u(x), \hat{p}(x))$$

where $u(x) = \max\{w(x) - B, 0\}$ is the amount of constraint violation of the bound B by the weight that should be minimized and $\hat{p}(x)$ is the discounted profit of solution x that should be maximized. We optimize f with respect to lexicographic order and have

$$f(x) \geq f(y)$$

iff

$$(u(x) < u(y)) \vee ((u(x) = u(y)) \wedge (\hat{p}(x) \geq \hat{p}(y))).$$

This implies a standard penalty approach where the weight $w(x)$ is reduced until it meets the constraint bound B, and the profit $\hat{p}(x)$ is maximized among all feasible solutions.

The key part if to develop formulations for $\hat{p}$ that take into account the stochastic part of the profits to make the formulations suitable for our chance constrained setting. Therefore, we will develop profit functions that reflect different stochastic settings in the following.

3.1   Chebyshev's Inequality

We give a formulation for $\hat{p}$ that can be applied in quite general settings, thereby providing only a lower bound on the value P for which a solution x still meets the profit chance constraint.

We assume that for a given solution only the expected value $\mu(x)$ and the variance $v(x)$ are known. The following lemma gives a condition for meeting the chance constraint based on Theorem 1.

Lemma 1. *Let x be a solution with expected profit $\mu(x)$ and variance $v(x)$. If*

$$\mu(x) - P \geq \sqrt{((1 - \alpha_p)/\alpha_p) \cdot v(x)} \text{ then } Pr(p(x) < P) \leq \alpha_p.$$

Proof. We have

$$Pr(p(x) \geq P) = Pr(p(x) \geq \mu(x) - (\mu(x) - P)) \geq 1 - \frac{v(x)}{v(x) + (\mu(x) - P)^2}$$

The chance constraint is met if

$$1 - \frac{v(x)}{v(x) + (\mu(x) - P)^2} \geq 1 - \alpha_p$$

$$\Longleftrightarrow \alpha_p \geq \frac{v(x)}{v(x) + (\mu(x) - P)^2}$$

$$\Longleftrightarrow \alpha_p \cdot (v(x) + (\mu(x) - P)^2) \geq v(x)$$

$$\Longleftrightarrow \alpha_p \cdot v(x) + \alpha_p \cdot (\mu(x) - P)^2 \geq v(x)$$

$$\Longleftrightarrow \alpha_p(\mu(x) - P)^2 \geq (1 - \alpha_p)v(x)$$

$$\Longleftrightarrow (\mu(x) - P)^2 \geq ((1 - \alpha_p)/\alpha_p) \cdot v(x)$$

$$\Longleftarrow \mu(x) - P \geq \sqrt{(1 - \alpha_p)/\alpha_p \cdot v(x)}$$

□

Given the last expression, P is maximal for

$$P = \mu(x) - \sqrt{((1 - \alpha_p)/\alpha_p) \cdot v(x)}.$$

We use the following profit function based on Chebyshev's inequality:

$$\hat{p}_{Cheb}(x) = \mu(x) - \sqrt{(1 - \alpha_p)/\alpha_p} \cdot \sqrt{v(x)} \qquad (8)$$

3.2 Hoeffding Bound

We now assume that each element i takes on a profit $p_i \in [\mu_i - \delta_p, \mu_i + \delta_p]$ independently of the other items. Let $\mu(x) = \sum \mu_i x_i$. We have $p(x) = \mu(x) - \delta_p |x|_1 + p'(x)$ where $p'(x)$ is the sum of $|x|_1$ independent random variables in $[0, 2\delta_p]$. We have $E[p'(x)] = |x|_1 \delta_p$ and

$$Pr(p(x) \le \mu(x) - \lambda) = Pr(p'(x) \le |x|_1 \delta_p - \lambda) \le e^{-2\lambda^2/(4\delta_p^2 |x|_1)} = e^{-\lambda^2/(2\delta_p^2 |x|_1)}$$

based on Theorem 2. The chance constraint is met if

$$e^{-\lambda^2/(2\delta_p^2 |x|_1)} \le \alpha_p$$
$$\iff -\lambda^2/(2\delta_p^2 |x|_1) \le \ln(\alpha_p)$$
$$\iff \lambda^2 \ge \ln(1/\alpha_p) \cdot (2\delta_p^2 |x|_1)$$
$$\iff \lambda \ge \delta_p \cdot \sqrt{\ln(1/\alpha_p) \cdot 2|x|_1}$$

Therefore, we get the following profit function based on the additive Hoeffding bound from Theorem 2:

$$\hat{p}_{Hoef}(x) = \mu(x) - \delta_p \cdot \sqrt{\ln(1/\alpha_p) \cdot 2|x|_1} \qquad (9)$$

3.3 Comparison of Chebyshev and Hoeffding Based Fitness Functions

The fitness functions $\hat{p}_{Hoef}$ and $\hat{p}_{Cheb}$ give a conservative lower bound on the value of P to be maximized. We now consider the setting investigated for the Hoeffding bound and compare it to the use of Chebyshev's inequality. If each element is chosen independently and uniformly at random from an interval of length $2\delta_p$ as done in Sect. 3.2, then we have $v(x) = |x|_1 \cdot \delta_p^2/3$. Based on this we can establish a condition on α_p which shows when $\hat{p}_{Hoef}(x) \le \hat{p}_{Cheb}(x)$ holds.
We have

$$\hat{p}_{Hoef}(x) \ge \hat{p}_{Cheb}(x)$$
$$\iff \sqrt{\ln(1/\alpha_p) \cdot 2 \cdot |x|_1} \le \sqrt{\frac{(1 - \alpha_p)|x|_1}{3\alpha_p}}$$
$$\iff \ln(1/\alpha_p) \cdot 2 \cdot |x|_1 \le \frac{(1 - \alpha_p)|x|_1}{3\alpha_p}$$
$$\iff \ln(1/\alpha_p) \cdot \alpha_p/(1 - \alpha_p) \le 1/6$$

Algorithm 1: (1+1) EA

1: Choose $x \in \{0,1\}^n$ to be a decision vector.
2: **while** *stopping criterion not met* **do**
3:    $y \leftarrow$ flip each bit of x independently with
    probability of $\frac{1}{n}$;
4:    **if** $f(y) \geq f(x)$ **then**
5:      $x \leftarrow y$;
6:    **end if**
7: **end while**

Note that the last inequality depends only on α_p but not on δ_p or $|x|_1$.

We will use values of $\alpha_p \in \{0.1, 0.01, 0.001\}$ in our experiments and have $\ln(1/\alpha_p) \cdot \alpha_p/(1-\alpha_p) > 1/6$ for $\alpha_p = 0.1$ and $\ln(1/\alpha_p) \cdot \alpha_p/(1-\alpha_p) < 1/6$ for $\alpha_p = 0.01, 0.001$. This means that the fitness function based on Chebyshev's inequality is preferable to use for $\alpha_p = 0.1$ as it gives a better (tighter) value for any solution x and the fitness function based on Hoeffding bounds is preferable for $\alpha_p = 0.01, 0.001$. Dependent on the given instance, it might still be useful to use the less tighter fitness function as the fitness functions impose different fitness landscapes.

4 Evolutionary Algorithms

We examine the performance of the (1+1) EA, the (1+1) EA with heavy-tail mutation and the (μ+1) EA. The (μ+1) EA uses a specific crossover operator for the optimization of the chance-constrained knapsack problem with stochastic profits together with heavy-tailed mutation. All algorithms use the fitness function f introduced in Sect. 3 and we will examine different choices of $\hat{p}$ in our experimental investigations.

4.1 (1+1) EA

We consider a simple evolutionary algorithm called the (1+1) EA that has been extensively studied in the area of runtime analysis. The approach is given in Algorithm 1. The (1+1) EA starts with an initial solution $x \in \{0,1\}^n$ chosen uniformly at random. It generates in each iteration a new candidate solution by standard bit mutation, i.e. by flipping each bit of the current solution with a probability of $1/n$, where n is the length of the bit string. In the selection step, the algorithm accept the offspring if it is at least as good as the parent. The process is iterated until a given stopping criterion is fulfilled. While the (1+1) EA is a very simple algorithm, it produces good results in many cases. Furthermore, it has the ability to sample new solutions globally as each bit is flipped independently with probability $1/n$. In order to overcome large inferior neighborhoods larger mutation rates might be beneficial. Allowing larger mutation rates from time to time is the idea of heavy tail mutations.

Algorithm 2: The heavy-tail mutation operator

Input: Individual $x = (x_1, \ldots, x_n) \in \{0,1\}^n$ and value β;
1: Choose $\theta \in [1, .., n/2]$ randomly according to $D_{n/2}^{\beta}$;
2: **for** $i = 1$ to n **do**
3:    **if** $rand([0,1]) \leq \theta/n$ **then**
4:       $y_i \leftarrow 1 - x_i$;
5:    **else**
6:       $y_i \leftarrow x_i$;
7:    **end if**
8: **end for**
9: **return** $y = (y_1, \ldots, y_n)$;

4.2 Heavy Tail Mutations

We also investigate the (1+1) EA with heavy tail mutation instead of standard bit mutation. In each operation of the heavy tail mutation operator (see Algorithm 9, first a parameter $\theta \in [1..n/2]$ is chosen according to the discrete power law distribution $D_{n/2}^{\beta}$. Afterwards, each bit n is flipped with probability θ/n. Based on the investigations in [9], we use $\beta = 1.5$ for our experimental investigations.

The heavy-tail mutation operator allows to flip significantly more bits in some mutation steps than the standard bit mutation. The use of heavy tail mutations has been shown to be provably effective on the OneMax and Jump benchmark problems in theoretical investigations [1,9]. Moreover, in [26] has been shown in that the use of heavy tail mutation effective improves performance of single-objective and multi-objective evolutionary algorithms for the weight chance constrained knapsack problem. For details, on the discrete power law distribution and the heavy tail operator, we refer the reader to [9].

4.3 Population-Based Evolutionary Algorithm

We also consider the population-based $(\mu + 1)$-EA shown in Algorithm 3. The algorithm produces in each iteration an offspring by crossover and mutation with probability p_c and by mutation only with probability $1 - p_c$. We use $p_c = 0.8$ for our experimental investigations. The algorithm makes use of the specific crossover operator shown in Algorithm 4 and heavy tail mutation. The crossover operator chooses two different individuals x and y from the population P and produces an offspring z. All bit positions where x and y are the same are transferred to z. Positions i where x_i and y_i are different form the set I. They are treated in a greedy way according to the discounted expected value to weight ratio. Setting $p_i = \mu_i - u(z,i)$ discounts the expected profit by an uncertainty value (see Line 3 in Algorithm 4). This uncertain value is based on the solution z and the impact if element i is added to z.

There are different ways of doing this. In our experiments, where the profits of the elements are chosen independently and uniformly at random, we use the calculation based on Hoeffding bounds and set

Algorithm 3: $(\mu + 1)$ EA

1: Randomly generate μ initial solutions as the initial population P;
2: **while** *stopping criterion not meet* **do**
3:    Let x and y be two different individual from P chosen uniformly at random;
4:    **if** $rand([0,1]) \leq p_c$ **then**
5:        apply the discounted greedy uniform crossover operator to x and y to produce an offspring z.
6:    **else**
7:        Choose one individual x from P uniformly at random and let z be a copy of x.
8:    **end if**
9:    apply the heavy-tail mutation operator to z;
10:    **if** $f(z) \geq f(x)$ **then**
11:        $P \leftarrow (P \setminus \{x\}) \cup \{z\}$;
12:    **else**
13:        **if** $f(z) \geq f(y)$ **then**
14:            $P \leftarrow (P \setminus \{y\}) \cup \{z\}$;
15:        **end if**
16:    **end if**
17: **end while**

$$p_i' = \mu_i - \delta_p \cdot \left(\sqrt{\ln(1/\alpha_p) \cdot 2(|z|_1 + 1)} - \sqrt{\ln(1/\alpha_p) \cdot 2|z|_1} \right).$$

The expected profit μ_i is therefore discounted with the additional uncertainty that would be added according to the Hoeffding bound when adding an additional element to z. Once, the discounted values p_i', the elements are sorted according to p_i'/w_i. The final steps tries the elements of I in sorted order and adds element $i \in I$ if it would not violate the weight constraint.

5   Experimental Investigation

In this section, we investigate the (1+1) EA and the (1+1) EA with heavy-tailed mutation on several benchmarks with chance constraints and compare them to the $(\mu + 1)$ EA algorithm with heavy-tailed mutation and new crossover operator.

5.1   Experimental Setup

Our goal is to study different chance constraint settings in terms of the uncertainty level δ_p, and the probability bound α_p. We consider different well-known benchmarks from [12,21] in their profit chance constrained versions. We consider two types of instances, uncorrelated and bounded strong correlated ones, with $n = 100, 300, 500$ items. For each benchmark, we study the performance of (1+1) EA, (1+1) EA with heavy-tailed mutation and $(\mu + 1)$ EA with value of $\mu = 10$. We consider all combinations of $\alpha_p = 0.1, 0.01, 0.001$, and $\delta_p = 25, 50$ for the experimental investigations of the algorithms. We allow 1 000 000 fitness evaluations for each of these problem parameter combinations. For each tested

---

Algorithm 4: Discounted Greedy Uniform Crossover

Input: Individuals $x = (x_1, \ldots, x_n)$ and $y = (y_1, \ldots, y_n)$;
1: Create $z = (z_1, \ldots, z_n)$ by setting $z_i \leftarrow x_i$ iff $x_i = y_i$ and $z_i \leftarrow 0$ iff $x_i! = y_i$;
2: Let $I = \{i \in \{1, \ldots, n\} \mid x_i! = y_i\}$;
3: Set $p'_i = \mu_i - u(z, i)$ for all $i \in I$;
4: Sort the items $i \in I$ in decreasing order with respect to p'_i/w_i ratio;
5: **for** each $i \in I$ in sorted order **do**
6:   **if** $w(z) + w_i \leq B$ **then**
7:     $z_i \leftarrow 1$;
8:   **end if**
9: **end for**
10: **return** $z = (z_1, \ldots, z_n)$;

---

instance, we carry out 30 independent runs and report the average results, standard deviation and statistical test. In order to measure the statistical validity of our results, we use the Kruskal-Wallis test with 95% confidence. We apply the Bonferroni post-hoc statistical correction, which is used for multiple comparison of a control algorithm, to two or more algorithms [6]. $X^{(+)}$ is equivalent to the statement that the algorithm in the column outperformed algorithm X. $X^{(-)}$ is equivalent to the statement that X outperformed the algorithm given in the column. If algorithm X^* does appear, then no significant difference was determined between the algorithms.

6   Experimental Results

We consider now the results for the (1+1) EA, the (1+1) EA with heavy-tailed mutation and the $(\mu + 1)$ EA with heavy-tailed mutation and specific crossover algorithm based on Chebyshev's inequality and Hoeffding bounds for the benchmark set.

We first consider the optimization result obtained by the above mentioned algorithms using Chebyshev's inequality for the combinations of α_p and δ_p. The experimental results are shown in Table 1. The results show that the (1 + 1) EA with heavy-tailed mutation is able to achieve higher average results for the instances with $100, 300, 500$ items for type bounded strongly correlated in most of the cases for all α_p and δ_p combinations. It can be observed that for the instance with 100 uncorrelated items the (1 + 1) EA with heavy-tailed mutation outperforms all algorithms for $\alpha_p = 0.1$ and $\delta_p = 50$ and for $\alpha_p = 0.01$ $\delta_p = 25, 50$, respectively. However, the $(\mu + 1)$ EA can improve on the optimization result for small α_p and high δ_p values, i.e. $\alpha_p = 0.001$, $\delta_p = 50$.

It can be observed that the $(\mu + 1)$ EA obtains the highest mean value for the instance with 300 and 500 items for the uncorrelated type. Furthermore, the statistical tests show that for all combinations of α_p and δ_p the $(\mu + 1)$ EA significantly outperforms the (1 + 1) EA and (1 + 1) EA with heavy-tailed mutation. For example, for the instance with $300, 500$ items uncorrelated and

Table 1. Experimental results for the Chebyshev based function $\hat{p}_{Cheb}$.

	B	α_p	δ_p	(1+1) EA $\hat{p}_{Cheb}$	std	stat	(1+1) EA-HT $\hat{p}_{Cheb}$	std	stat	(μ+1) EA $\hat{p}_{Cheb}$	std	stat
uncorr_100	2407	0.1	25	11073.5863	36.336192	$2^{(*)},3^{(*)}$	11069.0420	46.285605	$1^{(*)},3^{(*)}$	11057.4420	59.495722	$1^{(*)},2^{(*)}$
			50	10863.1496	85.210231	$2^{(*)},3^{(*)}$	10889.4840	37.175095	$1^{(*)},3^{(*)}$	10883.7163	53.635972	$1^{(*)},2^{(*)}$
		0.01	25	10641.9089	63.402329	$2^{(*)},3^{(*)}$	10664.5974	29.489838	$1^{(*)},3^{(*)}$	10655.7251	43.869265	$1^{(*)},2^{(*)}$
			50	10054.6427	49.184220	$2^{(*)},3^{(*)}$	10066.2854	36.689426	$1^{(*)},3^{(*)}$	10064.8734	39.556767	$1^{(*)},2^{(*)}$
		0.001	25	9368.33053	46.894877	$2^{(*)},3^{(*)}$	9368.2483	34.904933	$1^{(*)},3^{(*)}$	9365.5257	40.458098	$1^{(*)},2^{(*)}$
			50	7475.44948	50.681386	$2^{(*)},3^{(*)}$	7490.6387	27.819516	$1^{(*)},3^{(*)}$	7497.5054	14.098629	$1^{(*)},2^{(*)}$
strong_100	4187	0.1	25	8638.0428	68.740095	$2^{(-)},3^{(-)}$	8698.2592	64.435352	$1^{(+)},3^{(*)}$	8707.9271	49.633473	$1^{(+)},2^{(*)}$
			50	8441.9311	80.335771	$2^{(-)},3^{(-)}$	8483.1151	45.284814	$1^{(+)},3^{(*)}$	8481.0022	55.979520	$1^{(+)},2^{(*)}$
		0.01	25	8214.8029	56.705379	$2^{(-)},3^{(-)}$	8230.9642	42.084563	$1^{(+)},3^{(*)}$	8210.1448	55.148757	$1^{(+)},2^{(*)}$
			50	7512.3033	71.115520	$2^{(-)},3^{(-)}$	7563.5495	37.758812	$1^{(+)},3^{(*)}$	7554.7382	53.030592	$1^{(+)},2^{(*)}$
		0.001	25	6771.7849	58.314395	$2^{(-)},3^{(-)}$	6797.0376	42.944371	$1^{(+)},3^{(*)}$	6793.0387	43.492135	$1^{(+)},2^{(*)}$
			50	4832.2084	88.887119	$2^{(-)},3^{(-)}$	4929.1483	52.858392	$1^{(+)},3^{(*)}$	4902.0006	44.976733	$1^{(+)},2^{(*)}$
uncorr_300	6853	0.1	25	34150.7224	167.458986	$2^{(*)},3^{(-)}$	34218.9806	164.65331	$1^{(*)},3^{(*)}$	34319.8500	177.580430	$1^{(+)},2^{(*)}$
			50	33749.8625	202.704754	$2^{(*)},3^{(-)}$	33827.9115	158.675094	$1^{(*)},3^{(*)}$	33992.7669	157.059148	$1^{(+)},2^{(*)}$
		0.01	25	33298.9369	215.463952	$2^{(*)},3^{(-)}$	33482.2230	186.361325	$1^{(*)},3^{(*)}$	33584.5679	129.781221	$1^{(+)},2^{(*)}$
			50	32326.5299	203.976688	$2^{(*)},3^{(-)}$	32332.5785	190.826414	$1^{(*)},3^{(*)}$	32504.2005	178.815508	$1^{(+)},2^{(*)}$
		0.001	25	30989.2470	242.861056	$2^{(*)},3^{(-)}$	31150.1989	187.329891	$1^{(*)},3^{(*)}$	31281.7283	181.280416	$1^{(+)},2^{(*)}$
			50	27868.2812	180.822780	$2^{(*)},3^{(-)}$	27923.1672	148.146917	$1^{(*)},3^{(*)}$	28024.3756	144.125407	$1^{(+)},2^{(*)}$
strong_300	13821	0.1	25	24795.3122	143.413600	$2^{(-)},3^{(-)}$	24939.0678	94.941141	$1^{(+)},3^{(*)}$	24850.2784	135.783162	$1^{(*)},2^{(*)}$
			50	24525.1204	161.185000	$2^{(-)},3^{(*)}$	24585.2993	112.692219	$1^{(+)},3^{(*)}$	24589.7315	125.724850	$1^{(*)},2^{(*)}$
		0.01	25	24047.9634	147.055910	$2^{(-)},3^{(-)}$	24138.6765	103.635233	$1^{(+)},3^{(*)}$	24121.8843	132.985469	$1^{(*)},2^{(*)}$
			50	22982.7691	169.377913	$2^{(-)},3^{(*)}$	23088.9710	81.229946	$1^{(+)},3^{(*)}$	23057.3537	160.481591	$1^{(*)},2^{(*)}$
		0.001	25	21689.9288	168.324844	$2^{(-)},3^{(-)}$	21824.5028	77.615607	$1^{(+)},3^{(*)}$	21786.4256	126.077269	$1^{(*)},2^{(*)}$
			50	18445.0866	125.747992	$2^{(-)},3^{(*)}$	18545.0084	98.512038	$1^{(+)},3^{(*)}$	18543.0067	96.526569	$1^{(*)},2^{(*)}$
uncorr_500	11243	0.1	25	58309.8801	266.319166	$2^{(-)},3^{(-)}$	58454.4069	295.624416	$1^{(+)},3^{(*)}$	58708.9818	157.245339	$1^{(+)},2^{(+)}$
			50	57783.7554	316.155254	$2^{(-)},3^{(-)}$	57927.2459	299.811063	$1^{(+)},3^{(*)}$	58267.9737	204.854052	$1^{(+)},2^{(+)}$
		0.01	25	57262.7885	330.683000	$2^{(-)},3^{(-)}$	57538.1166	260.869372	$1^{(+)},3^{(*)}$	57770.6524	178.217884	$1^{(+)},2^{(+)}$
			50	55916.4463	260.392742	$2^{(-)},3^{(-)}$	56086.6031	224.647105	$1^{(+)},3^{(*)}$	56321.8437	197.704397	$1^{(+)},2^{(+)}$
		0.001	25	54149.7603	364.823822	$2^{(-)},3^{(-)}$	54406.8517	249.217045	$1^{(+)},3^{(*)}$	54806.6815	170.082092	$1^{(+)},2^{(+)}$
			50	50124.9811	265.408552	$2^{(-)},3^{(-)}$	50312.3993	286.632525	$1^{(+)},3^{(*)}$	50672.0950	197.712768	$1^{(+)},2^{(+)}$
strong_500	22223	0.1	25	41104.1611	321.324820	$2^{(*)},3^{(*)}$	41523.8952	222.691441	$1^{(*)},3^{(*)}$	41458.8477	238.463764	$1^{(*)},2^{(*)}$
			50	40834.8213	243.308935	$2^{(*)},3^{(*)}$	41067.8559	229.706142	$1^{(*)},3^{(*)}$	41043.6296	173.586544	$1^{(*)},2^{(*)}$
		0.01	25	40248.7094	289.114488	$2^{(*)},3^{(*)}$	40567.8724	133.387473	$1^{(*)},3^{(*)}$	40448.5671	206.754226	$1^{(*)},2^{(*)}$
			50	38831.0336	298.888606	$2^{(*)},3^{(*)}$	39123.3879	120.110352	$1^{(*)},3^{(*)}$	38984.3118	169.701352	$1^{(*)},2^{(*)}$
		0.001	25	37201.8768	273.119842	$2^{(*)},3^{(*)}$	37490.7767	118.382846	$1^{(*)},3^{(*)}$	37395.7375	164.601365	$1^{(*)},2^{(*)}$
			50	32880.2003	272.672330	$2^{(*)},3^{(*)}$	33013.4535	172.524052	$1^{(*)},3^{(*)}$	32951.6884	206.900731	$1^{(*)},2^{(*)}$

for 100 items bounded strongly correlated the statistical tests show that the (μ + 1) EA and (1 + 1) EA with heavy-tailed mutation outperforms the (1 + 1) EA. For the other settings there is no statistical significant difference in terms of the results between all algorithms.

Table 2 shows the results obtained by the above mentioned algorithms using Hoeffding bounds for the combinations of α_p and δ_p and statistical tests. The results show that the (1+1) EA with heavy-tailed mutation obtains the highest mean values compared to the results obtained by (1+1) EA and (μ + 1) EA for each setting for the instance with 100 items for both types, uncorrelated and bounded strongly correlated. Similar to the previous investigation in the case for the instances with 300 items, the (1+1) EA with heavy-tailed mutation obtains the highest mean values compared to the results obtained by other algorithms in most of the cases. However, the solutions obtained by (μ+1) EA are significantly better performance than in the case for $\alpha_p = 0.1, 0.001$, $\delta_p = 25$.

The use of the heavy-tailed mutation when compared to the use of standard bit mutation in the (1+1) EA achieves a better performance for all cases.

Table 2. Experimental results for the Hoeffding based function $\hat{p}_{Hoef}$.

	B	α_p	δ_p	(1+1) EA			(1+1) EA-HT			(μ+1) EA		
				$\hat{p}_{Hoef}$	std	stat	$\hat{p}_{Hoef}$	std	stat	$\hat{p}_{Hoef}$	std	stat
uncorr_100	2407	0.1	25	10948.7292	90.633230	$2^{(-)}, 3^{(*)}$	11016.8190	49.768932	$1^{(+)}, 3^{(+)}$	10981.3880	37.569308	$1^{(*)}, 2^{(-)}$
			50	10707.1094	43.869094	$2^{(-)}, 3^{(*)}$	10793.1175	58.150646	$1^{(+)}, 3^{(+)}$	10708.6094	44.384035	$1^{(*)}, 2^{(-)}$
		0.01	25	10836.0906	91.332983	$2^{(-)}, 3^{(*)}$	10928.3054	45.464936	$1^{(+)}, 3^{(+)}$	10866.9831	45.408500	$1^{(*)}, 2^{(-)}$
			50	10482.6216	46.444510	$2^{(-)}, 3^{(*)}$	10611.1895	69.341044	$1^{(+)}, 3^{(+)}$	10477.2328	47.065426	$1^{(*)}, 2^{(-)}$
		0.001	25	10765.3289	68.565293	$2^{(-)}, 3^{(*)}$	10862.7124	49.091526	$1^{(+)}, 3^{(+)}$	10784.7286	38.187390	$1^{(*)}, 2^{(-)}$
			50	10263.9426	90.504901	$2^{(-)}, 3^{(*)}$	10487.5621	32.625499	$1^{(+)}, 3^{(+)}$	10309.8572	44.811326	$1^{(*)}, 2^{(-)}$
strong_100	4187	0.1	25	8553.1744	74.046187	$2^{(-)}, 3^{(*)}$	8640.05156	39.413105	$1^{(+)}, 3^{(+)}$	8588.4894	53.878268	$1^{(*)}, 2^{(-)}$
			50	8264.8129	63.309264	$2^{(-)}, 3^{(*)}$	8398.4354	46.013234	$1^{(+)}, 3^{(+)}$	8273.9670	41.403505	$1^{(*)}, 2^{(-)}$
		0.01	25	8422.9258	70.464985	$2^{(-)}, 3^{(*)}$	8540.2095	63.072560	$1^{(+)}, 3^{(+)}$	8447.8489	59.841707	$1^{(*)}, 2^{(-)}$
			50	7996.0193	65.822419	$2^{(-)}, 3^{(*)}$	8181.2980	45.667034	$1^{(+)}, 3^{(+)}$	8013.1724	56.445427	$1^{(*)}, 2^{(-)}$
		0.001	25	8338.5159	57.880350	$2^{(-)}, 3^{(*)}$	8460.7513	53.402755	$1^{(+)}, 3^{(+)}$	8362.9405	51.607219	$1^{(*)}, 2^{(-)}$
			50	7794.1245	80.411946	$2^{(-)}, 3^{(*)}$	8017.8843	53.266120	$1^{(+)}, 3^{(+)}$	7833.5575	37.293481	$1^{(*)}, 2^{(-)}$
uncorr_300	6853	0.1	25	33831.9693	181.485453	$2^{(-)}, 3^{(*)}$	34118.7631	200.095911	$1^{(+)}, 3^{(+)}$	34129.8891	172.788856	$1^{(+)}, 2^{(*)}$
			50	33380.4952	157.014552	$2^{(-)}, 3^{(*)}$	33715.2964	199.074378	$1^{(+)}, 3^{(+)}$	33662.2668	124.206823	$1^{(+)}, 2^{(*)}$
		0.01	25	33655.5737	234.136500	$2^{(-)}, 3^{(*)}$	34014.2848	200.488072	$1^{(+)}, 3^{(+)}$	33962.8643	161.560953	$1^{(+)}, 2^{(*)}$
			50	32933.5174	291.623690	$2^{(-)}, 3^{(*)}$	33327.8984	235.915481	$1^{(+)}, 3^{(+)}$	33277.4015	142.387738	$1^{(+)}, 2^{(*)}$
		0.001	25	33515.7445	219.707660	$2^{(-)}, 3^{(*)}$	33806.1572	184.532069	$1^{(+)}, 3^{(+)}$	33835.4528	149.327823	$1^{(+)}, 2^{(*)}$
			50	32706.4466	176.599463	$2^{(-)}, 3^{(*)}$	33112.7494	177.218747	$1^{(+)}, 3^{(+)}$	32940.4397	173.836538	$1^{(+)}, 2^{(*)}$
strong_300	13821	0.1	25	24602.1254	171.596469	$2^{(-)}, 3^{(*)}$	24848.3209	100.078545	$1^{(+)}, 3^{(+)}$	24734.7210	127.268428	$1^{(+)}, 2^{(-)}$
			50	24184.8938	125.755762	$2^{(-)}, 3^{(*)}$	24457.7279	118.679623	$1^{(+)}, 3^{(+)}$	24205.9660	116.049342	$1^{(+)}, 2^{(-)}$
		0.01	25	24476.1412	159.274566	$2^{(-)}, 3^{(*)}$	24638.0751	105.088783	$1^{(+)}, 3^{(+)}$	24538.4199	101.959196	$1^{(+)}, 2^{(-)}$
			50	23653.3561	225.087307	$2^{(-)}, 3^{(*)}$	24060.0806	87.242862	$1^{(+)}, 3^{(+)}$	23830.8655	85.829604	$1^{(+)}, 2^{(-)}$
		0.001	25	24256.4468	173.293324	$2^{(-)}, 3^{(*)}$	24558.9506	105.253206	$1^{(+)}, 3^{(+)}$	24345.4340	144.094192	$1^{(+)}, 2^{(-)}$
			50	23377.6774	143.350899	$2^{(-)}, 3^{(*)}$	23843.7258	114.231223	$1^{(+)}, 3^{(+)}$	23520.1166	112.403711	$1^{(+)}, 2^{(-)}$
uncorr_500	11243	0.1	25	57995.2668	285.959899	$2^{(-)}, 3^{(*)}$	58286.1443	253.880622	$1^{(+)}, 3^{(-)}$	58527.7062	179.624520	$1^{(+)}, 2^{(+)}$
			50	57331.7069	319.089163	$2^{(-)}, 3^{(*)}$	57825.9426	227.649351	$1^{(+)}, 3^{(+)}$	57899.9614	167.585846	$1^{(+)}, 2^{(+)}$
		0.01	25	57757.1719	290.254639	$2^{(-)}, 3^{(*)}$	58023.1930	277.702516	$1^{(+)}, 3^{(-)}$	58224.2474	211.715398	$1^{(+)}, 2^{(+)}$
			50	56787.0897	411.706381	$2^{(-)}, 3^{(*)}$	57367.9869	206.916491	$1^{(+)}, 3^{(+)}$	57309.9927	227.397029	$1^{(+)}, 2^{(-)}$
		0.001	25	57519.6613	379.930530	$2^{(-)}, 3^{(*)}$	57910.4812	250.540248	$1^{(+)}, 3^{(+)}$	58052.4481	182.866780	$1^{(+)}, 2^{(-)}$
			50	56446.5408	273.663433	$2^{(-)}, 3^{(*)}$	57018.3566	253.684943	$1^{(+)}, 3^{(+)}$	56942.4016	183.464200	$1^{(+)}, 2^{(-)}$
strong_500	22223	0.1	25	41060.1634	306.686391	$2^{(-)}, 3^{(*)}$	41397.7895	146.844521	$1^{(+)}, 3^{(+)}$	41186.7266	213.577571	$1^{(*)}, 2^{(-)}$
			50	40244.7545	272.646652	$2^{(-)}, 3^{(*)}$	40897.9183	231.639926	$1^{(+)}, 3^{(+)}$	40543.1279	221.615657	$1^{(*)}, 2^{(-)}$
		0.01	25	40800.7084	271.459688	$2^{(-)}, 3^{(*)}$	41204.2676	179.999423	$1^{(+)}, 3^{(+)}$	40967.2373	229.232904	$1^{(*)}, 2^{(-)}$
			50	39839.2235	271.298804	$2^{(-)}, 3^{(*)}$	40445.3621	157.093438	$1^{(+)}, 3^{(+)}$	40012.5861	189.516720	$1^{(*)}, 2^{(-)}$
		0.001	25	40561.9235	348.722449	$2^{(-)}, 3^{(*)}$	41038.9246	136.670185	$1^{(+)}, 3^{(+)}$	40768.4056	206.509572	$1^{(*)}, 2^{(-)}$
			50	39404.7836	249.449911	$2^{(-)}, 3^{(*)}$	40087.0447	167.453651	$1^{(+)}, 3^{(+)}$	39561.6572	216.629134	$, 1^{(*)}, 2^{(-)}$

Furthermore, the statistical tests show that for most combinations of α_p and δ_p, the (1+1) EA with heavy-tailed mutation significantly outperforms the other algorithms. This can be due to the fact that a higher number of bits can be flipped than in the case of standard bit mutations flipping every bit with probability $1/n$.

7 Conclusions

Stochastic problems play an important role in many real-world applications. Based on real-world problems where profits in uncertain environments should be guaranteed with a good probability, we introduced the knapsack problem with chance constrained profits. We presented fitness functions for different stochastic settings that allow to maximize the profit value P such that the probability of obtaining a profit less than P is upper bounded by α_p. In our experimental study, we examined different types of evolutionary algorithms and compared their performance on stochastic settings for classical knapsack benchmarks.

Acknowledgements. We thank Simon Ratcliffe, Will Reid and Michael Stimson for very useful discussions on the topic of this paper and an anonymous reviewer for comments that significantly helped to improve the presentation of the paper. This work has been supported by the Australian Research Council (ARC) through grant FT200100536, and by the South Australian Government through the Research Consortium "Unlocking Complex Resources through Lean Processing".

References

1. Antipov, D., Buzdalov, M., Doerr, B.: Fast mutation in crossover-based algorithms. In: Proceedings of the Genetic and Evolutionary Computation Conference, GECCO 2020, pp. 1268–1276. ACM (2020)
2. Assimi, H., Harper, O., Xie, Y., Neumann, A., Neumann, F.: Evolutionary bi-objective optimization for the dynamic chance-constrained knapsack problem based on tail bound objectives. In: ECAI 2020–24th European Conference on Artificial Intelligence. Frontiers in Artificial Intelligence and Applications, vol. 325, pp. 307–314. IOS Press (2020)
3. Capponi, L.N., Peroni, R.d.L.: Mine planning under uncertainty. Insights Mining Sci. Technol. **2**(1), 17–25 (2020)
4. Charnes, A., Cooper, W.W.: Chance-constrained programming. Manage. Sci. **6**(1), 73–79 (1959)
5. Chiong, R., Weise, T., Michalewicz, Z. (eds.): Variants of evolutionary algorithms for real-world applications. Springer (2012)
6. Corder, G.W., Foreman, D.I.: Nonparametric Statistics for Non-statisticians: A Step-by-Step Approach. Wiley, Hoboken (2009)
7. Doerr, B.: Probabilistic tools for the analysis of randomized optimization heuristics. In: Doerr, B., Neumann, F. (eds.) Theory of Evolutionary Computation. NCS, pp. 1–87. Springer, Cham (2020). https://doi.org/10.1007/978-3-030-29414-4_1
8. Doerr, B., Doerr, C., Neumann, A., Neumann, F., Sutton, A.M.: Optimization of chance-constrained submodular functions. In: The Thirty-Fourth AAAI Conference on Artificial Intelligence, AAAI 2020, pp. 1460–1467. AAAI Press (2020)
9. Doerr, B., Le, H.P., Makhmara, R., Nguyen, T.D.: Fast genetic algorithms. In: Proceedings of the Genetic and Evolutionary Computation Conference, GECCO 2017, pp. 777–784. ACM (2017)
10. Eiben, A.E., Smith, J.E.: Introduction to evolutionary computing, 2nd edn. Springer, Natural Computing Series (2015)
11. Jana, R.K., Biswal, M.P.: Stochastic simulation-based genetic algorithm for chance constraint programming problems with continuous random variables. Int. J. Comput. Math. **81**(9), 1069–1076 (2004)
12. Kellerer, H., Pferschy, U., Pisinger, D.: Knapsack Problems. Springer, Berlin (2004)
13. Liu, B., Zhang, Q., Fernández, F.V., Gielen, G.G.E.: An efficient evolutionary algorithm for chance-constrained bi-objective stochastic optimization. IEEE Trans. Evol. Comput. **17**(6), 786–796 (2013)
14. Loughlin, D.H., Ranjithan, S.R.: Chance-constrained genetic algorithms. In: Proceedings of the 1st Annual Conference on Genetic and Evolutionary Computation - Volume 1, GECCO 1999, pp. 369–376. Morgan Kaufmann Publishers Inc. (1999)
15. Marcotte, D., Caron, J.: Ultimate open pit stochastic optimization. Comput. Geosci. **51**, 238–246 (2013)
16. Morton, D.P., Wood, R.K.: On a stochastic knapsack problem and generalizations, pp. 149–168. Springer, US, Boston, MA (1998)

17. Myburgh, C., Deb, K.: Evolutionary algorithms in large-scale open pit mine scheduling. In: Proceedings of the Genetic and Evolutionary Computation Conference, GECCO 2010, pp. 1155–1162. ACM (2010)

18. Neumann, A., Neumann, F.: Optimising monotone chance-constrained submodular functions using evolutionary multi-objective algorithms. In: Bäck, T., Preuss, M., Deutz, A., Wang, H., Doerr, C., Emmerich, M., Trautmann, H. (eds.) PPSN 2020. LNCS, vol. 12269, pp. 404–417. Springer, Cham (2020). https://doi.org/10.1007/978-3-030-58112-1_28

19. Neumann, F., Sutton, A.M.: Runtime analysis of the $(1 + 1)$ evolutionary algorithm for the chance-constrained knapsack problem. In: Proceedings of the 15th ACM/SIGEVO Conference on Foundations of Genetic Algorithms, FOGA 2019, pp. 147–153. ACM (2019)

20. Osada, Y., While, R.L., Barone, L., Michalewicz, Z.: Multi-mine planning using a multi-objective evolutionary algorithm. In: IEEE Congress on Evolutionary Computation, pp. 2902–2909 (2013)

21. Pisinger, D.: Where are the hard knapsack problems? Comput. Oper. Res. **32**(9), 2271–2284 (2005)

22. Rakshit, P., Konar, A., Das, S.: Noisy evolutionary optimization algorithms - a comprehensive survey. Swarm Evol. Comput. **33**, 18–45 (2017)

23. Roostapour, V., Neumann, A., Neumann, F.: Single- and multi-objective evolutionary algorithms for the knapsack problem with dynamically changing constraints. Theoret. Comput. Sci. **924**, 129–147 (2022)

24. Roostapour, V., Neumann, A., Neumann, F., Friedrich, T.: Pareto optimization for subset selection with dynamic cost constraints. Artif. Intell. **302**, 103597 (2022)

25. Xie, Y., Harper, O., Assimi, H., Neumann, A., Neumann, F.: Evolutionary algorithms for the chance-constrained knapsack problem. In: Proceedings of the Genetic and Evolutionary Computation Conference, GECCO 2019, pp. 338–346. ACM (2019)

26. Xie, Y., Neumann, A., Neumann, F.: Specific single- and multi-objective evolutionary algorithms for the chance-constrained knapsack problem. In: Proceedings of the Genetic and Evolutionary Computation Conference, GECCO 2020, pp. 271–279. ACM (2020)

27. Xie, Y., Neumann, A., Neumann, F., Sutton, A.M.: Runtime analysis of RLS and the $(1+1)$ EA for the chance-constrained knapsack problem with correlated uniform weights. In: Proceedings of the Genetic and Evolutionary Computation Conference, GECCO 2021, pp. 1187–1194. ACM (2021)

Self-adaptation via Multi-objectivisation: An Empirical Study

Xiaoyu Qin$^{(\boxtimes)}$ (ID) and Per Kristian Lehre$^{(\boxtimes)}$ (ID)

University of Birmingham, Birmingham B15 2TT, UK
{xxq896,p.k.lehre}@cs.bham.ac.uk

Abstract. Non-elitist evolutionary algorithms (EAs) can be beneficial in optimisation of noisy and or rugged fitness landscapes. However, this benefit can only be realised if the parameters of the non-elitist EAs are carefully adjusted in accordance with the fitness function. Self-adaptation is a promising parameter adaptation method that encodes and evolves parameters in the chromosome. Existing self-adaptive EAs often sort the population by first preferring higher fitness and then the mutation rate. A previous study (Case and Lehre, 2020) proved that self-adaptation can be effective in certain discrete problems with unknown structure. However, the population can be trapped on local optima, because individuals in "dense" fitness valleys which survive high mutation rates and individuals on "sparse" local optima which only survive with lower mutation rates cannot be simultaneously preserved.

Recently, the Multi-Objective Self-Adaptive EA (MOSA-EA) (Lehre and Qin, 2022) was proposed to optimise single-objective functions, treating parameter control via multi-objectivisation. The algorithm maximises the fitness and the mutation rates simultaneously, allowing individuals in "dense" fitness valleys and on "sparse" local optima to co-exist on a non-dominated Pareto front. The previous study proved its efficiency in escaping local optima with unknown sparsity, where some fixed mutation rate EAs become trapped. However, the performance is unknown in other settings.

This paper continues the study of MOSA-EA through an empirical study. We find that the MOSA-EA has a comparable performance on unimodal functions, and outperforms eleven randomised search heuristics considered on a bi-modal function with "sparse" local optima. For NP-hard problems, the MOSA-EA increasingly outperforms other algorithms for harder NK-LANDSCAPE and k-SAT instances. Notably, the MOSA-EA outperforms a problem-specific MAXSAT solver on several hard k-SAT instances. Finally, we show that the MOSA-EA self-adapts the mutation rate to the noise level in noisy optimisation. The results suggest that self-adaptation via multi-objectivisation can be adopted to control parameters in non-elitist EAs.

Keywords: Evolutionary algorithms · Self-adaptation · Local optima · Combinatorial optimisation · Noisy optimisation

All appendices can be found at https://research.birmingham.ac.uk/en/publications/self-adaptation-via-multi-objectivisation-an-empirical-study.

G. Rudolph et al. (Eds.): PPSN 2022, LNCS 13398, pp. 308–323, 2022.
https://doi.org/10.1007/978-3-031-14714-2_22

1 Introduction

Non-elitism is widely adopted in continuous EAs, and has recently shown to be promising also for combinatorial optimisation. Several runtime analyses have shown that non-elitist EAs can escape certain local optima efficiently [9,10] and can be robust to noise [11,32]. There exist a few theoretical results to investigate how non-elitist EAs can cope with local optima [9,10,14]. SPARSELOCALOPT [10] is a tunable problem class that describes a kind of fitness landscapes with *sparse deceptive regions* (local optima) and *dense fitness valleys*. Informally, search points in a dense fitness valley have many Hamming neighbours in the fitness valley, while search points in sparse deceptive regions have few neighbours within the deceptive region. Non-elitist EAs with *non-linear selection mechanisms* are proven to cope with sparse local optima [9,10]. In non-linear selection mechanisms [30,34], the selection probability is decreasing with respect to the rank of the individual, e.g., tournament and linear ranking selection [22]. The fitter individual has a higher probability to be selected, but worse individuals still have some chance to be selected. Thus, while individuals on sparse local optimum have higher chance of being selected, fewer of their offspring stay on the peak under a sufficiently high mutation rate. In contrast, even if individuals in dense fitness valley individuals have less chance of being selected, a larger fraction of their offspring stay within the fitness valley. However, it is critical to set the "right" mutation rate, which should be sufficiently high but below the *error threshold*. Non-elitist EAs with mutation rate above the error threshold will "fail" to find optima in expected polynomial time, assuming the number of global optima is polynomially bounded [29]. Finding the appropriate mutation rate which allows the algorithm to escape not too sparse local optima is non-trivial. In noisy environments, non-elitist EAs using the "right" mutation rate beat the current state of the art results for elitist EAs [32] on several settings of problems and noise models. However, the "right" mutation rate depends on the noise level, which is usually unknown in real-world optimisation.

Self-adaptation is a promising method to automate parameter configuration [4,36,41]. It encodes the parameters in the chromosome of each individual, thus the parameters are subject to variation and selection. Some self-adaptive EAs are proven efficient on certain problems, e.g., the ONEMAX function [17], the simple artificial two-peak PEAKEDLO function [12] and the unknown structure version of LEADINGONES [7]. These self-adaptive EAs sort the population by preferring higher fitness and then consider the mutation rate. They might be trapped in sparse local optima, because individuals in dense fitness valleys which survive high mutation rates and individuals on sparse local optima which only survive with lower mutation rates cannot be simultaneously preserved.

Recently, a new self-adaptive EA, the *multi-objective self-adaptive EA* (MOSA-EA) [33], was proposed to optimise single-objective functions, which treats parameter control from multi-objectivisation. The algorithm maximises the fitness and the mutation rates simultaneously, allowing individuals in dense fitness valleys and on sparse local optima to co-exist on a non-dominated Pareto front. The previous study showed its efficiency in escaping a local optimum with unknown sparsity, where some fixed mutation rate EAs including non-linear selection EAs become trapped. However, it is unclear whether the benefit of the MOSA-EA can also be observed for more complex problems, such as NP-hard combinatorial optimisation problems and noisy fitness functions.

This paper continues the study of MOSA-EA through an empirical study of its performance on selected combinatorial optimisation problems. We find that the MOSA-EA not only has a comparable performance on unimodal functions, e.g., ONEMAX and LEADINGONES, but also outperforms eleven randomised search heuristics considered on a bi-modal function with a sparse local optimum, i.e., FUNNEL. For NP-hard combinatorial optimisation problems, the MOSA-EA increasingly outperforms other algorithms for harder NK-LANDSCAPE and k-SAT instances. In particular, the MOSA-EA outperforms a problem-specific MAXSAT solver on some hard k-SAT instances. Finally, we demonstrate that the MOSA-EA can self-adapt the mutation rate to the noise level in noisy optimisation.

2   Multi-Objective Self-Adaptive EA (MOSA-EA)

This section introduces a general framework for self-adaptive EAs (Algorithm 1), then defines the algorithm MOSA-EA as a special case of this framework by specifying a self-adapting mutation rate strategy (Algorithm 2).

Algorithm 1. Framework for self-adaptive EAs [33]

Require: Fitness function $f : \{0,1\}^n \to \mathbb{R}$. Population size $\lambda \in \mathbb{N}$. Sorting mechanism Sort. Selection mechanism P_{sel}. Self-adapting mutation rate strategy D_{mut}. Initial population $P_0 \in \mathcal{Y}^\lambda$.
1: **for** t in $0, 1, 2, \ldots$ until termination condition met **do**
2:    Sort(P_t, f)
3:    **for** $i = 1, \ldots, \lambda$ **do**
4:       Sample $I_t(i) \sim P_{sel}([\lambda])$; Set $(x, \chi/n) := P_t(I_t(i))$.
5:       Sample $\chi' \sim D_{\text{mut}}(\chi)$.
6:       Create x' by independently flipping each bit of x with probability χ'/n.
7:       Set $P_{t+1}(i) := (x', \chi'/n)$.

Let $f : \mathcal{X} \to \mathbb{R}$ be any pseudo-Boolean function, where $\mathcal{X} = \{0,1\}^n$ is the set of bitstrings of length n. For self-adaptation in non-elitist EAs, existing algorithms [7,12,17] optimise f on an extended search space $\mathcal{Y} := \mathcal{X} \times [\varepsilon, 1/2]$ which includes $\mathcal{X}$ and an interval of mutation rates. Algorithm 1 [33] shows a framework for self-adaptive EAs. In each generation t, they first sort the population P_t using an order that depends on the fitness and the mutation rate. To sort the population, existing self-adaptive EAs [7,12,17] often prefer higher fitness and then consider the mutation rate. Then, each individual in the next population P_{t+1} is produced via selection and mutation. The selection mechanism is defined in terms of the ranks of the individuals in the sorted population. Then, the selected individual changes its mutation rate based on a self-adapting mutation rate strategy and each bit is flipped with the probability of the new mutation rate. For example, the *fitness-first sorting mechanism* in the self-adaptive (μ, λ) EA (SA-EA) [7] ensures that a higher fitness individual is ranked strictly higher than a lower fitness individual as illustrated in Fig. 1(a).

Different from the existing algorithms, the MOSA-EA sorts the population by the *multi-objective sorting mechanism* (Alg. 3 in Appendix 1). It first applies the *strict non-dominated sorting method* (Alg. 4 in Appendix 1) to divide the population into several Pareto fronts, then sorts the population based on the ranks of the fronts and then the fitness values. Unlike the non-dominated fronts used in multi-objective EAs, e.g., NSGA-II [13,42], each strict non-dominated front only contains a limited number of individuals, i.e., no pair of individuals have the same objective values. Alg. 5 in Appendix 1 shows an alternative way to do multi-objective sorting. Figure 1(b) [33] illustrates an example of the order of a population after multi-objective sorting.

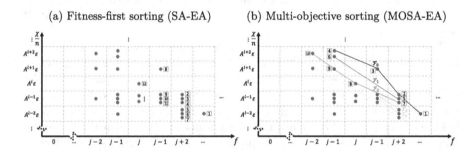

(a) Fitness-first sorting (SA-EA)   (b) Multi-objective sorting (MOSA-EA)

Fig. 1. Illustration of population sorting in (a) SA-EA and (b) MOSA-EA [33]. The points in the same cell have the same fitness and the same mutation rate.

In this paper, we consider comma selection (Alg. 6 in Appendix 1) which selects parents from the fittest μ individuals of the sorted population uniformly at random. To self-adapt the mutation rate, we apply the same strategy as in [33] (Algorithm 2), where the mutation rate is multiplied by $A > 1$ with probability $p_{\mathrm{inc}} \in (0,1)$, otherwise it is divided by A. The range of mutation rates is from $\varepsilon > 0$ to $1/2$. We sample the initial mutation rate of each individual from $\{\varepsilon A^i \mid i \in \left[0, \lfloor \log_A(\frac{1}{2\varepsilon}) \rfloor \right]\}$ uniformly at random, where i is an integer.

Most programming languages represent floats imprecisely, e.g., multiplying a number by A, and then dividing by A is not guaranteed to produce the original number. In the MOSA-EA, it is critical to use precise mutation rates to limit the number of distinct mutation rates in the population. We therefore recommend to implement the self-adaptation strategy as follows: build an indexed list containing all mutation rates $\chi/n = \varepsilon A^i$ for all integers $i \in \left[0, \lfloor \log_A(\frac{1}{2\varepsilon}) \rfloor \right]$, and encode the index into each individual. Then mutating the mutation rate can be achieved by adding or subtracting 1 to the index instead of the mutation rate multiplying A or dividing by A.

Algorithm 2. Self-adapting mutation rate strategy [33]

Require: Parameters $A > 1$, $\varepsilon > 0$ and $p_{\text{inc}} \in (0,1)$. Mutation parameter χ.

1: $\chi' = \begin{cases} \min(A\chi, \varepsilon n A^{\lfloor \log_A(\frac{1}{2\varepsilon}) \rfloor}) & \text{with probability } p_{\text{inc}}, \\ \max(\chi/A, \varepsilon n) & \text{otherwise.} \end{cases}$

2: **return** χ'.

---

3 Parameter Settings in MOSA-EA

One of the aims of self-adaptation is to reduce the number of parameters that must be set by the user. MOSA-EA has three parameters ε, p_{inc} and A, in addition to the population sizes λ and μ. We will first investigate how sensitive the algorithm is to these parameters. Adding three new parameters to adapt one parameter seems contradictory to the aim of self-adaptation. However, as we will see later in this paper, these parameters need not to be tuned carefully. We use the same parameter setting of the MOSA-EA for all experiments in this paper to show that the MOSA-EA does not require problem-specific tuning of the parameters.

The parameter ε is the lower bound of the mutation rate in the MOSA-EA. In fixed mutation rate EAs, we usually set a constant mutation parameter χ. To cover the range of all possible mutation rates χ/n, we recommend to set the lowest mutation rate $\varepsilon = c/(n \ln(n))$, where c is some small constant. In this paper, we set $\varepsilon = 1/(2n \ln(n))$. As mentioned before, $A > 0$ and $p_{\text{inc}} \in (0,1)$ are two self-adapting mutation rate parameters in Algorithm 2. We use some simple functions as a starting point to empirically analyse the effect of setting the parameters of A and p_{inc}. We run the MOSA-EA with different parameters A and p_{inc} on ONEMAX, LEADINGONES and FUNNEL ($n = 100$, the definitions can be found in Sect. 4.2) which represent single-modal and multi-modal functions. For each pair of A and p_{inc}, we execute the algorithm 100 times, with population sizes $\lambda = 10^4 \ln(n)$ and $\mu = \lambda/8$. Figures 2(a), (b) and (c) show the medians of the runtimes of the MOSA-EA for different parameters A and p_{inc} on ONEMAX, LEADINGONES and FUNNEL, respectively. The maximal number of fitness evaluations is 10^9.

From Figs. 2, the algorithm finds the optimum within 10^7 function evaluations for an extensive range of parameter settings. The algorithm is slow when A and p_{inc} are too large. Therefore, we recommend to set $p_{\text{inc}} \in (0.3, 0.5)$ and $A \in (1.01, 1.5)$. For the remainder of the paper, we will choose $p_{\text{inc}} = 0.4$ and $A = 1.01$. We also recommend to use a sufficiently large population size $\lambda = c \ln(n)$ for some large constant c. We will state the population sizes λ and μ later.

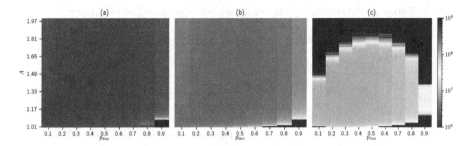

Fig. 2. Median runtimes of the MOSA-EA for different parameters A and p_{inc} on (a) ONEMAX, (b) LEADINGONES and (c) FUNNEL over 100 independent runs ($n = 100$).

4   Experimental Settings and Methodology

We compare the performance of the MOSA-EA with eleven other heuristic algorithms on three classical pseudo-Boolean functions and two more complex combinatorial optimisation problems. We also empirically study the MOSA-EA in noisy environments. In this section, we will first introduce the other algorithms and their parameter settings. We will then describe the definitions of benchmarking functions and problems. We will also indicate the statistical approach applied in the experiments.

4.1   Parameter Settings in Other Algorithms

We consider eleven other heuristic algorithms, including three single-individual elitist algorithms, *random search* (RS), *random local search* (RLS) and (1+1) EA, two population-based elitist algorithms, $(1 + (\lambda, \lambda))$ GA [15] and FastGA [16], two *estimation of distribution algorithms* (EDAs), cGA [24] and UMDA [37], two non-elitist EAs, 3-tournament EA and (μ, λ) EA, and two self-adjusting EAs, SA-EA [7] and self-adjusting population size $(1, \{F^{1/s}\lambda, \lambda/F\})$ EA (SA-$(1, \lambda)$ EA) [27], and a problem-specific algorithm, Open-WBO [35]. These heuristic algorithms are proved to be efficient in many scenarios, e.g., in multi-modal and noisy optimisation [3,6,7,9,10,16,18,19,26, 31]. Open-WBO is a MAXSAT solver that operates differently than randomised search heuristics. It was one of the best MAXSAT solvers in *MaxSAT Evaluations* 2014, 2015, 2016 and 2017 [2].

It is essential to set proper parameters for each algorithm for a comparative study [5]. In the experiments, we use parameter recommendations from the existing theoretical and empirical studies, which are summarised in Table 1.

Note that to investigate the effect of self-adaptation via multi-objectivisation, the SA-EA applied the same self-adapting mutation rate strategy and initialisation method as the MOSA-EA, instead of the strategy used in [7]. The only difference between the SA-EA and the MOSA-EA in the experiments is the sorting mechanism: The SA-EA uses the fitness-first sorting mechanism [7] (Alg. 7 in Appendix 1) and the MOSA-EA uses the multi-objective sorting mechanism.

Table 1. Parameter settings of algorithms considered in this paper

Category	Algorithm	Parameter settings
Elitist EAs	RS	-
	RLS	-
	(1+1) EA	Mutation rate $\chi/n = 1/n$
	$(1 + (\lambda, \lambda))$ GA [15]	Mutation rate $p = \lambda/n$; Crossover bias $c = 1/\lambda$;
		Population size $\lambda = 2\ln(n)$ [6]
	FastGA [16]	$\beta = 1.5$ [16]
EDAs	cGA [24]	$K = 7\sqrt{n}\ln(n)$ [44]
	UMDA [37]	$\mu = \lambda/8$
Non-Elitist EAs	3-tournament EA	Mutation rate $\chi/n = 1.09812/n$ [9,10]
	(μ, λ) EA	Mutation rate $\chi/n = 2.07/n$; Population size $\mu = \lambda/8$ [29,34]
Self-adjusting EAs	SA-$(1, \lambda)$ EA [27]	Population size $\lambda_{\text{init}} = 1$, $\lambda_{\max} = enF^{1/s}$; $F = 1.5$, $s = 1$ [27]
	SA-EA	Population size $\mu = \lambda/8$; $A = 1.01$, $p_{\text{inc}} = 0.4$, $\varepsilon = 1/(2\ln(n))$
	MOSA-EA	Population size $\mu = \lambda/8$; $A = 1.01$, $p_{\text{inc}} = 0.4$, $\varepsilon = 1/(2\ln(n))$
MAXSAT solver	Open-WBO [35][a]	Default

[a] We use version 2.1: https://github.com/sat-group/open-wbo.

4.2 Classical Functions

We first consider two well-known unimodal functions, ONEMAX and LEADING-GONES, i.e., $OM(x) := \sum_{i=1}^{n} x(i)$ and $LO(x) := \sum_{i=1}^{n} \prod_{j=1}^{i} x(j)$. One would not expect to encounter these functions in real-world optimisation. However, they serve as a good starting point to analyse the algorithms. We cannot expect good performance from an algorithm which performs poorly on these simple functions. We also consider FUNNEL which was proposed by Dang et al. [9], It is a bi-modal function with sparse local optima and a dense fitness valley which belongs to the problem class SPARSELOCALOPT$_{\alpha,\varepsilon}$ [10]. The parameters u, v, w in the FUNNEL function describe the sparsity of the deceptive region and the density of the fitness valley. Dang et al. [9] proved that the $(\mu + \lambda)$ EA and the (μ, λ) EA are inefficient on FUNNEL if $v - u = \Omega(n)$ and $w - v = \Omega(n)$, while the 3-tournament EA with the mutation rate $\chi/n = 1.09812/n$ can find the optimum in polynomial runtime. In the experiments, we test the FUNNEL function with the parameters $u = 0.5n$ $v = 0.6n$ and $w = 0.7n$ which satisfy the restrictions above.

For each problem, we independently run each algorithm 30 times for each problem size n from 100 to 200 with step size 10 and record the runtimes. For fair comparison, we set sufficiently large population sizes $\lambda = 10^4\ln(n)$ for the MOSA-EA, the 3-tournament EA, the (μ, λ) EA, the UMDA and the SA-EA.

4.3 Combinatorial Optimisation Problems

We consider two NP-hard problems, the random NK-LANDSCAPE problem and the random k-SAT problem, which feature many local optima [28,38]. We compare the performance of the MOSA-EA with other popular randomised search heuristics in a fitness evaluation budget. To further investigate the MOSA-EA, we compare it with the problem-specific algorithm Open-WBO [35] which is a MAXSAT solver in a fixed CPU time.

For fair comparisons, we set the population sizes $\lambda = 20000$ for the MOSA-EA, the 3-tournament EA, the (μ, λ) EA, the UMDA and the SA-EA. We also apply Wilcoxon rank-sum tests [45] between the results of each algorithm and the MOSA-EA.

Random NK-Landscape Problems. The NK-LANDSCAPE problem [28] can be described as: given $n, k \in \mathbb{N}$ satisfying $k \leq n$, and a set of sub-functions $f_i :$ $\{0,1\}^k \to \mathbb{R}$ for $i \in [n]$, to maximise NK-LANDSCAPE$(x) := \sum_{i=1}^{n} f_i \left(\Pi \left(x, i \right) \right)$, where the function $\Pi : \{0,1\}^n, [n] \to \{0,1\}^k$ returns a bit-string containing k right side neighbours of the i-th bit of x, i.e., $x_i, \ldots, x_{(i+k-1) \mod n}$. Typically, each sub-function is defined by a lookup table with 2^{k+1} entries, each in the interval $(0,1)$. The "difficulty" of instance can be varied by changing k [39]. Generally, the problem instances are considered to become harder for larger k. We generate 100 random NK-LANDSCAPE instances with $n = 100$ for each $k \in \{5, 10, 15, 20, 25\}$ by uniformly sampling values between 0 and 1 in the lookup table. We run each algorithm once on each instance and record the highest fitness value achieved in the fitness evaluation budget of 10^8.

Random k-Sat Problems. The k-SAT problem is an optimisation problem that aims to find an assignment in the search space $\{0,1\}^n$ which maximises the number of satisfied clauses of a given Boolean formula in conjunctive normal form [1,8,23]. For each random k-SAT instance, each of m clauses have k literals which are sampled uniformly at random from $[n]$ without replacement. We first generate 100 random k-SAT instances with $k = 5$, $n = 100$ and $m \in \{100i \mid i \in [30]\}$. Similarly with the NK-LANDSCAPE experiments, we run each algorithm on these random instances and record the smallest number of unsatisfied clauses during runs of 10^8 fitness function evaluations. Additionally, we run Open-WBO and the MOSA-EA on the same machine in one hour CPU time. The MOSA-EA is implemented in OCaml, while OpenWBO is implemented in C++, which generally leads to faster-compiled code than OCaml.

4.4 Noisy Optimisation

We consider the one-bit noise model (q) which are widely studied [18,20,21,31, 40,43]. Let $f^n(x)$ denote the noisy fitness value. Then the one-bit noise model (q) can be described as: given a probability $q \in [0,1]$, i.e., noise level, and a solution $x \in \{0,1\}^n$, then $f^n(x) = \begin{cases} f(x) & \text{with probability } 1 - q \\ f(x') & \text{with probability } q \end{cases}$, where x' is a uniformly sampled Hamming neighbour of x.

From the previous studies [11,32], non-elitist EAs can cope with the higher-level noise by reducing the mutation rate. However, we need to know the exact noise level to set a proper mutation rate. Our aim with the noisy optimisation experiments is to investigate whether the mutation rate self-adapts to the noise level when using the multi-objective self-adaptation mechanism. Therefore, we set the mutation rate of the (μ, λ) EA as $\ln(\lambda/\mu)/(2n)$ instead of the value close to the error threshold [29,34]. For a fair comparison, we set population sizes $\lambda = 10^4 \ln(n)$ and $\mu = \lambda/16$ for both the MOSA-EA and the (μ, λ) EA. The difference between the two EAs is only the parameter control method.

We test the algorithms on LEADINGONES in the one-bit noise model with noise levels $q = \{0.2, 0.6, 0.8, 0.9\}$. For each problem size $n = 100$ to 200 with step size 10, we execute 100 independent runs for each algorithm and record the runtimes. To track the behaviours of the MOSA-EA under different levels of noise, we also record the mutation rate of the individual with the highest real fitness value during the run.

5   Results and Discussion

5.1   Classical Functions

Figures 3(a), 3(b) and 3(c) show the runtimes of the MOSA-EA and nine other heuristic algorithms on ONEMAX, LEADINGONES and FUNNEL over 30 independent runs, respectively. Based on theoretical results [25], the expected runtimes of the (1+1) EA are $O\left(n \log(n)\right)$ and $O\left(n^2\right)$ on ONEMAX and LEADINGONES, respectively. We thus normalise the y-axis of Figs. 3(a) and (b) by $n \ln(n)$ and n^2, respectively. We also use the log-scaled y-axis for Figs. 3(a) and 3(b). The runtime of the 3-tournament EA with a mutation rate $\chi/n = 1.09812/n$ and a population size $c \log(n)$ for a sufficiently large constant c on FUNNEL is $O\left(n^2 \log(n)\right)$ [9]. We thus normalise the y-axis of Fig. 3(c) by $n^2 \ln(n)$. Note that (1+1) EA, RLS, (μ, λ) EA, cGA, FastGA, $(1 + (\lambda, \lambda))$ GA, SA-EA and SA-$(1, \lambda)$ EA cannot achieve the optimum of the FUNNEL function in 10^9 fitness evaluations. It is known that no elitist black-box algorithm can optimise FUNNEL in polynomial time with high probability [9,10].

Although the MOSA-EA is slower than EDAs and elitist EAs on the unimodal functions ONEMAX and LEADINGONES, it has comparable performance with the other non-elitist EAs and the SA-EA. Recall theoretical results on FUNNEL [9,10], elitist EAs and the (μ, λ) EA fail to find the optimum, while the 3-tournament EA is efficient. The results in Fig. 3 (c) are consistent with the theoretical results. In this paper, the (μ, λ) EA, the SA-EA and the MOSA-EA use the (μ, λ) selection. Compared with the (μ, λ) EA and the SA-EA, self-adaptation via multi-objectivisation can cope with sparse local optima and even achieve a better performance than the 3-tournament EA.

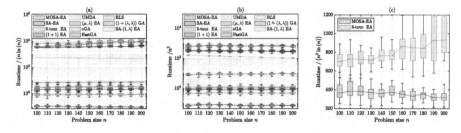

Fig. 3. Runtimes of nine algorithms on the (a) ONEMAX, (b) LEADINGONES, (c) FUNNEL ($u = 0.5n, v = 0.6n, w = 0.7n$) functions over 30 independent runs. The y-axis in sub-figures (a) and (b) are log-scaled. (1+1) EA, RLS, (μ, λ) EA, cGA, FastGA, $(1 + (\lambda, \lambda))$ GA, SA-EA and SA-$(1, \lambda)$ EA cannot find the optimum of the FUNNEL function in 10^9 fitness evaluations.

5.2 Combinatorial Optimisation Problems

Random NK-Landscape Problems. Figure 4 illustrates the experimental results of eleven algorithms on random NK-LANDSCAPE problems. From Wilcoxon rank-sum tests, the highest fitness values achieved by the MOSA-EA are statistically significantly higher than all other algorithms with significance level $\alpha = 0.05$ for all NK-LANDSCAPE with $k \in \{10, 15, 20, 25\}$. Furthermore, the advantage of the MOSA-EA is more significant for the harder problem instances.

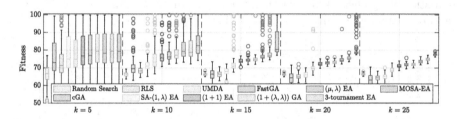

Fig. 4. The highest fitness values found in the end of runs in 10^8 fitness evaluations on 100 random NK-LANDSCAPE instances with different k ($n = 100$).

Figure 5 illustrates the highest fitness values found during the optimisation process on one random NK-LANDSCAPE instance ($k = 20$, $n = 100$). Note that the non-elitist algorithms, i.e., EDAs, (μ, λ) EA, 3-tournament EA, SA-$(1, \lambda)$ EA and MOSA-EA, do not always keep the best solution found. Therefore, the corresponding lines might fluctuate. In contrast, the elitist EAs, e.g., (1+1) EA, increase the fitness value monotonically during the whole run.

The elitist EAs converge quickly to solutions of medium quality, then stagnate. In contrast, the 3-tournament EAs, the (μ, λ) EA and the MOSA-EA improve the solution steadily. Most noticeably, the MOSA-EA improves the solution even after 10^7 fitness evaluations.

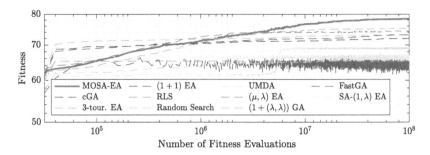

Fig. 5. The median of the highest fitness values found in every 2×10^4 fitness evaluations over 30 independent runs on one random NK-LANDSCAPE instance ($k = 20$, $n = 100$). The x-axis is log-scaled.

Random k-Sat Problems. Figure 6 illustrates the medians of the smallest number of unsatisfied clauses found in the 10^8 fitness evaluations budget among eleven algorithms on 100 random k-SAT instances ($k = 5$, $n = 100$) with different total number of clauses m. Coja-Oghlan [8] proved that the probability of generating a satisfiable instance drops from nearly 1 to nearly 0, if the ratio of the number of clauses m and the problem size n is greater than a threshold, $r_{k-\text{SAT}} = 2^k \ln(2) - \frac{1}{2}(1 + \ln(2)) + o_k(1)$, where $o_k(1)$ signifies a term that tends to 0 in the limit of large k. In this case, $r_{k-\text{SAT}}$ is roughly 2133 if we ignore the $o_k(1)$ term. We therefore call an instance with $m \geq 2133$ hard. The MOSA-EA is statistically significantly better than the other ten algorithms with significance level $\alpha = 0.05$ on hard instances from Wilcoxon rank-sum tests.

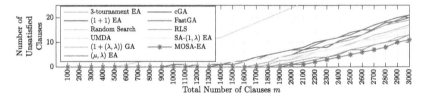

Fig. 6. The medians of the smallest number of unsatisfied clauses found in 10^8 fitness evaluations on 100 random k-SAT instances with different total numbers of clauses m ($k = 5$, $n = 100$).

Figure 7 illustrates the smallest number of unsatisfied clauses of the best solution found during the optimisation process on one random k-SAT instance ($k = 5$, $n = 100$, $m = 2500$). From Fig. 7, we come to similar conclusions with the experiments on NK-LANDSCAPE (Fig. 5).

Figure 8 illustrates the medians of the smallest number of unsatisfied clauses found in one hour CPU-time budget of Open-WBO and the MOSA-EA on 100 random k-SAT instances ($k = 5$, $n = 200$) with different total number of clauses m. For the instances with small numbers of clauses, i.e., $m \leq 1900$, Open-WBO

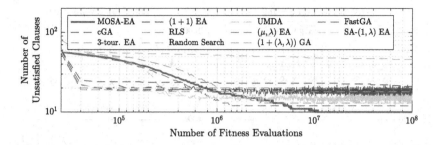

Fig. 7. The smallest number of unsatisfied clauses found over 2×10^4 fitness evaluations over 30 independent runs on one random k-SAT instance ($k = 5$, $n = 100$, $m = 2500$). The axis are log-scaled.

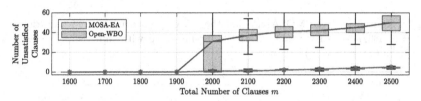

Fig. 8. The smallest number of unsatisfied clauses found in one hour CPU-time on 100 random k-SAT instances with different total numbers of clauses m ($k = 5$, $n = 100$).

returns all satisfied assignments within a few minutes, while the MOSA-EA takes up to one hour to find all satisfied assignments. However, the performance of the MOSA-EA is statistically significantly better than Open-WBO on hard instances in one hour of CPU time.

5.3 Noisy Optimisation

Figures 9 show the runtimes of the MOSA-EA and the (μ, λ) EA on LEADIN-GONES in the one-bit noise model. With a fixed mutation rate $\chi = \ln(\lambda/\mu)/(2n) = \ln(16)/(2n) \approx 1.386/n$, the runtimes of the (μ, λ) EA could be in $O(n^2)$ for low-level noise, while the runtimes rise sharply as problem size growing if the noise levels are $q = 0.9$. Based on the theoretical results [32], we could furthermore cope with the higher-level noise by using a lower mutation rate.

However, the exact noise level in real-world optimisation is usually unknown. Self-adaptation might help to configure the proper mutation rate automatically. From Fig. 9, the MOSA-EA handles the highest levels of one-bit noise, where the (μ, λ) EA encounters a problem. Figure 10 illustrates the relationships between mutation rates and real fitness values of the MOSA-EA. We observe a decrease in the mutation rate when the noise level increases. In particular, the MOSA-EA automatically adapts the mutation rate to below $1.386/n$, when using the (μ, λ) EA close to the optimum under the highest noise level. The lower mutation rate could be the reason for successful optimisation under noise.

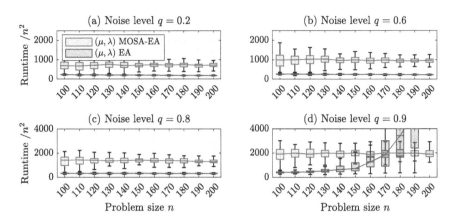

Fig. 9. Runtimes of the MOSA-EA and the (μ, λ) EA with the fixed mutation rate $\chi/n = 1.386/n$ on LEADINGONES under one-bit noise with different noise levels q. ($\lambda = 10^4 \ln(n)$, $\mu = \lambda/16$)

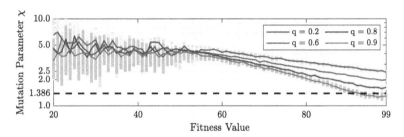

Fig. 10. Real fitness and mutation parameter of the highest real fitness individual per generation of the MOSA-EA on LEADINGONES under one-bit noise with different noise levels q. Lines show median value of mutation parameter χ. The corresponding shadows indicate the IQRs. The y-axis is log-scaled. ($n = 100$, $\lambda = 10^4 \ln(n)$, $\mu = \lambda/16$)

6   Conclusion

EAs applied to noisy or multi-modal problems can benefit from non-elitism. However, it is non-trivial to set the parameters of non-elitist EAs appropriately. Self-adaptation via multi-objectivisation, a parameter control method, is proved to be efficient in escaping local optima with unknown sparsity [33]. This paper continues the study of MOSA-EA through an empirical study of its performance on a wide range of combinatorial optimisation problems. We first empirically study the MOSA-EA on theoretical benchmark problems. The performance of the MOSA-EA is comparable with other non-elitist EAs on unimodal functions, i.e., ONEMAX and LEADINGONES. Self-adaption via multi-objectivisation can also help to cope with sparse local optima. For the NP-hard combinatorial optimisation problems, random NK-LANDSCAPE and k-SAT, the MOSA-EA is significantly better than the other nine heuristic algorithms. In particular, the MOSA-

EA can beat a state-of-the-art MAXSAT solver on some hard random k-SAT instances in a fixed CPU time. We then experimentally analyse the MOSA-EA in noisy environments. The results also demonstrate that self-adaptation can adapt mutation rates to given noise levels. In conclusion, the MOSA-EA outperforms a range of optimisation algorithms on several multi-modal and noisy optimisation problems.

Acknowledgements. This work was supported by a Turing AI Fellowship (EPSRC grant ref EP/V025562/1) and BlueBEAR/Baskerville HPC service.

References

1. Achlioptas, D., Moore, C.: Random k-SAT: two moments suffice to cross a sharp threshold. SIAM J. Comput. **36**(3), 740–762 (2006)
2. Ansótegui, C., Bacchus, F., Järvisalo, M., Martins, R.: MaxSAT Evaluation 2017. SAT (2017)
3. Antipov, D., Doerr, B., Karavaev, V.: A Rigorous Runtime Analysis of the $(1 + (\lambda, \lambda))$ GA on Jump Functions. Algorithmica, January 2022
4. Bäck, T.: Self-adaptation in genetic algorithms. In: Self Adaptation in Genetic Algorithms, pp. 263–271. MIT Press (1992)
5. Bartz-Beielstein, T., et al.: Benchmarking in Optimization: Best Practice and Open Issues. arXiv:2007.03488 [cs, math, stat], December 2020
6. Buzdalov, M., Doerr, B.: Runtime analysis of the $(1 + (\lambda, \lambda))$ genetic algorithm on random satisfiable 3-CNF formulas. In: Proceedings of the Genetic and Evolutionary Computation Conference, pp. 1343–1350. ACM, Berlin Germany (Jul 2017)
7. Case, B., Lehre, P.K.: Self-adaptation in non-Elitist Evolutionary Algorithms on Discrete Problems with Unknown Structure. IEEE Trans. Evol. Comput., 1 (2020)
8. Coja-Oghlan, A.: The asymptotic k-SAT threshold. In: Proceedings of the Forty-Sixth Annual ACM Symposium on Theory of Computing, pp. 804–813. ACM, New York, May 2014
9. Dang, D.C., Eremeev, A., Lehre, P.K.: Escaping local optima with non-elitist evolutionary algorithms. In: Proceedings of AAAI 2021. AAAI Press, Palo Alto, California USA (2020)
10. Dang, D.C., Eremeev, A., Lehre, P.K.: Non-elitist evolutionary algorithms excel in fitness landscapes with sparse deceptive regions and dense valleys. In: Proceedings of the Genetic and Evolutionary Computation Conference. ACM, Lille, France (2021)
11. Dang, D.C., Lehre, P.K.: Efficient optimisation of noisy fitness functions with population-based evolutionary algorithms. In: Proceedings of the 2015 ACM Conference on Foundations of Genetic Algorithms XIII - FOGA 2015, pp. 62–68. ACM Press, Aberystwyth, United Kingdom (2015)
12. Dang, D.-C., Lehre, P.K.: Self-adaptation of Mutation Rates in Non-elitist Populations. In: Handl, J., Hart, E., Lewis, P.R., López-Ibáñez, M., Ochoa, G., Paechter, B. (eds.) PPSN 2016. LNCS, vol. 9921, pp. 803–813. Springer, Cham (2016). https://doi.org/10.1007/978-3-319-45823-6_75
13. Deb, K., Pratap, A., Agarwal, S., Meyarivan, T.: A fast and elitist multiobjective genetic algorithm: NSGA-II. IEEE Trans. Evol. Comput. **6**(2), 182–197 (2002)

14. Doerr, B.: Does Comma Selection Help to Cope with Local Optima? Algorithmica, January 2022
15. Doerr, B., Doerr, C., Ebel, F.: From black-box complexity to designing new genetic algorithms. Theoret. Comput. Sci. **567**, 87–104 (2015)
16. Doerr, B., Le, H.P., Makhmara, R., Nguyen, T.D.: Fast genetic algorithms. In: Proceedings of the Genetic and Evolutionary Computation Conference, pp. 777–784. ACM, Berlin Germany, July 2017
17. Doerr, B., Witt, C., Yang, J.: Runtime analysis for self-adaptive mutation rates. Algorithmica **83**(4), 1012–1053 (2020). https://doi.org/10.1007/s00453-020-00726-2
18. Droste, S.: Analysis of the $(1 + 1)$ EA for a noisy ONEMAX. In: Deb, K. (ed.) GECCO 2004. LNCS, vol. 3102, pp. 1088–1099. Springer, Heidelberg (2004). https://doi.org/10.1007/978-3-540-24854-5_107
19. Friedrich, T., Kotzing, T., Krejca, M.S., Sutton, A.M.: The compact genetic algorithm is efficient under extreme gaussian noise. IEEE Trans. Evol. Comput., 1 (2016)
20. Friedrich, T., Kötzing, T., Krejca, M.S., Sutton, A.M.: Robustness of ant colony optimization to noise. Evol. Comput. **24**(2), 237–254 (2016), publisher: MIT Press
21. Gießen, C., Kötzing, T.: Robustness of populations in stochastic environments. Algorithmica **75**(3), 462–489 (2015). https://doi.org/10.1007/s00453-015-0072-0
22. Goldberg, D.E., Deb, K.: A comparative analysis of selection schemes used in genetic algorithms. In: Foundations of Genetic Algorithms, vol. 1, pp. 69–93. Elsevier (1991)
23. Gottlieb, J., Marchiori, E., Rossi, C.: Evolutionary algorithms for the satisfiability problem. Evol. Comput. **10**(1), 35–50 (2002)
24. Harik, G., Lobo, F., Goldberg, D.: The compact genetic algorithm. IEEE Trans. Evol. Comput. **3**(4), 287–297 (1999)
25. He, J., Yao, X.: A study of drift analysis for estimating computation time of evolutionary algorithms. Nat. Comput. **3**(1), 21–35 (2004)
26. Hevia Fajardo, M.A., Sudholt, D.: Self-adjusting offspring population sizes outperform fixed parameters on the cliff function. In: Proceedings of the 16th ACM/SIGEVO Conference on Foundations of Genetic Algorithms, pp. 1–15. ACM, Virtual Event Austria, September 2021
27. Hevia Fajardo, M.A.H., Sudholt, D.: Self-adjusting population sizes for non-elitist evolutionary algorithms: why success rates matter. In: Proceedings of the Genetic and Evolutionary Computation Conference, pp. 1151–1159. ACM, Lille France, June 2021
28. Kauffman, S.A., Weinberger, E.D.: The NK model of rugged fitness landscapes and its application to maturation of the immune response. J. Theoretical Biol. **141**(2), 211–245 (1989)
29. Lehre, P.K.: Negative drift in populations. In: Schaefer, R., Cotta, C., Kołodziej, J., Rudolph, G. (eds.) PPSN 2010. LNCS, vol. 6238, pp. 244–253. Springer, Heidelberg (2010). https://doi.org/10.1007/978-3-642-15844-5_25
30. Lehre, P.K.: Fitness-levels for non-elitist populations. In: Proceedings of the 13th Annual Conference on Genetic and Evolutionary Computation - GECCO 2011, pp. 2075. ACM Press, Dublin, Ireland (2011)
31. Lehre, P.K., Nguyen, P.T.H.: Runtime analyses of the population-based univariate estimation of distribution algorithms on LeadingOnes. Algorithmica **83**(10), 3238–3280 (2021). https://doi.org/10.1007/s00453-021-00862-3

32. Lehre, P.K., Qin, X.: More precise runtime analyses of non-elitist EAs in uncertain environments. In: Proceedings of the Genetic and Evolutionary Computation Conference, p. 9. ACM, Lille, France (2021)
33. Lehre, P.K., Qin, X.: Self-adaptation to multi-objectivisation: a theoretical study. In: Proceedings of the Genetic and Evolutionary Computation Conference. ACM (2022)
34. Lehre, P.K., Yao, X.: On the impact of mutation-selection balance on the runtime of evolutionary algorithms. IEEE Trans. Evol. Comput. **16**(2), 225–241 (2012)
35. Martins, R., Manquinho, V., Lynce, I.: Open-WBO: a modular MaxSAT solver'. In: Sinz, C., Egly, U. (eds.) SAT 2014. LNCS, vol. 8561, pp. 438–445. Springer, Cham (2014). https://doi.org/10.1007/978-3-319-09284-3_33
36. Meyer-Nieberg, S.: Self-adaptation in evolution strategies. Ph.D. thesis, Dortmund University of Technology (2007)
37. Mühlenbein, H., Paaß, G.: From recombination of genes to the estimation of distributions I. Binary parameters. In: Voigt, H.-M., Ebeling, W., Rechenberg, I., Schwefel, H.-P. (eds.) PPSN 1996. LNCS, vol. 1141, pp. 178–187. Springer, Heidelberg (1996). https://doi.org/10.1007/3-540-61723-X_982
38. Ochoa, G., Chicano, F.: Local optima network analysis for MAX-SAT. In: Proceedings of the Genetic and Evolutionary Computation Conference Companion, pp. 1430–1437. ACM, Prague Czech Republic, July 2019
39. Ochoa, G., Tomassini, M., Vérel, S., Darabos, C.: A study of NK landscapes' basins and local optima networks. In: Proceedings of the 10th Annual Conference on Genetic and Evolutionary Computation - GECCO 2008, pp. 555. ACM Press, Atlanta, GA, USA (2008)
40. Qian, C., Bian, C., Jiang, W., Tang, K.: Running time analysis of the (1+1)-EA for OneMax and LeadingOnes under bit-wise noise. Algorithmica **81**(2), 749–795 (2019)
41. Smith, J.E.: Self-adaptation in evolutionary algorithms for combinatorial optimisation. In: Adaptive and Multilevel Metaheuristics, pp. 31–57. Springer, Heidelberg (2008)
42. Srinivas, N., Deb, K.: Muiltiobjective optimization using nondominated sorting in genetic algorithms. Evol. Comput. **2**(3), 221–248 (1994)
43. Sudholt, D.: Analysing the robustness of evolutionary algorithms to noise: refined runtime bounds and an example where noise is beneficial. Algorithmica **83**(4), 976–1011 (2020). https://doi.org/10.1007/s00453-020-00671-0
44. Sudholt, D., Witt, C.: Update strength in EDAs and ACO: how to avoid genetic drift. In: Proceedings of the Genetic and Evolutionary Computation Conference 2016, pp. 61–68. ACM, Denver Colorado USA, July 2016
45. Wilcoxon, F.: Individual comparisons by ranking methods. In: Breakthroughs in Statistics, pp. 196–202. Springer, New York (1992)

The Combined Critical Node and Edge Detection Problem. An Evolutionary Approach

Tamás Képes, Noémi Gaskó$^{(\boxtimes)}$, and Géza Vekov

Babeş-Bolyai University, Cluj-Napoca, Romania
{tamas.kepes,noemi.gasko,geza.vekov}@ubbcluj.ro

Abstract. Studying complex networks has received a great deal of attention in recent years. A relevant problem is detecting critical nodes - nodes which, based on some measures, are more important than others in a certain network. In this paper, we propose a new optimization problem: the critical node and edge detection problem, which combines two well-known problems. A simple genetic algorithm is proposed to solve this problem, with numerical experiments having shown the potential of the method. As an application, we analyze several real-world networks and use the introduced problem as a new network robustness measure.

Keywords: Critical nodes · Critical edges · Genetic algorithm · Complex networks

1 Introduction

The study of complex networks has gained increased attention in recent years due to its applicability in different research fields (e.g. biology [1], ecology [7], telecommunication [12]). A relevant problem in networks study is the identification of a set of nodes, which, based on some properties, can be considered more important than others. If this property is a network measure, the problem is called the critical node detection problem.

The critical node detection problem (CNDP) has several application possibilities in different research fields, e.g. social network analysis [8,14], network risk management [5] and network vulnerability studies [10].

Generally, the CNDP consists of finding a set of k nodes in a given graph $G = (V, E)$, which, if deleted, maximally degrades the graph according to a given measure σ (σ can be for example, betweenness centrality, closeness centrality or page rank [18,24]).

This work was supported by a grant of the Romanian National Authority for Scientific Research and Innovation, CNCS - UEFISCDI, project number PN-III-P1-1.1-TE-2019-1633.

G. Rudolph et al. (Eds.): PPSN 2022, LNCS 13398, pp. 324–338, 2022.
https://doi.org/10.1007/978-3-031-14714-2_23

The definition of critical edge detection is almost the same. Given a graph $G = (V, E)$, the goal is to find a set of l edges in order to optimize a certain network property.

We propose a new combinatorial optimization problem, the critical node and edge detection problem. Although the critical node detection and the critical edge detection problems exist separately, the unification of the two problems is essential in some applications (e.g. road networks, computer networks, etc.) as it can model real-world situations better. In a certain network not only nodes but also edges can be deleted.

The next section of the paper presents related work about critical node and edge detection variants and algorithms. The third section describes the new proposed critical node and edge detection problem. In the fourth section, the designed algorithms are described, while section five presents the numerical results. The article ends with conclusions and further work.

2   Related Work

The CNDP can be considered as a special case of the node deletion problem [22]. Based on the survey [21] the CNDP can be divided in two main classes. The first class is the k-vertex-CNDP, where in the given graph $G = (V, E)$ and a connectivity metric σ and a given number k and the goal is to minimize the objective function $f(\sigma)$ of deleting k nodes. Some problems from this class are the MaxNum problem [32] (maximizing the number of connected components), MinMaxC (minimizing the size of the largest components) [33], and CNP (critical node problem - minimizing pairwise connectivity) [26]. The most studied variant is the CNP, with several algorithm proposals, for example, using integer linear programming [4], iterated local search algorithms [23], and greedy randomized adaptive search procedures [29]. In [3] an evolutionary framework is proposed which can deal with several variants of the CNDP.

The other main class is the β-connectivity-CNDP, where for a given $G = (V, E)$ graph, a connectivity metric σ and an integer β, the main goal is to limit the objective function $f(\sigma)$ to β, while minimizing the number of deleted nodes. Examples from this class are the Cardinality-Constrained-CNP (CC-CNP) [6] or the Component-Cardinality-Constrained CNP (3C-CNP) [20].

In an early work [40], edge deletion was studied in the case of maximum flow networks. In [37], the edge interdiction clique problem is introduced, where edges need to be removed so that the size of the maximum clique in the remaining graph is minimized and an exact algorithm is proposed to solve it for small graphs. In [15], a branch-and-cut algorithm is proposed to solve the problem.

In the matching interdiction problem [43] the weight of the maximum matching in the remaining graph after deleting edges or nodes needs to be minimized. In the same article, a pseudo-polynomial algorithm is proposed.

A recent article [9] proposes the online node and edge detection problem, where there are discussed and analyzed some online edge and node deletion problems. In [27] vertex and edge protection is proposed to stop a spreading process.

326 T. Képes et al.

Simultaneous deletion of nodes and links, for the best of our knowledge, appeared only in two variants: in [39] a joint region identification is proposed and in [11] the β-connectivity CNDP is studied where both nodes and edges can be deleted.

3   Combined Critical Node and Edge Detection Problem

The critical node and edge detection problem (CNEDP) consists of finding a set $W \subseteq V$ containing k nodes and a set $F \subseteq E$ having l edges in a given graph $G = (V, E)$, which deleted maximally degrades the graph according to a given measure σ. We denote this introduced problem by (k, l)-CNEDP.

In this article we study as a network connectivity measure the pairwise connectivity. The objective function, which needed to be minimized is the following:

$$f(A) = \sum_{C_i \in G[V \setminus A]} \frac{\delta_i(\delta_i - 1)}{2}, \tag{1}$$

where $A \subseteq V$, C_i is the set of connected components in the remaining graph, after the deletion of nodes and edges, and δ_i is the size of the connected component C_i.

Remark 1. It is obvious that $(k, 0)$-CNEDP reduces the CNDP, and $(0, l)$-CNEDP reduces the critical edge detection problem (CEDP).

Example 1. Let us consider the graph presented in Fig. 1. Considering $(1,1)$-CNDEP, if deleting the sixth node and the edge between node 3 and 4, $A = \{1, 2, 3, 4, 5, 7\}$, $C_1 = \{1, 2, 3\}$, $C_2 = \{4\}$, $C_3 = \{5, 7\}$, $\delta_1 = \{3\}$, $\delta_2 = \{1\}$, $\delta_3 = \{2\}$,

$$f(A) = \frac{3(3 - 1)}{2} + \frac{1(1 - 1)}{2} + \frac{2(2 - 1)}{2} = 4$$

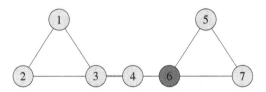

Fig. 1. A simple graph with 7 nodes

Remark 2. The complexity of the (k, l)-CNEDP is NP-complete. In [4] it is proved that the decision variant of the CNP problem is NP-complete, CNP is a subtask of the (k, l)-CNDEP problem.

4 Methods

We propose two algorithms to solve the (k, l)-CNEDP presented in the following.

4.1 Greedy Algorithm

In the framework proposed in [3] three non-evolutionary greedy solutions are presented to solve the CNDP. We adapted the second algorithm to fit the CNEDP, using the pairwise connectivity in the greedy decision-making process. The greedy selection is made based on the following function:

$$GR2(S_X) = argmax\{(f(S_X) - f(S_X \cup \{t\}) : t \in X \setminus S_X\}$$

where S_X is one of S_{nodes} or S_{edges}, and X represents respectively the original set of nodes or edges of the network.

The algorithm selects the best nodes and edges that minimize the objective function and, depending on the values of k and l, selects a single node or edge randomly, which will be removed from the network. The procedure is repeated until the maximal number of removed nodes and edges is reached. Since the selection is made randomly from the pre-selected components that have maximal impact on pairwise connectivity, the algorithm should be repeated several times to achieve the best possible results. According to [3] the number of iterations to perform should be set to $|S_{nodes}|^2$, thus we can achieve a feasible solution by setting this value to $(|S_{nodes}| + |S_{edges}|)^2$. The original framework recommends the execution of the other greedy selection rule to minimize the removed network component count after reaching a solution.

$$GR1(S_X) = argmax\{(f(S_X \setminus t) - f(S_X) : t \in S_X\}$$

Since the pairwise connectivity is driven by both the removed nodes and edges, and because of the specific structure of $(k, l) - CNEDP$, we consider that this step is not mandatory. We can argue that the complexity of the reintroduction of the removed components is too resource-intensive, but in case it is required, it can be executed. The outline of the algorithm is presented in Algorithm 1.

The number of total fitness function execution can be approximated for the algorithm, since the method is set to stop when k edges and l nodes are reached in the removal process. In each iteration the $GR2$ method computes the fitness of the network if any of the remaining nodes and edges are removed, one at a time, selecting the $Best*$ items those resulting in the best fitness value. Let us suppose that $D_{average}$ is the average node degree, and that with the removal of every node the number of fitness calculations in the next iteration will decrease with $1 + D_{average}$, and with the removal of an edge it decreases with 1. Then the approximate number of fitness execution will be $l(V + E) + \frac{1}{2}l(l + 1)(1 + D_{average}) + \frac{1}{2}k(2V - k)$.

Algorithm 1. Greedy algorithm
___
Parameters:

- G the network
- k the number of edges to remove
- l the number of nodes to remove
- S_{edges} and S_{nodes}, the set of nodes and edges to remove.
- $GR2(S_X, 1, a)$ - greedy selection algorithm notation based on [3]

$S_{edges} = \{\emptyset\}$
$S_{nodes} = \{\emptyset\}$
while $(|S_{edges}| < l)$ **or** $(|S_{nodes}| < k)$ **do**
$\quad Best_{edges} = \{GR2(S_{edges}, 1, a)\}$
$\quad Best_{nodes} = \{GR2(S_{nodes}, 1, a)\}$
$\quad [S_{nodes}, S_{edges}] = [S_{nodes}, S_{edges}] \cup Select(Best_{edges}, Best_{nodes})$
end while
return S_{nodes}, S_{edges}

4.2 Genetic Algorithm

We designed a simple genetic algorithm, the encoding, fitness evaluation and the operators used are presented in the next. The outline of the genetic algorithm is described in Algorithm 2.

Encoding: An individual is represented by two lists, one for nodes and the other for edges.

Fitness: Each individual is evaluated according to the pairwise connectivity of the connected components in the network, after the removal of the individual's nodes and edges.

Crossover and Parent Selection: This is realized using a tournament-based selection. For each round of the crossover tournament, the algorithm randomly chooses a set number of individuals from the population after which a selection is made, keeping only the two best individuals according to their fitness. They will then reproduce, by combining their node and edge lists. The algorithm will then split these lists randomly and evenly, keeping some restrictions, such as uniqueness. This way we generate two children for each round of the tournament. At the end of the crossover tournament, a set of new individuals is created (Algorithm 3).

Mutation: Two types of mutation are used. The first one is done by randomly replacing either a node or an edge in the offspring. The chance of either selection is 50%. Our new node or edge selection takes into account uniqueness inside an individual.

The second mutation is a time-varying operator. In the first step 50% of the $k + l$ nodes and edges are changed. The number of changes decreases linearly until half of the maximum generation number is reached, after which it will equal one.

While we have strict restrictions on each individual in our population, a repair operator is not necessary, since both the crossover and mutation operators self-repairs the potentially caused damage to any given individual.

Selection of Survivors: The algorithm combines the original population and any newly-created child, including the possible mutations, into a larger set of individuals, after which we trim this new set to the original population size using elitism (keeping the best individuals), this will become the new population for the next iteration of the algorithm (a $(\mu + \lambda)$ selection scheme is used).

Algorithm 2. Genetic algorithm

Parameters:

- G the network
- pop_size the number of individuals in the population
- k and l, the number of nodes and edges in an individual.
- p_{mut} the chance of mutation

Randomly initialize *pop*;
repeat
    Evaluate current population based on fitness value;
    Create child population using tournament based crossover;
    if random chance $== p_{mut}$ **then**
        Choose a random child from list of children.
        Mutate child by randomly replacing either a node or an edge with a new one;
    end if
    Elitist selection of *pop_size* number of individuals from combined parent and children population;
until Maximum number of generations;
return Best individual from final *pop*;

4.3   Experimental Set-Up

Benchmarks The synthetic benchmark set proposed in [38] contains three types of graphs, with different basic properties: Barabási-Albert (BA) - scale-free networks, Erdős-Rényi (ER) - random networks, and Forest-fire (FF) graphs, which simulate how fire spreads through a forest. Table 1 describes basic network measures of the benchmarks employed: number of nodes ($|V|$), number of edges ($|E|$), average degree ($\langle d \rangle$), density of the graph (ρ), and average path length (l_G).

In Table 2 the set of real networks used for numerical experiments is presented, including the source of the network.

Algorithm 3. Parent selection and Crossover

Parameters:

- *pop* the current population in the genetic algorithm.
- *tournament_size* the number of selected individuals to partake in the tournament.
- *max_round_number* the maximum number of rounds for the tournament, equal to half the size of newly generated child population

repeat
    Select *tournament_size* number of individuals to participate;
    Select the two best individuals according to evaluation from tournament contenders;
    Unite and then split evenly the two parents' node and edge lists.
    Append new children to the result children population.
until Maximum number of tournament rounds;
return *child_pop*

Table 1. Synthetic benchmark test graphs and basic properties.

Graph	$|V|$	$|E|$	$\langle d \rangle$	ρ	l_G
BA500	500	499	1.996	0.004	5.663
BA1000	1000	999	1.998	0.002	6.045
ER250	235	350	2.979	0.013	5.338
ER500	466	700	3.004	0.006	5.973
FF250	250	514	4.112	0.017	4.816
FF500	500	828	3.312	0.007	6.026

Parameter Setting. To find a good parameter configuration 16 parameter settings were tested on four networks: two synthetic (ER250 and FF250) and two real world networks (dolphins and karate). Table 3 presents the tested parameter settings, and Fig. 2 presents the obtained results. Based on a Wilcoxon non-parametric statistical test the configuration S11 was chosen for the further experiments.

The number of critical nodes (k) is 5% of the total nodes, while the number of critical edges (l) is set to 3% of the total number of edges (proportions are set general, to emphasize critical nodes and edges on different type of networks). The maximum generation number for both GA variants was set to 5000.

4.4   Results and Discussion

An example of the evolution of the fitness value of the genetic algorithm is presented in Fig. 3, we can observe the change of the values in each step.

For better understanding, Fig. 4 presents the smallest network, the Zebra network, and critical nodes and edges detected with the genetic algorithm.

Table 4 presents the results obtained from the genetic algorithm (GA_1), genetic algorithm with time-varying mutation (GA_2) and from the greedy algo-

Table 2. Real graphs and basic properties.

Graph	$|V|$	$|E|$	$\langle d \rangle$	ρ	l_G	Ref
Bovine	121	190	3.140	0.026	2.861	[30]
Circuit	252	399	3.167	0.012	5.806	[28]
Dolphins	62	159	5.1290	0.0841	3.3570	[19]
Ecoli	328	456	2.780	0.008	4.834	[25,41]
Football	115	613	10.6609	0.0935	2.5082	[16,19]
Hamsterster	2426	16631	13.7106	0.0057	2.4392	[31]
HumanDis	516	1188	4.605	0.008	6.509	[17]
Karate	34	78	4.5882	0.1390	2.4082	[19,42]
Zebra	27	111	8.2222	0.3162	1.3590	[19,36]

Table 3. Parameter setting used for parameter tuning

	S1	S2	S3	S4	S5	S6	S7	S8	S9	S10	S11	S12	S13	S14	15	S16
Pop. size	50	50	50	50	50	50	50	50	100	100	100	100	100	100	100	100
p_c	0.8	0.9	0.8	0.9	0.8	0.9	0.8	0.9	0.8	0.9	0.8	0.9	0.8	0.9	0.8	0.9
p_m	0.02	0.02	0.05	0.05	0.02	0.02	0.05	0.05	0.02	0.02	0.05	0.05	0.02	0.02	0.05	0.05
Tournament s.	3	3	3	3	5	5	5	5	3	3	3	3	5	5	5	5

rithm. GA_1 and GA_2 outperformed the greedy algorithm in most cases. Only in the case of the Football network did both algorithms perform in the same way (with standard deviation equal to 0). However, analyzing the results, in the case of the Football network the values for k and l were too small, because there was no change from the initial population in GA_1 and GA_2. The incorporation of time-varying mutation did not significantly improve the results.

Regarding the running time of both methods (GA_1 and greedy), in small networks, as expected, greedy runs faster (e.g. in the case of dolphins network 2.47 ± 0.02 s running time has the greedy algorithm, and 183.64 ± 0.48 s the GA_1), but in a larger network the GA_1 has better running time (e.g. for the FF500 network the greedy runs in average 1420.66 ± 63.48 s and the GA_1 1294.3 ± 25.15 s).

4.5   Application: New Network Robustness Measure Proposal

As an application of critical node and edge detection, we introduce a new network robustness measure. In the literature several robustness measure exist, trying to capture different properties of the networks. For example [13] describes different measures to characterize network robustness: k_v - vertex connectivity - the minimal number of vertices which need to be removed to disconnect the graph, k_e- edge connectivity - the same measure for edges, diameter of the graph (d), average distance ($d-$), average efficiency (E) - considering shortest paths, maximum edge betweenness (b_em), average vertex betweenness (b_v), average edge

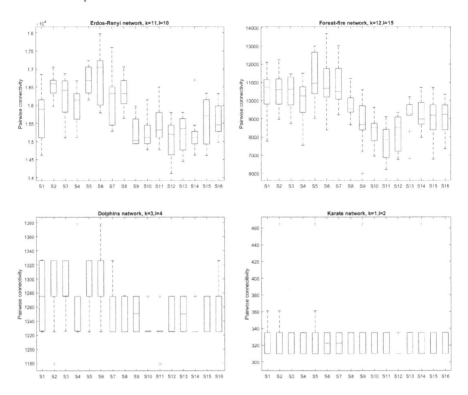

Fig. 2. Pairwise connectivity values over ten independent runs for four networks for 16 different parameter configurations

betweenness (b_e) - these measures considering shortest paths. The average clustering coefficient (C) is a proportion of triangles and connected triples. Algebraic connectivity (λ_2) is the second smallest eigenvalue of the Laplacian matrix of G, a number of spanning trees (ϵ) counts the possible different spanning trees of the graph, while effective graph resistance (Kirchhoff index) (R) investigates the graph as a network circuit.

We study several real-world networks from different application fields: two infrastructure networks - UsAir97 [31] a weighted undirected network with 332 nodes and 2126 edges (we do not take into account the weights) containing flights in the US in the year 1997 (nodes are airports and edges represent direct flight between them) and a road network [19,35] containing international E-roads, nodes representing cities (1174 nodes) and edges representing direct E-road connections between them (1417 edges). Two brain networks are studied: a mouse visual cortex network [2] with 123 nodes and 214 edges, and a cat-mixed-brain-species-1 network [2] with 65 nodes and 1100 edges (we will use the abbreviations Mouse_cortex and Cat_brain in the next). Two power networks are studied: 494-bus [31] (494 nodes and 586 edges) and 662-bus [31] (662 nodes and 906 edges), two interaction networks (Infect-dublin [34] having 410 nodes

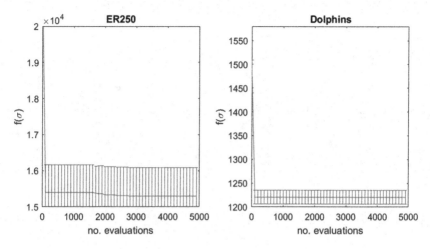

Fig. 3. Errorbar of 10 independent runs representing fitness values over evaluations on two example graphs: an Erdos-Renyi graph and Dolphins network

and 2800 edges and Infect-hyper [34] with 113 nodes and 2200 edges), and a computer network - Route_views [19] (6474 nodes and 13895 edges).

The above mentioned network measures are calculated for the studied networks, as presented in Table 5. As we can see, the majority of the indices cannot be used for disconnected networks (in this example, the E-road network is disconnected), this is one of the motivations to introduce the new measure to analyze the network robustness, based on the (k,l)-CNEDP.

The introduced measure ($NE_{k,l}$) has the following form:

$$NE_{k,l} = \frac{2 \cdot (k,l)\text{-CNEDP}}{(n-k-1)(n-k-2)}$$

$\frac{(n-k-1)(n-k-2)}{2}$ is the worst possible value of pairwise connectivity, after deleting k nodes, n is the number of nodes in the original network, $NE_{k,l} \in [0,1]$.

In the case of the USAir97 network, for example:

$$NE_{21,6} = \frac{2 \cdot 35264}{(332 - 21 - 1)(332 - 21 - 2)} = 0.66.$$

The $NE_{k,l}$ can be seen as a measure which based on the number of deletion of nodes and edges quantifies the network robustness. To analyse the results a correlation matrix was built (without the results of the E-road network), the new measure - $NE_{k,l}$ (new m) was compared with d, $d-$, E, $b_e m$, b_v and C. As presented in Fig. 5 a weak correlation exists between the new measure and the clustering coefficient (C).

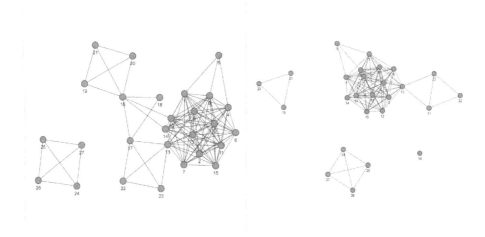

Fig. 4. The smallest real-world network, the zebra network (left). The remaining network after node and edge deletion - after (1, 3)-CNEDP (right)

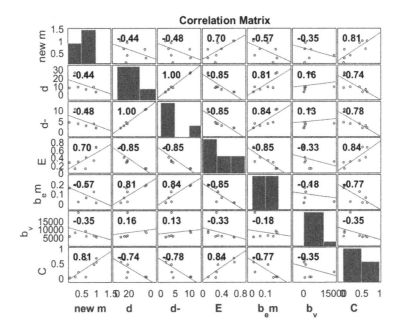

Fig. 5. Pearson's correlation coefficients between network robustness mesures

Table 4. Results for synthetic and real graphs. Mean values and standard deviation over 10 independent runs are presented. A (*) indicates best result based on a Wilcoxon sign-rank test

Graph	GA_1	GA_2	Greedy
BA500	$650.70 \pm 82.62^*$	835.00 ± 259.57	5722.90 ± 33.20
BA1000	$1947.20 \pm 235.93^*$	$2021.10 \pm 255.62^*$	362555.90 ± 924.58
ER250	$15296.30 \pm 785.33^*$	$14976.70 \pm 623.52^*$	25066.70 ± 156.40
ER500	$61664.60 \pm 1532.08^*$	$62929.20 \pm 2215.60^*$	100357.40 ± 568.82
FF250	$7905.50 \pm 1212.81^*$	$7909.70 \pm 619.53^*$	25508.20 ± 1052.20
FF500	$12303.30 \pm 4858.39^*$	$12342.90 \pm 4493.16^*$	105350.50 ± 4396.26
Bovine	$16.50 \pm 20.66^a{}^*$	$29.20 \pm 25.50^*$	1337.10 ± 31.78
Circuit	$274.00 \pm 332.22^*$	$274.70 \pm 417.62^*$	24110.10 ± 1448.70
Dolphins	$1220.40 \pm 14.55^*$	$1230.20 \pm 15.75^*$	1711.00 ± 0.00
Ecoli	$447.60 \pm 465.91^*$	$1338.40 \pm 1467.07^*$	47649.30 ± 576.69
Football	$5995.00 \pm 0.00^*$	$5995.00 \pm 0.00^*$	$5995.00 \pm 0.00^*$
HumanDis	$531.70 \pm 859.40^*$	3927.30 ± 9258.98	113625.90 ± 1521.87
Karate	$355.90 \pm 41.98^*$	$315.00 \pm 10.54^*$	411.10 ± 80.67
Zebra	$165.20 \pm 7.32^*$	$166.40 \pm 8.98^*$	237.00 ± 0.00

Table 5. Network robustness measures for studied networks

Measure	Networks								
	USAir97	E-road	Mouse_cortex	Cat_brain	494-bus	662-bus	Infect-dublin	Infect-hyper	Route_views
k_v	1	0	1	3	1	1	1	1	1
k_e	1	0	1	3	1	1	1	1	1
d	6	∞	8	3	26	25	9	3	9
$d-$	2.73	∞	4.27	1.69	10.47	10.24	3.63	1.65	3.70
E	0.40	0	0.27	0.66	0.11	0.11	0.32	0.67	0.29
$b_e m$	0.06	–	0.16	0.01	0.19	0.18	0.12	0.01	0.02
b_v	618.65	–	506.01	86.38	2827.38	3716.30	947	148.75	15227.74
b_e	70.76	–	369.78	4.84	1180.51	1429.46	110.10	4.77	5586.99
C	0.62	0	0.02	0.66	0.04	0.04	0.45	0.53	0.25
λ_2	0.12	0	0.03	2.88	-4.75	-5.14	0.19	0.99	-25.84
ϵ	3.37e+234	0	6.41e+10	1.45e+81	0	1.004	∞	1.7e+169	∞
R	45538.19	∞	47450.21	262.08	$-4.56e+18$	1.78e+18	30563.17	527.78	1587664.19
$NE^a_{k,l}$	0.66	0.33	0.09	1.00	0.28	0.64	0.92	1	0.39

[a] k and l are chosen as 3% of the nodes and 1% of edges

5   Conclusions

A new combinatorial optimization problem, the combined CNEDP, is defined and analyzed. Two methods are proposed to solve this problem: a greedy approach and a simple GA. Numerical experiments on both synthetic and real-world networks show the effectiveness of the proposed algorithm. As a direct application this newly-introduced problem is used as a new network centrality measure for network robustness testing. Further work will address other network measures (for example maximum components size, network centrality measures) and the refinement of the GA.

References

1. Albert, R.: Scale-free networks in cell biology. J. Cell Sci. **118**(21), 4947–4957 (2005)
2. Amunts, K., et al.: Bigbrain: an ultrahigh-resolution 3d human brain model. Science **340**(6139), 1472–1475 (2013)
3. Aringhieri, R., Grosso, A., Hosteins, P., Scatamacchia, R.: A general evolutionary framework for different classes of critical node problems. Eng. Appl. Artif. Intell. **55**, 128–145 (2016)
4. Arulselvan, A., Commander, C.W., Elefteriadou, L., Pardalos, P.M.: Detecting critical nodes in sparse graphs. Comput. Oper. Res. **36**(7), 2193–2200 (2009)
5. Arulselvan, A., Commander, C.W., Pardalos, P.M., Shylo, O.: Managing network risk via critical node identification. Risk management in telecommunication networks. Springer (2007)
6. Arulselvan, A., Commander, C.W., Shylo, O., Pardalos, P.M.: Cardinality-constrained critical node detection problem. In: Gülpınar, N., Harrison, P., Rüstem, B. (eds.) Performance Models and Risk Management in Communications Systems. SOIA, vol. 46, pp. 79–91. Springer, New York (2011). https://doi.org/10.1007/978-1-4419-0534-5_4
7. Bascompte, J.: Networks in ecology. Basic Appl. Ecol. **8**(6), 485–490 (2007)
8. Borgatti, S.P.: Identifying sets of key players in a social network. Comput. Math. Organ. Theory **12**(1), 21–34 (2006)
9. Chen, L.H., Hung, L.J., Lotze, H., Rossmanith, P.: Online node-and edge-deletion problems with advice. Algorithmica **83**(9), 2719–2753 (2021)
10. Dinh, T.N., Thai, M.T.: Precise structural vulnerability assessment via mathematical programming. In: 2011-MILCOM 2011 Military Communications Conference, pp. 1351–1356. IEEE (2011)
11. Dinh, T.N., Thai, M.T.: Network under joint node and link attacks: Vulnerability assessment methods and analysis. IEEE/ACM Trans. Networking **23**(3), 1001–1011 (2014)
12. Dzaferagic, M., Kaminski, N., McBride, N., Macaluso, I., Marchetti, N.: A functional complexity framework for the analysis of telecommunication networks. J. Complex Networks **6**(6), 971–988 (2018)
13. Ellens, W., Kooij, R.E.: Graph measures and network robustness. arXiv preprint arXiv:1311.5064 (2013)
14. Fan, N., Pardalos, P.M.: Robust optimization of graph partitioning and critical node detection in analyzing networks. In: Wu, W., Daescu, O. (eds.) COCOA 2010. LNCS, vol. 6508, pp. 170–183. Springer, Heidelberg (2010). https://doi.org/10.1007/978-3-642-17458-2_15
15. Furini, F., Ljubić, I., San Segundo, P., Zhao, Y.: A branch-and-cut algorithm for the edge interdiction clique problem. Eur. J. Oper. Res. **294**(1), 54–69 (2021)
16. Girvan, M., Newman, M.E.: Community structure in social and biological networks. Proc. Natl. Acad. Sci. **99**(12), 7821–7826 (2002)
17. Goh, K.I., Cusick, M.E., Valle, D., Childs, B., Vidal, M., Barabási, A.L.: The human disease network. Proc. Natl. Acad. Sci. **104**(21), 8685–8690 (2007)
18. Iyer, S., Killingback, T., Sundaram, B., Wang, Z.: Attack robustness and centrality of complex networks. PLoS ONE **8**(4), e59613 (2013)
19. Kunegis, J.: Konect: The koblenz network collection. In: Proceedings of the 22nd International Conference on World Wide Web, WWW 2013, pp. 1343–1350. Companion, Association for Computing Machinery, New York (2013). https://doi.org/10.1145/2487788.2488173

20. Lalou, M., Tahraoui, M.A., Kheddouci, H.: Component-cardinality-constrained critical node problem in graphs. Discret. Appl. Math. **210**, 150–163 (2016)
21. Lalou, M., Tahraoui, M.A., Kheddouci, H.: The critical node detection problem in networks: a survey. Comput. Sci. Rev. **28**, 92–117 (2018)
22. Lewis, J.M., Yannakakis, M.: The node-deletion problem for hereditary properties is np-complete. J. Comput. Syst. Sci. **20**(2), 219–230 (1980)
23. Lourenço, H.R., Martin, O.C., Stützle, T.: Iterated local search: framework and applications. In: Gendreau, M., Potvin, J.-Y. (eds.) Handbook of Metaheuristics. ISORMS, vol. 272, pp. 129–168. Springer, Cham (2019). https://doi.org/10.1007/978-3-319-91086-4_5
24. Lozano, M., García-Martínez, C., Rodriguez, F.J., Trujillo, H.M.: Optimizing network attacks by artificial bee colony. Inf. Sci. **377**, 30–50 (2017)
25. Lusseau, D., Schneider, K., Boisseau, O.J., Haase, P., Slooten, E., Dawson, S.M.: The bottlenose dolphin community of doubtful sound features a large proportion of long-lasting associations. Behav. Ecol. Sociobiol. **54**(4), 396–405 (2003)
26. Marx, D.: Parameterized graph separation problems. Theoret. Comput. Sci. **351**(3), 394–406 (2006)
27. Michalak, K.: Evolutionary graph-based V+E optimization for protection against epidemics. In: Bäck, T., Preuss, M., Deutz, A., Wang, H., Doerr, C., Emmerich, M., Trautmann, H. (eds.) PPSN 2020. LNCS, vol. 12270, pp. 399–412. Springer, Cham (2020). https://doi.org/10.1007/978-3-030-58115-2_28
28. Milo, R., et al.: Superfamilies of evolved and designed networks. Science **303**(5663), 1538–1542 (2004)
29. Purevsuren, D., Cui, G., Win, N.N.H., Wang, X.: Heuristic algorithm for identifying critical nodes in graphs. Adv. Comput. Sci. Int. J. **5**(3), 1–4 (2016)
30. Reimand, J., Tooming, L., Peterson, H., Adler, P., Vilo, J.: Graphweb: mining heterogeneous biological networks for gene modules with functional significance. Nucleic Acids Res. **36**, 452–459 (2008)
31. Rossi, R.A., Ahmed, N.K.: The network data repository with interactive graph analytics and visualization. In: AAAI (2015)
32. Shen, S., Smith, J.C.: Polynomial-time algorithms for solving a class of critical node problems on trees and series-parallel graphs. Networks **60**(2), 103–119 (2012)
33. Shen, S., Smith, J.C., Goli, R.: Exact interdiction models and algorithms for disconnecting networks via node deletions. Discret. Optim. **9**(3), 172–188 (2012)
34. SocioPatterns: Infectious contact networks. http://www.sociopatterns.org/datasets/
35. Šubelj, L., Bajec, M.: Robust network community detection using balanced propagation. The European Physical Journal B **81**(3), 353–362 (2011)
36. Sundaresan, S.R., Fischhoff, I.R., Dushoff, J., Rubenstein, D.I.: Network metrics reveal differences in social organization between two fission-fusion species, grevy's zebra and onager. Oecologia **151**(1), 140–149 (2007)
37. Tang, Y., Richard, J.P.P., Smith, J.C.: A class of algorithms for mixed-integer bilevel min-max optimization. J. Global Optim. **66**(2), 225–262 (2016)
38. Ventresca, M.: Global search algorithms using a combinatorial unranking-based problem representation for the critical node detection problem. Comput. Oper. Res. **39**(11), 2763–2775 (2012)
39. Wang, S., Zhang, T., Feng, C.: Nodes and links jointed critical region identification based network vulnerability assessing. In: 2016 IEEE International Conference on Network Infrastructure and Digital Content (IC-NIDC), pp. 66–71 (2016). https://doi.org/10.1109/ICNIDC.2016.7974537

40. Wollmer, R.: Removing arcs from a network. Oper. Res. **12**(6), 934–940 (1964)
41. Yang, R., Huang, L., Lai, Y.C.: Selectivity-based spreading dynamics on complex networks. Phys. Rev. E **78**(2), 026111 (2008)
42. Zachary, W.W.: An information flow model for conflict and fission in small groups. J. Anthropol. Res. **33**(4), 452–473 (1977)
43. Zenklusen, R.: Matching interdiction. Discret. Appl. Math. **158**(15), 1676–1690 (2010)

(Evolutionary) Machine Learning
and Neuroevolution

Attention-Based Genetic Algorithm for Adversarial Attack in Natural Language Processing

Shasha Zhou[1], Ke Li[2(✉)] ⬤, and Geyong Min[2]

[1] College of Computer Science and Engineering, University of Electronic Science and Technology of China, Chengdu 611731, China
[2] Department of Computer Science, University of Exeter, Exeter EX4 5DS, UK
ke.li@ieee.org

Abstract. Many recent studies have shown that deep neural networks (DNNs) are vulnerable to adversarial examples. Adversarial attacks on DNNs for natural language processing tasks are notoriously more challenging than that in computer vision. This paper proposes an attention-based genetic algorithm (dubbed AGA) for generating adversarial examples under a black-box setting. In particular, the attention mechanism helps identify the relatively more important words in a given text. Based on this information, bespoke crossover and mutation operators are developed to navigate AGA to focus on exploiting relatively more important words thus leading to a save of computational resources. Experiments on three widely used datasets demonstrate that AGA achieves a higher success rate with less than 48% of the number of queries than the peer algorithms. In addition, the underlying DNN can become more robust by using the adversarial examples obtained by AGA for adversarial training.

Keywords: Attention mechanism · Adversarial attack · Genetic algorithm · Natural language processing

1 Introduction

Deep neural networks (DNNs) have become pervasive tools for solving problems that have resisted the best attempts of the artificial intelligence community for decades. Since they have shown to be capable of discovering highly complex structures embedded in high-dimensional data, DNNs are applicable to various domains of science, engineering, business and government. For example, DNNs have achieved state-of-the-art performance in image recognition [23], computer vision [15] and natural language processing (NLP) [19]. However, many recent studies demonstrated that DNNs can be highly vulnerable, even being perturbed

This work was supported by UKRI Future Leaders Fellowship (MR/S017062/1), EPSRC (2404317), NSFC (62076056), Royal Society (IES/R2/212077) and Amazon Research Award.

by some hardly detectable noises [6,11,20], a.k.a. adversarial examples. The purpose of an adversarial attack is to find adversarial examples that fool the DNNs. There have been a wealth of studies of adversarial attacks in computer vision (e.g., [3,6]). However, due to intrinsic differences between images and textual data, the approaches for adversarial attacks in computer vision are not directly applicable for natural language processing (NLP) [26]. In a nutshell, there are two major reasons. First, the images are usually represented as continuous pixel values while the textual data are usually discrete by nature. Second, a small perturbation of images on the pixel values can hardly be recognized by a human wheres the perturbations on the text are relatively easy to be perceptible.

The existing approaches for adversarial attack for NLP can be divided into three categories. The first one is called the gradient-based attack [13,17] of which an attacker generates adversarial examples by leveraging the gradient of text vectors in a model. The second approach is called the importance-based attack [5,16]. This type of method believes that each word in the text has a different importance on the classification result of DNNs. The third one is called the population-based attack [1,7,25] that uses a population-based meta-heuristic to generate semantically and grammatically similar adversarial examples through exploiting synonyms. Comparing to the previous two approaches, the population-based attacks have shown to be powerful especially under a black-box setting. However, due to the use of a population of solutions and some intrinsic characteristics such as the importance of different words are largely ignored, they are usually less efficient.

Bearing these above considerations in mind, this paper develops a new GA based on an attention mechanism (dubbed AGA) for the adversarial example generation. In particular, the attention mechanism is designed to analyze and understand the importance of different tokens in the underlying text. Based on the extracted attention scores, our proposed AGA is able to strategically allocate the computational resources to the most influential tokens for offspring reproduction. In addition, AGA applies a $(\mu + \lambda)$-selection mechanism to achieve an accelerated convergence. Extensive experiments on three widely used datasets have fully validated the effectiveness of our proposed AGA for generating adversarial examples. In a nutshell, AGA is able to achieve a higher success rate with less than 48% of the number of queries compared against the peer methods. Moreover, our experiments also demonstrate that the adversarial examples found by AGA can be used in an adversarial training thus leading to a more robust DNNs.

The rest of this paper is organized as follows. Section 2 starts from a formal problem definition followed by description of the implementation of our proposed AGA. The experimental setup is introduced in Sect. 3 and the results are discussed in Sect. 4. At the end, Sect. 5 concludes this paper.

2   Proposed Method

In this section, we start from a formal problem definition of the adversarial attack considered in this paper. Then, we delineate the implementation of our proposed AGA step by step.

2.1   Problem Formulation

Given a text input $X = \{\mathbf{x}_i\}_{i=1}^{n} \in \mathcal{X}$, $h : \mathcal{X} \rightarrow \mathcal{Y}$ is defined as a classifier that predicts the label $y \in \mathcal{Y}$ of X, where $\mathbf{x} = (x_1, \cdots, x_{\tilde{n}})^\top$ is a word vector, $\mathcal{X}$ and $\mathcal{Y}$ are the input and output domains respectively. More specifically, the output of h is a logit vector $\phi_h(X) \in \mathbb{R}^K$ such that $y = \mathrm{argmax}_k \phi_h(X)_k$, where $k \in \{1, \cdots, K\}$ and $K > 1$ is the number of classes. An (untargeted) adversarial example is a data instance $X' \in \mathcal{X}$ such that $h(X') \neq y$ but X' and X are imperceptibly close to each other. In other words, X' and X have the same meaning to a human. In practice, the process of search for adversarial examples, a.k.a. adversarial attack, can be formulated as the following optimization problem.

$$\min_{X' \in \mathcal{X}, y \in \mathcal{Y}} \quad \mathcal{L}(X', y; h) \atop \text{subject to} \quad \rho(X, X') \leq \epsilon \tag{1}$$

where $\mathcal{L}(X', y; h)$ is defined as an adversarial loss that promotes the misclassification of the given text input $\mathbf{X}$:

$$\mathcal{L}(X, y; h) = \max\left(\phi_h(X)_y - \max_{k \neq y} \phi_h(X)_k, 0\right). \tag{2}$$

$\rho(X', X) : \mathcal{X} \times \mathcal{X} \rightarrow \mathbb{R}^+$ evaluates the difference between the X' and X, and $\epsilon > 0$ is a predefined threshold.

2.2   Implementation of Our Proposed AGA

The pseudo code of our proposed AGA is given in Algorithm 1. It follows the routine of a conventional GA but is featured with an evaluation of the attention scores of different words of the input text X at the outset. In the following paragraphs, we first introduce the working mechanism of the main crux of this paper, i.e., how to obtain the attention scores based on the attention mechanism. Then, we delineate the implementation of AGA step by step.

Attention for Score. One of the most attractive advantages of the attention mechanism is its ability to identify the information in an input most pertinent to accomplishing a task. Inspired by this, we expect to use the attention mechanism to help us identify the most important token(s) for perturbation. More specifically, this paper considers the hierarchical attention network (HAN) [24] to calculate the attention scores of different tokens in an input text (denoted as $\mathbf{s}^a$). As the overall architecture of HAN shown in Fig. 1, we can see that there are two levels of attention mechanisms in HAN. One is at the word level while the other is at the sentence level. An encoder and attention network are two building blocks of the attention model. The encoder learns the meaning behind those sequences of words and returns the vector corresponding to each word. The attention network returns weights corresponding to each token vector by using the corresponding shallow network. Afterwards, it aggregates the representation of those words to constitute a sentence vector, i.e., a weighted sum of

---

Algorithm 1: Pseudo code of AGA

Input: original text X, original label y, classifier h, maximum number of iterations T,
population size N, attention network HAN;

Output: adversarial example $\hat{X}^a$;

1 $t \leftarrow 1$, $\mathcal{P} \leftarrow \emptyset$;
2 Apply the HAN on X to obtain the attention scores $\mathbf{s}^a$;
3 Sort $\mathbf{s}^a$ in a descending order and return the sorted indices as $\mathbf{i}$;
4 **for** $i \leftarrow 1, \cdots, N$ **do**
5   Apply the mutation operation upon the i-th word of X to obtain a perturbed text $\hat{X}^i$;
6       $\mathcal{P}^t \leftarrow \mathcal{P}^t \bigcup \{\hat{X}^i\}$;

7 **while** *stopping criterion is not met* **do**
8       $\hat{X}^a = \underset{\hat{X}^i \in \mathcal{P}^t}{\arg\min} \mathcal{L}(\hat{X}^i, y; h)$;
9   **if** $h(\hat{X}^a) \neq y$ **then**
10   **return** $\hat{X}^a$;

11      $\mathcal{Q} \leftarrow \emptyset$;
12   **for** $i \leftarrow 1, \cdots, N$ **do**
13   Randomly pick up $\hat{X}^1$ and $\hat{X}^2$ from $\mathcal{P}^t$;
14   Apply the crossover operation upon $\hat{X}^1$ and $\hat{X}^2$ to obtain an offspring $\hat{X}^i_{new}$;
15   Select a site $1 \leq j \leq N$ with the probability proportional to the attention scores;
16   Apply the mutation operation upon the j-th word of $\hat{X}^i_{new}$ to obtain a further perturbed text $\hat{X}^i_{new}$;
17          $\mathcal{Q} \leftarrow \mathcal{Q} \bigcup \{\hat{X}^i_{new}\}$;

18      $\mathcal{Q} \leftarrow \mathcal{Q} \bigcup \mathcal{P}^t$;
19   Pick up the first N solutions in $\mathcal{Q}$ with the smallest adversarial loss to constitute $\mathcal{P}^{t+1}$;
20      $t \leftarrow t + 1$;

21 **return** NULL;

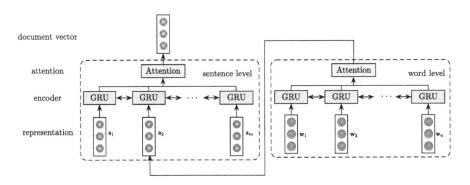

Fig. 1. The working mechanism of HAN.

those vectors that embrace the entire sentence. Analogously, we apply the same procedure to the sentence vectors so that the final vector is expected to embrace the gist of the whole document. In this paper, AGA mainly utilizes the weight vector of each token and sentence learned by HAN.

As shown in line 1 of Algorithm 1, the input text X is at first fed into HAN. By doing so, we can get the attention scores at both the word and the sentence levels. Thereafter, the importance of a word in the text X is evaluated as the product of its two attention scores. Let us use an example shown in Fig. 2 to illustrate this idea. Considering the word "loved", its sentence- and word-level attention scores are 0.1 and 0.78, respectively. Accordingly, its attention score is

$0.1 \times 0.78 = 0.078$. In practice, the higher the attention score, the larger impact on the output. By this means, considering the example shown in Fig. 2, we prefer perturbing "interesting" and "loved". Note that since the HAN is pretrained before being used in AGA, it does not cost any extra computational resource.

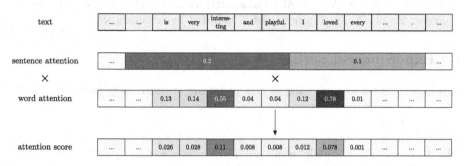

Fig. 2. An example on how to calculate the attention scores $\mathbf{s}^a$.

Optimization Procedure. As shown in Algorithm 1, AGA starts with using the HAN to obtain the attention score of each word of the input text X (line 2). Then, we sort the attention scores $\mathbf{s}^a$ in a descending order while the sorted indices is stored as a vector $\mathbf{i}$ for latter operations (line 3). For the first N words with the largest attention scores, we prepare a set of synonyms constituted by the closest 50 words in the word vector trained by GloVe [14]. GloVe is an unsupervised learning algorithm for obtaining vector representations for words. In particular, these synonyms will be used as the building blocks for the mutation operation. In lines 4 to 6, we apply the mutation operation upon each of the N important words spotted by the attention mechanism. Accordingly, we come up with a population of initialized solutions $\mathcal{P}^1 = \{\hat{X}^i\}_{i=1}^N$. During the main while loop, we first identify the solution $\hat{X}^a$ having the minimal adversarial loss as defined in Eq. (2) (line 8). Thereafter, if $\hat{X}^a$ can already achieve an attack, i.e., $h(\hat{X}^a) \neq y$, it is returned as an adversarial example we are looking for (lines 9 and 10). Otherwise, we will apply crossover and mutation operators together to generate another N offspring (lines 11 to 17). Afterwards, these offspring will be combined with the parent population $\mathcal{P}^t$ and the first N solutions having the smallest adversarial loss are used to constitute the parent population for the next iteration, a.k.a. $(\mu + \lambda)$-selection (lines 18 and 19). It is worth noting that if there is no successful adversarial example found after the computational resources are exhausted, the adversarial attack is treated as a failure and we return a NULL instead. At the end, we briefly introduce the working mechanisms of the crossover and mutation operations used in AGA.

- Crossover operation: Given a pair of texts $X = \{\mathbf{x}^i\}_{i=1}^n$ and $\bar{X} = \{\bar{\mathbf{x}}^i\}_{i=1}^n$ and a crossover position $1 \leq j \leq n$, the new texts are generated by swapping the words between the crossover position, i.e., $\hat{X}^1 = \{\mathbf{x}^i\}_{i=1}^j \bigcup \{\bar{\mathbf{x}}^i\}_{i=j+1}^n$ and $\hat{X}^2 = \{\bar{\mathbf{x}}^i\}_{i=1}^j \bigcup \{\mathbf{x}^i\}_{i=j+1}^n$. In particular, the one having a smaller adversarial loss is used as the offspring after the crossover operation, i.e., $\hat{X}_{new} = \underset{X \in \{\hat{X}^1, \hat{X}^2\}}{\text{argmin}} \ \mathcal{L}(X, y; h)$.

- Mutation operation: For a given text $X = \{\mathbf{x}^i\}_{i=1}^n$ and a mutation position $1 \le j \le N$ where $N < n$, AGA randomly pick a synonym $\tilde{\mathbf{x}}^j$ from the set of synonyms of $\mathbf{x}^j$. Then, a new text is mutated by replacing $\mathbf{x}^j$ by $\tilde{\mathbf{x}}^j$.

3 Experimental Setup

The experimental settings used in this paper are outlined as follows.

3.1 Datasets

The following three widely used public datasets including IMDB [10][1], AG's News [27][2], and SNLI [2][3] are considered in our empirical study.

- IMDB: This dataset considers binary classification tasks used for sentiment analysis. It consists of 50,000 movie reviews collected from IMDB, each of which is assigned with a binary sentiment, i.e., either positive or negative. More specifically, this dataset contains 9,998 unique words. The average length is 234 words and the standard deviation is 173 words. In our experiments, half of the data are used as the training data while the other half is used for the testing purpose.
- AG's News: This is a subset of AG's corpus. Its news articles collected from four largest classes of AG's corpus including 'World', 'Sports', 'Business', and 'Sci/Tech'. In particular, they are constructed by assembling titles and description fields. In our experiments, the AG's News dataset contains 30,000 training data and 1,900 test data for each class.
- SNLI: This dataset considers text entailment (TE) tasks for three classes including entailment, contradiction, and neutral. In particular, the TE in NLP is a directional relation between text fragments. The relation holds whenever the truth of one text fragment (text) follows from another text (hypothesis). In our experiments, the SNLI dataset consists of 570,000 human-written English sentence pairs.

Note that both AG's News and SNLI datasets are multi-class classification tasks. The average text length of IMDB is longer than the other two.

3.2 Target Model

For the IMDB ad AG's News dataset, we choose WordCNN [8], LSTM [22] and BERT [4] as the classification models to validate the effectiveness of the generated adversarial examples. For the SNLI dataset, since its inputs are sentence pairs, we

---

[1] https://www.kaggle.com/lakshmi25npathi/imdb-dataset-of-50k-movie-reviews.
[2] https://www.kaggle.com/amananandrai/ag-news-classification-dataset.
[3] https://nlp.stanford.edu/projects/snli/.

choose three variants of Transformer including BERT [4], DistilBERT [18], and ALBERT [9] as the target models. More specifically, we apply bert-base-uncased, distilbert-base-cased and albert-base-v2 published in HuggingFace[4]. They are fine tuned with five epochs while the batch size is set to 16 and the learning rate is set to 2×10^{-5}. As for the WordCNN and LSTM, we apply the models trained by TextAttack [12].

3.3 Evaluation Metrics

To quantitatively evaluate the performance different methods, we consider the following three metrics in our empirical study.

– Attack success rate (ASR): the percentage of the number of successful adversarial attacks examples (denoted as N_{succ}) w.r.t. the total number of examples generated by the corresponding algorithm (denoted as N_{total}):

$$ASR = \frac{N_{\mathrm{succ}}}{N_{\mathrm{total}}} \times 100\%. \tag{3}$$

The higher the ASR is, the better performance of the algorithm achieves.

– Average perturbation word rate (APWR): the average ratio of the perturbation words w.r.t. the total number of words:

$$APWR = \frac{1}{N_{succ}} \sum_{i=1}^{N_{succ}} \frac{|\mathbf{x}_{adv} - \mathbf{x}_i|}{|\mathbf{x}_i|} \times 100\%, \tag{4}$$

where $|*|$ indicates the number of tokens in the set. The lower the APWR is, the less perturbations caused by the adversarial example.

– Average query count (AQC): the average number of queries (denoted as N_{query}) w.r.t. the target model to find a successful adversarial example:

$$AQC = \frac{1}{N_{succ}} \sum_{i=1}^{N_{succ}} N_{\mathrm{query}}. \tag{5}$$

The larger the AQC is, the more resources required by the corresponding algorithm.

4 Experimental Results

We seek to answer the following three research questions (RQs) through our experimental evaluation:

– *RQ1:* How is the performance of our proposed AGA against the selected peer algorithms?
– *RQ2:* What are the benefits of the attention mechanism and $(\lambda + \mu)$-selection of AGA?
– *RQ3:* What are the effectiveness of adversarial examples generated by AGA?

---

[4] https://huggingface.co/.

Table 1. Comparision results of ASR, APWR and AQC values obtained by `AGA` and the other three selected peer algorithms on `IMDB`, `AG's News` and `SNLI`.

Dataset	Model	Metric	Genetic attack	IBP-certified	IGA	AGA
IMDB	WordCNN	ASR(%)	9.594E+1(3.16E-2)†	8.634E+1(6.58E-1)†	**9.881E+1(2.84E-1)**	9.735E+1(2.27E-1)
		APWR(%)	4.660E+0(1.49E-1)†	4.643E+0(1.84E-2)†	3.335E+0(1.06E-1)†	**3.033E+0(1.06E-1)**
		AQC	3.688E+3(2.45E+3)†	8.829E+3(1.34E+6)†	2.248E+5(5.40E+8)†	**3.324E+3(2.23E+3)**
	LSTM	ASR(%)	9.298E+1(1.50E-1)†	8.151E+1(2.81E-1)†	**9.638E+1(2.43E-1)‡**	9.415E+1(1.85E-1)
		APWR(%)	4.800E+0(1.14E+0)†	4.640E+0(3.08E+0)†	**2.980E+0(1.29E+0)**	3.080E+0(1.50E-1)
		AQC	3.987E+3(2.63E+2)†	8.653E+3(4.00E+5)†	1.975E+5(1.63E+8)†	**3.517E+3(1.08E+4)**
	BERT	ASR(%)	8.507E+1(2.12E+0)†	7.397E+1(2.46E+0)†	8.737E+1(3.56E+0)†	**9.053E+1(2.73E+0)**
		APWR(%)	7.940E+0(1.55E+0)†	7.425E+0(3.89E+0)†	**4.870E+0(8.70E-1)**	4.980E+0(1.02E+0)
		AQC	5.947E+3(9.34E+4)†	6.058E+3(3.14E+5)†	1.738E+5(3.73E+3)†	**4.667E+3(5.77E+5)**
AG's News	WordCNN	ASR(%)	5.565E+1(3.89E-1)†	2.722E+1(4.15E+0)†	7.685E+1(4.37E-1)†	**8.441E+1(7.52E-3)**
		APWR(%)	1.570E+1(5.33E-2)†	**1.358E+1(2.63E-1)‡**	1.561E+1(2.12E-2)†	1.457E+1(8.67E-2)
		AQC	3.090E+3(5.57E+2)†	4.936E+3(8.21E+3)†	5.056E+3(5.56E+4)†	**1.702E+3(3.56E+3)**
	LSTM	ASR(%)	5.208E+1(3.38E+0)†	2.875E+1(5.75E+0)†	6.981E+1(1.33E+0)†	**7.682E+1(1.50E-1)**
		APWR(%)	1.465E+1(1.86E+1)†	**1.332E+1(2.21E-1)‡**	1.592E+1(8.00E-4)†	1.472E+1(3.47E-1)
		AQC	3.287E+3(8.19E+2)†	4.792E+3(1.49E+3)†	5.529E+3(2.44E+3)†	**1.831E+3(1.15E+3)**
	BERT	ASR(%)	3.822E+1(4.67E+0)†	2.338E+1(4.26E+0)†	5.670E+1(2.46E+0)†	**5.996E+1(1.28E-1)**
		APWR(%)	1.447E+1(4.18E-2)†	**1.320E+1(1.01E-1)**	1.519E+1(2.64E-2)†	1.401E+1(4.03E-3)
		AQC	3.649E+3(1.84E+4)†	4.095E+3(1.40E+3)†	6.189E+3(3.35E+4)†	**2.259E+3(9.79E+3)**
SNLI	DistilBERT	ASR(%)	8.772E+1(4.26E-1)†	7.686E+1(1.40E+0)†	9.871E+1(3.33E-5)	**9.918E+1(3.00E-2)**
		APWR(%)	9.170E+1(5.19E-2)†	1.017E+1(1.16E-1)†	**8.88E+0(2.43E-2)‡**	8.550E+0(1.20E-1)
		AQC	7.664E+2(1.55E+1)†	1.059E+3(1.92E+2)†	4.047E+2(3.38E+2)†	**2.324E+2(1.71E+2)**
	ALBERT	ASR(%)	8.997E+1(3.05E-2)†	7.867E+1(4.90E-1)†	9.932E+1(3.98E-2)	**9.940E+1(1.76E-2)**
		APWR(%)	9.330E+0(3.01E-2)†	9.850E+0(3.92E-2)†	8.010E+0(3.84E-2)†	**7.827E+0(3.33E-5)**
		AQC	7.502E+2(1.12E+2)†	1.064E+3(1.17E+2)†	3.700E+2(4.40E+2)†	**2.108E+2(9.78E+1)**
	BERT	ASR(%)	8.692E+1(1.14E+0)†	7.723E+1(3.08E+0)†	**9.922E+1(2.24E-1)**	9.887E+1(1.50E-1)
		APWR(%)	8.910E-0(5.04E-2)†	1.011E+1(7.05E-2)†	8.350E+0(5.00E-5)†	**8.120E+0(8.33E-4)**
		AQC	8.623E+2(7.22E+2)†	1.082E+3(2.80E+3)†	3.811E+2(5.80E+2)†	**2.200E+2(9.98E+1)**

† denotes the performance of `AGA` is significantly better than other peers according to the Wilcoxon's rank sum test at a 0.05 significance level;

‡ denotes the corresponding algorithm significantly outperforms `AGA`.

4.1 Performance Comparison with the Selected Peer Algorithms

Method. In this experiment, `Genetic attack` [1], `IBP-certified` [7], and `IGA` [21] are chosen as the peer algorithms. All these algorithms apply a genetic algorithm to attack neural networks and are under a black-box attack setting. For each dataset introduced in Sect. 3.1, we randomly pick up 1,000 examples in our experiment. In view of the stochastic characteristics of the selected peer algorithms, each experiment is independently repeated 10 times with a different random seed. For a fair comparison, the population size is set to be 60, and the maximum number of iterations is set to be 20 for all peer algorithms.

Results and Analysis. From the comparison results shown in Table 1, it is clear to see the superiority of our proposed `AGA`. It obtains better ASR, APWR and AQC values in 18 out of 27 comparisons compared against the other three selected peer algorithms. More specifically, for the sentiment analysis task given in `IMDB`, the number of queries cost by `IGA` is at least 3× larger than the other peer algorithms. This can be partially attributed to the longer text length of the data in `IMDB`. Since the basic idea of `IGA` is to perturb the tokens in a sequential manner, it thus requires a larger number of queries in `IMDB`. The ASR obtained by AGA significantly outperforms the other peer algorithms on the AG's News whereas the superiority of the APWR is not that evident accordingly. This can be explained as none of the other peer algorithms can lead to a successful attack while our proposed AGA perturbs more words to achieve so.

Table 2. Selected adversarial attack examples for BERT on IMDB, AG's News, and SNLI. Note that the modified words are highlighted blue and red, respectively, for the original and adversarial texts.

Dataset	Attacker	Prediction	Text
IMDB	original	Pos	This movie is one of the funniest I have seen in years. A movie which deals with death and funerals without being depressing, or irreverant. Christopher Walken provides much of the comedy in this charming romance and I could hardly breathe for laughing so hard. I saw the movie a preview, and when it was over, the audience not only applauded, but cheered. I am telling all my friends to watch for it's arrival in the USA. I definitely plan on seeing it again in the theater and purchasing it on DVD as soon as it's available.
	IGA	Neg	Cette movie is one of the funniest I have seen in yr. A movie which deals with death and mortuary without being depressing, or irreverant. Christopher Walken provides much of the sitcom in this charming romance and I could hardly breathe for laugh so cumbersome. I saw the movie a preview, and when it was over, the audience not only applauded, but cheered. I am telling all my friends to watch for it's arrival in the US. I admittedly plan on seeing it again in the theater and purchasing it on DVD as soon as it's accessible.
	AGA	Neg	Cette movie is one of the funniest I have seen in yr. A movie which deals with death and funerals unless being depressing, or irreverant. Christopher Walken provides much of the sitcom in this charming romance and I could hardly breathe for smile so cumbersome. I saw the movie a preview, and when it was over, the audience not only applauded, but cheered. I am telling all my friends to watch for it's arrival in the USA. I definitely plan on seeing it again in the theater and purchasing it on DVDS as soon as it's available.
AG's News	original	World	Around the world Ukrainian presidential candidate Viktor Yushchenko was poisoned with the most harmful known dioxin, which is contained in Agent Orange, a scientist who analyzed his blood said Friday.
	IGA	Sci/Tech	Around the universe Ukrainian chair candidate Viktor Yushchenko was venomous with the most harmful known dioxin, which is contained in Actor Orange, a scientist who exploring his blood said Tuesday.
	AGA	Sci/Tech	Around the universe Ukrainian chairmen candidate Viktor Yushchenko was venomous with the most harmful known dioxin, which is contained in Agent Orange, a scientist who analyzed his blood said Friday.
SNLI	original	neutral	**premise:** A blond-haired woman squinting and wearing a bright yellow shirt. **hypothesis:** A woman is brushing her hair with a fork.
	IGA	entailment	**premise:** A blond-haired woman squinting and wearing a bright amber sweater. **hypothesis:** A nana is brushing her headgear with a fork.
	AGA	entailment	**premise:** A blond-haired woman frowning and wearing a radiant yellow shirt. **hypothesis:** A nana is brushing her headgear with a fork.

To have a better understanding of the adversarial examples generated by our proposed AGA and IGA, we pick up some selected examples for BERT on IMDB, AG's News, and SNLI, respectively, in Table 2. From these examples, we can see that AGA perturbs less words than IGA to achieve a successful attack, especially for the IMDB dataset. In addition, we notice that many of the words perturbed by IGA. This observation indicates that AGA is able to capture the words essential to the underlying text. Thus, the corresponding adversarial examples are more likely to be closer to the decision boundary.

> Response to *RQ1*: From the experimental results shown in this subsection, it is clear to see the outstanding performance of our proposed AGA compared against the other three selected peer algorithms for generating adversarial examples. In particular, it is interesting to note that AGA is able to find adversarial examples with nearly half of the computational cost (i.e., the amount of queries to the target models) of the peer algorithms.

4.2   Ablation Study of Attention Mechanism and $(\mu + \lambda)$-Selection

Method. To address the RQ2, we develop three variants to investigate the usefulness of the attention mechanism and the $(\mu+\lambda)$-selection mechanism therein.

- AGA-$v1$: This variant considers a vanilla genetic algorithm without using both the attention and the $(\mu+\lambda)$-selection mechanisms. Instead, AGA-$v1$ first picks up the best solution from the parent population to survive to the next iteration. Thereafter, it uses a uniform mutation upon each parent to generation $N-1$ offspring to directly survive to the next iteration. In particular, the uniform mutation simply picks up a token for a perturbation.
- AGA-$v2$: This variant is similar to AGA-$v1$ except the offspring reproduction is kept the same as AGA.
- AGA-$v3$: This variant is similar to AGA except the offspring reproduction is replaced by the uniform mutation as in AGA-$v1$.

Note that the other parameter settings are kept the same as in Sect. 4.1.

Results and Analysis. From the comparison results shown in Table 3, we find that although the ASR obtained by AGA-$v2$ is worse than that of AGA-$v1$, the number of queries incurred by AGA-$v2$ is reduced by 58%. This can be explained as the use of attention mechanism is able to narrow down the search space thus leading to a loss of population diversity. On the other hand, we find that the ASR obtained by AGA-$v3$ is improved on around 67% comparisons against AGA-$v1$. This can be attributed to the accelerated convergence provided by the $(\mu+\lambda)$-selection mechanism towards the decision boundary. In contrast, it is as anticipated that our proposed AGA is the most competitive algorithm in almost all cases, except for BERT on the IMDB dataset. Due to the use of both the attention mechanism and the $(\mu+\lambda)$-selection mechanism, AGA is able to achieve a successful with the least amount of queries.

> Response to *RQ2*: From the experimental results obtained in this subsection, we appreciate the usefulness of both the attention mechanism and the $(\mu+\lambda)$-selection mechanism in AGA. In particular, the attention mechanism is able to narrow down the search space while the $(\mu + \lambda)$-selection mechanism is able to provide a stronger selection pressure towards the decision boundary.

4.3   Comparison of the Generated Adversarial Examples

Method. To address RQ3, this subsection aims to investigate the effectiveness of the generated adversarial examples from the following two aspects.

Table 3. Comparison results of ASR, APWR and AQC values obtained by AGA and its three variants on IMDB, AG's News and SNLI.

	IMDB								
Attacker	WordCNN			LSTM			BERT		
	ASR(%)	APWR(%)	AQC	ASR(%)	APWR(%)	AQC	ASR(%)	APWR(%)	AQC
AGA-$v1$	**98.51**	3.36	203870.01	96.14	**3.02**	197649.08	88.17	4.82	173762.83
AGA-$v2$	95.56	3.14	5466.94	**96.73**	3.04	5665.29	88.85	**4.77**	6694.37
AGA-$v3$	98.37	3.78	3977.12	95.89	3.04	3999.88	89.52	4.84	**4393.78**
AGA	96.05	**3.03**	**3350.92**	94.15	3.08	**3577.11**	**90.75**	4.93	4722.83

	AG's News								
Attacker	WordCNN			LSTM			BERT		
	ASR(%)	APWR(%)	AQC	ASR(%)	APWR(%)	AQC	ASR(%)	APWR(%)	AQC
AGA-$v1$	76.21	15.63	4914.31	68.99	15.90	5419.71	57.85	15.30	6055.97
AGA-$v2$	55.40	**12.12**	2526.81	54.01	**12.13**	2568.27	56.47	14.73	3310.33
AGA-$v3$	78.28	13.73	3513.37	**77.37**	15.96	4813.49	**61.37**	14.65	2570.82
AGA	**84.36**	14.74	**1668.02**	77.04	15.07	**1811.27**	60.17	**14.05**	**2201.93**

	SNLI								
Attacker	DistilBERT			ALBERT			BERT		
	ASR(%)	APWR(%)	AQC	ASR(%)	APWR(%)	AQC	ASR(%)	APWR(%)	AQC
AGA-$v1$	98.71	8.47	415.35	99.32	8.01	**379.63**	**98.88**	8.34	395.04
AGA-$v2$	95.44	8.19	319.58	97.51	**7.65**	267.02	97.19	**7.89**	268.61
AGA-$v3$	**99.24**	8.38	342.82	**99.43**	8.58	234.59	98.79	8.19	323.68
AGA	99.18	**8.55**	**239.89**	99.32	7.83	216.49	98.65	8.14	**255.73**

- Transferability of the adversarial examples: It evaluates the usefulness of an adversarial example for attacking the models other than the target model. To this end, we first randomly pick up $1,000$ instances from the testing set of SNLI dataset. Then, for a given neural network, we apply Genetic attack, IBP-certified, IGA and AGA to find adversarial examples, respectively. Thereafter, the generated adversarial examples are used as inputs to feed into the other two neural networks. To evaluate the transferability, we evaluate the success rates of the generated examples that successfully attack the victim models. Note that each experiment is repeated 5 times and each of BERT, ALBERT and DistilBERT is used as the base model for adversarial example generation in a round-robin manner.
- Adversarial training: We randomly pick up $\tilde{N} \in \{1,000, 2,000, 3,000\}$ examples from the training set of the AG's News and SNLI datasets, respectively. Then, we apply IGA and AGA to generate adversarial examples accordingly. Thereafter, the generated adversarial examples are used to augment the training set for fine-tuning the BERT-base-uncased where the epoch is set to be 5 and the learning rate is set to be 5×10^{-5}. The testing accuracy of the fine-tuned BERT-base-uncased is used as the measure of the effectiveness of the fine-tuning. To investigate the robustness coming out of the adversarial training, we apply IGA and AGA again to conduct adversarial attacks on the

model fine-tuned with the generated adversarial examples and evaluate the ASR accordingly.

Results and Analysis. From the results shown in Fig. 3, we find that the generalization performance of the adversarial examples generated by our proposed AGA is similar to that of IGA. This can be explained as the adversarial examples generated by AGA are closer to the corresponding decision boundary. As for the effect of adversarial training, as shown in Fig. 4(a) and Fig. 4(b), we can see that the testing accuracy is improved after the fine-tuning by using the adversarial examples generated by AGA. Furthermore, as shown in Fig. 4(c) and Fig. 4(d), we can see that the ASR decreases with the increase of the number of adversarial examples. This indicates that the model thus becomes more robust.

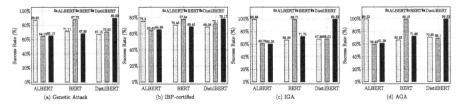

Fig. 3. Bar charts of the success rate of transferring adversarial examples generated from one model to the other on the SNLT dataset.

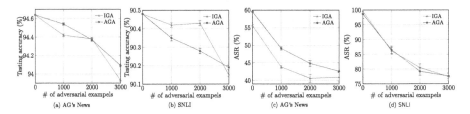

Fig. 4. Comparison of the impact of the number of adversarial examples used in the adversarial training on the testing accuracy (a) and (b) and the robustness (c) and (d) after fine-tuning.

Response to *RQ3*: We have the following takeaways from our experiments. The transferability of the adversarial examples generated by AGA is similar to the other peer algorithms on the SNLI dataset. Moreover, the adversarial examples generated by AGA have a better chance to improve the robustness of the victim model via adversarial training.

5 Conclusion

In this paper, we propose an efficient adversarial attack method based on genetic algorithm and attention, i.e. AGA. We analyze the shortcomings of the existing genetic algorithm attacks and improve these shortcomings. Inspired by the importance-based attacks, we use the attention mechanism to quickly find the most important words to reduce the number of queries. Then, we find the convergence rate of the population is slow, so we use the $(\lambda + \mu)$-selection strategy to accelerate the search procedure. Moreover, compared with the baseline methods, the proposed methods not only accelerate the speed to find adversarial examples but are also more successful in finding adversarial examples.

References

1. Alzantot, M., Sharma, Y., Elgohary, A., Ho, B., Srivastava, M.B., Chang, K.: Generating natural language adversarial examples. In: EMNLP'18: Proceedings of the 2018 Conference on Empirical Methods in Natural Language Processing, pp. 2890–2896. Association for Computational Linguistics (2018)
2. Bowman, S.R., Angeli, G., Potts, C., Manning, C.D.: A large annotated corpus for learning natural language inference. In: EMNLP'15: Proceedings of the 2015 Conference on Empirical Methods in Natural Language Processing, pp. 632–642. The Association for Computational Linguistics (2015). https://doi.org/10.18653/v1/d15-1075
3. Carlini, N., Wagner, D.A.: Towards evaluating the robustness of neural networks. In: 2017 IEEE Symposium on Security and Privacy, SP, pp. 39–57. IEEE Computer Society (2017). https://doi.org/10.1109/SP.2017.49
4. Devlin, J., Chang, M., Lee, K., Toutanova, K.: BERT: pre-training of deep bidirectional transformers for language understanding. In: NAACL'19: Proceedings of the 2019 Conference of the North American Chapter of the Association for Computational Linguistics: Human Language Technologies, pp. 4171–4186. Association for Computational Linguistics (2019). https://doi.org/10.18653/v1/n19-1423
5. Garg, S., Ramakrishnan, G.: BAE: bert-based adversarial examples for text classification. In: Webber, B., Cohn, T., He, Y., Liu, Y. (eds.) EMNLP'20: Proceedings of the 2020 Conference on Empirical Methods in Natural Language Processing, pp. 6174–6181. Association for Computational Linguistics (2020). https://doi.org/10.18653/v1/2020.emnlp-main.498, https://doi.org/10.18653/v1/2020.emnlp-main.498
6. Goodfellow, I.J., Shlens, J., Szegedy, C.: Explaining and harnessing adversarial examples. In: ICLR'15: Proceedings of the 2019 International Conference on Learning Representations (2015). http://arxiv.org/abs/1412.6572
7. Jia, R., Raghunathan, A., Göksel, K., Liang, P.: Certified robustness to adversarial word substitutions. In: EMNLP-IJCNLP'19 : Proceedings of the 2019 Conference on Empirical Methods in Natural Language Processing and the 9th International Joint Conference on Natural Language Processing, pp. 4127–4140. Association for Computational Linguistics (2019). https://doi.org/10.18653/v1/D19-1423
8. Kim, Y.: Convolutional neural networks for sentence classification. In: EMNLP'14: Proceedings of the 2014 Conference on Empirical Methods in Natural Language Processing, pp. 1746–1751. ACL (2014). https://doi.org/10.3115/v1/d14-1181
9. Lan, Z., Chen, M., Goodman, S., Gimpel, K., Sharma, P., Soricut, R.: ALBERT: a lite BERT for self-supervised learning of language representations. In: ICLR'20: Proceedings of the 2020 International Conference on Learning Representations. OpenReview.net (2020). https://openreview.net/forum?id=H1eA7AEtvS

10. Maas, A.L., Daly, R.E., Pham, P.T., Huang, D., Ng, A.Y., Potts, C.: Learning word vectors for sentiment analysis. In: ACL'11: Proc. of the 2011 Association for Computational Linguistics: Human Language Technologies. pp. 142–150. The Association for Computer Linguistics (2011), https://aclanthology.org/P11-1015/

11. Maheshwary, R., Maheshwary, S., Pudi, V.: Generating natural language attacks in a hard label black box setting. In: AAAI'21: Proc. of the Thirty-Fifth AAAI Conference on Artificial Intelligence, AAAI 2021, Thirty-Third Conference on Innovative Applications of Artificial Intelligence, IAAI 2021, The Eleventh Symposium on Educational Advances in Artificial Intelligence, EAAI, pp. 13525–13533. AAAI Press (2021). https://ojs.aaai.org/index.php/AAAI/article/view/17595

12. Morris, J.X., Lifland, E., Yoo, J.Y., Grigsby, J., Jin, D., Qi, Y.: Textattack: a framework for adversarial attacks, data augmentation, and adversarial training in NLP. In: EMNLP'20: Proceedings of the 2020 Conference on Empirical Methods in Natural Language Processing: System Demonstrations, pp. 119–126. Association for Computational Linguistics (2020). https://doi.org/10.18653/v1/2020.emnlp-demos.16

13. Papernot, N., McDaniel, P.D., Swami, A., Harang, R.E.: Crafting adversarial input sequences for recurrent neural networks. In: Brand, J., Valenti, M.C., Akinpelu, A., Doshi, B.T., Gorsic, B.L. (eds.) MILCOM'16: Proceedings of the 2016 IEEE Military Communications Conference, pp. 49–54. IEEE (2016). https://doi.org/10.1109/MILCOM.2016.7795300

14. Pennington, J., Socher, R., Manning, C.D.: Glove: global vectors for word representation. In: EMNLP'14: Proceedings of the 2014 Conference on Empirical Methods in Natural Language Processing, pp. 1532–1543. ACL (2014). https://doi.org/10.3115/v1/d14-1162

15. Ren, S., He, K., Girshick, R.B., Sun, J.: Faster R-CNN: towards real-time object detection with region proposal networks. In: NIPS'15: Proceedings of the 2015 Advances in Neural Information Processing Systems, pp. 91–99 (2015). https://proceedings.neurips.cc/paper/2015/hash/14bfa6bb14875e45bba028a21ed38046-Abstract.html

16. Ren, S., Deng, Y., He, K., Che, W.: Generating natural language adversarial examples through probability weighted word saliency. In: ACL'19: Proceedings of the 2019 Association for Computational Linguistics, pp. 1085–1097. Association for Computational Linguistics (2019). https://doi.org/10.18653/v1/p19-1103

17. Samanta, S., Mehta, S.: Towards crafting text adversarial samples. CoRR abs/1707.02812 (2017). http://arxiv.org/abs/1707.02812

18. Sanh, V., Debut, L., Chaumond, J., Wolf, T.: Distilbert, a distilled version of BERT: smaller, faster, cheaper and lighter. CoRR abs/1910.01108 (2019). http://arxiv.org/abs/1910.01108

19. Sutskever, I., Vinyals, O., Le, Q.V.: Sequence to sequence learning with neural networks. In: NIPS'14: Proc. of the 2014 Advances in Neural Information Processing Systems. pp. 3104–3112 (2014), https://proceedings.neurips.cc/paper/2014/hash/a14ac55a4f27472c5d894ec1c3c743d2-Abstract.html

20. Szegedy, C., Zaremba, W., Sutskever, I., Bruna, J., Erhan, D., Goodfellow, I.J., Fergus, R.: Intriguing properties of neural networks. In: Bengio, Y., LeCun, Y. (eds.) ICLR'14: Proc. of the 2014 International Conference on Learning Representations (2014), http://arxiv.org/abs/1312.6199

21. Wang, X., Jin, H., He, K.: Natural language adversarial attacks and defenses in word level. CoRR abs/1909.06723 (2019). http://arxiv.org/abs/1909.06723

22. Wang, Y., Huang, M., Zhu, X., Zhao, L.: Attention-based LSTM for aspect-level sentiment classification. In: EMNLP'16: Proceedings of the 2016 Conference on Empirical Methods in Natural Language Processing, pp. 606–615. The Association for Computational Linguistics (2016). https://doi.org/10.18653/v1/d16-1058

23. Wang, Y., Huang, R., Song, S., Huang, Z., Huang, G.: Not all images are worth 16x16 words: Dynamic transformers for efficient image recognition. In: NIPS'21: Proc. of the 2021 Advances in Neural Information Processing Systems, vol. 34 (2021). https://proceedings.neurips.cc/paper/2021/hash/64517d8435994992e682b3e4aa0a0661-Abstract.html

24. Yang, Z., Yang, D., Dyer, C., He, X., Smola, A.J., Hovy, E.H.: Hierarchical attention networks for document classification. In: NAACL'16: Proc. of the 2016 Conference of the North American Chapter of the Association for Computational Linguistics: Human Language Technologies, pp. 1480–1489. The Association for Computational Linguistics (2016). https://doi.org/10.18653/v1/n16-1174, https://doi.org/10.18653/v1/n16-1174

25. Zang, Y., Qi, F., Yang, C., Liu, Z., Zhang, M., Liu, Q., Sun, M.: Word-level textual adversarial attacking as combinatorial optimization. In: ACL'20: Proceedings of the 2020 Annual Meeting of the Association for Computational Linguistics, pp. 6066–6080. Association for Computational Linguistics (2020). https://doi.org/10.18653/v1/2020.acl-main.540

26. Zhang, W.E., Sheng, Q.Z., Alhazmi, A., Li, C.: Adversarial attacks on deep-learning models in natural language processing: a survey. ACM Trans. Intell. Syst. Technol. 11(3), 24:1–24:41 (2020). https://doi.org/10.1145/3374217

27. Zhang, X., Zhao, J.J., LeCun, Y.: Character-level convolutional networks for text classification. In: NIPS'15: Proceedings of the 2015 Advances in Neural Information Processing Systems, pp. 649–657 (2015). https://proceedings.neurips.cc/paper/2015/hash/250cf8b51c773f3f8dc8b4be867a9a02-Abstract.html

Deep Reinforcement Learning with Two-Stage Training Strategy for Practical Electric Vehicle Routing Problem with Time Windows

Jinbiao Chen, Huanhuan Huang, Zizhen Zhang, and Jiahai Wang[✉]

School of Computer Science and Engineering, Sun Yat-sen University,
Guangzhou, China
{chenjb69,huanghh29}@mail2.sysu.edu.cn,
{zhangzzh7,wangjiah}@mail.sysu.edu.cn

Abstract. Recently, it is promising to apply deep reinforcement learning (DRL) to the vehicle routing problem (VRP), which is widely employed in modern logistics systems. A practical extension of VRP is the electric vehicle routing problem with time windows (EVRPTW). In this problem, the realistic traveling distance and time are non-Euclidean and asymmetric, and the constraints are more complex. These characteristics result in a challenge when using the DRL approach to solve it. This paper proposes a novel end-to-end DRL method with a two-stage training strategy. First, a graph attention network with edge features is designed to tackle the graph with the asymmetric traveling distance and time matrix. The node and edge features of the graph are effectively correlated and captured. Then, a two-stage training strategy is proposed to handle the complicated constraints. Some constraints are allowed to be violated to enhance exploration in the first stage, while all the constraints are enforced to be satisfied to guarantee a feasible solution in the second stage. Experimental results show that our method outperforms the state-of-the-art methods and can be generalized well to different problem sizes.

Keywords: Deep reinforcement learning · Electric vehicle routing problem with time windows · Graph attention network · Two-stage training

1 Introduction

Vehicle routing problem (VRP) [2], as a classic combinatorial optimization problem, aims at dispatching a fleet of vehicles to serve a set of customers so as to minimize the total traveling cost. Recently, electric vehicles (EVs) have been extensively popularized. Compared with traditional vehicles, EVs can be beneficial to sustainable transportation systems and environmental protection. They

G. Rudolph et al. (Eds.): PPSN 2022, LNCS 13398, pp. 356–370, 2022.
https://doi.org/10.1007/978-3-031-14714-2_25

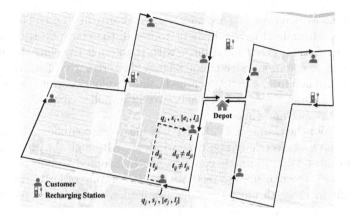

Fig. 1. An illustration of the practical EVRPTW.

will undoubtedly become the mainstream of vehicles in the future. By incorporating EVs into VRP, an interesting problem called the electric VRP (EVRP) attracts researchers' attention [25]. In addition to basic properties of VRP, EVRP further considers limited electricity of EVs. This implies that EVs need to be recharged at recharging stations before running out of battery. A natural and practical variant of EVRP is to impose a specific time window for each customer, known as EVRP with time windows (EVRPTW). Figure 1 illustrates an example of EVRPTW.

To solve VRP and its variants, traditional methods include exact algorithms and heuristic algorithms. Exact algorithms [6] usually employ the branch-and-bound framework to produce an optimal solution, but they can only handle small-scale problems in general. Heuristic algorithms [26] can obtain acceptable solutions in reasonable time, but they require problem-specific experience and knowledge. In addition, heuristic methods, which independently address problem instances and iteratively perform search in solution space, also suffer from long computation time for large-scale problems.

Recently, a novel framework based on deep reinforcement learning (DRL) is developed to rapidly obtain near-optimal solutions of combinatorial optimization problems [22], especially routing problems [21,28]. By using an end-to-end learning paradigm, a deep network model is trained offline with numerous instances. It can automatically learn underlying features from the data and generalize to unseen instances. Then it is leveraged to rapidly construct a solution by a direct forward inference without any iterative search.

The DRL methods have achieved some success for traditional VRPs. However, when these methods are applied to the practical EVRPTW, additional issues need to be addressed.

1) Most of the existing works only focus on routing problems where the symmetric Euclidean traveling distance and time are calculated by the given coordinates of nodes. In the practical EVRPTW, the traveling distance and

time are non-Euclidean and asymmetric, as shown in Fig. 1. The traveling distance is determined by the realistic routes in the transportation network. The traveling time is not only linear with the actual traveling distance but also related to comprehensive traffic factors like terrain, weather, and crowding degree. Therefore, existing methods that exploit coordinates as inputs may fail to extract the edge features between two nodes containing the actual traveling distance and time.

2) Different from the classic VRP, EVRPTW has more complicated constraints including limited electricity and time windows. Although the masking mechanism in most DRL methods can be directly adopted to ensure a feasible solution, such mechanism also restricts the exploration of DRL, which makes it difficult to cross through the infeasible region to learn the relationships among solutions, constraints and objectives.

To tackle the aforementioned issues, this paper proposes a novel DRL method with a two-stage training strategy (DRL-TS) to solve the practical EVRPTW. Our contributions can be summarized as follows.

1) A deep neural network based on a graph attention network (GAT) with edge features is proposed to effectively capture both node and edge features of a graph input. The network correlates the node embeddings with the edge embeddings, which can produce a high-quality solution.
2) A two-stage training strategy is proposed to deal with complex constraints. It is a general strategy, as it does not rely on a specific form of constraints. Specifically, in the first stage, some constraints are treated as soft and allowed to be violated to enhance the exploration, but penalties are appended to the objective if constraints are violated. In the second stage, all the constraints must be satisfied by the masking mechanism to guarantee a feasible solution.
3) Computational experiments conducted on real-world asymmetric datasets show that the proposed method outperforms conventional methods and state-of-the-art DRL methods.

2   Literature Review

This section first briefly reviews traditional methods for EVRPTW, including exact and heuristic algorithms. EVRPTW was first introduced by Schneider et al. [26] and solved by a hybrid heuristic algorithm combining a variable neighborhood search heuristic with a tabu search heuristic. Exact branch-and-price-and-cut algorithms relying on customized labeling algorithms [6] were proposed for four variants of EVRPTW. A few conventional methods for EVRPTW have been proposed. Other heuristic algorithms are further developed for some extensions of EVRPTW considering various recharging policies [4,11,12].

Next, we review recent DRL methods for routing problems. A sequence-to-sequence model based on a pointer network [1] was first proposed to solve routing problems. The recurrent neural network (RNN) and attention mechanism [23] were used to improve the policy network for VRP. The framework of [23] was adapted to solve EVRPTW [19], which is named as DRL-R. Inspired by the Transformer architecture [27], the multi-head self-attention mechanism [14] was adopted to improve the policy network for routing problems. A dynamic attention model with dynamic encoder-decoder architecture [24] was developed to improve the performance for VRP. Policy optimization with multiple optima [15] was introduced to further improve the performance by utilizing the symmetries in the representation of a solution. A structural graph embedded pointer network [10] was presented to iteratively produce tours for the online VRP. A deep model based on the graph convolutional network with node and edge features [7] was proposed for the practical VRP. These methods were further developed for variants of routing problems [16–18,29,33] and their multi-objective versions [30,32,34]. Different from the above end-to-end methods that learn to directly construct solutions, a learn-to-improve framework [5,20,31] that learns local search operators was developed to iteratively improve an initial solution.

In summary, most of the DRL methods were focused on VRP and its simple variants. Only one work, DRL-R [19], applied the DRL method to EVRPTW. This method treating coordinates as inputs cannot be directly applied to the practical EVRPTW with asymmetric traveling distance and time. In addition, it cannot well handle the complex constraints. These facts motivate us to develop a more effective DRL method for EVRPTW.

3   The Proposed DRL-TS for Practical EVRPTW

3.1   Problem Statements and Reinforcement Learning Formulation

EVRPTW is defined on a complete directed graph $G = (N, E)$, where N is the node set and E is the edge set. The node set $N = V \cup F \cup \{0\}$, where V is the set of customers, F is the set of recharging stations, and 0 is the depot. Each node $i \in V$ has a positive demand q_i and a service time s_i. Each node $i \in F$ also has a demand $q_i = 0$ and a service time s_i representing its full recharging time. Also, each node $i \in N$ has a time window $[e_i, l_i]$. Each edge is associated with a traveling distance d_{ij}, a traveling time t_{ij}, and a battery consumption λd_{ij}, where λ is a battery consumption rate. In the practical EVRPTW, the traveling distance and time are both possibly asymmetric, i.e., $d_{ij} \neq d_{ji}$ and $t_{ij} \neq t_{ji}$.

A fleet of homogeneous EVs with identical loading capacity C and battery capacity Q is initially placed at the depot with full battery power. EVs must serve all the customers and finally return to the depot. The objective is to minimize the total traveling distance. The following constraints must be satisfied.

1) *Capacity constraints*: The remaining capacity of EVs for serving node $i \in V$ must be no less than the demand q_i.
2) *Time window constraints*: EVs need to wait if they arrive at node $i \in N$ before e_i. The service after l_i is not allowed.
3) *Electricity constraints*: The remaining electricity of EVs arriving at each node $i \in N$ must be no less than 0.

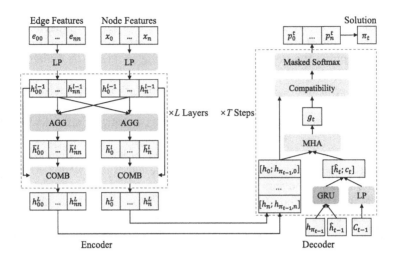

Fig. 2. The GAT-based policy network. LP denotes a linear projection, AGG denotes the aggregation sub-layer, COMB denotes the combination sub-layer, GRU denotes a gated recurrent unit, MHA denotes a multi-head attention layer, and $N = \{0, 1, ..., n\}$.

To formulate EVRPTW as a reinforcement learning (RL) form, it can be naturally deemed as a sequential decision problem by constructing the routes step-by-step. Especially, a solution is represented by a sequence $\boldsymbol{\pi} = \{\pi_0, ..., \pi_T\}$, where $\pi_t \in N$, $\pi_0 = \pi_T = 0$. Note that the sequence length T is unfixed, because there are multiple routes and the recharging stations can be visited any times.

In RL, the state is defined as a partial solution $\boldsymbol{\pi}_{0:t-1}$. The action is defined as visiting a node $\pi_t \in V \backslash \{\boldsymbol{\pi}_{0:t-1}\} \cup F \cup \{0\}$ satisfying all the constraints. The state transition is defined as $\boldsymbol{\pi}_{0:t} = \{\boldsymbol{\pi}_{0:t-1}, \pi_t\}$. The reward is defined as $R = \sum_{t=1}^{T} -d_{\pi_{t-1}, \pi_t}$, where d_{π_{t-1}, π_t} represents the traveling distance at step t. A stochastic policy $p(\boldsymbol{\pi}|G)$ generating a solution $\boldsymbol{\pi}$ from the graph G of an instance is calculated as $p(\boldsymbol{\pi}|G) = \prod_{t=1}^{T} p_\theta(\pi_t|\boldsymbol{\pi}_{0:t-1}, G)$, where $p_\theta(\pi_t|\boldsymbol{\pi}_{0:t-1}, G)$ is the probability of the node selection parameterized by $\boldsymbol{\theta}$.

3.2 GAT-based Policy Network with Edge Features

To learn the policy p_θ, a GAT-based policy network with edge features is designed, which follows the encoder-decoder architecture (see Fig. 2). It can extract both the node and edge features including the traveling distance and time matrix of the practical EVRPTW. The encoder produces the correlated node and edge embeddings of the graph. At each time step, the decoder aggregates the embeddings with the context to generate a probability vector and selects a node accordingly. The process is iteratively repeated until all the customers are served.

Encoder. The encoder first computes initial node and edge embeddings of the graph input with node features $x_i = (q_i, e_i, l_i, z_i)$ and edge features $e_{ij} = (d_{ij}, t_{ij}, a_{ij})$, $i, j \in N$, where z_i denotes the node type (depot, customer or recharging station) and a_{ij} denotes whether i and j are adjacent as follows.

$$z_i = \begin{cases} 0, & i \in \{0\} \\ 1, & i \in V \\ -1, & i \in F \end{cases} , \tag{1}$$

$$a_{ij} = \begin{cases} 1, & i \text{ and } j \text{ are } r\text{-nearest neighbors} \\ -1, & i = j \\ 0, & \text{otherwise} \end{cases} , \tag{2}$$

where r is empirically set to 10 as in [7]. The inputs are linearly projected to initial node embeddings h_i^0 and edge embeddings h_{ij}^0 with the same dimension d_h.

$$\begin{aligned} h_i^0 &= W_N^0 x_i + b_N, \\ h_{ij}^0 &= W_E^0 e_{ij} + b_E, \end{aligned} \tag{3}$$

where W_N, W_E, b_N, and b_E are all trainable parameters. Let h_i^l and h_{ij}^l respectively denote the embeddings of node i and edge (i, j) produced by layer $l \in \{1, ..., L\}$. The final embeddings h_i^L and h_{ij}^L are obtained by the encoder. Each layer l contains an aggregation and a combination sub-layer, which aggregate each embedding with its neighbors and combine itself with the aggregated embeddings [7], respectively.

In the aggregation sub-layer, the node and edge embeddings are simultaneously computed as follows.

$$\begin{aligned} \tilde{h}_i^l &= \text{MHA}^l(h_i^{l-1}, \{[h_j^{l-1}; h_{ij}^{l-1}] \mid j \in N\}), \\ \bar{h}_i^l &= \sigma(\text{BN}^l(W_N^l \tilde{h}_i^l)), \\ \tilde{h}_{ij}^l &= W_{E1}^l h_{ij}^{l-1} + W_{E2}^l h_i^{l-1} + W_{E3}^l h_j^{l-1}, \\ \bar{h}_{ij}^l &= \sigma(\text{BN}^l(W_E^l \tilde{h}_{ij}^l)), \end{aligned} \tag{4}$$

where [;] is the concatenation of two vectors, BN is a batch normalization layer [9], σ is the ReLU activation, and W_N^l, W_E^l, W_{E1}^l, W_{E2}^l, W_{E3}^l are all trainable parameters. MHA is a multi-head attention layer with M heads [27], in which $q_i^{m,l-1} = W_Q^m h_i^{l-1}$, $k_{ij}^{m,l-1} = W_K^m[h_j^{l-1}; h_{ij}^{l-1}]$, $v_{ij}^{m,l-1} = W_V^m[h_j^{l-1}; h_{ij}^{l-1}]$, and $m \in \{1, ..., M\}$.

In the combination sub-layer, the node and edge embeddings are both combined by a skip-connection [8], as follows.

$$\begin{aligned} h_i^l &= \sigma(h_i^{l-1} + \text{BN}^l(\text{FF}^l(\bar{h}_i^l))), \\ h_{ij}^l &= \sigma(h_{ij}^{l-1} + \text{BN}^l(\text{FF}^l(\bar{h}_{ij}^l))), \end{aligned} \tag{5}$$

where FF is a fully connected feed-forward layer.

Decoder. The decoder sequentially selects a node according to a probability distribution obtained by the node embeddings h_i^L and edge embeddings h_{ij}^L from the encoder (the superscript L is omitted for readability in the following).

Specifically, at the decoding step t, the *context* c_t, previous partial tour, node embeddings, and edge embeddings are firstly aggregated as the *glimpse* g^t [1].

$$
\begin{aligned}
c_t &= W_C C_{t-1} + b_C, \\
\hat{h}_t &= \mathrm{GRU}(h_{\pi_{t-1}}, \hat{h}_{t-1}), \\
g^t &= \mathrm{MHA}([\hat{h}_t; c_t], \{[h_j; h_{\pi_{t-1},j}] \mid j \in N\}),
\end{aligned}
\tag{6}
$$

where W_C and b_C are trainable parameters, GRU is a gated recurrent unit, which can better handle the partial solution generated by sequential steps. $C_t = (\mathcal{T}_t, D_t, B_t)$ is composed of the traveling time $\mathcal{T}_t$, remaining capacity D_t, and remaining electricity B_t when leaving node π_t, which are updated as follows.

$$
\begin{aligned}
\mathcal{T}_t &= \begin{cases} \mathcal{T}_{t-1} + t_{\pi_{t-1},\pi_t} + s_{\pi_t}, & \pi_t \in V \cup F \\ 0, & \pi_t \in \{0\} \end{cases}, \\
D_t &= \begin{cases} D_{t-1} - q_{\pi_t}, & \pi_t \in V \cup F \\ C, & \pi_t \in \{0\} \end{cases}, \\
B_t &= \begin{cases} B_{t-1} - \lambda d_{\pi_{t-1},\pi_t}, & \pi_t \in V \\ Q, & \pi_t \in F \cup \{0\} \end{cases}.
\end{aligned}
\tag{7}
$$

Then, the *compatibility* u^t is calculated by the *query* $q^t = W_Q g^t$ and *key* $k_i^t = W_K[h_i; h_{\pi_{t-1},i}]$ with trainable parameters W_Q and W_K.

$$
u_i^t = \begin{cases} \zeta \cdot \tanh(\frac{(q^t)^T k_i^t}{\sqrt{d_h}}), & \mathrm{mask}_i^t = 1 \\ -\infty, & \text{otherwise} \end{cases},
\tag{8}
$$

where ζ is used to clip the result. $\mathrm{mask}_i^t = 1$ represents that the feasible node i is unmasked at step t. A node is called feasible or unmasked, if upon its arrival, the capacity, time window and electricity constraints are not violated. Finally, the probability distribution to select a node at step t is computed using the softmax function as follows.

$$
p_\theta(\pi_t | \boldsymbol{\pi}_{0:t-1}, G) = \mathrm{softmax}(u^t).
\tag{9}
$$

Two common decoding strategies, which are sampling decoding and greedy decoding, are adopted to choose a node at each step. A node is chosen according to the probability distribution in the sampling decoding, or according to the maximum probability in the greedy decoding.

3.3   Two-Stage Training Strategy

The masking mechanism in DRL methods can be directly applied to EVRPTW by including the capacity, time window, and electricity constraints [19]. However, such mechanism would limit the exploration of those infeasible regions, thereby impairing the seeking of global-best solutions. In order to enrich the search space, we propose a two-stage training strategy, which tries to well balance the exploration and feasibility of the search.

Specifically, in the first stage, the capacity, time window and electricity constraints are treated as the soft constraints. Only the constraints ensuring a tour are retained hard. In this case, node i is masked, i.e., $\text{mask}_i^t = 0$, if one of the following conditions is satisfied.

1) The customer has already been visited before, i.e., $i \in V$ and $i \in \{\boldsymbol{\pi}_{0:t-1}\}$.
2) The depot has been visited in the last step, i.e., $i \in \{0\}$ and $\pi_{t-1} = 0$.
3) The recharging station is visited by the EV with full electricity, i.e., $i \in F, \pi_{t-1} \in F \cup \{0\}$.

In the second stage, all the original constraints must be respected. In this case, $\text{mask}_i^t = 0$ if one of the following conditions in addition to above three conditions in the first stage is satisfied.

1) The capacity constraint: $D_{t-1} < q_i$.
2) The time window constraint: $T_{t-1} + t_{\pi_{t-1},i} > l_i$.
3) The electricity constraint: $B_{t-1} < \lambda d_{\pi_{t-1},i} + \min_{j \in F \cup \{0\}} \lambda d_{ij}$.

The loss is defined as $\mathcal{L}(\boldsymbol{\theta}|G) = \mathbf{E}_{p_{\boldsymbol{\theta}}(\boldsymbol{\pi}|G)}[y(\boldsymbol{\pi})]$, where $y(\boldsymbol{\pi})$ is defined as an uniform form for two training stages as follows.

$$
\begin{aligned}
y(\boldsymbol{\pi}) = & \sum_{t=1}^{T} d_{\pi_{t-1},\pi_t} + \alpha \sum_{t=1}^{T} \max(T_{t-1} + t_{\pi_{t-1},\pi_t} - l_{\pi_t}, 0) \\
& + \beta \sum_{t=1}^{T} \max(q_{\pi_t} - D_{t-1}, 0) + \gamma \sum_{t=1}^{T} \max(\lambda d_{\pi_{t-1},\pi_t} - B_{t-1}, 0)
\end{aligned}
\tag{10}
$$

For the first stage, α, β, and γ are respectively three penalties for the violation of the capacity, time window, and electricity constraints. For the second stage, $y(\boldsymbol{\pi})$ is degenerated to the total traveling distance $\sum_{t=1}^{T} d_{\pi_{t-1},\pi_t}$, since the hard constraints are considered. $\mathcal{L}(\boldsymbol{\theta}|G)$ is optimized by gradient descent using the well-known REINFORCE algorithm with a rollout baseline $b(G)$, as follows.

$$
\nabla \mathcal{L}(\boldsymbol{\theta}|G) = \mathbf{E}_{p_{\boldsymbol{\theta}}(\boldsymbol{\pi}|G)}[(y(\boldsymbol{\pi}) - b(G))\nabla \log p_{\boldsymbol{\theta}}(\boldsymbol{\pi}|G)],
\tag{11}
$$

$$
\boldsymbol{\theta} \leftarrow \text{Adam}(\boldsymbol{\theta}, \nabla \mathcal{L}),
\tag{12}
$$

where Adam is the Adam optimizer [13]. The training algorithm is similar to that in [14], and $b(G)$ is a greedy rollout produced by the current model.

The proportions of the epochs of the first and second stage are respectively controlled by η and $1 - \eta$, where η is a user-defined parameter.

3.4 Characteristics of DRL-TS

The characteristics of DRL-TS are summarized as follows.

1) For the policy network, DRL-TS uses GAT with edge features to effectively tackle the graph input with the asymmetric traveling distance and time matrix, while DRL-R [19] exploiting coordinates as inputs can only deal with the symmetric Euclidean traveling distance. In addition, unlike DRL-R adopting a dynamic T-step encoder based on RNN, which consumes enormous memory for large-scale cases, DRL-TS uses an efficient one-step encoder.
2) For the training, like most of DRL-based methods, DRL-R [19] directly adopts the masking mechanism to EVRPTW to ensure a feasible solution. However, DRL-TS uses a two-stage training strategy. The capacity, time window, and electricity constraints are all allowed to be violated to enhance exploration in the first stage, while the original constraints must be satisfied by the masking mechanism to guarantee feasibility in the second stage.

4   Experimental Results

In this section, we conduct computational experiments to evaluate the proposed method on practical EVRPTW instances. All the experiments are implemented on a computer with an Intel Xeon 4216 CPU and an RTX 3090 GPU. The source code of the proposed algorithm is available on request.

4.1   Experimental Settings

Benchmarks. The practical EVRPTW instances are generated from the real-world data of the Global Optimization Challenge competition[1] held by JD Logistics, where the traveling distance and time are both asymmetric. For each instance, each customer i is randomly selected from the entire customer set of the data. The time window $[e_i, l_i]$, service time (or recharging time) s_i, traveling distance d_{ij}, and traveling time t_{ij} are all directly obtained from the original data. The time window of the depot $[e_0, l_0]$ is [8:00,20:00], as the working time is 720 min. Each demand q_i is randomly chosen from $\{1,...,9\}$. Following the previous work [14,19], we conduct experiments on the instances with different customer sizes $|V| = 10/20/50/100$, denoted as C10/C20/C50/C100. The capacities of EVs Q are correspondingly set to 20/30/40/50. The number of recharging stations $|F|$ is set to $|V|/10$. The battery capacity Q is set to 1.6×10^5. The battery consumption rate λ is fixed to 1.

Baselines. The following representative methods, including the state-of-the-art DRL method, exact method, and heuristic method, are considered as baselines for the comparisons.

---

[1] https://jdata.jd.com/html/detail.html?id=5.

1) SCIP[2]: an open-source exact solver for combinatorial optimization problems, which solves EVRPTW using the mathematical model described in [26]. SCIP(180s) indicates running the SCIP solver for 180 s.
2) DRL-R [19]: the state-of-the-art DRL method for EVRPTW, which adopts an RNN-based policy network and uses an ordinary masking mechanism to train the network.
3) SA-VNS: the combination of simulated annealing (SA) and variable neighborhood search (VNS), which is an especially designed meta-heuristic method based on [26] for EVRPTW. To efficiently explore the solution space, different neighborhood operators are employed, including 2-opt, 2-opt*, or-opt, cross-exchange, merge, and stationInRe. The last operator is specially designed for EVRPTW [26], while the others are widely used in routing problems [3]. The SA framework is adopted to avoid local optima. The number of outer iterations is 1000. The linear annealing rate is 0.99. The initial temperature is 10. The number of inner iterations is 10. The six operators are successively executed in a random order in each inner iteration.

Hyper-parameters. In the encoder, $L = 2$ GAT-layers are used as in [7]. d_h, M, and ζ are respectively set to 128, 8, and 10 as in [14]. The Adam optimizer with a constant learning rate 10^{-4} is adopted to train the model. The penalties α, β, and γ are all set to 1. Regarding the training, 200 epochs are run. In each epoch, 250 batches are generated. The batch size is set to 128 for C10/C20/C50 and 64 for C100 due to memory constraints. η is set to 0.5, i.e., 100 epochs are run both for the first and second stage.

4.2   Comparison Analysis

The experimental results of SCIP, SA-VNS, DRL-R [19] and DRL-TS (ours) are recorded in Table 1, where the objective value, optimality gap, and average computing time of an instance are shown. The objective values are normalized by a scale of 10^5. The gap of a method is calculated by $(O_m - O_b)/O_b \times 100\%$, where O_m is the objective value of the compared method and O_b is the best objective value among all methods. The performance of our method and that of other methods are all statistically different by the Wilcoxon rank-sum test with a significance level 1% for each experiment.

From Table 1, it can be seen that our method outperforms DRL-R with both greedy and sampling decoding. DRL-R is first trained on C20 and then tested on instances of various sizes like that in [19], since it is a dynamic encoding model with T steps consuming enormous memory. DRL-TS20 uses the same training and testing form as DRL-R, but it still outperforms DRL-R. Regarding sampling decoding, it achieves smaller objective values and gaps than greedy

---

[2] https://scip.zib.de/.

Table 1. The average performance of our method and baselines on 1000 random instances. DRL-TS20/50 means that our proposed model trained on the C20/C50 is tested on instances of various sizes. G denotes greedy decoding and S denotes sampling decoding that chooses the best solution from 1280 sampled solutions.

Method	C10			C20		
	Obj.	Gap	Time/s	Obj.	Gap	Time/s
SCIP(180s)	**3.665**	**0.00%**	4.01	**5.626**	**0.00%**	146.20
SA-VNS	3.687	0.58%	20.23	5.771	2.58%	27.79
DRL-R(G)	4.172	13.81%	0.01	6.554	16.49%	0.01
DRL-TS20(G)	4.038	10.16%	<0.01	6.247	11.04%	<0.01
DRL-TS50(G)	4.184	14.15%	<0.01	6.425	14.19%	<0.01
DRL-TS(G)	3.950	7.76%	<0.01	6.247	11.04%	<0.01
DRL-R(S)	3.930	7.23%	0.71	6.112	8.63%	1.99
DRL-TS20(S)	3.819	4.18%	0.95	5.903	4.91%	1.67
DRL-TS50(S)	3.818	4.17%	1.24	5.975	6.19%	1.94
DRL-TS(S)	3.796	3.57%	0.91	5.903	4.91%	1.67
Method	C50			C100		
	Obj.	Gap	Time/s	Obj.	Gap	Time/s
SCIP(180s)	14.499	26.22%	180.00	62.910	217.62%	181.06
SA-VNS	11.693	1.80%	49.87	20.470	3.35%	105.22
DRL-R(G)	13.147	14.46%	0.01	23.197	17.12%	0.03
DRL-TS20(G)	12.642	10.06%	<0.01	22.008	11.11%	0.01
DRL-TS50(G)	12.187	6.10%	<0.01	21.084	6.45%	0.01
DRL-TS(G)	12.187	6.10%	<0.01	20.847	5.25%	0.01
DRL-R(S)	12.155	5.82%	5.26	21.515	8.63%	30.91
DRL-TS20(S)	11.604	1.03%	3.72	20.879	5.41%	12.85
DRL-TS50(S)	11.486	0.00%	3.77	19.844	0.19%	14.97
DRL-TS(S)	**11.486**	**0.00%**	3.77	**19.806**	**0.00%**	13.10

decoding despite using slightly more computing time, which shows that the sampling strategy can effectively improve the solution quality. SCIP can engender the optimal solutions for all of C10 instances and 400 instances of C20, but only feasible solutions for all of C50 and C100 instances, as it suffers from exponentially growing computing time. SA-VNS can produce near-optimal solutions within reasonable computing time. Our method can rapidly achieve better performance in terms of objective values and gaps for large problem sizes, i.e., C50 and C100, and achieve acceptable performance for small problem sizes. Moreover, compared with results of DRL-TS that performs best correspondingly trained on each problem size, the results of DRL-TS20 and DRL-TS50 demonstrate that the model has a desirable ability of generalization for different problem sizes.

In summary, our method has fast solving speed and high generalization ability, and it is superior to the state-of-the-art DRL method. For small-scale cases, compared with the exact and heuristic methods carefully calibrated for problems, our method can still produce promising solutions with acceptable gaps.

4.3 Ablation Study

To verify the significance of the two-stage training strategy and the GAT-based network considering the feature edges, the results of ablation experiments are shown in Table 2. DRL-H/DRL-S is our proposed GAT-based network trained only with purely hard/soft constraints, i.e., the same as the second/first stage, which is used to evaluate the effects of the proposed two-stage training strategy. DRL-TS w/o E is our proposed policy network without edge features, which is used to evaluate the effects of the edge features. The results indicate that our model performs better for both greedy and sampling decoding than that trained only with the hard or soft constraints and without edge features.

Furthermore, for the two-stage training strategy, the sensitivity of the proportion of the epochs of the first stage η is studied. $\eta = 0$ and $\eta = 1$ represent DRL-H and DRL-S, respectively. The results on C50 instances are shown in Fig. 3. Except $\eta = 0$ and $\eta = 1$, η has a little influence on the objective value and $\eta = 0.5$ is always a robust strategy. This again verifies the effectiveness of the two-stage training strategy.

Table 2. Different DRL methods on 1000 random instances.

Method	C10		C20		C50		C100	
	Obj.	Gap	Obj.	Gap	Obj.	Gap	Obj.	Gap
DRL-TS(S)	**3.796**	**0.00%**	**5.903**	**0.00%**	**11.486**	**0.00%**	**19.806**	**0.00%**
DRL-H(S)	3.833	0.97%	5.931	0.48%	11.551	0.56%	19.965	0.80%
DRL-S(S)	3.819	0.60%	5.990	1.48%	11.729	2.11%	20.298	2.48%
DRL-TS w/o E(S)	4.210	10.91%	6.875	16.47%	15.817	37.70%	30.437	53.67%
DRL-TS(G)	**3.950**	**4.05%**	**6.247**	**5.83%**	**12.187**	**6.10%**	**20.847**	**5.25%**
DRL-H(G)	3.971	4.60%	6.257	6.01%	12.248	6.63%	21.121	6.64%
DRL-S(G)	4.024	6.00%	6.483	9.83%	12.769	11.17%	21.515	8.62%
DRL-TS w/o E(G)	4.781	25.95%	8.545	44.76%	18.841	64.03%	34.417	73.77%

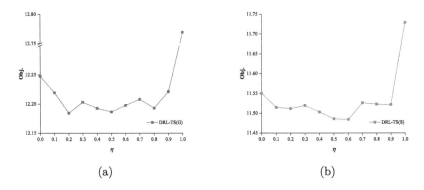

Fig. 3. Effect of the parameter η. (a) Greedy. (b) Sampling.

5   Conclusion

This paper proposes a novel DRL method for the practical EVRPTW. A GAT-based policy network with edge features is designed to cope with the non-Euclidean and asymmetric traveling distance and time. A two-stage training strategy is presented to handle the complicated constraints. The experimental results based on the real-world asymmetric data verify the effectiveness of the proposed approach. In the future, the proposed method can be extended to other VRP variants with practical characteristics and their multi-objective versions.

Acknowledgements. This work is supported by the National Key R&D Program of China (2018AAA0101203), the National Natural Science Foundation of China (62072483), and the Guangdong Basic and Applied Basic Research Foundation (2022A1515011690, 2021A1515012298).

References

1. Bello, I., Pham, H., Le, Q.V., Norouzi, M., Bengio, S.: Neural combinatorial optimization with reinforcement learning. In: International Conference on Learning Representations (2017)
2. Braekers, K., Ramaekers, K., Van Nieuwenhuyse, I.: The vehicle routing problem: state of the art classification and review. Comput. Ind. Eng. **99**, 300–313 (2016)
3. Bräysy, O., Gendreau, M.: Vehicle routing problem with time windows, Part I: route construction and local search algorithms. Transp. Sci. **39**(1), 104–118 (2005)
4. Bruglieri, M., Pezzella, F., Pisacane, O., Suraci, S.: A variable neighborhood search branching for the electric vehicle routing problem with time windows. Electron. Notes Discret. Math. **47**, 221–228 (2015)
5. Chen, X., Tian, Y.: Learning to perform local rewriting for combinatorial optimization. In: Advances in Neural Information Processing Systems (2019)
6. Desaulniers, G., Errico, F., Irnich, S., Schneider, M.: Exact algorithms for electric vehicle-routing problems with time windows. Oper. Res. **64**(6), 1388–1405 (2016)

7. Duan, L., et al.: Efficiently solving the practical vehicle routing problem: a novel joint learning approach. In: ACM SIGKDD Conference on Knowledge Discovery & Data Mining, pp. 3054–3063 (2020)
8. He, K., Zhang, X., Ren, S., Sun, J.: Deep residual learning for image recognition. In: IEEE Conference on Computer Vision and Pattern Recognition, pp. 770–778 (2016)
9. Ioffe, S., Szegedy, C.: Batch normalization: accelerating deep network training by reducing internal covariate shift. In: International Conference on Machine Learning, vol. 37, pp. 448–456 (2015)
10. James, J., Yu, W., Gu, J.: Online vehicle routing with neural combinatorial optimization and deep reinforcement learning. IEEE Trans. Intell. Transp. Syst. **20**(10), 3806–3817 (2019)
11. Keskin, M., Çatay, B.: Partial recharge strategies for the electric vehicle routing problem with time windows. Transp. Res. Part C Emerg. Technol. **65**, 111–127 (2016)
12. Keskin, M., Çatay, B.: A matheuristic method for the electric vehicle routing problem with time windows and fast chargers. Comput. Oper. Res. **100**, 172–188 (2018)
13. Kingma, D.P., Ba, J.: Adam: a method for stochastic optimization. In: International Conference on Learning Representations (2015)
14. Kool, W., Van Hoof, H., Welling, M.: Attention, learn to solve routing problems! In: International Conference on Learning Representations (2019)
15. Kwon, Y.D., Choo, J., Kim, B., Yoon, I., Gwon, Y., Min, S.: POMO: policy optimization with multiple optima for reinforcement learning. In: Advances in Neural Information Processing Systems (2020)
16. Li, J., et al.: Deep reinforcement learning for solving the heterogeneous capacitated vehicle routing problem. IEEE Trans. Cybern. (2021)
17. Li, J., Xin, L., Cao, Z., Lim, A., Song, W., Zhang, J.: Heterogeneous attentions for solving pickup and delivery problem via deep reinforcement learning. IEEE Trans. Intell. Transp. Syst. **23**(3), 2306–2315 (2022)
18. Li, K., Zhang, T., Wang, R., Wang, Y., Han, Y., Wang, L.: Deep reinforcement learning for combinatorial optimization: covering salesman problems. IEEE Trans. Cybern. (2021)
19. Lin, B., Ghaddar, B., Nathwani, J.: Deep reinforcement learning for the electric vehicle routing problem with time windows. IEEE Trans. Intell. Transp. Syst. (2021)
20. Lu, H., Zhang, X., Yang, S.: A learning-based iterative method for solving vehicle routing problems. In: International Conference on Learning Representations (2020)
21. Mazyavkina, N., Sviridov, S., Ivanov, S., Burnaev, E.: Reinforcement learning for combinatorial optimization: a survey. Comput. Oper. Res. **134**, 105400 (2021)
22. Mirhoseini, A., et al.: A graph placement methodology for fast chip design. Nature **594**, 207–212 (2021)
23. Nazari, M., Oroojlooy, A., Takáč, M., Snyder, L.V.: Reinforcement learning for solving the vehicle routing problem. In: Advances in Neural Information Processing Systems (2018)
24. Peng, B., Wang, J., Zhang, Z.: A deep reinforcement learning algorithm using dynamic attention model for vehicle routing problems. In: Li, K., Li, W., Wang, H., Liu, Y. (eds.) ISICA 2019. CCIS, vol. 1205, pp. 636–650. Springer, Singapore (2020). https://doi.org/10.1007/978-981-15-5577-0_51
25. Qin, H., Su, X., Ren, T., Luo, Z.: A review on the electric vehicle routing problems: variants and algorithms. Front. Eng. Manag. **8**(3), 370–389 (2021). https://doi.org/10.1007/s42524-021-0157-1

26. Schneider, M., Stenger, A., Goeke, D.: The electric vehicle-routing problem with time windows and recharging stations. Transp. Sci. **48**(4), 500–520 (2014)
27. Vaswani, A., et al.: Attention is all you need. In: Advances in Neural Information Processing Systems (2017)
28. Wang, Q., Tang, C.: Deep reinforcement learning for transportation network combinatorial optimization: a survey. Knowl.-Based Syst. **233**, 107526 (2021)
29. Wu, G., Zhang, Z., Liu, H., Wang, J.: Solving time-dependent traveling salesman problem with time windows with deep reinforcement learning. In: IEEE International Conference on Systems, Man, and Cybernetics (2021)
30. Wu, H., Wang, J., Zhang, Z.: MODRL/D-AM: multiobjective deep reinforcement learning algorithm using decomposition and attention model for multiobjective optimization. In: Li, K., Li, W., Wang, H., Liu, Y. (eds.) ISICA 2019. CCIS, vol. 1205, pp. 575–589. Springer, Singapore (2020). https://doi.org/10.1007/978-981-15-5577-0_45
31. Wu, Y., Song, W., Cao, Z., Zhang, J., Lim, A.: Learning improvement heuristics for solving routing problems. IEEE Trans. Neural Netw. Learn. Syst. (2021)
32. Zhang, Y., Wang, J., Zhang, Z., Zhou, Y.: MODRL/D-EL: multiobjective deep reinforcement learning with evolutionary learning for multiobjective optimization. In: International Joint Conference on Neural Networks (2021)
33. Zhang, Z., Liu, H., Zhou, M., Wang, J.: Solving dynamic traveling salesman problems with deep reinforcement learning. IEEE Trans. Neural Netw. Learn. Syst. (2021)
34. Zhang, Z., Wu, Z., Zhang, H., Wang, J.: Meta-learning-based deep reinforcement learning for multiobjective optimization problems. IEEE Trans. Neural Netw. Learn. Syst. (2022)

Evolving Through the Looking Glass: Learning Improved Search Spaces with Variational Autoencoders

Peter J. Bentley[1,2]([✉]), Soo Ling Lim[1], Adam Gaier[2], and Linh Tran[2]

[1] Department of Computer Science, University College London (UCL), London, UK
p.bentley@cs.ucl.ac.uk
[2] Autodesk Research, London, UK

Abstract. Nature has spent billions of years perfecting our genetic representations, making them evolvable and expressive. Generative machine learning offers a shortcut: learn an evolvable latent space with implicit biases towards better solutions. We present SOLVE: Search space Optimization with Latent Variable Evolution, which creates a dataset of solutions that satisfy extra problem criteria or heuristics, generates a new latent search space, and uses a genetic algorithm to search within this new space to find solutions that meet the overall objective. We investigate SOLVE on five sets of criteria designed to detrimentally affect the search space and explain how this approach can be easily extended as the problems become more complex. We show that, compared to an identical GA using a standard representation, SOLVE with its learned latent representation can meet extra criteria and find solutions with distance to optimal up to two orders of magnitude closer. We demonstrate that SOLVE achieves its results by creating better search spaces that focus on desirable regions, reduce discontinuities, and enable improved search by the genetic algorithm.

Keywords: Variational autoencoder · Latent variable evolution · Generative machine learning · Genetic algorithm

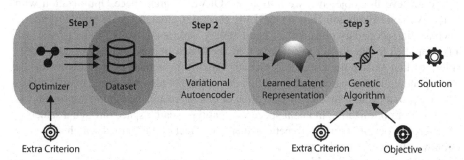

Fig. 1. *Search space Optimization with Latent Variable Evolution* (SOLVE). An optimizer produces a dataset of random solutions satisfying an extra criterion (e.g., constraint or secondary objective). A variational autoencoder learns this dataset and produces a learned latent representation biased towards the desired region of the search space. This learned representation is then used by a genetic algorithm to find solutions that meet the objective and extra criterion together.

© The Author(s) 2022
G. Rudolph et al. (Eds.): PPSN 2022, LNCS 13398, pp. 371–384, 2022.
https://doi.org/10.1007/978-3-031-14714-2_26

1 Introduction

In nature, the mapping from gene to phenotypic effect is hugely complex. Any new human genetic trait will be propagated through trillions of cells, which via a complex developmental process involving gene regulatory networks, intercellular communication, pattern formation and differentiation, results in altered phenotypic characteristics: an improved ability to taste bitter substances; an increased likelihood of developing an immunity to certain diseases; a reduced propensity to be creative. There is pervasive pleiotropy in the human genome [1] and yet like all living organisms, we have evolved and continue to do so, with most of our offspring remaining viable as healthy functioning organisms. Somehow, nature has learned a genetic representation that is astonishingly expressive, searchable, and despite constant genetic innovation, maps to viable living creatures that satisfy the multiple criteria of survival.

In evolutionary computation our hand-designed genetic representations are usually mapped directly to phenotypic effects so that we have minimal pleiotropy. Yet if we introduce any additional criterion or constraint, our evolutionary algorithms still struggle. From an optimization perspective, the additional criteria distort the search space, adding discontinuities and deceptive regions that result in ineffective optimization [2]. Typical solutions involve modifying the search operators or the optimization algorithm to overcome problems in the search space [3–5]. These specialized algorithms may need tuning for each problem and require expertise, which may not always be available.

Recent work has used autoencoders to learn representations when performing black box optimization. We extend our previous work [6] and propose a variation of this idea: *learn a better search space*. In contrast to previous work which aims to reduce the problem dimensionality, here we investigate the idea that a generative machine learning approach could map a difficult-to-search genotype space into an easier-to-search latent space. This new space would be biased towards solutions that satisfy the additional criteria to the problem, while at the same time smoothing out discontinuities in the space, effectively achieving evolution of evolvability [9] by using deep learning as a shortcut.

To achieve this objective, we introduce SOLVE: Search space Optimization with Latent Variable Evolution (Fig. 1). SOLVE generates a dataset from problem criteria (a constraint, secondary objective, or heuristic). A Variational Autoencoder (VAE) [10, 11] is applied to the dataset to generate a learned latent representation biased towards solutions that satisfy the criteria. A Genetic Algorithm is then used to evolve in the corresponding latent search space and find solutions that also satisfy the overall objective. We provide a step-by-step investigation of this approach, examining improvements provided for optimization in latent space vs. genotype search space through a selection of different types of criterion. To better isolate the effect of the learned search space, we employ only a very simple optimizer.

SOLVE is not a multi-objective optimization algorithm – it aims to find single solutions that meet one objective and one or more extra criteria. SOLVE is also not a constraint satisfaction approach – while constraints can be recast as additional criteria [6], it cannot guarantee that they will always be met. Instead, SOLVE is a search space optimizer. It is suitable for difficult problems that can be broken down into separate objectives, criteria and/or constraints, or that may have important domain knowledge available in the form of heuristics or required features. SOLVE shows for the first time that a VAE can be used to map a difficult-to-search space into a latent space that is easier to search, without relying on parameter reduction.

2 Background

2.1 Variational Autoencoders

An autoencoder [7] is a neural network originally used for feature learning or dimensionality reduction, but its concept became widely popular for learning generative models of the data. An autoencoder consists of two parts - an encoder p and a decoder q. The encoder maps the observations x to a (lower dimensional) embedding space z, whereas the decoder maps the embeddings back to the original observation space. We denote the reconstructed data as x'. The autoencoder is trained to minimize the reconstruction error between observation and decoded output and simultaneously project the observations into the lower dimensional "bottleneck".

The Variational Autoencoder (VAE) is a probabilistic autoencoder proposed concurrently by Kingma et al. [8] and Rezende et al. [9]. The architecture of a VAE is similar to the one of autoencoders described above. However, instead of encoding an observation as a single point, VAEs encode it as a distribution over the latent space. Due to its simplicity and the resulting analytical solution for the regularization, the distribution is set to be an isotropic Gaussian distribution. From a Bayesian perspective, VAEs maximize a lower bound on the log-marginal likelihood, which is given by

$$\log p(x) \geq \underbrace{\mathbb{E}_{q_\phi}\big[\log p_\theta(x|z)\big]}_{\text{log likelihood}} - \underbrace{D_{KL}(q(z|x)||p(z))\big]}_{\text{latent space regularization}} =: -\mathcal{L}_{VAE}(x; \theta, \phi) \qquad (1)$$

where θ are the model parameters of encoder p and ϕ are the model parameters of decoder q. The expected log-likelihood or "reconstruction", the first term of Eq. (1), is proportional to the mean squared loss between decoded output and input observation if the output distribution is Gaussian. The second term of Eq. (1) denotes the Kullback-Leibler (KL) divergence and measures the similarity between the latent variable distribution q and a chosen prior p. In the common VAE, the latent variable distribution is isotropic Gaussian and parameterized through the neural network q_ϕ and the prior distribution is a Gaussian distribution with zero mean and diagonal unit variance $\mathcal{N}(0, I)$. The KL divergence has an analytical solution. The final minimization objective is

$$\mathcal{L}_{VAE}(x; \theta, \phi) = -\mathbb{E}_{q_\phi}\big[\log p_\theta(x|z)\big] + D_{KL}(q(z|x)||p(z))\big] \propto \|x' - x\|^2$$
$$+ D_{KL}(q(z|x)||p(z))\big] \qquad (2)$$

2.2 Evolving Latent Variables

The ability of deep learning systems like VAEs to learn representations has not gone unnoticed by the evolutionary optimization community. Latent variable evolution (LVE) techniques first train generative models on existing datasets, such as video game levels, fingerprints, or faces and then use evolution to search the latent spaces of those models. Game levels can be optimized for a high or low number of enemies [10], fingerprints to defeat biometric security [11], celebrity look-alike faces can be generated with varied hair and eye colors [12], and VAEs can learn alternative search spaces for GP [13]. When no existing dataset of solutions is available, these solution sets must be generated. In [14, 15], solutions were collected by saving the champion solutions found after repeatedly running an optimizer on the problem. A representation learned from this set of champions can then be effective in solving similar sets of problems.

Quality-diversity [16, 17] approaches have been blended with LVE to improve optimization in high dimensional spaces. DDE-Elites [18] learns a 'data-driven encoding' by training a VAE based on the current collection, and uses that encoding together with the direct encoding to accelerate search. The related Policy Manifold Search [19] uses a VAE trained in the same way as part of a mutation operator. Standard and surrogate-assisted GAs have also been shown to benefit from having learned encodings "in-the-loop" in order to better tackle high-dimensional search spaces [20, 21].

A learned representation does more than reduce the dimensionality of the search space – it reduces the range of solutions which can be generated [22]. A model trained on Mario levels will never produce Pac-man levels, a model trained on predominantly white faces will only produce white faces. In this work we subvert the biases of models to limit search to desirable regions. Our previous work introduced this idea for constraint handling [6], here we expand the concept to problems with additional criteria.

3   Method

The SOLVE approach comprises three steps: Dataset Generation, Representation Learning, and Optimization.

3.1   Step 1: Dataset Generation

SOLVE decomposes the problem of optimization into stages, similar to [23]: the first step (Fig. 1 left) is the generation of a set of solutions that meet a criterion, without regard to their performance on the objective. When the only requirement is to satisfy a single criterion out of several, the search problem becomes much easier. Where it is feasible to calculate random values that satisfy the criterion using a dedicated algorithm, e.g., a constraint solver, this is typically the fastest method. However, it is sufficient to use a simple genetic algorithm for many criteria.

To produce the dataset of solutions we use the simple genetic algorithm defined in the DEAP framework [22]. We use real encoding, with a real-valued gene corresponding to each variable of the overall problem including variables that may be in the objective function and not in the extra criterion. Fitness is defined only by the criterion with a threshold used to determine acceptability for that criterion. The criterion fitness is zero if the value is under the threshold, otherwise it is set to a linear distance value representing the degree to which the criterion has been met, e.g., for $45 - x < = 0$: if $x < 45$ then fitness $f(x) = 45 - x$; otherwise $f(x) = 0$. If the best individual in the population achieves a fitness of 0 it is added to the dataset and the run terminates.

The genetic algorithm is run repeatedly until a dataset of d values has been found (typical execution times for C1 were no more than 10 ms per run, with a dataset of 5000 values taking less than 30 s on a MacBook Air 2020 M1, 16 GB memory). Each run used different initial populations, producing a distribution of values in the feasible region. These solutions provide a sampling of the valid region that serves as the basis for the learned representation. Should coverage be insufficient, or should a more efficient method be required, alternative algorithms which explicitly search for diversity such as Clustering [24], Clearing [25], Novelty Search [26], or MAP-Elites [27] could be employed for this step. Specialized constraint satisfaction algorithms could be used if the criterion takes the form of a constraint [28, 29].

3.2 Step 2: Representation Learning

Redesigning representations offers an alternative approach to optimization: a representation that has a bias towards the expression of useful or desirable solutions can enable simple optimizers to find good solutions, removing the requirement of needing expert tuning or development of specialized optimization algorithms.

In the second step of SOLVE (Fig. 1 middle), we use a simple VAE[1] with a standard loss function from [8]. We use one latent variable for every variable in the problem. We transform the input from the dataset to -1.0 to 1.0 (although the data-generation GA was limited to this range, its output data can focus on a smaller area of the search space and thus the dataset range may be smaller) and learn for E epochs. We refer the reader to the Supplemental Material[2] for full details.

3.3 Step 3: Optimization

In the third step (Fig. 1 right), we use the objective function for the first time and use a GA to search for optimal solutions that also satisfy the extra criterion in the learned latent space. Again, we use DEAP with real encoding, each gene corresponding to a latent variable with range -2.0 to 2.0 (typically sufficient to express the full learned range in the latent representation) with the results of all operators bound to this range. The same crossover and mutation operators are used as described in step 1.

Individuals are evaluated by decoding the latent values using the learned VAE model, mapping back to the range of the problem, and applying the fitness function. Fitness is a combination of the criterion, as formulated in Subsect. 3.1, and objective function. To ensure both are treated equally, we use tournament fitness, which awards individuals an average fitness score based on how many times it beats another individual for each objective and criterion over a series of smaller tournaments, similar to [3], chosen to encourage diversity akin to [30]. This approach avoids the need for summing and weighting separate criteria or using penalties for constraints [2, 31] and preliminary experiments using this method on benchmark problems with a standard GA resulted in significant

---

[1] https://github.com/pytorch/examples/tree/master/vae.

[2] http://www.cs.ucl.ac.uk/staff/p.bentley/solvesupplemental.pdf.

improvements to optimization with the settings used here. See the Supplemental Material for the tournament fitness algorithm. Parent selection is then performed using the same tournament selection (t size = 3) as described in step 1. A small population size evolving for few generations is sufficient.

4 Experiments

For all experiments we use a simple objective function to minimize for D variables:

$$f(\overline{x}) = \sum_{i=0}^{D-1} x_i^2 \tag{3}$$

Table 1 provides the additional criteria used to modify the search space of Eq. (3). These represent several commonly observed constraint and optimization functions inspired by the analysis in [34]: the range limit (C1), correlated variables in the form

Table 1. Extra criteria designed to conflict with the objective optimal.

Criterion	Equation	Search space
C1: Range limit	$\sum_{i=0}^{D-1}(45 - x_i) \leq 0$	
C2: Dependency	$\forall i \in \{0,..,D-2\}$ $8 \leq (x_{i+1} - x_i) \leq 10$	
C3: Nonlinear (Shifted, generalised Rosenbrock [32])	$\forall i \in \{0,..,D-2\}$ $100((x_i - 3)^2 - x_{i+1} - 3)^2 + (x_i - 4) <= 0$	
C4: Multiple dependencies and range limits over different subsets of problem variables	**C4.1**: $80 - x_0 - x_1 \leq 0$ **C4.2**: $x_2 + 45 \leq 0$ **C4.3**: $60 - x_1/2 - x_3 \leq 0$ **C4.4**: $-7 \leq (x_2 + x_3) \leq -5$	
C5: Discontinuous, $D=2$ (Shifted, rotated DeJong [33])	$1/[\sum_{j=1}^{25}((x_0' - A[1,j])^6$ $-(x_1' - A[2,j])^6 + j)^{-1} + \epsilon] - 445 < 0$ where: $A \in R^{2\times25}$ and $\epsilon = 0.002$ $\begin{bmatrix} x_0' \\ x_1' \end{bmatrix} = \begin{bmatrix} \cos(d) & -\sin(d) \\ \sin(d) & \cos(d) \end{bmatrix} \begin{bmatrix} ax_0 \\ ax_1 \end{bmatrix} + b$ $d = -33, a = 0.39, b = [9, -3]^T$	

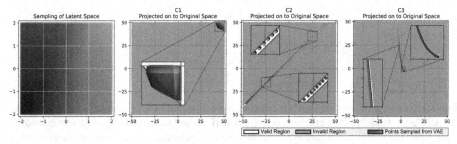

Fig. 2. Visualizing the Learned Landscapes. We can observe how the VAE shapes the space to be searched. Left: x_0 and x_1 values from -2 to 2 are color-coded to their locations in the learned space. *Others*: when these points in the learned space are projected back to the original space they are confined closely to the valid regions as defined by C1, C2 and C3.

of chained inequality (C2), the more complex Rosenbrock function [32] used in optimization competitions (C3) [35], multiple criteria (C4), and a discontinuous, multimodal function (C5). For all experiments, the range of the problem (objective and criteria) is -50 to 50.

4.1 Single Criterion (C1, C2, C3)

We initially focus on problems with one extra criterion. We investigate C1, C2, and C3, each designed to create a search space that cannot be solved reliably using a standard GA. The criteria achieve this by conflicting with the true optimal of Eq. (3) forcing the optimizer to compromise to meet the objective and criterion equally. We apply each step of SOLVE: generating data, learning new representation, using a GA to find optimal solutions with this representation. First, we examine the VAE and its learned representation.

Visualizing Learned Latent Representations. To understand how the learned latent representation may be beneficial for the GA in SOLVE, we plot the distribution of values returned by the latent representation as we vary the latent variables from -2 to 2 for C1 and C2. Figure 2 shows how the search space is compressed and folded into the small valid region for C1. This gives the dual advantage of reducing the search space to focus on the valid region and improving evolvability through duplication, with different values for the latent variables mapping to the same valid region. For C2 the space is compressed into a straight line, showing that the VAE has learned the correlation between the two variables.

Comparing Learned Latent Representation with Standard Representation. We examine criteria C1 to C3, using SOLVE to generate valid, optimal solutions and compare results to a standalone GA. The number of parameters D for the objective and criterion were varied from 1 to 10 to examine the effects of scaling the problem. The simple GA within step 1 of SOLVE generated data for C1 and C3 (5000 points).

This GA was unable to generate data for C2 for higher dimensions in feasible time, so 5000 random datapoints were calculated directly from the criterion in this case. The algorithm is provided in Supplementary Material.

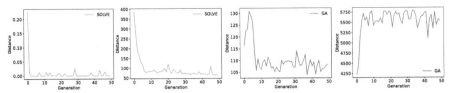

Fig. 3. Example run showing SOLVE criterion error (far left), SOLVE objective error over time (middle left), standalone GA criterion error (middle right) and standalone GA objective error over time (right) for 3 variable C1, and averaged over entire population. Note difference in *y*-axis scales for SOLVE and standalone GA results.

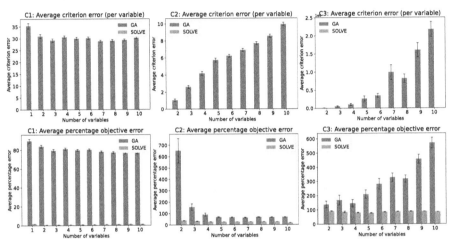

Fig. 4. Average error per criteria and average percentage error from best solution, over 100 runs, for different numbers of variables ($D = 1..10$). SOLVE always met C1 and C2; almost always C3. Error bars: mean ± SE.

We compare SOLVE with a standalone GA evolving a direct representation of the same problem (each gene real coded and bound to between -1.0 to 1.0, and using the same fitness calculation, selection, population size of 20 and 50 generations). Every run for both representations was repeated 100 times and average (mean) results reported. In order to enable comparison of achieving different optimal solutions, we calculate objective error as difference between $f(\overline{x})$ (Eq. 3) and optimal, divided by optimal $\times 100$, and criterion error is distance to criterion divided by number of variables.

Figure 3 shows a representative sample run comparing SOLVE evolution vs. standalone GA for C1 with D = 3 variables. While the GA struggles to evolve solutions that satisfy C1 and meet the objective, with evolution becoming stuck in a poor local optimal, SOLVE is consistently able to evolve solutions that meet C1 and are of high quality according to the objective. (Similar findings are presented in [6]).

The learned latent representation enables dramatic improvement (two orders of magnitude better) for the evolution of solutions meeting the objective and criteria. Figure 4 shows results for the three criteria with the number of variables increased from 1 to 10.

In C1 the standalone GA performs equally poorly for criteria and objective in all cases, while SOLVE achieves all criteria and a good objective for all numbers of variables. In C2, where variables are correlated, the standalone GA performs worse for larger numbers of variables, but objective values improve as the constrained problem reduces the number of possible good solutions. In contrast, SOLVE consistently nearly always meets all criteria and achieves solutions that reach the objective more closely for all numbers of variables. In the more difficult C3, the standalone GA performs worse for criteria and objectives as the number of variables increases. SOLVE meets the criteria better and consistently achieves good objective values.

4.2 Multiple Criteria (C4)

We next consider problems with multiple criteria (C4 in Table 1). This common form of optimization comprises several criteria and an objective. Each criterion applies only to a subset of the variables in the problem.

While SOLVE may be successfully applied to problems with a single extra criterion as shown for C1 to C3, for multiple criteria the creation of datasets that comprise examples of valid solutions for each criterion is incompatible with the VAE. This incompatibility is caused by the fact that by considering one criterion at a time to keep the problem easily solvable we can only generate valid values for the subsets of the variables belonging to that criterion. If we were to create a dataset for each criterion, we would generate valid values for variables belonging to the criterion in question, and random values for the free variables. For example, C4.1 applies only to x_0 and x_1, leaving x_2 and x_3 free to take any values; conversely, C4.4 applies only to x_2 and x_3, leaving x_0 and x_1 free. The result would be datasets comprising, in part, random values for all variables – effectively training the VAE that all values for all variables are valid and removing or even harming its ability to learn a useful representation. (While a GA could generate one dataset for all criteria simultaneously, this partially solves the difficult problem that we wish to transform by the VAE, defeating the objective of the work.)

The solution is to layer SOLVEs. We generate a latent representation for the first criterion, and then use the GA with the learned latent representation to create a dataset for the second criterion, which is used to learn a new latent representation that encapsulates both criteria, and so on, until all criteria have been learned in turn. The final latent representation is then used with all criteria and the objective to evolve optimal solutions that meet all criteria together, Fig. 5.

To assess the advantages of using a learned latent representation for multiple criteria, we used C4 with the same objective and compared a standalone GA against the original SOLVE and the LayeredSOLVE. We grow the difficulty incrementally, first trying C4.1 and C4.2 with a 2-layer SOLVE, then C4.1, C4.2, C4.3 with a 3-layer SOLVE, and finally C4.1, C4.2, C4.3, C4.4 with a 4-layer SOLVE (using 4 variables for all). To determine whether the order of the criteria alters the results, we also perform the same experiment, this time ordered: C4.1, C4.2, C4.4, C4.3. Figure 6 shows the results.

The results show that the original SOLVE offers little advantage compared to the standalone GA, both showing poor performance and large variance over the 100 runs. In contrast, the LayeredSOLVE can meet the criteria better, and enable solutions close to the optimum to be found, both with high consistency. The experiments also show that the

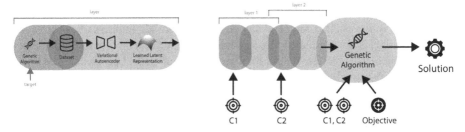

Fig. 5. LayeredSOLVE for two criteria over different subsets of problem variables. The number of stacks = number of criteria in the problem.

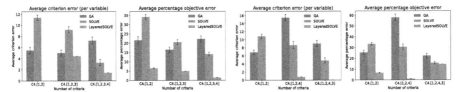

Fig. 6. Average error per criterion (average degree to which each criterion was broken) and average percentage error from best solution, over 100 runs, for a standalone GA, SOLVE and LayeredSOLVE, for 2, 3 and 4 criteria on 4 variables. Left two charts: C4.1, C4.2, C4.3, C4.4; Right two charts: C4.1, C4.2, C4.4, C4.3. Error bars: mean ± SE.

order in which criteria are presented to LayeredSOLVE has an effect. When presenting the "stricter" C4.4 before C4.3, the criteria are better satisfied. (Although the objective error increases, analysis of the solutions shows the better objective error observed for cases where criteria are not met are caused by the standalone GA cheating – criteria are conflicted to have invalid but deceptively good objective scores.) The notion that criteria more difficult to satisfy should be presented first to an algorithm was exploited in [36] where constraints are ordered according to "the sum of the constraints violations of all solutions in the initial population from the least violated to the most violated". Our results suggest that using the same strategy in the LayeredSOLVE will also improve its ability to satisfy all criteria.

4.3 Discontinuous Criteria (C5)

Finally, we focus on discontinuous problems. While SOLVE has been successfully applied to single (C1–C3) and multiple (C4) criteria problems, discontinuous criteria provide a challenge for the SOLVE model used in Sect. 4.1. We use an intuitive two-dimensional discontinuous constrained optimization problem (Table 1 C5). The problem is based on the well-known DeJong #5 function [33, 37], designed as a multimodal optimization benchmark, adjusted to create discontinuous valid regions and rotated to make it non-trivial for our optimizer, akin to [35].

This problem is difficult for a standalone GA, but its discontinuous nature is also incompatible with the use of a simple VAE, which attempts to fit a single Gaussian distribution to the set of valid points, i.e., it assumes a Gaussian distribution $N(0, I)$ as

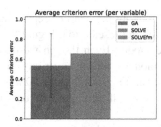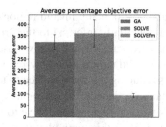

Fig. 7. Average error per criterion (average degree to which each criterion was broken) and average percentage error from best solution, over 100 runs, for original SOLVE, SOLVE with flow model (SOLVEfm), and a standalone GA for criterion C5. Average criterion error for SOLVEfm is zero. Error bars: mean ± SE.

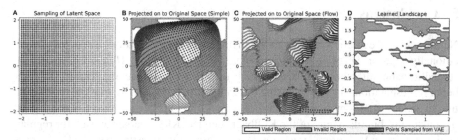

Fig. 8. Visualizing the learned latent representation. **A**: Sampling space from -2 to 2. **B**: Points projected onto space learned by a simple VAE. **C**: Points projected onto space learned by VAE with normalizing flow prior (SOLVEfm). **D**: Latent space colored by valid and invalid regions – this is the transformed landscape searched by SOLVEfm.

prior. The solution to improve expressivity in SOLVE is to increase the complexity of the VAE prior (i.e., "the structure of the sampling space"). While a Gaussian distribution allows for easy sampling and the KL divergence term can be calculated in closed form, for many datasets the representation is much more complicated than Gaussian distribution. For discontinuous criterion C5, we use a **normalizing flow** [38] as prior. A normalizing flow transforms a simple Gaussian distribution into a complex one by applying a sequence of invertible transformations. Given the chain of transformations, we change the latent variable to obtain a more complex target variable according to the change of variables theorem.

We compare the standalone GA with SOLVE and SOLVE using a normalizing flow VAE on C5 plus the usual objective function, reporting the average results of 100 runs. Figure 7 shows that the standalone GA and original SOLVE find valid solutions with large variance but fail to locate the optimum much of the time. SOLVE with normalizing flow model (denoted as SOLVEfm in Fig. 7) finds valid solutions with high consistency and finds better solutions for the objective with smaller variance.

We can understand how this result is achieved by examining Fig. 8. When using a simple VAE within SOLVE, the VAE fails to learn a good representation for the nine unbalanced and separated data modes for valid solutions (Fig. 8B). The generated data from the simple VAE tries to cover the center of the data, however, it misses some data modes and covers many invalid regions, making this learned representation little different (and potentially inferior) to the original representation. Figure 8C shows the resulting expressed values for the flow-based model. Even though we observe some samples in the invalid regions, the flow-based model is able to cover most modes of the criterion, thus reducing the search space to the valid regions. Figure 8D illustrates how this representation "connects the disconnected" – transforming the discontinuous space into a more connected region, conducive to search.

The advantages of this method still need to be weighed against the generation of datasets, which even when the problem has been simplified by considering just one criteria at a time, requires computation time, see [6] for discussion. A comparison of the quality of SOLVE solutions should also be made with state-of-the-art optimizers.

5 Conclusions

Nature achieved evolution of evolvability in its genetic representations through a computationally expensive process that is infeasible for us to duplicate. Here we have demonstrated a viable alternative: SOLVE, which uses generative machine learning to learn better representations for search. Using this method, not only can we bias the representation so that it focusses mainly on desirable regions of the space according to extra criteria or constraints, but the nature of the Kullback-Leibler (KL) divergence used for regularization within the VAE provides a natural "smoothing" effect on the resulting latent space akin to the notion of evolvability, which can change a discontinuous space into a continuous space enabling highly effective search by the optimizer. We have demonstrated that with zero dimensionality reduction (i.e., using the same number of latent variables as problem variables), SOLVE can map different forms of hard-to-search spaces onto improved latent spaces. These spaces enable even a simple optimizer to achieve substantive performance increases in terms of quality of solution found and effort required to find that solution, as evidenced by the small population sizes and low number of generations required for the GA within SOLVE.

This work used a GA for data generation and optimization and a VAE for representation learning, but other equivalent approaches could be employed within SOLVE. The field of generative machine learning continues to advance at a great pace, so we anticipate that the integration of these newer techniques into optimization for the purposes of generating improved search spaces will be a fruitful area of research going forwards.

Source Code. The source code necessary to reproduce the experiments in this paper is available at: https://github.com/writingpeter/SOLVE.

References

1. Watanabe, K., et al.: A global overview of pleiotropy and genetic architecture in complex traits. Nat. Genet. **51**, 1339–1348 (2019)
2. Homaifar, A., Qi, C.X., Lai, S.H.: Constrained optimization via genetic algorithms. SIMULATION **62**, 242–253 (1994)
3. Deb, K.: An efficient constraint handling method for genetic algorithms. Comput. Methods Appl. Mech. Eng. **186**, 311–338 (2000)

4. Yu, T., Bentley, P.: Methods to evolve legal phenotypes. In: Eiben, A.E., Bäck, T., Schoenauer, M., Schwefel, H.-P. (eds.) PPSN 1998. LNCS, vol. 1498, pp. 280–291. Springer, Heidelberg (1998). https://doi.org/10.1007/BFb0056871

5. Deb, K., Pratap, A., Agarwal, S., Meyarivan, T.: A fast and elitist multiobjective genetic algorithm: NSGA-II. IEEE Trans. Evol. Comput. **6**, 182–197 (2002)

6. Bentley, P., Lim, S.L., Gaier, A., Tran, L.: COIL: Constrained optimization in learned latent space. Learning representations for valid solutions. In: ACM Genetic and Evolutionary Computation Conference Companion, p. 8 (2022)

7. Hinton, G.E., Salakhutdinov, R.R.: Reducing the dimensionality of data with neural networks. Science **313**, 504–507 (2006)

8. Kingma, D.P., Welling, M.: Auto-encoding variational bayes. In: International Conference on Learning Representation, p. 14 (2014)

9. Rezende, D.J., Mohamed, S., Wierstra, D.: Stochastic backpropagation and approximate inference in deep generative models. In: International Conference on Machine Learning, pp. 1278–1286 (2014)

10. Volz, V., Schrum, J., Liu, J., Lucas, S.M., Smith, A., Risi, S.: Evolving Mario levels in the latent space of a deep convolutional generative adversarial network. In: ACM Genetic and Evolutionary Computation Conference, pp. 221–228 (2018)

11. Bontrager, P., Roy, A., Togelius, J., Memon, N., Ross, A.: DeepMasterPrints: generating masterprints for dictionary attacks via latent variable evolution. In: IEEE 9th International Conference on Biometrics Theory, Applications and Systems, pp. 1–9 (2018)

12. Fontaine, M.C., Nikolaidis, S.: Differentiable quality diversity. In: 35th Conference on Neural Information Processing Systems, pp. 10040–10052 (2021)

13. Liskowski, P., Krawiec, K., Toklu, N.E., Swan, J.: Program synthesis as latent continuous optimization: Evolutionary search in neural embeddings. In: ACM Genetic and Evolutionary Computation Conference, pp. 359–367 (2020)

14. Scott, E.O., De Jong, K.A.: Toward learning neural network encodings for continuous optimization problems. In: ACM Genetic and Evolutionary Computation Conference Companion, pp. 123–124 (2018)

15. Moreno, M.A., Banzhaf, W., Ofria, C.: Learning an evolvable genotype-phenotype mapping. In: ACM Genetic and Evolutionary Computation Conference, pp. 983–990 (2018)

16. Pugh, J.K., Soros, L.B., Stanley, K.O.: Quality diversity: a new frontier for evolutionary computation. Front. Robot. AI **3**, 40 (2016)

17. Cully, A., Demiris, Y.: Quality and diversity optimization: a unifying modular framework. IEEE Trans. Evol. Comput. **22**, 245–259 (2017)

18. Gaier, A., Asteroth, A., Mouret, J.-B.: Discovering representations for black-box optimization. In: ACM Genetic and Evolutionary Computation Conference, pp. 103–111 (2020)

19. Rakicevic, N., Cully, A., Kormushev, P.: Policy manifold search: exploring the manifold hypothesis for diversity-based neuroevolution. In: ACM Genetic and Evolutionary Computation Conference, pp. 901–909 (2021)

20. Cui, M., Li, L., Zhou, M.: An Autoencoder-embedded Evolutionary Optimization Framework for High-dimensional Problems. In: IEEE International Conference on Systems, Man, and Cybernetics, vol. 2020-October, pp. 1046–1051 (2020)

21. Cui, M., Li, L., Zhou, M., Abusorrah, A.: Surrogate-assisted autoencoder-embedded evolutionary optimization algorithm to solve high-dimensional expensive problems. IEEE Trans. Evol. Comput. (2021). https://doi.org/10.1109/TEVC.2021.3113923

22. Hagg, A., Berns, S., Asteroth, A., Colton, S., Bäck, T.: Expressivity of parameterized and data-driven representations in quality diversity search. In: ACM Genetic and Evolutionary Computation Conference, pp. 678–686 (2021)

23. Venkatraman, S., Yen, G.G.: A generic framework for constrained optimization using genetic algorithms. IEEE Trans. Evol. Comput. **9**, 424–435 (2005)

24. Yin, X., Germay, N.: A fast genetic algorithm with sharing scheme using cluster analysis methods in multimodal function optimization. In: Artificial Neural Nets and Genetic Algorithms, pp. 450–457 (1993)
25. Pétrowski, A.: A clearing procedure as a niching method for genetic algorithms. In: IEEE International Conference on Evolutionary Computation, pp. 798–803 (1996)
26. Lehman, J., Stanley, K.O.: Abandoning objectives: Evolution through the search for novelty alone. Evol. Comput. **19**, 189–223 (2011)
27. Mouret, J.-B., Clune, J.: Illuminating search spaces by mapping elites. arXiv preprint arXiv: 1504.04909 (2015)
28. Coello, C.A.C.: Theoretical and numerical constraint-handling techniques used with evolutionary algorithms: a survey of the state of the art. Comput. Methods Appl. Mech. Eng. **191**, 1245–1287 (2002)
29. Tsang, E.: Foundations of constraint satisfaction: the classic text. BoD–Books on Demand (2014)
30. Forrest, S., Hightower, R., Perelson, A.: The Baldwin effect in the immune system: learning by somatic hypermutation. Individual Plasticity in Evolving Populations: Models and Algorithms (1996)
31. Yeniay, Ö.: Penalty function methods for constrained optimization with genetic algorithms. Math. Comput. Appl. **10**, 45–56 (2005)
32. Biscani, F., Izzo, D.: A parallel global multiobjective framework for optimization: pagmo. J. Open Source Softw. **5**, 2338 (2020)
33. De Jong, K.A.: An Analysis of the Behavior of a Class of Genetic Adaptive Systems. University of Michigan. Ph.D. thesis (1975)
34. Hellwig, M., Beyer, H.-G.: Benchmarking evolutionary algorithms for single objective real-valued constrained optimization–a critical review. Swarm Evol. Comput. **44**, 927–944 (2019)
35. Suganthan, P.N., et al.: Problem definitions and evaluation criteria for the CEC 2005 special session on real-parameter optimization. KanGAL Report **2005005**, 51 (2005)
36. Sallam, K.M., Elsayed, S.M., Chakrabortty, R.K., Ryan, M.J.: Improved multi-operator differential evolution algorithm for solving unconstrained problems. In: 2020 IEEE Congress on Evolutionary Computation, pp. 1–8 (2020)
37. Molga, M., Smutnicki, C.: Test functions for optimization needs. Test Functions Optim. Needs **101**, 43 (2005)
38. Rezende, D., Mohamed, S.: Variational inference with normalizing flows. In: International Conference on Machine Learning, pp. 1530–1538 (2015)

Generalization and Computation for Policy Classes of Generative Adversarial Imitation Learning

Yirui Zhou[1], Yangchun Zhang[1], Xiaowei Liu[1], Wanying Wang[1],
Zhengping Che[2], Zhiyuan Xu[2], Jian Tang[2], and Yaxin Peng[1(✉)]

[1] Department of Mathematics, School of Science, Shanghai University,
Shanghai 200444, China
{zyr050798,zycstatis,davidlau,wywang,yaxin.peng}@shu.edu.cn
[2] AI Innovation Center, Midea Group, Shanghai 201702, China
{chezp,xuzy70,tangjian22}@midea.com

Abstract. Generative adversarial imitation learning (GAIL) learns an optimal policy by expert demonstrations from the environment with unknown reward functions. Different from existing works that studied the generalization of reward function classes or discriminator classes, we focus on policy classes. This paper investigates the generalization and computation for policy classes of GAIL. Specifically, our contributions lie in: 1) We prove that the generalization is guaranteed in GAIL when the complexity of policy classes is properly controlled. 2) We provide an off-policy framework called the two-stage stochastic gradient (TSSG), which can efficiently solve GAIL based on the soft policy iteration and attain the sublinear convergence rate to a stationary solution. The comprehensive numerical simulations are illustrated in MuJoCo environments.

Keywords: Generative adversarial imitation learning ·
Generalization · Computation · Policy classes

1 Introduction

Imitation learning (IL) [1,16,23,30,33,36], a powerful and practical alternative to reinforcement learning (RL) [24,32], aims at recovering expert policies from limited demonstration data. It has been widely applied to decision-making in many complex fields, including robotics [17], autonomous driving [6], and recommendation systems [31].

Promoted by increasingly sophisticated sequential decision tasks, IL mainly includes two kinds of algorithms: behavioral cloning (BC) [5,26,34] and adversarial imitation learning (AIL) [1,33,39]. BC attempts to minimize the difference between the agent and expert policies, and converts the IL task to ordinary regression or classification [26,34]. AIL mimics the expert policy through the state-action distribution matching, which mitigates the compounding error [26,27]. For large and high-dimensional environments, traditional BC and AIL

G. Rudolph et al. (Eds.): PPSN 2022, LNCS 13398, pp. 385–399, 2022.
https://doi.org/10.1007/978-3-031-14714-2_27

are not good at imitating complex expert policies. As a generalization of AIL, generative adversarial imitation learning (GAIL) [16] formulates the IL problem as a min-max optimization based on dual representation, which can be efficiently solved by stochastic gradient algorithms. Many empirical studies with the complicated environment based on GAIL have been successfully conducted [9,17,31].

Unfortunately, theories on GAIL are less complete. Chen et al. [8] analyzed the generalization capability of $\mathcal{R}$-distance, which is essentially an integral probability metric (IPM) [22]. Then Xu et al. [36] utilized the neural network distance [3] to link the generalization capability with the expected return. Existing works, however, mainly focus on the properties of reward functions or discriminator classes. $\mathcal{R}$-distance and neural network distance can analyze the generalization capability of reward function classes or discriminator classes well. The discriminator aims to maximize the difference between the expert and the learned policies, while the learned policy contributes to making itself as close to the expert policy as possible. As a result, $\mathcal{R}$-distance and neural network distance are not suitable for depicting the generalization property of policy classes.

In this paper, we introduce a novel state-action distribution error, which investigates the infimum of the expected return gap between the expert policy and the policy class. Furthermore, this new definition can normalize the reward function as a distribution. Simultaneously, we prove that the generalization is guaranteed in GAIL when the complexity of policy classes is properly controlled. Moreover, GAIL easily suffers from the phenomenon that the discriminator learns faster than the policy. Such an imbalance has a bad impact on the convergence of GAIL [37]. Our policy generalization ensures that the computation of GAIL can be performed using reproducing kernel policy functions. Based on reproducing kernel reward functions [8] and reproducing kernel policy functions, we consider a slightly modified constrained min-max optimization problem from the primal problem for more sufficient explorations to enhance the policy in both performance and sample efficiency. Inspired by the success of the updating strategy of soft policy iteration in soft actor-critic (SAC) [10,13–15], we leverage the two-stage stochastic gradient (TSSG) with the automatic entropy tuning to our framework. TSSG can efficiently solve GAIL and attain sublinear convergence to a stationary solution.

2 Preliminaries

2.1 Markov Decision Process

We take an infinite-horizon Markov decision process (MDP) [24,32,36] into consideration. It is formalized by the tuple $(\mathcal{S}, \mathcal{A}, p(s'|s,a), r(s,a), \gamma, p(s_0))$, i.e., the finite state space $\mathcal{S}$, the finite action space $\mathcal{A}$, the transition distribution $p(s'|s,a) : \mathcal{S} \times \mathcal{A} \times \mathcal{S} \to [0,1]$, the reward function $r(s,a) : \mathcal{S} \times \mathcal{A} \to \mathbb{R}$, the discount factor $\gamma \in (0,1)$, and the initial state distribution $p(s_0) : \mathcal{S} \to [0,1]$.

We define $d^\pi(s) = (1-\gamma) \sum_{t=0}^{\infty} \gamma^t \mathrm{Pr}(s_t = s; \pi)$ as *the discounted stationary state distribution*. It measures the overall "frequency" of visiting a state under the policy π. Similarly, *the discounted stationary state-action distribution* is defined as $\rho^{\pi}(s,a) = (1-\gamma)\sum_{t=0}^{\infty}\gamma^t \Pr(s_t = s, a_t = a; \pi)$, measuring the overall "frequency" of visiting a state-action under the policy π.

The goal of RL is to find a policy π^* that maximizes the expected return, i.e., $\max_{\pi \in \Pi} V_{\pi} = \max_{\pi \in \Pi} \mathbb{E}_{\pi}\left[\sum_{t=0}^{\infty}\gamma^t r(s_t, a_t)\right]$, where $s_{t+1} \sim p(\cdot|s_t, a_t)$ and Π is the policy class. Another equivalent formulation of V_{π} is

$$V_{\pi} = \frac{1}{1-\gamma}\mathbb{E}_{(s,a)\sim\rho^{\pi}}\left[r(s,a)\right]. \tag{1}$$

2.2 Generative Adversarial Imitation Learning

IL [1,16,23,30,33,36] requires expert demonstrations to train a policy. The purpose of IL is to minimize the expected return gap between π_E and $\pi \in \Pi$, i.e., $V_{\pi_E} - V_{\pi}$, where π_E is the expert policy.

GAIL [16] uses a discriminator to provide rewards for the agent. To distinguish whether a state-action is from the expert demonstration or generated by the agent, the optimization objective of the discriminator $D(s,a)$ from the discriminator class $\mathcal{D}$ can be written as

$$\max_{D \in \mathcal{D}} \mathbb{E}_{(s,a)\sim\rho^{\pi_E}}\left[\log(D(s,a))\right] + \mathbb{E}_{(s,a)\sim\rho^{\pi}}\left[\log(1 - D(s,a))\right].$$

The target of the agent is to minimize the discrepancy between the distributions generated by the agent and the expert. With the reward function $r(s,a) = -\log(1 - D(s,a))$, the optimization objective of the agent is formalized as maximizing its expected return:

$$\max_{\pi \in \Pi} \mathbb{E}_{(s,a)\sim\rho^{\pi}}\left[-\log(1 - D(s,a))\right].$$

3 Related Work

IL [1,16,23,30,33,36] trains a policy by learning from expert demonstrations. GAIL [16], a famous IL algorithm, has achieved vital empirical progress [6,17,31].

In the theoretical aspect, $\mathcal{R}$-distance $d_{\mathcal{R}}(\pi, \pi')$ [8] is defined as $d_{\mathcal{R}}(\pi, \pi') = \sup_{r \in \mathcal{R}}\{\mathbb{E}_{\pi}[r(s,a)] - \mathbb{E}_{\pi'}[r(s,a)]\}$ for two policies π and π', as well as the reward function class $\mathcal{R}$. The generalization of $\mathcal{R}$-distance can be guaranteed as long as the class of the reward functions is properly controlled for GAIL with general reward parameterization [8]. In [8], the authors also designed convergent stochastic first-order optimization algorithms to solve the min-max optimization in GAIL, but they set the entropy regularizer as an unchangeable constant. Our paper regards the entropy regularizer as dynamically updated to capture the change in the policy. Zhang et al. [38] studied a gradient-based algorithm with alternating updates and established its sublinear convergence to the globally optimal solution of GAIL. Then Guan et al. [12] characterized the global convergence with a sublinear rate for a broad range of commonly used policy gradient algorithms. In terms of generalization, the closest work to ours is [36], where the authors considered the generalization capability of GAIL. In particular, they analyzed the generalization capability of discriminator classes but they did not provide the result for policy classes, which is another way of learning the generalization of GAIL. Besides, they also focused on the value discrepancy of BC and GAIL respectively [35,36].

4   Generalization for Policy Classes of GAIL

4.1   State-Action Distribution Error

First, we take a classic example.

Example 1. Here we set a toy environment to elaborate on the gap between the expert policy and the policy class. In an adversarial environment, the state s_i can be viewed as the opponent in a two-player game and the action a_j as our response. Note that for a state-action pair (s_i, a_j), if $j \geq i$, the player loses the game. Likewise, if $j < i$, the player wins the game. The policy class Π contains six policies.

The scoring mechanism states the score for each win or loss in the game, which can be determined by a discriminator D. In Fig. 1, we discover that the score for D_1 is 1 per win and -2 per loss. Similarly, the score for D_2 is 3 per win and -1 per loss. The total score gap between the expert and the policy class for all games in the discriminators D_1 and D_2 are $1 - (-2) = 3$ and $8 - 4 = 4$ respectively, which implies that the total score gaps in various discriminators are different.

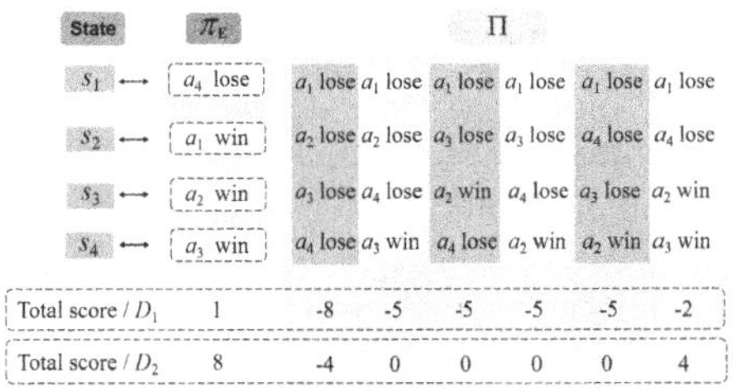

An error is introduced to measure

State	π_E	\multicolumn{6}{c}{Π}
s_1	a_4 lose	a_1 lose	a_1 lose	a_1 lose	a_1 lose	a_1 lose	a_1 lose
s_2	a_1 win	a_2 lose	a_2 lose	a_3 lose	a_3 lose	a_4 lose	a_4 lose
s_3	a_2 win	a_3 lose	a_4 lose	a_2 win	a_4 lose	a_3 lose	a_2 win
s_4	a_3 win	a_4 lose	a_3 win	a_4 lose	a_2 win	a_2 win	a_3 win
Total score / D_1	1	-8	-5	-5	-5	-5	-2
Total score / D_2	8	-4	0	0	0	0	4

Fig. 1. An example depicting the gap between the expert policy and the policy class. The scoring mechanism is determined by a discriminator D.

the discrepancy between the distributions generated by the expert and agent, which aims to learn the closest policy in the policy class Π to the expert policy. The error essentially describes the infimum of the expected return gap between the expert policy and the policy class.

Definition 1. *(State-action distribution error) Given an expert policy π_E, a policy class Π and a **fixed discriminator** $D \in (0,1)^{S \times A}$, the state-action distribution error between π_E and Π is*

$$e(C_D \rho^{\pi_E}, C_D \rho^{\Pi})$$
$$= \inf_{\pi \in \Pi} \{ \mathbb{E}_{(s,a) \sim \rho^{\pi_E}} [-\log(1 - D(s,a))] - \mathbb{E}_{(s,a) \sim \rho^{\pi}} [-\log(1 - D(s,a))] \},$$

where $C_D = -\sum_{(s,a) \in S \times A} \log(1 - D(s,a))$ denotes the discriminant coefficient and ρ^{Π} is the state-action distribution functional class induced by Π.

Remark 1. The reward function $r(s,a) = -\log(1 - D(s,a))$ is not a distribution, and we normalize it by $\mathbb{D}(s,a) = r(s,a)/C_D$. Intuitively, this state-action distribution measures the overall "frequency" of visiting a state-action according to the discriminator D. Then Definition 1 can be rewritten as

$$e(C_D \rho^{\pi_E}, C_D \rho^{\Pi}) = C_D \inf_{\pi \in \Pi} \{\mathbb{E}_{(s,a) \sim \mathbb{D}}[\rho^{\pi_E}(s,a) - \rho^{\pi}(s,a)]\}. \tag{2}$$

Actually, Eq. (2) is equivalent to Definition 1:

$$e(C_D \rho^{\pi_E}, C_D \rho^{\Pi}) = C_D \inf_{\pi \in \Pi} \{\mathbb{E}_{(s,a) \sim \mathbb{D}}[\rho^{\pi_E}(s,a) - \rho^{\pi}(s,a)]\}$$

$$= C_D \inf_{\pi \in \Pi} \{\sum_{(s,a) \in \mathcal{S} \times \mathcal{A}} \rho^{\pi_E}(s,a) \frac{-\log(1 - D(s,a))}{C_D}$$

$$- \sum_{(s,a) \in \mathcal{S} \times \mathcal{A}} \rho^{\pi}(s,a) \frac{-\log(1 - D(s,a))}{C_D}\}$$

$$= \inf_{\pi \in \Pi} \{\sum_{(s,a) \in \mathcal{S} \times \mathcal{A}} \rho^{\pi_E}(s,a)[-\log(1 - D(s,a))]$$

$$- \sum_{(s,a) \in \mathcal{S} \times \mathcal{A}} \rho^{\pi}(s,a)[-\log(1 - D(s,a))]\}.$$

In addition, the learned policy is fixed to investigate the generalization capability of the reward function class or discriminator class [8,36]. The discriminator D is required to be fixed while dealing with the generalization capability of policy classes in comparison.

Remark 2. In Example 1, $e(C_D \rho^{\pi_E}, C_D \rho^{\Pi})$ is the total score gap between the expert and the policy class for all games.

The purpose of a discriminator is to maximize the difference between the expert policy and the learned policy. Accordingly, GAIL maximizes the empirical state-action distribution error

$$e\left(\left(-\sum_{i=1}^{m} \log(1 - D(s_D^{(i)}, a_D^{(i)}))\right)\rho^{\pi_E}, \left(-\sum_{i=1}^{m} \log(1 - D(s_D^{(i)}, a_D^{(i)}))\right)\rho^{\Pi}\right),$$

where $\{(s_D^{(i)}, a_D^{(i)})\}_{i=1}^{m}$ is the demonstration samples collected by $\mathbb{D}$. Similarly, *the empirical state-action distribution error* is defined as follows.

Definition 2. *Given an expert policy π_E, a policy class Π and a fixed discriminator $D \in (0,1)^{\mathcal{S} \times \mathcal{A}}$, the empirical state-action distribution error between π_E and Π is*

$$e(\hat{c}_D^{(m)} \rho^{\pi_E}, \hat{c}_D^{(m)} \rho^{\Pi}) = \hat{c}_D^{(m)} \inf_{\pi \in \Pi} \{\mathbb{E}_{(s,a) \sim \hat{\mathbb{D}}}[\rho^{\pi_E}(s,a) - \rho^{\pi}(s,a)]\},$$

where $\hat{c}_D^{(m)} = -\sum_{i=1}^{m} \log(1 - D(s_D^{(i)}, a_D^{(i)}))$ denotes the empirical discriminant coefficient and $\hat{\mathbb{D}}$ is the empirical version of $\mathbb{D}$ with m samples, i.e., $\hat{\mathbb{D}}(s_D^{(i)}, a_D^{(i)}) = -\log(1 - D(s_D^{(i)}, a_D^{(i)}))/\hat{c}_D^{(m)}$ means the empirical "frequency" of visiting a state-action according to D.

4.2 Generalization Properties for Policy Classes

To facilitate later analysis, we present the upper bound with respect to the state-action distribution and the discriminant coefficient in Assumption 1.

Assumption 1. *Given π_E and a fixed discriminator D_I, suppose that there is an upper bound B_Π on the product of the discriminant coefficient and the state-action distribution functional class induced by (i) the policy class Π, (ii) the expert policy π_E, i.e., for all $\pi \in \Pi$, $\max_{(s,a)\in\mathcal{S}\times\mathcal{A}}\{C_{D_I}\rho^\pi(s,a), C_{D_I}\rho^{\pi_E}(s,a)\} \leq B_\Pi$.*

We now show the result on the generalization property for policy classes in the view of distribution error. We denote the empirical Rademacher complexity [21,29,36] of $C_D\rho^\Pi$ on m samples $Z = \{(s_D^{(i)}, a_D^{(i)})\}_{i=1}^m$ as $\hat{\mathfrak{R}}_{\mathbb{D}}^{(m)}(C_D\rho^\Pi) = \mathbb{E}_\sigma\left[\frac{1}{m}\sup_{\pi\in\Pi}\{\sum_{i=1}^m \sigma_i C_D\rho^\pi(s_D^{(i)}, a_D^{(i)})\}\right]$, where σ_i is the independent and identically distributed (i.i.d.) Rademacher random variable for $i = 1, ..., m$.

Theorem 1. *(Generalization for policy classes of distribution error) Suppose Assumption 1 holds, and for any $\hat{\epsilon} > 0$, given D_I and π_E with*

$$e(\hat{c}_{D_I}^{(m)}\rho^{\pi_E}, \hat{c}_{D_I}^{(m)}\rho^\Pi) \geq \sup_{D\in\mathcal{D}}\{e(\hat{c}_D^{(m)}\rho^{\pi_E}, \hat{c}_D^{(m)}\rho^\Pi)\} - \hat{\epsilon},$$

then

$$e(C_{D_I}\rho^{\pi_E}, C_{D_I}\rho^\Pi)$$

$$\geq \underbrace{\sup_{D\in\mathcal{D}}\{e(\hat{c}_D^{(m)}\rho^{\pi_E}, \hat{c}_D^{(m)}\rho^\Pi)\}}_{\mathrm{Appr}(\mathcal{D},m)} \underbrace{-2\hat{\mathfrak{R}}_{\mathbb{D}_I}^{(m)}(C_{D_I}\rho^\Pi) - 8B_\Pi\sqrt{\frac{\log(3/\delta)}{2m}}}_{\mathrm{Estm}(\Pi,m,\delta)} -\hat{\epsilon}$$

for all $\delta \in (0,1)$, with a probability of at least $1 - \delta$.

Proof Sketch. To prove Theorem 1, we decompose the state-action distribution error $e(C_{D_I}\rho^{\pi_E}, C_{D_I}\rho^\Pi)$ into three parts: the approximation error $\mathrm{Appr}(\mathcal{D}, m)$, the estimation error and the optimization error $-\hat{\epsilon}$, where $\mathrm{Appr}(\mathcal{D}, m)$ corresponds to the approximation error induced by the limited discriminator class. Then we only need to study the lower bound $\mathrm{Estm}(\Pi, m, \delta)$ of the estimation error. Here $\mathrm{Estm}(\Pi, m, \delta)$ denotes the estimation error of GAIL regarding the complexity of the policy class and the number of samples. For detailed proof, please refer to the Supplementary Material[1]. □

Theorem 1 implies that the generalization of state-action distribution error can be guaranteed as long as the complexity of policy class Π is properly controlled. Concretely, a simpler policy class increases the estimation error, then tends to increase the state-action distribution error.

---

[1] The Supplementary Material is released at
https://github.com/MDM-shu/GAIL-Policy-Generalization-and-TSSG.

Additionally, applying Dudley's entropy integral [8], we can further connect $\hat{\mathfrak{R}}_{\mathbb{D}_I}^{(m)}(C_{D_I}\rho^{\Pi})$ with the covering number [21]. Then we study the generalization property of state-action distribution error in terms of the covering number, which is a direct consequence of Theorem 1. For detailed proof, please refer to the Supplementary Material.

Example 2. **(Reproducing kernel policy function)** One available choice to parameterize the product of the state-action distribution functional class and the discriminant coefficient is the reproducing kernel Hilbert space (RKHS) [18,19]. We consider *the feature mapping* approach. To be specific, $h : \mathbb{R}^n \times \mathbb{R}^m \to \mathbb{R}^p$ is taken into account, and the product of the state-action functional class and the discriminant coefficient can be written as $C_{D_I}\rho^{\pi}(s,a) = C_{D_I}\rho^{\pi_\theta}(\psi_s,\psi_a) = \theta^{\top}h(\psi_s,\psi_a)$, where h satisfies Assumption 2 and $\theta \in \mathbb{R}^p$.

Assumption 2. *The feature mapping h satisfies $h(0,0) = 0$, where h is Lipschitz continuous with respect to (ψ_s,ψ_a), i.e., there exists a positive constant L_h such that for any $\psi_s, \psi_a, \psi_s', \psi_a'$, we have*

$$\|h(\psi_s,\psi_a) - h(\psi_s',\psi_a')\|_2 \le L_h\sqrt{\|\psi_s - \psi_s'\|_2^2 + \|\psi_a - \psi_a'\|_2^2}.$$

Moreover, for all $s \in \mathcal{S}$ and $a \in \mathcal{A}$, we have $\|\psi_s\|_2 \le 1$ and $\|\psi_a\|_2 \le 1$.

Assumption 2 is a standard condition for studying the properties of popular feature mappings [4,8,25]. A step further, we investigate the generalization of GAIL using feature mapping (see Corollary 1).

Corollary 1. *Suppose Assumption 2 holds and $\|\theta\|_2 \le B_\theta$. Then*

$$e(C_{D_I}\rho^{\pi_E}, C_{D_I}\rho^{\Pi}) \ge \sup_{D \in \mathcal{D}}\{e(\hat{c}_D^{(m)}\rho^{\pi_E}, \hat{c}_D^{(m)}\rho^{\Pi})\}$$

$$- \frac{8}{m} - \frac{24B_{\Pi}}{\sqrt{m}}\sqrt{p\log(1 + 2\sqrt{2m}B_\theta L_h)} - 8B_{\Pi}\sqrt{\frac{\log(3/\delta)}{2m}} - \hat{\epsilon}$$

with a probability of at least $1 - \delta$.

Proof Sketch. We first exploit the Lipschitz continuity of $C_{D_I}\rho^{\pi}(s,a)$ with respect to the parameter θ. Then applying the standard argument of the volume ratio, we bound the covering number. For detailed proof, please refer to the Supplementary Material.   □

Example 2 shows that, with respect to a class of properly normalized reproducing kernel policy functions, GAIL generalizes in terms of the state-action distribution error.

Based on Theorem 1, we can obtain the result of the generalization capability of GAIL from the perspective of the expected return between the expert policy and the best-learned policy. This best-learned policy is based on the given discriminator D_I in the policy class Π.

Theorem 2. *(GAIL Generalization for policy classes)* *Under the same assumption of Theorem 1, we have*

$$V_{\pi_E} - \sup_{\pi \in \Pi} V_\pi \geq \frac{1}{1-\gamma}\left(\mathrm{Appr}(\mathcal{D}, m) + \mathrm{Estm}(\Pi, m, \delta) - \hat{\epsilon}\right)$$

with a probability of at least $1 - \delta$.

Proof Sketch. Combining the definition of V_π in Eq. (1) with the reward function of GAIL, we get the relationship between $V_{\pi_E} - \sup_{\pi \in \Pi} V_\pi$ and the state-action distribution error. Then plugging the result into the conclusion of Theorem 1, we complete the proof of Theorem 2. For detailed proof, please refer to the Supplementary Material.   □

Theorem 2 discloses that the generalization is guaranteed in GAIL when the complexity of policy classes is properly controlled. Moreover, Theorem 2 implies a necessary condition (the right-hand side less than or equal to zero) for the learned policy to outperform the expert policy. Future work could be sufficient conditions for the learned policy to outperform the expert policy.

5   Two-Stage Stochastic Gradient Algorithm

In this section, we show the computational properties of GAIL, which are the direct application of Example 2 and reproducing kernel reward functions [8]. Particularly, the reward function can be parameterized by $r(s, a) = r_\phi(\psi_s, \psi_a) = \phi^\top g(\psi_s, \psi_a)$, $\phi \in \mathbb{R}^q$, where g is based on the following assumption.

Assumption 3. [8] *The feature mapping g satisfies $g(0, 0) = 0$, and there exists a constant L_g such that for any $\psi_s, \psi_a, \psi_s', \psi_a'$, we have*

$$\|g(\psi_s, \psi_a) - g(\psi_s', \psi_a')\|_2 \leq L_g\sqrt{\|\psi_s - \psi_s'\|_2^2 + \|\psi_a - \psi_a'\|_2^2}.$$

Assumption 3 is consistent with Assumption 2. For computational convenience, a slightly modified min-max optimization problem from the primal problem with a constraint condition

$$\min_\theta \max_{\|\phi\|_2 \leq \kappa} \mathbb{E}_{(s,a)\sim\rho^{\pi_E}}[r_\phi(s, a)] - \mathbb{E}_{(s,a)\sim\rho^{\pi_\theta}}[r_\phi(s, a)] - \frac{\mu}{2}\|\phi\|_2^2$$

$$\text{s.t. } \mathbb{E}_{(s_t,a_t)\sim\rho^{\pi_\theta}}\left[-\log(\pi_\theta(a_t|s_t))\right] \geq H_0, \tag{3}$$

is considered. Here $\mathbb{E}_{(s_t,a_t)\sim\rho^{\pi_\theta}}\left[-\log(\pi_\theta(a_t|s_t))\right] = \mathbb{E}_{s_t\sim d^{\pi_\theta}}\left[\mathbb{H}(\pi_\theta(\cdot|s_t))\right]$ is the entropy regularizer for the policy, H_0 is a desired minimum expected entropy [15] and $\mu > 0$ is a tuning parameter.

Based on SAC, the constraint in Eq. (3) is usually tight. It is unnecessary to impose an upper bound on the entropy [15]. Compared with the primal problem, this optimization problem with constraint conditions supports:

(i) Any useful action is not omitted in the learned policy that corresponds with the idea of maximum entropy RL.

(ii) Computational stability is improved in practice.

Then the optimization problem of Eq. (3) can be transformed into two sub-problems:

(i) For a fixed ϕ, the sub-problem is

$$\min_{\theta} -\mathbb{E}_{(s,a)\sim\rho^{\pi_\theta}}[r_\phi(s,a)] \quad \text{s.t.} \quad \mathbb{E}_{(s_t,a_t)\sim\rho^{\pi_\theta}}\big[-\log(\pi_\theta(a_t|s_t))\big] \geq H_0. \quad (4)$$

(ii) For a fixed θ, the sub-problem is

$$\max_{\|\phi\|_2\leq\kappa} \mathbb{E}_{(s,a)\sim\rho^{\pi_E}}[r_\phi(s,a)] - \mathbb{E}_{(s,a)\sim\rho^{\pi_\theta}}[r_\phi(s,a)] - \frac{\mu}{2}\|\phi\|_2^2. \quad (5)$$

By the fact that $\min_\theta -\mathbb{E}_{(s,a)\sim\rho^{\pi_\theta}}[r_\phi(s,a)] = \max_\theta \mathbb{E}_{\pi_\theta}[\sum_{t=0}^{T}\gamma^t r_\phi(s_t,a_t)]$, Eq. (4) is equivalent to:

$$\max_{\theta} \min_{\alpha\geq 0} \mathbb{E}_{\pi_\theta}\big[\sum_{t=0}^{T}\gamma^t r_\phi(s_t,a_t)\big] + \alpha(\mathbb{E}_{(s_t,a_t)\sim\rho^{\pi_\theta}}\big[-\log(\pi_\theta(a_t|s_t))\big] - H_0). \quad (6)$$

To solve the optimization problem in Eq. (3), the inner and outer layers are updated alternately. The framework of TSSG is shown in the Supplementary Material.

In the inner layer, we introduce the soft alternating mini-batch (SAM) submodule. Specifically,

Update of θ: we follow the soft policy improvement in SAC to solve the min-max optimization problem in Eq. (6). Particularly, the optimized objective functions of the value network and the policy network are denoted by

$$J_Q(w;\theta,\phi,\alpha) = \mathbb{E}_{(s_t,a_t)\sim\mathcal{D}_I}\Big[\frac{1}{2}\big(Q_w^{\text{soft}}(s_t,a_t;\theta,\phi) - (r_\phi(s_t,a_t)$$
$$+ \gamma\mathbb{E}_{s_{t+1}\sim p(\cdot|s_t,a_t)}[V_w^{\text{soft}}(s_{t+1})])\big)^2\Big],$$
$$F(\theta;w,\phi,\alpha) = J_\pi(\theta) + \mathbb{E}_{(s,a)\sim\rho^{\pi_E}}[r_\phi(s,a)] - \frac{\mu}{2}\|\phi\|_2^2,$$

respectively, where $\mathcal{D}_I$ denotes the experience replay buffer [20] of SAC. The value function is implicitly parameterized through the soft Q-function parameters via:

$$V_w^{\text{soft}}(s_t) = \mathbb{E}_{a_t\sim\pi_\theta}[Q_w^{\text{soft}}(s_t,a_t;\theta,\phi) - \alpha\log(\pi_\theta(a_t|s_t))]. \quad (7)$$

$J_\pi(\theta)$ is the objective function in SAC:

$$J_\pi(\theta) = \mathbb{E}_{s_t\sim\mathcal{D}_I}\big[\mathbb{E}_{a_t\sim\pi_\theta}[\alpha\log(\pi_\theta(a_t|s_t)) - Q_w^{\text{soft}}(s_t,a_t;\theta,\phi)]\big]. \quad (8)$$

Both Eq. (7) and Eq. (8) are referred to [14,15]. For the reason that $\nabla_\theta F = \nabla_\theta J_\pi$, the convergence of SAC is not affected. Therefore, we take

$$w_{t+1} = w_t - \frac{\eta_w}{n_w}\sum_{j\in D_w^t}\nabla_w \tilde{J}_Q^{(j)}(w_t;\theta_t,\hat{\phi}(\theta^{(k)}),\alpha_t), \quad (9)$$
$$\theta_{t+1} = \theta_t - \frac{\eta_\theta}{n_\theta}\sum_{j\in D_\theta^t}\nabla_\theta \tilde{f}^{(j)}(\theta_t;w_{t+1},\hat{\phi}(\theta^{(k)}),\alpha_t) \quad (10)$$

at the $(t+1)$-th iteration, where η_w and η_θ are learning rates of w and θ respectively. $\nabla_w \tilde{J}_Q^{(j)}, \nabla_\theta \tilde{f}^{(j)}$ are independent stochastic approximations of $\nabla_w J_Q$ and $\nabla_\theta F$, and D_w^t, D_θ^t are mini-batches with sizes n_w and n_θ respectively. Note that $\hat{\phi}(\theta^{(k)})$ is the unbiased estimator of the sub-problem Eq. (5):

$$\mathbb{E}[\hat{\phi}(\theta^{(k)})] = \phi^\star(\theta^{(k)})$$
$$= \arg\max_\phi \mathbb{E}_{(s,a)\sim\rho^{\pi_E}}[r_\phi(s,a)] - \mathbb{E}_{(s,a)\sim\rho^{\pi_\theta}}[r_\phi(s,a)] - \frac{\mu}{2}\|\phi\|_2^2, \quad (11)$$

which is the direct consequence of [8].

Update of α: we claim that α is not only a Lagrange multiplier but also an entropy temperature parameter. Thus, α can be self-updated directly by the loss function $J(\alpha;\theta) = \mathbb{E}_{a_t\sim\pi_\theta}\left[-\alpha\log(\pi_\theta(a_t|s_t)) - \alpha H_0\right]$ introduced in SAC [15]. Therefore, we take the $(t+1)$-th iteration as

$$\alpha_{t+1} = \alpha_t - \eta_\alpha \nabla_\alpha \tilde{J}_t^{(k)}(\alpha_t; \theta_{t+1}), \quad (12)$$

where η_α is the learning rate of α and $\nabla_\alpha \tilde{J}_t^{(k)}$ is a stochastic approximation of $\nabla_\alpha J$.

Moreover, two Q-functions in the SAM submodule are used to restrain the overestimation of the value function.

The pseudocode for the SAM submodule is presented in Algorithm 1.

Algorithm 1 SAM submodule: $\theta_T = \text{SAM}(\theta_0; w_0, \alpha_0, \phi)$

Input: $\theta_0 \in \mathbb{R}^p, w_0 \in \mathbb{R}^d, \alpha_0 \in \mathbb{R}$, fixed $\phi \in \mathbb{R}^q$;
Output: θ_T.
1: **for** $t = 0, ..., T-1$ **do**
2:     Apply Eq. (9) and Eq. (10) to update w_{t+1} and θ_{t+1} respectively.
3:     Apply Eq. (12) to automate entropy adjustment for α_{t+1}.
4: **end for**

In the outer layer, we introduce the TSSG algorithm. ϕ is updated by Eq. (11) and θ is updated by the SAM submodule, where $w_T^{(k)} = w_0^{(k+1)}$, $\theta_0^{(k)} = \theta^{(k)}$, $\theta_T^{(k)} = \theta^{(k+1)}$ and $\alpha_T^{(k)} = \alpha_0^{(k+1)}$.

The pseudocode for TSSG is presented in Algorithm 2.

Algorithm 2 TSSG

Input: Expert demonstrations $\mathcal{D}^\star$, an empty dataset $\mathcal{D}_I$, $\theta^{(0)} \in \mathbb{R}^p$, $w_0^{(0)} \in \mathbb{R}^d$, $\alpha_0^{(0)} \in \mathbb{R}$;

Output: $\theta^{(N)}$.

1: **for** $k = 0, ..., N-1$ **do**
2:     Collect samples with $\pi_{\theta^{(k)}}$. Add them to $\mathcal{D}_I$.
3:     Apply Eq. (11) using data from $\mathcal{D}_I$ and $\mathcal{D}^\star$ to update $\hat{\phi}(\theta^{(k)})$.
4:     Apply the SAM submodule (Alg. 1) using data from $\mathcal{D}_I$ to update the policy:

$$\theta^{(k+1)} = \mathrm{SAM}(\theta^{(k)}; w_0^{(k)}, \alpha_0^{(k)}, \hat{\phi}(\theta^{(k)})).$$

5: **end for**

The convergence analysis of TSSG is in the Supplementary Material. The result shows that TSSG attains sublinear convergence to a stationary solution. Compared with the alternating mini-batch stochastic gradient algorithm and the greedy stochastic gradient algorithm in [8], TSSG involves a dynamic entropy regularizer rather than regarding it as an unchangeable constant.

6 Experiments

In the sequel, we compare the TSSG algorithm with some AIL variations. The sensitivity of inner layer steps T is shown in Subsect. 6.2.

6.1 Evaluation and Results

The purpose of IL is to recover the expert policy based on limited expert demonstrations. These demonstrations are typically collected from human experts or static datasets in practice. In our experiment settings, to create such an expert for generating demonstrations, we utilize a trained model-free SAC [15] agent to simulate the expert policy in Hopper-v2, Walker2d-v2 and HalfCheetah-v2 respectively. The demonstration data containing a buffer size of 10^6 is obtained with 0.01 standard deviation in each environment. The imitator aims to achieve expert-level control under the limited buffers. The mean return of the demonstration data in each environment is listed in Table 1.

When training imitation learning approaches, we compare two contenders in all the aforementioned environments. The first is adversarial inverse reinforcement learning (AIRL) [11] where the policy parameters are updated by the proximal policy optimization (PPO) [28]. The second is GAIL with PPO, where the parameters of the kernel function are updated by the greedy stochastic gradient algorithm [8].

Table 1. Mean return of the demonstration data in MuJoCo environments.

Environment	Mean return
Hopper-v2	3433
Walker2d-v2	3509
HalfCheetah-v2	9890

We use the same neural network architecture in all experiments. We establish a multi-layer perceptron (MLP) structure to build the policy network, the Q-functions and the discriminator network. To implement the kernel function, we use two layers of the neural network to mimic the random feature mapping [2,7].

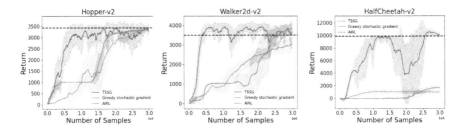

Fig. 2. Performances of the three algorithms.

We set $T = 32$ in Hopper-v2, $T = 64$ in Walker2d-v2 and $T = 32$ in HalfCheetah-v2 for the TSSG algorithm. Figure 2 shows the performance of all methods. The dashed lines indicate the mean return of expert demonstrations. More details and experiments are displayed in the Supplementary Material. These results show that the TSSG algorithm yields significant improvements and reduces the sample complexity by an order of magnitude. It is due to:

(i) The framework of TSSG achieves higher efficiency in learning the expert policy by making full use of the samples from the experience replay buffer of SAC, compared with the previous algorithms in [8,11].

(ii) The phenomenon that the discriminator learns faster than the policy in GAIL [37] can be alleviated by sufficient exploration of the agent training in TSSG.

6.2   Hyper-parameter Sensitivity

Without loss of generality, we choose Hopper-v2 as an example for sensitivity analysis. In this subsection, we explore the performance of the TSSG algorithm under different T. Figure 3 visualizes that increasing T increases the sample efficiency of TSSG. However, T should not be set too large owing to two aspects:

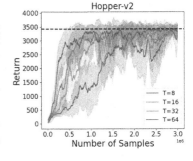

(i) $T = 64$ takes much longer time than $T = 32$, while the effect does not witness more improvement.

(ii) $T = 64$ has the risk of instability caused by overfitting.
From the above discussion, T dropping in [16, 32] is the best choice for Hopper-v2.

Fig. 3. Learning curves from Hopper-v2 for TSSG at different T's.

7 Conclusion

In this paper, we have explored the generalization and computation for policy classes of GAIL. In terms of generalization, we normalize the reward function as a distribution. We also propose the state-action distribution error to study the infimum of the expected return gap between the expert policy and the policy class. We prove that the generalization can be guaranteed when the complexity of policy classes is properly controlled. Our TSSG algorithm can efficiently solve GAIL and attain sublinear convergence to a stationary solution. Numerical experiments are provided to support our analysis.

We investigate the necessary condition of the learned policy outperforming the expert policy. The sufficient conditions are left for future works.

References

1. Abbeel, P., Ng, A.Y.: Apprenticeship learning via inverse reinforcement learning. In: International Conference on Machine Learning, pp. 1–8 (2004)
2. Arora, S., Du, S.S., Hu, W., Li, Z., Salakhutdinov, R.R., Wang, R.: On exact computation with an infinitely wide neural net. In: Advances in Neural Information Processing Systems, vol. 32, pp. 8139–8148 (2019)
3. Arora, S., Ge, R., Liang, Y., Ma, T., Zhang, Y.: Generalization and equilibrium in generative adversarial nets (GANs). In: International Conference on Machine Learning, pp. 224–232 (2017)
4. Bach, F.: On the equivalence between kernel quadrature rules and random feature expansions. J. Mach. Learn. Res. **18**(1), 714–751 (2017)
5. Bain, M., Sammut, C.: A framework for behavioural cloning. Mach. Intell. **15**, 103–129 (1995)
6. Bhattacharyya, R.P., Phillips, D.J., Wulfe, B., Morton, J., Kuefler, A., Kochenderfer, M.J.: Multi-agent imitation learning for driving simulation. In: 2018 IEEE/RSJ International Conference on Intelligent Robots and Systems (IROS), pp. 1534–1539. IEEE (2018)
7. Bietti, A., Mairal, J.: On the inductive bias of neural tangent kernels. In: Advances in Neural Information Processing Systems, vol. 32, pp. 12873–12884 (2019)
8. Chen, M., et al.: On computation and generalization of generative adversarial imitation learning. In: International Conference on Learning Representations (2020)
9. Chi, W., et al.: Collaborative robot-assisted endovascular catheterization with generative adversarial imitation learning. In: 2020 IEEE International Conference on Robotics and Automation (ICRA), pp. 2414–2420 (2020)
10. Dally, K., Van Kampen, E.J.: Soft actor-critic deep reinforcement learning for fault tolerant flight control. In: AIAA SCITECH 2022 Forum, pp. 2078–2097 (2022)
11. Fu, J., Luo, K., Levine, S.: Learning robust rewards with adversarial inverse reinforcement learning. arXiv preprint arXiv:1710.11248 (2017)
12. Guan, Z., Xu, T., Liang, Y.: When will generative adversarial imitation learning algorithms attain global convergence? In: International Conference on Artificial Intelligence and Statistics, pp. 1117–1125 (2021)
13. Haarnoja, T., Tang, H., Abbeel, P., Levine, S.: Reinforcement learning with deep energy-based policies. In: International Conference on Machine Learning, pp. 1352–1361 (2017)

14. Haarnoja, T., Zhou, A., Abbeel, P., Levine, S.: Soft actor-critic: off-policy maximum entropy deep reinforcement learning with a stochastic actor. In: International Conference on Machine Learning, pp. 1861–1870 (2018)

15. Haarnoja, T., et al.: Soft actor-critic algorithms and applications. arXiv preprint arXiv:1812.05905 (2018)

16. Ho, J., Ermon, S.: Generative adversarial imitation learning. In: Advances in Neural Information Processing Systems, vol. 29, pp. 4565–4573 (2016)

17. Jabri, M.K.: Robot manipulation learning using generative adversarial imitation learning. In: Proceedings of the Thirtieth International Joint Conference on Artificial Intelligence, IJCAI-21, pp. 4893–4894 (2021)

18. Kim, K.E., Park, H.S.: Imitation learning via kernel mean embedding. In: Proceedings of the AAAI Conference on Artificial Intelligence, pp. 3415–3422 (2018)

19. Li, S., Xiao, S., Zhu, S., Du, N., Xie, Y., Song, L.: Learning temporal point processes via reinforcement learning. arXiv preprint arXiv:1811.05016 (2018)

20. Lin, L.J.: Self-improving reactive agents based on reinforcement learning, planning and teaching. Mach. Learn. 8(3–4), 293–321 (1992)

21. Mohri, M., Rostamizadeh, A., Talwalkar, A.: Foundations of Machine Learning. MIT Press, Cambridge (2018)

22. Müller, A.: Integral probability metrics and their generating classes of functions. Adv. Appl. Probab. 29(2), 429–443 (1997)

23. Ng, A.Y., Russell, S.J., et al.: Algorithms for inverse reinforcement learning. In: International Conference on Machine Learning, pp. 663–670 (2000)

24. Puterman, M.L.: Markov Decision Processes: Discrete Stochastic Dynamic Programming. John Wiley & Sons, Hoboken (2014)

25. Rahimi, A., Recht, B.: Random features for large-scale kernel machines. In: Advances in Neural Information Processing Systems, vol. 20, pp. 1177–1184 (2007)

26. Ross, S., Bagnell, D.: Efficient reductions for imitation learning. In: International Conference on Artificial Intelligence and Statistics, pp. 661–668 (2010)

27. Ross, S., Gordon, G., Bagnell, D.: A reduction of imitation learning and structured prediction to no-regret online learning. In: International Conference on Artificial Intelligence and Statistics, pp. 627–635 (2011)

28. Schulman, J., Wolski, F., Dhariwal, P., Radford, A., Klimov, O.: Proximal policy optimization algorithms. arXiv preprint arXiv:1707.06347 (2017)

29. Shalev-Shwartz, S., Ben-David, S.: Understanding Machine Learning: From Theory to Algorithms. Cambridge University Press, Cambridge (2014)

30. Shani, L., Zahavy, T., Mannor, S.: Online apprenticeship learning. arXiv preprint arXiv:2102.06924 (2021)

31. Shi, J.C., Yu, Y., Da, Q., Chen, S.Y., Zeng, A.X.: Virtual-Taobao: virtualizing real-world online retail environment for reinforcement learning. In: Proceedings of the AAAI Conference on Artificial Intelligence, pp. 4902–4909 (2019)

32. Sutton, R.S., Barto, A.G.: Reinforcement Learning: An Introduction. MIT Press, Cambridge (1998)

33. Syed, U., Schapire, R.E.: A game-theoretic approach to apprenticeship learning. In: Advances in Neural Information Processing Systems, vol. 20, pp. 1449–1456 (2007)

34. Syed, U., Schapire, R.E.: A reduction from apprenticeship learning to classification. In: Advances in Neural Information Processing Systems, vol. 23, pp. 2253–2261. Citeseer (2010)

35. Xu, T., Li, Z., Yu, Y.: On value discrepancy of imitation learning. arXiv preprint arXiv:1911.07027 (2019)

36. Xu, T., Li, Z., Yu, Y.: Error bounds of imitating policies and environments. In: Advances in Neural Information Processing Systems, vol. 33, pp. 15737–15749 (2020)
37. Zhang, Y.F., Luo, F.M., Yu, Y.: Improve generated adversarial imitation learning with reward variance regularization. Mach. Learn. **111**(3), 977–995 (2022)
38. Zhang, Y., Cai, Q., Yang, Z., Wang, Z.: Generative adversarial imitation learning with neural network parameterization: global optimality and convergence rate. In: International Conference on Machine Learning, pp. 11044–11054 (2020)
39. Ziebart, B.D., et al.: Maximum entropy inverse reinforcement learning. In: Proceedings of the AAAI Conference on Artificial Intelligence, vol. 8, pp. 1433–1438 (2008)

Generative Models over Neural Controllers for Transfer Learning

James Butterworth$^{(\boxtimes)}$ ⓘ, Rahul Savani ⓘ, and Karl Tuyls ⓘ

University of Liverpool, Liverpool L69 3BX, UK
{j.butterworth2,rahul.savani}@liverpool.ac.uk

Abstract. We introduce a technique that leverages the power of indirect encodings (IE) from the field of evolutionary computation to improve the speed of evolution in transfer learning control tasks. Although generative models have previously been used to construct IEs, their potential in transfer learning, specifically in reinforcement learning domains, has not yet been utilised. We train three types of generative models: an autoencoder (AE), a variational autoencoder (VAE) and a generative adversarial network (GAN) on the neural network weights of well-performing solutions of a set of paramaterised source domains. The decoder of the AE and VAE or the generator of the GAN is then used as the IE in an evolutionary run on unseen, but related, target domains. We compare against two baselines: a direct encoding (DE) and a DE *starting* evolution from a controller pre-trained to maximise the average fitness over the set of source domains. We show that, by using these IEs, the speed of learning on the target domains is greatly increased with respect to the baselines.

Keywords: Indirect encodings · Evolutionary algorithms · Generative models · Neuroevolution

1 Introduction

Transfer learning refers to the situation where what has been learned on one task can be exploited in a different but related task [7]. An example from the supervised learning setting would be to use the initial layers of a classifier trained to detect pictures of cats in order to initialise or pretrain a classifier used to detect pictures of dogs. One would assume that the features learned in the first layers to detect cats such as edges, corners, changes in lighting etc. will also be useful in detecting dogs. Pretraining a model with a large auxiliary dataset that differs from the target dataset enhances object detection performance compared to a model trained *solely* on the target image dataset [5]; suggesting the features that were useful in classifying objects in the first dataset were also useful for classification in the second dataset.

In reinforcement learning (RL) tasks, transfer learning consists of leveraging prior knowledge from a set of source domains to improve learning on some target domain [15]. For example, the ability to walk is a prerequisite for many different tasks such as hunting, building and evading predators, however, when learning each of these tasks, animals do not have to relearn how to walk each time. It would be inefficient for RL agents to relearn primitive actions every time they learn a new, but related, task. To avoid this, there needs to be some mechanism by which *invariant* knowledge over the source domains is gathered and stored.

Evolutionary algorithms (EA) are a set of gradient-free search algorithms that can be used to find and optimise policies for RL tasks. EAs also have the ability to store invariant domain knowledge via an indirect encoding (IE). An IE is a mapping from a genotype space to a phenotype space, where the phenotype space *is* the solution space, and the genotype space is that on which evolutionary operators are applied. One way of producing a domain-dependent IE is by learning a distribution over the parameters of previously found solutions using a generative model, this is known as a data-driven encoding (DDE) [4]. Data-driven encodings have been explored in a number of works referenced in Sect. 2. However, despite the proven link between evolutionary processes and learning theory suggesting evolution has the ability to generalise via their IEs [11,14], to the authors knowledge, no work has yet applied DDEs to transfer learning in RL tasks.

Consequently, in this work we use indirect encodings to capture the similarities in neural network controller parameter spaces for a set of source domains, and then reuse these IEs to evolve solutions on a set of unseen target domains with much greater speed. Three different IEs are explored: the decoders derived from a trained autoencoder (AE) and a variational autoencoder (VAE), as well as the generator derived from a trained generative adversarial network (GAN). Hence, we address the following questions, which are central to transfer learning:

1. What knowledge is relevant for generalisation to future tasks?
2. Which storage mechanism should be used for this knowledge?

For 1., we model the distribution of well-performing solutions in parameter space over a set of source domain; for 2., we store relevant knowledge in the IE. We demonstrate the ability of this technique on three OpenAI gym environments: Continuous Mountain Car, Frozen Lake and Bipedal Walker and give evidence that IEs trained in this way perform much better than two other baselines.

2   Related Work

An autoencoder and a VAE are used in [13] and [2], respectively, to learn a distribution over neural network controller parameters for benchmark RL tasks, such as Bipedal Walker and Cart Pole. Unlike our work, the focus is not on transferring gained knowledge to unseen target domains, rather, the enumeration and analysis of behaviours on a single domain instance. Furthermore, the fitness of discovered solutions is not used as a metric of evaluation in either [13] or [2], whereas our work considers the fitness of produced solutions of paramount importance.

DDE-Elites [4] uses a VAE to learn a distribution over joint angles for well-performing solutions to a 2D planar arm inverse kinematics environment. Similar to our work, evolutionary search is subsequently performed in the latent space of the VAE decoder, which acts as an IE, and is compared to a direct encoding. Also, optimisation on novel but similar tasks is performed, akin to the transfer learning experiments in our work. However, unlike our work, the technique is not evaluated on RL tasks. Furthermore, in [4], the parameters of optimisation are joint angles as opposed to the parameters of neural network controllers.

Conditional GANs (cGAN) have also been used to construct IEs. The optimisation of *high-level* control policies for a robotic arm, and the design of buildings, an energy plant and the respective energy distribution network for an urban neighbourhood is assisted by a cGAN in [9] and [10] respectively. Similar to our work, [9] tests the generalisation capabilities of the trained generative model on unseen but related domains. Unlike our work, neither [9] or [10] consider neural network controllers, nor do they perform any form of search in the latent space of their respective generative models.

COIL [1] applies the ideas of [4] to the concept of constrained mathematical optimisation problems. AutoMap [12] uses the decoder of an autoencoder as an IE to re-evolve on a rugged fitness landscape resulting in *much* faster learning on the same task.

All of the works cited in Sect. 2 consider only a single type of IE, whereas our work compares three different types of IE.

3   Conceptual Overview

3.1   Indirect Encodings

A typical evolutionary algorithm takes a genotype space, Γ, and a fitness function, $f : \Gamma \to \mathbb{R}$ and aims to find

$$\mathbf{g}^* \in \Gamma \quad s.t.$$

$$f(\mathbf{g}^*) \geq f(\mathbf{g}) \quad \forall \mathbf{g} \in \Gamma$$

by applying evolutionary operators such as mutation, crossover and selection directly to Γ.

An indirect encoding $\delta : \Gamma \to \Phi$ introduces the concept of a phenotype space, Φ, and in turn reformulates the concept of an EA as aiming to find

$$\mathbf{g}^* \in \Gamma \quad s.t.$$

$$f(\delta(\mathbf{g}^*; \boldsymbol{\theta})) \geq f(\delta(\mathbf{g}; \boldsymbol{\theta})) \quad \forall \mathbf{g} \in \Gamma$$

where δ may have some additional function parameters, θ. Γ and Φ are vectors of elements, these elements can take a number of different forms such as: real numbers, integers, binary values, characters, etc.

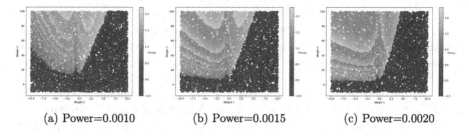

(a) Power=0.0010   (b) Power=0.0015   (c) Power=0.0020

Fig. 1. Fitness function plots for three different engine power settings in CMC. Plots were generated by producing 1000 random neural networks and assessing their fitnesses. The axes are the weight values of the neural networks and the colour represents fitness.

In the above formulation, a direct encoding would be a special case of an indirect encoding where $\Gamma = \Phi$ and δ is the identity function resulting in no change to the genotype, in other words, the phenotype is the same as the genotype.

In this work, we use IEs in neuroevolution - the optimisation of neural networks using evolutionary algorithms - therefore, $\Phi \in \mathbb{R}^n$ where n is the number of neural network weights. We use $\Gamma \in \mathbb{R}^m$, however, Γ does not have to be restricted to real numbers when using IEs in neuroevolution.

3.2   Related Fitness Functions

In RL, each domain typically has a reward function that gives a set reward for each state entered by an agent; an RL agent will try to maximise this cumulative reward by taking actions resulting in higher rewards. An agent that is optimised by an EA typically receives a singular reward at the end of an environmental run, which is used in the selection procedure. Step-wise RL rewards can be converted into a singular end reward by simply summing the rewards for all the steps of the episode.

One can imagine that for completely different domains, the fitness functions are vastly different. For example, a neural network trained on Continuous Mountain Car (CMC) would not work at all on Bipedal Walker. Aside from the fact that the number of inputs and number of outputs of the control networks are different, even if they were not, the likelihood is that a good solution on one domain would not be a good solution on another. However, for similar domains it might be the case that the fitness functions are somewhat related.

For example, CMC can be solved with a very simple neural controller; there are 2 inputs and 1 output and a solution can be found using a linear controller with only 2 weights (no bias). The more complicated fitness function used in our work encourages the car to get as close to the goal as possible but also rewards the car for getting there in a faster time *and* for having a lower speed at the goal. There are a number of adjustable domain parameters within CMC, such as gravity and engine power. For each of these parameterised domains, the resulting fitness function will be different.

Figure 1 illustrates the fitness functions for three different engine power settings. It can be observed that despite the fitness functions being different, they contain structural similarities. For example, no matter what the power value, all networks with a second weight value of less than 0, or a first weight value greater than 5, have poor fitness scores. Information of this kind could be very useful in future search because we can assume that for domains with similar power values, it will be fruitless to search within these low valued fitness areas. Similarly, we can see that the higher fitness solutions are located in a *ridge* on the top left of the fitness function. This ridge is at a different place for each power setting but still in the same area of the weight space.

Without prior information about where good solutions in the weight space are located, there would not be much reason to start search anywhere other than at 0 with an unknown starting distribution variance. If one were therefore going to constrain the search space to $[-100, 100]$ (which is reasonable when searching over neural network weight space) the areas of interest in Fig. 1 are a small proportion of the space. Therefore, without integrating previous information into search, a substantial amount time and computational power might be required to locate this region.

The main proposition of this work is that we can capture the area of good solutions common to a number of different parameterised domains using generative models. These generative models will then act as the indirect encoding, thereby mapping arbitrary genotype values to high fitness areas of the phenotype space. We can then evolve again on some unseen domains using this IE, resulting in a much more informed search. However, this proposition can be extended much further than the neuroevolution experiments ran here and can be used for any parameterised optimisation problem where there are commonalities amongst the different search spaces.

4   Methodology

We prepared training data for the generative models using CMA-ES [8] to evolve solutions on a set of parameterised domains (source domains) using a direct encoding. A solution was defined as a neural network controller that achieves some minimal fitness value. Next, the generative models were trained using the neural network weights of the training data. We then performed evolution using CMA-ES over the latent space of the decoder, in the case of the AE and VAE, or the generator, in the case of the GAN. For the AE, evolution was started at 0.5 as the learnt code was bound in the range $[0, 1]$ due to the sigmoid activation function in the pre-code layer. For the VAE and GAN, evolution was started at 0.0 because their training procedures are such that the typical code processed by the latent space is that derived from a unit normal distribution.

For evaluation, we tested the speed of evolution and the fitness of the best winner found for the three IEs and two baselines on a set of target domains. The two baselines were: (1) a direct encoding, and (2) a direct encoding *starting* evolution from a controller pretrained to maximise the average fitness over the full set of source domains. We call this pretrained controller a universal controller (UC), which has been previously used as a baseline for MAML [3][1].

5   Experiments

5.1   Continuous Mountain Car

Continuous Mountain Car (CMC) is a variation of Mountain Car in which a continuous valued force is applied to the car instead of a singular valued discrete force. The car begins in the trough of a valley, and aims to reach a flag atop one of the mountains. The simulation ends once 1000 time steps have elapsed or the car reaches the goal position.

In order to make the environment more difficult, we modified the reward function such that the car achieves a greater reward for having a smaller speed at the goal position. This requires more precise control over the acceleration, making a solution more difficult to come by. There are three aspects of the reward function: a higher reward for the car getting closer to the goal, a higher reward for finding the goal in a smaller amount of time, and a higher reward for a slower speed at the goal location.

In CMC one can modify the engine power of the car such that the force applied to the car by the controller, and thereby the acceleration of the car, will be different given the same control signal. We use this engine power as the modifiable domain parameter in these experiments. Each instance of engine power will result in a similar but related fitness function, as in Fig. 1. A controller trained on one engine power instance will either overshoot or undershoot the goal, thereby reducing the overall fitness.

In order to collect training data for the IE, a DE was used to evolve 333 solutions for each of the three source domain instances with an engine power in the following set: $\{0.0008, 0.0012, 0.0016\}$, giving a total of 999 solutions. These solutions were evolved using CMA-ES with an initial sigma of 1.0 and the initial distribution centroid at $[0, 0]$. A population size of 100 was ran for 100 generations. Each element in the search space was bound between $[-100, 100]^2$. No IE was used in the collection of the training data.

Each of the three generative models were trained five times. It should be noted that the decoder (in the case of the AE and VAE) or the generator (in the case of the GAN) had a hidden layer with a non-linear activation function. Therefore, the resulting output space was non-linear.

---

[1] Both the training of the IEs and the initialisation of the UC have additional preparatory overheads compared to evolution using a DE only. To compare techniques according to the total number of FLOPS, including pretraining, would be particularly meticulous and, more importantly, implementation dependent. We have therefore decided to evaluate with respect to the number of generations inline with evaluation methods used in [4] (no. of generations) and [3] (no. of gradient steps).

[2] Often CMA-ES can discover solutions with very large values making it more difficult to train a generative model over.

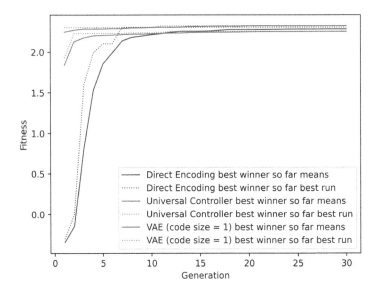

Fig. 2. Ten evolutionary runs for the DE and five evolutionary runs for the UC and a VAE with a code size of 1 on CMC with an engine power of 0.0014. The fitnesses plotted are those of the best winner so far, this is the best solution found so far in the evolutionary run. The solid lines are the mean fitnesses of the runs and the dotted line is the best run according to the final generation fitness.

Once trained, we used the decoder or generator as the IE for another evolutionary process on target CMC domains with test engine power values in the set: $\{0.0010, 0.0014\}$. We ran the evolutionary process 5 times for each test parameter and for each of the three generative models. We compared these results with 10 runs of a DE and 5 runs of a DE starting from a UC trained on the source domains.

Figure 2 shows the fitnesses of the best winner so far of the evolutionary runs for an engine power of 0.0014. The plotted IE is derived from a VAE with a code size of 1. The VAE begins the evolutionary run with a much higher fitness than both the DE and the UC, and repeatedly finds a solution that beats the baselines after 30 generations. The mean of the best winner so far for this IE starts above 2.0 in the first generation, which is considered to be a full solution to this environment. These results show that the IE and UC have the ability to integrate information from their own respective training procedures to bootstrap their evolutionary procedures.

Figure 3 highlights some key information about the controller's weight space. The illustrated training data, which represents the weights of solutions to the source domains, is shown as lying in regular parabolic shapes for each particular engine power. The grey dotted line, which represents the enumeration over the latent space of the IE, shows how the decoder of the VAE learnt to map values from its latent space to the weight space such that it could reconstruct

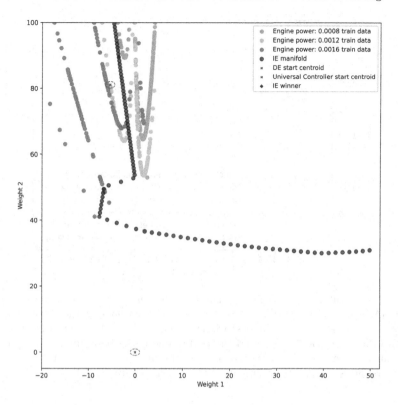

Fig. 3. The weight space of the neural network controller for the CMC domain. The thistle, gold and dark orange points represent the training data used to train the generative models. The grey dotted line labelled 'IE manifold' represents an enumeration over the one dimensional latent space of the decoder derived from the VAE in Fig. 2 mapped into the two dimensional weight space. The enumeration is over the range $[-3, 3]$ with increments of 0.05. The blue diamond at $(-3.68, 89.58)$ represents the best winner found by the decoder. The red and green crosses represent the initial centroids of search for the DE and the UC, respectively, with the dotted circles representing the initial sigma of the search distributions. (Color figure online)

the training data as accurately as possible using a single dimension. It intersects the weight space near the center of the source domain solutions, which evidently happens to coincide with high fitness solutions of the target domains. It is interesting to see how this enumerated weight manifold is more sparse in areas of least interest and more dense in areas with a higher likelihood of finding a good solution.

Figure 3 also shows the starting positions of evolutionary search for the DE and the UC as red and green crosses, respectively. Without previous knowledge of the search space, there is no other information to suggest that the best place to start search is anything other than the origin; for this reason we start evolutionary search of the DE at $[0, 0]$. However, starting search at the origin results

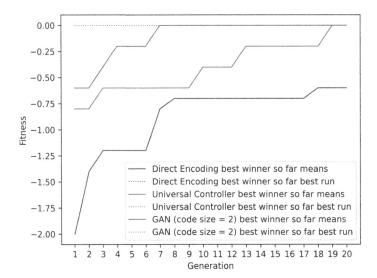

Fig. 4. Ten evolutionary runs for the DE and 5 evolutionary runs for both the UC and a GAN with code size = 2 on Frozen Lake with an goal position of (1, 3). The fitnesses plotted are those of the best winner so far in the evolutionary run. The solid lines are the means of the runs and the dotted line is the best run according to the final generation fitness. All three dotted lines are on top of each other, however this renders as showing only the green dotted line. (Color figure online)

in a significant amount of time and compute expended as the search distribution maneuvers into the area of good solutions for every single new domain instance. Alternatively, the UC has been trained to maximise the average fitness over the three source domains. Even though the UC does not achieve perfect fitness on any of the source domains individually *or* the newly tested 0.0014 target domain, it allows search to start in a much more informed position, which leads to much faster convergence to a solution on the new target domain.

5.2   Frozen Lake

Frozen Lake (FL) is a simple, text-based, maze-like environment with a discrete state and action space. An agent aims to move from a start location to a goal location without falling through a hole into the lake within a designated time limit. By default, FL is a stochastic environment, however we modify it to be deterministic for these experiments. The action space consists of four actions: north, east, south, west. The state space consists of one integer representing the current tile that the agent is located at.

A neural network controller with 1 input, 4 hidden units with a ReLU activation function and 4 output units (one for each of the 4 discrete actions) with a sigmoid activation was used. In order to select the action of the agent, the argument of the largest value output was used. Not all positions in the maze

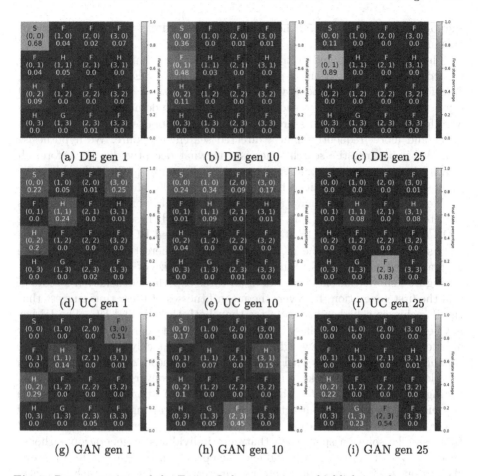

Fig. 5. Representations of the Frozen Lake environment highlighting the percentage of the population in a single generation that ends the episode in a particular tile. Each tile in the 4×4 FL environment is labeled by one of the following types: S, the starting location; F, frozen tile (traversable); H, hole; and G, the goal location. The target domain with goal position $(1, 3)$ is shown. The coordinates of the tile and the aforementioned percentage are also shown. Each subfigure highlights the state of the environment at different generations for the DE, UC and GAN plotted in Fig. 4.

can be located by a controller with zero hidden layers, it was for this reason that a hidden layer was included in the controller. FL is therefore an appropriate domain to demonstrate the ability of our learnt IEs to find neural network controllers with a hidden layer and with a larger number of weights than in the CMC experiments.

For FL, we use goal position as the modifiable domain parameter. The source domains consists of those with goal positions in the set: $\{(1,2), (3,2), (3,3)\}$. The reward given is inversely proportional to the manhattan distance between the end location of the agent and the goal, with a reward of -10 given if the agent falls in a hole.

The UCs were prepared by training solutions to maximise the average of the reward over the three training goal positions using a DE. Then, the training data for the generative models was collected in the same way as in CMC: 999 solutions were found, 333 for each of the three source domains. These solutions were found using a random search. In practice, random search found solutions faster and more frequently than a directed search procedure. We hypothesise that this is because the search space is somewhat deceptive [6]. As before, all weights in these experiments were bound in the range $[-100, 100]$. All three generative models were trained five times on these 999 solutions.

We then ran 5 evolutionary runs using the three IEs on the target domains with test goal positions: $\{(1,3), (3,0)\}$. We compared these with 10 runs using a DE and 5 runs using a DE starting at a pre-trained UC. Figure 4 shows the results of these runs on the $(1, 3)$ goal position where the IE plotted is the generator derived from a GAN trained with a code size of 2.

Figure 4 shows that the best run of all three experiments achieved a score of 0 in the first generation, however, the mean fitnesses of these runs suggest that the IE achieves a higher fitness faster than both the DE and UC. The IE has not been trained on controllers that locate this target goal but it is successfully able to interpolate over the weight space in order to quickly generate controllers capable of finding this new goal location. The DE encodes no previous domain knowledge and therefore takes a much longer time to converge to a good solution.

Figure 5 illustrates that without any information about the solution space the DE chooses from the four actions equally in the starting tile resulting in half of the controllers in generation 1 not moving at all. Alternatively, the trained IE biases the solution space such that *zero* individuals in generation 1 choose either the action north or west and end the episode in the start state. Figure 5c shows that as the number of generations increases the DE gets stuck in a local minima at the deceptive $(0, 1)$ location. In contrast, 23% of the controllers that the GAN derived IE produces find the goal (Fig. 5i).

5.3 Bipedal Walker

Bipedal Walker (BPW) is a simulation that requires the design of a controller that allows a two legged robot to walk as far as possible in a fixed time without falling over. It is more complicated than the previously tested domains due to the fact that there are 24 state inputs and 4 action outputs. In these experiments we use a controller network with no hidden layers resulting in 100 tunable weights. This demonstrates the capability of our technique to scale to a one hundred dimensional problem.

For this domain, we use the knee speed as the modifiable domain parameter. We collected training data for the IEs for the knee speeds: $\{2, 4, 6\}$. The default reward function for BPW is used. Target domains with knee speeds of $\{3, 5\}$ were used to evaluate performance.

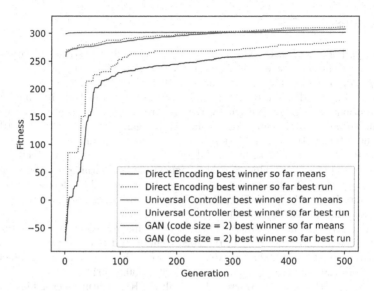

Fig. 6. Ten evolutionary runs for the DE and 5 evolutionary runs for both the UC and a GAN with code size = 2 on Bipedal Walker with aknee speed of 5. The fitnesses plotted are those of the best winner so far. The solid lines are the means of the runs and the dotted line is the best run according to the final generation fitness.

Figure 6 shows that the best winners for both the IE and UC start the evolutionary run with a much higher fitness than the DE. Due to the difficulty of this domain, it takes a significant number of generations for the DE to generate a solution with fitness greater than 250, however, the IE and UC already achieve this in the first generation. The GAN finds solutions with a much higher fitness faster than the UC, however, it begins to plateau after a short amount of time, and is eventually overtaken by the UC.

6   Conclusion and Future Work

In summary, we have demonstrated the ability of three generative models, namely autoencoders, VAEs and GANs, to produce indirect encodings that successfully evolve neural controller solutions for unseen target domains in transfer learning RL tasks. We have also highlighted certain settings in which these IEs significantly outperform two baseline techniques with respect to speed of search.

In future work, we wish to scale these techniques to domains that require controllers with a larger number of parameters than those used in this work. We would also like to apply the techniques to more significant cross-domain transfer examples.

References

1. Bentley, P.J., Lim, S.L., Gaier, A., Tran, L.: Coil: Constrained optimization in learned latent space - learning representations for valid solutions. CoRR **abs/2202.02163** (2022). https://doi.org/10.48550/arXiv.2202.02163
2. Chang, O., Kwiatkowski, R., Chen, S., Lipson, H.: Agent embeddings: a latent representation for pole-balancing networks. In: Proceedings of the 18th International Conference on Autonomous Agents and MultiAgent Systems, AAMAS 2019, pp. 656–664. International Foundation for Autonomous Agents and Multiagent Systems, Richland, SC (2019). https://dl.acm.org/doi/10.5555/3306127.3331753
3. Finn, C., Abbeel, P., Levine, S.: Model-agnostic meta-learning for fast adaptation of deep networks. In: Proceedings of the 34th International Conference on Machine Learning, ICML 2017, vol. 70, pp. 1126–1135. JMLR.org (2017). https://dl.acm.org/doi/10.5555/3305381.3305498
4. Gaier, A., Asteroth, A., Mouret, J.B.: Discovering representations for black-box optimization. In: Proceedings of the 2020 Genetic and Evolutionary Computation Conference, GECCO 2020, pp. 103–111. Association for Computing Machinery, New York (2020). https://doi.org/10.1145/3377930.3390221
5. Girshick, R., Donahue, J., Darrell, T., Malik, J.: Rich feature hierarchies for accurate object detection and semantic segmentation. In: 2014 IEEE Conference on Computer Vision and Pattern Recognition, pp. 580–587 (2014). https://doi.org/10.1109/CVPR.2014.81
6. Goldberg, D.E.: Simple genetic algorithms and the minimal, deceptive problem. In: Davis, L. (ed.) Genetic Algorithms and Simulated Annealing, pp. 74–88. Research Notes in Artificial Intelligence, Pitman, London (1987)
7. Goodfellow, I., Bengio, Y., Courville, A.: Deep Learning. MIT Press (2016). http://www.deeplearningbook.org
8. Hansen, N., Ostermeier, A.: Completely derandomized self-adaptation in evolution strategies. Evol. Comput. **9**, 159–195 (2001). https://doi.org/10.1162/106365601750190398
9. Jegorova, M., Doncieux, S., Hospedales, T.: Behavioral repertoire via generative adversarial policy networks. IEEE Trans. Cogn. Dev. Syst., 1 (2020). https://doi.org/10.1109/TCDS.2020.3008574
10. Kalehbasti, P.R., Lepech, M.D., Pandher, S.S.: Augmenting high-dimensional nonlinear optimization with conditional gans. In: Proceedings of the Genetic and Evolutionary Computation Conference Companion, GECCO 2021, pp. 1879–1880. Association for Computing Machinery, New York (2021). https://doi.org/10.1145/3449726.3463675
11. Kouvaris, K., Clune, J., Kounios, L., Brede, M., Watson, R.A.: How evolution learns to generalise: using the principles of learning theory to understand the evolution of developmental organisation. PLoS Comput. Biol. **13**(4), 1–20 (2017). https://doi.org/10.1371/journal.pcbi.1005358
12. Moreno, M.A., Banzhaf, W., Ofria, C.: Learning an evolvable genotype-phenotype mapping. In: Proceedings of the Genetic and Evolutionary Computation Conference, GECCO 2018, pp. 983–990. Association for Computing Machinery, New York (2018). https://doi.org/10.1145/3205455.3205597
13. Rakicevic, N., Cully, A., Kormushev, P.: Policy manifold search: Exploring the manifold hypothesis for diversity-based neuroevolution. In: Proceedings of the Genetic and Evolutionary Computation Conference, pp. 901–909 (2021). https://doi.org/10.1145/3449639.3459320

14. Watson, R.A., Szathmáry, E.: How can evolution learn? Trends Ecol. Evol. **31**, 147–157 (2016). https://doi.org/10.1016/j.tree.2015.11.009
15. Zhu, Z., Lin, K., Zhou, J.: Transfer learning in deep reinforcement learning: a survey. CoRR abs/2009.07888 (2020). https://arxiv.org/abs/2009.07888

HVC-Net: Deep Learning Based Hypervolume Contribution Approximation

Ke Shang, Weiduo Liao, and Hisao Ishibuchi[✉]

Guangdong Provincial Key Laboratory of Brain-inspired Intelligent Computation,
Department of Computer Science and Engineering, Southern University of Science
and Technology, Shenzhen, China
{shangk,hisao}@sustech.edu.cn, liaowd@mail.sustech.edu.cn

Abstract. In this paper, we propose HVC-Net, a deep learning based
hypervolume contribution approximation method for evolutionary multi-
objective optimization. The basic idea of HVC-Net is to use a deep neural
network to approximate the hypervolume contribution of each solution
in a non-dominated solution set. HVC-Net has two characteristics: (1) It
is permutation equivalent to the order of solutions in the input solution
set, and (2) a single HVC-Net can handle solution sets of various size
(e.g., solution sets with 20, 50 and 100 solutions). The performance of
HVC-Net is evaluated through computational experiments by comparing it with two commonly-used hypervolume contribution approximation
methods (i.e., point-based method and line-based method). Our experimental results show that HVC-Net outperforms the other two methods in
terms of both the runtime and the ability to identify the smallest (largest)
hypervolume contributor in a solution set, which shows the superiority
of HVC-Net for hypervolume contribution approximation.

Keywords: Hypervolume contribution · Approximation ·
Evolutionary multi-objective optimization · Deep learning

1 Introduction

Hypervolume [11,15] is a popular performance indicator in the field of evolutionary multi-objective optimization (EMO). It possesses rich theoretical properties
(e.g., Pareto compliance [16], submodularity [13]), which make it attractive to
use in practice. For example, it has been used to design EMO algorithms (e.g.,
SMS-EMOA [2,6]) and subset selection algorithms (e.g., greedy hypervolume
subset selection [4,7]).

In SMS-EMOA, the population evolves in a steady-state manner. In each
generation, one offspring is generated and added to the population, then one
solution is removed from the population so that the hypervolume of the remaining population is maximized. In greedy hypervolume subset selection, in each
step, one solution is selected from a candidate set and added to the subset so
that the hypervolume of the subset is maximized. In these two cases, in order

© The Author(s), under exclusive license to Springer Nature Switzerland AG 2022
G. Rudolph et al. (Eds.): PPSN 2022, LNCS 13398, pp. 414–426, 2022.
https://doi.org/10.1007/978-3-031-14714-2_29

to remove (select) the correct solution, we need to calculate the hypervolume contribution of each solution. Hypervolume contribution is an important concept which describes the amount of hypervolume contributed by one solution to a solution set. In SMS-EMOA, we need to identify the solution with the smallest hypervolume contribution to the population. In greedy hypervolume subset selection, we need to identify the solution with the largest hypervolume contribution to the subset.

However, the calculation of the hypervolume contribution is #P-hard [3], which limits its applicability in many-objective optimization. In order to overcome this drawback, some hypervolume contribution approximation methods have been proposed [1,5,12]. Two representative methods are the point-based method [1] and the line-based method [12]. The point-based method is also known as the Monte Carlo sampling method. In this method, in order to approximate the hypervolume contribution of a solution, a sampling space is firstly determined. After that, a large number of samples are uniformly drawn in the sampling space to do the approximation. The line-based method is also known as the R2 indicator variant. In this method, a large number of line segments are uniformly drawn in the hypervolume contribution region of a solution to do the approximation.

In this paper, we propose a new hypervolume contribution approximation method. The proposed method, named HVC-Net, uses a deep neural network to do the approximation. The input of HVC-Net is a non-dominated solution set, and the output of HVC-Net is the approximated hypervolume contribution of each solution in the input solution set. HVC-Net has two characteristics. One is that it is permutation equivalent [14]. That is, a change of the order of solutions in the input solution set will cause the same change of the order of the outputs (i.e., the same results are obtained for any permutation of solutions in the input solution set). The other is that it can handle solution sets of different size (e.g., 20, 50, 100 solutions). That is, a single HVC-Net is trained using solution sets of various size. These two characteristics guarantee high applicability of HVC-Net for hypervolume contribution approximation.

The rest of this paper is organized as follows. Section 2 presents the preliminaries of the study. Section 3 introduces a new hypervolume contribution approximation method, HVC-Net. Section 4 conducts experimental studies. Section 5 concludes the paper.

2   Preliminaries

In this section, we present the preliminaries of this paper, including the definitions of hypervolume and hypervolume contribution, and two representative hypervolume contribution approximation methods.

2.1   Hypervolume and Hypervolume Contribution

Hypervolume. The hypervolume is a widely used performance indicator in the field of evolutionary multi-objective optimization. Formally, for a solution set S in the objective space, the hypervolume of S is defined as

$$HV(S, \mathbf{r}) = \mathcal{L}\left(\bigcup_{\mathbf{s} \in S} \{\mathbf{s}' | \mathbf{s} \prec \mathbf{s}' \prec \mathbf{r}\}\right), \tag{1}$$

where $\mathcal{L}(\cdot)$ is the Lebesgue measure of a set, $\mathbf{r}$ is a reference point which is dominated by all solutions in S, and $\mathbf{s} \prec \mathbf{s}'$ denotes that $\mathbf{s}$ dominates $\mathbf{s}'$ (i.e., $s_i \leq s_i'$ for all $i = 1, ..., m$ and $s_j < s_j'$ for at least one $j = 1, ..., m$ in the minimization case, where m is the number of objectives).

Figure 1(a) gives an illustration of the hypervolume of a solution set $\{\mathbf{a}^1, \mathbf{a}^2, \mathbf{a}^3\}$ in a two-dimensional objective space, where each objective is to be minimized.

Hypervolume Contribution. The hypervolume contribution is an important concept based on the hypervolume indicator. It describes the amount of the hypervolume value contributed by a solution to the solution set. Formally, for a solution $\mathbf{s} \in S$, the hypervolume contribution of $\mathbf{s}$ to S is defined as

$$HVC(\mathbf{s}, S, \mathbf{r}) = HV(S, \mathbf{r}) - HV(S \setminus \{\mathbf{s}\}, \mathbf{r}). \tag{2}$$

Figure 1(b) gives an illustration of the hypervolume contribution of each solution to the solution set $\{\mathbf{a}^1, \mathbf{a}^2, \mathbf{a}^3\}$.

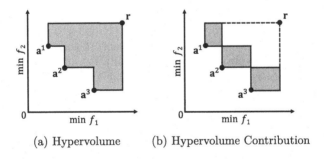

(a) Hypervolume   (b) Hypervolume Contribution

Fig. 1. Illustrations of the hypervolume and the hypervolume contribution. The shaded area in (a) is the hypervolume of the solution set $\{\mathbf{a}^1, \mathbf{a}^2, \mathbf{a}^3\}$, and each shaded area in (b) is the hypervolume contribution of the corresponding solution to the solution set $\{\mathbf{a}^1, \mathbf{a}^2, \mathbf{a}^3\}$.

2.2   Hypervolume Contribution Approximation

Two representative hypervolume contribution approximation methods are the point-based method and the line-based method. These two methods are briefly explained as follows.

Point-Based Method. The point-based method is also known as the Monte Carlo sampling method [1]. Figure 2(a) illustrates this method. The basic idea is as follows. To approximate the hypervolume contribution of a solution $\mathbf{s} \in S$, a sampling space (i.e., a hyperrectangle) which contains $\mathbf{s}$'s hypervolume contribution region is firstly determined (e.g., the rectangle bounded by $\mathbf{a}^2$ and $\mathbf{p}$ in Fig. 2(a)). Then a large number of samples are uniformly drawn in the sampling space (e.g., k samples). Suppose there are k' samples uniquely dominated by $\mathbf{s}$ (e.g., the red samples in Fig. 2(a)), then the hypervolume contribution of $\mathbf{s}$ is approximated as

$$HVC(\mathbf{s}, S, \mathbf{r}) \approx \frac{k'}{k}V, \tag{3}$$

where V is the volume of the sampling space (i.e., the hyperrectangle).

In practice, the sampling space should be as tight as possible. In [1], the tightest sampling space is theoretically derived. The lower bound of the sampling space is the solution itself (e.g., $\mathbf{a}^2$ in Fig. 2(a)). The upper bound (e.g., $\mathbf{p}$ in Fig. 2(a)) is determined as follows:

$$p_i = \min\left\{s_i'|\mathbf{s}' \in S \setminus \{\mathbf{s}\} \text{ and } \mathbf{s}' \prec_i \mathbf{s}\right\}, i = 1, ..., m, \tag{4}$$

where $\mathbf{s}' \prec_i \mathbf{s}$ denotes that $\mathbf{s}'$ dominates $\mathbf{s}$ in all but the ith objective.

Therefore, the tightest sampling space in Fig. 2(a) is exactly the hypervolume contribution region of $\mathbf{a}^2$ (i.e., $\mathbf{p} = (a_1^3, a_2^1)$). We did not put $\mathbf{p}$ exactly at position (a_1^3, a_2^1) (i.e., the red point) in Fig. 2(a) for easy illustration.

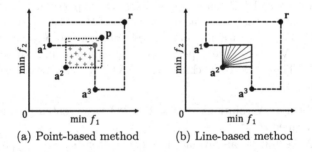

(a) Point-based method   (b) Line-based method

Fig. 2. Illustrations of the point-based and line-based methods for hypervolume contribution approximation.

Line-Based Method. The line-based method is also known as the R2 indicator variant [12]. Figure 2(b) illustrates this method. The basic idea is as follows. To approximate the hypervolume contribution of a solution $\mathbf{s} \in S$, a set of line segments starting from $\mathbf{s}$ and with different directions are drawn in its hypervolume contribution region. Suppose there are n line segments and the length of each line segment is $l_i, i = 1, ..., n$, then the hypervolume contribution of $\mathbf{s}$ is approximated as

$$HVC(\mathbf{s}, S, \mathbf{r}) \approx \frac{1}{n} \sum_{i=1}^{n} (l_i)^m, \qquad (5)$$

where m is the number of objectives.

The directions of the line segments can be defined using a direction vector set $\Lambda = \{\boldsymbol{\lambda}^1, ..., \boldsymbol{\lambda}^n\}$ where each direction vector satisfies $\|\boldsymbol{\lambda}^i\|_2 = 1$, $\lambda_j^i \geq 0$, $i = 1, ..., n$, $j = 1, ..., m$. The length of each line segment can be calculated as

$$l_i = \min \left\{ \min_{\mathbf{s}' \in S \setminus \{\mathbf{s}\}} \left\{ g^{*2\text{tch}}(\mathbf{s}'|\boldsymbol{\lambda}^i, \mathbf{s}) \right\}, g^{\text{mtch}}(\mathbf{r}|\boldsymbol{\lambda}^i, \mathbf{s}) \right\}, i = 1, ..., n, \qquad (6)$$

where $g^{*2\text{tch}}(\cdot)^1$ and $g^{\text{mtch}}(\cdot)$ are defined as follows:

$$g^{*2\text{tch}}(\mathbf{s}'|\boldsymbol{\lambda}^i, \mathbf{s}) = \max_{j \in \{1, ..., m\}} \left\{ \frac{s_j' - s_j}{\lambda_j^i} \right\}, \qquad (7)$$

$$g^{\text{mtch}}(\mathbf{r}|\boldsymbol{\lambda}^i, \mathbf{s}) = \min_{j \in \{1, ..., m\}} \left\{ \frac{|s_j - r_j|}{\lambda_j^i} \right\}. \qquad (8)$$

3   HVC-Net

Motivated by DeepSets [14], which is a deep neural network for dealing with a set as its input, we design HVC-Net to approximate the hypervolume contribution of each solution in a solution set. The architecture of HVC-Net is shown in Fig. 3. The input of HVC-Net is a non-dominated solution set $S = \{\mathbf{s}^1, \mathbf{s}^2, ..., \mathbf{s}^N\}$. The output of HVC-Net is the hypervolume contribution approximation of each solution in S. The working mechanism of HVC-Net can be described in the following three steps.

- **Step 1**: Each of N solutions $\mathbf{s}^i$ ($i = 1, ..., N$) is presented to the network ϕ and transformed to $\mathbf{a}^i = \phi(\mathbf{s}^i)$.
- **Step 2**: N vectors $\mathbf{a}^i$ are averaged as one vector $\mathbf{b} = \frac{1}{N} \sum_{i=1}^{N} \mathbf{a}^i$. Vector $\mathbf{b}$ is further presented to network η and transformed to $\mathbf{c} = \eta(\mathbf{b})$. Vector $\mathbf{c}$ is added to each of N vectors $\mathbf{a}^i$ and N vectors $\mathbf{d}^i = \mathbf{c} + \mathbf{a}^i$ are obtained.
- **Step 3**: Each of N vectors $\mathbf{d}^i$ is presented to network ρ and N outputs $\widehat{HVC}_\theta(\mathbf{s}^i, S, \mathbf{r}) = \rho(\mathbf{d}^i)$ are obtained, where $\boldsymbol{\theta}$ is the parameter vector of HVC-Net.

It should be noted that **Step 2** can be stacked multiple times (i.e., K) as shown in Fig. 3. **Step 2** is used to learn the relation between a solution and the whole solution set. The two main characteristics of HVC-Net are as follows:

---

[1] The $g^{*2\text{tch}}(\cdot)$ function defined in Eq. (7) is used in minimization case. For maximization case, s_j' and s_j should be swapped in Eq. (7). Please refer to [12] for more detailed explanations.

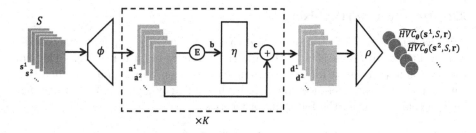

Fig. 3. The architecture of HVC-Net.

1. It is permutation equivalent. That is, for any permutation π of the input solutions (i.e., $S = \{\mathbf{s}^{\pi(1)}, ..., \mathbf{s}^{\pi(N)}\}$), the outputs of HVC-Net are $\widetilde{HVC}_\theta(\mathbf{s}^{\pi(1)}, S, \mathbf{r}), ..., \widetilde{HVC}_\theta(\mathbf{s}^{\pi(N)}, S, \mathbf{r})$. This means that the approximated hypervolume contribution value for each solution is not affected by the order of the input solutions. This is because the change of the order of the input solutions will lead to the change of the order of the approximated values consistently. This characteristic guarantees the robustness of HVC-Net.
2. A single HVC-Net can handle solution sets of various size. For example, we can use a trained HVC-Net to handle a solution set with 10 solutions. We can also use the same HVC-Net to handle a solution set with 100 solutions. This characteristic guarantees high applicability of HVC-Net.

3.1   How to Train HVC-Net

In HVC-Net, we implicitly assume the minimization case where the reference point for hypervolume contribution calculation is set to $\mathbf{r} = (1, ..., 1)$ and all solutions in S are located in $[0, 1]^m$. For the training of HVC-Net, we prepare the training data as follows. First, we prepare L non-dominated solution sets $\{S_1, S_2, ..., S_L\}$ where each solution set is located in $[0, 1]^m$. Then, we calculate the hypervolume contribution of each solution in each solution set based on the reference point $\mathbf{r} = (1, ..., 1)$. That is, we obtain the target output $HVC(\mathbf{s}^i, S_j, \mathbf{r})$ for each solution $\mathbf{s}^i \in S_j$ $(i = 1, ..., |S_j|, j = 1, ..., L)$.

Based on the training data (i.e., the pairs of the solution sets and the corresponding hypervolume contributions), we define the loss function of HVC-Net as follows:

$$\mathcal{L}(\boldsymbol{\theta}) = \frac{1}{L} \sum_{j=1}^{L} \frac{1}{|S_j|} \sum_{i=1}^{|S_j|} \left(\log \widetilde{HVC}_\theta(\mathbf{s}^i, S_j, \mathbf{r}) - \log HVC(\mathbf{s}^i, S_j, \mathbf{r}) \right)^2. \quad (9)$$

The loss function defined in Eq. (9) is similar to the mean squared error (MSE) loss function. The only difference is that we add log function to the hypervolume contribution (approximation) values. This is because the hypervolume contribution values are usually very small (e.g., in the magnitude of 10^{-4}). Using log values can make the training easier and better. More details about the training of HVC-Net are described in Sect. 4.1.

3.2   How to Use HVC-Net

After we train a HVC-Net, we can use it to approximate the hypervolume contribution if the solution set is in $[0,1]^m$ and the reference point is $\mathbf{r} = (1,...,1)$. The question is how to use it for hypervolume contribution approximation when the solution set and the reference point are both arbitrarily given. Before answering this question, we need the following properties.

Property 1. For any positive vector $\boldsymbol{\alpha} \in \mathbb{R}_{>0}^m$, $HVC(\mathbf{s},S,\mathbf{r}) = \frac{1}{\prod_{i=1}^m \alpha_i} HVC(\boldsymbol{\alpha} \odot \mathbf{s}, \boldsymbol{\alpha} \odot S, \boldsymbol{\alpha} \odot \mathbf{r})$, where $\odot$ denotes the element-wise multiplication[2].

Property 2. For any real vector $\boldsymbol{\beta} \in \mathbb{R}^m$, $HVC(\mathbf{s},S,\mathbf{r}) = HVC(\mathbf{s}+\boldsymbol{\beta}, S+\boldsymbol{\beta}, \mathbf{r}+\boldsymbol{\beta})$.

Property 3. $HVC(\mathbf{s},S,\mathbf{r}) = HVC(-\mathbf{s},-S,-\mathbf{r})$ where $HVC(-\mathbf{s},-S,-\mathbf{r})$ is calculated for maximization problems whereas $HVC(\mathbf{s},S,\mathbf{r})$ is calculated for minimization problems.

The above three properties can be easily obtained from the properties of the hypervolume indicator obtained in [10]. Based on these properties, we can first transform the solution set and the reference point so that the reference point is $\mathbf{r} = (1,...,1)$ and the solution set is located in $[0,1]^m$. Then we use HVC-Net to approximate the hypervolume contribution for the transformed solution set. Lastly, we calculate the hypervolume contribution approximation for the original solution set based on the output of HVC-Net. The last step is not needed when our task is to find the solution with the smallest or largest hypervolume contribution in a solution set.

Thus, although HVC-Net is trained based on solution sets in $[0,1]^m$ and reference point $\mathbf{r} = (1,...,1)$, it can be used for any solution set with any reference point.

4   Experiments

In this section, we conduct computational experiments to examine the performance of HVC-Net by comparing it with the point-based and line-based methods for hypervolume contribution approximation.

4.1   Experimental Settings

HVC-Net Specifications. In HVC-Net in Fig. 3, three networks ϕ, η, and ρ need to be specified. In our experiments, all three networks are specified as feedforward neural networks. Figure 4 shows the structures of ϕ, η, and ρ used in our experiments. All the networks have three hidden layers. In the hidden layers of the three networks, we use ReLU activation function for efficient training.

---

[2] For two vectors $\mathbf{a} = (a_1,...,a_m)$ and $\mathbf{b} = (b_1,...,b_m)$, $\mathbf{a} \odot \mathbf{b} = (a_1 b_1,...,a_m b_m)$. For a set B, $\mathbf{a} \odot B$ means $\mathbf{a} \odot \mathbf{b}$ for all $\mathbf{b} \in B$.

In the output layer of network ρ, we use Sigmoid activation function since the hypervolume contribution values are in $[0, 1]$ for the training solution sets. In HVC-Net, **Step 2** can be stacked K times as shown in Fig. 3. In our experiments, we set $K = 10$. That is, we have 10 different η networks in HVC-Net.

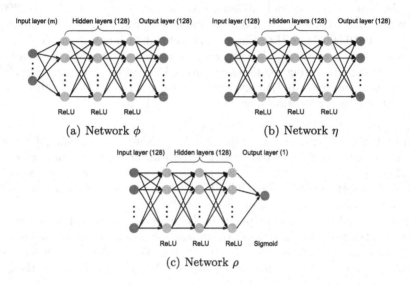

(a) Network ϕ (b) Network η

(c) Network ρ

Fig. 4. Networks ϕ, η, and ρ in HVC-Net. The number in the parentheses indicates the number of neurons in each layer. The activation function used in each layer is shown under each layer.

Training and Testing Data Generation. We examine 5, 8, and 10-objective cases (i.e., $m = 5, 8, 10$). We generate training data with L solution sets $\{S_1, ..., S_L\}$ for each case where $L = 1,000,000$. Each solution set is generated using the following procedure:

- Step 1: Randomly sample an integer $num \in [1, 100]$ where num denotes the number of solutions in the solution set.
- Step 2: Randomly sample 1000 solutions in $[0, 1]^m$ as candidate solutions.
- Step 3: Apply non-dominated sorting to these 1000 solutions and obtain different fronts $\{F_1, ..., F_l\}$ where F_1 is the first front (i.e., the set of non-dominated solutions in the 1000 solutions) and F_l is the last front.
- Step 4: Identify all the fronts F_i with $|F_i| \geq num$. If no front satisfies this condition, go back to Step 2.
- Step 5: Randomly pick one front F_i with $|F_i| \geq num$ and randomly select num solutions from this front to construct one solution set.

This procedure is used in order to select a wide variety of non-dominated solution sets for training. On average, about 10,000 solution sets with the same size are generated (1,000,000 solution sets in total for 100 different sizes).

We generate two types of testing solution sets. Type-I testing solution sets are generated using exactly the same procedure as described above. We generate 10,000 Type-I testing solution sets for each case (i.e., $m = 5, 8, 10$). These 10,000 solution sets form one group. We generate 10 different groups of Type-I testing solution sets for each case of m. Type-II testing solution sets are generated using a similar procedure. The only difference is that an integer $num \in [101, 200]$ is randomly sampled in Step 1 and 10,000 solutions are randomly sampled in Step 2. We generate 10,000 Type-II testing solution sets for each case of m. These 10,000 solution sets form one group. We generate 10 different groups of Type-II for each case of m. Type-II testing solution sets are used to test the generalization ability of HVC-Net since we train HVC-Net using solution sets with 1–100 solutions and test HVC-Net using solution sets with 101–200 solutions.

Parameter Settings. For the training of HVC-Net, we use Adam [8], an effective gradient-based optimization method with an adaptive learning rate. The initial learning rate is set to 10^{-4}. For all the other parameters in Adam, we use their default settings in PyTorch [9]. The batch size during training is set to 100. The number of epochs for training is set to 100.

For the number of sampling points in the point-based method, we examine 20 different settings: 5, 10, ..., 100. For the number of lines in the line-based method, we examine 20 different settings: 5, 10, ..., 100. We use the unit normal vector method [5] to generate the direction vector set Λ in the line-based method.

Performance Metrics. To compare the performance of different hypervolume contribution approximation methods, we use the correct identification rate (CIR). CIR is a metric which can evaluate the ability of a method to identify the smallest (largest) hypervolume contributor in a solution set. For example, suppose we have P solution sets. If a method can correctly identify the smallest (largest) hypervolume contributor for Q out of P solution sets, then CIR is calculated as Q/P. In our experiments, we use $\text{CIR}_{\min}$ (i.e., CIR for identifying the smallest hypervolume contributor) and $\text{CIR}_{\max}$ (i.e., CIR for identifying the largest hypervolume contributor). A larger CIR value means better approximation quality of a method.

We also record the runtime of the three methods to compare their efficiency. Here the runtime of HVC-Net means its evaluation time on the testing solution sets, not the training time.

Platforms. All the methods are coded in Python and tested on a server with Intel(R) Xeon(R) Gold 6130 CPU @ 2.10 GHz, GeForce RTX 2080 GPU, and Ubuntu 18.04.6 LTS. HVC-Net is implemented based on PyTorch version 1.9.0.

4.2 The Training of HVC-Net

Figure 5 shows the training curve of HVC-Net in each case of m. We can see that the loss decreases sharply in the first 20 epochs. This is mainly because we use a batch size of 100 for training 1M solutions, which means that the parameters of HVC-Net can be updated 10K times in each epoch. We can also observe that the loss becomes very small at the end of the training process in each figure, which shows the success of the training of HVC-Net.

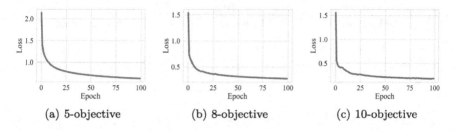

(a) 5-objective   (b) 8-objective   (c) 10-objective

Fig. 5. The training curve of HVC-Net in each case.

Table 1 shows the time used for training HVC-Net. Although the training of HVC-Net needs quite a substantial time, the trained HVC-Net models can be saved and are ready to use at any time. That is, once we obtain a well trained HVC-Net model, we can save it and use it directly in the future without spending a lot of time to retrain it.

Table 1. The time (GPU hours) used for training HVC-Net in each case.

5-objective	8-objective	10-objective
369.13	370.47	383.79

Next, we will examine the performance of our trained HVC-Net models on the testing solution sets.

4.3 Testing on Type-I Solution Sets

We apply the three hypervolume contribution approximation methods to Type-I testing solution sets, i.e., solution sets with 1–100 solutions. For fair comparison, all the methods are tested on CPU. That is, we disable GPU when using HVC-Net for evaluation.

Figure 6 shows the results of CIR_{min} and CIR_{max} for each case. We can see that HVC-Net clearly dominates the other two methods in most cases in terms of both the correct identification rate and the runtime, which shows the advantage of using HVC-Net for identifying the smallest (largest) hypervolume contributor in a solution set. It is worth noting that the point-based and line-based methods are very time-consuming compared with HVC-Net. HVC-Net is able to process the testing solution sets in less than 10 s and achieves a good CIR value.

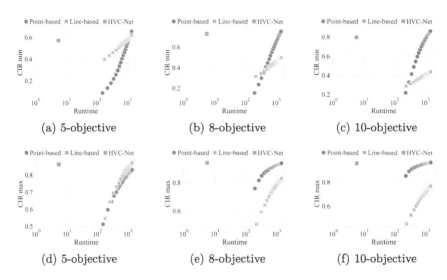

Fig. 6. Experimental results of the three hypervolume contribution approximation methods on Type-I testing solution sets in each case. The runtime (in seconds) means the total time for evaluating 10,000 testing solution sets. The CIR (CIR_{min} in (a)–(c) and CIR_{max} in (d)–(f)) means the correct identification rate over 10,000 testing solution sets. All the results are the average over 10 groups of Type-I testing solution sets.

The other two methods even perform worse (i.e., a smaller CIR value) than HVC-Net by consuming 1000 s. Of course, the other two methods can achieve better CIR values than HVC-Net using more points or lines in some cases (e.g., in Fig. 6(a) the other two methods achieve better CIR values than HVC-Net by consuming more than 1000 s). However, the runtime cost is too high to realize this goal for the point-based and line-based methods.

4.4 Testing on Type-II Solution Sets

We also apply the three hypervolume contribution approximation methods to Type-II testing solution sets, i.e., solution sets with 101–200 solutions. We use Type-II testing solution sets to examine the generalization ability of HVC-Net.

Figure 7 shows the results of CIR_{min} and CIR_{max} for each case. We can observe that HVC-Net performs well in general. It can still dominate the other two methods in terms of the correct identification rate and the runtime in most cases. There is almost no runtime increase for HVC-Net compared with the results on Type-I solution sets. However, there is a significant runtime increase for the point-based and line-based methods. These results show the strong generalization ability and the efficiency of HVC-Net. Although HVC-Net is trained using solution sets with 1–100 solutions, it is able to handle solution sets with 101–200 solutions effectively and efficiently.

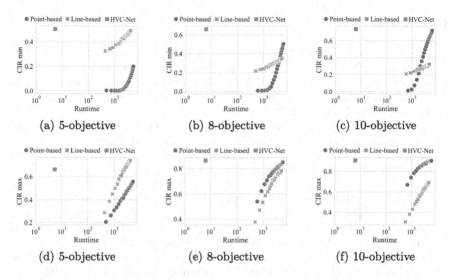

Fig. 7. Experimental results of the three hypervolume contribution approximation methods on Type-II testing solution sets in each case. The runtime (in seconds) means the total time for evaluating 10,000 testing solution sets. The CIR (CIR$_{min}$ in (a)–(c) and CIR$_{max}$ in (d)–(f)) means the correct identification rate over 10,000 testing solution sets. All the results are the average over 10 groups of Type-II testing solution sets.

5   Conclusions

In this paper, we proposed HVC-Net, a deep learning based method for hypervolume contribution approximation. We compared HVC-Net with the point-based method and the line-based method. The experimental results showed that HVC-Net outperformed the other two methods in terms of both the correct identification rate and the runtime, which showed the potential of using deep learning technique for hypervolume contribution approximation.

Our future work is the development of a hypervolume-based EMO algorithm and a hypervolume subset selection algorithm based on HVC-Net for many-objective optimization. Of course, we will also try to further improve the performance of HVC-Net by improving its structure, parameter settings, training method, and so on.

All the source codes and the trained HVC-Net models are available at https://github.com/HisaoLabSUSTC/HVC-Net.

Acknowledgements. This work was supported by National Natural Science Foundation of China (Grant No. 62002152, 61876075), Guangdong Provincial Key Laboratory (Grant No. 2020B121201001), the Program for Guangdong Introducing Innovative and Enterpreneurial Teams (Grant No. 2017ZT07X386), The Stable Support Plan Program of Shenzhen Natural Science Fund (Grant No. 20200925174447003), Shenzhen Science and Technology Program (Grant No. KQTD2016112514355531).

References

1. Bader, J., Deb, K., Zitzler, E.: Faster hypervolume-based search using Monte Carlo sampling. In: Ehrgott, M., Naujoks, B., Stewart, T., Wallenius, J. (eds.) Multiple Criteria Decision Making for Sustainable Energy and Transportation Systems. Lecture Notes in Economics and Mathematical Systems, vol. 634, pp. 313–326. Springer, Heidelberg (2010)
2. Beume, N., Naujoks, B., Emmerich, M.: SMS-EMOA: multiobjective selection based on dominated hypervolume. Eur. J. Oper. Res. **181**(3), 1653–1669 (2007)
3. Bringmann, K., Friedrich, T.: Approximating the least hypervolume contributor: NP-hard in general, but fast in practice. In: Ehrgott, M., Fonseca, C.M., Gandibleux, X., Hao, J.-K., Sevaux, M. (eds.) EMO 2009. LNCS, vol. 5467, pp. 6–20. Springer, Heidelberg (2009). https://doi.org/10.1007/978-3-642-01020-0_6
4. Chen, W., Ishibuchi, H., Shang, K.: Fast greedy subset selection from large candidate solution sets in evolutionary multi-objective optimization. IEEE Trans. Evol. Comput. (2021). https://ieeexplore.ieee.org/document/9509298
5. Deng, J., Zhang, Q.: Approximating hypervolume and hypervolume contributions using polar coordinate. IEEE Trans. Evol. Comput. **23**(5), 913–918 (2019)
6. Emmerich, M., Beume, N., Naujoks, B.: An EMO algorithm using the hypervolume measure as selection criterion. In: International Conference on Evolutionary Multi-Criterion Optimization, pp. 62–76 (2005)
7. Guerreiro, A.P., Fonseca, C.M., Paquete, L.: Greedy hypervolume subset selection in low dimensions. Evol. Comput. **24**(3), 521–544 (2016)
8. Kingma, D.P., Ba, J.: Adam: a method for stochastic optimization. arXiv preprint arXiv:1412.6980 (2014)
9. Paszke, A., et al.: Pytorch: an imperative style, high-performance deep learning library. Adv. Neural. Inf. Process. Syst. **32**, 8026–8037 (2019)
10. Shang, K., Chen, W., Liao, W., Ishibuchi, H.: HV-Net: hypervolume approximation based on deepsets. IEEE Trans. Evol. Comput. (2022). https://ieeexplore.ieee.org/document/9790869
11. Shang, K., Ishibuchi, H., He, L., Pang, L.M.: A survey on the hypervolume indicator in evolutionary multiobjective optimization. IEEE Trans. Evol. Comput. **25**(1), 1–20 (2021)
12. Shang, K., Ishibuchi, H., Ni, X.: R2-based hypervolume contribution approximation. IEEE Trans. Evol. Comput. **24**(1), 185–192 (2020)
13. Ulrich, T., Thiele, L.: Bounding the effectiveness of hypervolume-based (μ + λ)-Archiving Algorithms. In: Hamadi, Y., Schoenauer, M. (eds.) LION 2012. LNCS, pp. 235–249. Springer, Heidelberg (2012). https://doi.org/10.1007/978-3-642-34413-8_17
14. Zaheer, M., Kottur, S., Ravanbhakhsh, S., Póczos, B., Salakhutdinov, R., Smola, A.J.: Deep Sets. Advances in Neural Information Processing Systems, pp. 3394–3404 (2017)
15. Zitzler, E., Thiele, L., Laumanns, M., Fonseca, C., da Fonseca, V.: Performance assessment of multiobjective optimizers: an analysis and review. IEEE Trans. Evol. Comput. **7**(2), 117–132 (2003)
16. Zitzler, E., Brockhoff, D., Thiele, L.: The hypervolume indicator revisited: on the design of pareto-compliant indicators via weighted integration. In: Obayashi, S., Deb, K., Poloni, C., Hiroyasu, T., Murata, T. (eds.) EMO 2007. LNCS, vol. 4403, pp. 862–876. Springer, Heidelberg (2007). https://doi.org/10.1007/978-3-540-70928-2_64

Multi-objective Evolutionary Ensemble Pruning Guided by Margin Distribution

Yu-Chang Wu, Yi-Xiao He, Chao Qian$^{(\boxtimes)}$, and Zhi-Hua Zhou

State Key Laboratory for Novel Software Technology, Nanjing University,
Nanjing 210023, China
{wuyc,heyx,qianc,zhouzh}@lamda.nju.edu.cn

Abstract. Ensemble learning trains and combines multiple base learners for a single learning task, and has been among the state-of-the-art learning techniques. Ensemble pruning tries to select a subset of base learners instead of combining them all, with the aim of achieving a better generalization performance as well as a smaller ensemble size. Previous methods often use the validation error to estimate the generalization performance during optimization, while recent theoretical studies have disclosed that margin distribution is also crucial for better generalization. Inspired by this finding, we propose to formulate ensemble pruning as a three-objective optimization problem that optimizes the validation error, margin distribution, and ensemble size simultaneously, and then employ multi-objective evolutionary algorithms to solve it. Experimental results on 20 binary classification data sets show that our proposed method outperforms the state-of-the-art ensemble pruning methods significantly in both generalization performance and ensemble size.

Keywords: Machine learning · Ensemble pruning · Multi-objective optimization · Margin distribution · Multi-objective evolutionary algorithm

1 Introduction

For one machine learning task, ensemble methods [31] train and combine multiple base learners, which can achieve a better generalization performance than a single base learner, and has been one of the most successful learning algorithms. Based on the way how the base learners are generated, ensemble methods can be generally classified into two categories: sequential methods such as Boosting [26], and parallel methods such as Bagging [4]. After generating a set of trained base learners, ensemble pruning [31] selects and combines a subset of base learners instead of combining them all, which can not only save the storage space and accelerate the prediction speed, but also lead to a better generalization performance than the whole ensemble [7,20,24,34].

This work was supported by the National Science Foundation of China (62022039, 61921006).

In the past twenty-five years, a number of effective ensemble pruning methods have been proposed, which can be roughly classified into two groups, ordering-based pruning and optimization-based pruning. Ordering-based methods are usually based on greedy strategies. Given a set of trained base learners, this kind of method iteratively selects the base learner with the largest marginal gain on some specially designed evaluation criterion. Representative criteria include minimizing the error on the validation set (i.e., validation error) [10,21], maximizing the diversity [2], maximizing the complementarity [20], or combining different evaluation criteria [17]. It has been shown that compared with combining all base learners, ordering-based methods can often achieve a smaller error on the test set (i.e., test error) by selecting only a subset of base learners [20].

Different from ordering-based methods, optimization-based pruning methods formulate ensemble pruning as an optimization problem explicitly, and then apply optimization techniques to search for the optimal subset of base learners that constitutes the final pruned ensemble. As evolutionary algorithms (EAs) [1] inspired by natural evolution are a kind of general-purpose optimization algorithms, they have been naturally used for ensemble pruning. Indeed, the first work which opened the direction of optimization-based pruning [34] used a standard genetic algorithm to select a subset of base learners minimizing the validation error. Compared with the ordering-based methods, the generated pruned ensemble has a competitive test error, but also has a much larger ensemble size.

In order to obtain not only a good generalization performance but also a small ensemble size, Qian et al. [24] formulated ensemble pruning as an explicit bi-objective optimization problem that minimizes the validation error and ensemble size simultaneously, and proposed the Pareto Ensemble Pruning (PEP) method, which employs a simple MOEA [16,23] combined with a local search operator to solve the bi-objective problem. It has been shown [24] that PEP can be significantly better on both test error and ensemble size than various ordering-based methods [2,10,17,20,21] as well as the single-objective optimization-based method that minimizes the validation error only [34].

Ensemble pruning naturally has two goals: maximizing the generalization performance and minimizing the ensemble size. The above-mentioned works (e.g., [20,24,34]) mainly measured the generalization performance by the validation error during the optimization process. However, it has been revealed that the generalization performance depends on not only the error on a sampled data set, but also the margin, i.e., the distance from a sampled data to the decision boundary. Margin theory for Boosting was first presented by Schapire et al. [3] to explain the success of AdaBoost. Soon after, Breiman [5] proved that the minimum margin is crucial to the margin theory, but optimizing the minimum margin led to poor empirical generalization performance; this sentenced margin theory to death. Later, Reyzin and Schapire conjectured that it is margin distribution rather than minimum margin concerns [25]. Gao and Zhou [14] finally proved that it is crucial to optimize margin distribution, characterized by maximizing margin mean and minimizing margin variance simultaneously. Later, Grønlund et al. [15] proved that one cannot hope for much stronger upper bounds than Gao and Zhou's result. Gao and Zhou's result has inspired many advanced machine learning algorithms to maximize margin mean and minimize margin variance simultaneously [29,30], generally by taking one of them as an objective whereas the other as a constraint. Lyu et al. [19] tried to take margin ratio, defined by the standard deviation of margin over margin mean, and applied it to improve deep forest. But to the best of our knowledge, the margin distribution has not been exploited for ensemble pruning.

In this paper, we propose a Margin Distribution guided multi-objective evolutionary Ensemble Pruning (MDEP) method, which formulates ensemble pruning as a three-objective optimization problem that minimizes the validation error, margin ratio [19] and ensemble size simultaneously, and then applies advanced multi-objective EAs (MOEAs) to solve it. Experiments have been conducted on 20 binary classification data sets. We first examine the performance of MDEP equipped with three typical MOEAs, i.e. NSGA-II [9], MOEA/D [28] and NSGA-III [8], suggesting that NSGA-III leads to the best performance. Then, we compare MDEP using NSGA-III against all the state-of-the-art pruning methods introduced before, showing that MDEP can achieve a better test error with a significantly smaller ensemble size. Finally, we also perform an ablation study to show that introducing the objective of minimizing the margin ratio (i.e., optimizing the margin distribution) really contributes to the advantage of MDEP.

2   MDEP Method

In this section, we first introduce the three-objective formulation (i.e., validation error, margin distribution and ensemble size) of the ensemble pruning problem, and then show how to solve this three-objective problem by MOEAs.

2.1   Three-Objective Formulation with Margin Distribution

Given a set of n trained base learners $H = \{h_t\}_{t=1}^n$, where $h_t : \mathcal{X} \to \mathcal{Y}$ maps the instance space $\mathcal{X}$ to the label space $\mathcal{Y}$. Let H_s denote a pruned ensemble with the selector vector $s \in \{0,1\}^n$, where $\forall t \in \{1, 2, \ldots, n\}$, $s_t = 1$ and $s_t = 0$ mean that the base learner h_t is selected and unselected, respectively. Considering using voting to combine the base learners, the output of H_s on an instance $x \in \mathcal{X}$ is calculated by taking an average of the selected base learners, i.e.,

$$H_s(x) = \frac{1}{|s|} \sum_{t=1}^n s_t h_t(x), \tag{1}$$

where $|s| = \sum_{t=1}^n s_t$ represents the ensemble size. The goal of ensemble pruning is to select a pruned ensemble H_s that optimizes the generalization performance (i.e., the expected prediction error under the unknown data distribution $\mathcal{D}$ over $\mathcal{X} \times \mathcal{Y}$) while containing as few base learners as possible.

Ensemble pruning can be naturally formulated as a bi-objective optimization problem that optimizes the generalization performance of H_s and minimizes the ensemble size $|s|$, simultaneously. Previous work [24,34] measured the generalization performance by the validation error only. However, it has been proved by Gao and Zhou [14] that the generalization performance depends on not only the error on a sampled data set, but also the margin distribution.

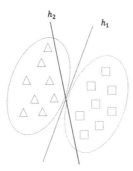

Fig. 1. A simple illustration of two linear classifiers h_1, h_2 with the same validation error but different margin distributions. Dotted ellipses are two underlying distributions, from which blue triangles and green squares are validation instances sampled for two classes. This illustration takes the idea from [32]. (Color figure online)

To intuitively show that the generalization performance of a learner is related to the margin distribution, we consider an example of binary classification in Fig. 1. The margin of an instance with respect to a learner is the distance from the instance to the learner's decision boundary, which can also be viewed as a measure of confidence in classification. The larger the margin, the better it is. Figure 1 illustrates the importance of margin distribution. h_1 and h_2 are different linear classifiers with equal validation errors which cannot be distinguished if we only consider the validation error. But when we also consider the margin distribution, h_1 has larger margins on most sampled instances and will be selected, which is the true better classifier that separates the two classes perfectly.

For achieving a better generalization performance, it is thus required to optimize both the error and the margin distribution on the validation set. Let $D = \{(\boldsymbol{x}_i, y_i)\}_{i=1}^m$ denote the given validation set. Considering binary classification, i.e., $\mathcal{Y} = \{+1, -1\}$, the validation error of a pruned ensemble H_s can be represented as

$$\text{error}_D(H_s) = \frac{1}{m} \sum_{i=1}^m \left(I(y_i H_s(\boldsymbol{x}_i) < 0) + \frac{I(y_i H_s(\boldsymbol{x}_i) = 0)}{2} \right), \qquad (2)$$

where $I(\cdot)$ is the indicator function that is 1 if the inner expression is true and 0 otherwise. Note that $y_i H_s(\boldsymbol{x}_i) < 0$ implies that the pruned ensemble H_s makes the wrong prediction; $y_i H_s(\boldsymbol{x}_i) = 0$ implies that $H_s(\boldsymbol{x}_i)$ in Eq. (1) is equal to 0, and the pruned ensemble will make a random guess, resulting in an error with probability $1/2$. The margin of the labeled instance $(\boldsymbol{x}_i, y_i)$ with respect to a pruned ensemble H_s is

$$\rho_{H_s}(\boldsymbol{x}_i, y_i) = y_i H_s(\boldsymbol{x}_i) = \frac{1}{|s|} \left(\sum_{t: y_i = h_t(\boldsymbol{x}_i)} s_t - \sum_{t: y_i \neq h_t(\boldsymbol{x}_i)} s_t \right). \qquad (3)$$

Gao and Zhou [14] have revealed that a smaller margin variance and a larger margin mean will lead to a better margin distribution, and Lyu et al. [19] have further proved that margin distribution can be characterized by margin ratio related to the margin standard deviation against the margin mean. For a pruned ensemble H_s, its margin ratio on the validation set D can be calculated as

$$\rho_D^{\text{ratio}}(H_s) = \sqrt{\frac{\text{Var}_D(\rho_{H_s})}{\text{Mean}_D^2(\rho_{H_s})}} = \sqrt{\frac{m \sum_{i \neq j} \left(\rho_{H_s}(\boldsymbol{x}_i, y_i) - \rho_{H_s}(\boldsymbol{x}_j, y_j) \right)^2}{2(m-1)(\sum_{i=1}^m \rho_{H_s}(\boldsymbol{x}_i, y_i))^2}}, \quad (4)$$

where $\text{Var}_D(\rho_{H_s})$ and $\text{Mean}_D(\rho_{H_s})$ denote the margin variance and mean, respectively, of the instances in D with respect to H_s, and $\rho_{H_s}(\boldsymbol{x}_i, y_i)$ is the margin of $(\boldsymbol{x}_i, y_i)$ with respect to H_s, as calculated in Eq. (3). The smaller the margin ratio, the better the margin distribution and thus the generalization performance.

Based on the above analysis, we formulate ensemble pruning as a three-objective minimization problem

$$\arg\min_{s \in \{0,1\}^n} \left(\text{error}_D(H_s), \rho_D^{\text{ratio}}(H_s), |s| \right). \quad (5)$$

That is, the validation error, the margin ratio and the ensemble size are minimized simultaneously. Note that minimizing the first two objectives corresponds to optimizing the generalization performance. To the best of our knowledge, this is the first time that margin distribution is utilized for ensemble pruning.

By solving the three-objective problem formulated in Eq. (5) by MOEAs, we propose the Margin Distribution guided multi-objective evolutionary Ensemble Pruning method, briefly called MDEP. Though the margin in Eq. (3) is defined for binary classification, it can be adapted to multi-class classification and regression accordingly [11,22], and thus MDEP can also be applied to these tasks.

2.2 Multi-objective Evolutionary Algorithms

Next, we will show how MDEP applies MOEAs to solve the three-objective problem in Eq. (5). The input of MDEP is a set of trained base learners $H = \{h_t\}_{t=1}^n$ and a validation data set D. As introduced before, a pruned ensemble can be naturally represented by a Boolean vector $s \in \{0,1\}^n$, where the t-th bit $s_t = 1$ if and only if the base learner h_t is selected. Note that the solution with all 0s is excluded during optimization. The procedure of MDEP is presented in Algorithm 1. In fact, MDEP can be equipped with any existing MOEA, e.g., NSGA-II [9], MOEA/D [28] and NSGA-III [8]. Here, we mainly introduce the special initialization, crossover and mutation operations that MDEP adopts.

Initialization. With the goal of improving the search efficiency of MDEP in the solution space with a small ensemble size, we evaluate all the solutions with size 1 in line 1 of Algorithm 1, and select the non-dominated solutions among them as initial solutions in line 2. Note that these solutions must be Pareto optimal, because solutions with size larger than 1 cannot dominate them. To

---

Algorithm 1. MDEP Method

Input: Original ensemble $H = \{h_t\}_{t=1}^n$, validation data set $D = \{(\boldsymbol{x}_i, y_i)\}_{i=1}^m$
Output: Pruned ensemble H_s
 1: Evaluate all the solutions $\{\boldsymbol{s}^i\}_{i=1}^n$ with size 1, where $\boldsymbol{s}^i$ has value 1 on the i-th bit,
    and 0 otherwise;
 2: Select the non-dominated solutions among $\{\boldsymbol{s}^i\}_{i=1}^n$, and add them into the initial
    population P_1;
 3: For the remaining required initial solutions, randomly select them from $\{0,1\}^n$;
 4: **for** $t = 1$: maximum #generations **do**
 5:     Select solutions from P_t to compose the mating pool;
 6:     Generate offspring population P' by uniform crossover and bit-wise mutation;
 7:     **for** each offspring solution $\boldsymbol{s}' \in P'$ **do**
 8:         **if** $|\boldsymbol{s}'| \leq 1$ **then**
 9:             **repeat**
10:                 Apply the bit-wise mutation operator to update $\boldsymbol{s}'$
11:             **until** $|\boldsymbol{s}'| > 1$
12:         **end if**
13:         Evaluate $\boldsymbol{s}'$
14:     **end for**
15:     Select next population P_{t+1} from $P_t \cup P'$
16: **end for**
17: Select a non-dominated solution $\boldsymbol{s}$ from the final population

---

fill in the initial population P_1, the remaining initial solutions are randomly selected from the whole solution space $\{0,1\}^n$ in line 3. Note that this setting implicitly requires that the population size is at least the number of Pareto optimal solutions with size 1. In our experiments, the population size will be set to n, which obviously satisfies this requirement.

Reproduction. To reproduce offspring solutions from the selected parent solutions, we employ the common operators over Boolean vector representation: uniform crossover and bit-wise mutation [13], as shown in line 6 of Algorithm 1. The uniform crossover operator generates the first offspring solution by inheriting each bit from the first parent solution independently with probability $1/2$, and otherwise from the second parent. The second offspring is created using inverse mapping. The bit-wise mutation operator flips each bit of a solution independently with probability $1/n$. Since we have explored all the solutions with size 1 in the initialization procedure, when an offspring solution $\boldsymbol{s}'$ with $|\boldsymbol{s}'| \leq 1$ is generated, the bit-wise mutation operator is applied repeatedly to update $\boldsymbol{s}'$ until $|\boldsymbol{s}'| > 1$ (i.e., lines 8–12).

Though the above settings are simple, they have been sufficient to lead to a good performance of MDEP, which will be shown in our experiments. More careful designs may further improve the performance. Note that the uniform crossover and bit-wise mutation operators are usually applied with some probabilities, denoted as P_c and P_m, respectively. They are treated as two hyperparameters. The parent selection strategy for reproduction in line 5 as well as the survival selection strategy for updating the population in line 15 depends on the concrete MOEA employed by MDEP. For example, if NSGA-II [9] is employed, binary tournament selection is used to select parent solutions and the survival selection strategy is based on non-dominated sorting and crowding distance.

MDEP will continue to run until a predefined number of generations (i.e., maximum #generations in line 4 of Algorithm 1) is reached. After MDEP terminates, we will get a set of solutions, and the final output solution can be selected according to the user's preference. Here we propose to select the solution with the smallest validation error from the final population. If such a solution is not unique, we select the solution with the smallest ensemble size among them. This strategy of selecting the final solution will be used in our experiments.

3 Experiments

In this section, we empirically examine the performance of MDEP. Section 3.1 introduces the general experimental settings. As MDEP can be equipped with any MOEA, we compare the performance of MDEP using three typical MOEAs, i.e., NSGA-II [9], MOEA/D [28] and NSGA-III [8], in Sect. 3.2, showing that NSGA-III is the best choice. Next, we compare MDEP equipped with NSGA-III against state-of-the-art pruning methods in Sect. 3.3. Finally, Sect. 3.4 performs an ablation study to examine whether considering the margin distribution in problem formulation, i.e., introducing the objective of minimizing the margin ratio in Eq. (5), really contributes to the advantage of MDEP.

3.1 Settings

We conduct experiments on 20 binary classification data sets from the UCI repository [12]. Some of the binary classification data sets are generated from multi-class data sets: *letter-ah* is based on the *letter* data and classifies 'a' against 'h', and alike *letter-br* and *letter-oq*; *optdigits* classifies '01234' against '56789'; *satimage-12v57* is based on the *satimage* data and classifies labels '1' and '2' against '5' and '7', and alike *satimage-2v5*; *vehicle-bo-vs* is based on the *vehicle* data and classifies 'bus' and 'opel' against 'van' and 'saab', and alike *vehicle-b-v*.

To evaluate each method on each data set, a data set is evenly and randomly split into three parts: training set, validation set and test set. We use Bagging [4] to train 100 C4.5 decision trees [6] on the training set as the original ensemble $H = \{h_t\}_{t=1}^{n}$, and then prune the ensemble by a pruning method on the validation set. Finally, we report the performance of the pruned ensemble on the test set. In order to reduce the influence of randomness, each data set is randomly partitioned 30 times independently, and each method will be performed on each partition of the data set and the average performance will be reported.

3.2 Comparison of MDEP Using Various MOEAs

Since MDEP can employ various MOEAs to solve the three-objective problem in Eq. (5), we first compare the performance of MDEP equipped with NSGA-II [9], MOEA/D [28] and NSGA-III [8]. Because the optimization process of an

MOEA is inherently stochastic, for each partition of each data set, the MOEA is repeated 5 times further. That is, each MOEA on each data set is repeated 150 times (30 partitions × 5 times per partition). For fairness of comparison, we use the same hyperparameter setting for each MOEA. The population size is 100, the number of generations is 500. The probabilities P_c and P_m of applying crossover and mutation are set arbitrarily to 0.7 and 1, respectively. The more careful setting may achieve better results.

The comparative methods also include two baselines: Bagging which uses the original ensemble (i.e., all 100 trained base learners), and Best Individual (BI) which selects the base classifier with the smallest validation error. Table 1 gives the detailed results, i.e., the mean and standard deviation of test error and ensemble size of each method on each data set. To save space, MDEP equipped with a specific MOEA is denoted by the name of the MOEA in Table 1. For example, NSGA-III actually means MDEP equipped with NSGA-III. Among all the comparison methods, BI has the worst test error on all data sets, which is consistent with the fact that an ensemble of multiple classifiers usually achieves better generalization performance than a single classifier. From the row of "w/t/l to Bagging", we can observe that MDEP using any MOEA achieves a smaller test error than Bagging on at least 80% (16/20) data sets. Furthermore, by the Wilcoxon rank-sum test [27] with confidence level 0.1, MDEP using any MOEA can be significantly better than Bagging on 60% (12/20) of the data sets.

By the row of "count of the best", we can observe that MDEP using NSGA-III achieves the smallest test error on 75% (15/20) data sets, which is better than using other MOEAs. This may be because NSGA-III is proposed to improve the performance of NSGA-II for problems with more objectives. Though using NSGA-II most often achieves the smallest ensemble size, the average ensemble size of NSGA-III, NSGA-II, and MOEA/D on 20 data sets is similar, which is 8.66, 8.42 and 8.76, respectively. That is, MDEP using any MOEA will reduce the original ensemble size greatly.

In conclusion, MDEP using any MOEA can result in better generalization performance with significantly reduced ensemble size. Furthermore, using NSGA-III leads to the best performance of MDEP, which achieves a smaller test error with a similar ensemble size, compared with using other MOEAs.

3.3   MDEP vs. State-of-the-Art Pruning Methods

Next, we compare MDEP with state-of-the-art ensemble pruning methods. Note that MDEP uses NSGA-III here, which has been shown to be the best choice in Sect. 3.2. We implement seven state-of-the-art pruning methods, including five ordering-based methods: Reduce-Error (RE) [7], Kappa [2], ComPlementarity (CP) [20], Margin Distance (MD) [21] and DREP [17]; two optimization-based methods: EA [33,34] that employs a standard genetic algorithm to minimize the validation error only, and PEP [24] that employs a simple MOEA [16] combined with a local search operator to minimize the validation error and ensemble size simultaneously. Note that EA and PEP output the pruned ensemble with

Table 1. The test errors and ensemble sizes (mean+std.) of the compared methods on 20 binary data sets. The smallest error and size on each data set are bolded. In the row of "count of the best", the largest values are bolded. The "w/t/l to Bagging" denotes the number of data sets where the test error of MDEP using a specific MOEA is smaller, same, or larger, compared to Bagging.

Data set	Test error					Ensemble size		
	NSGA-III	NSGA-II	MOEA/D	Bagging	BI	NSGA-III	NSGA-II	MOEA/D
australian	**.143 ± .020**	.144 ± .021	**.143 ± .020**	**.143 ± .017**	.152 ± .023	8.2 ± 3.4	**7.5 ± 3.3**	**7.5 ± 3.1**
breast-cancer	**.273 ± .035**	.279 ± .038	.278 ± .035	.279 ± .037	.298 ± .044	7.4 ± 2.7	6.9 ± 1.6	**6.8 ± 2.2**
liver-disorders	.312 ± .033	.313 ± .033	**.310 ± .035**	.327 ± .047	.365 ± .047	11.2 ± 3.8	10.6 ± 3.7	10.9 ± 3.3
heart-statlog	**.192 ± .037**	.197 ± .040	.195 ± .040	.195 ± .038	.235 ± .049	**7.7 ± 2.4**	7.9 ± .2.7	**7.7 ± 2.1**
house-votes-84	.044 ± .018	.045 ± .019	.043 ± .020	**.041 ± .013**	.047 ± .016	**3.0 ± 1.4**	3.1 ± 1.8	**3.0 ± 1.9**
ionosphere	**.083 ± .022**	.085 ± .025	**.083 ± .023**	.092 ± .025	.117 ± .022	5.0 ± 1.6	**4.9 ± 1.7**	5.1 ± 1.7
kr-vs-kp	**.009 ± .003**	.010 ± .003	**.009 ± .003**	.015 ± .007	.011 ± .004	**3.8 ± 1.4**	4.0 ± 1.2	4.4 ± 1.8
letter-AH	**.012 ± .006**	.014 ± .006	**.012 ± .006**	.021 ± .006	.023 ± .008	5.1 ± 2.0	**4.9 ± 1.9**	5.1 ± 1.7
letter-BR	**.045 ± .011**	.047 ± .012	.048 ± .010	.059 ± .013	.078 ± .012	9.8 ± 2.2	**9.4 ± 2.5**	10.7 ± 2.9
letter-OQ	**.041 ± .009**	.042 ± .010	.043 ± .009	.049 ± .012	.078 ± .017	9.9 ± 2.5	**9.8 ± 2.7**	10.7 ± 2.9
optdigits-b	.035 ± .005	**.034 ± .005**	.037 ± .005	.038 ± .007	.095 ± .008	**21.1 ± 4.1**	21.7 ± 4.5	21.5 ± 5.3
satimage-12v57	**.028 ± .004**	**.028 ± .004**	**.028 ± .004**	.029 ± .004	.052 ± .006	**13.7 ± 3.1**	14.3 ± 4.6	14.7 ± 4.2
satimage-25	.022 ± .006	**.021 ± .007**	**.021 ± .006**	.023 ± .009	.033 ± .010	**5.4 ± 1.3**	5.6 ± 1.9	5.7 ± 1.9
sick	**.015 ± .003**	**.015 ± .003**	.016 ± .003	.018 ± .004	.018 ± .004	5.8 ± 2.2	**5.6 ± 2.7**	6.2 ± 1.8
sonar	**.244 ± .052**	.257 ± .057	.257 ± .040	.266 ± .052	.310 ± .051	10.9 ± 3.5	**9.9 ± 2.7**	10.9 ± 3.5
spambase	**.065 ± .006**	.066 ± .007	.066 ± .006	.068 ± .007	.093 ± .008	14.0 ± 4.9	**13.7 ± 3.7**	14.0 ± 3.4
tic-tac-toe	**.128 ± .024**	.131 ± .021	**.128 ± .022**	.164 ± .028	.212 ± .028	12.4 ± 3.2	**11.2 ± 3.2**	12.0 ± 3.1
vehicle-bo-vs	.226 ± .022	**.223 ± .021**	.229 ± .021	.228 ± .026	.257 ± .025	13.1 ± 4.6	**11.9 ± 4.1**	12.6 ± 3.6
vehicle-b-v	**.019 ± .011**	.020 ± .012	**.019 ± .013**	.027 ± .014	.024 ± .013	**2.8 ± 1.0**	**2.8 ± 1.1**	2.9 ± 1.5
vote	**.044 ± .018**	.046 ± .019	.046 ± .020	.047 ± .018	.046 ± .016	2.9 ± 1.5	**2.7 ± 1.1**	2.8 ± 1.3
count of the best	**15**	5	9	2	0	7	**13**	4
w/t/l to Bagging	18/1/1	16/1/3	16/2/2	-	-	-	-	-

the smallest validation error from the final population [24,33,34]. The hyperparameter p of MD is set to 0.075 [21], and the hyperparameter ρ of DREP is selected from $\{0.2, 0.25, \ldots, 0.5\}$ [17]. As suggested by [24], the total number of fitness evaluations used by EA and PEP is set to $n^4 \log n$, which is much greater than that (i.e., population size 100×500 #generations $= 50,000$) of MDEP as $n = 100$. Though this is unfair MDEP, better performance on test error and ensemble size can still be achieved by MDEP, and will be shown later.

The average test error and ensemble size are shown in Table 2. In terms of test error, MDEP performs the best on 65% (13/20) of the data sets, while the other methods are at most 40% (8/20). Compared with any other method, MDEP is better on at least 55% (11/20) of the data sets, and is never significantly worse since no 'o' appears in the upper half of Table 2. In terms of ensemble size, MDEP and PEP perform the best on 85% (17/20) and 20% (4/20) of the data sets, respectively, while the other methods never achieve the smallest size. This may be because only MDEP and PEP minimize the ensemble size explicitly. EA minimizes the validation error only, and generates ensembles with the largest size on all data sets, which is consistent with previous observation [18,34]. Compared with the runner-up PEP, MDEP achieves a smaller ensemble size on 80% (16/20) of the data sets, and is significantly better on 45% (9/20) of the data sets. To sum up, MDEP can achieve better generalization performance than other pruning methods, while with a significantly smaller ensemble size.

Table 2. The test errors and ensemble sizes (mean+std.) of the compared methods on 20 binary data sets. The smallest error and size on each data set are bolded, and '•/○' denotes that MDEP is significantly better/worse than the corresponding method by the Wilcoxon rank-sum test with confidence level 0.1. In the rows of "count of the best", the largest values are bolded. The "MDEP: w/t/l" denotes the number of data sets where the test error (or ensemble size) of MDEP is smaller, same or larger, compared to the corresponding method.

Test error										
Data set	MDEP	DREP	Kappa	CP	MD	RE	EA	PEP	Bagging	BI
australian	**.143**±.020	.144±.019	**.143**±.021	.145±.022	.148±.022	.144±.020	**.143**±.020	.144±.020	**.143**±.017	.152±.023•
breast-cancer	**.273**±.035	.275±.036	.287±.037•	.282±.043	.295±.044•	.277±.031	.275±.032	.275±.041	.279±.037	.298±.044•
liver-disorders	.312±.033	.316±.045	.326±.042•	.306±.039	.337±.035•	.320±.044	.317±.046	**.304**±.039	.327±.047•	.365±.047•
heart-statlog	.192±.037	.194±.044	.201±.038	.199±.044	.226±.048•	**.187**±.044	.196±.032	.197±.037	.195±.038	.235±.049•
house-votes-84	.044±.018	.045±.017	.044±.017	.045±.017	.048±.018	.043±.018	**.041**±.012	.045±.019	**.041**±.013	.047±.016
ionosphere	**.083**±.022	.085±.021	.084±.020	.089±.021•	.100±.026•	.086±.021	.093±.026•	.088±.021•	.092±.025•	.117±.022•
kr-vs-kp	**.009**±.003	.011±.003	.010±.003	.011±.003	.011±.005	.010±.004	.012±.004•	.010±.003	.015±.007•	.011±.004
letter-AH	**.012**±.006	.014±.005•	**.012**±.006	.015±.006•	.017±.007•	.015±.006•	.017±.006•	.013±.005	.021±.006•	.023±.008•
letter-BR	**.045**±.011	.048±.009	.048±.014	.048±.012	.057±.014•	.048±.012	.053±.011•	.046±.008	.059±.013•	.078±.012•
letter-OQ	.041±.009	**.041**±.010	.042±.011	.042±.010	.046±.011•	.046±.011•	.044±.011	.043±.009	.049±.012•	.078±.017•
optdigits-b	**.035**±.005	.035±.006	**.035**±.005	.036±.005	.037±.006	.036±.006	.035±.006	.035±.006	.038±.007•	.095±.008•
satimage-12v57	**.028**±.004	.029±.004	**.028**±.004	.029±.004	.029±.004	.029±.004	.029±.004	.028±.004	.029±.004	.052±.006•
satimage-25	.022±.006	.022±.008	.022±.007	**.021**±.008	.026±.010•	.023±.007	.021±.008	.021±.007	.023±.009	.033±.010•
sick	**.015**±.003	.016±.003	.017±.003	.016±.003	.017±.003•	.016±.003	.017±.004•	**.015**±.003	.018±.004•	.018±.004•
sonar	**.244**±.052	.257±.056	.249±.059	.250±.048	.268±.055	.267±.053•	.251±.041	.248±.056	.266±.052•	.310±.051•
spambase	**.065**±.006	**.065**±.006	.066±.006	.066±.006	.068±.007•	.066±.006	.066±.006	**.065**±.006	.068±.007•	.093±.008•
tic-tac-toe	**.128**±.024	.129±.026	.132±.023	.132±.026	.145±.022•	.135±.026	.138±.020•	.131±.027	.164±.028•	.212±.028•
vehicle-bo-vs	.226±.022	.234±.026	.233±.024	.234±.024	.244±.024•	.226±.022	.230±.024	**.224**±.023	.228±.026	.257±.025•
vehicle-bus-van	.019±.011	.019±.013	.019±.012	.020±.011	.021±.011	.020±.011	.026±.013•	**.018**±.011	.027±.014•	.024±.013
vote	.044±.018	.043±.019	**.041**±.016	.043±.016	.045±.014	.044±.017	.045±.015	.044±.018	.047±.018	.046±.016
count of the best	**13**	3	5	1	0	1	4	8	2	0
MDEP: w/t/l	—	14/5/1	12/7/1	17/0/3	20/0/0	16/2/2	16/2/2	11/5/4	18/1/1	20/0/0
Ensemble size										
australian	**8.2**±3.4	11.7±4.7•	14.7±12.6•	11.0±9.7	8.5±14.8	12.5±6.0•	41.9±6.7•	10.6±4.2•	—	—
breast-cancer	**7.4**±2.7	9.2±3.7•	26.1±21.7•	8.8±12.3	7.8±15.2	8.7±3.6•	44.6±6.6•	8.4±3.5•	—	—
liver-disorders	**11.2**±3.8	13.9±5.9•	24.7±16.3•	15.3±10.6	17.7±20.0	13.9±4.2•	42.0±6.2•	14.7±4.2•	—	—
heart-statlog	**7.7**±2.4	11.3±2.7•	17.9±11.1•	13.2±8.2	13.6±21.1	11.4±5.0•	44.2±5.1•	9.3±2.3	—	—
house-votes-84	3.0±1.4	4.1±2.7•	5.5±3.3•	4.7±4.4	5.9±14.1	3.9±4.0	46.5±6.1	**2.9**±1.7	—	—
ionosphere	**5.0**±1.6	8.4±4.3•	10.5±6.9•	8.5±6.3•	10.7±14.6	7.9±5.7•	48.8±5.1•	5.2±2.2	—	—
kr-vs-kp	**3.8**±1.4	7.1±3.9•	10.6±9.1•	9.6±8.6•	7.2±15.2	5.8±4.5	45.9±5.8	4.2±1.8	—	—
letter-AH	5.1±2.0	7.8±3.6•	7.1±3.8•	8.7±4.7•	11.0±10.9	7.3±4.4•	42.5±6.5•	**5.0**±1.9	—	—
letter-BR	**9.8**±2.2	11.3±3.5•	13.8±6.7•	12.9±6.8•	23.2±17.6•	15.1±7.3•	38.3±7.8•	10.9±2.6	—	—
letter-OQ	**9.9**±2.5	13.7±4.9•	13.9±6.0•	12.3±4.9•	23.5±15.6•	13.6±5.8•	39.3±8.2•	12.0±3.7•	—	—
optdigits-b	**21.1**±4.1	25.0±8.0•	25.2±8.1•	21.4±7.5•	46.8±23.9•	25.0±9.3•	41.4±7.6•	22.7±3.1•	—	—
satimage-12v57	**13.7**±3.1	18.1±4.9•	22.1±10.3•	21.2±10.0•	37.6±24.3•	20.8±9.2•	42.7±5.2•	17.1±5.0•	—	—
satimage-25	**5.4**±1.3	7.7±3.5•	7.6±4.2•	10.9±7.0•	26.2±28.1•	6.8±3.2•	44.1±4.8	5.7±1.7	—	—
sick	**5.8**±2.2	11.6±6.7•	10.9±6.0•	11.5±10.0•	8.3±13.6	7.5±3.9•	44.7±8.2•	6.9±2.8	—	—
sonar	10.9±3.5	14.4±5.9•	20.6±9.3•	13.9±7.1	20.6±20.7	**11.0**±4.1	43.1±6.4	11.4±4.2	—	—
spambase	**14.0**±4.9	16.7±4.6•	20.0±8.1•	19.0±9.9•	28.8±17.0•	18.5±5.0•	39.7±6.4•	17.5±4.5•	—	—
tic-tac-toe	**12.4**±3.2	13.6±3.4	17.4±6.5•	15.4±6.3	28.0±22.6•	16.1±5.4•	39.8±8.2•	14.5±3.8•	—	—
vehicle-bo-vs	**13.1**±4.6	13.2±5.0	16.5±8.2•	11.2±5.7	21.6±20.4	15.7±5.7•	41.9±5.6•	16.5±4.5•	—	—
vehicle-bus-van	**2.8**±1.0	4.0±3.9	4.5±1.6•	5.3±7.4	2.8±3.8	3.4±2.1	48.0±5.6	2.8±1.1	—	—
vote	2.9±1.5	3.9±2.5	5.1±2.6•	5.4±5.2	6.0±9.8	3.2±2.7	47.8±6.1	**2.7**±1.1	—	—
count of the best	**17**	0	0	0	0	0	0	4	—	—
MDEP: w/t/l	—	20/0/0	20/0/0	20/0/0	20/0/0	20/0/0	20/0/0	16/1/3	—	—

We further make a more comprehensive comparison between MDEP and the runner-up PEP [24]. We map all the solutions in their final population into the space of test error and ensemble size. Figure 2(a) shows the results on the data

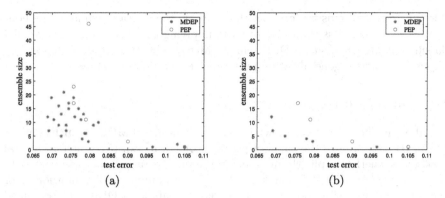

Fig. 2. The final solution sets of MDEP (blue stars) and PEP (red dots) in the space of test error and ensemble size on the data set *spambase*. (a) All solutions. (b) Non-dominated solutions in (a). (Color figure online)

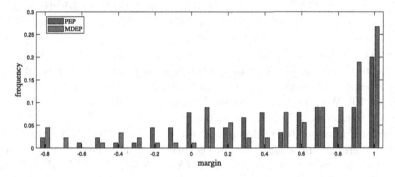

Fig. 3. The histogram of the margin distributions (i.e., the frequency on each margin) obtained by MDEP and PEP on the data set *heart-statlog*.

set *spambase*. It can be observed that MDEP obtains a much larger solution set than PEP, with more solutions in the lower-left corner of the figure. Note that PEP does not maintain a fixed-size population, and thus may obtain few final solutions, as observed. Figure 2(b) shows the non-dominated solutions in Fig. 2(a). It can be more clearly observed that for each solution obtained by PEP, MDEP has at least one solution that can dominate it.

Since MDEP optimizes the margin distribution explicitly, we also visualize the margin distribution of the final pruned ensemble by plotting the histogram of the frequency on each margin. Figure 3 shows the results of MDEP and PEP on the data set *heart-statlog*. It can be seen that MDEP obtains larger margins (e.g., margins greater than 0.7). Although MDEP also gets more very negative margins (e.g., margins no greater than -0.7), the overall frequency of non-positive margins is less than that of PEP, implying that fewer samples are misclassified by MDEP. Thus, MDEP achieves an overall better margin distribution, suggesting a better generalization performance as observed before.

3.4   Ablation Study

The above experiments have shown the clear advantage of MDEP, which employs NSGA-III to minimize the three-objective problem in Eq. (5). Then, a natural question is whether explicitly minimizing the margin ratio really contributes to the advantage of MDEP. Though PEP [24] is to minimize the validation error and ensemble size simultaneously, it employs a simple MOEA [16] combined with local search for optimization, and thus the superiority of MDEP over PEP cannot answer the question due to the difference of the employed optimizer.

To answer the question, we next compare the performance of the same MOEA (i.e., NSGA-III, NSGA-II or MOEA/D) optimizing the three objectives and two objectives (i.e., only the validation error and ensemble size), respectively. The hyper-parameters of all MOEAs are the same as in the previous experiments. The results are shown in Table 3. We can observe that for the same MOEA, minimizing the margin ratio additionally (corresponding to the columns of '3-obj') usually results in a smaller test error, which also supports the margin distribution theory [14,19]. We note that the ensemble size obtained by optimizing the three objectives is relatively larger, which may be because a solution with a larger ensemble size is easier to be dominated under the bi-objective formulation. In fact, the difference in the ensemble size is very small. For the three-objective formulation, the average ensemble size of NSGA-III, NSGA-II and MOEA/D on the 20 data sets is 8.66, 8.42 and 8.76, respectively; while for the bi-objective formulation, the average size is 8.15, 7.66 and 8.20, respectively. Furthermore, as shown in Sect. 3.3, the ensemble size achieved by NSGA-III under the three-objective formulation is still significantly smaller than other state-of-the-art pruning methods. Thus, these results give a positive answer, i.e., confirm that optimizing the margin distribution explicitly brings advantages.

Table 3. The test errors and ensemble sizes (mean+std.) of each MOEA optimizing three or two objectives on 20 binary data sets. For each MOEA on each data set, the smaller error and size are bolded. The "3-obj vs. 2-obj: w/t/l" denotes the number of data sets where the test error (or ensemble size) of an MOEA optimizing three objectives is smaller, same or larger, compared to that of the MOEA optimizing two objectives.

Data set	Test error						Ensemble size					
	NSGA-III		NSGA-II		MOEA/D		NSGA-III		NSGA-II		MOEA/D	
	3-obj	2-obj	3-obj	2-obj	3-obj	2-obj	3-obj	2-obj	3-obj	2-obj	3-obj	2-obj
australian	.143±.020	.143±.019	.144±.021	.147±.022	.143±.020	.144±.022	8.2±3.4	7.0±3.1	7.5±3.3	6.7±2.6	7.8±3.1	7.9±3.2
breast-cancer	.273±.035	.283±.037	.279±.038	.275±.038	.278±.035	.280±.039	7.4±2.7	5.5±2.2	6.9±1.6	5.9±1.6	6.8±2.2	5.6±2.3
liver-disorders	.312±.033	.325±.043	.313±.033	.325±.040	.310±.035	.313±.040	11.2±3.8	11.5±4.2	10.6±3.7	10.2±4.1	10.9±3.3	11.6±3.9
heart-statlog	.192±.037	.193±.042	.197±.040	.209±.039	.195±.040	.202±.033	7.7±2.4	6.6±2.4	7.9±2.7	6.9±2.2	7.7±2.1	7.2±2.3
house-votes-84	.044±.018	.045±.019	.045±.019	.046±.018	.043±.020	.044±.018	3.0±1.4	2.9±1.4	3.1±1.8	2.9±1.3	3.0±1.9	2.7±1.3
ionosphere	.083±.022	.092±.021	.085±.025	.095±.025	.083±.023	.090±.023	5.0±1.6	4.5±1.5	4.9±1.7	4.0±1.2	5.1±1.7	5.1±2.2
kr-vs-kp	.009±.003	.010±.004	.010±.003	.010±.003	.009±.003	.010±.003	3.8±1.4	3.7±1.4	4.0±1.2	3.6±1.2	4.4±1.8	3.7±1.4
letter-AH	.012±.006	.013±.006	.014±.006	.012±.006	.012±.006	.013±.005	5.1±2.0	4.7±1.7	4.9±1.9	4.9±1.8	5.1±1.7	5.1±1.5
letter-BR	.045±.011	.049±.010	.047±.012	.048±.010	.048±.010	.047±.011	9.8±2.2	9.2±3.5	9.4±2.5	8.3±3.1	10.7±2.9	8.9±3.5
letter-OQ	.041±.009	.046±.010	.042±.010	.044±.011	.043±.009	.045±.011	9.9±2.5	8.9±2.2	9.8±2.7	9.0±3.0	10.7±2.9	11.1±2.6
optdigits-b	.035±.005	.036±.006	.034±.005	.036±.006	.037±.005	.037±.006	21.1±4.1	20.2±4.9	21.7±4.5	20.3±5.6	21.5±5.3	18.9±5.0
satimage-12v57	.028±.004	.029±.004	.028±.004	.029±.004	.028±.004	.030±.004	13.7±3.1	13.2±4.3	14.3±4.6	12.5±3.7	14.7±4.2	13.7±3.3
satimage-25	.022±.006	.022±.008	.021±.007	.022±.006	.021±.006	.022±.007	5.4±1.3	5.7±2.4	5.6±1.9	4.7±1.2	5.7±1.9	5.1±1.9
sick	.015±.003	.016±.003	.015±.003	.016±.003	.016±.003	.017±.003	5.8±2.2	5.7±2.2	5.6±2.7	5.1±2.0	6.2±1.8	5.3±2.3
sonar	.244±.052	.267±.071	.287±.057	.263±.064	.257±.040	.255±.048	10.9±3.5	8.5±3.5	9.9±2.7	8.6±3.4	10.9±3.5	8.8±3.3
spambase	.065±.006	.067±.006	.066±.007	.066±.005	.066±.006	.067±.007	14.0±4.9	15.1±4.8	13.7±3.7	12.2±3.7	14.0±3.4	14.2±3.9
tic-tac-toe	.128±.024	.133±.024	.131±.021	.135±.024	.128±.022	.137±.020	12.4±3.2	11.0±3.5	11.2±3.2	10.7±3.6	12.0±3.1	12.4±3.2
vehicle-bo-vs	.226±.022	.228±.023	.223±.021	.225±.019	.229±.021	.233±.024	13.1±4.6	13.7±5.0	11.9±4.1	11.1±4.1	12.6±3.6	12.1±5.1
vehicle-b-v	.019±.011	.021±.013	.020±.012	.018±.012	.019±.013	.019±.013	2.8±1.0	2.7±1.1	2.8±1.1	2.8±1.1	2.9±1.5	2.8±1.2
vote	.044±.018	.045±.020	.046±.019	.045±.019	.046±.020	.045±.017	2.9±1.5	2.7±1.2	2.7±1.1	2.7±1.1	2.8±1.3	2.5±1.1
3-obj vs. 2-obj: w/t/l	18/2/0		14/2/4		15/2/3		4/0/16		0/3/17		5/2/13	

4 Conclusion

In this paper, we introduce the three-objective (i.e., validation error, margin ratio and ensemble size) formulation of ensemble pruning, and propose a new optimization-based ensemble pruning method MDEP, which employs MOEAs to solve the three-objective problem. Experimental results show that MDEP using NSGA-III is better than using NSGA-II and MOEA/D, and more importantly, it can outperform state-of-the-art pruning methods significantly in both generalization performance and ensemble size. In the future, it would be interesting to perform theoretical analysis [35], as well as to design the components of MDEP more carefully or apply more advanced MOEAs, which may bring further performance improvement.

References

1. Back, T.: Evolutionary Algorithms in Theory and Practice: Evolution Strategies, Evolutionary Programming. Genetic Algorithms. Oxford University Press, Oxford, UK (1996)
2. Banfield, R.E., Hall, L.O., Bowyer, K.W., Kegelmeyer, W.P.: Ensemble diversity measures and their application to thinning. Inf. Fusion **6**(1), 49–62 (2005)

3. Bartlett, P., Freund, Y., Lee, W.S., Schapire, R.E.: Boosting the margin: a new explanation for the effectiveness of voting methods. Ann. Stat. **26**(5), 1651–1686 (1998)
4. Breiman, L.: Bagging predictors. Mach. Learn. **24**(2), 123–140 (1996)
5. Breiman, L.: Prediction games and arcing algorithms. Neural Comput. **11**(7), 1493–1517 (1999)
6. Breiman, L., Friedman, J., Olshen, R.A., Stone, C.J.: Classification and Regression Trees. Wadsworth and Brooks, Monterey (1984)
7. Caruana, R., Niculescu-Mizil, A., Crew, G., Ksikes, A.: Ensemble selection from libraries of models. In: Proceedings of the 21st International Conference on Machine Learning (ICML 2004), pp. 18–25. Banff, Canada (2004)
8. Deb, K., Jain, H.: An evolutionary many-objective optimization algorithm using reference-point-based nondominated sorting approach, Part I: Solving problems with box constraints. IEEE Trans. Evol. Comput. **18**(4), 577–601 (2013)
9. Deb, K., Pratap, A., Agarwal, S., Meyarivan, T.: A fast and elitist multiobjective genetic algorithm: NSGA-II. IEEE Trans. Evol. Comput. **6**(2), 182–197 (2002)
10. Dietterich, T., Margineantu, D.: Pruning adaptive boosting. In: Proceedings of the 14th International Conference on Machine Learning (ICML 1997), pp. 211–218. Nashville, TN (1997)
11. Drucker, H., Burges, C.J.C., Kaufman, L., Smola, A.J., Vapnik, V.: Support vector regression machines. In: Advances in Neural Information Processing Systems 9 (NIPS 1996), pp. 155–161. Denver, CO (1996)
12. Dua, D., Graff, C.: UCI machine learning repository (2017). http://archive.ics.uci.edu/ml
13. Eiben, A., Smith, J.: Introduction to Evolutionary Computing. Springer, Bering (2015)
14. Gao, W., Zhou, Z.H.: On the doubt about margin explanation of boosting. Artif. Intell. **203**, 1–18 (2013)
15. Grønlund, A., Kamma, L., Larsen, K.G., Mathiasen, A., Nelson, J.: Margin-based generalization lower bounds for boosted classifiers. In: Advances in Neural Information Processing Systems 32 (NeurIPS 2019), Vancouver, Canada, pp. 11940–11949 (2019)
16. Laumanns, M., Thiele, L., Zitzler, E.: Running time analysis of multiobjective evolutionary algorithms on pseudo-Boolean functions. IEEE Trans. Evol. Comput. **8**(2), 170–182 (2004)
17. Li, N., Yu, Y., Zhou, Z.H.: Diversity regularized ensemble pruning. In: Proceedings of the 23rd European Conference on Machine Learning (ECML 2012), Bristol, UK, pp. 330–345 (2012)
18. Li, N., Zhou, Z.H.: Selective ensemble under regularization framework. In: Proceedings of the 8th International Workshop on Multiple Classifier Systems (MCS 2009), Reykjavik, Iceland, pp. 293–303 (2009)
19. Lyu, S.H., Yang, L., Zhou, Z.H.: A refined margin distribution analysis for forest representation learning. In: Advances in Neural Information Processing Systems 32 (NeurIPS 2019), Vancouver, Canada, pp. 5531–5541 (2019)
20. Martínez-Muñoz, G., Hernández-Lobato, D., Suárez, A.: An analysis of ensemble pruning techniques based on ordered aggregation. IEEE Trans. Pattern Anal. Mach. Intell. **31**(2), 245–259 (2008)
21. Martínez-Muñoz, G., Suárez, A.: Pruning in ordered bagging ensembles. In: Proceedings of the 23rd International Conference on Machine Learning (ICML 2006), Pittsburgh, PA, pp. 609–616 (2006)

22. Mohri, M., Rostamizadeh, A., Talwalkar, A.: Foundations of Machine Learning. MIT Press, Cambridge (2018)
23. Qian, C., Yu, Y., Zhou, Z.H.: An analysis on recombination in multi-objective evolutionary optimization. Artif. Intell. **204**, 99–119 (2013)
24. Qian, C., Yu, Y., Zhou, Z.H.: Pareto ensemble pruning. In: Proceedings of the 29th AAAI Conference on Artificial Intelligence (AAAI 2015), Austin, TX, pp. 2935–2941 (2015)
25. Reyzin, L., Schapire, R.E.: How boosting the margin can also boost classifier complexity. In: Proceedings of the 23rd International Conference on Machine Learning (ICML 2006), Pittsburgh, PA , pp. 753–760 (2006)
26. Schapire, R.E.: The strength of weak learnability. Mach. Learn. **5**(2), 197–227 (1990)
27. Wilcoxon, F.: Individual comparisons by ranking methods. Biometrics Bull. **1**(6), 80–83 (1945)
28. Zhang, Q., Li, H.: MOEA/D: a multiobjective evolutionary algorithm based on decomposition. IEEE Trans. Evol. Comput. **11**(6), 712–731 (2007)
29. Zhang, T., Zhou, Z.H.: Optimal margin distribution clustering. In: Proceedings of the 32nd AAAI Conference on Artificial Intelligence (AAAI 2018), New Orleans, LA, pp. 4474–4481 (2018)
30. Zhang, T., Zhou, Z.H.: Optimal margin distribution machine. IEEE Trans. Knowl. Data Eng. **32**(6), 1143–1156 (2019)
31. Zhou, Z.H.: Ensemble Methods: Foundations and Algorithms. Chapman & Hall/CRC Press, Boca Raton, FL (2012)
32. Zhou, Z.H.: Large margin distribution learning. In: Proceedings of the 6th International Workshop on Artificial Neural Networks in Pattern Recognition (ANNPR 2014), Montreal, Canada, pp. 1–11 (2014)
33. Zhou, Z.H., Tang, W.: Selective ensemble of decision trees. In: Proceddings of the 9th International Conference on Rough Sets, Fuzzy Sets, Data Mining, and Granular Computing (RSFDGrC 2003), Chongqing, China, pp. 476–483 (2003)
34. Zhou, Z.H., Wu, J., Tang, W.: Ensembling neural networks: many could be better than all. Artif. Intell. **137**(1–2), 239–263 (2002)
35. Zhou, Z.H., Yu, Y., Qian, C.: Evolutionary Learning: Advances in Theories and Algorithms. Springer, Singapore (2019)

Revisiting Attention-Based Graph Neural Networks for Graph Classification

Ye Tao[1], Ying Li[2(✉)], and Zhonghai Wu[2]

[1] School of Software and Microelectronics, Peking University, Beijing, China
`tao.ye@pku.edu.cn`
[2] National Engineering Center of Software Engineering, Peking University, Beijing, China
{`li.ying,wuzh`}`@pku.edu.cn`

Abstract. The attention mechanism is widely used in GNNs to improve performances. However, we argue that it breaks the prerequisite for a GNN model to obtain the maximum expressive power of distinguishing different graph structures. This paper performs theoretical analyses of attention-based GNN models' expressive power on graphs with both node and edge features. We propose an enhanced graph attention network (EGAT) framework based on the analysis to deal with this problem. We add a degree-related scale term to the attention coefficients and adjust the message extraction function to enhance the expressive power, which is critical in the graph classification task. Furthermore, we introduce a virtual node connected with all nodes to augment the node representation update process with global information. To prove the effectiveness of our EGAT framework, we first construct synthetic datasets to validate our theoretical proposal, then we apply EGAT to two Open Graph Benchmark (OGB) graph classification tasks to empirically demonstrate that our model also performs well in real applications.

Keywords: Graph neural network · Attention mechanism · Expressive power

1 Introduction

Graph-structured data is ubiquitous across application domains ranging from science to engineering [1,4,11,23,27,29,40]. One of the major tasks in graph analysis is graph classification. Traditional approaches often use hand-crafted graph features to deal with this task, which is not necessarily suitable for all datasets. Recently, a significant research effort is to develop Graph Neural Networks (GNNs) to learn graph representations from raw features, and use these learned representations for the download tasks [5,10,12,19,32].

GNN models follow a neighborhood aggregation scheme. Generally speaking, GNNs are formed using a composition of i) an *aggregate* function, which collects messages from the target node's neighbors to capture local structure information, ii) a *combine* function, which fuses the neighborhood information

G. Rudolph et al. (Eds.): PPSN 2022, LNCS 13398, pp. 442–458, 2022.
https://doi.org/10.1007/978-3-031-14714-2_31

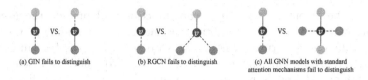

(a) GIN fails to distinguish   (b) RGCN fails to distinguish   (c) All GNN models with standard attention mechanisms fail to distinguish

Fig. 1. Examples of local structures that some classical GNNs cannot distinguish.

with the target node embedding, and iii) a *readout* function, which gets the representation of the entire graph based on the representations of all nodes. Most GNN models follow the above framework and only differ in how these operations are defined and composed. The attention mechanism has been widely used in GNN models. By assigning different weights to nodes within the neighborhood, attention-based GNN models can focus on the most relevant parts of the input to make decisions and have achieved success in real applications. Although extensive studies conducted on attention-based GNN models have achieved unprecedented results, there is little understanding of why these models are successful in practice. Recently, studies [28,38] have proved that any GNN model that follows the above framework is at most as powerful as the Weisfeiler-Lehman (WL) test in distinguishing different graph structures. However, their theoretical results are limited to graphs that only consider node features, and they did not discuss GNN models with attention mechanisms. Because of the above restrictions, the existing theoretical analyses cannot well meet the needs of practical applications.

In order to adapt to a broader range of application scenarios, we propose to extend the existing theoretical analyses. We first generalize the WL test algorithm to graphs with both node features and edge features. When considering edge features, we observed that many classical GNN models could not distinguish some different structures that the generalized WL test decides as non-isomorphic, even for models that have the maximum expressive power on graphs with only node features. For example, GIN [38] and RGCN [33] cannot distinguish the different local structures around v in Fig. 1(a) and Fig. 1(b) respectively. Further, we argue that the widely used attention mechanism breaks the prerequisite for a GNN model to obtain the maximum expressive power. This means that the existing attention-based GNN model has some inherent limitations that restrict the upper bound of its theoretical representation ability. For example, all GNN models with standard attention mechanisms cannot distinguish Fig. 1(c). Besides, GNN models lack the ability to propagate messages across remote parts. In most GNN models, the *combine* function only aggregates information collected from one-hop neighbors so that a k-layer GNN can only obtain information within k-hop neighbors. In practice, we usually use shallow structures because piling up too many GNN layers does not improve the predictive performance [31]. This means that each node cannot obtain information from distant nodes.

To deal with the above problems, we propose an enhanced graph attention network (EGAT) framework. The standard attention mechanism performs *softmax* to normalize the attention coefficients, which makes the *aggregate* function not injective. We add a degree-related scale term to attention coefficients to enhance the attention-based *aggregate* function and provide the necessary and sufficient conditions for it to obtain the maximum expressive power to discriminate different structures. Further, we also revise the *combine* function by introducing a virtual node connected with all nodes to guide the fusion process between the target node's representation and the neighborhood information. Therefore, each node can determine how to employ neighborhood information to obtain a higher-level node representation based on the global state.

We first construct synthetic datasets to validate our theoretical proposal experimentally. Then, in order to illustrate that strong theoretical expressive power will bring benefit in practical applications, we compare our EGAT with many classical competitors on Open Graph Benchmark (OGB) [16]. The experimental results demonstrate that our model achieves state-of-the-art performance on two OGB graph property prediction tasks. Our main contributions are as follows:

- We generalize the theoretical analysis of GNNs' expressive power to graphs with both edge features. And we discussed how the attention mechanism influences the expressive power of GNN models. Besides, we explain why GNNs lack the ability to propagate messages across remote parts.
- We propose a new attention-based GNN framework EGAT to solve the above two problems. We add a degree-related scale term to the attention coefficients to enhance the theoretical expressive power and prove it is as powerful as the generalized WL test. We introduce a virtual node connected with all nodes to enhance the information propagation process with global status.
- We conducted comprehensive experiments on both synthetic datasets and real-world datasets. The experimental results show that our model has good theoretical and practical performance, outperforming state-of-the-art GNN based competitors in graph classification tasks.

2 Preliminaries

2.1 Graph Neural Networks Framework

We denote a graph as a tuple $\mathcal{G} = (\mathcal{V}, \mathcal{E}, X, A)$. X_v is the node feature vector for $v \in \mathcal{V}$ and A_{uv} is the edge feature vector for $e_{uv} \in \mathcal{E}$, where e_{uv} denotes the edge from node u to node v. We use $\mathcal{N}_v$ to denote the neighbor nodes set of $v \in \mathcal{V}$, i.e., $\mathcal{N}_v = \{u \mid \exists e_{uv} \in \mathcal{E}\}$. A node may have multiple neighbors with the same representation, so we regard the set of feature vectors of a given node's neighbors as a multiset, i.e., a generalization of a set that allows repeated elements. We use $\{\{\cdot\}\}$ to denote a multiset, $\vec{h}_v^k$ to denote the embedding of node v at the k-th layer, and $\vec{h}_{uv}$ to denote the embedding of edge e_{uv}. The k-th layer of the classical GNN models can be written as follows

$$\vec{a}_v^k = \text{Aggregate}^k \left(\left\{ \left\{ \phi \left(\vec{h}_u^{k-1}, \vec{h}_{uv}, \vec{h}_v^{k-1} \right) \mid u \in \mathcal{N}_v \right\} \right\} \right) \tag{1}$$

$$\vec{h}_v^k = \text{Combine}^k \left(\vec{h}_v^{k-1}, \vec{a}_v^k \right) \tag{2}$$

For the graph classification task, the representative approach for obtaining the final graph representation is to exploit different levels of graph level embeddings, i.e., using the *read* function to obtain the graph representation at each layer and then aggregate all layers graph representations for prediction.

$$\vec{h}_G^k = \text{Read}^k \left(\left\{ \vec{h}_v^k \mid v \in \mathcal{V} \right\} \right), \quad \vec{h}_G = f \left(\left\{ \vec{h}_G^k \right\} \right), \quad \hat{y}_G = \text{Prediction} \left(\vec{h}_G \right) \quad (3)$$

where $\vec{h}_G^k$ and $\vec{h}_G$ are the k-th layer and final graph representation.

2.2 Weisfeiler-Lehman Test

The Weisfeiler-Lehman (WL) test is an algorithm to distinguish whether two graphs are isomorphic. Given a labeled graph, the WL test proceeds recursively to update node labels. At the k-th iteration, a new label is computed for each node from the current label and its neighbors' labels, i.e.,

$$l_v^k = \text{Hash}^k \left(l_v^{k-1}, \{\{ l_u^{k-1} \mid u \in \mathcal{N}_v \}\} \right) \quad (4)$$

where Hash is an injective function that maps different inputs to a unique label. The WL test algorithm decides that two graphs are non-isomorphic if at some iteration, the labels of the nodes between the two graphs differ. Although there are some corner cases that the WL test can not distinguish [8], this algorithm is a successful isomorphism test for a broad class of graphs [3].

2.3 Theoretical Analysis

Compare Eqs. 1–2 with Eq. 4, we observe that a GNN model can be viewed as replacing the Hash function of the WL test algorithm with a neural network so that it can handle node/edge representations in continuous space. Without considering edge attributes, it has been proved that a GNN model can be as most powerful as the WL test in distinguishing non-isomorphic graphs [28,38]. We generalize the WL test to graphs with edge features,

$$l_v^k = \text{Hash}^k \left(l_v^{k-1}, \{\{ l_{uv}, l_u^{k-1} \mid u \in \mathcal{N}_v \}\} \right) \quad (5)$$

where l_{uv} is the label obtained from the attribute of the edge e_{uv}.

Our first theoretical result shows that a GNN following Eqs. 1–3 is at most as powerful as the generalized WL test in distinguishing graph structures.

Theorem 1. *Let $\mathcal{G}_1$ and $\mathcal{G}_2$ be any two graphs. If a GNN maps $\mathcal{G}_1$ and $\mathcal{G}_2$ to different embeddings, the generalized WL also decides they are not isomorphic.*

Our second theoretical result shows that a GNN can obtain the same expressive power as the generalized WL test when the feature extraction is injective.

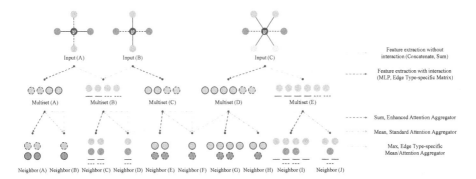

Fig. 2. Expressive power comparison for GNN models with different aggregate functions. The top panel shows the local input structure for the target node v. Then the message extraction function constructs the neighbor nodes' representation multiset, which is illustrated in the middle panel. Finally, the aggregate function gets the uniform neighborhood representation. Different inputs should result in different outputs to obtain the maximum expressive power. If two different elements on the previous layer point to the same element on the next layer, the mapping process is not injective, which means that it restricts the model's expressive power.

Theorem 2. *Let* $\mathcal{A} : \mathcal{G} \to d \in \mathbb{R}^d$ *be a GNN. With sufficient layers,* $\mathcal{A}$ *maps any graphs* $\mathcal{G}_1$ *and* $\mathcal{G}_2$ *that the generalized Weisfeiler-Lehman test decides as non-isomorphic to different embeddings if the following conditions hold:*

a) $\mathcal{A}$ aggregates and updates node features iteratively with

$$\vec{h}_v^k = f_2\left(\vec{h}_v^{k-1}, f_1\left(\left\{\left\{\left(\vec{h}_{uv}, \vec{h}_u^{k-1}\right) \mid u \in \mathcal{N}_v\right\}\right\}\right)\right)$$

where the function f_1 and f_2 are injective.

b) $\mathcal{A}$'s graph-level readout function is injective.

3   Enhanced Graph Attention Networks

3.1   Enhancing the Aggregate Function

The *aggregate* function collects neighborhood messages to capture local structure. Since graph data is generated from non-Euclidean domains, a node's neighbors have no natural ordering. Thus, the messages transmitted to a target node from all neighbors can be regarded as a multiset. It has been proved that a function $f(X)$ operating on a set X that has elements from a countable universe is invariant to the permutation of instances in X *iff* it can be decomposed in the form $\rho\left(\Sigma_{x\in X}\xi(x)\right)$, for suitable transformations ρ and ξ [39]. Thus, we formulate our attention-based *aggregate* function framework as follows,

$$\vec{a}_v^k = \sum_{u\in\mathcal{N}_v} \alpha_{uv}\vec{m}_{uv}^k, \quad \vec{m}_{uv}^k = \phi^k\left(\vec{h}_u^{k-1}, \vec{h}_{uv}, \vec{h}_v^{k-1}\right) \tag{6}$$

$$\alpha_{uv}^{k} = \frac{D_{v}{}^{\beta} \exp\left(c_{uv}^{k}\right)}{\Sigma_{u \in \mathcal{N}_{v}} \exp\left(c_{uv}^{k}\right)}, \quad c_{uv}^{k} = \psi^{k}\left(\vec{h}_{u}^{k-1}, \vec{h}_{uv}, \vec{h}_{v}^{k-1}\right) \tag{7}$$

where α_{uv} is the attention coefficient for u, $\vec{m}_{uv}$ is the message transmitted to v from u, ϕ is the message extraction function, c_{uv} is the importance score for node u, D_{v} is the degree of target node v and β is a hyper-parameter.

If we employ concatenation or addition to map the edge and node feature into a single neighbor message $\vec{m}_{uv}$ (like most existing works), and set the attention coefficient of each neighbor to the same value (such as sum and mean aggregator), we cannot distinguish the structures with the same multiset of neighbor node features and edge features but a different neighborhood. This explains why GIN has the most expressive power for graphs without edge features but cannot distinguish the different graphs in Fig. 1(a) and Fig. 1(c), i.e., we have $f\left((1+\epsilon)\vec{h}_{v} + [(\vec{a}_{1} + \vec{b}_{1}) + (\vec{a}_{2} + \vec{b}_{2})]\right) = f\left((1+\epsilon)\vec{h}_{v} + [(\vec{a}_{1} + \vec{b}_{2}) + (\vec{a}_{2} + \vec{b}_{1})]\right)$, while the multiset $\left\{\left\{(\vec{a}_{1}, \vec{b}_{1}), (\vec{a}_{2}, \vec{b}_{2})\right\}\right\} \neq \left\{\left\{(\vec{a}_{1}, \vec{b}_{2}), (\vec{a}_{2}, \vec{b}_{1})\right\}\right\}$.

Compared with the standard attention mechanism, we simply add a degree-related scale term $D_{v}{}^{\beta}$. This modification strategy is compatible and can be combined with any specific attention coefficient calculation methods under the standard attention mechanism framework. Although simple, the following theorem shows this item is essential to the theoretical expressive power.

Theorem 3. *Suppose node and edge features are from a countable universe. Iff $\beta \neq 0$, for multiset $\left\{\left\{\left(\vec{h}_{uv}, \vec{h}_{u}\right)\right\}\right\}$ of bounded size, there exist a function ϕ and a function ψ so that $\sum_{u \in \mathcal{N}_{v}} \frac{D_{v}{}^{\beta} \exp\left(\psi^{k}\left(\vec{h}_{u}, \vec{h}_{uv}, \vec{h}_{v}\right)\right)}{\Sigma_{u \in \mathcal{N}_{v}} \exp\left(\psi\left(\vec{h}_{u}, \vec{h}_{uv}, \vec{h}_{v}\right)\right)} \phi\left(\vec{h}_{u}, \vec{h}_{uv}, \vec{h}_{v}\right)$ is injective .*

Suppose $\beta = 0$, the new framework will degenerate into a traditional attention-based GNN model. If so, according to Theorems 1–3, it cannot distinguish some different graph structures that the generalized WL test decides as non-isomorphic. Take Fig. 1(c) as an example, the GNN model cannot distinguish these two structures if we set $\beta = 0$. No matter how we implement ψ, after normalization with *softmax*, the ratio of the attention coefficients in the left panel to these values in the right panel is always 2:1. After aggregation by weighted sum, the same representation is obtained for the two different neighborhood multisets. Without the degree-related scale term, attention-based GNNs only capture the (proportions) of elements in a multiset but not the exact multiset, as shown in Fig. 1. Using node degrees as extra node input features is another way to solve the above problem. In this way, GNNs need to learn how the node degree affects node representation through neural networks. By contrast, using $D_{v}{}^{\beta}$ means that we have introduced the prior knowledge into the aggregation process: the node representation is scaled in each dimension according to its degree. We compare these two methods in experiments to prove the effectiveness of our design.

3.2   Enhancing the Combine Function

In the traditional *combine* function, the node representation update process only depends on local information, which restricts the expressive power. GNNs should determine the node representation update strategy based on the global state or a specific remote substructure to alleviate this problem. Therefore, we also integrate the global status in addition to the neighborhood information.

To learn the representation of global status at different levels, we introduce a virtual node that is connected with all nodes in a graph. At first, the virtual node representation is initialized according to the initial node and edge embeddings. We denotes the virtual node embedding at k-th layer as $\vec{h}_g^k$. Formally,

$$\vec{h}_v^k = \text{Combine}\left(\vec{h}_v^{k-1}, \vec{a}_v^k, \vec{h}_g^{k-1}\right) \tag{8}$$

To obtain higher-level global status, we aggregate node representations to obtain a new global state $\vec{g}^k$, and use it to update the virtual node embedding.

$$\vec{g}^k = \text{Readout}\left(\vec{h}_g^{k-1}, \left\{\left\{\vec{h}_v^k \mid v \in \mathcal{V}\right\}\right\}\right), \quad \vec{h}_g^k = \text{Update}\left(\vec{h}_g^{k-1}, \vec{g}^k\right) \tag{9}$$

Although the Read function in Eq. 3 and the Readout function in Eq. 9 both obtain the global status, they have different goals. The Readout function focuses on determining the update strategy to capture higher-level node representations more effectively, while the Read function focuses on capturing features that can be directly used for downstream graph-level tasks.

3.3   Implementation

We first discuss the *aggregate* function. The hyper-parameter β depends on the prior knowledge that we want to introduce. We set $\beta > 0$, so that D_v^{β} is monotonically increasing, which means the node representation becomes more prominent as the degree increase. This is usually more in line with the actual application scenario, where the core nodes have a greater impact on the overall property of the whole graph. For the message extraction function ϕ, we concatenate the input $\vec{h}_u^{k-1}$, $\vec{h}_{uv}$ and $\vec{h}_v^{k-1}$, and process it with a MLP. For the attention function ψ, we use a single-layer FFN parameterized by a weight vector $\vec{a}$. It is obvious that concatenating operation is injective. Thanks to the universal approximation theorem [14,15], we can use an MLP and $\vec{a}$ to model ϕ and ψ in Theorem 3. Thus our *aggregate* function is injective, which corresponds to the f_1 in Theorem 2.

$$\vec{m}_{uv}^k = \text{MLP}_1^k\left([\vec{h}_u^{k-1}\|\vec{h}_{uv}\|\vec{h}_v^{k-1}]\right), \quad c_{uv}^k = \vec{a}_k^T \vec{m}_{uv}^k \tag{10}$$

Then we discuss the *combine* function. For simplicity, we set the initial virtual node representation to $\vec{0}$. When updating the node representations, we concatenate the input $\vec{h}_v^{k-1}$, $\vec{a}_v^k$, $\vec{h}_g^{k-1}$, and process it with another MLP.

$$\vec{h}_v^k = \text{MLP}_2\left([\vec{h}_v^{k-1}\|\vec{a}_v^k\|\vec{h}_g^{k-1}]\right) \tag{11}$$

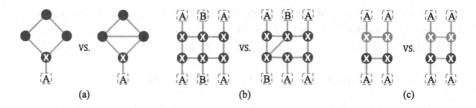

Fig. 3. Examples of different graphs that some classical GNNs cannot distinguish. The structures illustrated in detail determine the label of a graph, which is called *decisive structures*, The dotted box with the letter A/B represents the rest of the graph that obeys a certain distribution A/B. Nodes labeled as X represent the connection nodes that bridge these two parts. In (b), the ratio of the average number of nodes in B to the average number of nodes in A is 3:2.

For the same reason as Eq. 10, we can use MLP_2 to model any function over the tuple $(\vec{h}_u^{k-1}, \vec{h}_{uv}, \vec{h}_v^{k-1})$, which corresponds to the f_2 in Theorem 2.

We use attention sum as the Readout function to get the global state. A linear function is used to calculate the attention coefficients for each node, and we use residual connection with another MLP to update the virtual node embedding.

$$\alpha_v^k = \text{softmax}\left(W^k \vec{h}_v^k + \vec{b}^k\right), \quad \vec{g}^k = \sum_{v \in \mathcal{V}} \alpha_v^k \vec{h}_v^k \tag{12}$$

$$\vec{h}_g^k = \text{MLP}_3\left(\vec{h}_g^{k-1} + \vec{g}^k\right) \tag{13}$$

According to Theorem 2, since our *aggregate* and *combine* functions are both injective, our implementation of EGAT is as powerful as the generalized WL-test if the Read function in Eq. 3 is injective. To consider structural information at different levels, we concatenate the output of Read function across all layers to get the final graph representations for graph-level tasks.

4 Experimental Setup

4.1 Synthetic Datasets

To the best of our knowledge, existing studies only provide examples that some classical GNNs cannot distinguish at the node-level. Although we will get the same representation for v, GNN models can still distinguish these different graph structures with an effective graph readout function. To verify our analyses, we provide three representative structures that cannot be distinguished by some classical GNN models in Fig. 3. In some cases, we assume a graph's property can be determined by a specific substructure (called *decisive structure*).

In Fig. 3(a), the *decisive structure* is a relatively independent part that is connected to the rest of the graph through a small number of nodes and edges. In Fig. 3(b), the *decisive structure* constitutes the skeleton of the whole graph, and the rest is generated around the skeleton. In Fig. 3(c), each node in the *decisive structure* has a different neighborhood multiset $\left\{\left\{\left(\vec{h}_{uv}, \vec{h}_u\right)\right\}\right\}$, but they have the same multiset of neighbor node features and edge features. We constructed three synthetic datasets based on the above three situations illustrated in Fig. 3(a)(b)(c) and set the graph label accordingly.

Table 1. Statistics of OGB graph-level datasets.

Dataset	ogbg-molpcba	ogbg-molpcba
#Graphs	437,929	452,741
Average #Nodes	26.0	125.2
Average #Edges	28.1	124.2
Split Scheme	Scaffold	Project
Split Ratio	80/10/10	90/5/5
Task Type	Binary class	Sub-token prediction
Metric	Average Precision	F1 score

Baselines. To validate our theoretical proposal, we compare our model with classical GNN models GCN [21], GAT [34], GraphSAGE [13] and GIN [38].

Table 2. Experimental results on synthetic datasets. We use -x to denote EGAT with $\beta = x$, + d to denote using node degrees as extra features, ✓ to denote a model can make perfect predictions and ✗ to denote it fails to give any effective predictions.

Method	structure-a	structure-b	structure-c
GCN/GraphSAGE	✗	✗	✗
GCN/GraphSAGE + d	✓	✓	✗
GAT	✗	✗	✓
GAT + d	✓	✓	✓
GIN	✓	✓	✗
EGAT-0	✗	✗	✓
EGAT-0 + d	✓	✓	✓
EGAT-1	✓	✓	✓

4.2   Real World Datasets

To prove that our EGAT also has better prediction performance in practice, we conducted experiments on two datasets in Open Graph Benchmark (OGB) [16]. The statistics of datasets are summarized in Table 1.

ogbg-code2 is a collection of ASTs obtained from approximately 450 thousand Python method definitions. *ogbg-code2* originates from CodeSearchNet [18]. The task is to predict the method name, given the method body represented by AST and its node features. *Project split* [2] is used for dataset splitting, where the methods in the training set are obtained from projects that do not appear in the

Table 3. Experimental results for *ogbg-molpcba*.

Method	Test Average Precision	#Params
GINE+ w/virtual nodes	0.2917 ± 0.0015	6,147,029
DGN	0.2885 ± 0.0030	6,732,696
PNA	0.2838 ± 0.0035	6,550,839
DeeperGCN+virtual node	0.2781 ± 0.0038	5,550,208
GIN+virtual node	0.2703 ± 0.0023	3,374,533
GCN+virtual node	0.2424 ± 0.0034	2,017,028
GIN	0.2266 ± 0.0028	1,923,433
GAT*	0.2032 ± 0.0027	1,213,068
GCN	0.2020 ± 0.0024	565,928
GraphSAGE*	0.1987 ± 0.0030	1,657,068
EGAT	**0.2966 ± 0.0028**	6,766,844

test set. **ogbg-molpcba** is a molecular dataset adopted from MOLECULENET [37]. Each graph represents a molecule, where nodes are atoms and edges are chemical bonds. All molecules are preprocessed using RDKIT [22]. The task is to predict the target molecular properties. OGB uses a scaffold splitting procedure to separate structurally different molecules.

Baselines. OGB provides public leaderboards to keep track of recent advances, so we compare EGAT with these recent public results. For *ogbg-code2*, the methods include GCN, GCN + virtual node, GIN and GIN + virtual node. For *ogbg-pcba*, in addition to the above methods, the methods also include DeeperGCN + virtual node [25], PNA [9], DGN [6] and GINE+ w/virtual nodes [7]. We also implemented GAT and GraphSAGE for comparison on OGB datasets.

4.3   Implementation Details

We implement all GNN methods with Deep Graph Library (DGL) [35] and run experiments on an NVIDIA RTX TITAN. We use Adam [20] optimizer with learning rate 0.001. We set the embedding size to 300 (the same as the OGB example models). We use the same edge encoder as EGAT to get edge embedding and sum it up with node embedding during message processing for methods that do not consider edge features. For EGAT, GAT and GraphSAGE, we use the validing data to select suitable hyper-parameters (the number of GNN layers $\in \{4, 5, 6\}$, the dropout value $\in \{0.4, 0.5, 0.6\}$, the value of $\beta \in \{0.1, 0.5, 1, 2, 10\}$). For other baseline models, we quote the experimental results on the leaderboards provided by OGB. The data and source code are publicly available at here.

Table 4. Experimental results for *ogbg-code2*.

Method	Test F1	#Params
GCN+virtual node	0.1595 ± 0.0018	12,484,310
GIN+virtual node	0.1581 ± 0.0026	13,841,815
GraphSAGE*	0.1578 ± 0.0019	11,487,710
GAT*	0.1564 ± 0.0020	11,487,710
GCN	0.1507 ± 0.0018	11,033,210
GIN	0.1495 ± 0.0023	12,390,715
EGAT	**0.1783 ± 0.0014**	14,825,335

Table 5. Ablation study for EGAT on *ogbg-molpcba* and *ogbg-code2*.

Method	ogbg-pcba Test AP	ogbg-code2 Test F1
EGAT	0.2966 ± 0.0028	0.1783 ± 0.0014
w/o $D_v{}^\beta$	0.2814 ± 0.0027	0.1754 ± 0.0017
w/o $D_v{}^\beta + d$	0.2778 ± 0.0031	0.1743 ± 0.0015
old v-node	0.2749 ± 0.0032	0.1698 ± 0.0014
w/o v-node	0.2324 ± 0.0022	0.1632 ± 0.0018
w/o $D_v{}^\beta$, v-node	0.2198 ± 0.0028	0.1596 ± 0.0012

5   Experimental Results

5.1   Synthetic Datasets Performance

We first validate our theoretical analyses of GNN methods' representational power by comparing models' performances on synthetic datasets. Since each graph's label in the dataset is determined by a certain *decisive structure*, the model can either make perfect predictions or fail to give any effective prediction.

We show the results in Table 2, which are in line with our expectations. GCN, GraphSAGE, GAT, and EGAT-0 can only capture the distribution of a multiset so they fail to give any effective predictions with (a) and (b). These situations are illustrated in Fig. 2 as the *Mean, Standard Attention Aggregator* and the *Max, Edge Type-specific Mean/Attention Aggregator*. By adding the degree-related scale term or using node degree as extra features, models can represent the exact multiset of neighbor node features and edge features. Assuming that a GNN model only concatenates or sums the edge and node features to obtain a single neighbor message and does not use the attention mechanism, it still cannot process (c). These situations are illustrated in Fig. 2 as the *Feature extraction without interaction(Concatenate, Sum)*. To summarize, without using node degrees as extra node features, EGAT-1 is the only one that can deal with all three situations among these GNN models.

5.2 Real World Datasets Performance

To prove the proposed EGAT also performs well in real applications, we apply EGAT to two OGB graph-level datasets. The experimental results in Tables 3 and 4 show that EGAT significantly outperform the state-of-the-art GNN methods on both datasets. It is worth noting that in addition to modifying the *aggregate*, *combine* and *readout* functions, some recent related work did some extra work to improve model performance. For example, DGN [6] performs pre-computed steps to compute eigenvectors and GINE+ [7] includes higher-order neighbors in a single aggregation step. Since our EGAT is modified on the general framework of attention-based GNNs, as future work, we would like to combine their methodology into our EGAT framework to further improve the performance.

5.3 Comparison Among EGAT Variants

Further, we compare the variants of EGAT concerning the following two aspects to demonstrate the effectiveness of our design: (1) whether adding the degree-related term to the attention coefficients, we use w/o $D_v{}^\beta$ to denote removing the degree-related term $D_v{}^\beta$. Using node degrees as extra node features is another way to restore the theoretical representative capacity, which is denoted as + d. (2) whether using the virtual node to enhance the local node representation update process with global status, we use w/o v-node to denote removing the virtual node. We also tried to use virtual nodes in the same way as GCN/GIN+virtual node in [16], which is denoted as old v-node.

The results in Table 5 shows that: (a) For *ogbg-molpcba*, no matter whether introducing the virtual node, adding the degree-related scale term always improves the performance. We also try to employ D_v^β in GAT, and the Test AP rises from 0.2032 to 0.2178. It proves that our extension of the classical attention mechanism not only guarantees the theoretical expressive power but also improves the predictive performance. On the contrary, using node degrees as extra node features worsens the performance, which illustrates that it is difficult for the model to use this information effectively. For *ogbg-code2*, there are large differences between node features. Therefore, the distribution of a multiset is sufficient to determine its property. The performance improvement brought by adding D_v^β is relatively small, showing that our technology is more suitable for graphs with many nodes with similar features. (b) On both datasets, the introduction of the virtual node has significantly improved the performance. It proves that enhancing the local node representation update process with global status is necessary. We compared two strategies for using virtual nodes. Our strategy is better than the existing one, proving that sufficient interaction between local information and global information is essential to improve performance.

6   Related Work

GNN extends the deep learning technique to deal with graph data. Graph attention networks (GAT) [34] firstly introduced the attention mechanism into GNNs.

Later, many attention-based GNNs have been proposed to perform improvements and extensions on GAT [17,24,36]. The attention mechanism allows GNNs to focus on the most informative parts and avoid noise, enhancing the robustness and generalization ability. However, existing work does not discuss the impact of the attention mechanism on model representation capabilities.

Two recent papers [28,38] have started exploring the expressive power of GNN models by establishing a close connection between GNN models and the WL test in distinguishing different graph structures. However, many classical GNNs do not satisfy the requirement, such as GCN, GraphSAGE, and most attention-based GNN models. Moreover, although the attention mechanism has been widely used in GNN models, studies have not yet explored its impact on GNN models' expressive ability. Therefore, we think it is necessary to generalize the theoretical analyses to cover more general application scenarios.

Besides, studies have illustrated that GNN models lack the ability to capture long-range dependencies. Introducing a virtual node connected to all nodes to aggregate and propagate messages is a common solution to this problem. However, most of the existing works divide combining neighbor information and applying virtual node embedding into two steps with simple network structures [26,30], which makes the interaction between the virtual node and other nodes insufficient. Therefore we need to design a new structure to get, update and employ the virtual node representation to improve the performance further.

7   Conclusion

In this paper, we theoretically analyze the expressive power of attention-based GNN models. Based on the analyses, we propose a new attention-based aggregation framework EGAT by adding a degree-related scale term and introducing a virtual node to enhance the local node representation update process with global information. The experimental results show that our model outperforms the state-of-the-art GNN methods on two OGB real-world datasets.

Acknowledgement. The work is partly supported by Delta Research Program.

References

1. Aggarwal, C.C., Bar-Noy, A., Shamoun, S.: On sensor selection in linked information networks. Comput. Networks **126**, 100–113 (2017). https://doi.org/10.1016/j.comnet.2017.05.024
2. Allamanis, M.: The adverse effects of code duplication in machine learning models of code. In: Masuhara, H., Petricek, T. (eds.) Proceedings of the 2019 ACM SIGPLAN International Symposium on New Ideas, New Paradigms, and Reflections on Programming and Software, Onward! 2019, Athens, Greece, 23–24 October 2019, pp. 143–153. ACM (2019). https://doi.org/10.1145/3359591.3359735
3. Babai, L., Kucera, L.: Canonical labelling of graphs in linear average time. In: 20th Annual Symposium on Foundations of Computer Science, San Juan, Puerto Rico, 29–31 October 1979, pp. 39–46. IEEE Computer Society (1979). https://doi.org/10.1109/SFCS.1979.8

4. Backstrom, L., Leskovec, J.: Supervised random walks: predicting and recommending links in social networks. In: King, I., Nejdl, W., Li, H. (eds.) Proceedings of the Forth International Conference on Web Search and Web Data Mining, WSDM 2011, Hong Kong, China, 9–12 February, 2011, pp. 635–644. ACM (2011). https://doi.org/10.1145/1935826.1935914

5. Battaglia, P.W., Pascanu, R., Lai, M., Rezende, D.J., Kavukcuoglu, K.: Interaction networks for learning about objects, relations and physics. In: Lee, D.D., Sugiyama, M., von Luxburg, U., Guyon, I., Garnett, R. (eds.) Advances in Neural Information Processing Systems 29: Annual Conference on Neural Information Processing Systems 2016, 5–10 December 2016, Barcelona, Spain, pp. 4502–4510 (2016). https://proceedings.neurips.cc/paper/2016/hash/3147da8ab4a0437c15ef51a5cc7f2dc4-Abstract.html

6. Beaini, D., Passaro, S., Létourneau, V., Hamilton, W.L., Corso, G., Liò, P.: Directional graph networks. CoRR abs/2010.02863 (2020). https://arxiv.org/abs/2010.02863

7. Brossard, R., Frigo, O., Dehaene, D.: Graph convolutions that can finally model local structure. CoRR abs/2011.15069 (2020). https://arxiv.org/abs/2011.15069

8. Cai, J., Fürer, M., Immerman, N.: An optimal lower bound on the number of variables for graph identifications. Comb. 12(4), 389–410 (1992). https://doi.org/10.1007/BF01305232

9. Corso, G., Cavalleri, L., Beaini, D., Liò, P., Velickovic, P.: Principal neighbourhood aggregation for graph nets. In: Larochelle, H., Ranzato, M., Hadsell, R., Balcan, M., Lin, H. (eds.) Advances in Neural Information Processing Systems 33: Annual Conference on Neural Information Processing Systems 2020, NeurIPS 2020, 6–12 December 2020, virtual (2020). https://proceedings.neurips.cc/paper/2020/hash/99cad265a1768cc2dd013f0e740300ae-Abstract.html

10. Defferrard, M., Bresson, X., Vandergheynst, P.: Convolutional neural networks on graphs with fast localized spectral filtering. In: Lee, D.D., Sugiyama, M., von Luxburg, U., Guyon, I., Garnett, R. (eds.) Advances in Neural Information Processing Systems 29: Annual Conference on Neural Information Processing Systems 2016, 5–10 December, 2016, Barcelona, Spain, pp. 3837–3845 (2016). https://proceedings.neurips.cc/paper/2016/hash/04df4d434d481c5bb723be1b6df1ee65-Abstract.html

11. Deng, S., Huang, L., Xu, G., Wu, X., Wu, Z.: On deep learning for trust-aware recommendations in social networks. IEEE Trans. Neural Networks Learn. Syst. 28(5), 1164–1177 (2017). https://doi.org/10.1109/TNNLS.2016.2514368

12. Duvenaud, D., et al.: Convolutional networks on graphs for learning molecular fingerprints. In: Cortes, C., Lawrence, N.D., Lee, D.D., Sugiyama, M., Garnett, R. (eds.) Advances in Neural Information Processing Systems 28: Annual Conference on Neural Information Processing Systems 2015, 7–12 December 2015, Montreal, Quebec, Canada, pp. 2224–2232 (2015). https://proceedings.neurips.cc/paper/2015/hash/f9be311e65d81a9ad8150a60844bb94c-Abstract.html

13. Hamilton, W.L., Ying, Z., Leskovec, J.: Inductive representation learning on large graphs. In: Guyon, I., von Luxburg, U., Bengio, S., Wallach, H.M., Fergus, R., Vishwanathan, S.V.N., Garnett, R. (eds.) Advances in Neural Information Processing Systems 30: Annual Conference on Neural Information Processing Systems 2017, 4–9 December 2017, Long Beach, CA, USA, pp. 1024–1034 (2017). https://proceedings.neurips.cc/paper/2017/hash/5dd9db5e033da9c6fb5ba83c7a7ebea9-Abstract.html

14. Hornik, K.: Approximation capabilities of multilayer feedforward networks. Neural Networks 4(2), 251–257 (1991). https://doi.org/10.1016/0893-6080(91)90009-T

15. Hornik, K., Stinchcombe, M.B., White, H.: Multilayer feedforward networks are universal approximators. Neural Networks **2**(5), 359–366 (1989). https://doi.org/10.1016/0893-6080(89)90020-8

16. Hu, W., et al.: Open graph benchmark: Datasets for machine learning on graphs. In: Larochelle, H., Ranzato, M., Hadsell, R., Balcan, M., Lin, H. (eds.) Advances in Neural Information Processing Systems 33: Annual Conference on Neural Information Processing Systems 2020, NeurIPS 2020, 6–12 December, 2020, virtual (2020). https://proceedings.neurips.cc/paper/2020/hash/fb60d411a5c5b72b2e7d3527cfc84fd0-Abstract.html

17. Hu, Z., Dong, Y., Wang, K., Sun, Y.: Heterogeneous graph transformer. In: Huang, Y., King, I., Liu, T., van Steen, M. (eds.) WWW '20: The Web Conference 2020, Taipei, Taiwan, 20–24 April, 2020, pp. 2704–2710. ACM / IW3C2 (2020). https://doi.org/10.1145/3366423.3380027

18. Husain, H., Wu, H., Gazit, T., Allamanis, M., Brockschmidt, M.: Codesearchnet challenge: evaluating the state of semantic code search. CoRR abs/1909.09436 (2019). http://arxiv.org/abs/1909.09436

19. Kearnes, S.M., McCloskey, K., Berndl, M., Pande, V.S., Riley, P.: Molecular graph convolutions: moving beyond fingerprints. J. Comput. Aided Mol. Des. **30**(8), 595–608 (2016). https://doi.org/10.1007/s10822-016-9938-8

20. Kingma, D.P., Ba, J.: Adam: A method for stochastic optimization. In: Bengio, Y., LeCun, Y. (eds.) 3rd International Conference on Learning Representations, ICLR 2015, San Diego, CA, USA, 7–9 May, 2015, Conference Track Proceedings (2015). http://arxiv.org/abs/1412.6980

21. Kipf, T.N., Welling, M.: Semi-supervised classification with graph convolutional networks. In: 5th International Conference on Learning Representations, ICLR 2017, Toulon, France, April 24–26, 2017, Conference Track Proceedings. OpenReview.net (2017). https://openreview.net/forum?id=SJU4ayYgl

22. Landrum: Rdkit: Open-source cheminformatics (2006)

23. Lee, J.B., Kong, X., Bao, Y., Moore, C.M.: Identifying deep contrasting networks from time series data: application to brain network analysis. In: Chawla, N.V., Wang, W. (eds.) Proceedings of the 2017 SIAM International Conference on Data Mining, Houston, Texas, USA, 27–29 April, 2017, pp. 543–551. SIAM (2017). https://doi.org/10.1137/1.9781611974973.61

24. Lee, J.B., Rossi, R.A., Kong, X.: Graph classification using structural attention. In: Guo, Y., Farooq, F. (eds.) Proceedings of the 24th ACM SIGKDD International Conference on Knowledge Discovery & Data Mining, KDD 2018, London, UK, 19–23 August 2018, pp. 1666–1674. ACM (2018). https://doi.org/10.1145/3219819.3219980

25. Li, G., Xiong, C., Thabet, A.K., Ghanem, B.: Deepergcn: All you need to train deeper gcns. CoRR abs/2006.07739 (2020). https://arxiv.org/abs/2006.07739

26. Li, J., Cai, D., He, X.: Learning graph-level representation for drug discovery. CoRR abs/1709.03741 (2017). http://arxiv.org/abs/1709.03741

27. Liu, Q., Xiang, B., Yuan, N.J., Chen, E., Xiong, H., Zheng, Y., Yang, Y.: An influence propagation view of pagerank. ACM Trans. Knowl. Discov. Data **11**(3), 30:1–30:30 (2017). https://doi.org/10.1145/3046941

28. Morris, C., Ritzert, M., Fey, M., Hamilton, W.L., Lenssen, J.E., Rattan, G., Grohe, M.: Weisfeiler and leman go neural: Higher-order graph neural networks. In: The Thirty-Third AAAI Conference on Artificial Intelligence, AAAI 2019, The Thirty-First Innovative Applications of Artificial Intelligence Conference, IAAI 2019, The Ninth AAAI Symposium on Educational Advances in Artificial Intelligence, EAAI 2019, Honolulu, Hawaii, USA, January 27 - February 1, 2019, pp. 4602–4609. AAAI Press (2019). https://doi.org/10.1609/aaai.v33i01.33014602

29. Pei, J., Jiang, D., Zhang, A.: On mining cross-graph quasi-cliques. In: Grossman, R., Bayardo, R.J., Bennett, K.P. (eds.) Proceedings of the Eleventh ACM SIGKDD International Conference on Knowledge Discovery and Data Mining, Chicago, Illinois, USA, 21–24 August 2005, pp. 228–238. ACM (2005). https://doi.org/10.1145/1081870.1081898

30. Pham, T., Tran, T., Dam, K.H., Venkatesh, S.: Graph classification via deep learning with virtual nodes. CoRR abs/1708.04357 (2017). http://arxiv.org/abs/1708.04357

31. Rong, Y., Huang, W., Xu, T., Huang, J.: Dropedge: towards deep graph convolutional networks on node classification. In: 8th International Conference on Learning Representations, ICLR 2020, Addis Ababa, Ethiopia, 26–30, April 2020. OpenReview.net (2020). https://openreview.net/forum?id=Hkx1qkrKPr

32. Scarselli, F., Gori, M., Tsoi, A.C., Hagenbuchner, M., Monfardini, G.: Computational capabilities of graph neural networks. IEEE Trans. Neural Networks **20**(1), 81–102 (2009). https://doi.org/10.1109/TNN.2008.2005141

33. Schlichtkrull, M., Kipf, T.N., Bloem, P., van den Berg, R., Titov, I., Welling, M.: Modeling relational data with graph convolutional networks. In: Gangemi, A., Navigli, R., Vidal, M.-E., Hitzler, P., Troncy, R., Hollink, L., Tordai, A., Alam, M. (eds.) ESWC 2018. LNCS, vol. 10843, pp. 593–607. Springer, Cham (2018). https://doi.org/10.1007/978-3-319-93417-4_38

34. Velickovic, P., Cucurull, G., Casanova, A., Romero, A., Liò, P., Bengio, Y.: Graph attention networks. In: 6th International Conference on Learning Representations, ICLR 2018, Vancouver, BC, Canada, 30 April–3 May 2018, Conference Track Proceedings. OpenReview.net (2018). https://openreview.net/forum?id=rJXMpikCZ

35. Wang, M., et al.: Deep graph library: towards efficient and scalable deep learning on graphs. CoRR abs/1909.01315 (2019). http://arxiv.org/abs/1909.01315

36. Wang, X., et al.: Heterogeneous graph attention network. In: Liu, L., White, R.W., Mantrach, A., Silvestri, F., McAuley, J.J., Baeza-Yates, R., Zia, L. (eds.) The World Wide Web Conference, WWW 2019, San Francisco, CA, USA, 13–17 May 2019, pp. 2022–2032. ACM (2019). https://doi.org/10.1145/3308558.3313562

37. Wu, Z., et al.: Moleculenet: a benchmark for molecular machine learning. CoRR abs/1703.00564 (2017). http://arxiv.org/abs/1703.00564

38. Xu, K., Hu, W., Leskovec, J., Jegelka, S.: How powerful are graph neural networks? In: 7th International Conference on Learning Representations, ICLR 2019, New Orleans, LA, USA, 6–9 May 2019. OpenReview.net (2019). https://openreview.net/forum?id=ryGs6iA5Km

39. Zaheer, M., Kottur, S., Ravanbakhsh, S., Póczos, B., Salakhutdinov, R., Smola, A.J.: Deep sets. In: Guyon, I., et al. (eds.) Advances in Neural Information Processing Systems 30: Annual Conference on Neural Information Processing Systems 2017, 4–9 December 2017, Long Beach, CA, USA, pp. 3391–3401 (2017). https://proceedings. neurips.cc/paper/2017/hash/f22e4747da1aa27e363d86d40ff442fe-Abstract.html
40. Zheng, Y., Capra, L., Wolfson, O., Yang, H.: Introduction to the special section on urban computing. ACM Trans. Intell. Syst. Technol. **5**(3), 37:1–37:2 (2014). https://doi.org/10.1145/2642650

Robust Neural Network Pruning
by Cooperative Coevolution

Jia-Liang Wu[1], Haopu Shang[1], Wenjing Hong[2], and Chao Qian[1(✉)]

[1] State Key Laboratory for Novel Software Technology, Nanjing University,
Nanjing 210023, China
{wujl,shanghp,qianc}@lamda.nju.edu.cn

[2] Department of Computer Science and Engineering, Southern University of Science
and Technology, Shenzhen 518055, China
hongwj@sustech.edu.cn

Abstract. Convolutional neural networks have achieved success in various tasks, but often lack compactness and robustness, which are, however, required under resource-constrained and safety-critical environments. Previous works mainly focused on enhancing either compactness or robustness of neural networks, such as network pruning and adversarial training. Robust neural network pruning aims to reduce computational cost while preserving both accuracy and robustness of a network. Existing robust pruning works usually require expert experiences and trial-and-error to design proper pruning criteria or auxiliary modules, limiting their applications. Meanwhile, evolutionary algorithms (EAs) have been used to prune neural networks automatically, achieving impressive results but without considering the robustness. In this paper, we propose a novel robust pruning method CCRP by cooperative coevolution. Specifically, robust pruning is formulated as a three-objective optimization problem that optimizes accuracy, robustness and compactness simultaneously, and solved by a cooperative coevolution pruning framework, which prunes filters in each layer by EAs separately. The experiments on CIFAR-10 and SVHN show that CCRP can achieve comparable performance with state-of-the-art methods.

Keywords: Model compression · Neural network pruning ·
Robustness · Evolutionary algorithm · Cooperative coevolution

1 Introduction

Recently, convolutional neural networks (CNNs) have achieved great success in the field of computer vision, such as image classification [9] and object detection [6]. Despite the impressive performance, the high computational cost of CNNs inhibits their deployments in resource-limited scenarios. The CNNs are

This work was supported by the NSFC (62022039, 62106098) and the Jiangsu NSF (BK20201247).

G. Rudolph et al. (Eds.): PPSN 2022, LNCS 13398, pp. 459–473, 2022.
https://doi.org/10.1007/978-3-031-14714-2_32

also vulnerable to malicious attacks, challenging their reliability in safety-critical scenarios. Therefore, in many real-world applications like autonomous driving [4,5], enhancing the compactness and robustness of CNNs simultaneously is essential.

Most previous works, however, only focus on enhancing either compactness or robustness of networks. On the one hand, various model compression methods have been proposed to reduce the computational cost of neural networks, such as neural network pruning [8] and quantization [30]. Among them, neural network pruning aims to remove the redundant parameters in networks while preserving accuracy, which has achieved impressive success. On the other hand, methods like adversarial training [7], which aims to minimize the training loss on adversarial examples, can significantly improve the robustness of neural networks.

Recently, several works [21,22,27] took the network robustness into consideration when pruning neural networks. They usually use criteria designed by experts to measure the importance of network weights and prune the networks accordingly. However, the designing and tuning of such criteria require plenty of expertise and tiring trials, making them difficult to be applied to the practical scenarios where the data sets and neural network architectures can be various. Meanwhile, these works mainly focus on unstructured neural network pruning [2], which can hardly reduce the computation cost in practical applications, since the consequent irregular structures are incompatible with the mainstream software and hardware frameworks. Therefore, an automatic structured robust pruning method is essential in real-world applications.

Robust neural network pruning can be naturally formulated as an optimization problem that aims to search for a sub-net of the original network which still maintains high accuracy and robustness but has less computation cost. Evolutionary algorithms (EAs) [1] are a kind of heuristic randomized optimization algorithms inspired by natural evolution, which have been used for pruning neural networks automatically since the 1990s [26]. However, unlike artificial neural networks in the last century, modern CNNs usually consist of dozens of layers and millions of parameters, implying a huge search space. It is difficult for EAs to obtain satisfactory solutions within a limited computational overhead. Recently, Shang et al. have proposed CCEP [23], an evolutionary pruning method inspired by cooperative coevolution, which has achieved encouraging results on the large-scale pruning problem. However, they only focused on the accuracy but did not take robustness into consideration.

In this paper, we propose a novel Cooperative Coevolution method for Robust Pruning (CCRP). The robust pruning problem is formulated as an explicit three-objective optimization problem, i.e., optimizing accuracy, robustness and compactness simultaneously. A cooperative coevolution framework is adopted to solve the formulated robust pruning problem, which divides the search space by layer and applies an EA to optimize each group. Besides, since the process of generating adversarial examples for each pruned network is time-consuming, we design an adversarial example generating method to improve the efficiency of robustness evaluation.

The contributions of this paper are summarized as follows.

1. We propose a novel framework, CCRP, that considers network robustness during the pruning process and automatically solves the three-objective robust pruning problem by cooperative coevolution. To the best of our knowledge, this is the first application of EAs to robust neural network pruning.
2. We propose an adversarial example generating method to improve the efficiency of evaluating the robustness of pruned networks.
3. We compare CCRP with previous methods through experiments on three network architectures and two data sets. Experimental results show that CCRP can achieve comparable performance with state-of-the-art methods.

2   Related Work

2.1   Neural Network Pruning

Neural network pruning aims to enhance the efficiency of a network by removing redundant components. Existing methods can be generally classified into two categories, i.e., unstructured pruning and structured pruning [2]. Unstructured pruning methods directly prune the weights in the parameter matrices of the network. Even though such methods can achieve impressive theoretical acceleration, the resulted sparse matrices and broken structures are incompatible with the mainstream software and hardware platforms, which can hardly obtain actual acceleration. Instead, structured pruning methods focus on pruning structured components such as filters in convolution layers, which has shown better overall performance in real-world application, and thus has prevailed and attracted more attention nowadays.

Based on how to identify the redundant component, previous structured pruning methods can be generally categorized into criteria-based and learning-based methods. Criteria-based methods (e.g., [15,16]) use expert-designed criteria to identify unimportant components and prune them while learning-based methods (e.g., [18]) use auxiliary modules to measure the importance of components and conduct pruning accordingly. However, both of these methods heavily rely on the expertise, limiting their application and extensibility.

To get rid of the reliance on expertise, it is natural to use EAs to search for the good pruned networks automatically, which has been studied since the 1990s [26]. However, the huge search space of a deep neural network brings severe challenge to EAs [13,14]. Recently, a novel pruning method inspired by cooperative coevolution named CCEP [23] is proposed, which employs the idea of divide-and-conquer to settle the huge search space and has achieved impressive performance, showing the great potential of EA-based methods for neural network pruning. But the previous EA-based pruning methods never considered the network robustness, which is important to many application scenarios [4,5].

2.2   Robustness of Neural Network

In application scenarios [4,5], neural networks are typically vulnerable to adversarial attacks [7]. Generally, adversaries utilize the model information to generate adversarial examples for attack. An adversarial example can be defined as

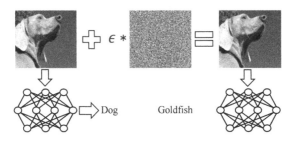

Fig. 1. Illustration of an adversarial example in the image classification task.

$$\hat{x} = x + \eta \qquad s.t. \quad \|\eta\|_\infty \leq \epsilon, \tag{1}$$

where x is the original example and η is the perturbation subject to budget ϵ. As shown in Fig 1, an adversarial example in the image classification task is generated by adding perturbations to the original image. The perturbations are imperceptible to the human eyes, but will mislead the neural network to incorrect prediction. When facing adversarial attacks, the robustness of a neural network $\mathcal{N}$ is usually measured by the robust accuracy on an adversarial data set D_a, i.e.,

$$\mathrm{ACC}_r(\mathcal{N}) = \frac{1}{|D_a|} \sum_{\hat{x},y \in D_a} \mathbb{I}(\mathcal{N}(\hat{x}) = y), \tag{2}$$

where $|D_a|$ denotes the size of D_a (i.e., the number of adversarial examples), $\mathcal{N}(\hat{x})$ denotes the prediction of the neural network $\mathcal{N}$ on the adversarial example $\hat{x}$, and $\mathbb{I}(\cdot)$ is the indicator function that is 1 if the inner expression is true and 0 otherwise.

Adversarial training [7,19,29] is one of the primary defense methods against adversarial attacks. The main idea is to minimize the training loss on adversarial examples generated by adversarial attacks, such as fast gradient sign method (FGSM) [7]. Thus, the objective of adversarial training can be formulated as

$$\min_{\theta} \mathbb{E}_{(x,y) \sim \mathcal{D}} \left[\max_{\|\eta\|_\infty \leq \epsilon} L(x + \eta, y, \theta) \right], \tag{3}$$

where θ denotes the parameters of the neural network, and L is a loss function. Previous empirical results have indicated that adversarial training requires the networks owning a larger capacity for better overall performance. Therefore, neural networks with robustness are usually too computationally intensive to be deployed on resource-constrained applications.

2.3   Robust Neural Network Pruning

Recently, some works [10,25] have studied on the relationship between robustness and network capacity, revealing that a sub-net of the original network can have similar or even better robustness than the original one, and different subnets can have quite different robustness. This finding has inspired robust neural network pruning, which aims to find a compact neural network and retain the robustness. The few existing methods usually train a network by adversarial training and conduct unstructured pruning based on expert-designed criteria. For example, ADV-LWM [21] prunes weights with small l_1-norm, and fine-tunes the obtained network by adversarial training to recover robustness. Ye *et al.* [27] adopts the ADMM pruning framework by replacing the original training loss with an adversarial one, and prunes weights with small l_1-norm. HYDRA [22] adds importance scores to all the weights in the network, and optimizes the adversarial loss by adjusting importance scores while freezing weights. Then the weights with small importance scores are pruned. DNR [12] chooses the feature matrices with small Frobenius norm and prunes the corresponding filters. Furthermore, these methods need proper pruning ratios of each layer, which also often require a lot of expert experience and trial-and-error.

3   CCRP Method

Let $\mathcal{N}$ denote a well-trained neural network with n convolution layers $\{\mathcal{L}_1, \mathcal{L}_2, \cdots, \mathcal{L}_n\}$, where $\mathcal{L}_i$ denotes the ith layer which has l_i filters and $\mathcal{L}_{ij}$ denotes the jth filter in the ith layer. Robust neural network pruning can be formulated as an optimization problem, with the aim of searching for a subset of filters in $\mathcal{N}$, which can maximize the accuracy and robustness while minimizing computational cost simultaneously. Let the mask vector $\mathcal{M} = \{m_{ij} \mid m_{ij} \in \{0,1\}, i \in \{1, 2, ..., n\}, j \in \{1, 2, ..., l_i\}\}$, where $m_{ij} = 1$ if and only if $\mathcal{L}_{ij}$ is retained. Thus, a pruned network can be represented by the mask $\mathcal{M}$ as

$$\mathcal{N}_{\mathcal{M}} = \bigcup_{i=1}^{n} \bigcup_{j=1}^{l_i} m_{ij} \mathcal{L}_{ij}. \tag{4}$$

Let $\mathrm{ACC}(\mathcal{N}_{\mathcal{M}})$ denote the accuracy of the pruned network $\mathcal{N}_{\mathcal{M}}$ on the clean data sets, $\mathrm{ACC}_r(\mathcal{N}_{\mathcal{M}})$ denote the robust accuracy on the generated adversarial examples, and $\mathrm{FLOPs}(\mathcal{N}_{\mathcal{M}})$ denote the number of FLoating point OPerations, which is a common metric to measure the computational cost. The robust neural network pruning problem can be formulated as

$$\underset{\mathcal{M} \in \{0,1\}^{\sum_{i=1}^{n} l_i}}{\arg\max} \ (\mathrm{ACC}(\mathcal{N}_{\mathcal{M}}), \mathrm{ACC}_r(\mathcal{N}_{\mathcal{M}}), -\mathrm{FLOPs}(\mathcal{N}_{\mathcal{M}})) \tag{5}$$

Because the number of alternative filters to be pruned in a CNN, i.e. $\sum_{i=1}^{n} l_i$, can be very large, this is essentially a challenging large-scale optimization problem. To solve this problem, we propose a novel robust pruning method named CCRP. Inspired by our previous work CCEP [23], we adopt a cooperative coevolution pruning framework which divides the search space by layer and conducts an EA on each layer independently. Robust accuracy is set as an optimization objective to directly guide pruned neural networks towards robustness. Note that the evaluation of $\mathrm{ACC}_r(\mathcal{N}_{\mathcal{M}})$ is time-consuming since specialized adversarial examples need to be generated for each pruned network. To settle this, we propose an adversarial example generating method that needs to generate adversarial samples pnly once in each iteration of CCRP. The generated examples can be used to evaluate all the pruned networks in this iteration.

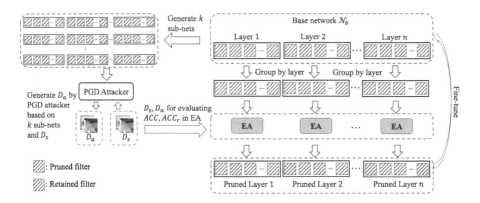

Fig. 2. Illustration of the framework of CCRP.

3.1   Framework of CCRP

The framework of CCRP is shown in Algorithm 1. It prunes a well-trained network iteratively and finally outcomes a set of pruned networks with robustness for user selection. Each iteration works as follows. Firstly, a mask $\mathcal{M}$ is generated based on the network to be pruned in line 4. Then the mask $\mathcal{M}$ will be split into n groups by layer in line 5. After that, in line 6, a set D_a of adversarial examples is generated for the evaluation of the pruned networks, which is shown in Algorithm 3 in detail. For each group, an EA is employed to optimize it and obtain $\boldsymbol{m}_i'$ representing the pruned result of the ith layer. The process of EA in each group is described in Algorithm 2. By collecting all the $\boldsymbol{m}_i'$ of n layers and applying it to $\mathcal{N}_b$, the corresponding pruned network $\mathcal{N}'$ is obtained, as presented in line 8. After pruning, to recover the accuracy and robust accuracy, the pruned network $\mathcal{N}'$ will be fine-tuned by adversarial training in line 9. The fine-tuned model will be used as the new base network $\mathcal{N}_b$ to be pruned in the next iteration, and added into archive H in line 11. After T iterations, the CCRP method will stop and return the pruned networks in archive H. An illustration of the framework of CCRP is also shown in Fig. 2.

3.2   EA in Each Group

For EA in each group, we use a typical evolutionary process: generate an initial subpopulation randomly, breed new individuals by applying reproductive

---

Algorithm 1. CCRP Framework

Input: A well trained CNN $\mathcal{N}$ with n layers, maximum number T of iterations, training set D_t, a randomly sampled part D_s of the training set D_t
Output: A set of pruned networks with different sizes

1: Let $H = \emptyset$, $i = 0$;
2: Set base network $\mathcal{N}_b = \mathcal{N}$;
3: **while** $i < T$ **do**
4:     Generate a mask $\mathcal{M}$ based on $\mathcal{N}_b$ and initialize it with all bits equal to 1;
5:     $m_1, m_2, \cdots, m_n = \text{Decompose}(\mathcal{M})$;
6:     $D_a = $ Generate adversarial samples based on $(\mathcal{N}_b, D_s)$;
7:     $m'_1, m'_2, \cdots, m'_n = \text{EA}(m_1, D_s, D_a), \text{EA}(m_2, D_s, D_a), \cdots, \text{EA}(m_n, D_s, D_a)$;
8:     $\mathcal{N}' = \bigcup_{i=1}^{n} \bigcup_{j=1}^{l_i} m'_{ij} \mathcal{L}_{ij}$, where $\mathcal{L}_{ij}$ is a filter of $\mathcal{N}_b$;
9:     $\mathcal{N}_b = \text{Fine-tune}(\mathcal{N}', D_t)$;
10:     $H = H \cup \mathcal{N}_b$;
11:     $i = i + 1$
12: **end while**
13: **return** H

---

operators, evaluate the fitness of each individual, and select better individuals to remain in the next generation. When the termination condition is reached, it selects an individual from the final subpopulation, which represents the corresponding pruned layer.

The detailed description of EA in each group is shown in Algorithm 2. At the beginning, it generates the initial subpopulation P with d individuals, as shown in line 2. An individual m_0 with all bits equal to 1 is created and added into P to encourage conservative pruning. The rest $d - 1$ individuals are generated by applying a modified bit-wise mutation operator with mutation rate p_1. In each generation of EA, it generates d new offspring individuals by uniformly and randomly choosing d individuals from the subpopulation and applying the bit-wise mutation operator with mutation rate p_2. Following the prior work CCEP [23], we make a modification to the standard bit-wise mutation operator to prevent the pruning process from being too violent. That is, a ratio bound r is introduced to limit the number of filters to be pruned. Specifically, a standard bit-wise mutation operator with mutation rate p is first performed on an individual m; if the number of 0s in the mutated m, denoted by $|m|_0$, is larger than $\text{Len}(m) \times r$, it will randomly select $|m|_0 - \text{Len}(m) \times r$ bits from where $m_i = 0$, and flip them to 1.

When evaluating the fitness of an offspring individual, noting that each individual only corresponds to a single pruned layer, it first obtains a complete network by splicing this single layer with the other layers obtained from the base network $\mathcal{N}_b$. Then, we can evaluate the accuracy, robust accuracy, and FLOPs of offspring individuals. More specifically, ACC is evaluated on the clean data set D_s, which is randomly sampled from the training set D_t; ACC_r is evaluated on the adversarial data set D_a, which is generated by Algorithm 3 and will be introduced in Sect. 3.3; FLOPs can be calculated directly. After evaluation in lines 6–7, the d offspring individuals and d individuals in the current subpopu-

---

Algorithm 2. EA in each group

Input: A mask m_i of the ith layer, population size d, a randomly sampled part D_s of the training set D_t, adversarial data set D_a, mutation rate p_1, p_2, ratio bound r, maximum number G of generations

Output: Mask vector m'_i

1: Let $j = 0$, $m_0 = m_i$;
2: Initialize a subpopulation P with m_0 and $d - 1$ individuals generated from m_0 by applying the bit-wise mutation operator with p_1 and r;
3: **while** $j < G$ **do**
4:     Uniformly randomly select d individuals from P with replacement as the parent individuals;
5:     Generate d offspring individuals by applying the bit-wise mutation operator with p_2 and r on each parent individual;
6:     Calculate the ACC and ACC_r of d offspring individuals by using D_s and D_a;
7:     Calculate the FLOPs of d offspring individuals;
8:     Set Q as the union of P and d offspring individuals;
9:     Rank the $2d$ individuals in Q in descending order by $\frac{\text{ACC}+\text{ACC}_r}{2}$; for two individuals with the same value, the one with less FLOPs is ranked ahead;
10:    Replace the individuals in P with the top d individuals in Q;
11:    $j = j + 1$
12: **end while**
13: Select the rank one individual in P as m'_i
14: **return** m'_i

---

lation P will be merged into a collection Q. Since we consider three objectives as in Eq. (5), it is not easy to find a proper ranking of the individuals. For simplicity, the individuals in Q are ranked by the average value of ACC and ACC_r in descending order. As for two individuals with the same average value, the one with less FLOPs is ranked ahead. Other techniques (e.g., non-dominated sorting [3]) may also be employed, and will be investigated in our future work. After the evolution of G generations, the individual that ranks first in the final subpopulation is chosen as the pruned result of the corresponding group.

3.3 Robustness Evaluation

Typically, the robustness of a neural network is based on its ability against adversarial attacks. In this paper, we use the robust accuracy on the generated adversarial examples as the metric of robustness, which is denoted as ACC_r. CCRP applies, the state-of-the-art white-box attack algorithm PGD [19] to generate the adversarial examples. PGD is designed to attack a specialized network in an iterative style, which is time-consuming. If we use PGD to generate specialized adversarial examples when evaluating each pruned network, the computational overhead will be prohibitive. To settle this issue, we design an adversarial example generating method shown in Algorithm 3 to generate an adversarial data set D_a, which can be shared in one iteration of CCRP. The method samples a sub-net $\mathcal{N}'$ of base network $\mathcal{N}_b$ by randomly selecting $\lceil n/k \rceil$ layers in $\mathcal{N}_b$ and applying bit-wise mutation operator with p_1 and r to them in lines 4-5, and

---

Algorithm 3. Adversarial Example Generating

Input: Base network $\mathcal{N}_b$ with n layers, a randomly sampled part D_s of the training set D_t, number k of sampled sub-nets.
Output: Adversarial data set D_a

1: Let $D_a = \emptyset$, $i = 0$;
2: **while** $i < k$ **do**
3:      Randomly select $\lceil n/k \rceil$ layers from $\mathcal{N}_b$;
4:      Apply mutation with p_1 and r on these layers to obtain a sub-net $\mathcal{N}'$;
5:      $A = \text{PGD}(\mathcal{N}', D_s)$;
6:      $D_a = D_a \cup A$;
7:      $i = i + 1$
8: **end while**
9: **return** D_a

---

then employs PGD on $\mathcal{N}'$ to generate adversarial examples based on D_s in line 6. This process will be repeated k times independently, and all the generated adversarial examples constitutes the adversarial data set D_a, which will be used to evaluate the robustness of all the pruned networks in the current iteration of CCRP. Note that the goal of generating adversarial examples from diverse sub-nets of $\mathcal{N}_b$ is to better measure the robustness of a pruned network.

3.4 Comparison with CCEP

In this subsection, we make a comparison between CCEP and CCRP. CCEP applies cooperation coevolution to neural network pruning and achieves impressive results. CCRP is inspired by CCEP and extended to robust neural network pruning, i.e., by taking the robustness of networks into consideration. These two methods use a similar decomposition strategy that splits the search space by layer. The most significant difference between them is the problem formulation. CCRP introduces robustness as an optimization objective while CCEP concerns accuracy and compactness only. An adversarial example generation method has been introduced into CCRP, which can reduce the cost of robustness evaluation. In addition, adversarial training is applied in fine-tuning to retain the robustness of the pruned network.

4 Experiments

We conduct experiments from three aspects. First, we compare CCRP with the state-of-the-art unstructured robust pruning methods. Second, we extend several popular structured pruning methods to robust pruning and compare CCRP with them. In the third aspect, we visualize the architecture of pruned networks and conduct repeated experiments to examine the stability of CCRP.

Two popular image classification data sets CIFAR-10 [11] and SVHN [20], and three typical neural networks VGG [24], ResNet [9] and WRN [28] are used for examination. Following the common filter pruning settings, CCRP prunes all

Table 1. Comparison in terms of ACC drop, ACC_r drop, and inference speed with unstructured robust pruning methods on CIFAR-10 and SVHN. The best results of each objective are shown in bold.

Data set	Architecture	Method	Base ACC (%)	Base ACC_r (%)	$ACC\downarrow$ (%)	$ACC_r \downarrow$ (%)	Speed (images/s)
CIFAR-10	VGG-16	ADV-LWM	82.70	51.90	3.90	4.20	2082.13
		ADV-ADMM	78.36	47.07	3.50	3.76	2114.77
		HYDRA	82.70	51.90	2.20	2.40	2077.57
		CCRP	81.57	61.71	−1.39	**0.63**	**6842.39**
	WRN-28-4	ADV-LWM	85.60	57.20	2.80	3.40	4142.74
		ADV-ADMM	78.22	51.56	2.46	2.50	4375.58
		HYDRA	85.60	57.20	1.90	2.00	4016.55
		CCRP	85.91	53.42	**−0.05**	**−9.12**	**4737.09**
SVHN	VGG-16	ADV-LWM	90.50	53.50	1.30	2.00	2308.65
		ADV-ADMM	89.35	54.61	−0.23	4.10	2322.72
		HYDRA	90.50	53.50	1.30	**1.10**	2334.29
		CCRP	86.86	53.18	**−1.58**	2.36	**11124.56**
	WRN-28-4	ADV-LWM	93.50	60.10	1.20	0.70	5259.51
		ADV-ADMM	92.14	59.07	1.32	4.53	5482.91
		HYDRA	93.50	60.10	−0.90	**−2.70**	5294.31
		CCRP	90.07	57.47	**−1.63**	−0.18	**6467.55**

convolution layers for VGG and the first convolution layer of the residual blocks for ResNet and WRN. The popular adversarial training method, TRADES [29], is used in the pre-train and fine-tune processes. The settings of CCRP are described as follows. It runs for 16 iterations, i.e., $T = 16$. For EA in each group, the population size m is 5, the mutation rate p_1 and p_2 are 0.05 and 0.1, respectively, the ratio bound r is 0.1, the maximum generation G is 10, and D_s is generated by selecting 10% of the training set randomly. When generating adversarial examples, the number k of sampled sub-nets is 5.

For adversarial training by TRADES [29], the common settings are used. The optimizer is SGD with an initial learning rate 0.1, and a Cosine Annealing scheduler [17] is employed to adjust the learning rate during fine-tuning. The weight decay is 0.0001 and the momentum is 0.9. The number of epochs in each process of fine-tuning is 30. The batch size for training is 128. For adversarial attack by PGD, the l_∞ perturbation budget, number of steps, and perturbation per step are set as 8/255, 10, 2/255 respectively in adversarial training and 8/255, 40, 2/255 for evaluation and testing.

We compare CCRP with various methods, including three state-of-the-art unstructured robust pruning methods: ADV-LWM [21], ADV-ADMM [27] and HYDRA [22], as well as two structured pruning methods L1 [15] and HRank [16] extended to robust pruning. The results of HYDRA and ADV-LWM are obtained from their released models. All the experiments are realized based on PyTorch and carried out on a single Nvidia GeForce RTX-3090 GPU.

Comparison with Unstructured Robust Pruning Methods: We first compare CCRP with state-of-the-art unstructured robust pruning methods in terms of accuracy drop, robust accuracy drop, and inference speed, as shown in Table 1.

Inference speed is used to measure the computation cost here since FLOPs drop of unstructured models cannot reflect the actual computational performance in applications. The inference speed is tested on 100,000 32 × 32 images with a batch size of 128. For CCRP, the solution in the 10th iteration is presented in Table 1 for comparison. CCRP achieves a smaller drop in accuracy and robust accuracy with faster inference speed in most cases. On SVHN, HYDRA [22] and ADV-LWM [21] achieve a smaller drop in robust accuracy than CCRP but more drop in accuracy and slower inference speed. It is worth noting that CCRP achieves the fastest inference speed in all cases.

Comparison with Structured Robust Pruning Methods: For a more comprehensive comparison, two structured pruning methods, L1 [15] and HRank [16], are extended to the scenario of robust pruning by introducing adversarial training in the pre-train and fine-tune steps. For CCRP, we select the solution in the 16th iteration for comparison. The results in Table 2 show that compared with L1 and HRank, CCRP can always achieve better performance on at least two of the three metrics. Sometimes CCRP prevails on two metrics but only with minor disadvantage on the third metric.

Table 2. Comparison in terms of ACC drop, ACC$_r$ drop, and pruning ratio with structured robust pruning methods on CIFAR-10 and SVHN. The best results of each objective are shown in bold.

Data set	Architecture	Method	Base ACC (%)	Base ACC$_r$ (%)	ACC↓ (%)	ACC$_r$ ↓ (%)	FLOPs ↓ (%)
CIFAR-10	VGG-16	L1	81.57	61.71	2.00	3.37	69.23
		HRank	81.91	61.11	7.05	**3.01**	65.85
		CCRP	81.57	61.71	**0.05**	6.22	**77.95**
	ResNet-56	L1	80.31	48.95	2.47	−4.62	68.53
		HRank	80.31	48.95	**0.13**	−2.22	50.02
		CCRP	80.31	48.95	0.23	**−8.31**	**72.30**
	WRN-28-4	L1	85.91	53.61	2.00	3.37	**69.23**
		CCRP	85.91	53.61	**−0.88**	**−8.06**	66.92
SVHN	VGG-16	L1	86.86	53.18	2.17	4.35	**85.88**
		HRank	86.06	54.53	0.40	5.03	65.85
		CCRP	86.86	53.18	**−0.98**	**2.68**	80.44
	ResNet-56	L1	85.91	52.24	**−2.04**	−1.78	60.08
		HRank	87.09	55.57	−1.53	3.25	50.02
		CCRP	85.91	52.24	−1.95	**−2.01**	**70.71**
	WRN-28-4	L1	90.07	57.47	1.45	4.19	**72.11**
		CCRP	90.07	57.47	**−1.25**	**1.46**	70.99

Further Studies: Because experiments of network pruning are very time-consuming and may require dozens of hours, previous works [15,16,21,22] usually conducted experiments only once. However, considering the stochastic characteristic of EAs, we conduct a repeated test on a relatively small data set CIFAR-10, to prune ResNet-56 and VGG-16 for ten independent runs. The ACC, ACC$_r$, and ACC$_a$ (i.e., the average of ACC and ACC$_r$), are recorded and shown in

Fig. 3. The solid line is the mean value, and the shadow area represents the 95% confidence interval. We can observe that the ACC and ACC_r are even better than the base model when the pruning ratio is low and get a slight drop as the pruning ratio increases; the ACC_a is always better than the base model. The 95% confidence interval implies the good stability of CCRP.

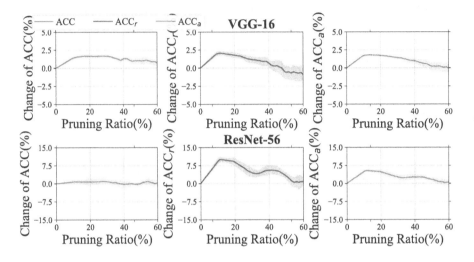

Fig. 3. Repeated test of CCRP on CIFAR-10.

In Fig. 4, we visualize the architecture (i.e., the number of filters in each layer or residual block) of pruned networks on CIFAR-10. The results show that CCRP can choose different pruning ratios at different layers (or residual blocks) automatically. For ResNet-56, CCRP prunes fewer filters around the expansion of channels, while on VGG-16, CCRP prunes more filters after the 6th layer. As for WRN-28-4, more filters at 9th block are preserved. Note that previous robust pruning methods may require lots of trial and error to design the proper pruning ratios at each layer manually.

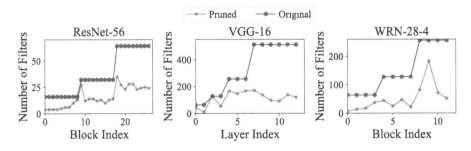

Fig. 4. Visualization of the original networks and the pruned networks obtained by CCRP on CIFAR-10.

5   Conclusion

This paper proposes the automatic robust neural network pruning method, which formulates robust pruning as a three-objective optimization problem considering robustness, and solves it by an adapted cooperative coevolution framework. To the best of our knowledge, this is the first application of EAs to robust neural network pruning. Experiments show that CCRP can achieve a comparable performance with the state-of-the-art methods. In the future, we will try to perform theoretical analysis [31], as well as incorporate more advanced multi-objective optimization techniques to improve the performance of CCRP.

References

1. Bäck, T.: Evolutionary Algorithms in Theory and Practice: Evolution Strategies, Evolutionary Programming. Genetic Algorithms. Oxford University Press, Oxford, UK (1996)
2. Cheng, J., Wang, P., Li, G., Hu, Q., Lu, H.: Recent advances in efficient computation of deep convolutional neural networks. Front. Inf. Technol. Electron. Eng. **19**(1), 64–77 (2018). https://doi.org/10.1631/FITEE.1700789
3. Deb, K., Agrawal, S., Pratap, A., Meyarivan, T.: A fast and elitist multiobjective genetic algorithm: NSGA-II. IEEE Trans. Evol. Comput. **6**(2), 182–197 (2002)
4. Duan, R., Ma, X., Wang, Y., Bailey, J., Qin, A.K., Yang, Y.: Adversarial camouflage: hiding physical-world attacks with natural styles. In: Proceedings of the 2020 IEEE Conference on Computer Vision and Pattern Recognition (CVPR), Seattle, WA, pp. 997–1005 (2020)
5. Eykholt, K., et al.: Robust physical-world attacks on deep learning visual classification. In: Proceedings of the 2018 IEEE Conference on Computer Vision and Pattern Recognition (CVPR), Salt Lake City, UT, pp. 1625–1634 (2018)
6. Girshick, R.B., Donahue, J., Darrell, T., Malik, J.: Rich feature hierarchies for accurate object detection and semantic segmentation. In: Proceedings of the 2014 IEEE Conference on Computer Vision and Pattern Recognition (CVPR), Columbus, OH, pp. 580–587 (2014)
7. Goodfellow, I.J., Shlens, J., Szegedy, C.: Explaining and harnessing adversarial examples. In: Proceedings of the 3rd International Conference on Learning Representations (ICLR), San Diego, CA (2015)
8. Han, S., Mao, H., Dally, W.J.: Deep compression: Compressing deep neural network with pruning, trained quantization and huffman coding. In: Proceedings of the 4th International Conference on Learning Representations (ICLR), San Juan, Puerto Rico (2016)
9. He, K., Zhang, X., Ren, S., Sun, J.: Deep residual learning for image recognition. In: Proceedings of the 2016 IEEE Conference on Computer Vision and Pattern Recognition (CVPR), Las Vegas, NV, pp. 770–778 (2016)
10. Huang, H., Wang, Y., Erfani, S., Gu, Q., Bailey, J., Ma, X.: Exploring architectural ingredients of adversarially robust deep neural networks. In: Advances in Neural Information Processing Systems (NeurIPS), vol. 34, New Orleans, LA, pp. 5545–5559 (2021)
11. Krizhevsky, A., Hinton, G.: Learning multiple layers of features from tiny images. Technical report, University of Toronto, Toronto, Canada (2009)

12. Kundu, S., Nazemi, M., Beerel, P.A., Pedram, M.: DNR: a tunable robust pruning framework through dynamic network rewiring of DNNs. In: Proceedings of the 26th Asia and South Pacific Design Automation Conference (ASPDAC), Tokyo, Japan, pp. 344–350 (2021)

13. Li, G., Qian, C., Jiang, C., Lu, X., Tang, K.: Optimization based layer-wise magnitude-based pruning for DNN compression. In: Proceedings of the 27th International Joint Conference on Artificial Intelligence (IJCAI), Stockholm, Sweden, pp. 2383–2389 (2018)

14. Li, G., Yang, P., Qian, C., Hong, R., Tang., K.: Magnitude-based pruning for recurrent neural networks. IEEE Trans. Neural Networks Learn. Syst. (in press)

15. Li, H., Kadav, A., Durdanovic, I., Samet, H., Graf, H.P.: Pruning filters for efficient convnets. In: Proceedings of the 5th International Conference on Learning Representations (ICLR), Toulon, France (2017)

16. Lin, M., et al.: HRank: filter pruning using high-rank feature map. In: Proceedings of the 2020 IEEE Conference on Computer Vision and Pattern Recognition (CVPR), Los Alamitos, CA, pp. 1526–1535 (2020)

17. Loshchilov, I., Hutter, F.: SGDR: stochastic gradient descent with warm restarts. In: Proceedings of the 5th International Conference on Learning Representations (ICLR), Toulon, France (2017)

18. Luo, J., Wu, J.: Autopruner: an end-to-end trainable filter pruning method for efficient deep model inference. Pattern Recogn. 107(107461), 107461 (2020)

19. Madry, A., Makelov, A., Schmidt, L., Tsipras, D., Vladu, A.: Towards deep learning models resistant to adversarial attacks. In: Proceedings of the 6th International Conference on Learning Representations (ICLR), Vancouver, Canada (2018)

20. Netzer, Y., Wang, T., Coates, A., Bissacco, A., Wu, B., Ng, A.Y.: Reading digits in natural images with unsupervised feature learning. In: Advances in Neural Information Processing Systems, Workshop (NeurIPS) (2011)

21. Sehwag, V., Wang, S., Mittal, P., Jana, S.: Towards compact and robust deep neural networks. CoRR p. abs/1906.06110 (2019)

22. Sehwag, V., Wang, S., Mittal, P., Jana, S.: HYDRA: pruning adversarially robust neural networks. In: Advances in Neural Information Processing Systems (NeurIPS), vol. 33, Vancouver, Canada, pp. 19655–19666 (2020)

23. Shang, H., Wu, J.L., Hong, W., Qian, C.: Neural network pruning by cooperative coevolution. In: Proceedings of the 31st International Joint Conference on Artificial Intelligence (IJCAI), Vienna, Austria (2022)

24. Simonyan, K., Zisserman, A.: Very deep convolutional networks for large-scale image recognition. In: Proceedings of the 3rd International Conference on Learning Representations (ICLR), San Diego, CA (2015)

25. Wu, B., Chen, J., Cai, D., He, X., Gu, Q.: Do wider neural networks really help adversarial robustness? In: Advances in Neural Information Processing Systems (NeurIPS), vol. 34, New Orleans, LA New Orleans, LA, pp. 7054–7067 (2021)

26. Yao, X.: Evolving artificial neural networks. Proc. IEEE 87(9), 1423–1447 (1999)

27. Ye, S., et al.: Adversarial robustness vs. model compression, or both? In: Proceedings of the 2019 IEEE International Conference on Computer Vision (ICCV), Seoul, Korea (South), pp. 111–120 (2019)

28. Zagoruyko, S., Komodakis, N.: Wide residual networks. In: Proceedings of the 2016 British Machine Vision Conference (BMVC), York, UK (2016)

29. Zhang, H., Yu, Y., Jiao, J., Xing, E.P., Ghaoui, L.E., Jordan, M.I.: Theoretically principled trade-off between robustness and accuracy. In: Proceedings of the 36th International Conference on Machine Learning (ICML), Long Beach CA, pp. 7472–7482 (2019)

30. Zhou, A., Yao, A., Guo, Y., Xu, L., Chen, Y.: Incremental network quantization: Towards lossless CNNs with low-precision weights. In: Proceedings of the 5th International Conference on Learning Representations (ICLR), Toulon, France (2017)
31. Zhou, Z., Yu, Y., Qian, C.: Evolutionary Learning: Advances in Theories and Algorithms. Springer, Singapore (2019)

SemiGraphFL: Semi-supervised Graph Federated Learning for Graph Classification

Ye Tao[1], Ying Li[2(✉)], and Zhonghai Wu[2]

[1] School of Software and Microelectronics, Peking University, Beijing, China
`tao.ye@pku.edu.cn`
[2] National Engineering Center of Software Engineering, Peking University, Beijing, China
{`li.ying,wuzh`}`@pku.edu.cn`

Abstract. GNNs have achieved remarkable performance on graph classification tasks. It can be attributed to the accessibility of abundant graph data, which are usually isolated by different data owners. Graph Federated Learning (GraphFL) allows multiple clients to collaboratively build GNN models without explicitly sharing data. However, all existing works assume that all clients have fully labeled data, which is impractical in reality. This work focuses on the graph classification task with partially labeled data. (1) Enhancing the collaboration processes: We propose a new personalized FL framework to deal with Non-IID data. Clients with more similar data have greater mutual influence, where the similarities can be evaluated via unlabeled data. (2) Enhancing the local training process: We introduce auxiliary loss for unlabeled data that restrict the training process. We propose a new pseudo-label strategy for our SemiGraphFL framework to make more effective predictions. Extensive experimental results prove the effectiveness of our design.

Keywords: Graph Neural Network · Federated Learning · Semi-supervised learning

1 Introduction

The availability of amount graph data has shed interest into the topic of graph analysis. Recently, Graph Neural Networks (GNNs) have demonstrated remarkable performance in modelling graph data and derived various researches and applications [11,23]. The success of GNNs relies on abundant data, which usually exists in an isolated manner in real applications. Due to privacy or commercial concerns, it gives rise to challenges on centrally training GNNs. Given this situation, Graph Federated Learning (GraphFL) has been proposed [2,9]. FL is a paradigm where multiple clients collaboratively train machine learning models via coordinated communication while keeping the data decentralized [10,12,19].

To the best of our knowledge, all existing GraphFL methods designed for the graph classification task assumes that all clients have fully labeled data.

G. Rudolph et al. (Eds.): PPSN 2022, LNCS 13398, pp. 474–487, 2022.
https://doi.org/10.1007/978-3-031-14714-2_33

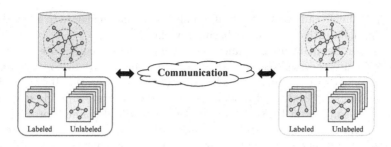

Fig. 1. An example of semi-supervised graph federated learning.

However, large labeled datasets are scarce in many practical cases. Creating labeled datasets requires a considerable amount of resources, limiting the adoption of these methods. To deal with the problem in Fig. 1, we propose a new data-efficient Semi-Supervised Graph Federated Learning framework SemiGraphFL to overcome the need for large annotated datasets. Since the GraphFL framework consists of multi-client collaboration and single-client local training, we focus on how to take advantage of unlabeled data to enhance these two processes.

To enhance the multi-client collaboration process, we propose a decentralized framework where each client performs local training and aggregates messages from others. In classical FL methods, a server aggregates the locally learned model parameters to obtain a unified global model [12]. However, FL applications generally face the data Non-IID problem, making it challenging to ensure good performance across different clients. Intuitively, clients with more similar data should have more significant mutual influence. To evaluate the similarities between clients, we design a new communication strategy. Each client collects parameters from other clients and tests them on the local unlabeled data. They use these predictions to compute the similarities and perform a weighted sum to aggregate parameters from other clients to update the local parameters. Using labeled data to do the same thing cannot achieve good results because clients perform local supervised learning on these data and over-fit them.

To enhance the local training process, we borrow ideas for general semi-supervised learning methods. According to the low-density assumption, we restrict predictions on unlabeled data to be sharper, i.e., the decision boundary should not pass through unlabeled data. The auxiliary supervision makes training more stable when there are little training data in each client. We also introduce a new pseudo-label strategy to alleviate the problem of insufficient labeled data. We propose integrating the output of multiple models for unlabeled data to obtain more effective and robust pseudo-labeled data.

Since there is no semi-supervised graph federated learning benchmarks, we employ the widely used graph datasets in Open Graph Benchmark (OGB) [3] and split data in different ways to simulate various Non-IID scenarios in reality. To summarize, our main contributions are as follows:

- To the best of our knowledge, we are the first to deal with semi-supervised graph federated learning. We propose a data-efficient framework SemiGraphFL that multiple clients can employ both the labeled and unlabeled data to train a GNN model collaboratively without sharing raw data.

- Under our SemiGraphFL framework, we design strategies to enhance the FL training process. We propose to use unsupervised data to evaluate the similarity of different clients, which enhances the collaboration between clients to train personalized local models. And we design semi-supervised learning strategies suitable in FL settings to enhance the local training process.
- To evaluate our SemiGraphFL framework, we provide semi-supervised GraphFL datasets. The experimental results show that our SemiGraphFL performs well on universal scenarios with different data distributions. And the ablation studies prove the effectiveness of our design for each module.

2   Related Work and Preliminaries

2.1   Federated Learning

Federated Learning (FL) has gained increasing attention as a decentralized learning technique where many clients collaboratively train a model under the orchestration of a central server while protecting data privacy. FedAvg [12] is currently the most widely adopted FL framework. Most existing FL frameworks rely on the optimization by SGD. However, data of Non-IID distribution will not guarantee the stochastic gradients to be an unbiased estimation of the full gradients, thus hurting the convergence of FL. A research trend is to accommodate personalized local models for Non-IID data, e.g., by integrating FL with meta-learning [5], assisted learning [20] and multi-task learning [15]. Besides, there are many improvement efforts devoted to addressing other problems, including reducing the communication cost [8] and protecting data privacy [10].

Federated Learning collaboratively trains machine learning models via coordinated communication with multiple clients. Suppose there are N clients, let $\mathcal{D} = \{\mathcal{D}_1, \ldots, \mathcal{D}_N\}$ be a given dataset with n samples, where $\mathcal{D}_i$ is privately collected local dataset at i-th client. For a machine learning problem, we typically take $f_i(\theta) = l(x; \theta)$, i.e., the loss of the prediction on example x with parameter θ. Generally, the objective function of a FL method can be written as follows

$$f(\theta) = \sum_{i=1}^{N} \frac{n_i}{n} F_i(\theta), \quad F_i(\theta) = \frac{1}{n_i} \sum_{j \in \mathcal{D}_i} f_j(\theta) \tag{1}$$

If $\{\mathcal{D}_1, \ldots, \mathcal{D}_N\}$ was formed by distributing examples over $\mathcal{D}$ at random, we would have $\mathbb{E}_{\mathcal{D}_i}[F_i(\theta)] = f(\theta)$, where the expectation is over the set of examples assigned to a fixed client i. This is the IID scenario. We refer to the case where this does not hold as the Non-IID setting.

2.2   Semi-supervised Learning

Semi-Supervised Learning (SSL) is a machine learning paradigm that uses partially labeled data. SSL algorithms only work under some assumptions about the structure of the data need to hold [13,17]. If sufficient unlabeled data is available and under certain assumptions about the distribution, this data can help construct a better classifier. The *low-density assumption* assumes that the decision boundary of a classifier should preferably pass through low-density regions in the input space. The *smoothness assumption* assumes that for two input points $x, x' \in \mathcal{X}$ that are close by in the input space, the corresponding labels y, y' should be the same. And the *manifold assumption* assumes that high-dimensional data are likely to be located in a low-dimensional manifold.

Semi-supervised learning employs labeled as well as unlabeled data to perform certain learning tasks. Usually, the examples of unlabeled data are much more than those of the labeled data. Let $\mathcal{S} = \{(x_i, y_i)\}_{i=1}^{n_s}$ be a set of n_s labeled data instances and $\mathcal{U} = \{(x_i)\}_{i=1}^{n_u}$ be a set of n_u of unlabeled instances without corresponding label. Given a neural network parameterized by weights θ with these two datasets, the objective is to minimize the following loss function

$$\ell_{\text{final}}(\theta) = \ell_s(x, y; \theta) + \ell_u(x; \theta) \qquad (2)$$

where ℓ_s is loss term for supervised learning on $\mathcal{S}$ and ℓ_u is loss term for unsupervised learning on $\mathcal{U}$. Some recent works have studied the problems of Semi-Supervised Federated Learning (SSFL) [4,6,26], but none of them paid attention to the GNN-based graph classification problem.

2.3 Graph Neural Network

Graph Neural Networks (GNNs) extend the deep learning technique to deal with graph-structure data. Early studies define the graph convolution operation on the spectral domain [1]. GCN [7] generalizes the graph convolution technique to the spatial domain, which has become a prevalent formulation of GNNs in current graph representation learning methods. GIN [22] theoretically proves that standard GNNs are at most as powerful as the Weisfeiler-Lehman graph isomorphism test in distinguishing graph structures. Generally, GNNs consist of (i) *aggregate* function, which collects messages from neighbors to capture local structure information, (ii) *combine* function, which fuses the neighborhood information with the node embedding, and (iii) *readout* function, which performs graph-level readout to get the representation of the entire graph.

GNN-Based Graph Classification. GNNs broadly follow a neighborhood aggregation scheme. They iteratively update the representation of a node by aggregating representations of its neighbors. We denote a graph as a tuple $\mathcal{G} = (\mathcal{V}, \mathcal{E}, X, A)$, where X_v is the node feature vector for $v \in \mathcal{V}$ and A_{uv} is the edge attribute vector for $e_{uv} \in \mathcal{E}$. Let $\vec{h}_v^k$ denotes the embedding of node v at the k-th layer, and $\vec{h}_{uv}$ denotes the embedding of edge e_{uv}. We use N_v denotes the neighbor nodes of node v, the *aggregate* and *combine* function at k-th layer of the a GNN model can be written as follows

$$\vec{a}_v^k = \text{Aggregate}^k \left(\left\{ \left\{ \phi \left(\vec{h}_u^{k-1}, \vec{h}_{uv}, \vec{h}_v^{k-1} \right) \mid u \in \mathcal{N}_v \right\} \right\} \right) \qquad (3)$$

$$\vec{h}_v^k = \text{Combine}\left(\vec{h}_v^{k-1}, \vec{a}_v^k\right) \tag{4}$$

For graph-level tasks, sometimes different levels of graph representations are needed. To deal with the graph classification tasks, GNN models employ the *readout* function to obtain the graph-level representation for prediction. Then a final layer makes predictions based on graph representation.

$$\vec{g} = \text{Readout}\left(\left\{\vec{h}_v \mid v \in \mathcal{V}\right\}\right), \quad \hat{y} = \text{Pre}(\vec{g}) \tag{5}$$

3 SemiGraphFL: A Semi-supervised Graph Federated Learning Framework

We borrow the ideas from the Graph Attention Networks (GAT) [18]. First, we construct a fully connected communication graph, where each client is regarded as a node, and the model parameters are considered as the node representation. Then, each FL training round is accomplished as a communication step in GAT that consists of the (1) local feature extraction, which corresponds to the model local training process, (2) message aggregation between neighbors, which corresponds to the multi-client communication, and (3) local representation updates, which corresponds to updating the local model parameters. In Sect. 3.1, we mainly concentrated on the last two parts, while the first part will be illustrated in detail in Sect. 3.2. Figure 2 is an overview of our strategies.

3.1 Multi-client Collaboration

In FL tasks, one key problem is how multiple clients collaboratively train deep learning models. Most FL structures iteratively cycle between downloading model parameters from server to client, performing local training, and sending back the updated models for future rounds. In a Traditional FL framework such as FedAvg, the server average the model parameters collected from clients and use the result as the result parameters, which means that multiple clients share the same model. However, to deal with the Non-IID problem, we propose a new decentralized framework that does not compute a single global model.

For each FL training round, all clients receive GNN model parameters from other clients. We denote the parameters at client i as θ_i and the corresponding model as f_{θ_i}, they construct a parameter set $\{\theta_i\}$ and a model set $\{f_{\theta_i}\}$. Assuming N federating clients, we update each local model parameters θ_i at round t as the weighted sum the parameter set $\{\theta_i\}$

$$\theta_i^{t+1} = \sum_{n=1}^{N} w_{ij}^{model,t} \cdot \theta_j^t \tag{6}$$

where the value of $w_{ij}^{model,t}$ indicates how important client j is to client i at round t. Intuitively, clients with more similar data should have greater mutual

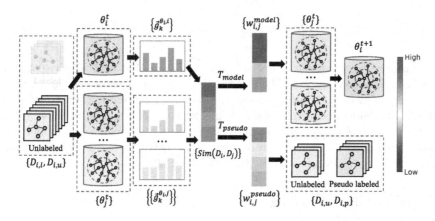

Fig. 2. An illustration of how SemiGraphFL framework employs the unlabeled data to perform personalized parameters aggregation and data pseudo-labeling. Parameter aggregation executes every FL round, while pseudo-labeling only executes once.

influence, i.e., $w_{ij}^{model,t}$ should be positively correlated with $Sim(\mathcal{D}_i, \mathcal{D}_j)$. However, to protect data privacy, the local data are not available to other clients. Therefore, one should estimate $Sim(\mathcal{D}_i, \mathcal{D}_j)$.

Recently, some related works have also proposed to enhance the training process of FL by evaluating the similarity between different clients [14,21,25]. Most of them evaluate $Sim(\mathcal{D}_i, \mathcal{D}_j)$ directly by defining the distance between model parameters, e.g., [21] computes the distance between the local model and the global model and divides all clients into little clusters with smaller intra-cluster distance. However, models with different parameters can perform a similar feature extraction process. The randomness of mini-batch data sampling, parameter dropout, and other model components makes parameter-level distances unstable. Thus we argue that the similarity evaluation of model parameters, such as the Euclidean distance or cosine similarity, does not well reflect the data distribution similarity. Thus, we propose to evaluate the similarity of models by their behavior rather than the values of their items.

Under our SemiGraphFL framework, each client receives model parameters from other client and use these parameters to perform predictions on its local unsupervised dataset. It is worth noting that GNN-based graph classification model first get the graph representations and then perform the final prediction with Eqs. 3–5. We argue that the hidden representations are more effective and stable than final predictions to evaluate these similarities. We denote the representation with the j-th client's parameter on i-th client's k-th unlabeled data sample as $\vec{g}_k^{\theta_j,i}$. Then we compute the cosine similarities between graph representations. Formally, we have

$$Sim(\mathcal{D}_i, \mathcal{D}_j) = \frac{1}{n_i} \sum_{k=1}^{n_i} \frac{\vec{g}_k^{\theta_i,i} \cdot \vec{g}_k^{\theta_j,i}}{||\vec{g}_k^{\theta_i,i}|| \cdot ||\vec{g}_k^{\theta_j,i}||} \qquad (7)$$

We employ *softmax* to normalize weights for all clients.

$$w_{ij}^{model} = \frac{e^{\mathrm{Sim}(\mathcal{D}_i, \mathcal{D}_j)/T_{model}}}{\sum_{k=1}^{N} e^{\mathrm{Sim}(\mathcal{D}_i, \mathcal{D}_k)/T_{model}}} \tag{8}$$

We introduce a temperature hyper-parameter T_{model} that controls the parameter synchronization process. There are significant randomness and fluctuations at the beginning of the training process. So we use a high temperature to enhance mutual constraints between multiple clients. As the training progresses, we gradually reduce the T_{model} so that each client can obtain a personalized model that is not constrained by clients with different data distributions. The personalized parameters aggregation process is illustrated in Fig. 2.

3.2   Single-Client Training

Since we have explained how multiple clients work together, now we turn our attention to how each client performs its local training with both labeled data.

For labeled data, we perform supervised learning with the standard cross entropy loss. Therefore, our SemiGraphFL framework can enhance most existing GNN-based graph classification methods. Formally, we have

$$\ell_l = -\frac{1}{n_l} \sum_{i=1}^{n_l} \sum_{c=1}^{C} w_c \cdot y_{i,c} \log \hat{y_{i,c}} \tag{9}$$

where ℓ_l is the supervision loss for labeled data, C is the number of classes, $\hat{y_{i,c}}$ represents the probability of predicting sample x_i with label c, and $\{w_c\}$ is the weight set to control the loss weights for different classes.

For unlabeled data, we first design a supervision signal according to the low-density assumption, i.e., the decision boundary should preferably not pass through unlabeled data. Specifically, for GNN-based graph classification tasks, we restrict the predictions for unlabeled data to be sharp. Formally, we restrict the output of the classifier to have low entropy.

$$\ell_i = -\frac{1}{n_u} \sum_{i=1}^{n_u} \sum_{c=1}^{C} \hat{y_{i,c}} \log \hat{y_{i,c}} \tag{10}$$

Since each client has little labeled data, introducing this auxiliary loss will restrict the training process to be stable.

Another simple approach to extending existing GNN methods to the semi-supervised setting is to first train GNN models on labeled graphs, and then use the predictions for unlabeled data of the resulting classifiers to generate additional labeled data [17]. Some existing methods design multiple classifiers that mutually enhance each other. The classifiers are supposed to be sufficiently diverse, which is usually achieved by operating on different subsets of the given objects or features. In our SemiGraphFL framework, each client can obtain other clients' models, which are trained under different data distribution. Therefore, we propose a new pseudo-label strategy that can take the advantages of this feature to get more effective pseudo-labeled data. Formally,

$$\hat{y}_{i,k}^{pseudo} = \sum_{j=1}^{N} \frac{e^{\mathrm{Sim}(\mathcal{D}_i, \mathcal{D}_j)/T_{pseudo}}}{\sum_{k=1}^{N} e^{\mathrm{Sim}(\mathcal{D}_i, \mathcal{D}_k)/T_{pseudo}}} \hat{y}_{j,k} \qquad (11)$$

where $\hat{y}_{i,k}^{pseudo}$ is the pseudo prediction for unlabeled data x_k on client i, $\hat{y}_{j,k}$ is the prediction for x_k at client i with the j-th client's model, $\mathrm{Sim}(\mathcal{D}_i, \mathcal{D}_j)$ is obtained with Eq. 8, and T_{pseudo} is a temperature hyper-parameter. We select top $n_{p,i} = p \cdot n_i$ samples with the highest confidence as supplement for labeled data, where p is the pseudo ratio. And we use y^{pseudo} to denote the one-hot value for $\hat{y}^{pseudo}$, the pseudo-label loss is as follows

$$\ell_p = -\frac{1}{n_p} \sum_{i=1}^{n_p} \sum_{c=1}^{C} w_c \cdot y_{i,k}^{pseudo} \log \hat{y_{i,c}} \qquad (12)$$

To get more effective pseudo-labeled data, we should set $T_{pseudo} > T_{model}$. Therefore, the pseudo-label process will obtain more restrictions from other clients than the model parameter aggregation, which is illustrated in Fig. 2.

To further enhance the pseudo-label procedure, we propose to employ a two-stage training strategy. At the first stage, the target is set to

$$\ell_{first} = \ell_l^{first} + \lambda_u \cdot \ell_u \qquad (13)$$

We set $w_c = \frac{n_p}{|C| \cdot \sum_{i=1}^{n_p} \mathbf{1}_{(y_{i,c}=1)}}$ for ℓ_l^{first}. Although the labeled data distributions of multiple clients are different, we can get pseudo-label data with a uniform and consistent distribution across datasets. Besides, since the distribution of training data samples are not balanced, without employing w_c to normalize the loss weights, the pseudo-labeled data with high confidence will only concentrated in a few categories with a large amount of labeled data. We use the model to perform pseudo label. For the second stage, the target is set to

$$\ell = \ell_l^{second} + \lambda_u \cdot \ell_u + \lambda_p \cdot \ell_p \qquad (14)$$

We set $w_c = 1$ for ℓ_l^{second} so that each client will train their optimum personalized model. Our framework can alleviate the confirmation bias with these techniques, which is a common problem with pseudo-labeling strategies.

4 Experimental Setup

4.1 Datasets

Since there is no semi-supervised graph federated learning benchmarks, we adopt and process the widely used datasets in Open Graph Benchmark (OGB) [3]. The dataset ogbg-ppa is a set of undirected protein association graphs extracted from the protein-protein association networks of 1,581 different species [16] that cover 37 broad taxonomic groups. Each species contains 100 samples, so there are 158,100 graphs in total. Given a graph, the task is a 37-way multi-class classification to predict what taxonomic group the graph originates from.

Table 1. Statistics of Non-IID distribution for *ogbg-ppa*. For example, in NIID-1 dataset, the Client-1 distribution is 70%/10%/10%/10%, meaning that client-1 hold 70% data samples in the data cluster and hold 10% data in the 2/3/4 data cluster.

Clint name	NIID-1	NIID-2	NIID-3
Client-1	70%/10%/10%/10%	40%/20%/20%/20%	40%/40%/10%/10%
Client-2	10%/70%/10%/10%	20%/40%/20%/20%	40%/40%/10%/10%
Client-3	10%/10%/70%/10%	20%/20%/40%/20%	10%/10%/40%/40%
Client-4	10%/10%/10%/70%	20%/20%/20%/40%	10%/10%/40%/40%

To simulate different Non-IID degrees in practical application scenarios, we first split the *ogbg-ppa* dataset into 4 clusters. Each cluster consists of graphs of multiple species, and species in different clusters do not contain intersections. Then, these data are distributed to 4 clients. Each client holds different proportions of data in each cluster. We first construct an IID dataset by randomly dividing data in all clusters into 4 clients, which is denoted as *IID-data*. Then, we construct three None-IID datasets by dividing data in different clusters into 4 clients with different ratios/proportions. Detailed data distributions are illustrated in Table 1. After divide data into different clients, we randomly divide the data at each client into the labeled/unlabeled/test set with the proportion of 1%/50%/49% to perform the following experiments.

4.2 Baseline Methods and Model Settings

To illustrate our SemiGraphFL framework can be easily applied to existing GNN-based graph classification methods, we select GCN [7] and GAT [18] to be the base models. We compare our framework with: (1) the single-training method, (2) classical FL frameworks FedAvg [12] and FedProx [10], (3) a personalized FL framework FeSEM [21] that split clients into multiple cluster with little intra-cluster differences, and (4) a graph contrastive learning method GraphCL [24], we add the contrastive learning loss under the FedAvg framework.

For all methods, we employ the SGD optimizer. We follow [3] to employ a GNN model consisting of 5 GNN layers and a linear prediction layer to perform graph classification. The batch size for labeled, unlabeled, and pseudo-labeled data are all set to 32, the embedding size is set to 300, and the drop-out ratio is set to 0.5. The loss weight parameters λ_u and λ_p are both set to 1. T_{model} is decreasing gradually as the model training, and the final value are searched from $\{1, 0.5, 0.25, 0.1\}$. For the pseudo-labeling process, we set $T_{pseudo} = 2.0$, pseudo ratio $p = 0.1$. Since we have 4 clients, we report the mean Accuracy for all clients as the final result.

5 Experimental Results

5.1 Comparison with Baseline Methods

The results in Table 2 show that our SemiGraphFL framework significantly and consistently outperforms all baseline methods on all experimental scenarios.

Table 2. Experimental results under different data distribution scenarios.

Model name	IID-data	NIID-1	NIID-2	NIID-3
Single-GCN	22.14%	36.08%	23.95%	28.80%
Single-GAT	22.51%	35.98%	24.51%	29.01%
FedAvg-GCN	35.46%	35.56%	35.66%	35.89%
FedAvg-GAT	35.45%	35.47%	35.53%	35.70%
FedProx-GCN	35.57%	35.81%	24.15%	36.41%
FedProx-GAT	35.72%	35.73%	24.48%	36.51%
FeSEM-GCN	33.86%	36.18%	35.08%	36.58%
FeSEM-GAT	33.94%	36.07%	35.18%	36.78%
GraphCL-GCN	35.28%	35.38%	34.98%	35.47%
GraphCL-GAT	35.37%	35.47%	35.02%	35.60%
Ours-GCN	**38.68%**	**40.31%**	**38.98%**	**39.88%**
Ours-GAT	**38.80%**	**40.62%**	**39.02%**	**40.18%**

For the IID-data scenario, all FL methods significantly outperform the single-training method. And we also came to a similar conclusion for scenarios where there is a slight data distribution difference (NIID-2/3). The above results prove that FL methods can indeed improve the performance of GNN-based graph classification tasks while protecting data privacy in many application scenarios. However, things are different when significant differences in data distributions (NIID-1). In this scenario, models trained with only local data can achieve better results than federated methods. A unified global model restricts each client from making effective predictions based on locally distributed prior knowledge. Therefore, it is necessary to design a personalized FL strategy.

FeSEM is a personalized FL framework that divides clients into multiple small clusters with slight data distribution differences. It only outperforms the classical FL methods FedAvg and FedProx under NIID-3. Under this scenario, clients are naturally divided into two clusters: {Client 1, Client 2} and {Client 3, Client 4}. FeSEM reduces the mutual impact between clients with large differences. However, results in other scenarios illustrate the limitations of this approach. By contrast, our framework can adapt to various scenarios and bring benefits. Recently, Graph Contrastive Learning (GCL) has received extensive attention. We have tried the feature-drop/edge-perturb/sub-graph methods proposed by GraphCL, but none of them improves the final results. We also tried to use GCL to pre-train the model, and the results were not satisfactory. Perhaps a more suitable GCL method for a specific dataset can benefit, and we can add it to the local training part of our framework. This is reserved for future work.

Single-training can be regarded as an extreme case for personalized FL strategies, while FedAvg can be regarded as an extreme case for non-personalized FL strategies. Actually, our SemiGraphFL framework is a more universal framework that generalizes the above two methods. If we only concentrated on multi-client

Table 3. Experimental results of ablation study.

Model name	IID-data	NIID-1	NIID-2	NIID-3
SemiGraphFL	38.68%	40.31%	38.98%	39.88%
Param-sim	38.66%	39.61%	38.56%	39.28%
Label-sim	38.63%	39.42%	38.47%	39.14%
w/o λ_u	36.53%	37.11%	36.48%	36.97%
w/o λ_p	38.13%	38.92%	38.02%	38.78%
Single-pseudo	38.52%	39.51%	38.44%	39.26%

collaboration, SemiGraphFL will result in FedAvg with $T_{model} \rightarrow +\infty$, i.e., $w_{ij}^{model} = 1/N$. And SemiGraphFL will result in single-training with $T_{model} \rightarrow 0$, i.e., $w_{ij}^{model} = 1$ where $i = j$ while $w_{ij}^{model} = 0$ where $i \neq j$.

5.2   Ablation Study and Hyper-Parameter Analysis

We perform ablation studies to illustrate the effectiveness of each module. (1) We compare using the model parameters or the final predictions of labeled data to compute the $Sim(\mathcal{D}_i, \mathcal{D}_j)$, which are denoted as *param-sim* and *label-sim*; (2) We analyze whether the unlabeled supervision loss can bring benefits to the final results, which are denoted as *w/o λ_u*; (3) We remove the pseudo-labeling step, which is denoted as *w/o λ_p*. And we also tried a pseudo-labeling strategy that uses the local model after model aggregation, which is denoted as *single-pseudo*.

The ablation results are included in Table 3. Due to space limitations, we only show results with GCN as the base model, and models based on GAT have similar results. The results show that each module in SemiGraphFL contributes to the final result: (1) Employing hidden representations of unlabeled samples to evaluate the similarities between clients is better than using the model parameters or the hidden representations of labeled data. This allows the model to have better generalization performance; (2) unlabeled supervision restricts the classifier to sharp output is the most critical part of enhancing the predictive performance. It proves that the underlying marginal data distribution $p(x)$ over the input space contains information about the posterior distribution $p(y|x)$; (3) Pseudo label is a general method for semi-supervised tasks. However, if we employ the local model to perform pseudo-labeling, a personalized model tends to select samples similar to local labeled data. This procedure will lead to the accumulation of training bias that hurts the generalization performance. Our new pseudo-label strategy alleviates this problem and enhances the final performance.

For the personalized parameter aggregation, T_{model} is an important hyper-parameter. We conduct experiments to verify how its value affects performance. The results in Fig. 3 show that NIID-1 achieves the best performances when $T_{model} = 0.25$ and NIID-2/3 achieves the best performances when $T_{model} = 0.5$. If T_{model} is too small, the mutual constraints between clients are too weak.

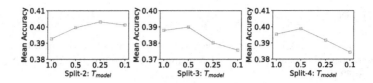

Fig. 3. The parameters sensitivity analyses about T_{model}.

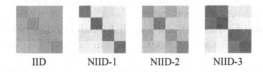

Fig. 4. The parameter aggregation weights w_{ij}^{model} under different Non-IID scenarios.

Each client will over-fit the local labeled data quickly. On the contrary, large T_{model} may lead to a low level of model personalization, which may not be the optimum solution. Generally, scenarios with high data Non-IID level need small T_{model} value and high-level personalization. To verify whether the personalization parameter aggregation is executed as expected, we compute the mean value of w_{ij}^{model} for all communication steps during the training process, which is visualized in Fig. 4. It shows that our method can well evaluate the similarity of data distribution between different clients without exposing raw data.

5.3 Complexity Analysis

Assume there are N clients, K communication rounds, and the model parameter size is d. The communication cost for each client in FedAvg is $O(K \cdot d)$, and for all clients are $O(N \cdot K \cdot d)$. SemiGraphFL is a decentralized framework, the communication cost for each client is $O(N \cdot K \cdot d)$, and the overall cost is $O(N^2 \cdot K \cdot d)$. During local training, the computation cost of additional supervision for unlabeled data is similar to that of labeled data. So the computational cost is about twice as much as before. For most cross-silo FL tasks, N is small (usually $N \leq 10$). And the size d of GNN-based graph classification models is usually small compared with other DNN models, e.g., CV/NLP. Therefore, the additional communication and computation cost is acceptable.

6 Conclusion

This paper proposes a new semi-supervised graph federated learning framework SemiGraphFL. We propose to employ the unlabeled data to enhance the multi-client collaboration and single-client training process, including personalized parameter aggregation, unlabeled data supervision, and FL pseudo-labeling strategy. Extensive experimental results illustrate the effectiveness of our design.

Acknowledgements. The work is partly supported by Delta Research Program.

References

1. Defferrard, M., Bresson, X., Vandergheynst, P.: Convolutional neural networks on graphs with fast localized spectral filtering. Adv. Neural Inf. Process. Syst. **29**, 3844–3852 (2016)
2. He, C., et al.: Fedgraphnn: a federated learning system and benchmark for graph neural networks. arXiv preprint arXiv:2104.07145 (2021)
3. Hu, W., et al.: Open graph benchmark: datasets for machine learning on graphs. arXiv preprint arXiv:2005.00687 (2020)
4. Jeong, W., Yoon, J., Yang, E., Hwang, S.J.: Federated semi-supervised learning with inter-client consistency & disjoint learning. arXiv preprint arXiv:2006.12097 (2020)
5. Jiang, Y., Konečný, J., Rush, K., Kannan, S.: Improving federated learning personalization via model agnostic meta learning. arXiv preprint arXiv:1909.12488 (2019)
6. Jin, Y., Wei, X., Liu, Y., Yang, Q.: Towards utilizing unlabeled data in federated learning: a survey and prospective. arxiv. Learning (2020)
7. Kipf, T.N., Welling, M.: Semi-supervised classification with graph convolutional networks. arXiv preprint arXiv:1609.02907 (2016)
8. Konečný, J., McMahan, H.B., Yu, F.X., Richtárik, P., Suresh, A.T., Bacon, D.: Federated learning: strategies for improving communication efficiency. arXiv preprint arXiv:1610.05492 (2016)
9. Lalitha, A., Kilinc, O.C., Javidi, T., Koushanfar, F.: Peer-to-peer federated learning on graphs. arXiv preprint arXiv:1901.11173 (2019)
10. Li, T., Sahu, A.K., Zaheer, M., Sanjabi, M., Talwalkar, A., Smith, V.: Federated optimization in heterogeneous networks. Proc. Mach. Learn. Syst. **2**, 429–450 (2020)
11. Liu, Z., Chen, C., Yang, X., Zhou, J., Li, X., Song, L.: Heterogeneous graph neural networks for malicious account detection. In: Proceedings of the 27th ACM International Conference on Information and Knowledge Management, pp. 2077–2085 (2018)
12. McMahan, B., Moore, E., Ramage, D., Hampson, S., y Arcas, B.A.: Communication-efficient learning of deep networks from decentralized data. In: Artificial Intelligence and Statistics, pp. 1273–1282. PMLR (2017)
13. Ouali, Y., Hudelot, C., Tami, M.: An overview of deep semi-supervised learning. arXiv preprint arXiv:2006.05278 (2020)
14. Sattler, F., Müller, K.R., Samek, W.: Clustered federated learning: model-agnostic distributed multitask optimization under privacy constraints. IEEE Trans. Neural Netw. Learn. Syst. **32**(8), 3710–3722 (2021)
15. Smith, V., Chiang, C.K., Sanjabi, M., Talwalkar, A.: Federated multi-task learning. arXiv preprint arXiv:1705.10467 (2017)
16. Szklarczyk, D., et al.: STRING v11: protein-protein association networks with increased coverage, supporting functional discovery in genome-wide experimental datasets. Nucleic Acids Res. **47**(D1), D607–D613 (2019)
17. van Engelen, J.E., Hoos, H.H.: A survey on semi-supervised learning. Mach. Learn. **109**(2), 373–440 (2019). https://doi.org/10.1007/s10994-019-05855-6
18. Veličković, P., Cucurull, G., Casanova, A., Romero, A., Lio, P., Bengio, Y.: Graph attention networks. arXiv preprint arXiv:1710.10903 (2017)
19. Wang, H., Yurochkin, M., Sun, Y., Papailiopoulos, D., Khazaeni, Y.: Federated learning with matched averaging. arXiv preprint arXiv:2002.06440 (2020)

20. Xian, X., Wang, X., Ding, J., Ghanadan, R.: Assisted learning: a framework for multi-organization learning. arXiv preprint arXiv:2004.00566 (2020)
21. Xie, M., et al.: Multi-center federated learning. arXiv preprint arXiv:2108.08647 (2021)
22. Xu, K., Hu, W., Leskovec, J., Jegelka, S.: How powerful are graph neural networks? arXiv preprint arXiv:1810.00826 (2018)
23. Ying, R., He, R., Chen, K., Eksombatchai, P., Hamilton, W.L., Leskovec, J.: Graph convolutional neural networks for web-scale recommender systems. In: Proceedings of the 24th ACM SIGKDD International Conference on Knowledge Discovery & Data Mining, pp. 974–983 (2018)
24. You, Y., Chen, T., Sui, Y., Chen, T., Wang, Z., Shen, Y.: Graph contrastive learning with augmentations. Adv. Neural Inf. Process. Syst. **33**, 5812–5823 (2020)
25. Zhang, M., Sapra, K., Fidler, S., Yeung, S., Alvarez, J.M.: Personalized federated learning with first order model optimization. arXiv preprint arXiv:2012.08565 (2020)
26. Zhang, Z., et al.: Improving semi-supervised federated learning by reducing the gradient diversity of models. In: 2021 IEEE International Conference on Big Data (Big Data), pp. 1214–1225. IEEE (2021)

Evolvable Hardware and Evolutionary Robotics

Evolutionary Design of Reduced Precision Preprocessor for Levodopa-Induced Dyskinesia Classifier

Martin Hurta$^{(\boxtimes)}$ (ID), Michaela Drahosova (ID), and Vojtech Mrazek (ID)

Faculty of Information Technology, Brno University of Technology,
Brno, Czech Republic
{ihurta,drahosova,mrazek}@fit.vut.cz

Abstract. The aim of this work is to design a hardware-efficient implementation of data preprocessing in the task of levodopa-induced dyskinesia classification. In this task, there are three approaches implemented and compared: 1) evolution of magnitude approximation using Cartesian genetic programming, 2) design of preprocessing unit using two-population coevolution (2P-CoEA) of cartesian programs and fitness predictors, which are small subsets of training set, and 3) a design using three-population coevolution (3P-CoEA) combining compositional coevolution of preprocessor and classifier with coevolution of fitness predictors. Experimental results show that all of the three investigated approaches are capable of producing energy-saving solutions, suitable for implementation in hardware unit, with a quality comparable to baseline software implementation. Design of approximate magnitude leads to correctly working solutions, however, more energy-demanding than other investigated approaches. 3P-CoEA is capable of designing both preprocessor and classifier compositionally while achieving smaller solutions than the design of approximate magnitude. Presented 2P-CoEA results in the smallest and the most energy-efficient solutions along with producing a solution with significantly better classification quality for one part of test data in comparison with the software implementation.

Keywords: Cartesian genetic programming · Compositional coevolution · Adaptive size fitness predictors · Levodopa-induced dyskinesia · Approximate magnitude · Energy-efficient

1 Introduction

Parkinson's disease (PD) is one of the most common neurological conditions affecting the motor system. Treatment of symptoms usually involves the administration of a drug containing levodopa. The proper dosage of levodopa is crucial for reducing its side effects, which include levodopa-induced dyskinesia (LID). Lones et al. [4] developed a non-invasive wearable monitoring system for assessing LID in people with PD. In our previous work [3], we followed up their work with a goal to evolve LID-classifiers with respect to hardware (HW) implementation

© The Author(s), under exclusive license to Springer Nature Switzerland AG 2022
G. Rudolph et al. (Eds.): PPSN 2022, LNCS 13398, pp. 491–504, 2022.
https://doi.org/10.1007/978-3-031-14714-2_34

and thus enable LID-classifier to be implemented directly in a home wearable device. We successfully applied *coevolution of cartesian programs and adaptive size fitness predictors* (CGPcoASFP) [1] in order to design LID-classifiers working with fixed-point arithmetic with reduced precision, which is suitable for implementation in application-specific integrated circuits (ASIC).

However, evolved HW-efficient LID-classifier still employed data preprocessing in the form of magnitude calculation, which is considerably complex to be implemented directly in HW as the square root has to be calculated. In the most effective and accurate HW implementations, the square root is typically computed iteratively in $n/2$ cycles for n-bit inputs using additions and subtractions. That makes the calculation complex and power-inefficient. To achieve a good throughput, the operation can be pipelined but the final area on the chip significantly grows up. The square root can also be approximated by a lookup table, but for 12-bit inputs such a table would be enormous. Therefore the goal of this paper is to replace this preprocessor with HW-effective implementation while keeping a reasonable accuracy of the classifier and thus to produce a solution that could be implemented directly into a home wearable device. A small low-power solution would enable long-term continuous monitoring of people with PD in their own homes and allow clinicians accurate assessment of their patient's condition and the advised adjustment of levodopa dosage.

We propose and evaluate three evolutionary approaches to design HW-efficient preprocessor and classifier. The first approach represents the state-of-the-art approximation of magnitude in terms of *symbolic regression* with the use of HW-friendly functions performed by *Cartesian genetic programming* (CGP) [7]. The second approach designs a preprocessor that is connected to an existing LID-classifier [3] using CGPcoASFP. The third approach is *compositional coevolution* [8] of preprocessor and classifier supplemented with the coevolution of adaptive size fitness predictors in order to accelerate the design process. The proposed evolutionary approaches are compared with a baseline software implementation in terms of classification quality. Selected solutions evolved using proposed approaches are then compared in terms of their HW characteristics.

2   LID-Classifier Model

Lones et al. [4] developed a system for monitoring LID in people with PD's own homes, which included non-invasive wearable units and a model of LID-classifier. Sensing modules comprised a tri-axial accelerometer and a tri-axial gyroscope and stored data in local memory.

Baseline Preprocessing: Data used for classification of LID contain data from the three-axial accelerometer, recorded with a sample rate 100 Hz. Each sample contains three unsigned 12-bit numbers corresponding to axes a_1, a_2 and a_3, which need to be normalized by some preprocessing algorithm. Baseline data preprocessing for LID-classifier, as introduced by Lones et al. [4], consists in calculating a magnitude a from a_1, a_2 and a_3 values, using the formula:

$$a = \sqrt{(a_1^2 + a_2^2 + a_3^2)}. \tag{1}$$

Classification of LID: According to [4], a *fitness case* for LID-classifier consists of L preprocessed values (i.e. $L = 100$ represents a record of length 1 s). The fitness case is processed through classifier in the form of $L - 31$ overlapping windows of length 32 (0.32 s). The classifier produces an output for each of the $L - 31$ windows (i.e. input vectors), and the resulting response from the classifier is then expressed as the mean of output values. A target class is then decided by applying a threshold to the response range of the classifier. The threshold is not included in the classifier model, as for determining the quality of classifiers during evolution we use AUC (Area Under the receiver operating characteristics Curve), which allows accurate assessment of the ability to distinguish classes without defining a threshold value. The procedure of LID classification is illustrated in Fig. 1.

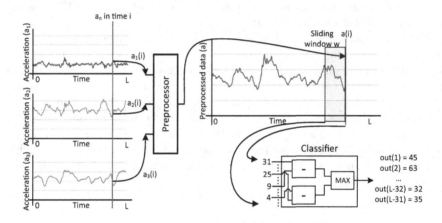

Fig. 1. LID-classifier model. Figure also shows classifier C0 proposed in [3].

In our previous work [3], we presented a HW-efficient LID-classifier operating with unsigned 8-bit data representation and including only a few function blocks, evolved using CGPcoASFP. The resulting classifier is shown in Fig. 1 and is labelled as C0 in the rest of this paper. To adjust magnitude calculated according to Eq. 1 to the range of 0 to 255, a logical shift by five bits to the right has shown as the most suitable from investigated techniques in this task [3].

3   Evolutionary Design of LID-Preprocessor

Hardware implementation of data preprocessing using Eq. 1 is considerably complex. Our goal is to replace expensive magnitude calculation with a simpler solution evolved using CGP which was successful in various related topics, such as the design of efficient digital circuits [5], solving problems of classification and prediction [6], or medical applications [10].

3.1 Model of Magnitude Approximation

The first approach replaces computationally expensive magnitude calculation with a fixed-point magnitude approximation, evolved in terms of symbolic regression performed using standard CGP [7]. A preprocessor is then represented using a cartesian grid of functional nodes operating over three primary inputs, taken from three axes a_1, a_2 and a_3, and giving one primary output representing the magnitude approximation.

3.2 Model of LID-Preprocessor

The aim of this approach is to evolve a preprocessor without a need of defining a specific mathematical expression, that preprocessor should accomplish. In this model, data from clinical studies (described below in Sect. 4.1) that are employed in LID-classifier design, are directly used during LID-preprocessor evolution. Fitness of LID-preprocessor is then evaluated in connection with LID-classifier, i.e. in terms of classification quality.

At first, we used the same preprocessor model as for magnitude approximation, i.e. the cartesian grid operating over the three axes a_1, a_2 and a_3, and giving one primary output. Our initial experiments have shown, that evolution of this model produces preprocessors that prefer to operate over one or two of three axes, which does not lead to correctly working solutions in general.

Thus, we propose to evolve one component to be used for operating each of the three axes and then aggregated. In this model, we propose to aggregate component outputs in the following way: Each component output shift by 2 bits to the right and summed. This model affords the important property of the same weighting of all axes.

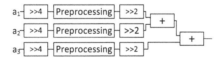

Fig. 2. The layout of proposed preprocessing enforcing equal weight of all axes.

As we aim to use 8-bit unsigned integer representation in LID-classifier, it is fundamental to use this representation in preprocessor. For this reason, acceleration values, which are originally in 12-bit range, are at first fitted to a range of 0–255 by a logical shift of 4 bits to the right. The LID-preprocessor model is illustrated in Fig. 2.

For the automated design of the LID-preprocessor, we propose to use two-population coevolution of cartesian programs and adaptive size fitness predictors (CGPcoASFP). Next, we propose a three-population coevolutionary algorithm where two types of candidate components, one of LID-preprocessors and one of LID-classifiers, are evolved in separate populations and composed to evaluate their fitness together using fitness predictors (the third population).

3.3   Two-Population Coevolutionary Algorithm

There are two concurrently evolving populations in the proposed two-population coevolutionary algorithm (2P-CoEA), one of candidate programs (LID-prepro-cessors) evolving using CGP and one containing fitness predictors (FP, i.e. small subsets of training set). FPs are evolving using a simple genetic algorithm accompanied with a specific heuristic allowing to change the size of fitness predictor dynamically in response to the evolutionary process [1]. The variable size of the fitness predictor helps to evaluate candidate programs on the proper amount of input data, i.e. to find a good trade-off between the time and quality of evaluation. Fitness predictors with variable size reduce evolution time, allow leaving local optima and help to prevent overfitting [1,3]. Both populations evolve simultaneously and interact through the fitness function. Two archives, one of fitness trainers (A_{FT}) and one containing the best-evolved fitness predictors (A_{FP}), supplement these populations. A_{FT} is used by the predictor population for the evaluation of evolved fitness predictors. It contains copies of selected candidate programs obtained during the evolution. The fitness predictor from A_{FP} is used to evaluate candidate programs. The method is depicted in Fig 3. The detailed approach description is summarised by Drahosova et al. [1].

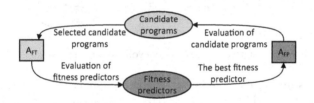

Fig. 3. Two-population coevolutionary algorithm.

3.4   Three-Population Coevolutionary Algorithm

Inspired by the previous work of Sikulova et al. [9], we propose to employ a compositional coevolution in this task. Compositional coevolution sprang from cooperative coevolutionary algorithms, wherein the originally stated aim was to attack the problem of evolving complex objects by explicitly breaking them into parts, evolving these parts separately and then assembling the parts into a working whole [8]. When designing LID-preprocessor, we naturally use two components: 1) the preprocessor and 2) the classifier; see Fig. 1. Our target is to replace the evolution of non-interacting components with a coevolutionary algorithm, in which the fitness of a component depends on fitness of other components, i.e. the components are adapted to work together. Two populations, one of preprocessors and one of classifiers, are accompanied with three archives: a) the archive of preprocessors (A_{prep}) used by the classifiers population for their evaluation, b) the archive of classifiers (A_{class}) used by the preprocessors population for their evaluation, and c) the archive of the top-ranked composition (A_{comp}). The detailed approach description for use with CGP is summarised by Sikulova et al. [9]. Fitness of each candidate preprocessor is thus determined as a fitness of the whole module composed of the candidate preprocessor and a classifier in the A_{class}, and vice-versa.

Moreover, we propose to supplement this compositional coevolutionary approach with the coevolution of fitness predictors to evaluate the fitness of candidate components and thus include its benefits in the compositional approach. Then, this approach involves three populations (i.e. three-population coevolutionary algorithm; 3P-CoEA), one of preprocessor components, one of classifier components and one containing FPs. These three populations are supplemented with five archives: a) A_{prep}, b) A_{class}, c) A_{comp}, d) A_{FT} - i.e. compositions used by the predictor population for the evaluation of fitness predictors, and e) A_{FP}.

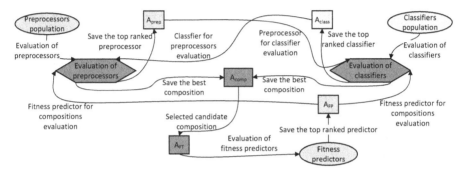

Fig. 4. Three-population coevolutionary algorithm.

3P-CoEA is initialised as follows: All three populations are initialised using random individuals while *complete mixing* interaction of populations [8] is utilised to evaluate initial generation. That means, all individuals of preprocessors are composed with all individuals of classifiers and their fitness is evaluated. Fitness predictors are evaluated using randomly initialized A_{FT}. The top-ranked preprocessor is then copied to A_{prep}, the top-ranked classifier to A_{class}, the top-ranked fitness predictor to A_{FP}, A_{FT} is filled with selected compositions, while the top-ranked composition is placed to A_{comp}. For further generations, the complete mixing approach is not utilised (as it is time demanding) – only the top-ranked preprocessor from A_{prep} is utilised to evaluate classifiers and the top-ranked classifier from A_{class} is used to evaluate preprocessors. When a new top-ranked preprocessor occurs in preprocessor population, it replaces an old one in A_{prep}, and vice-versa. A_{comp} is used as evolutionary process memory. After the coevolution is terminated, a final solution, i.e. the best-evolved composition, is found in A_{comp}. Interactions of fitness predictors are utilised in terms of CGPcoASFP [1]. The overall methodology is depicted in Fig. 4.

4   Experimental Results

This section presents experimental data, experimental setup and experimental evaluation of the proposed coevolutionary approaches and their comparisons.

4.1 Clinical Study Data

In our work, we adopt two clinical studies (from [4]) conducted at Leeds Teaching Hospitals NHS Trust, UK, with granted ethics approval and written informed consent given by all participants. Collected movement records were split into time periods, of different lengths, according to the severity of LID it character-izes. Each time period (time-series) is graded by the standard UDysRS (Unified Dyskinesia Rating Scale) scoring system from 0 (no dyskinesia) to 4 (severe dyskinesia) and contains information, among others, about the patient's activ-ity during a set moment, e.g. sitting at rest, walking, drinking.

Data from clinical studies are used the same as by Lones et al. [4] and Hurta et al. [3] in their previous works. Due to the sliding window size in the proposed model, only time-series of at least 32 samples (i.e. 0.32 s) are considered. As described in [4], training data is constructed of Clinical study 1 [4] using 2939 time series of LID grade 0, understood to be LID negative (N) and 745 time series of merged grades 3 and 4, understood to be LID positive (P). The reason for excluding grades 1 and 2 is that it is easier to generate robust classifiers when these are not involved during training, as described in [4].

Evolved solutions are re-evaluated using the movement samples from Clinical study 2 [4] to obtain a more robust measure of generality. Six test groups are created to allow detailed measurement of the LID-classifier quality, see Table 1. Besides four test groups representing classification of LID severity (LID 0 to LID 4), two more groups are created to classify severe LID (i.e. LID grade 3 and 4) during specific movement activities of sitting at rest and walking.

Table 1. Number of positive (P) and negative (N) samples in training data used for fitness calculation and in test groups used for re-evaluation of solutions in order to obtain a more robust measure of generality.

	Training data (LID 3 + LID 4)	Test groups					
		LID 1	LID 2	LID 3	LID 4	*Walking*	*Sitting*
Number of (P)	745	895	628	179	361	21	170
Number of (N), i.e. LID 0	2939	1588	1588	1588	1588	90	733

4.2 Test Scenarios

In order to evaluate the proposed approach, three scenarios and thus three approaches to LID-preprocessor design are examined:

Scenario S1: The first scenario replaces computationally expensive magnitude calculation with a fixed-point *magnitude approximation*, searched in terms of symbolic regression performed using standard CGP. In order to form a training set representing LID-preprocessor function (as used in [3]), 68 921 equidistant distributed samples were taken from function

$$f_{magnitude}(a_1, a_2, a_3) = \sqrt{(a_1{}^2 + a_2{}^2 + a_3{}^2)} >> 5, \tag{2}$$

with sampling $U[0, 100, 4095]$ for training and $U[0, 1, 4095]$ for testing. A fitness function in Scenario S1 is represented by the mean square error (MSE).

Scenario S2: The second scenario involves the 2P-CoEA design of LID-prepro-cesor that is trained using training data set from clinical studies Sect. 4.1. The preprocessor is then evaluated in composition with the fixed C0 classifier [3] in terms of AUC (Area Under the receiver operating characteristics Curve). This allows us to find such a preprocessor that is adapted to work with the C0 classifier without finding a magnitude function. The training set consists of 2939 fitness cases of class (N) and 745 fitness cases of class (P), as described in Sect. 4.1.

Scenario S3: The third scenario employs 3P-CoEA to evolve LID-preprocessor together with the evolution of LID-classifier to adapt them to work together directly during evolution. AUC fitness for each component evaluation is used as same as in Scenario S2. The detailed AUC-based fitness calculation in the task of LID-classifier design is summarised in Hurta et al. [3]. Presented scenarios are illustrated in Fig. 5.

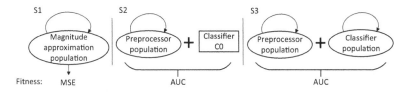

Fig. 5. Presented scenarios.

4.3   Experimental Setup

The experimental setup of CGP uses some shared parameter settings for all scenarios S1–S3 as follows (based on our previous work [3]): The initial populations are randomly seeded, the $(1 + 4)$ evolutionary strategy and the Goldman mutation operator [2] are used to produce a new generation.

The remaining CGP parameters are based on our initial experiments. The set of functions was altered to reflect our intention to design HW efficient solution. Computationally expensive multiply and divide functions used in [3] were removed. The set of functions has been supplemented with functions suitable for use in HW and successfully used for magnitude approximation search by Wiglasz and Sekanina [11]. The used function set is shown in Table 2. The remaining CGP parameters differ for surveyed scenarios.

Table 2. CGP node functions.

Function	Description	Function	Description	Function	Description
255	Constant	$\overline{i_1 \wedge i_2}$	Bit. NAND	$i_1 >> 1$	Right shift by 1
i_1	Identity	$i_1 \oplus i_2$	Bit. XOR	$i_1 >> 2$	Right shift by 2
$255 - i_1$	Inversion	$\overline{i_1} \vee i_2$	Bit. $\overline{i_1}$ OR i_2	$i_1 - i_2$	$-$ (subtraction)
$i_1 \vee i_2$	Bit. OR	$swap(i_1, i_2)$	Swap nibbles	$i_1 -^S i_2$	$-$ with saturation
$i_1 \wedge i_2$	Bit. AND	$i_1 + i_2$	+ (addition)	$i_1 +^S i_2$	+ with saturation
$min(i_1, i_2)$	Minimum	$max(i_1, i_2)$	Maximum	$(i_1 >> 1) + (i_2 >> 1)$	Mean

CGP used to train an approximate magnitude according to scenario S1 comprised of up to 96 function instances laid out on a 12×8 Cartesian plane. CGP takes inputs from 3 terminal nodes fed from the accelerometer axes (a_1, a_2 and a_3). We used a generation limit of 100 000.

To train a preprocessor according to scenarios S2 and S3, CGP comprises up to 32 function instances laid out on a 4×8 Cartesian plane. CGP takes inputs from 1 terminal node fed from the accelerometer axis, i.e. one of a_1, a_2 and a_3 as an aggregation is done according to Fig. 2. A generation limit is set to 20 000, where significant changes are no longer observed in evolved solutions. CGP used to train a classifier according to scenario S3 differs in having 32 inputs from 32 terminal nodes fed from the preprocessor outputs (see Fig. 1). Function set is also given by Table 2.

CGPcoASFP employed in scenarios S2 and S3 is used according to [3], i.e. 6 fitness trainers in the A_{FT}, 8 FPs in the predictor population, and evolution of FPs is conducted using a simple GA, where one-point crossover and mutation with probability 0.01 per gene operators are used. The size of FP is initialized with 300 fitness cases, which is around 8 % of the original training set. The detailed setup of CGPcoASFP for the classifier design task is surveyed in [3].

4.4   The Quality and the Size of Evolved Solutions

The quality of solutions evolved using proposed approaches evaluated using test groups (Table 1) is shown in Fig. 6. Preprocessors evolved using S1 and S2 are composed with C0 classifier in order to evaluate AUC, preprocessors evolved using S3 are composed with corresponding classifiers coevolved with them according to S3. It can be seen that all three investigated scenarios are capable of producing solutions with AUC comparable to baseline implementation, i.e. exact magnitude calculation composed with classifier C0.

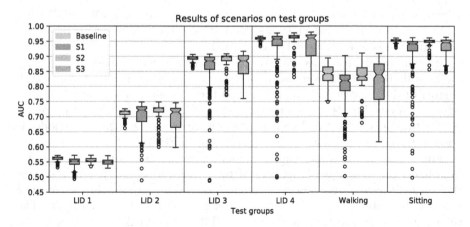

Fig. 6. AUC of baseline solution and solutions evolved using proposed scenarios on the test groups. Baseline represents evolution with the use of exact magnitude [3].

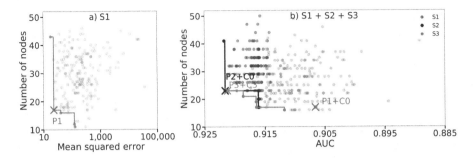

Fig. 7. Trade-off between fitness and number of nodes in evolved solutions composed with corresponding classifiers. Figure **a)** shows MSE fitness vs. number of nodes for scenario S1. Figure **b)** shows AUC fitness vs. number of nodes for scenarios S2 and S3 supplemented with the selected solution from S1 re-evaluated in terms of AUC. Lines represent the Pareto frontiers and cross highlight the selected solutions. (Color figure online)

It can be noticed that the LID 1 and LID 2 cases were much more difficult to fit. It might be caused by negligible signs of lower LID grades in some body parts.

As our target is a small solution (i.e. highly optimized energy-efficient implementation) along with the classification accuracy as close as possible to a baseline software implementation, we analyse evolved solutions in terms of the number of active nodes and the AUC fitness. Figure 7 shows the fitness and the number of nodes for all evolved solutions.

In S1, five Pareto optimal solutions were found spanning their fitness (MSE) on test data from 18.25 to 126.99. Their number of active nodes, including the C0 classifier, ranges from 11 to 43, see Fig. 7a. The preprocessor (P1) with MSE = 22.95 (AUC = 0.91 for P1 in composition with C0) using 17 active nodes is selected for further evaluation of HW characteristics. It should be noted that a lower MSE does not lead to a solution with a better classification AUC – see the red cross in Fig. 7b, which shows the AUC and the number of used nodes of selected P1 in composition with classifier C0.

In scenario S2, only three Pareto optimal solutions with fitness (AUC) ranging from 0.916 to 0.922 were found, see the green Pareto frontier in Fig. 7b. A number of their active nodes multiplied by three axes and with the addition of aggregation (see Fig. 2) and C0 classifier spans from 17 to 41. The preprocessor (P2) with AUC = 0.92 (for P2 in composition with C0) using 23 active nodes is selected for further evaluations (the green cross in Fig. 7b).

Five Pareto optimal solutions were found in scenario S3 with fitness (AUC) from 0.912 to 0.921 and the number of total nodes including preprocessing of three axes, their aggregation and composition with corresponding co-evolved classifier from 16 to 23 nodes, see the orange Pareto frontier in Fig. 7b. The composition of preprocessor and classifier (P3+C3) with AUC = 0.92 using 23 active nodes is selected for further evaluations (the orange cross in Fig. 7b).

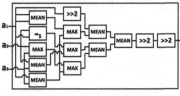

(a) Magnitude approximation P1.

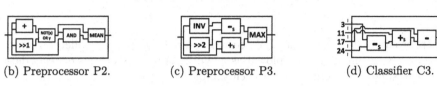

(b) Preprocessor P2.   (c) Preprocessor P3.   (d) Classifier C3.

Fig. 8. Selected solutions of scenarios S1 (P1), S2 (P2) and S3 (P3 and C3).

The number of nodes in solutions designed by individual scenarios tends to be the lowest in solutions designed by scenario S3, with a mean of 27 nodes, an increase in size by two nodes in scenario S2 and an additional node in S1.

4.5   HW Characteristics of Selected Preprocessors

For the evaluation of HW characteristics, one solution from each of the scenarios is selected. To select precise enough and, at the same time, space-saving solution, a solution from the Pareto frontier with fitness in the top ten percent of the fitness span with the lowest number of used nodes is selected. In Fig. 7, the selected solutions are marked by the cross.

Selected magnitude approximation (preprocessor P1), evolved according to scenario S1, is shown in Fig. 8a. Figure 8b shows the preprocessor P2 evolved according to scenario S2 for composition with the baseline classifier C0. Coevolutionary designed preprocessor P3 and classifier C3, evolved according to scenario S3, i.e. using 3P-CoEA, are shown in Fig. 8c and Fig. 8d.

AUC of the presented solutions are shown in Table 3. All three proposed solutions achieve comparable AUC to the baseline implementation (P0+C0) on all investigated test groups. Moreover, the solution evolved using scenario S2, i.e. P2+C0, achieved significantly better AUC on the test group "walking".

Table 3. AUC of presented preprocessors and classifiers on training data and on test groups according to Table 1. The highest AUC of each group is marked in bold font. A significant improvement of AUC on test group *Walking* is marked in italic bold font.

	Training data	LID 1	LID 2	LID 3	LID 4	*Walking*	*Sitting*
P0+C0 (baseline solution)	0.91	**0.57**	0.72	0.89	0.96	0.82	**0.95**
P1+C0 (S1: magnitude approximation)	0.91	0.55	**0.73**	0.89	0.96	0.82	**0.95**
P2+C0 (S2: 2P-CoEA)	**0.92**	0.56	**0.73**	0.89	**0.97**	*0.87*	**0.95**
P3+C3 (S3: 3P-CoEA)	**0.92**	0.55	**0.73**	0.90	**0.97**	0.83	**0.95**

Table 4. HW characteristics of synthesized solutions for a 45 nm technology on 100 MHz frequency.

LID-Classifier	Area on chip [µm²]	Number of cycles	Power [mW]	Energy [pJ]
P0+C0 (baseline solution)	7115	13	1.0459	135.967
P1+C0 (S1: magnitude approx.)	5014	1	0.8593	8.593
P2+C0 (S2: 2P-CoEA)	2435	1	0.4645	4.645
P3+C3 (S3: 3P-CoEA)	2629	1	0.5131	5.131

In order to determine the HW cost, Synopsys Design Compiler targeting 45 nm ASIC technology is employed as a synthesis tool. Synthesis expects data preprocessing to be done once for each sample in a stream of acceleration data. This is in contrast to placing preprocessor before each of the primary inputs of a classifier that would generally result in four to five times more calculations in preprocessing stage and require three times more storage.

A combination of magnitude approximation preprocessor P1 and classifier C0 are synthesised with 16-bit data representation necessary for preprocessor P1. Combination P2+C0 and P3+C3 use a more efficient 8-bit representation. Preprocessors P2 and P3 are in both cases presented in the solution three times together with the aggregation unit shown in Fig. 2.

Investigated hardware characteristics are shown in Table 4. The baseline solution P0+C0 is the largest LID-classifier in terms of area on chip. Iterative calculation of square roots also requires 13 clock cycles (i.e. 130 ns), resulting in an energy of 136 pJ. All proposed solutions result in better characteristics. Solution P1+C0 (Scenario 1) reduces area on chip by 30% with the main benefit of requiring only one clock cycle for calculation and thus almost 16 times reduction of energy. Solutions P2+C0 (Scenario 2) and P3+C3 (Scenario 3) both result in an additional reduction of area on chip compared to solution P1+C0 due to 8-bit data representation. Energy of P2+C0 and P3+C3 also achieve an additional reduction of energy in comparison with P1+C0 by 46%, 40% respectively.

5   Conclusion

In this paper, we proposed and compared three methods for the automatic design of an energy-efficient variant of data preprocessing for the levodopa-induced dyskinesia classifier. We have shown that all of three investigated methods, including compositional coevolution of preprocessor and classifier, are capable of producing space-saving solutions with AUC comparable to the baseline software implementation.

Evolutionary design of magnitude approximation using Cartesian genetic programming (CGP) (scenario S1) leads to larger and more energy-demanding solutions. Compositional coevolution of preprocessor and classifier supplemented with coevolution of adaptive size fitness predictors (scenario S3) has resulted on an average in the smallest solutions while evolution of preprocessor for the baseline classifier implementation (scenario S2) has resulted in a solution with significantly better AUC for "walking" test-group in comparison with other investigated approaches. All of the proposed solutions significantly improve investigated hardware characteristics compared to the baseline solution, which is the most evident in up to 29 times lower energy consumption.

Our future work will be devoted to utilisation of compositional evolution and inclusion of multi-objective optimization into the design of HW-efficient solutions, whereas HW parameters of candidate solutions will be concerned during the evolutionary process.

Acknowledgements. This work was supported by the Czech science foundation project 21-13001S. The computational resources were supported by the MSMT project e-INFRA CZ (ID:90140). We also acknowledge Prof. Lukas Sekanina, Prof. Stephen Smith and Dr. Michael Lones for their advice; Dr. Jane Alty, Dr. Jeremy Cosgrove, and Dr. Stuart Jamison, for their contribution to the clinical study that generated the data, and the UK National Institute for Health Research (NIHR) for adopting the study within in its Clinical Research Network Portfolio.

References

1. Drahosova, M., Sekanina, L., Wiglasz, M.: Adaptive fitness predictors in coevolutionary cartesian genetic programming. Evol. Comput. **27**(3), 497–523 (2019). https://doi.org/10.1162/evco_a_00229
2. Goldman, B.W., Punch, W.F.: Reducing wasted evaluations in cartesian genetic programming. In: Krawiec, K., Moraglio, A., Hu, T., Etaner-Uyar, A.Ş, Hu, B. (eds.) EuroGP 2013. LNCS, vol. 7831, pp. 61–72. Springer, Heidelberg (2013). https://doi.org/10.1007/978-3-642-37207-0_6
3. Hurta, M., Drahosova, M., Sekanina, L., Smith, S.L., Alty, J.E.: Evolutionary design of reduced precision levodopa-induced dyskinesia classifiers. In: Medvet, E., Pappa, G., Xue, B. (eds.) EuroGP 2022: Proceedings of the 25th European Conference on Genetic Programming, LNCS, pp. 85–101. Springer Verlag, Madrid (2022). https://doi.org/10.1007/978-3-031-02056-8_6
4. Lones, M.A., et al.: A new evolutionary algorithm-based home monitoring device for Parkinson's dyskinesia. J. Med. Syst. **41**(11), 1–8 (2017). https://doi.org/10.1007/s10916-017-0811-7
5. Manazir, A., Raza, K.: Recent developments in cartesian genetic programming and its variants. ACM Comput. Surv. **51**(6), 1–29 (2019). https://doi.org/10.1145/3275518
6. Miller, J.F.: Cartesian genetic programming. In: Cartesian Genetic Programming, pp. 17–34. Springer, Heidelberg (2011). https://doi.org/10.1007/978-3-642-17310-3_2
7. Miller, J.F.: Cartesian genetic programming. In: Natural Computing Series, vol. 43, 1 edn. Springer, Heidelberg (2011). https://doi.org/10.1007/978-3-642-17310-3
8. Popovici, E., Bucci, A., Wiegand, R., De Jong, E.: Coevolutionary principles. In: Handbook of Natural Computing, pp. 987–1033. Springer, Heidelberg (2012). https://doi.org/10.1007/978-3-540-92910-9_31

9. Sikulova, M., Komjathy, G., Sekanina, L.: Towards compositional coevolution in evolutionary circuit design. In: 2014 IEEE International Conference on Evolvable Systems, pp. 157–164. Institute of Electrical and Electronics Engineers, Piscataway (2014). https://doi.org/10.1109/ICES.2014.7008735

10. Smith, S.L., Lones, M.A.: Medical applications of cartesian genetic programming. In: Stepney, S., Adamatzky, A. (eds.) Inspired by Nature, pp. 247–266. Springer International Publishing, Cham (2018). https://doi.org/10.1007/978-3-319-67997-6_12

11. Wiglasz, M., Sekanina, L.: Cooperative coevolutionary approximation in hog-based human detection embedded system. In: 2018 IEEE Symposium Series on Computational Intelligence (SSCI 2018), pp. 1313–1320. Institute of Electrical and Electronics Engineers (2018). https://doi.org/10.1109/SSCI.2018.8628910

In-Materio Extreme Learning Machines

Benedict. A. H. Jones[1]([✉]) [iD], Noura Al Moubayed[2] [iD], Dagou A. Zeze[1] [iD],
and Chris Groves[1] [iD]

[1] Department of Engineering, Durham University, Durham DH1 3LE, UK
{benedict.jones,chris.groves}@durham.ac.uk
[2] Department of Computer Science, Durham University, Durham DH1 3LE, UK

Abstract. Nanomaterial networks have been presented as a building
block for unconventional in-Materio processors. Evolution in-Materio
(EiM) has previously presented a way to configure and exploit physical
materials for computation, but their ability to scale as datasets get larger
and more complex remains unclear. Extreme Learning Machines (ELMs)
seek to exploit a randomly initialised single layer feed forward neural network by training the output layer only. An analogy for a physical ELM
is pro0duced by exploiting nanomaterial networks as material neurons
within the hidden layer. Circuit simulations are used to efficiently investigate diode-resistor networks which act as our material neurons. These
in-Materio ELMs (iM-ELMs) outperform common classification methods
and traditional artificial ELMs of a similar hidden layer size. For iM-
ELMs using the same number of hidden layer neurons, leveraging larger
more complex material neuron topologies (with more nodes/electrodes)
leads to better performance, showing that these larger materials have a
better capability to process data. Finally, iM-ELMs using virtual material neurons, where a single material is re-used as several virtual neurons,
were found to achieve comparable results to iM-ELMs which exploited
several different materials. However, while these Virtual iM-ELMs provide significant flexibility, they sacrifice the highly parallelised nature of
physically implemented iM-ELMs.

Keywords: Evolution in-Materio · Evolvable processors · Extreme
learning machines · Material neurons · Virtual neurons · Classification

1 Introduction

The inevitable slowdown in traditional CMOS technology improvement [6] has
led to a growing interest in alternative and unconventional computing techniques.
Evolution in-Materio (EiM) is one such paradigm which attempts to leverage a
material's inherent properties and exploit them for computation, these materials
were initially referred to as Configurable Analogue Processors [29]. EiM uses an
Evolutionary Algorithm (EA) to optimise external stimuli and other parameters such that the material can perform a target application. Materials which
might have limited uses in conventional electronic devices, may in fact provide

G. Rudolph et al. (Eds.): PPSN 2022, LNCS 13398, pp. 505–519, 2022.
https://doi.org/10.1007/978-3-031-14714-2_35

sufficiently complex and interesting characteristics to be exploited as an EiM device. EiM processors have been used to achieve a range of applications such as logic gates [2,4,23,27] and for classification [5,28,40]. Notably, EiM attempts to exploit a system's intrinsic physics with only a few configurable parameters, which is important for the complexity engineering approach [15] and helps prevent an over parametrised system. If this relative computational simplicity is paired with low power materials, then in-Materio processors present a possible candidate for edge case computing [30]. Recent analysis of EiM systems has established fundamental good practices [21] and enhancements [20] to the EiM paradigm. However, even with these improvements, methods to scale in-Materio processing systems to larger and more complex datasets remains lacking. Without this, any real-world adoption remains unlikely.

Generally, within a 'material' or 'in-Materio' processor, some nanomaterial is placed on a microelectrode such that stimuli and data can be applied as a voltage. Some form of hardware interface is necessary to apply and read voltages to the material, often controlled from a PC (which in the case of EiM hosts the EA that optimises the system). Devices operating in such a manner include liquid crystals [17,18], metallic nanoparticles [2,16], single walled carbon nanotubes [27,28], and dopant networks [4,35]. However, in-Materio processors could be configured using any external stimuli such as light [39] or radio waves [25]. In fact, any medium with complex intrinsic characteristics which can be interfaced with and leveraged could be used as an in-Materio processor. Networks of common electronic devices (resistors, diodes, etc.) can provide interesting tunable dynamics [22] and can be realised physically or investigated using reliable and fast SPICE (Simulation Program with Integrated Circuit Emphasis) simulations [20,21].

Extreme Learning Machines (ELM) and Reservoir Computing (RC) present a good analogy for in-Materio processors since both involve the exploitation of random networks. These systems depend on the underlying assumption that the randomised network/reservoir will produce useful and often higher dimensional output states that are used to process the data more successfully. Notably, within these fields of research it is generally assumed that the network/reservoir remains fixed after its inception. However, previous work has shown that a small amount of stochastic optimisation can improve a systems performance [7,13,43]. RC was developed from recurrent neural networks and is generally employed to process temporal data. Physical RCs [38] could lead to low power, efficient and fast systems which can operate at 'the edge'. Examples include the use of circuit (anti-parallel diode) based non-linear neuron [22], memristive network [1,12] and magnetic spintronic [8] based reservoirs. ELMs were developed from single layer feed forward neural networks (SLFN) and are generally employed to process non-temporal data [19]. Examples of physical implementations of ELM remain sparse but include memristor based networks [1] and photonic systems [26,32]. Their remains significant opportunity to develop both classical and quantum substrates [31] for both RC and ELM.

Here we present a method of exploiting nanomaterial networks as 'material neurons', grouping them into a SLFN's hidden layer and training them as an ELM. To enable efficient analysis, random diode-resistor networks are used as a proxy for physical nanomaterials, which are solved using fast, reliable SPICE simulations. The performance of these 'in-Materio ELMs' on several common machine learning datasets is examined for various hidden layer sizes and physical material network topologies. They are found to outperform other common classification techniques and traditional (artificial) ELMs of a similar size. Finally, drawing from the work showing EiM processors can be configured via external voltage stimuli and other parameters, we implement a material 're-use' system, whereby a single material neuron is used to create several virtual material neurons. These physical neuron based ELMs provide a scalable in-Materio unconventional computing method whereby the intrinsic properties of a material (or medium) can be exploited in a highly parallelisable way.

2 Background

2.1 Evolution In-Materio Processors

EiM exploits nanomaterials using an optimisation algorithm such that they can perform useful tasks. Since in-Materio processors are analogue and generally lack an analytical model to describe their electrical characteristics, derivative-free optimisation algorithms are used, rather than gradient based algorithms [28]. EAs are a subset of evolutionary computing [36], consisting of population-based metaheuristic search algorithms, making them ideal for EiM. Many types of EAs have been used for EiM such as Evolutionary Strategies [9], Genetic Algorithms [2], Differential Evolution [28,40]. In particular, Differential Evolution (DE) is an easily implemented and effective optimisation algorithm [36,37] which only requires a few robust control variables [33] and is attractive for real parameter optimisation [10].

Such a DE algorithm can be combined with a material simulation (developed in [21]) to allow for significantly faster testing and analysis of EiM processors than physical manufacturing and experimentations would allow. Full details are available elsewhere [10,37], but briefly, the DE algorithm uses the greedy criterion that involves evaluating the fitness of each member of a generation's population, with those members of the population with better fitness being more likely to proceed to the next generation. The characteristics of the population therefore change gradually over time due to the random mutation of characteristics and cross-over with other population members. Every member of the population is represented by a vector of decision variables $\mathbf{X}$. This decision vector contains configuration parameters which the EA optimises each generation. A basic EiM processor might commonly have a decision vector defined as:

$$\mathbf{X} = [\, V_{c1}, \; V_{c2}, \; ..., \; V_{cP}, \; l_1, \; l_2, \; ..., \; l_R \,]^T, \tag{1}$$

where T is the vector transpose. Here, the included configuration parameters are as follows: Input "configuration" voltage stimuli $V_{cp} \in [-5, 5] \, V$ applied to a

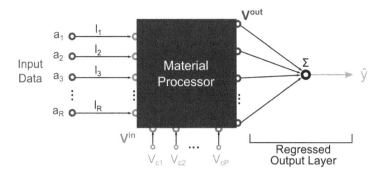

Fig. 1. Illustration of a typical monolithic EiM processor's structure. Input data is applied to the material as voltages. The output voltages (i.e., material processor output states) are regressed to produce an output layer which predicts the class ($\hat{y}$) of the processed data instances. Input weights (l_r) can be used to scale the input data voltages, and configurable voltage stimuli (V_c) can manipulate the processor's behaviour [21].

node p, where the total number of configuration nodes is P. These configuration voltages have been shown to both introduce a bias but also alter the decision boundary of an EiM classifier [21]; therefore, these effect how the materials IV characteristic is exploited which could be analogous to altering the material's 'activation function'. Input weights $l_r \in [-1, 1]$, are used to scale the input voltages V_r^{in} applied at the data driven input electrodes r due to a corresponding input attribute a_r, such that:

$$V_r^{in}(k) = l_r \times a_r(k), \tag{2}$$

where k is a given data instance and the total number of data driven input electrodes is R. The structure of this type of typical EiM processor is shown in Fig. 1, and illustrates how the configurable parameters effect the material processor's operation.

While an evolved output layer and threshold can be used to interpret the materials output voltages and assign a label to a particular data instance which has been processed [21], it has been found that a regressed output layer is more successful at evaluating and exploiting a material's output voltage states [20]. This regressed layer is generated when evaluating a population member on the training dataset and must be maintained and updated in tandem with the EA's population.

Generally, for classification, a dataset D, containing R attributes a_1, $a_2, \ldots, a_R$, is split into two subsets: a training set D^{train} and a test set D^{test}. During each generation, every member of the population is evaluated using the training data and an associated fitness is calculated using the EA's objective function. The objective function Φ has commonly been the classification error, but other types of fitnesses may be desirable, such as binary cross entropy [20]. The best population member p_{best} is tracked during training and the test set is used to evaluate the final best population member.

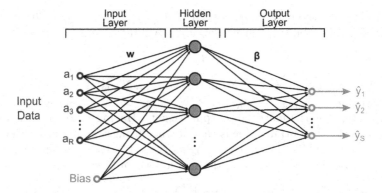

Fig. 2. Basic structure of an artificial single-hidden layer feed forward network used as an extreme learning machine.

2.2 Extreme Learning Machines

ELMs generally consist of a SLFN, as seen in Fig. 2. They operate by assigning random weights and biases to the input and hidden layers respectively [19]. These parameters are fixed and remain unchanged during training. Here, the hidden layer neurons use the sigmoid activation function. Whilst many activation functions exist [34], the sigmoid function is widely used [42] and can achieve good performance in most cases [3]. The only parameters learned are the weights (and sometimes biases) associated with the output layer, done during the training phase. Therefore, ELMs converge significantly faster than traditional artificial neural network algorithms, such as back propagation. ELMs have been shown to perform well and are more likely to reach a global optimum than systems with networks which have all parameters trained [42]. Specifically, ELM systems achieve fast training speeds with good generalisation capability.

Keeping our nomenclature consistent with Sect. 2.1, consider K data instances, where a particular data instance k is defined by its inputs $\mathbf{a}_k$ and its target outputs $\mathbf{y}_k$. Here, we define a particular instances' input containing R attributes as $\mathbf{a}_k = [a_{k1}, a_{k2}, ..., a_{kR}]^T$, and its corresponding target containing S outputs as $\mathbf{y}_k = [y_{k1}, y_{k2}, ..., y_{kS}]^T$. The predicted outputs $\hat{y}$ from an ELM with N hidden neurons can be expressed as:

$$\hat{\mathbf{y}}_k = \sum_{n=1}^{N} \boldsymbol{\beta}_n g(\mathbf{w}_n \cdot \mathbf{a}_k + b_n) = \sum_{n=1}^{N} \boldsymbol{\beta}_n h_{kn} . \qquad k = 1, ..., K \qquad (3)$$

where $\mathbf{w}_n = [w_{n1}, w_{n2}, ..., w_{nR}]^T$ is the weight vector connecting the nth hidden neuron and the input neurons, $\boldsymbol{\beta}_n = [\beta_{n1}, \beta_{n2}, ..., \beta_{nS}]^T$ is the weight vector connecting the nth hidden neuron and the output neurons, b_n is the bias of the nth hidden neuron, $g(x)$ is the activation function of the hidden layer neurons, and h_{kn} is a hidden layer neurons' output. A SLFN with enough hidden neurons can approximate these K samples such that $\sum_{n=1}^{N} \|\hat{\mathbf{y}}_n - \mathbf{y}_k\| = 0$ (universal approximation capability), hence a set of β_n, $\mathbf{w}_n$ and b_n must exist so that [19]:

$$\mathbf{H}\beta = \mathbf{Y}, \tag{4}$$

where $\mathbf{H} = \{h_{kn}\}$ ($k = 1, ..., K$ and $n = 1, ..., N$) is the hidden layer output matrix, $\mathbf{Y} = [\mathbf{y}_1, \mathbf{y}_2, ..., \mathbf{y}_K]^T$ is the matrix of target outputs, and $\beta = [\beta_1, \beta_2, ..., \beta_N]^T$ is the matrix of output weights. Having randomised and fixed the input layer, the output layer is then learnt during training using the training data subset D^{train}. The output weights β are traditionally obtained by the Moore-Penrose inverse. Therefore, the smallest norm least-squares solution is:

$$\hat{\beta} = \mathbf{H}^\dagger \mathbf{Y}, \tag{5}$$

where $\mathbf{H}^\dagger$ is the Moore-Penrose inverse of matrix $\mathbf{H}$. The final solution is then tested on the test set D^{test} to provide a uniform evaluation of the system.

Often, many randomly initialised networks are considered and the network size is incrementally increased. Various methods of calculating the output layer, adjusting network structure, and increasing convergence speed have been proposed [42]. Specifically, we highlight work producing an RR-ELM algorithm [24] which optimises the output layer using ridge regression rather than the Moore-Penrose method described above. The RR-ELM algorithm is shown to have good generalisation and stability, while also reducing adverse effects caused by perturbation or multicollinearity - properties likely to be useful in physical systems.

3   In-Materio Extreme Learning Machines

Traditionally, as described in Sect. 2.1, EiM processors have used one-to-one mapping. Here, we refer to this as a typical monolithic EiM processor, where each attribute is applied as a voltage to a corresponding input node. However, as a dataset becomes more complex, with more attributes, the size of the material network would need to physically grow. We postulate that in real microelectrode-based nanomaterial processors, larger networks might lead to fewer 'interactions' between distance electrodes, leading to poorer performance i.e., material processors may struggle to scale as the data does.

In order to overcome this problem we can take inspiration from SLFNs, as shown in Fig. 2. Specifically, the Configurable Analogue Processors or 'material processors' used in previous EiM work can instead be viewed as a complex physical neuron. These 'material neurons' can then be grouped into a Hidden Layer (HL) to produce a typical SLFN like structure. The output voltages from these material neurons are the HL's output states; we note that a material generally projects the applied input data voltages to a higher dimensional number of output voltages. The remaining question then becomes how to translate the input data into usable voltage which can be fed into the material neurons' input data electrodes/nodes i.e., an input layer.

We propose a *directly connected* input layer network structure as shown in Fig. 3. Consider a network with M material neurons, each of which contains J

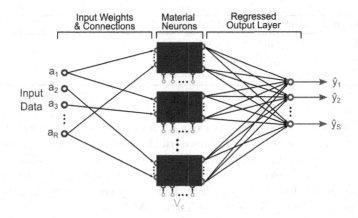

Fig. 3. Diagram of a structured network of material processors which are exploited as hidden layer neurons for an ELM using a *directly connected* input layer, where M material neurons make up the hidden layer.

input data electrodes/nodes. Each data voltage input V^{in} is the product of a weight and a selected attribute. Therefore, the voltage applied to a particular material m and node j is defined as:

$$V_{mj}^{in} = l_j^m \times a_{C_j^m}, \tag{6}$$

where $C_j^m \in \{1, 2, ..., R\}$ defines which attribute a_r is being passed to a particular material's data input node, and $l_j^m \in [-1, 1]$ is that connection's associated weight. This is a relatively simple input layer scheme, not requiring the introduction of an activation function, ensuring that the computation within the system is carried out by the material neuron only and that the hardware voltage limits are note exceeded. The relatively few number of parameters helps the system comply with the complexity engineering approach [15] and potentially benefit from concepts such as weight agnostic and minimal neural network topologies which have been found to be beneficial in ANNs [14].

The system can be defined using a vector of decision variables $\mathbf{X}$ as discussed in Eq. 1. Expanding this to include all the discussed adjustable parameters, the new structured network's decision vector can be defined as:

$$\mathbf{X} = [V_{c1}^1, ..., V_{cP}^1, l_1^1, ..., l_J^1, C_1^1, ..., C_J^1, ..., V_{c1}^M, ..., V_{cP}^M, l_1^M, ..., l_J^M, C_1^M, ..., C_J^M]^T. \tag{7}$$

Now, any single material neuron based SLFN or population p of material neuron based SLFNs (i.e., multiple initialisations of $\mathbf{X}$) can be randomly generated and trained as an ELM network using Algorithm 1. We refer to this method of combining a physical neuron based SLFN and ELM training as an in-Materio ELM (iM-ELM). In this paper, the output layer of an iM-ELM is trained using ridged regression rather then the Moore-Penrose inverse detailed in Sect. 2.2.

Finally, we highlight the possibility of re-using a single nanomaterial network as several 'virtual' neurons. The basis of this method stems from EiM processors

---

Algorithm 1: Method for in-Materio ELM.

Initialise random population of solutions p;

Train population p on D^{train};

Evaluate population using the test data D^{test};

---

whereby a wide variety of operations can be discovered by configuring the external stimuli of a single nanomaterial network. Therefore, by randomly initialising the different configurable parameters, but using only a single material, several virtual material neurons can be produced. Each of these will manifest their own unique internal IV characteristics which the ELM system will attempt to exploit. Here, we refer to such a network as a Virtual iM-ELM.

4 Problem Formulation

4.1 Simulated Material Networks

A circuit SPICE model is used to generate Diode Random Networks (DRNs) which behaves as a complex and exploitable network, and acts as a proxy for a non-linear nanomaterial. These networks contain: voltage driven input data nodes (*in-nodes*), voltage driven configuration stimuli (*c-nodes*), and measured output voltage nodes (*out-nodes*) calculated using a DC analysis. The DRN consists of randomly orientated diodes and series resistors between every node pair. The rapid changes in conductivity when a diode turns on allows for complex non-linear IV characteristics which can be exploited for classification. The DRN is physically realisable using discrete circuit components and its properties are common in nanomaterials. An example of a five node (i.e., electrode) material is given in Fig. 4. We note that other IV characteristics or circuit components could be used to create a variety of 'material networks' for different types of analysis. 1N4148PH Signal Diodes are used, and the resistance of the interconnecting resistors are uniformly randomly selected between $\in [10, 100]$ kΩ.

4.2 Classification Tasks

The performance of the iM-ELM systems are compared against several common machine learning datasets which can be found on the UCI repository [11]. These include: the Pima Indians Diabetes Database (diabetes) dataset containing 8 attributes, 768 data instances and 2 classes. The wine (wine) dataset containing 13 attributes, 178 instances and 3 classes. The Australian Credit Approval (aca) dataset containing 14 attributes, 690 instances and 2 classes. The Wisconsin Diagnostic Breast Cancer (wdbc) dataset containing 30 attributes, 569 instances and 2 classes. These datasets are more complex (i.e., contain more attributes) than have been previously used, specifically within Evolution in-Materio based literature [20,27,41], but remain small enough (i.e., relatively few data instances) to ensure comprehensive analysis without excessive simulation times.

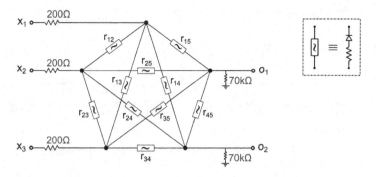

Fig. 4. An example five node DRN material, where each node is connected to every other node via a resistor and diode. In this example, two nodes are behaving as outputs (o_1, o_2) and three nodes as inputs (x_1, x_2, x_3) which could be allocated as either data driven or configuration stimuli voltages.

The datasets are randomly split into a training (D^{train}) and test (D^{test}) subset using a 70%–30% split respectively. The datasets are normalised (using D^{train}) and then scaled to the max/min allowed hardware voltages $\in [-5, 5]$ V. The performance of these datasets was considered using some simple (default) python sklearn classification methods and the results are shown in Table 1.

5   Results and Discussion

Considering the model developed in Sect. 3, iM-ELMs of incremental sizes were analysed. Specifically, the number of material neurons in the HL was increased from one to fifteen (beyond which the results plateau). For each HL size, thirty different random seeds were used to generate the material neurons within thirty iM-ELMs. The same thirty seeds are used for each network size incrementation, meaning that each system continues to include the same material neurons that were used in its corresponding previous smaller networks. Therefore, we can consider the change in performance of the iM-ELM networks as they are

Table 1. Test accuracy results for the datasets when using several common classification methods. The best accuracy achieved for each dataset is highlighted in bold.

Dataset	Classification method				
	Ridge Regression	Logistic Regression	Random Forest	SVM (linear)	SVM (poly)
diabetes	0.7576	0.7619	0.7403	**0.7662**	0.6970
wine	0.9815	0.9815	**1.000**	0.9630	0.9815
aca	0.8357	0.8502	**0.8599**	0.8213	0.8261
wdbc	0.9357	**0.9591**	0.9298	0.9532	**0.9591**

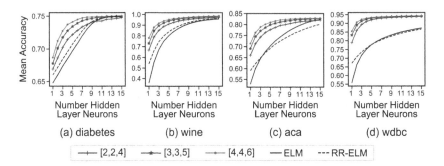

Fig. 5. Mean test accuracy of all the (30 systems, each with 100 parameter initialisations) in-Materio ELMs for each hidden layer size increment used to classify the (a) diabetes, (b) wine, (c) aca and (d) wdbc datasets. Three different material neuron topologies are considered ([No. *in-nodes*, No. *c-nodes*, No. *out-nodes*]), and these are compared to the mean accuracy of 3000 traditional artificial ELMs and RR-ELMs.

enlarged. For each iM-ELM, a 'population' of 100 randomly generated decision vectors are considered (i.e., randomly initialised input layer and configuration parameters), which was observed to provide a good insight into performance and maintain reasonable simulation times. The mean test accuracy from these thirty systems each with 100 parameter initialisation is shown for each HL size in Fig. 5. Recall from Sect. 4.1 that these material neurons' consist of a fully interconnected network containing three main classes of nodes: input voltage nodes for data (*in-nodes*), input voltage nodes for configuration/stimuli altering the material neurons behaviour (*c-nodes*), and output voltage nodes (*out-nodes*). Notably, the *directly connected* input layer connects each data input node to only a single data attribute; so, if too few neurons are in use, then not all data attributes may be 'connected'. The experiment is performed with three increasingly larger material neuron network topologies: (i) materials containing two *in-nodes*, two *c-nodes* and four *out-nodes* denoted by [2,2,4], (ii) materials containing three *in-nodes*, three *c-nodes* and five *out-nodes* denoted by [3,3,5], (iii) materials containing four *in-nodes*, four *c-nodes* and six *out-nodes* denoted by [4,4,6]. This will provide some initial insight on the effect of scaling the size of (well connected) materials. The performance of these iM-ELMs is plotted against the mean accuracy of 3000 artificial (Moore-Penrose) ELMs and (Ridge Regression) RR-ELMs, selected because it matches the total 'computational expense' (i.e., total number of data instances) used over the 30 iM-ELMs systems.

On average, the iM-ELMs considered outperformed the artificial ELM and RR-ELM systems of equivalent network sizes. The simulated 'material' neurons are successfully generating useful, higher dimensional output states which the ELM algorithm can exploit, and these are out-performing their artificial neuron counterparts. As more material neurons are operated in parallel, the mean classification accuracy improves. Notably, the larger and more complex material neuron topologies (i.e., when using more input, configuration, and output nodes/electrodes) achieve higher mean accuracies for iM-ELMs with the same

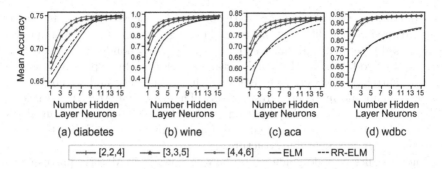

Fig. 6. Mean test accuracy of all the (30 systems, each with 100 parameter initialisations) Virtual in-Materio ELMs for each hidden layer size increment used to classify the (a) diabetes, (b) wine, (c) aca and (d) wdbc datasets. Three different material neuron topologies are considered ([No. *in-nodes*, No. *c-nodes*, No. *out-nodes*]), and these are compared to the mean accuracy of 3000 traditional artificial ELMs and RR-ELMs.

size of HL. This in turn means that fewer neurons are required within the SLFN HL to achieve comparable results with networks leveraging less capable neurons.

The best accuracy achieved, across all HL sizes, for the different material neuron topologies and datasets, is shown in Table 2. The iM-ELMs discussed can significantly outperform some of the common classification methods presented in Table 1. Indeed, the best iM-ELMs also compare favourably with the traditional artificial ELM networks, and in the case of the wdbc dataset the iM-ELMs can achieve a 1.76% increase in the best obtained accuracy.

As discussed in Sect. 3 any single material could be re-used as a virtual material neuron. Ideally, the different randomised parameters (input weights, connections and configuration voltages) enable the different virtual neurons to behave in a sufficiently independent manner. To investigate this, the previous analysis is repeated i.e., generating thirty SLFN for each HL size, each with 100 random initialisations. However, now each HL contains several virtual neurons which are generated from only a single material. The mean accuracy of these Virtual iM-ELMs is plotted against the size of the SLFN in Fig. 6, and the best ever achieved accuracies are shown in Table 2. The Virtual iM-ELMs have a

Table 2. Best accuracy achieved from the different systems, from across the different hidden layer sizes. The best accuracy for each dataset is highlighted in bold.

Dataset	iM-ELM			Virtual iM-ELM			ELM	RR-ELM
	[2,2,4]	[3,3,5]	[4,4,6]	[2,2,4]	[3,3,5]	[4,4,6]		
diabetes	0.7922	0.7922	**0.7965**	0.7879	0.7922	**0.7965**	**0.7965**	0.7922
wine	**1.000**	**1.000**	**1.000**	**1.000**	**1.000**	**1.000**	**1.000**	**1.000**
aca	0.8792	0.8792	**0.8889**	0.8841	0.8841	0.8792	**0.8889**	0.8744
wdbc	**0.9708**	**0.9708**	**0.9708**	**0.9708**	**0.9708**	0.9649	0.9532	0.9532

near identical average performance to the previously discussed iM-ELMs systems that exploited several different materials. This suggests that the type of material neuron used here (i.e., the simulated circuit based DRN non-linear network) can successfully produce several virtual instances, achieved by exploiting the wide range of current-voltage characteristics which can be tuned and selected by the voltage stimuli and input layer respectively. These Virtual iM-ELMs are significantly more flexible, only requiring one material substrate to create an SLFN. However, by 're-using' a single material, the systems loses its ability to benefit from the highly parallelisable structure.

These results provide guidance on how to operate in-Materio processors in parallel to process much more complex datasets then would have previously been possible with only a single monolithic material. However, further work is needed to consider how well these systems can scale to the much larger datasets commonly found in state of the art machine learning problems.

6 Conclusion

In this paper, material networks are exploited as physical material neurons to implement a single hidden layer feed forward network (SLFN) which was trained as an Extreme Learning Machine (ELM). The input data was passed to the physical neurons using a *directly connected* input layer which ensured physical hardware limits were obeyed and that 'computation' within the system was carried out by the materials only. Complex diode-resistor networks were simulated to provide a convenient, fast and reliable proxy to nanomaterial based Configurable Analogue Processors, used as the physical neurons. This enabled the efficient investigation of these in-Materio ELMs (iM-ELMs) when classifying several datasets of varying complexity. It was found that these iM-ELMs could significantly outperform other common classification methods, as well as traditional (artificial) ELMs. The complex current-voltage characteristics of the materials are successfully being exploited to leverage them as physical neurons, which outperform traditional artificial neurons. As the size of the material topology increases (i.e., the number of nodes/electrodes used in the material network), the performance of iM-ELMs with similar hidden layer sizes improves; showing the capability of the material neurons to process data is increasing. As more material neurons are used in parallel, the average classification performance improves rapidly before plateauing.

Drawing from previous work with Evolution in-Materio processors, which show that a single nanomaterial network can be tuned for a range of operations, we present a method to re-use a single material as several 'virtual' physical neurons. These Virtual iM-ELMs which leveraged only a single physical material performed comparably to the iM-ELMs which used several different physical material neurons. This suggests that our circuit based 'materials' can achieve a wide range of physical properties which are successfully exploited as different virtual neurons. While this forgoes the benefits of parallelised operation, it grants more flexibility when creating larger SLFN and when designing the required physical hardware interface.

These physical analogue neurons have the potential to produce efficient in-Materio ELMs which can exploit the non-differentiable, complex characteristics presented by a nanomaterial. We anticipate that, when implemented using physical substrates, highly parallelisable and fast ELMs will be produced.

Acknowledgements. This work was supported by the Engineering and Physical Sciences Research Council [EP/R513039/1].

References

1. Bennett, C., Querlioz, D., Klein, J.O.: Spatio-temporal learning with arrays of analog nanosynapses. In: Proceedings of the IEEE/ACM International Symposium on Nanoscale Architectures, NANOARCH 2017, pp. 125–130 (2017). https://doi.org/10.1109/NANOARCH.2017.8053708
2. Bose, S.K., et al.: Evolution of a designless nanoparticle network into reconfigurable Boolean logic. Nat. Nanotechnol. **10**(12), 1048–1052 (2015). https://doi.org/10.1038/nnano.2015.207
3. Cao, W., Gao, J., Ming, Z., Cai, S.: Some tricks in parameter selection for extreme learning machine. IOP Conf. Ser. Mater. Sci. Eng. **261**, 012002 (2017). https://doi.org/10.1088/1757-899X/261/1/012002
4. Chen, T., et al.: Classification with a disordered dopant-atom network in silicon. Nature **577**(7790), 341–345 (2020). https://doi.org/10.1038/s41586-019-1901-0
5. Clegg, K.D., Miller, J.F., Massey, M.K., Petty, M.C.: Practical issues for configuring carbon nanotube composite materials for computation. In: 2014 IEEE International Conference on Evolvable Systems, pp. 61–68, December 2014. https://doi.org/10.1109/ICES.2014.7008723
6. Conte, T.M., DeBenedictis, E.P., Gargini, P.A., Track, E.: Rebooting computing: the road ahead. Computer **50**(1), 20–29 (2017). https://doi.org/10.1109/MC.2017.8
7. Dale, M., Stepney, S., Miller, J., Trefzer, M.: Reservoir computing in materio: an evaluation of configuration through evolution. In: 2016 IEEE Symposium Series on Computational Intelligence (SSCI) (2016). https://doi.org/10.1109/SSCI.2016.7850170
8. Dale, M., et al.: Reservoir computing with thin-film ferromagnetic devices. arXiv:2101.12700 [cond-mat] (January 2021)
9. Dale, M., Stepney, S., Miller, J.F., Trefzer, M.: Reservoir computing in materio: a computational framework for in materio computing. In: 2017 International Joint Conference on Neural Networks (IJCNN), pp. 2178–2185, May 2017. https://doi.org/10.1109/IJCNN.2017.7966119
10. Das, S., Suganthan, P.N.: Differential evolution: a survey of the state-of-the-art. IEEE Trans. Evol. Comput. **15**(1), 4–31 (2011). https://doi.org/10.1109/TEVC.2010.2059031
11. Dheeru Dua, E.K.T.: UCI machine learning repository (2017). http://archive.ics.uci.edu/ml
12. Du, C., Cai, F., Zidan, M., Ma, W., Lee, S., Lu, W.: Reservoir computing using dynamic memristors for temporal information processing. Nat. Commun. **8**(1) (2017). https://doi.org/10.1038/s41467-017-02337-y
13. Eshtay, M., Faris, H., Obeid, N.: Metaheuristic-based extreme learning machines: a review of design formulations and applications. Int. J. Mach. Learn. Cybern. **10**(6), 1543–1561 (2018). https://doi.org/10.1007/s13042-018-0833-6

14. Gaier, A., Ha, D.: Weight agnostic neural networks. arXiv:1906.04358 [cs, stat] (September 2019)

15. Ganesh, N.: Rebooting neuromorphic hardware design–a complexity engineering approach. arXiv:2005.00522 [cs] (September 2020)

16. Greff, K., et al.: Using neural networks to predict the functionality of reconfigurable nano-material networks. Int. J. Adv. Intell. Syst. **9**, 339–351. IARIA (2017)

17. Harding, S., Miller, J.: Evolution in materio: a tone discriminator in liquid crystal. In: Proceedings of the 2004 Congress on Evolutionary Computation (IEEE Cat. No. 04TH8753), vol. 2, pp. 1800–1807, June 2004. https://doi.org/10.1109/CEC. 2004.1331114

18. Harding, S., Miller, J.F.: Evolution in materio: evolving logic gates in liquid crystal. Int. J. Unconv. Comput. **3**, 243–257 (2007)

19. Huang, G.B., Zhu, Q.Y., Siew, C.K.: Extreme learning machine: theory and applications. Neurocomputing **70**(1), 489–501 (2006). https://doi.org/10.1016/j. neucom.2005.12.126

20. Jones, B.A.H., Al Moubayed, N., Zeze, D.A., Groves, C.: Enhanced methods for evolution in-materio processors. In: 2021 International Conference on Rebooting Computing (ICRC), pp. 109–118, November 2021. https://doi.org/10.1109/ ICRC53822.2021.00026

21. Jones, B.A.H., et al.: Towards intelligently designed evolvable processors. Evolut. Comput. 1–23 (2022). https://doi.org/10.1162/evco_a_00309

22. Kan, S., Nakajima, K., Takeshima, Y., Asai, T., Kuwahara, Y., Akai-Kasaya, M.: Simple reservoir computing capitalizing on the nonlinear response of materials: theory and physical implementations. Phys. Rev. Appl. **15**(2), 024030 (2021). https:// doi.org/10.1103/PhysRevApplied.15.024030

23. Kotsialos, A., Massey, M.K., Qaiser, F., Zeze, D.A., Pearson, C., Petty, M.C.: Logic gate and circuit training on randomly dispersed carbon nanotubes. Int. J. Unconv. Comput. **10**(5–6), 473–497 (2014)

24. Li, G., Niu, P.: An enhanced extreme learning machine based on ridge regression for regression. Neural Comput. Appl. **22**(3), 803–810 (2013). https://doi.org/10. 1007/s00521-011-0771-7

25. Linden, D.: A system for evolving antennas in-situ. In: Proceedings Third NASA/DoD Workshop on Evolvable Hardware. EH-2001, pp. 249–255, July 2001. https://doi.org/10.1109/EH.2001.937968

26. Lupo, A., Butschek, L., Massar, S.: Photonic extreme learning machine based on frequency multiplexing. Opt. Express **29**(18), 28257 (2021). https://doi.org/10. 1364/OE.433535

27. Massey, M.K., et al.: Computing with carbon nanotubes: optimization of threshold logic gates using disordered nanotube/polymer composites. J. Appl. Phys. **117**(13), 134903 (2015). https://doi.org/10.1063/1.4915343

28. Massey, M.K., et al.: Evolution of electronic circuits using carbon nanotube composites. Sci. Rep. **6**(1), 32197 (2016). https://doi.org/10.1038/srep32197

29. Miller, J., Downing, K.: Evolution in materio: looking beyond the silicon box. In: Proceedings 2002 NASA/DoD Conference on Evolvable Hardware, pp. 167–176. IEEE Comput. Soc, Alexandria, VA, USA (2002). https://doi.org/10.1109/EH. 2002.1029882

30. Morán, A., et al.: Hardware-optimized reservoir computing system for edge intelligence applications. Cognit. Comput. (2021). https://doi.org/10.1007/s12559-020-09798-2

31. Mujal, P., et al.: Opportunities in quantum reservoir computing and extreme learning machines. Adv. Quantum Technol. **4**(8), 2100027 (2021). https://doi.org/10.1002/qute.202100027
32. Ortín, S., et al.: A unified framework for reservoir computing and extreme learning machines based on a single time-delayed neuron. Sci. Rep. **5**(1), 14945 (2015). https://doi.org/10.1038/srep14945
33. Pedersen, M.E.H.: Good parameters for differential evolution (2010)
34. Ratnawati, D.E., Marjono, Widodo, Anam, S.: Comparison of activation function on extreme learning machine (ELM) performance for classifying the active compound. In: AIP Conference Proceedings, vol. 2264, no. 1, p. 140001, September 2020. https://doi.org/10.1063/5.0023872
35. Ruiz-Euler, H.C., Alegre-Ibarra, U., van de Ven, B., Broersma, H., Bobbert, P.A., van der Wiel, W.G.: Dopant network processing units: towards efficient neural-network emulators with high-capacity nanoelectronic nodes. arXiv:2007.12371 [cs, stat] (July 2020)
36. Sloss, A.N., Gustafson, S.: 2019 evolutionary algorithms review. arXiv:1906.08870 [cs] (June 2019)
37. Storn, R., Price, K.: Differential evolution–a simple and efficient heuristic for global optimization over continuous spaces. J. Glob. Optim. **11**(4), 341–359 (1997). https://doi.org/10.1023/A:1008202821328
38. Tanaka, G., et al.: Recent advances in physical reservoir computing: a review. Neural Netw. **115**, 100–123 (2019). https://doi.org/10.1016/j.neunet.2019.03.005
39. Viero, Y., et al.: Light-stimulatable molecules/nanoparticles networks for switchable logical functions and reservoir computing. Adv. Func. Mater. **28**(39), 1801506 (2018). https://doi.org/10.1002/adfm.201801506
40. Vissol-Gaudin, E., Kotsialos, A., Groves, C., Pearson, C., Zeze, D., Petty, M.: Computing based on material training: application to binary classification problems. In: 2017 IEEE International Conference on Rebooting Computing (ICRC), pp. 1–8. IEEE, Washington, DC, November 2017. https://doi.org/10.1109/ICRC.2017.8123677
41. Vissol-Gaudin, E., et al.: Confidence measures for carbon-nanotube/liquid crystals classifiers. In: 2018 IEEE Congress on Evolutionary Computation (CEC), pp. 1–8, July 2018. https://doi.org/10.1109/CEC.2018.8477779
42. Wang, J., Lu, S., Wang, S.H., Zhang, Y.D.: A review on extreme learning machine. Multimed. Tools Appl. (2021). https://doi.org/10.1007/s11042-021-11007-7
43. Zhu, Q.Y., Qin, A.K., Suganthan, P.N., Huang, G.B.: Evolutionary extreme learning machine. Pattern Recognit. **38**(10), 1759–1763 (2005). https://doi.org/10.1016/j.patcog.2005.03.028

On the Impact of the Duration of Evaluation Episodes on the Evolution of Adaptive Robots

Larissa Gremelmaier Rosa[1(✉)], Vitor Hugo Homem[1], Stefano Nolfi[2]⊙, and Jônata Tyska Carvalho[1,2]⊙

[1] Federal University of Santa Catarina, Florianópolis, Santa Catarina, Brazil
`larissa.gremelmaier.rosa@grad.ufsc.br`
[2] Institute of Cognitive Science and Technologies National Research Council (CNR-ISTC), Rome, Italy

Abstract. We investigate the impact of the duration of evaluation episodes and of the way in which the duration is varied during the course of the evolutionary process in evolving robots. The results obtained demonstrates that these factors can have drastic effects on the performance of the evolving robots and on the characteristics of the evolved behaviors. Indeed, the duration of the evaluation episodes do not alter simply the accuracy of the fitness estimation but also the quality of the estimation. The comparison of the results indicates that the best results are obtained by starting with short evaluation episodes and by increasing their duration during the course of the evolutionary process, or by using shorter evaluation episodes during the first part of the evolutionary process.

Keywords: Evolutionary robotics · Fitness evaluation · Experimental parameter tuning

1 Introduction

Evolutionary robotics [16,17] permits to develop robots capable to perform a desired function automatically, with limited human intervention. Indeed, it frees the experimenter from the need to identify the behavior that should be exhibited by the robot and the control rules which enable the robot to exhibit such behavior.

The experimenter, however, should specify manually the fitness function, i.e. the criterion which is used to evaluate the performance of alternative robots. Despite designing the fitness function, which is used to evolve a robot automatically, is much less demanding than directly designing the characteristics of the robot, it can be still challenging. For this reason, the problem of the fitness function design has been investigate by several authors [6,7,12].

The problem is further complicated by the fact that the fitness assigned to robots depends also on other factors, such as the termination criteria, the way in which the initial properties of the robot and of the environment are varied at the beginning of evaluation episodes, and the number and duration of evaluation episodes.

The termination criteria are rules that determine the premature termination of evaluation episodes. They are introduced to minimize the cost of the evaluation of the robots. An example of termination criterion is a rule that terminate the evaluation episode when a robot which is evolved for the ability to walk fall to the ground. The rationale, is that keep evaluating a fallen robot is pointless since it will not be able to keep improving its fitness once it is fallen down. On the other hand, the criterion which is used to determine whether a robot is fallen down or not might influence the behavior of the evolving robots. The effect of termination conditions, however, have not been sufficiently studied yet.

The way in which the initial state of the robot and of the environment are varied at the beginning of evaluation episodes is also crucial to evolve robots which are robust with respect to variations of the environmental conditions [10,19] and to evolve robots which can cross the reality gap [9,11].

In this article we will demonstrate that also the duration of the evaluation episode and eventually the way in which the duration of the episodes vary during the course of the evolutionary process has a strong influence on the quality of the evolved solutions.

As far as we know, the impact of the duration of evaluation episodes on the efficacy of the evolved robots have not been investigated before. Several works investigated methods for optimizing the number of evaluation episodes [2–4,8]. The utility of evaluating the robots for multiple episodes originates from the fact the fitness measure is stochastic in the presence of environmental variations. Indeed, the fitness obtained by an individual depends both on the ability of the individual and on the difficulty of the environmental conditions experienced by that individual. The easier the conditions experienced, the higher the fitness obtained by the individual. The stochasticity of the fitness measure can be reduced by evaluating the robots for multiple episodes and by using the fitness averaged over the episodes. On the other hand, carrying multiple episodes increases the computation cost. For this reason, the works referenced above investigated mechanisms for setting automatically the number of episodes with the objective of improving the estimation of the fitness while maintaining the computation cost as lower as possible.

The only work, we are aware of, that investigated the impact of the duration of evaluation episodes is [5]. In this work, the authors analyzed the problem of adapting the duration of the episodes to the characteristics of the environment with the objective of identifying the minimal duration which permit to estimate the performance of the evolving robot with sufficient accuracy.

In this work, instead, we will demonstrate that the duration of evaluation episodes alters not only the accuracy of the estimation but also the quality of the estimation. The duration of evaluation episodes can have drastic effects on the behavior and on the performance of evolving robots. Consequently, the duration of evaluation episodes should be considered as a fundamental characteristics of the fitness function and, overall, of the evaluation process.

2   Method

To investigate the impact of the length of evaluation episodes we carried a set of experiments by using the PyBullet locomotor environments [1], a popular set of problems which are widely used to benchmark evolutionary and reinforcement learning algorithms.

The Pybullet environments involve multi-segments robots with varying morphologies which are evaluated for the ability to locomote as fast as possible toward a given destination. More specifically, we used the Hopper, Ant, Walker2D, and Humanoid environments (Fig. 1). The Hopper agent is formed by 4 segments which form a single-leg system which can locomote by jumping. The Ant agent is constituted by four legs attached to a central body element. The Walker2D agent is constituted of biped with two legs attached to a hip. Finally, the Humanoid agent has a humanoid morphology which include two legs, two arms, a torso and a head.

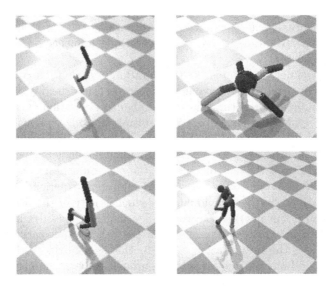

Fig. 1. The Hopper (top-left), Ant (top-right), Walker2D (bottom-left) and Humanoid (bottom-right) agents.

The agents are controlled by 3-layered feedforward neural networks including 50 internal neurons in the case of the Hopper, Ant and Walker2d environments and 250 internal neurons in the case of the Humanoid environment.

The connection weights of the neural networks are evolved. We used the OpenAI evolutionary algorithm [21] which is a state-of-the-art evolutionary strategy. Connection weights are normalized with weight-decay [13]. Observation vectors are normalized with virtual batch normalization [20,21]. This setup was chosen based on previous good results reported on [21]. The evolutionary process is continued up to a total of 50 million evaluation steps in the case of the Hopper, Ant, Walker environments and up to a total of 100 million evaluation steps in the case of the Humanoid environment.

The experiments were performed with the open-source Evorobotpy2 software [14,15]. The software can also be used to easily replicate the experiments reported in this paper. The agents are rewarded for their ability to locomote on the basis of their speed toward the destination. In addition, they receive: (1) a positive bonus for staying upright, (2) a punishment for the electricity cost which corresponds to the average of the dot product of the action vector and of the joint speed vector, (3) a punishment for the stall cost which corresponds to the average of the squared action vector, (4) a punishment proportional to the number of joints that reached a joint limit, and (5) a punishment of –1 for falling down. The bonus for staying upright is set to 2.0 in the case of the Humanoid and to 1.0 in the case of the other problems. The electricity cost, stall cost, and joint at limit cost are weighted by −8.5, −0.425, and −0.1 in the case of the Humanoid, and for −2.0, −0.1, and −0.1 in the case of the other problems [1]. The five additional components have been introduced to facilitate the development of the locomotion behavior. The fitness of the agents is computed by summing the reward received during each step of the evaluation episodes. When the agents are evaluated for multiple evaluation episodes, the fitness is computed by averaging the sum of the rewards received during each episode.

As demonstrated by [18], the efficacy of the reward functions depends on the algorithm. The standard reward functions described above and implemented in the Pybullet environment is optimized for reinforcement learning algorithm but can produce sub-optimal solutions in combination with evolutionary algorithms. We will thus also carry on experiments with the modified reward function proposed by [18] which are optimized for evolutionary algorithms. These functions are simpler with respect to the standard version. Indeed they reward the agents on the basis of their speed toward the destination only in the case of the hopper, and include fewer additional components in the case of the other environments.

At the beginning of an evaluation episode, the agents are initialized at the center of the surface slightly elevated over the plane. Evaluation episodes are terminated prematurely when the agents fall down, e.g. when certain part of the agents' body touch the ground and/or when the inclination of the agents' body exceed a certain threshold. The effect of the actions performed by the agents are perturbed by adding a random value selected within the range [−0.01, 0.01] the value of the motor neurons. The standard duration of evaluation episodes is 1000 steps. However, in our experiments, we study the effect of different durations.

Some of the agents are post-evaluated for multiple episodes. During these post-evaluations the initial posture of the agents and the noise added to the effects the motor neurons vary randomly. In particular, the best agent of each generation are post-evaluated for 3 episodes. The fitness obtained during these post-evaluations is used to identify the best individual obtained in an evolutionary experiment. The best agents of each evolucionary experiment are post-evaluated for 1000 episodes. The fitness obtained in this manner is used to compare the relative efficacy of the experimental conditions.

2.1 Experimental Conditions

To investigate the effect of the duration of learning episodes and the effect of using constant or variable durations we considered the following experimental conditions:

1. A standard experimental condition in which the duration of evaluation episodes is set to 1000 steps and remains constant. This is the modality implemented in the standard Pybullet locomotor environments. As mentioned above, the duration of evaluation episodes indicate the maximum duration. The episodes are terminated prematurely when the agents fall down.
2. A fixed-nsteps experimental condition in which the duration of evaluation episodes is constant and is set to nsteps. The experiments have been repeated by setting nsteps to 100, 200, 300, 400 and 500.
3. A fixed-nsteps-1000 experimental condition in which the duration of the evaluation episodes is constant and is set to nsteps during the first half of the evolutionary process and to 1000 during the remaining part. As in the cases of the previous condition, the experiments have been repeated by setting nsteps to 100, 200, 300, 400 and 500.
4. An incremental condition in which the duration of evaluation episodes is initially set to 100 and is incremented of 100 steps every 1/10 of the evolutionary process.
5. An mod-reward conditions in which the duration of evaluation episodes is set to 1000 and in which the agents are evaluated with the reward functions optimized for evolutionary algorithms described in [18].
6. A random-n experimental condition in which the duration of evaluation episodes is set randomly in the range [50, 1000] with an uniform distribution every generation.

In all experimental condition the performance of the evolve agents are evaluated by post-evaluating them for 1000 episodes with the standard reward function and by setting the duration of the episodes to 1000. This is necessary to compare the efficacy of the agents evolved in different conditions on the same settings.

3 Results and Discussion

Figures 2, 3, 4 and 5 display the results obtained. For the fixed-nsteps and fixed-nsteps-1000 conditions, in which the experiments have been repeated by setting the nsteps parameter to different values, the Figures show the results obtained in the best case. The performance of agents obtained with the random-n experimental condition are very poor and, consequently, are not shown in this section.

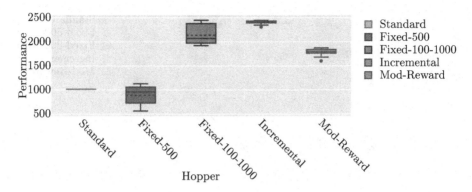

Fig. 2. Performance of the hopper agents evolved in the different experimental conditions. For the fixed-nsteps and fixed-nsteps-1000 conditions in which the experiments have been repeated by setting the nsteps parameter to different value we report the results obtained with the best parameter. Average results obtained by post-evaluating the best evolved agents for 1000 episodes. Each histogram shows the average result of 10 replications.

Overall, the duration of the episodes and the variation of the duration during the evolutionary process have a strong impact on performance.

In the case of the Hopper, Ant and Walker environments, the standard setting in which the length of the episodes is fixed and is set to 1000 is rather sub-optimal and is outperformed by all other conditions.

Simply reducing the duration of the episodes to 100/500 (see the fixed-nsteps condition) steps leads to better results. The best value of nsteps is task dependent and corresponds to 100, 500 and 200 steps in the case of the Hopper, Ant, and Walker, respectively.

Using shorter evaluation episodes during the first half of the evolutionary process leads to even better performance in the three environments. Also in this case, the best value of nsteps during the first part of the evolutionary process is task dependent and correspond to 100, 200, and 100 steps in the case of the Hopper, Ant, and Walker, respectively.

Linearly increasing the duration of episodes leads to the best performance in the three environments considered and to much better performance with respect to the standard condition in which the duration of evaluation episode is fixed and set to 1000 steps.

The usage of the modified reward function leads to better performance with respect to the standard condition in all cases. However, the advantage gained by using the reward function optimized for evolutionary algorithms is lower than the advantage gained in the incremental experimental condition, i.e. of the advantage that is obtained by starting with short evaluation episodes and by linearly increasing the length of the episodes during the evolutionary process.

In the case of the Humanoid problem, instead, the performances are low in all cases. The only experiments which lead to reasonably good performance

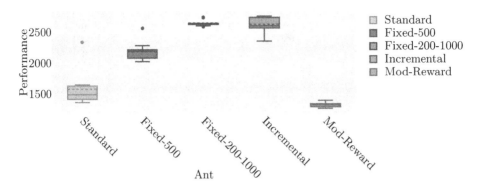

Fig. 3. Performance of the Ant agents evolved in the different experimental conditions. For the fixed-nsteps and fixed-nsteps-1000 conditions in which the experiments have been repeated by setting the nsteps parameter to different value we report the results obtained with the best parameter. Average results obtained by post-evaluating the best evolved agents for 1000 episodes. Each histogram shows the average result of 10 replications.

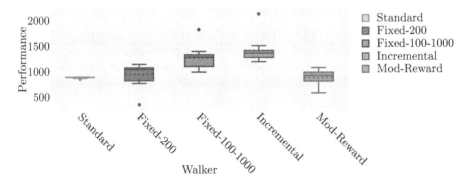

Fig. 4. Performance of the walker agents evolved in the different experimental conditions. For the fixed-nsteps and fixed-nsteps-1000 conditions in which the experiments have been repeated by setting the nsteps parameter to different value we report the results obtained with the best parameter. Average results obtained by post-evaluating the best evolved agents for 1000 episodes. Each histogram shows the average result of 10 replications.

are those performed by using the reward function optimized for evolutionary algorithms. Possibly, even better results can be obtained by running experiments which combine this reward function with the incremental condition.

All the performance differences were statistically significant ($p < .001$ for all experiments). We performed the Shapiro-wilk test for normality checking and Kruskal-Wallis H-test for checking statistical significance. It is noteworthy that the statistical difference for the Humanoid case was due to the Mod-Reward experiment, rather than the episode duration experiments.

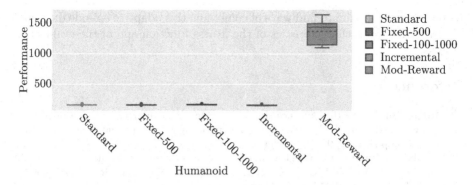

Fig. 5. Performance of the Humanoid agents evolved in the different experimental conditions. For the fixed-nsteps and fixed-nsteps-1000 conditions in which the experiments have been repeated by setting the nsteps parameter to different value we report the results obtained with the best parameter. Average results obtained by post-evaluating the best evolved agents for 1000 episodes. Each histogram shows the average result of 10 replications.

4   Conclusion

We investigated the impact of the duration of evaluation episodes and of the way in which the duration is varied during the course of the evolutionary process in the case of the Pybullet locomotors environments.

The results obtained demonstrate that the duration of evaluation episodes and the way in which the duration is varied have drastic effect on the performance of the evolving agents and on the characteristics of the evolved behavior. The duration of the evaluation episodes do not alter simply the accuracy of the estimation of the fitness but also the quality of the estimation. This qualitative effect can be explained by considering that the duration of evaluation episodes alter the relative strength of fitness components. In particular, in the case of the environments considered, the length of the evaluation episodes alter the adaptive pressure exhorted toward mediocre/prudent behaviors, which achieve low progress by taking low risks, versus high-performing/risky behaviors, which achieve high progress by taking high risk. A good example of this case is the Hopper robot that evolves a jumping behavior when a 100-step episode is used, instead of a standing still behavior when a 1000-step episode is used.

The comparison of the results obtained by varying the duration of the evaluation episodes and eventually the way in which the duration is varied during the course of the evolutionary process indicates that best results are obtained by starting with short evaluation episodes and by increasing the duration of the evaluation episodes during the course of the evolutionary process or by using shorter evaluation episodes during the first part of the evolutionary process.

Our results also show that, in some cases, the advantage which can be gained by optimizing the duration of evaluation episodes can be higher than the advantage which can be gained by optimizing the characteristics of the fitness function.

As future work, we aim to find ways of combining this adaptive episode duration strategy with other characteristics of the fitness function and of the evaluation process.

References

1. Admin: Bullet real-time physics simulation, May 2021. https://pybullet.org/
2. Aizawa, A.N., Wah, B.W.: Scheduling of genetic algorithms in a noisy environment. Evol. Comput. **2**(2), 97–122 (1994)
3. Branke, J., Schmidt, C.: Selection in the presence of noise. In: Cantú-Paz, E., et al. (eds.) GECCO 2003. LNCS, vol. 2723, pp. 766–777. Springer, Heidelberg (2003). https://doi.org/10.1007/3-540-45105-6_91
4. Cantú-Paz, E.: Adaptive sampling for noisy problems. In: Deb, K. (ed.) GECCO 2004. LNCS, vol. 3102, pp. 947–958. Springer, Heidelberg (2004). https://doi.org/10.1007/978-3-540-24854-5_95
5. Dinu, C.M., Dimitrov, P., Weel, B., Eiben, A.: Self-adapting fitness evaluation times for on-line evolution of simulated robots. In: Proceedings of the 15th Annual Conference on Genetic and Evolutionary Computation, pp. 191–198 (2013)
6. Divband Soorati, M., Hamann, H.: The effect of fitness function design on performance in evolutionary robotics: the influence of a priori knowledge. In: Proceedings of the 2015 Annual Conference on Genetic and Evolutionary Computation, pp. 153–160 (2015)
7. Doncieux, S., Mouret, J.-B.: Beyond black-box optimization: a review of selective pressures for evolutionary robotics. Evol. Intell. **7**(2), 71–93 (2014). https://doi.org/10.1007/s12065-014-0110-x
8. Hansen, N., Niederberger, A.S., Guzzella, L., Koumoutsakos, P.: A method for handling uncertainty in evolutionary optimization with an application to feedback control of combustion. IEEE Trans. Evol. Comput. **13**(1), 180–197 (2008)
9. Jakobi, N., Husbands, P., Harvey, I.: Noise and the reality gap: the use of simulation in evolutionary robotics. In: Morán, F., Moreno, A., Merelo, J.J., Chacón, P. (eds.) ECAL 1995. LNCS, vol. 929, pp. 704–720. Springer, Heidelberg (1995). https://doi.org/10.1007/3-540-59496-5_337
10. Milano, N., Carvalho, J.T., Nolfi, S.: Moderate environmental variation across generations promotes the evolution of robust solutions. Artif. Life **24**(4), 277–295 (2019)
11. Mouret, J.B., Chatzilygeroudis, K.: 20 years of reality gap: a few thoughts about simulators in evolutionary robotics. In: Proceedings of the Genetic and Evolutionary Computation Conference Companion, pp. 1121–1124 (2017)
12. Nelson, A.L., Barlow, G.J., Doitsidis, L.: Fitness functions in evolutionary robotics: a survey and analysis. Robot. Auton. Syst. **57**(4), 345–370 (2009)
13. Ng, A.Y.: Feature selection, L_1 vs. L_2 regularization, and rotational invariance. In: Proceedings of the Twenty-first International Conference on Machine Learning, p. 78 (2004)
14. Nolfi, S.: A tool for training robots through evolutionary and reinforcement learning methods (2020). https://github.com/snolfi/evorobotpy2
15. Nolfi, S.: Behavioral and cognitive robotics: an adaptive perspective, January 2021. https://bacrobotics.com/
16. Nolfi, S., Bongard, J., Husbands, P., Floreano, D.: Evolutionary robotics. In: Springer Handbook of Robotics, pp. 2035–2068. Springer, Berlin, Heidelberg (2016). https://doi.org/10.1007/978-1-4899-7502-7_94-1

17. Nolfi, S., Floreano, D.: Evolutionary Robotics: The Biology, Intelligence, and Technology of Self-Organizing Machines. MIT press, Cambridge (2000)
18. Pagliuca, P., Milano, N., Nolfi, S.: Efficacy of modern neuro-evolutionary strategies for continuous control optimization. Front. Robot. AI **7**, 98 (2020)
19. Pagliuca, P., Nolfi, S.: Robust optimization through neuroevolution. PloS one **14**(3), e0213193 (2019)
20. Salimans, T., Goodfellow, I., Zaremba, W., Cheung, V., Radford, A., Chen, X.: Improved techniques for training GANs. Adv. Neural Inf. Process. Syst. **29** (2016)
21. Salimans, T., Ho, J., Chen, X., Sidor, S., Sutskever, I.: Evolution strategies as a scalable alternative to reinforcement learning. arXiv preprint arXiv:1703.03864 (2017)

Fitness Landscape Modeling
and Analysis

Analysing the Fitness Landscape Rotation for Combinatorial Optimisation

Joan Alza[1,2][✉] [iD], Mark Bartlett[1,2] [iD], Josu Ceberio[3] [iD], and John McCall[1,2] [iD]

[1] National Subsea Centre, Aberdeen AB21 0BH, Scotland
[2] Robert Gordon University, Aberdeen AB10 7GJ, Scotland
{j.alza-santos,m.bartlett3,j.mccall}@rgu.ac.uk
[3] University of the Basque Country UPV/EHU, Donostia 20018, Spain
josu.ceberio@ehu.eus

Abstract. Fitness landscape rotation has been widely used in the field of dynamic combinatorial optimisation to generate test problems with academic purposes. This method changes the mapping between solutions and objective values, but preserves the structure of the fitness landscape. In this work, the rotation of the landscape in the combinatorial domain is theoretically analysed using concepts of discrete mathematics. Certainly, the preservation of the neighbourhood relationship between the solutions and the structure of the landscape are studied in detail. Based on the theoretical insights obtained, landscape rotation has been employed as a strategy to escape from local optima when local search algorithms get stuck. Conducted experiments confirm the good performance of the rotation-based local search algorithms to perturb the search towards unexplored local optima on a set of instances of the linear ordering problem.

Keywords: Landscape rotation · Combinatorial optimisation · Group theory

1 Introduction

In the field of dynamic optimisation, there still remains the issue of translating real-world applications to academia [20]. Researchers often design simplified and generalised benchmark problems for algorithm development in controlled-changing environments, although they often omit important properties of real-world problems [12,16]. In academia, dynamic problems are generally created using generators that introduce regulated changes to an existing static optimisation problem by means of adjustable parameters. The Moving Peaks Benchmark (MPB) [3] is probably the most popular generator, where a set of n parabolic peaks change in height, width and position in a continuous space $\mathbb{R}^n$.

In the combinatorial domain, the *landscape rotation* is presumably the most popular method to construct dynamic problems for academic purposes [2,11,18, 19,21]. Introduced by Yang and Yao, the XOR dynamic problem generator [18,19] periodically modifies the mapping between solutions and objective values by means of defined operators (the exclusive-or and the composition operators).

G. Rudolph et al. (Eds.): PPSN 2022, LNCS 13398, pp. 533–547, 2022.
https://doi.org/10.1007/978-3-031-14714-2_37

According to [16], the landscape rotation in the binary space progressively permutes the initial problem, preserving important properties of the problem, such as the landscape structure, stable. In fact, the wide use of this strategy comes from its preservation nature, as well as its simplicity for comparing algorithms in controlled-changing environments. However, despite its popularity, a theoretical analysis of the preservation of the landscape structure, and study further applications of this operation in combinatorial optimisation problems, beyond the binary space, are still lacking [16].

This work introduces an analysis of the fitness landscape rotation in combinatorial problems using notions of group and graph theory. The theoretical notations provided investigate the preservation of the neighbourhood relationship between solutions even when the landscape is rotated, and capture the repercussion of rotations in the permutation space. The study is supported by proofs and examples to demonstrate its validity.

Utilising the theoretical insights gained, we experimentally investigate different ways to employ the landscape rotation for the development of advanced local search algorithms. Particularly, the goal is to illustrate the applicability of the landscape rotation to perturb the search of the algorithm when it gets trapped. To that end, two rotation-based algorithms, obtained from [2], are employed and compared to study the exploratory profit of this strategy. Conducted experiments on a set of instances of the Linear Ordering Problem [5,10] reveal the good performance of rotation-based local search algorithms, and also show the ability of the landscape rotation strategy to reach unexplored local optima. The results obtained are supported by visualisations that illustrate the behaviour of these algorithms by means of Search Trajectory Networks [13].

The remainder of the paper is structured as follows. Section 2 introduces background on the fitness landscape in the combinatorial domain, and provides important properties of the group theory. Section 3 explores the landscape rotation under group actions, and studies the repercussion of rotations in combinatorial fitness landscapes. Section 4 presents two rotation-based algorithms that are studied in the experimentation. Section 5 describes the experimental study, and discusses the applicability of the landscape rotation from the observed results. Finally, Sect. 6 concludes the paper.

2   Background

This section introduces concepts to comprehend the rotation of the fitness landscape in the combinatorial domain, and presents some basics of group theory to analyse the rotation consequences in the next section.

2.1   Combinatorial Fitness Landscape

Formally, a combinatorial optimisation problem is a tuple $P = (\Omega, f)$, where Ω is a countable finite set of structures, called search space, and $f : \Omega \to \mathbb{R}$ is an objective function that needs to be maximised or minimised. Most combinatorial problems are categorised as NP-Hard [6], which means that there is no algorithm able to solve them in polynomial time. As a result, heuristic algorithms, and especially local search algorithms, have been widely used to solve combinatorial problems [8].

A key assumption about local search algorithms is the *neighbourhood* operator, which links solutions to each other through their similarity. Formally, a neighbourhood $\mathcal{N}$ is a mapping between a solution $x \in \Omega$ and a set of solutions $\mathcal{N}(x)$ after a certain operation in the encoding of x, such that

$$\mathcal{N} : \Omega \rightarrow \mathcal{P}(\Omega),$$

where $\mathcal{P}(\Omega)$ is the power set of Ω. In other words, two solutions x and y are neighbours when a modification in the encoding of x transforms it into y, so $x \in \mathcal{N}(y)$. The neighbourhood operator in combinatorial optimisation usually implies *symmetric* relations, meaning that any operation is reversible, i.e. $x \in \mathcal{N}(y) \Leftrightarrow y \in \mathcal{N}(x)$. This property naturally leads to define *regular* neighbourhoods, which implies the same cardinality of the neighbourhood of every solution in Ω, i.e. each solution has the same number of neighbours.

The fitness landscape in the combinatorial domain can be defined as the combination of combinatorial optimisation problems together with the neighbourhood operator [14]. Formally, the fitness landscape is a triple $(\Omega, f, \mathcal{N})$, where Ω is the search space, f is the objective function and $\mathcal{N}$ stands for the neighbourhood operator. The metaphor of the fitness landscape allows comprehending the behaviour of local search algorithms when solving a combinatorial problem, given a specific neighbourhood operator. In other words, the behaviour of local search algorithms, along with the suitability of different neighbourhood operators, can be studied based on properties of the fitness landscape, such as the number of local optima, global optima, basins of attraction or plateaus. These components are thoroughly described in the following paragraphs.

A local optimum is a solution $x^* \in \Omega$ whose objective value is better or equal than its neighbours' $\mathcal{N}(x^*) \in \Omega$, i.e. for any maximisation problem, $\forall y \in \mathcal{N}(x^*)$, $f(x^*) \geq f(y)$. The number of local optima of a combinatorial problem can be certainly associated to the difficulty of a local search algorithm to reach the global optimum (the local optimum with the best objective value) [7]. Nevertheless, there are other problem features, such as those explained in [15], that are also valid for understanding the dynamics of local search algorithms.

Some works in the combinatorial domain study the basins of attraction of local optima to shape the fitness landscape, and calculate the probability to reach the global optimum [7,8,17]. Formally, an attraction basin of a local optimum, $\mathcal{B}(x^*)$, is a set of solutions that lead to the local optimum x^* when a steepest-ascent hill-climbing algorithm is applied, so $\mathcal{B}(x^*) = \{x \in \Omega | a_x = x^*\}$, where a_x is the final solution obtained by the algorithm starting from x. Figuratively, an attraction basin $\mathcal{B}(x^*)$ can be seen as a tree-like directed acyclic graph, where the nodes are solutions, and the edges represent the steepest-ascent movement from a solution to a neighbour. This assumption leads to the following definition.

Definition 1 (Attraction graph). *Let us define an attraction graph to be a directed graph $\mathcal{G}_f(x^*) = (V, E)$, where f is the objective function, $V \subseteq \Omega$ is a set of solutions, and E is a set of directed edges representing the movement from a solution to a neighbour with a better, or equal, objective value. For every solution in the graph, there is an increasing path (sequence of solutions connected by directed edges) until reaching the local optima, such that $\forall x \in V, (x = a_1, a_2, \ldots, a_h = x^*)$, where $a_{i+1} \in \mathcal{N}(a_i)$, $(a_i, a_{i+1}) \in E$, and $f(a_i) \leq f(a_{i+1})$ for any maximisation problem.*

The fitness landscape can be represented as the collection of all the attraction graphs, such that $O_f = \cup_{x^* \in \Omega^*} \mathcal{G}_f(x^*)$, where x^* is a local optimum of the set composed by local optima $\Omega^* \subset \Omega$, given a triple $(\Omega, f, \mathcal{N})$. Note that a solution (node) may belong to multiple attraction graphs if some neighbours, that belong to different attraction graphs, share the same objective value.

In the case that neighbouring solutions have equal objective values, we say that the landscape contains flat structures, called *plateaus*. Formally, a plateau $\Gamma \subseteq \Omega$ is a set of solutions with the same objective value, such that for every pair of solutions $x, y \in \Gamma$, there is a path $(x = a_1, a_2, \ldots, a_k = y)$, where $a_i \in \Gamma, a_{i+1} \in \mathcal{N}(a_i)$ and $f(a_i) = f(a_{i+1})$. The authors in [8] demonstrated that combinatorial problems often contain plateaus, and remark the importance of considering plateaus when working with problems in the combinatorial domain. The authors also differentiate three classes of plateaus, and point out that a plateau composed by multiple local optima can be considered as a single local optimum when applying local search-based algorithms, as their basins of attraction lead to the same plateau.

2.2 Permutation Space

One of the most studied fields in combinatorial optimisation is the permutation space, where solutions of the problem are represented by permutations. Formally, a permutation is a bijection from a finite set, usually composed by natural numbers $\{1, 2, \ldots, n\}$, onto the same set. The search space Ω represents the set of all permutations of size n, called *symmetric group* and denoted as $\mathbb{S}_n$, whose size is $n!$. Permutations are usually denoted using $\sigma, \pi \in \mathbb{S}_n$, except for the identity permutation $e = 12...n$. We direct the interested reader to [4] for a deeper analysis of permutation-based problems.

The similarity between permutations can be specified by permutation distances. The distance between two permutations is the minimum number of operations to convert one permutation into another. Irurozki in [9] studies the Kendall's-τ, Cayley, Ulam and Hamming distance metrics, and suggests some methods to generate new permutations uniformly at random for each distance metric. It is worth mentioning that each distance metric has its own maximum and minimum distances, d_{min} and d_{max}. We direct the interested reader to [9] for more details on permutation distances.

2.3 Landscape Rotation

The fitness landscape rotation has been probably the most popular benchmark generator, for academic purposes, in the combinatorial domain. Introduced as the XOR dynamic problem generator [18,19], this method periodically applies the rotation operation to alter the mapping between solutions and objective values by means of the exclusive-or (rotation) operator. Formally, given a static binary problem, a rotation degree ρ and the frequency of change τ, the objective value of a solution $x \in \Omega$ is altered by

$$f_t(x) = f(x \oplus M_t),$$

where f_t is the objection function at instance $t = \lceil \frac{i}{\tau} \rceil$, i is the iteration of the algorithm, f is the original (static) objective function, '$\oplus$' is the exclusive-or operator and $M_t \in \Omega$ is a binary mask. The mask M_t is incrementally generated by $M_t = M_{t-1} \oplus T$, where T is a binary template randomly generated containing $\lfloor \rho \times n \rfloor$ number of ones. The initial mask is a zero vector, $M_1 = \{0\}^n$.

Some works in the literature extended the XOR dynamic problem generator to the permutation space [2,11,21]. The landscape rotation in the permutation space can be represented as

$$f_t(\sigma) = f(\Pi_t \circ \sigma),$$

where $\sigma \in \mathbb{S}_n$ is a solution, '$\circ$' is the composition operation and Π_t is a permutation mask. The permutation mask is incrementally generated by $\Pi_t = \Pi_{t-1} \circ \pi$, where π is a permutation template generated using the methods in [9], containing $\lfloor d_{max} \times \rho \rfloor$ operations from the identity permutation given a permutation distance. The permutation mask is initialised by the identity permutation, $\Pi_1 = e$.

According to Tínos and Yang [16], the XOR dynamic problem generator changes the fitness landscape according to a permutation matrix, where the neighbourhood relations between solutions are maintained over time. However, as far as we are concerned, these assumptions have never been studied in the permutation space. Hence, this is the motivation of this work.

2.4 Group Theory

The landscape rotation can be represented by group actions, where the search space along with the rotation operation satisfy certain properties. Formally, given a finite set of solutions Ω and a group operation '$\cdot$', $G = (\Omega, \cdot)$ is a group if the closure, associativity, identity, and invertibility properties are satisfied. Mathematically, these fundamental group properties (axioms) are defined as:

- **Closure:** $x, y \in G, x \cdot y \in G$.
- **Associativity:** $x, y, z \in G, (x \cdot y) \cdot z = x \cdot (y \cdot z)$.
- **Identity:** $i \in G, \forall x \in G, x \cdot i = i \cdot x = x$.
- **Invertibility:** $x, x^{-1} \in G, x \cdot x^{-1} = x^{-1} \cdot x = i$.

There is another property, the **commutativity**, that is not fundamental for the definition of a group. A group is said to be commutative when $x, y \in G$, $x \cdot y = y \cdot x$. It is worth mentioning that the commutation property holds in the binary space, but it does not in the permutation space.

3   Analysis of the Fitness Landscape Rotation

In this section, we aim to theoretically analyse some important consequences of rotating the landscape in the combinatorial domain using group properties. To that end, we will study the (i) the neighbourhood relation preservation after rotating the landscape, and (ii) the repercussion of rotations in landscapes encoded by permutations.

3.1   Neighbourhood Analysis

The properties of the fitness landscape rotation can be studied using notions of group theory, i.e. the rotation (exclusive-or '$\oplus$' and composition '$\circ$') operators can be generalised to the group operation '$\cdot$'. Following previous notations, we can demonstrate the preservation of neighbourhood relations between solution before and after a rotation without loss of generality.

Theorem 1. *Given a group G, let $\mathcal{N}(x) \in G$ be the neighbourhood of x and $t \in G$ the mask used to rotate the space. We say that the neighbourhood relations are preserved iff $\mathcal{N}(t \cdot x) \Leftrightarrow t \cdot \mathcal{N}(x)$.*

Proof. Let $x, y \in G$ be two neighbouring solution in the group, such that $x \in \mathcal{N}(y)$ is a neighbour of y (and vice versa). We can define the neighbourhood operation as $c(i,j) \cdot x \in \mathcal{N}(x)$, where $c(i,j)$ is an operation in the encoding of a solution. For example, $c(i,j)$ can represent the swap of the elements i and j, or the insertion of the element at position i into the position j.

Based on this notation, we can define the fundamental group properties (identity, invertibility and associativity) as

$$c(i,j) \cdot e = e \cdot c(i,j) = c(i,j) \tag{1}$$
$$c(i,j) \cdot c(i,j)^{-1} = c(i,j)^{-1} \cdot c(i,j) = e \tag{2}$$
$$x = c(i,j) \cdot x \cdot c(i,j)^{-1} \tag{3}$$

Note that $c(i,j)^{-1} = c(j,i)$ represents the inverse operation of $c(i,j)$. Assuming that the landscape is rotated using the template $t \in G$, we can say that Ω and $t \cdot \Omega$ are defined independently. In the following, we aim to prove that $t \cdot \mathcal{N}(x) \Leftrightarrow \mathcal{N}(t \cdot x)$.

First, we must ensure that the rotation of the landscape preserves the neighbourhood relation between solutions, so $\mathcal{N}(t \cdot x) \subset t \cdot \mathcal{N}(x)$. It can be demonstrated in the following way.

$$x \in \mathcal{N}(y)$$
$$x = c(i,j) \cdot y$$
$$t \cdot x = t \cdot c(i,j) \cdot y \qquad (4)$$
$$c(i,j)^{-1} \cdot t \cdot x = c(i,j)^{-1} \cdot t \cdot c(i,j) \cdot y$$
$$c(i,j)^{-1} \cdot t \cdot x = t \cdot y$$

Note that the last step in the equation is given by the Eq. 3. Considering that $c(i,j)^{-1} = c(j,i)$ and $c(i,j) \cdot x \in \mathcal{N}(x)$, we can say that $c(i,j)^{-1} \cdot t \cdot x \in \mathcal{N}(t \cdot x)$. Therefore, given the symmetry of the neighbourhood operator, we can prove that the rotation of the neighbourhood is a subset of the neighbourhood of the rotated $t \cdot x$, so $t \cdot y \in \mathcal{N}(t \cdot x)$.

Then, we must prove the inverse statement, i.e. the rotated neighbourhood derives in the rotation of the neighbourhood, $t \cdot \mathcal{N}(x) \subset \mathcal{N}(t \cdot x)$. It can be demonstrated in the following way.

$$x \in \mathcal{N}(t \cdot y)$$
$$x = c(i,j) \cdot t \cdot y$$
$$x = c(i,j) \cdot [c(i,j)^{-1} \cdot t \cdot c(i,j)] \cdot y \qquad (5)$$
$$x = e \cdot t \cdot c(i,j) \cdot y$$

In this case, after considering $c(i,j) \cdot x \equiv \mathcal{N}(x)$, we use the inverse property of the neighbourhood operation (Eq. 3) and the identity property (Eq. 2) to demonstrate that $x = t \cdot c(i,j) \cdot y$, so $x \in t \cdot \mathcal{N}(y)$. Therefore, we can prove that $x \in t \cdot G$ derives in $x \in G$.

In summary, by showing the symmetry of the rotation operation, we can confirm that the neighbourhood relations between solutions are preserved.

3.2 Repercussion of the Landscape Rotation

The authors in [16] mention that the topological structure of the fitness landscape must be analysed to comprehend the behaviour of the algorithms. In order to study the repercussion of changes in the space, we will use the definition of the attraction graphs (Definition 1) to represent the topological features of the fitness landscape. In the following, we will consider the permutation space ($\mathbb{S}_n$), the swap operation (2-exchange operator) and the steepest-ascent hill-climbing algorithm (saHC) to represent the fitness landscape, and more precisely, the collection of attraction graphs.

Note that the preservation of the neighbourhood relations (previous section) is independent of the algorithm and objective function. Thus, we can demonstrate that the landscape structure is preserved even if solutions are rearranged. Using graph theory notations, we could point out that attraction graphs are *isomorphic* (structurally equivalent) to themselves in the rotated environment, such that $O_f \cong O_{f_t}$, where O_{f_t} is the set of attraction graphs that composes the rotated fitness landscape f_t. Despite the conservation of the landscape structure,

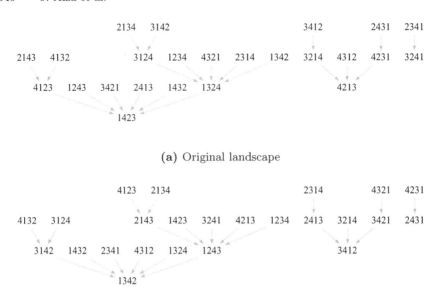

(a) Original landscape

(b) Landscape rotated by $t = 1342$

Fig. 1. Illustrative visualisation of the landscape structure before and after a rotation.

all solutions are mapped to different positions when the landscape is rotated. Let us illustrate these concepts with the following example.

Figure 1 displays the landscape of a permutation problem of size $n = 4$ as a collection of attraction graphs produced by a given objective function f. These images illustrate the preservation of the landscape structure. For example, the attraction graph on the right side of both images will always contain two solutions, such that $\mathcal{G}_f(3241)$ and $\mathcal{G}_{f_t}(1324 \circ 3241 = 2431)$ are isomorphic.

It is worth noting that the landscape rotation alters the mapping between solutions and objective values. Hence, we can conclude that, since solutions are rearranged at different positions in the fitness landscape, the objective values are preserved. In other words, the fitness landscapes before and after a rotation, O_f and O_{f_t}, are *equal* in terms of objective values, e.g. $f(3241) = f_t(2431)$.

In order to measure the impact of the rearrangement after a rotation, we can use the total number of solution exchanges between attraction graphs. This assumption is motivated by the fact that, for a local search-based algorithm, it is more likely to "escape" from an attraction graph when a rotation implies a big number of solution exchanges between graphs. Continuing with the previous example, Table 1 summarises the total number of solution exchanges between attraction graphs for all the possible rotations generated at a given Cayley distance (d_C). This distance metric uses the smallest number of swaps assumption to generate permutations uniformly at random [1]. The table entries demonstrate that the rotation degree, measured as the Cayley distance, is not necessarily pro-

Table 1. Example of the number of exchanges between attraction graphs of all possible rotation masks generated by the Cayley distance metric.

t (d_C)	Exchanges	t (d_C)	Exc.	t (d_C)	Exc.
1234 (0)	0 (original)	1423 (2)	10	4213 (2)	13
1243 (1)	4	2143 (2)	16	4321 (2)	14
1324 (1)	8	2341 (2)	14	2314 (3)	12
1432 (1)	8	2431 (2)	14	2413 (3)	12
2134 (1)	16	3124 (2)	12	3142 (3)	12
3214 (1)	12	3241 (2)	13	3421 (3)	13
4231 (1)	14	3412 (2)	12	4123 (3)	14
1342 (2)	10	4132 (2)	14	4312 (3)	13

portional to the number of solution exchanges between attraction graphs in the permutation space. For example, the average number of solution exchanges for each Cayley distance in Table 1 reflects that rotating to $d_C = 1$ produces 10.3 exchanges, $d_C = 2$ produces 12.9 exchanges and $d_C = 3$ produces 12.6 exchanges, on average, respectively. Hence, the use of permutation distances as a rotation degree should be used with caution, since medium rotations can be severe, in terms of the total number of solution exchanges between attraction graphs.

4   Rotation as a Perturbation Strategy

From the theoretical insights gained from the previous section analysis, we suggest the landscape rotation as a perturbation strategy for local search-based algorithms to react when algorithms get trapped in poor quality local optima. The rotation action can be used to relocate a stuck algorithm's search, ideally into a different attraction graph, by means of the permutation distance that controls the magnitude of the perturbation. Note that, generally, a local optimum in the original landscape is not mapped to a local optimum in the rotated landscape. This assumption motivates its usage to reach unexplored local optima. In short, rotation-based local search algorithms can be summarised as follows:

1: Run the local search algorithm until reaching a local optimum, x^*.
2: The rotation operation is applied to relocate the algorithm at the solution $t \cdot x^*$. Ideally, the algorithm will reach a new local optimum, such that $t \cdot x^* \subseteq \mathcal{G}_{f_t}(y^*)$.
3: This process is repeated until meeting the stopping criterion.

In [2], we presented two rotation-based perturbation strategies: a depth-first and a width-first strategy. These methods differ in the way they use the rotation operation. If we apply these strategies into the steepest-ascent hill climbing algorithm (saHC), the depth-first algorithm (saHC-R1) will use the rotation operation to move the search away, and continue the search from a new position until getting trapped again. Then, both local optima are compared, and the search is relocated to the best found solution. On the other hand, the width-first algorithm (saHC-R2) applies the rotation operation for some iterations, and then continues to search from a new position (undo the rotation) until it gets stuck again. Unlike saHC-R1, this strategy does not relocate the search to the best solution found, which encourages continuous exploration of different graphs. For more details, see [2].

5   Results and Discussion

This section evaluates the performance of the proposed strategies to solve the Linear Ordering Problem (LOP) [5,10]. This problem aims to maximise the entries above the main diagonal of a given matrix $B = [b_{i,j}]_{n \times n}$. The objective is to find a permutation σ that orders the rows and columns of B, such that

$$\arg \max_{\sigma \in \mathbb{S}_n} f(\sigma) = \sum_{i=1}^{n-1} \sum_{j=i+1}^{n} b_{\sigma_i, \sigma_j}.$$

The specific instances used in the experimentation of this work are obtained from the supplementary material web[1] presented in [8]. The web contains 12 LOP instances: eight instances of size 10, and four instances of size 50. The parameters employed in the rotation-based algorithms [1] are summarised in Table 2. Note that the rotation degree is designed to start with big rotations, and exponentially decreasing it to intensify the search. The motivation of this strategy is to balance the intensification-diversification trade off based on the search process of the algorithm.

Table 2. Parameter settings for the experimental study.

Parameter	Value
Rotation degree	$d = \left\lfloor d_{max} e^{\left(\frac{\ln\left(\frac{d_{min}}{d_{max}}\right)}{S} \right) i} \right\rfloor$
Distance metric	Cayley distance
Number of repetitions	30
Stopping criterion	$10^3 n$ iterations

---

[1] http://www.sc.ehu.es/ccwbayes/members/leticia/AnatomyOfAB/instances/ InstancesLOP.html.

Obtained results and instance properties are summarised in Table 3, where the best found objective values, the number of rotations and the number (and percentage) of visited local optima are shown for each algorithm. The total number of local optima for each instance has been obtained from [8]. However, due to the incompleteness of the number of local optima for instances of size $n = 50$, we only show the number of local optima explored for these instances, without showing the percentages of local optima explored.

The results show the good performance of rotation-based algorithms, as well as their explorability ability. Both algorithms are able to find the same optimal solutions (except for N-be75oi), in terms of objective values, but they differ in the number of rotations and local optima explored. The percentages represent the ability of the algorithms to explore the attraction graphs that compose the landscape. saHC-R2 always performs more rotations than saHC-R1, and thus, it can find a larger (or the same) number of local optima than saHC-R1. Therefore, we can say that saHC-R2 tends to be more exploratory than saHC-R1.

In order to illustrate the influence of the rotation degree on the search of the algorithms, Table 4 shows the number of rotations and local optima reached by each algorithm, over the 30 runs, for each Cayley distance, on Instance 8. Remember that the rotation degree describes an exponential decrease as the search progresses[2].

Table 3. Information of the instances, and results of the rotation-based algorithms on LOP instances. Percentages for instances of size $n = 50$ are not available, since their total number of local optima is unknown.

Instance	LO	saHC-R1			saHC-R2		
		Obj. value	Rotations	LO (%)	Obj. value	Rotations	LO (%)
Instance 1	13	1605	2015	13 (100%)	1605	2636	13 (100%)
Instance 2	24	1670	2011	24 (100%)	1670	2629	24 (100%)
Instance 3	112	4032	1956	104 (92.8%)	4032	2545	106 (94.6%)
Instance 4	129	3477	1988	121 (93.8%)	3477	2565	125 (96.9%)
Instance 5	171	32952	2093	169 (98.8%)	32952	2712	171 (100%)
Instance 6	226	40235	1954	215 (95.1%)	40235	2571	220 (97.3%)
Instance 7	735	22637	2138	683 (92.9%)	22637	2742	716 (97.4%)
Instance 8	8652	513	2528	6063 (70%)	513	3351	6887 (79.6%)
N-be75eec	>500	236464	8527	62143 (−%)	236464	10737	75404 (−%)
N-be75np	>500	716994	8306	36046 (−%)	716994	10364	58423 (−%)
N-be75oi	>500	111171	8507	80001 (−%)	111170	10698	96371 (−%)
N-be75tot	>500	980516	8437	31550 (−%)	980516	10606	53450 (−%)

---

[2] The repercussion of the rotation degree in other instances is available online.

Table 4. Number of rotations and reached local optima by each algorithm on Instance 8.

		Cayley distance								
		1	2	3	4	5	6	7	8	9
saHC-R1	Rotations	19594	19304	10851	7416	5623	4525	3778	3279	1452
	LO	678	1968	2500	2679	2725	2633	2438	2234	1227
saHC-R2	Rotations	27680	26090	14191	9432	7039	5633	4676	4027	1746
	LO	736	3809	4014	3719	3337	3037	2761	2528	1346

The table shows that, although the number of rotations exponentially decays, the highest exploratory behaviour of the algorithms holds on medium-small distances, i.e. both algorithms find more local optima when the rotation operates at $d_C = \{3, 4, 5\}$. This performance matches with the example in Table 1, where rotating to medium distances is sufficient to perturb the search of algorithms to different attraction graphs.

In order to visually represent and analyse the evolution of the algorithms, we use the Search Trajectory Networks (STNs) tool [13], a directed-graph-based model that displays search spaces in two or three dimensions. Figure 2 displays a single run of each algorithm on Instance 8 using STNs. The colours in the figures highlight the starting and ending points of the search (blue and green nodes), the best found solutions (yellow nodes) and the rotation operations (red edges), respectively. The entire experimentation is available online[3].

The plots show the behaviour of each algorithm in a two-dimensional space. The left plot shows the behaviour of saHC-R1, where the algorithm always applies the rotation operation from the best solution found. This visualisation gives an insight of the structure of the landscape, since the algorithm is able to explore the paths that compose the attraction graphs. On the other hand, Fig. 2b shows a continuous search of saHC-R2, where the algorithm moves through attraction graphs, meaning that it rarely gets stuck on the same local optimum after a rotation. This behaviour can be comprehended by the fact that saHC-R2 does not rotate from the best solution found, but from the last local optimum found. That said, we can conclude that saHC-R1 outperforms in instances with few but deep attraction graphs, while saHC-R2 outperforms in instances composed of many attraction graphs.

Finally, it is worth noting the presence of a plateau composed of local optima in Instance 8, i.e. multiple local optima have the same objective value, which turns out to be the optimal value. The figures confirm that both algorithms can detect and deal with plateaus. Interestingly, Fig. 2a shows that some local optima that form the plateau are visibly larger, which means that saHC-R1 visits them several times. From previous statements, we can deduce that Instance 8 is composed of neighbouring local optima with the same objective value, and also that saHC-R2 reaches more local optima than saHC-R1.

---

[3] https://zenodo.org/record/6406825#.YkcxaW7MI-Q.

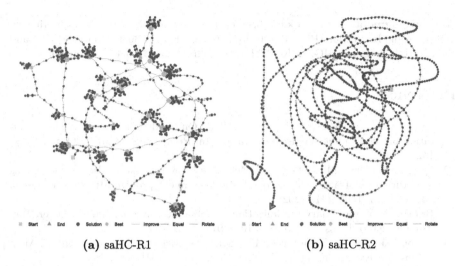

(a) saHC-R1   (b) saHC-R2

Fig. 2. Search Trajectory Networks of rotation-based algorithms on Instance 8.

6   Conclusions and Future Work

Landscape rotation has been widely used to generate dynamic problems for academic purposes due to its preserving nature, where important properties of the problem are maintained. In this article, we study the preservation of the landscape structure using group actions, and based on the insights gained, we suggest using the rotation operation to relocate the search of local search-based algorithms when they get stuck. The experiments carried out show the good application of the landscape rotation to perturb the search of the local search algorithm through unexplored local optima. Obtained results also illustrate that 'medium' rotations can cause a big repercussion, when it comes to the number of rearranged solutions.

This work can be extended in several ways. First and most obvious, there are some landscape properties that have been ignored in this manuscript, such as the number and size of attraction graphs, or the frontier and the centrality of local optima [8]. These assumptions, along with the consideration of problem properties, such as the symmetries of the problem instance, could lead to very different outcomes, where the landscape rotation may be less applicable. Finally, this work has considered the swap operator and the steepest-ascent hill-climbing algorithm to construct the attraction graphs in the Linear Ordering Problem. This study can be naturally extended to other combinatorial optimisation problems, as well as other ways to represent the landscape, such as the insertion operation or the first-improvement hill-climbing heuristic.

Acknowledgments. This work has been partially supported by the Spanish Ministry of Science and Innovation (PID2019-106453GA-I00/AEI/10.13039/501100011033), and the Elkartek Program (SIGZE, KK-2021/00065) from the Basque Government.

References

1. Alza, J., Bartlett, M., Ceberio, J., McCall, J.: On the definition of dynamic permutation problems under landscape rotation. In: Proceedings of GECCO, pp. 1518–1526 (2019)
2. Alza, J., Bartlett, M., Ceberio, J., McCall, J.: Towards the landscape rotation as a perturbation strategy on the quadratic assignment problem. In: Proceedings of GECCO, pp. 1405–1413 (2021)
3. Branke, J.: Evolutionary Optimization in Dynamic Environments. Springer, Heidelberg (2002). https://doi.org/10.1007/978-1-4615-0911-0
4. Ceberio, J.: Solving Permutation Problems with Estimation of Distribution Algorithms and Extensions Thereof. Ph.D. thesis, UPV/EHU (2014)
5. Ceberio, J., Mendiburu, A., Lozano, J.A.: The linear ordering problem revisited. EJOR **241**(3), 686–696 (2015)
6. Garey, M.R., Johnson, D.S.: Computers and Intractability: A Guide to the Theory of NP-Completeness. W. H. Freeman & Co., New York (1979)
7. Hernando, L., Mendiburu, A., Lozano, J.A.: An evaluation of methods for estimating the number of local optima in combinatorial optimization problems. Evol. Comput. **21**(4), 625–658 (2013)
8. Hernando, L., Mendiburu, A., Lozano, J.A.: Anatomy of the attraction basins: breaking with the intuition. Evol. Comput. **27**(3), 435–466 (2019)
9. Irurozki, E.: Sampling and learning distance-based probability models for permutation spaces. Ph.D. thesis, UPV/EHU (2014)
10. Martí, R., Reinelt, G.: The Linear Ordering Problem: Exact and Heuristic Methods in Combinatorial Optimization, vol. 175. Springer, Heidelberg (2011). https://doi.org/10.1007/978-3-642-16729-4
11. Mavrovouniotis, M., Yang, S., Yao, X.: A benchmark generator for dynamic permutation-encoded problems. In: PPSN, pp. 508–517 (2012)
12. Nguyen, T.T., Yang, S., Branke, J.: Evolutionary dynamic optimization: a survey of the state of the art. Swarm Evol. Comput. **6**, 1–24 (2012)
13. Ochoa, G., Malan, K.M., Blum, C.: Search trajectory networks: a tool for analysing and visualising the behaviour of metaheuristics. Appl. Soft Comput. **109**, 107492 (2021)
14. Reidys, C.M., Stadler, P.F.: Combinatorial Landscapes. Working Papers 01-03-014, Santa Fe Institute (2001)
15. Tayarani-N., M.H., Prügel-Bennett, A.: On the landscape of combinatorial optimization problems. IEEE TEVC **18**(3), 420–434 (2014)
16. Tinós, R., Yang, S.: Analysis of fitness landscape modifications in evolutionary dynamic optimization. Inf. Sci. **282**, 214–236 (2014)
17. Vérel, S., Daolio, F., Ochoa, G., Tomassini, M.: Local optima networks with escape edges. In: Hao, J.-K., Legrand, P., Collet, P., Monmarché, N., Lutton, E., Schoenauer, M. (eds.) EA 2011. LNCS, vol. 7401, pp. 49–60. Springer, Heidelberg (2012). https://doi.org/10.1007/978-3-642-35533-2_5
18. Yang, S.: Non-stationary problem optimization using the primal-dual genetic algorithm. In: Proceedings of CEC, vol. 3, pp. 2246–2253 (2003)

19. Yang, S., Yao, X.: Experimental study on population-based incremental learning algorithms for dynamic optimization problems. Soft Comput. **9**, 815–834 (2005)
20. Yazdani, D., Cheng, R., Yazdani, D., Branke, J., Jin, Y., Yao, X.: A survey of evolutionary continuous dynamic optimization over two decades-part b. IEEE TEVC **25**(4), 630–650 (2021)
21. Younes, A., Calamai, P., Basir, O.: Generalized benchmark generation for dynamic combinatorial problems. In: Proceedings of GECCO, pp. 25–31 (2005)

Analysis of Search Landscape Samplers for Solver Performance Prediction on a University Timetabling Problem

Thomas Feutrier[ID], Marie-Éléonore Kessaci[ID], and Nadarajen Veerapen[(⊠)][ID]

Univ. Lille, CNRS, Centrale Lille, UMR 9189 CRIStAL, 59000 Lille, France
{thomas.feutrier,mkessaci,nadarajen.veerapen}@univ-lille.fr

Abstract. Landscape metrics have proven their effectiveness in building predictive models, including when applied to University Timetabling, a highly neutral problem. In this paper, two Iterated Local Search algorithms sample search space to obtain over 100 landscape metrics. The only difference between the samplers is the exploration strategy. One uses neutral acceptance while the other only accepts strictly improving neighbors. Different sampling time budgets are considered in order to study the evolution of the fitness networks and the predictive power of their metrics. Then, the performance of three solvers, Simulated Annealing and two versions of a Hybrid Local Search, are predicted using a selection of landscape metrics. Using the data gathered, we are able to determine the best sampling strategy and the minimum sampling time budget for models that are able to effectively predict the performance of the solvers on unknown instances.

Keywords: University Timetabling · Performance prediction · Landscape analysis · Local search

1 Introduction

University Timetabling is an active research area in combinatorial optimization. Research into the topic is stimulated, notably, by the International Timetabling Competition (ITC) organized at PATAT. Many different heuristics [6] have been proposed to solve University Timetabling problems, including both local search [12] and crossover-based algorithms [18]. Given the complex nature of the problems, hybrid and hyper-heuristic approaches are also well suited [17].

Fitness landscapes [19] and their analysis can help to understand the nature of the search space. Over the last ten years or so, landscape analysis has moved from an admittedly mainly theoretical construct to being a more practical tool [8,10].

In general, the whole landscape cannot be enumerated and sampling is required. One such approach relies on gathering the traces of Iterated Local Search (ILS) runs. This is the approach taken in this paper, where we compare

G. Rudolph et al. (Eds.): PPSN 2022, LNCS 13398, pp. 548–561, 2022.
https://doi.org/10.1007/978-3-031-14714-2_38

two ILS samplers with different exploration strategies. We wish to investigate whether considering the neutrality of the landscape will provide a better sample or not. We rely on the predictive ability of performance models built from features derived from the samples as a proxy for the quality of the samples.

In the context of performance prediction, some papers have shown that landscape analysis and associated metrics can build accurate models [10]. In particular, when considering continuous optimization, Bisch et al. [1] have used support vector regression models, Muñoz et al. [11] considered neural networks to predict the performance of CMA-ES, Malan and Engelbrecht [9] used decision trees to predict failure in particle swarm optimization, while Jankovic and Doerr [4] considered random forests and CMA-ES. In the combinatorial context, Daolio et al. [2] and Liefooghe et al. [7] respectively applied mixed-effects multi-linear regression and random forest models to multiobjective combinatorial optimization, and Thomson et al. [20] considered random forests and linear regression on the Quadratic Assignment Problem, finding that random forests performed better.

In this paper, we consider the Curriculum-Based Course Timetabling problem (CB-CTT). We first use two variants of an ILS to explore and sample the search space for problem instances of the ITC 2007 competition [16] across 4 different time budgets. After a feature selection step, a model is built for each time budget and each sampler. Each model is evaluated, via cross-validation, according to its ability to predict the final fitness on 3 different solvers on unseen instances. Our results show that, on our instances, sampling the search space with 100 ILS runs of 5 s each allows us to build models that can accurately predict fitness across solvers.

The paper is organized as follows: Sect. 2 presents our problem; Sect. 3 introduces fitness landscapes and relevant definitions; we develop our experimental protocol in Sect. 4; the features used in our models are laid out in Sect. 5; Sect. 6 analyzes the effects of the sampler time budget on networks; Sect. 7 describes the preprocessing involved in building the models as well as the evaluation procedure; the models obtained are discussed in Sect. 8; finally, Sect. 9 concludes the paper and outlines potential for future research.

2   Curriculum-Based Course Timetabling

The paper focuses on a specific University Timetabling problem: Curriculum-Based Course Timetabling (CB-CTT). ITC 2007 was especially important for CB-CTT because it formalized several instances and imposed a runtime limit of 5 min, encouraging the use of metaheuristics instead of exact solvers.

CB-CTT is centered on the notion of a curriculum, that is a simple package of courses. This may be a simplification of the real-life problem since a student can choose only one curriculum. Curricula cluster students together, with each change of event impacting all students in the same way.

Courses are sets of lectures and are taught by one teacher. One course can belong to several different curricula. In this case, all students across curricula attend lectures together. The problem is scheduled over a limited number of days, divided into periods or timeslots. One period corresponds to the duration time of one lecture.

In CB-CTT, a solution consists of scheduling lectures in timeslots and available rooms following hard and soft constraints. Hard constraints must always be respected. A timetable is said to be feasible when all the hard constraints are met. An example of a hard constraint is preventing one teacher from teaching two lectures at once. On the contrary, soft constraints are optional and generally represent targets to strive for. The violations of each soft constraint are represented as a function to minimize.

The objective function to optimize for CB-CTT is a weighted sum of the constraint violations: $f(s) = \sum_{i=1}^{4} SoftConstraints_i(s) * \omega_i$, where S represents a timetable. The weights, as used for ITC 2007, are set to 1, 5, 2 and 1 for ω_1, ω_2, ω_3 and ω_4 respectively. $SoftConstraints_i(s)$ represents the number of violations for the soft constraints listed below.

1. *RoomCapacity*: All students can be sat in the room.
2. *MinWorkingDays*: A course has lectures which should be scheduled within a minimum number of days.
3. *CurriculumCompactness*: A student should have always two consecutive lectures before a gap.
4. *RoomStability*: Lectures of a course should be in the same room.

3   Search Landscape

Educational timetabling, and CB-CTT in particular, is known to be a very neutral problem [13]. This means that a large number of similar solutions share the same fitness value. Neutrality may hinder the solving process because finding a suitable trajectory in the search landscape becomes more difficult. In this context, landscape analysis can help to understand the nature of the search space, for instance by characterizing the ruggedness of the landscape or its connectivity patterns.

The notion of landscape is strongly tied to the algorithm and neighborhood operators used to explore the search space.

Landscape. A landscape [19] may be formally defined as a triplet (S, N, f) where

- S is a set of solutions, or search space,
- $N : S \longrightarrow \wp(S)$, the neighborhood structure, is a function that assigns, to every $s \in S$, a set of neighbors $N(s)$ ($\wp(S)$ is the power set of S), and
- $f : S \longrightarrow \mathbb{R}$ is a fitness function.

In CB-CTT, a problem instance establishes fixed relationships between curricula, courses, lectures and teachers. A solution then describes the tripartite graph that links lectures, rooms and timeslots together. We choose to implement this as an object-oriented representation where lecture, room and timeslot objects are connected together as appropriate.

We use 6 classic timetabling neighborhood operators: 3 basic ones and 3 designed to deal with specific soft constraints. A link between two solutions in our landscapes is the result of the application of any one of these operators. The 3 simplest operators focus, at each call, on a single lecture. RoomMove and TimeMove respectively change the room and the timeslot. LectureMove changes the room and timeslot at the same time. The 3 other operators attempt to lower the violation penalty of their associated soft constraint using a combination of the basic moves. For instance, the CurriculumCompactnessMove identifies an isolated lecture in a curriculum and performs a TimeMove to bring it closer to another lecture.

Besides exhaustive exploration that gives a perfect model of the landscape, any other sampling method will only provide an approximation. In our case, exhaustive exploration is computationally infeasible. We therefore rely on an ILS to sample the search space, as has been done for instance by Ochoa et al. [14]. We focus in particular on local optima.

Local Optimum. A local optimum is a solution $s^* \in S$ such that $\forall s \in N(s^*)$, $f(s^*) \leq f(s)$. In order to allow for plateaus and neutral landscapes, the inequality is not strict. Minimization is considered since we deal with constraint violations.

A number of landscape metrics can be measured by building Local Optima Networks (LONs) [21]. These provide compressed graph models of the search space, where nodes are local optima and edges are transitions between them according to some search operator. LONs for neutral landscapes have been studied before by Verel et al. [22]. The latter work introduces the concept of *Local Optimum Neutral Network* that considers that a neutral network is a local optimum if all the configurations of the neutral network are local optima. Another approach to neutrality in LONs is by Ochoa et al. [14] who develop the notion of *Compressed LONs* where connected LON nodes of equal fitness are aggregated together. For our purposes, we will consider two slightly different kinds of networks: Timeout Plateau Networks and Fitness Networks.

Plateau. A plateau, sometimes called a neutral network, is usually defined as a set of connected solutions with the same fitness value. Two solutions s_1 and s_2 are connected if they are neighbors, i.e., $s_2 \in N(s_1)$. Depending on the neighborhood function, we may also have $s_1 \in N(s_2)$. Plateaus are defined as sequences of consecutive solutions with the same fitness.

Timeout Plateau. The first ILS sampler we consider, $ILS_{neutral}$, contains a hill-climber that stops if it remains on the same plateau for too long (50,000 consecutive evaluations at the same fitness). Furthermore, the hill-climber used accepts the first non-deteriorating move (i.e., an improving or a neutral neighbor). We call the last plateaus thus found *timeout plateaus* because the hill-climber has not been able to escape from them within a given number of iterations. However they are not necessarily a set of local optima. Some exploratory analysis showed at least 1 % of these solutions were not actual local optima. The other sampler,

ILS$_{strict}$, has the same components but will only accept a strictly improving solution. Its timeout plateaus therefore trivially only contain one solution.

Timeout Plateau Network. A Timeout plateau network is a graph where each node represents one timeout plateau and an edge represents a transition between two such plateaus. Here this transition is an ILS perturbation followed by hill-climbing. Timeout Plateau Networks are a set of independent chains, where each represents one ILS run.

Fitness Network. This is a simplification of Timeout Plateau Networks where all nodes with the same fitness are contracted together. This provides a graph structure with much higher connectivity than a Timeout Plateau Network. While it is not meant as an accurate representation of the landscape, several different metrics related to connectivity between fitness levels can be computed. Note that this is an even greater simplification than *Compressed LONs* [14] which only aggregate nodes sharing the same fitness that are connected together at the LON level.

4   Experimental Protocol

Experiments use 19 of the 21 instances proposed for ITC 2007. Instances 01 and 11 are set aside they are very easy to solve. All instances contain between 47 and 131 courses, 138–434 lectures, 52–150 curricula, 25–36 timeslots and 9–20 rooms.

Our solver of choice is the Hybrid Local Search (HLS) proposed by Müller [12]. It combines Hill-Climbing (HC), Great Deluge and Simulated Annealing algorithms in an Iterated Local Search and won ITC 2007. Our experiments use two versions of the HLS. We will refer to the default version, that accepts equal or better solutions during HC, simply as HLS. The other, HLS$_{strict}$, uses a strict acceptance criterion for HC. In addition, iterated Simulated Annealing (SA) is tested. We reuse the SA component found within HLS and place it inside a loop that stops when runtime budget is reached. All executions run on Intel(R) Core(TM) i7-9700 CPU @ 3.00GHz, all solvers have a time budget of 5 min and were tested 100 times on each instance.

Sampling the Search Space. To sample our search space, we use one of two ILS, based on algorithmic components found in the HLS mentioned above. Both ILS include an Iterative Bounded Perturbation (IBP) and use a hill-climber (HC) following a first improvement strategy. HC stops when it finds a local optimum or when it has evaluated 50,000 solutions without strict improvement. The distinction between the two ILS samplers lies in the acceptance criterion. The first follows the same strategy as HLS, i.e., accepting any equal or better solution; it is called ILS$_{neutral}$. Second sampler, ILS$_{strict}$, follows the strategy of HLS$_{strict}$, accepting only strictly improving solutions.

IBP takes a baseline fitness, $Fit_{FirstSol}$, corresponding to the fitness of the first solution after construction. Then it deteriorates the final solution found by HC, $Fit_{LastSol}$, to reach a solution with fitness equal to $Bound = Fit_{LastSol} + 0.1(Fit_{FirstSol} - Fit_{LastSol})$.

Due to memory constraints, $ILS_{neutral}$ only records fitness and size for all timeout plateaus met. ILS_{strict} just saves the fitness of last solution of each HC, its timeout plateaus have a size of 1. Both ILS samplers are run 100 times with a time budget of $\{5, 10, 20, 30\}$ seconds per run on each instance. This budget was set in order to obtain enough predictive information on the landscape. Each (time budget, sampler) pair produces one network, so we have 8 networks for each instance.

Timeout Plateau Network. Each Timeout Plateau Network is built from the data gathered from 100 runs. Each node of the network is a timeout plateau and each directed edge is a transition corresponding to a perturbation followed by a hill-climber. At this stage the weight of a directed edge is 1 and there is no connectivity between the trajectories of the individual ILS sampler runs. Thus, a contraction step is required to obtain further information.

Fitness Network. Given the high level of neutrality of the problem, we have chosen to consider that all solutions with the same fitness belong to a single wide plateau. This hypothesis allows us to obtain a connected network very easily. The contraction process of the Timeout Plateau Network into a Fitness Network preserves all the data required to compute the metrics mentioned in Sect. 5.2. The new weights on the directed edges correspond to the sum of the contracted directed edges.

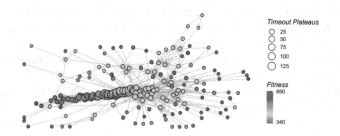

Fig. 1. Fitness network of instance 12 built by $ILS_{neutral}$ with 30 s.

5 Features

Gathering a large number of features ranging from instance features to global and local landscape metrics increases the probability of finding combinations of them that contain complementary information for prediction.

5.1 Instance Features

Instance features include descriptive data about the problem instance. The most basic ones count the courses, curricula, lectures, and days. Others quantify the complexity of the problem. LecturesByCourse counts the minimum, maximum and average number of lectures for one course. TeachersNeeded is the number of lectures divided by the number of timeslots. Finally, CourseByCurriculum is number of courses divided by the number of curricula. It measures the difficulty of scheduling without violations. In total, there are 24 instance features.

5.2 Landscape Metrics

We consider different metrics that are computed on the landscape.

Node-Level Metrics. A first set of metrics relates to what the nodes represent. Plateaus is the number of plateaus that have been contracted to form the node. It is used as the size property of nodes in Fig. 1. Size is the sum of number of accepted solutions in the contracted plateaus. Fit corresponds to the Fitness of all Timeout Plateaus represented by the current node. Loops is the number of consecutive loops on the same fitness, an estimator for attraction power.

A second set of metrics relates to the connectivity of the nodes within the network. Each connectivity metric has 9 variants by combining whether *all*, *ingoing* or *outgoing* edges are considered, together with whether we considered directed edges that reach *any* node or only select the ones that reach *better* (resp. *worse*) nodes. The Degree metric measures the number of different arcs connected to the nodes. The Weight metric represent the number of times the samplers have passed from one timeout plateau to another.

We also consider two variants of weight and degree metrics. For some given node, the *better* (resp. *worse*) variant only considers directed edges between this node and better (resp. *worse*) nodes.

The above metrics are computed for each node and five points corresponding to the quartiles (Q1, median and Q3) and the 10th and 90th percentiles of the distribution are used as features for our models. The number of features calculated with node-level metrics amounts to 65 features.

Network Metrics. To describe the networks themselves, we used additional metrics including the Mean fitness, the number of timeout plateaus, nodes and edges. Moreover, the number of sink nodes is stored, as well as the matching coefficients of assortivity and transitivity.

Sink nodes are ones that do not have any outgoing edges to nodes with better fitness. Assortivity is a measure of similarity between linked nodes [15]. The more numerous the connected nodes with the same attributes, the higher the coefficient. The transitivity coefficient, also called the clustering coefficient, is the probability of a link between adjacent neighbors and one chosen vertex [15]. In total, we consider 23 network metrics.

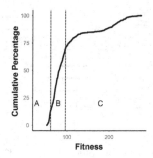

Fig. 2. Cumulative percentage of the number of plateaus as a function of fitness, and the associated groups, for $ILS_{neutral}$ with 30 s.

We also wish to identify and quantify the most and least promising regions of the network. To do so, we consider the cumulative percentage distribution of the number of plateaus with respect to fitness, as illustrated in Fig. 2 for a representative instance, and split the network into 3 groups:

- *Group A*: The first sub-network has a low local density. Its fitness values are little visited and are the best found.
- *Group B*: This set of nodes represents a big part of the networks. Nodes correspond to good fitness values, and solving methods often find them. Vertices are inter-connected and arcs are frequently traveled, with high weights.
- *Group C*: The nodes in this group are almost all of size one. They represent the worst fitness values found. They are not connected to each other because their arcs lead only to vertices belonging to Group B.

For all instances, the distributions are of the same shape. That implies that behaviors are similar. In order to automate the partitioning of nodes into the above groups, we identify two points of inflection on the curve as follows. The first point is found when a percentage difference of at least 1% point is observed. If the difference in percentage between two consecutive fitness values represents a variation greater than 1% point, the first fitness values are part of Group A and the following ones are from Group B. In the cases where this point is reached very early, the 10 best fitness values are assigned to Group A, as in Fig. 2. Afterwards, the second point is identified when the difference drops below 1% point, and the remaining fitness values are in Group C. For sub-networks A and B, the mean fitness, number of nodes, number of plateaus, and the number of sink nodes are computed. There are thus 8 features relative to sub-networks.

6   Effects of Sampler Choice on Sampled Fitness Values

Here we consider the effects of the (sampler, time budget) pairs on fitness distribution. Recall that the samplers are two Iterative Local Search algorithms,

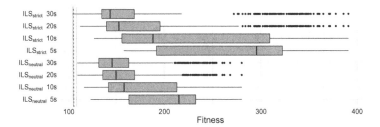

Fig. 3. Fitness distribution for each (sampler, time budget) pair on instance 21. Vertical lines represent the mean fitness values for each solver.

ILS_{strict} and $ILS_{neutral}$, differing in the neutral or strict acceptance strategy of neighboring solutions.

Figure 3 shows the fitness distribution of the timeout plateaus obtained by each sampler on one representative instance as boxplots. As might be expected, we can observe a clear relationship between the time budget and the skew of the distribution towards better solutions. This stands for both samplers. The best solutions found with 20 and 30 s sampling runs reach, or are very close to, the mean fitness value obtained by the solvers when run for 5 min. In addition, it stands to reason that as time increases, timeout plateaus are added but those found within shorter time budgets remain, only they represent a smaller proportion. The tighter time budgets also, naturally, sample fewer solutions. The study of network features shows that networks become sparser in terms of connectivity as time is reduced. This behavior is similar for both samplers.

The fact that some sampling scenarios reach, or are very close to, the mean solver fitness likely indicates that any feature that encodes some information about the best sample fitness will be very important to the model. In our case this is exactly what $\overline{Fit}_A$, the mean fitness of Group A, does. This is investigated further in Sect. 8.

The boxplots also show that the two samplers do not have the same behavior. It is clear that $ILS_{neutral}$ finds, within the same time, better solutions in terms of fitness. Furthermore, Fig. 3 reveals that as the time decreases, the gap between the two ILS widens. With the neutral acceptance policy, the first plateau from which the ILS needs to escape via a perturbation is further down the landscape. These observations imply that $ILS_{neutral}$ is the most efficient and the most robust in finding better solutions in the face of time. Thus, the neutral strategy improves ILS performance and sampling effectiveness on the most promising parts of the space.

7 Model Construction and Evaluation

A small but growing number of landscape analysis papers [2,7,20] have successfully shown that landscapes contain meaningful information that is linked to search algorithm performance. We employ a model building process consisting

Table 1. Selected features with respect to sampler and budget.

ILS_strict				ILS_neutral				Feature	Description
5s	10s	20s	30s	5s	10s	20s	30s		
✓	✓	✓	✓	✓	✓	✓	✓	Cu	Number of Curricula
✓	✓	✓	✓	✓	✓	✓	✓	$\overline{Fit}_A$	Mean fitness of Group A
✓	✓	✓	✓		✓	✓	✓	$\overline{Fit}_B$	Mean fitness of Group B
✓	✓	✓	✓	✓	✓	✓	✓	$\overline{Fit}$	Mean fitness
					✓			Sink_B	Number of sink nodes in Group B
					✓			CC	Number of connected components

of feature selection followed by linear regression to predict the fitness value. The objective will then be to assess how the sampler and its budget affect the quality of the prediction for different solvers.

Pre-processing. After merging all data, there is a total of 120 features, some of which may not be useful. We first remove 21 features with a constant value: most of them are 90th percentile features, i.e., they describe the top of the landscape. Features are then standardized. After these two steps, features have to be selected in order to improve the potential success of our models.

We use correlation preselection which computes the correlation value between the outcome variable and features. Here, the outcome variable corresponds to the fitness value, which is the result of one of the three solvers used. This step selects all the features correlated with fitness above a fixed threshold that we set to a relatively high value of 0.9. Table 1 summarizes the selected features depending on the version of the ILS sampler and the allotted budget. Between 3 and 6 features are selected, with the number of curricula, the mean fitness across all sampled timeout plateaus, and the mean fitness of groups A and B always being present. These last two features are not only about fitness but also indirectly encode some information about the proportion of plateaus since this is used to create the groups. It is interesting to note the relative homogeneity of the selected features across samplers and budgets, as well as the absence of more complex features.

One potential caveat of this restricted feature set relates to the different fitness features and the associated multicollinearity that is not usually recommended for linear regression. Multicollinearity makes it difficult to interpret regression coefficients, however in this work we are essentially interested in the models' predictions and their precision, so this is not a concerning problem.

Evaluation. In order to obtain a robust evaluation of the models, especially given that we have few instances, we use complete 5-fold cross-validation.

Cross-validation uses complementary subsets of the data for training and testing in order to assess the model's ability to predict unseen data. With a k-fold approach, data are partitioned into k subsets, one of which is retained for testing, the remaining $k-1$ being used for training the model. The cross-validation process is repeated k times such that each subset is used once for testing. The results are then averaged to produce a robust estimation. The specificity of the complete cross validation is to apply a k-fold on all possible cutting configurations [5]. Therefore $\binom{m}{m/k}$ configurations are considered instead of only k. In our case, with 19 instances and 5 folds, we have $\binom{19}{4} = 3876$ configurations. This complete 5-fold cross-validation algorithm has two main advantages. The first is to check whether the model can predict final fitness for new instances. The second smooths out the impact of how the data are split between training and test sets. When two problem instances are very similar and are not in the same fold, information about the first helps prediction. However, our objective is to obtain a robust model for all problem instances and not only very similar instances. Testing all combinations reduces this boosting effect.

The quality of the regression is assessed using the coefficient of determination, R^2, a well-known indicator for regression tasks.

8   Model Results and Discussion

Using the data collected and the selected features, we build linear regression models for each of the 2 ILS samplers, $ILS_{neutral}$ and ILS_{strict}, across 4 different time budgets to predict the performance of the 3 solvers considered, HLS, or HLS_{strict} and SA. Since we observed that the $\overline{Fit}_A$ feature was likely to have a major impact on the model, we also build a set of trivial models that incorporate this single feature and another set of models that exclude this feature. There are therefore 72 different models in total. The resulting R^2 values are plotted on Fig. 4 where each line represents a (sampler, solver) pair.

The first observation is that in all cases we obtain relatively good to very good models, with R^2 values ranging from 0.62 to 0.97. As was expected, $\overline{Fit}_A$ by itself is a very good predictor of the final fitness obtained by the different solvers. Nonetheless, models that use all the selected features come out on top, even if the advantage is somewhat marginal, indicating that it is worth using the extra features. Models that do not use $\overline{Fit}_A$ perform less well but remain competitive. The outliers to this general trend are the models built using $ILS_{neutral}$ and a 5 s budget per sampling run, where removing $\overline{Fit}_A$ causes a major decrease in R^2. In that specific scenario, however, the sampling did not reach the mean fitness of the solvers and so $\overline{Fit}_A$ is a non-trivial feature.

If we consider neutral versus strict acceptance sampling, $ILS_{neutral}$ is always better except in the scenario mentioned before. This is to be expected since the landscape is known to be neutral. What is more surprising, is that strict acceptance holds up nonetheless and provides decent models.

Next, we consider what happens w.r.t. the different solvers. The performance of all 3 solvers is adequately predicted, even though the sampling algorithm differs from the solvers, as has been observed in the literature [3,20]. The prediction for SA is slightly worse since it is the most different from the sampling algorithm,

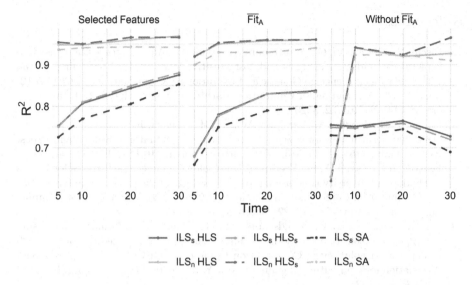

Fig. 4. R^2 values for models, where each point represents a (sampler, solver) pair.

whereas ILS is a component within HLS. Interestingly, there is no major difference between HLS with neutral (default version) and strict acceptance.

A general observable trend that holds for most cases, is that R^2 improves as
the sampling budget increases, which seems fairly intuitive. Perhaps surprisingly
however, predictive performance remains almost flat (but very good) when all
the selected features and neutral sampling are used. This robustness with respect
to time makes it easy to recommend using the smallest budget in that scenario.

9 Conclusion

In this paper, we considered the Curriculum-based Course Timetabling problem
known to be a very neutral problem where a large number of solutions share the
same fitness value. We proposed to characterized the search landscape taking into
account the specificity of neutrality and used relevant metrics to build predictive
models of solver performance. We compared two ILS samplers, ILS_{strict} and
$ILS_{neutral}$, based on hill-climbers that differ in their acceptance criterion. We
showed that the sampler that considers the neutral specificity of the search
space leads to better predictive models and is more time-robust.

In future work, we intend to consider different types of solvers, for example
evolutionary algorithms, in order to study further the notion of neutrality and
how it relates to predicting the performance of the solving algorithm. Moreover,
we plan to investigate other university timetabling problems to observe the similarities and differences, and especially to check whether the same models can
be used elsewhere.

References

1. Bischl, B., Mersmann, O., Trautmann, H., Preuß, M.: Algorithm selection based on exploratory landscape analysis and cost-sensitive learning. In: Proceedings of the 14th Annual Conference on Genetic and Evolutionary Computation, GECCO 2012, pp. 313–320. Association for Computing Machinery, New York (2012). https://doi.org/10.1145/2330163.2330209
2. Daolio, F., Liefooghe, A., Verel, S., Aguirre, H., Tanaka, K.: Problem features vs. algorithm performance on rugged multiobjective combinatorial fitness landscapes. Evol. Comput. **25**(4), 555–585 (2017). https://doi.org/10.1162/EVCO_a_00193
3. Feutrier, T., Kessaci, M.E., Veerapen, N.: Exploiting landscape features for fitness prediction in university timetabling. In: Proceedings of the Genetic and Evolutionary Computation Conference Companion, GECCO 2022. Association for Computing Machinery, New York (2022, [accepted as poster paper])
4. Jankovic, A., Doerr, C.: Landscape-aware fixed-budget performance regression and algorithm selection for modular cma-es variants. In: Proceedings of the 2020 Genetic and Evolutionary Computation Conference, GECCO 2020, pp. 841–849. Association for Computing Machinery, New York (2020). https://doi.org/10.1145/3377930.3390183
5. Kohavi, R.: A study of cross-validation and bootstrap for accuracy estimation and model selection. In: Proceedings of the 14th International Joint Conference on Artificial Intelligence, IJCAI 1995, vol. 2, pp. 1137–1143. Morgan Kaufmann Publishers Inc., San Francisco (1995)
6. Lewis, R.: A survey of metaheuristic-based techniques for University Timetabling problems. OR Spect. **30**(1), 167–190 (2008). https://doi.org/10.1007/s00291-007-0097-0
7. Liefooghe, A., Daolio, F., Verel, S., Derbel, B., Aguirre, H., Tanaka, K.: Landscape-aware performance prediction for evolutionary multiobjective optimization. IEEE Trans. Evol. Comput. **24**(6), 1063–1077 (2020). https://doi.org/10.1109/TEVC.2019.2940828
8. Malan, K.M., Engelbrecht, A.P.: A survey of techniques for characterising fitness landscapes and some possible ways forward. Inf. Sci. **241**, 148–163 (2013). https://doi.org/10.1016/j.ins.2013.04.015
9. Malan, K.M., Engelbrecht, A.P.: Particle swarm optimisation failure prediction based on fitness landscape characteristics. In: 2014 IEEE Symposium on Swarm Intelligence, pp. 1–9 (2014). https://doi.org/10.1109/SIS.2014.7011789
10. Malan, K.M.: A survey of advances in landscape analysis for optimisation. Algorithms **14**(2) (2021). https://doi.org/10.3390/a14020040
11. Muñoz, M.A., Kirley, M., Halgamuge, S.K.: A meta-learning prediction model of algorithm performance for continuous optimization problems. In: Coello, C.A.C., Cutello, V., Deb, K., Forrest, S., Nicosia, G., Pavone, M. (eds.) PPSN 2012. LNCS, vol. 7491, pp. 226–235. Springer, Heidelberg (2012). https://doi.org/10.1007/978-3-642-32937-1_23
12. Müller, T.: ITC2007 solver description: a hybrid approach. Ann. Oper. Res. **172**(1), 429–446 (2009). https://doi.org/10.1007/s10479-009-0644-y
13. Ochoa, G., Qu, R., Burke, E.K.: Analyzing the landscape of a graph based hyper-heuristic for timetabling problems. In: Proceedings of Genetic and Evolutionary Computation Conference (GECCO 2009) (2009)

14. Ochoa, G., Veerapen, N., Daolio, F., Tomassini, M.: Understanding phase transitions with local optima networks: number partitioning as a case study. In: Hu, B., López-Ibáñez, M. (eds.) EvoCOP 2017. LNCS, vol. 10197, pp. 233–248. Springer, Cham (2017). https://doi.org/10.1007/978-3-319-55453-2_16
15. Ochoa, G., Verel, S., Daolio, F., Tomassini, M.: Local optima networks: a new model of combinatorial fitness landscapes. In: Richter, H., Engelbrecht, A. (eds.) Recent Advances in the Theory and Application of Fitness Landscapes. ECC, vol. 6, pp. 233–262. Springer, Heidelberg (2014). https://doi.org/10.1007/978-3-642-41888-4_9
16. PATAT: International Timetabling Competition 2007 (2007). publication Title: International Timetabling Competition
17. Pillay, N.: A review of hyper-heuristics for educational timetabling. Ann. Oper. Res. **239**(1), 3–38 (2014). https://doi.org/10.1007/s10479-014-1688-1
18. Salwani, A.: On the use of multi neighbourhood structures within a Tabu-based memetic approach to university timetabling problems - ScienceDirect (2012)
19. Stadler, P.F.: Fitness landscapes. In: Lässig, M., Valleriani, A. (eds.) Biological Evolution and Statistical Physics, Lecture Notes in Physics, pp. 183–204. Springer, Heidelberg (2002). https://doi.org/10.1007/3-540-45692-9_10
20. Thomson, S.L., Ochoa, G., Verel, S., Veerapen, N.: Inferring future landscapes: sampling the local optima level. Evol. Comput. **28**(4), 621–641 (2020). https://doi.org/10.1162/evco_a_00271
21. Tomassini, M., Verel, S., Ochoa, G.: Complex-network analysis of combinatorial spaces: the nk landscape case. Phys. Rev. E **78**, 066114 (2008). https://doi.org/10.1103/PhysRevE.78.066114
22. Verel, S., Ochoa, G., Tomassini, M.: Local optima networks of NK landscapes with neutrality. IEEE Trans. Evol. Comput. **15**(6), 783–797 (2011). https://doi.org/10.1109/TEVC.2010.2046175

Fractal Dimension and Perturbation Strength: A Local Optima Networks View

Sarah L. Thomson[1]([✉]), Gabriela Ochoa[1], and Sébastien Verel[2]

[1] University of Stirling, Stirling, UK
{s.l.thomson,gabriela.ochoa}@stir.ac.uk
[2] Université du Littoral Côte d'Opale (ULCO), Dunkirk, France
verel@univ-littoral.fr

Abstract. We study the effect of varying perturbation strength on the fractal dimensions of Quadratic Assignment Problem (QAP) fitness landscapes induced by iterated local search (ILS). Fitness landscapes are represented as Local Optima Networks (LONs), which are graphs mapping algorithm search connectivity in a landscape. LONs are constructed for QAP instances and fractal dimension measurements taken from the networks. Thereafter, the interplay between perturbation strength, LON fractal dimension, and algorithm difficulty on the underlying combinatorial problems is analysed. The results show that higher-perturbation LONs also have higher fractal dimensions. ILS algorithm performance prediction using fractal dimension features may benefit more from LONs formed using a high perturbation strength; this model configuration enjoyed excellent performance. Around half of variance in Robust Taboo Search performance on the data-set used could be explained with the aid of fractal dimension features.

Keywords: Local Optima Network · Fractal dimension · Quadratic Assignment Problem · QAP · Iterated local search · Perturbation strength · Fitness landscapes

1 Introduction

Many systems can be characterised by their *fractal* geometry. Fractals are patterns which contain parts resembling the whole [1]. This kind of geometry is non-Euclidean in nature and a non-integer dimension can be computed for a pattern—the *fractal* dimension. This is an index of spatial complexity and captures the relationship between the amount of detail and the scale of resolution the detail is measured with. Not all systems can be characterised by a single fractal dimension, however [2] and multiple fractal dimensions—a spectrum— can be obtained through *multifractal* analysis. If there is diversity within the spectrum, this is an indication that the pattern is multifractal; i.e., the spatial complexity may be heterogeneous in nature.

Local Optima Networks (LONs) [3] are a tool to study fitness landscapes. The nodes are local optima, and the edges are transitions between local optima using a given search operator. Analysing the features of LONs can help explain algorithm search difficulty on the underlying optimisation problem. LONs have been subject to fractal analysis previously [4]; results have suggested that their fractal dimension, and extent of multifractality, may be linked to increased search difficulty.

The connection between perturbation strength and fractal dimension in LONs has not been studied before. We speculate that there may be some untapped knowledge concerning algorithm performance explanation in this area, and advance towards this aim in the present work.

The Quadratic Assignment Problem (QAP)—a benchmark combinatorial optimisation domain—is used for this study. We extract LONs with low and high perturbation strength, then compute fractal dimension features from them. Separately, two metaheuristics (iterated local search and robust taboo search) are executed on the QAP instances to collect algorithm performance information. The interplay between perturbation strength, fractal dimensions, and algorithm performance is then examined.

2   Methodology

2.1   Quadratic Assignment Problem

Definition. A solution to the QAP is generally written as a permutation s of the set $\{1, 2, ..., n\}$, where s_i gives the location of item i. Therefore, the search space is of size $n!$. The cost, or fitness function associated with a permutation s is a quadratic function of the distances between the locations, and the flow between the facilities, $f(s) = \sum_{i=1}^{n} \sum_{j=1}^{n} a_{ij} b_{s_i s_j}$, where n denotes the number of facilities/locations and $A = \{a_{ij}\}$ and $B = \{b_{ij}\}$ are the distance and flow matrices, respectively.

Instances. We consider the instances from the QAPLIB[1] [5] with between 25 and 50 facilities; these are of moderate size, and yet are not always trivial to solve. Some of the instances in this group have not been solved to optimality; for those, we use their best-known fitness as the stand-in global optimum. In the rest of this paper, for simplicity we **refer to these as the global optimum**. According to [6,7], most QAPLIB instances can be classified into four types: uniform random distances and flows, random flows on grids, real-world problems, and random real-world like problems. All of these are present in the instance set used in this work.

2.2   Monotonic Local Optima Networks

Monotonic LON. Is the directed graph $MLON = (L, E)$, where nodes are the local optima L, and edges E are the monotonic perturbation edges.

---

[1] http://www.seas.upenn.edu/qaplib/.

Local Optima. We assume a search space S with a fitness function f and a neighbourhood function N. A local optimum, which in the QAP is a minimum, is a solution l such that $\forall s \in N(l)$, $f(l) \leq f(s)$. Notice that the inequality is not strict, in order to allow the treatment of neutrality (local optima of equal fitness), which we found to occur in some QAP instances. The set of local optima, which corresponds to the set of nodes in the network model, is denoted by L.

Monotonic Perturbation Edges. Edges are directed and based on the perturbation operator (k-exchange, $k > 2$). There is an edge from local optimum l_1 to local optimum l_2, if l_2 can be obtained after applying a random perturbation (k-exchange) to l_1 followed by local search, and $f(l_2) \leq f(l_1)$. These edges are called *monotonic* as they record only non-deteriorating transitions between local optima. Edges are weighted with estimated frequencies of transition. We estimated the edge weights in a sampling process. The weight is the number of times a transition between two local optima basins occurred with a given perturbation. The set of edges is denoted by E.

2.3   Multifractal Dimensions

A fractal dimension is the logarithmic ratio between amount of detail in a pattern, and the scale used to measure the detail: $\frac{ln(detail)}{ln(scale)}$. *Multifractal* analysis [2] can be used for systems where a single fractal dimension may not be sufficient to characterise the spatial complexity. With this approach, a spectrum of dimensions is instead produced. Study of the spectrum can provide information about the multifractality (i.e., the heterogeneity of fractal complexity), as well as dimensionality. We approach multifractal analysis using the *sandbox* algorithm [8] where several nodes are randomly selected to be sandbox 'centres'. Members of the sandboxes are computed as nodes which are r edges apart from the centre c. After that the average sandbox size is calculated. The procedure is replicated for different values of r which is the sandbox radius. To facilitate the production of a dimension spectrum the whole process is repeated for several arbitrary real-valued numbers which supply a parameter we call q.

   The sandbox algorithm has been specialised and modified to suit LONs [4] and this is the process we use for our fractal analysis experiments. *Fitness* distance—as well as network edge distance—is considered. The comparison between two local optima fitness values is conducted through *logarithmic returns: fitness difference* $= |ln(f_1/f_2)|$ where f_1 and f_2 are the fitnesses of two local optima at the start and end of a LON edge. The resultant value can then be compared with a set fitness-distance maximum allowable threshold, ϵ. Pseudo-code for the multifractal algorithm we use on the LONs is given in Algorithm 1. Sandbox centre selection is at Line 7 of the Algorithm. A node n is included in the 'sandbox' of a central node c (Line 15 of the pseudo-code) if *either* the LON edge distance $d(n,c) = 1$ *or* $d(n,c) = r-1$ **and** the fitness-distance between the two local optima is less than a threshold: $|ln(\frac{f(n)}{f(c)})| < \epsilon$ (see Line 14).

Algorithm 1. Multifractal Analysis of a LON

    Input: *LON, q.values, radius.values, fitness.thresholds, number.centres*
    Output: *mean sandbox size*

1: Initialisation:
2: *centre.nodes* ← ∅, *noncentre.nodes* ← *all.nodes*
3: *mean.sandbox.sizes* ← ∅
4: **for** *q* in *q.values* **do**
5:     **for** *r* in *radius.values* **do**
6:         **for** ϵ in *fitness.thresholds* **do**
7:             *centre.nodes* ← RANDOM.SELECTION(*all.nodes, number.centres*)
8:             *sandbox.sizes* ← ∅
9:             **for** *c* in *centre.nodes* **do**
10:                *number.boxed* ← 0
11:                **for** *v* in *all.nodes* **do**
12:                    *d* ← DISTANCE(*c, v*)
13:                    *j* ← DIFFERENCE($f(c), f(v)$)
14:                    **if** (*d* == 1) **OR** (*d* == *r* - 1 **and** *j* < ϵ) **then:**
15:                        *number.boxed* ← *number.boxed* + 1
16:                    **end if**
17:                **end for**
18:                *sandbox.sizes* ← *sandbox.sizes* ∪ {[*number.boxed*]}
19:             **end for**
20:             *bs* ← MEAN(*sandbox.sizes*)
21:             *mean.sandbox.sizes*[*q*][*r*][ϵ] ← *bs*
22:         **end for**
23:     **end for**
24: **end for**

At the end of each 'sandboxing' iteration conducted with particular values for the parameters q, r and ϵ, the associated fractal dimension is calculated:

$$fractal\ dimension = \frac{ln(detail^{q-1})}{(q-1)*ln(scale)} \tag{1}$$

where *detail* is the average sandbox size (as a proportion of the network size), q is an arbitrary real-valued value, and *scale* is $\frac{r}{dm}$, with r being the radius of the boxes and dm the diameter of the network. The sandbox algorithm has a cubic time complexity and quadratic space complexity [9].

3 Experimental Setup

3.1 Iterated Local Search

We use Stützle's iterated local search (**ILS**) for both gathering performance data and as the foundation of LON construction [7]. The local search stage uses a first improvement hill-climbing variant with the pairwise (2-exch ange) neighbourhood. This operator swaps any two positions in a permutation.

The perturbation operator exchanges k randomly chosen items. We consider two perturbation strengths **for both constructing the LONs and computing the performance metrics**: $\frac{ND}{8}$ (we will henceforth refer to this as **low** perturbation) and $\frac{3ND}{4}$ (this will be referred to as **high** perturbation) with ND being the problem dimension. These perturbation magnitudes were chosen because they have been studied previously for the QAP and ILS [10]; in that work, $\frac{ND}{8}$ is the lowest strength considered, while $\frac{3ND}{4}$ is the second-strongest (the strongest was a total restart, which we decided was too extreme for our purposes). Only local optima which have improved or equal fitness to the current are accepted. Worsening local optima are never accepted.

3.2 Robust Taboo Search

Robust Taboo Search (**ROTS**) [11] is a competitive heuristic for the QAP and is also executed on the instances in this study. ROTS is a best-improvement pairwise exchange local search with a variable-length taboo list tail. For each facility-location combination, the most recent point in the search when the facility was assigned to the location is retained. A potential move is deemed to be 'taboo' (not allowed) if both facilities involved have been assigned to the prospective locations within the last y cycles. The value for y is changed randomly, but is always from the range $[0.9n, 1.1ND]$, where ND is the problem dimension.

Algorithm Performance Metric. We compute the *performance gap* to summarise ILS and ROTS performance on the instances. In the case of ILS, runs terminate when *either* the known best fitness is found or after 10,000 iterations with no improvement. For ROTS, runs complete when the best-known fitness is found or after 100,000 iterations. The performance gap is calculated over 100 runs for each, and is defined as the *mean* obtained fitness as a proportion of the best-known fitness.

3.3 LON Construction and Metrics

The LON models are constructed by aggregating the unique nodes and edges encountered during 100 independent ILS runs with the standard acceptance strategy (i.e. accepting improvements and equal solutions). Runs terminate after 10,000 non-improving iterations.

At this stage, **esc** instances are removed from the set: their local optima networks are uninteresting to study because there is a very high degree of LON neutrality. Removing these anomalies left us with the remaining moderate-size (between 25 and 50, inclusive) QAPLIB: 40 instances. There are two LONs per problem instance (for the two perturbation strengths), totalling 80 LONs.

For each LON, thousands of fractal dimensions are produced. The exact number depends on the diameter of the network: full parameter details are given in the next Section. The measurements we compute from the set of fractal dimensions for a given LON are: the median fractal dimension (simply the median of all the dimensions calculated); the maximum fractal dimension (maximum of all dimensions); the dimension variance; the multifractality (measured by taking the absolute value of a fractal dimension at the end of the spectrum divided by the absolute value of a dimension at the beginning), and an excerpt dimension (randomly chosen from the spectrum).

We consider some other LON metrics too: the flow towards global optima (computed as the incoming network edge strength to global optima in the LON); the number of local optima (simply the number of nodes in the LON); and the number of global optima (number of LON nodes with the best-known fitness).

3.4   Multifractal Analysis

We implement the multifractal analysis algorithm for LONs in C programming language and have made it publicly available for use; some of the code functionality was obtained from a monofractal analysis algorithm [12] available on Hernan A. Makse's webpage[2]. To generate *multifractal* spectra, a range of arbitrary real-valued numbers is needed. We set these as q in the range $[3.00, 8.90]$ in step sizes of 0.1. The number of 'sandbox' centres in each iteration is set at 50 and the choice of these centres is randomised. A range of ten values is used for the local optima fitness-distance threshold: $\epsilon \in [0.01, 0.19]$ in step sizes of 0.02. The sizes of sandboxes are integers in the range $r \in \{2..diameter - 1\}$ where *diameter* is the LON diameter. Note that in the interest of reducing computation, we constrain the maximum considered box radius to eleven—that is, when the LON diameter exceeds twelve ($diameter - 1 > 11$), then the upper limit for r is set to 11, to allow ten possible values $r \in \{2..11\}$.

3.5   Regression Models

Random Forest regression [13] is used. We separate LONs by the ILS perturbation strength which was applied during their construction; in this way, for modelling there are two distinct data-sets, each of them totalling 40 rows. Each observation is a set of LON features such as median fractal dimension (these are the independent variables) alongside performance metrics (the dependent variables). LONs formed using low perturbation are mapped to low-perturbation ILS performance runs, and high-perturbation LONs are mapped to high-perturbation ILS performance runs. The same Taboo Search performance metrics are used for both sets of LONs. The candidate independent variables are:

○ Number of local optima
○ Number of global optima
○ Search flow towards global optima
○ Median fractal dimension for the LON

---

[2] https://hmakse.ccny.cuny.edu/.

○ Variance of fractal dimension (proxy for *multifractality*)
○ Maximum fractal dimension
○ Variation in the multifractal spectrum (proxy for *multifractality*)

The manner of computing the metrics was described in Sect. 3.3. Iterated local search and Robust Taboo Search *performance gap* on the instances serve as response variables, making this a regression setting.

We aimed for models with as few independent variables as possible, owing to the limited number of eligible QAPLIB instances of moderate size. The *one-in-ten* rule [14] stipulates that roughly ten observations are required per independent variable. Our instance sets are each comprised of 40 instances—so we correspondingly set the maximum number of independents as four and conduct feature selection, as described now.

Recursive Feature Elimination. Backwards *recursive feature elimination* (RFE) was used to select model configurations with subsets of the predictors. We use Root Mean Squared Error (RMSE) as the quality metric for model comparisons. RMSE is the square-root of the MSE, which itself is the mean squared difference between the predicted values and true values. For the experiments, we configure RFE as follows. Random Forest is the modelling method. We consider feature subset sizes of one, two, three, and four from a set of eight candidates (listed earlier). The RFE cross-validation is set to 10-fold; model configurations are compared based on the mean RMSE over the 10 folds.

Models Using Selected Features. After feature selection, Random Forest regression is conducted using the selected features only. There are several separate model configurations owing to the different ILS perturbations under scrutiny and the two optimisation performance algorithms. To attempt to mitigate the effect of the limited training set size—which is due to the available quantity of moderate-size QAPLIB instances—we *bootstrap* the selection of the training and validation sets. We consider an 80–20 split for training and validation with 1000 iterations. Quality metrics are computed on both the training set and also from the predictions made on the validation set. The first included measurement is the R-Squared (RSQ, computed as $1 - \frac{MSE}{variance(t)}$, where t is the response variable). Also considered is the RMSE, as detailed already. The metrics are computed as the mean value over 1000 bootstrapping iterations, and their standard error is also included in the results. The standard error reported here is a measurement for how varied the means for RSQ and RMSE are across different random sub-samplings: it is the standard deviation of the means for these parameters.

Details. For all feature selection and subsequent modelling, the default hyperparameters for Random Forest in R are used, namely: 500 trees; minimum size of terminal nodes set to five; a sample size set to the number of observations; re-sampling with replacement; features considered per split set to one-third of the number of features. Independent variables are standardised as follows: $p = \frac{(p - E(p))}{sd(p)}$, with p being the predictor in question, E the expected value (mean), and sd the standard deviation.

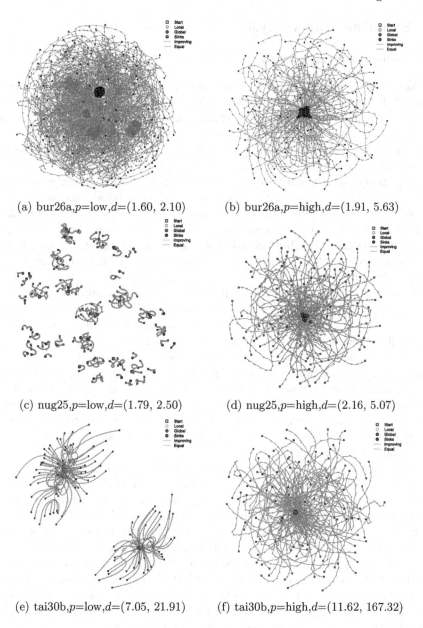

(a) bur26a,p=low,d=(1.60, 2.10)   (b) bur26a,p=high,d=(1.91, 5.63)

(c) nug25,p=low,d=(1.79, 2.50)   (d) nug25,p=high,d=(2.16, 5.07)

(e) tai30b,p=low,d=(7.05, 21.91)   (f) tai30b,p=high,d=(11.62, 167.32)

Fig. 1. Monotonic LONs for selected instances and the two perturbation strengths, $p = low$ (left) $p = high$ (right). The median and maximum fractal dimension are also indicated in the sub-captions as d = (median, maximum) (Color figure online)

4   Experimental Analysis

4.1   Network Visualisation

Visualisation is a powerful tool to get insight into the structure of networks. Figure 1 illustrates MLONs for three representative QAP instances: a real-world instance **bur26**, a random flows on grids instance **nug25**, and a random real-world like instance **tai30b**. The networks in Fig. 1 capture the whole set of sampled nodes and edges for each instance and perturbation strength. The two perturbation strengths, low and high, are shown. In the plots, each node is a local optimum and edges are perturbation transitions, either improving in fitness (visualised in grey) or leading to equal fitness nodes (visualised in orange). Node and edge decorations reflect features relevant to search. The edges colour reflect their transition type, to nodes with improving (grey) or equal fitness (orange). Global optima are highlighted in red. The start nodes (without incoming edges) are highlighted as yellow squares, while the sink nodes (without outgoing edges) are visualised in blue.

Figures 1a and 1b reflect the same problem instance (**bur26a**) but with LONs constructed using different perturbation strengths. Figure 1b has higher fractal dimensions and this is probably because of the lesser extent of neutrality at the local optima level (in the image, this can be seen through the amount of orange connections), as well as fewer connection patterns between local optima. There are also some long monotonic paths. All of these factors would result in higher fractal dimension, because they would lend to the fitness-distance and edge-distance boxing constraints in the multifractal analysis algorithm not being satisfied—and consequently, nodes remaining un-boxed, leading to a higher level of detail being computed and a higher fractal dimension (recall Sect. 2.3 for particulars on this process).

The LONs of instance **tai30b** (Fig. 1e and 1f) have the highest fractal dimensions shown. This is probably because of the lack of LON neutrality (lack of orange edges in the plot), as well as long and separate monotonic pathways (notice the number of edge-steps forming some of the paths). Additionally, compared to the networks associated with the other two instances, the LON fitness ranges are quite large for here: the minimum LON fitness is around 73–74% of the maximum in the **tai30b** LONs, while for the other two instances shown in Fig. 1a–1d, it is between approximately 92%–99%. This means that the fitness-distance boxing condition in the fractal algorithm will be satisfied much less for these LONs, resulting in higher fractal dimensions. This situation also implies that the monotonic pathways contain large fitness jumps.

4.2   Distributions

Figure 2 presents distributions for fractal dimension measurements, split by perturbation strength: *low* and *high*. In Fig. 2a are median fractal dimensions for the LONs. Notice that the dimensions are noticeably higher—and more varied—in the high-perturbation group (on the right) when compared to the low-perturbation group on the left. Next, in Fig. 2b, are the *maximum* fractal dimensions for the LONs. The same trend is evident here; that is, high-perturbation LONs (on the right) have higher and more varied dimensions than the low-perturbation LON group.

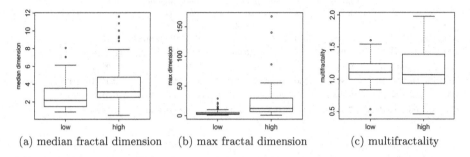

(a) median fractal dimension   (b) max fractal dimension   (c) multifractality

Fig. 2. Distributions of fractal dimension measurements taken from local optima networks. Note the different scales on the y-axes.

Figure 2c shows the amount of multifractality (heterogeneity of fractal geometry), computed as the absolute value of a fractal dimension at the end of the spectrum divided by the absolute value of a dimension at the beginning. This time, in the plot, it cannot confidently be said that one group contains more multifractality than another; however, it seems clear that the high-perturbation group have more varied multifractality values.

4.3   Predictive Modelling

Table 1 presents the configuration and quality of regression models for algorithm performance prediction. Each column (columns two to four) is a model setting. The first two rows are configuration information: the ILS perturbation strength used to form the LONs whose features are used as predictors (*LON perturbation*) and the features selected for the model by recursive feature elimination. All the remaining rows convey data about the quality of the models. Provided are the RSQ, RMSE, and RMSE as a percentage of the range of the target variable. Each of these are reported for the training and validation data. In the Table, abbreviations are used for feature names. *Multifractality* is computed as the absolute value of a fractal dimension at the end of the spectrum divided by the absolute value of a dimension at the beginning; *FD* is short for fractal dimension; *flow GO* is the combined strength of LON edges incoming to global optima; and *var FD* is the variance of the fractal dimension.

Recall from Sect. 3.5 that the number of local optima and the number of global optima were candidate predictors. Notice from Table 1, row two, that these are never selected from the pool. Instead, fractal dimension metrics and the incoming search flow to global optima (*flow GO*) are chosen by the RFE

Table 1. Information about models with features selected by recursive feature elimination in a Random Forest setting.

	Iterated local search		Robust taboo search	
LON perturbation	Low	High	Low	High
Selected features	[*multifractality, median FD, max FD*]	[*flow GO, multifractality, max FD, median FD*]	[*median FD, multifractality, flow GO, max FD*]	[*var FD, median FD*]
RSQ-*train* (SE)	0.1663 (0.2650)	0.7917 (0.2457)	**0.8610 (0.3081)**	0.7523 (0.2790)
RMSE-*train* (SE)	0.0001 (0.0001)	0.0000 (0.0000)	0.0005 (0.0042)	0.0010 (0.0044)
RMSE%range-*train*	0.1% (0.1%)	**0% (0%)**	0.1% (0.7%)	0.2% (0.8%)
RSQ-*validation* (SE)	0.8661 (0.4630)	**0.9891 (0.4147)**	0.5124 (1.1331)	0.4973 (1.0439)
RMSE-*validation* (SE)	0.0086 (0.0076)	0.0007 (0.0021)	0.0709 (0.0509)	0.0720 (0.0482)
RMSE%range-*validation*	7.9% (7.0%)	**0.3% (0.8%)**	12.2% (8.8%)	12.4% (8.3%)

algorithm. We particularly note that *multifractality*, which captures how varied the fractal complexity in a LON is, appears in three of the four model setups. The median fractal dimension appears in all four, and maximum dimension in three. Differing quantities of predictors are selected. In two cases, there is the maximum allowable amount (recall Sect. 3.5) chosen from eight candidates: four. The remaining models, however, contain less selected features: two and three, respectively.

Bold text in the Table draw the eye to the best value within a row. RMSE values are not highlighted in this way because they do not have a common range (owing to different response variable distributions). Instead, the RSQ and RMSE as a percentage of the range are emphasised with emboldened text. Notice that the model built using features of high perturbation LONs and which is modelling ILS performance gap as a response seems to be the best of the four models; this can be seen by comparing the second model column with the other three. RMSE is very low on both training and validation data, suggesting that this is a good model. While the RSQ-*train* is lower than for the ROTS response using low-perturbation LONs modelling (in the next column along), the RSQ-*validation* is superior to that—and indeed, the others—by a large margin.

Using features of low-perturbation LONs to model ILS performance response results in a much weaker model (view this in the first model column). The RSQ for training data is poor—only approximately 0.17. Even though the RSQ for validation data is significantly higher (approximately 0.87), the low RSQ on training data suggests that it does not accurately capture the patterns. Comparing this model (low-perturbation LONs) with its neighbour in the Table (high-perturbation LONs), we observe that using a higher perturbation strength to construct LONs may result in fractal dimension metrics which are more useful in predicting ILS performance.

Focusing now on the two models which consider ROTS performance as response variables (model columns three and four), we can see that—on validation data—each of them explains around 50% of variance (RSQ-*validation* row).

That being said, both RSQ means have very high standard errors (in brackets). This means that while the results hold true for this set of QAP instances, we would be cautious in extrapolating these specific results to other instance sets. A high standard error can occur with a limited sample size and with high diversity of training instances—both of which are present in our dataset. Nevertheless, the fact that some ROTS variance can be explained (at least for this specific dataset) is important because the LONs were not formed using a ROTS process; ILS was the foundation (Sect. 3.3).

The finding means that performance of a separate metaheuristic can be partially explained using ILS-built LON fractal dimension features, even when different perturbation strengths are used. Notice also that the low-perturbation model for ROTS is slightly better than the high-perturbation LON model. This might be because ROTS does not conduct dramatic perturbations on solutions. While RMSE is low on the training data (RMSE%range-*train*), it is much higher on validation data (although still not what might be considered 'high').

5   Conclusions

We have conducted a study of the relationship between Iterated Local Search (ILS) perturbation strength and fractal dimensions. The ILS perturbation strength is used when constructing Local Optima Networks (LONs), and fractal dimension can be computed from those LONs.

We found that higher-perturbation LONs also have higher fractal dimensions. Fractal dimension measurements drawn from LONs which were constructed using low and high perturbation strengths were related to algorithm performance on the underlying Quadratic Assignment Problems (QAPs). The results showed that ILS algorithm performance prediction using fractal dimension features may benefit more from LONs formed using a high perturbation strength; this model configuration enjoyed excellent performance. Around half of variance in Robust Taboo Search performance on the dataset used could be explained using predictors including fractal dimension features, and the model using the low-perturbation features was slightly stronger than the high-perturbation model.

The local optima networks are available online[3]; the fractal analysis algorithm for local optima networks is published here[4].

References

1. Mandelbrot, B.B.: Possible refinement of the lognormal hypothesis concerning the distribution of energy dissipation in intermittent turbulence. In: Rosenblatt, M., Van Atta, C. (eds.) Statistical Models and Turbulence. LNP, vol. 12, pp. 333–351. Springer, Heidelberg (1972). https://doi.org/10.1007/3-540-05716-1_20

---

[3] https://github.com/sarahlouisethomson/fractal-dimension-perturbation-strength.

[4] https://github.com/sarahlouisethomson/compute-fractal-dimension-local-optima-networks.

2. Mandelbrot, B.B., Fisher, A.J., Calvet, L.E.: A multifractal model of asset returns. In: Cowles Foundation Discussion Paper (1997)
3. Ochoa, G., Tomassini, M., Vérel, S., Darabos, C.: A study of NK landscapes' basins and local optima networks. In: Proceedings of the 10th Annual Conference on Genetic and Evolutionary Computation, pp. 555–562. ACM (2008)
4. Thomson, S.L., Ochoa, G., Verel, S.: The fractal geometry of fitness landscapes at the local optima level. Nat. Comput. **21**, 317–333 (2022)
5. Burkard, R.E., Karisch, S.E., Rendl, F.: QAPLIB - a quadratic assignment problem library. J. Global Optim. **10**(4), 391–403 (1997). https://doi.org/10.1023/A:1008293323270
6. Taillard, E.: Comparison of iterative searches for the quadratic assignment problem. Locat. Sci. **3**(2), 87–105 (1995)
7. Stützle, T.: Iterated local search for the quadratic assignment problem. Eur. J. Oper. Res. **174**(3), 1519–1539 (2006)
8. Liu, J.L., Yu, Z.G., Anh, V.: Determination of multifractal dimensions of complex networks by means of the sandbox algorithm. Chaos Interdisc. J. Nonlinear Sci. **25**(2), 023–103 (2015)
9. Ding, Y., Liu, J.L., Li, X., Tian, Y.C., Yu, Z.G.: Computationally efficient sandbox algorithm for multifractal analysis of large-scale complex networks with tens of millions of nodes. Phys. Rev. E **103**(4), 043303 (2021)
10. Ochoa, G., Herrmann, S.: Perturbation strength and the global structure of QAP fitness landscapes. In: Auger, A., Fonseca, C.M., Lourenço, N., Machado, P., Paquete, L., Whitley, D. (eds.) PPSN 2018, Part II. LNCS, vol. 11102, pp. 245–256. Springer, Cham (2018). https://doi.org/10.1007/978-3-319-99259-4_20
11. Taillard, É.: Robust taboo search for the quadratic assignment problem. Parallel Comput. **17**(4–5), 443–455 (1991)
12. Song, C., Gallos, L.K., Havlin, S., Makse, H.A.: How to calculate the fractal dimension of a complex network: the box covering algorithm. J. Stat. Mech. Theory Exp. **2007**(03), P03006 (2007)
13. Breiman, L.: Random forests. Mach. Learn. **45**(1), 5–32 (2001)
14. Harrell, F.E., Jr., Lee, K.L., Califf, R.M., Pryor, D.B., Rosati, R.A.: Regression modelling strategies for improved prognostic prediction. Stat. Med. **3**(2), 143–152 (1984)

HPO × ELA: Investigating Hyperparameter Optimization Landscapes by Means of Exploratory Landscape Analysis

Lennart Schneider[1]([✉])(ID), Lennart Schäpermeier[2](ID), Raphael Patrick Prager[3](ID),
Bernd Bischl[1](ID), Heike Trautmann[3,4](ID), and Pascal Kerschke[2](ID)

[1] Chair of Statistical Learning and Data Science, LMU Munich, Munich, Germany
{lennart.schneider,bernd.bischl}@stat.uni-muenchen.de
[2] Big Data Analytics in Transportation, TU Dresden, Dresden, Germany
{lennart.schaepermeier,pascal.kerschke}@tu-dresden.de
[3] Data Science: Statistics and Optimization, University of Münster,
Münster, Germany
{raphael.prager,heike.trautmann}@wi.uni-muenster.de
[4] Data Management and Biometrics Group, University of Twente,
Enschede, Netherlands

Abstract. Hyperparameter optimization (HPO) is a key component of machine learning models for achieving peak predictive performance. While numerous methods and algorithms for HPO have been proposed over the last years, little progress has been made in illuminating and examining the actual structure of these black-box optimization problems. Exploratory landscape analysis (ELA) subsumes a set of techniques that can be used to gain knowledge about properties of unknown optimization problems. In this paper, we evaluate the performance of five different black-box optimizers on 30 HPO problems, which consist of two-, three- and five-dimensional continuous search spaces of the XGBoost learner trained on 10 different data sets. This is contrasted with the performance of the same optimizers evaluated on 360 problem instances from the black-box optimization benchmark (BBOB). We then compute ELA features on the HPO and BBOB problems and examine similarities and differences. A cluster analysis of the HPO and BBOB problems in ELA feature space allows us to identify how the HPO problems compare to the BBOB problems on a structural meta-level. We identify a subset of BBOB problems that are close to the HPO problems in ELA feature space and show that optimizer performance is comparably similar on these two sets of benchmark problems. We highlight open challenges of ELA for HPO and discuss potential directions of future research and applications.

Keywords: Hyperparameter optimization · Exploratory landscape analysis · Machine learning · Black-box optimization · Benchmarking

L. Schneider, L. Schäpermeier and R. P. Prager—Equal contributions.

G. Rudolph et al. (Eds.): PPSN 2022, LNCS 13398, pp. 575–589, 2022.
https://doi.org/10.1007/978-3-031-14714-2_40

1   Introduction

In machine learning (ML), hyperparameter optimization (HPO) constitutes one of the most frequently used tools for improving the predictive performance of a model [3]. The goal of classical single-objective HPO is to find a hyperparameter configuration that minimizes the estimated generalization error. Generally, neither a closed-form mathematical representation nor analytic gradient information is available, making HPO a *black-box* optimization problem and evolutionary algorithms (EAs) and model-based optimizers good candidate algorithms. As a consequence, no prior information about the optimization landscape – which could allow comparisons of HPO and other black-box problems, or provide guidance regarding the choice of optimizer – is available. This also extends to automated ML (AutoML) [14], which builds upon HPO.

In contrast, in the domain of *continuous* black-box optimization, a sophisticated toolbox for landscape analysis and the characterization of their properties has been developed over the years. In exploratory landscape analysis (ELA), optimization landscape features are calculated from small samples of evaluated points from the original black-box problem. It has been shown in numerous studies that ELA feature sets capture relevant landscape characteristics and that they can be used for automated algorithm selection, improving upon the state-of-the-art selector [5,17]. Particularly well-studied are the functions from the black-box optimization benchmark (BBOB) [12].

Empirical studies [30,31] in the closely related area of algorithm configuration hint that performance landscapes often are rather benign, i.e., unimodal and convex, although this only holds for an aggregation over larger instance sets and their analysis does not allow further characterization of individual problem landscapes. There exists some work to circumvent HPO altogether, by automatically configuring an algorithm for a given problem instance [1,28]. However, these are limited to configuring optimization algorithms rather than ML models. In addition, they are often restricted in the number and type of variables they are able to configure. [26] apply fitness landscape analysis on AutoML landscapes, computing fitness distance correlations and neutrality ratios on various AutoML problems. They utilize these features only in an exploratory manner, characterizing the landscapes, without a link to optimizer performance, and cannot compare the analyzed landscapes to other black-box problems in a natural way. Similar work on fitness landscape analysis exists but focuses mostly on neural networks [6,35]. Some preliminary work [9] on the hyperparameters of a $(1+1)$-EA on a OneMax problem suggests that the ELA feature distribution of a HPO problem can be significantly different from other benchmark problems. Recently, [32] developed statistical tests for the deviation of loss landscapes from uni-modality and convexity and showed that loss landscapes of AutoML problems are highly structured and often uni-modal.

In this work, we characterize continuous HPO problems using ELA features, enabling comparisons between different black-box optimization problems and optimizers. Our main contributions are as follows:

1. We examine similarities and differences of HPO and BBOB problems by investigating the performance of different black-box optimizers.
2. We compute ELA features for all HPO and BBOB problems and demonstrate their usefulness in distinguishing between HPO and BBOB.
3. We demonstrate how HPO problems position themselves in ELA feature space on a meta-level by performing a cluster analysis on principle components derived from ELA features of HPO and BBOB problems and investigate performance differences of optimizers on HPO problems and BBOB problems that are close to the HPO problems in ELA feature space.
4. We discuss how ELA can be used for HPO in future work and highlight open challenges of ELA in the context of HPO.
5. We release code and data of all our benchmark experiments hoping to facilitate future research (which currently may be hindered due to the computationally expensive HPO black-box evaluations).

The remainder of this paper is structured as follows: Fundamentals for HPO and ELA are introduced in Sect. 2. The experimental setup is presented in Sect. 3, with the results regarding the algorithm performance and ELA feature space analysis in Sect. 4 and 5, respectively. Section 6 concludes this paper and offers future research directions.

2   Background

Hyperparameter Optimization. Hyperparameter optimization (HPO) methods aim to identify a well-performing hyperparameter configuration $\boldsymbol{\lambda} \in \tilde{\Lambda}$ for an ML algorithm $\mathcal{I}_{\boldsymbol{\lambda}}$ [3]. An ML *learner* or *inducer* $\mathcal{I}$ configured by hyperparameters $\boldsymbol{\lambda} \in \Lambda$ maps a data set $\mathcal{D} \in \mathbb{D}$ to a model $\hat{f}$, i.e., $\mathcal{I} : \mathbb{D} \times \Lambda \to \mathcal{H}, (\mathcal{D}, \boldsymbol{\lambda}) \mapsto \hat{f}$. $\mathcal{H}$ denotes the so-called hypothesis space, i.e., the function space to which a model belongs [3]. The considered search space $\tilde{\Lambda} \subset \Lambda$ is typically a subspace of the set of all possible hyperparameter configurations: $\tilde{\Lambda} = \tilde{\Lambda}_1 \times \tilde{\Lambda}_2 \times \cdots \times \tilde{\Lambda}_d$, where $\tilde{\Lambda}_i$ is a bounded subset of the domain of the i-th hyperparameter Λ_i. This $\tilde{\Lambda}_i$ can be either real, integer, or category valued, and the search space can contain dependent hyperparameters, leading to a possibly hierarchical search space. The classical (single-objective) HPO problem is defined as:

$$\boldsymbol{\lambda}^* \in \arg\min_{\boldsymbol{\lambda} \in \tilde{\Lambda}} \widehat{\mathrm{GE}}(\boldsymbol{\lambda}), \tag{1}$$

i.e., the goal is to minimize the estimated generalization error. This typically involves a costly resampling procedure that can take a significant amount of time, see [3] for further details. $\widehat{\mathrm{GE}}(\boldsymbol{\lambda})$ is a black-box function, as it generally has no closed-form mathematical representation, and analytic gradient information is generally not available. Therefore, the minimization of $\widehat{\mathrm{GE}}(\boldsymbol{\lambda})$ forms an *expensive black-box* optimization problem. In general, $\widehat{\mathrm{GE}}(\boldsymbol{\lambda})$ is only a stochastic estimate of the true unknown generalization error. Formally, $\widehat{\mathrm{GE}}(\boldsymbol{\lambda})$ depends on the concrete inducer, a resampling strategy (e.g., cross-validation) and a performance metric, for more details see [3]. In the following, we use the logloss as performance metric:

$$\frac{1}{n_{\text{test}}} \sum_{i=1}^{n_{\text{test}}} \left(-\sum_{k=1}^{g} \sigma_k \left(y^{(i)}\right) \log \left(\hat{\pi}_k \left(\mathbf{x}^{(i)}\right)\right) \right). \tag{2}$$

Here, g is the total number of classes, $\sigma_k \left(y^{(i)}\right)$ is 1 if y is class k, and 0 otherwise (multi-class one-hot encoding), and $\hat{\pi}_k \left(\mathbf{x}^{(i)}\right)$ is the estimated probability for observation $\mathbf{x}^{(i)}$ belonging to class k.

Exploratory Landscape Analysis. The optimization landscapes of black-box functions, by design, carry no prior problem information, beyond the definition of their search parameters, which can be used for their characterization. In the continuous domain, ELA [23] addresses this problem by computing features on a small sample of evaluated points, which can be used for better understanding optimizer performance [24], algorithm selection [17] and even algorithm configuration [28].

The original ELA features consist, e.g., of meta model features (`ela_meta`) such as adjusted R^2 values for quadratic and linear models and y-distribution features (`ela_distr`) such as the skewness and kurtosis of the objective values. Over time, researchers continued to propose further feature sets, including nearest better clustering (`nbc`) [16] and dispersion (`disp`) [22] features to measure multi-modality, and information content (`ic`) features [25], which extract features from random walks across the problem landscape. The R package `flacco` [18] and Python package `pflacco` [27] implement a collection of the most widely used ELA feature sets.

ELA studies often focus on the noiseless BBOB functions, as they offer diverse, well-understood challenges (such as conditioning and multimodality) and a wide range of algorithm performance data is readily available. BBOB consists of 24 minimization problems, which are identified by their function ID (FID) and scalable with respect to their dimensionality, which ranges from 2 to 40. Furthermore, different instances, identified by instance IDs (IIDs), are defined for each function, creating slightly different optimization problems with the same fundamental characteristics by means of randomized transformations in the decision and objective space. All D-dimensional BBOB problems share a decision space of $[-5, 5]^D$, which is guaranteed to contain the (known) optimum.

3   Experimental Setup

We compare the following optimizers: CMAES (a simple CMA-ES with $\sigma_0 = 0.5$ and no restarts), GENSA (a generalized simulated annealing approach as described in [37]), Grid (a grid search performed by generating a uniform sized grid over the search space and evaluating configurations of the grid in random order),

Random (random search performed by sampling configurations uniformly at random), and MBO (Bayesian optimization using a Gaussian process as surrogate model and expected improvement as acquisition function [15], similarly configured as in [20]). All optimizers were given a budget of $50D$ function evaluations in total (where D is the dimensionality of the problem). All optimizer runs were replicated 10 times. We choose these optimizers for the following reasons: (1) they cover a wide range of optimizers that can be used for a black-box problem, (2) Grid and especially Random are frequently used for HPO and Random often can be considered a strong baseline [2].

As HPO problems, we tune XGBoost[1] [8] on ten different OpenML [36] data sets (classification tasks) chosen from the OpenML-CC18 benchmarking suite [4]. The specific data sets were chosen to cover a variety of the number of classes, instances, and features (cf. Table 1). To reduce noise as much as possible, performance (logloss) is estimated via 10-fold cross-validation with a fixed instantiating per data set. On each data set, we create $2, 3$ and 5 dimensional XGBoost problems by tuning nrounds, eta ($2D$), lambda ($3D$), gamma and alpha ($5D$), resulting in 30 problems in total. We selected these hyperparameters because (1) they can be incorporated in a purely continuous search space which is generally required for the computation of ELA features, (2) they have been shown to be influential on performance [29] and (3) have a straightforward interpretation, i.e., nrounds controls the number of boosting iterations (typically increasing performance but also the tendency to overfit) while the other hyperparameters counteract overfitting and control various aspects of regularization. The full search space is described in Table 2. Note that nrounds is tuned on a logarithmic scale and therefore all parameters are treated as continuous during optimization. Missing values of numeric features were imputed using Histogram imputation (values are drawn uniformly at random between lower and upper histogram breakpoints with cells being sampled according to the relative frequency of points contained in a cell). Missing values of factor variables were imputed by adding a new factor level and factor variables were encoded using one-hot-encoding. While XGBoost is a practically relevant learner we do have to note that only considering a single learner is somewhat restrictive. We discuss this limitation in Sect. 6. In the following, individual HPO problems are abbreviated by <name>_<d>, i.e., wilt_2 for the $2D$ wilt problem.

As BBOB problems we select FIDs 1–24 with IIDs 1–5 with a dimensionality of $\{2, 3, 5\}$, resulting in 360 problems in total. We abbreviate individual BBOB problems by <fid>_<iid>_<dim>, i.e., 24_1_5 for FID 24 with IID 1 in the $5D$ setting. Experiments have been conducted in R [33], where the individual implementation of an optimizer is referenced in the mlr3 ecosystem [19]. The package smoof [7] provides the aforementioned BBOB problems. We release all data and code for running the benchmarks and analyzing results via the following GitHub repository: https://github.com/slds-lmu/hpo_ela. HPO benchmarks took around 2.2 CPU years on Intel Xeon E5-2670 instances, with optimizer overhead ranging from 10% (MBO for $5D$) to less than 1% (Random or Grid).

---

[1] Using a gbtree booster.

Table 1. OpenML data sets.

ID	Name	Number of		
		Cl.	Inst.	Feat.
40983	wilt	2	4839	5
469	analcatdata_dmft	6	797	4
41156	ada	2	4147	48
6332	cylinder-bands	2	540	37
23381	dresses-sales	2	500	12
1590	adult	2	48842	14
1461	Bank-marketing	2	45211	16
40975	car	4	1728	6
41146	sylvine	2	5124	20
40685	shuttle	7	58000	9

IDs correspond to OpenML data set IDs, which enable to query data set properties via https://www.openml.org/d/<id>.

Table 2. XGBoost search space.

Hyper-param.	Type	Range	Trafo
nrounds	int.	$[3, 2000]$	log
eta	cont.	$[\exp(-7), \exp(0)]$	log
lambda	cont.	$[\exp(-7), \exp(7)]$	log
gamma	cont.	$[\exp(-10), \exp(2)]$	log
alpha	cont.	$[\exp(-7), \exp(7)]$	log

"log" in the Trafo column indicates that this parameter is optimized on a (continuous) logarithmic scale, i.e., the range is given by [log(lower), log(upper)], and values are re-transformed via the exponential function prior to their evaluation. Parameters part of the full XGBoost search space that are not shown are set to their default.

4 Optimizer Performance

For each BBOB problem, we computed optimizer rankings based on the average final performance (best target value of an optimizer run averaged over replications). Figures 1a to 1c visualize the differences in rankings on the BBOB problems split for the dimensionality. Friedman tests indicated overall significant differences in rankings ($2D$: $\chi^2(4) = 154.55, p < 0.001$, $3D$: $\chi^2(4) = 219.16, p < 0.001$, $5D$: $\chi^2(4) = 258.69, p < 0.001$). We observe that MBO and CMAES perform well throughout all three dimensionalities, whereas GENSA only is significantly better than Grid or Random for dimensionalities 3 and 5. Moreover, Grid only falls behind Random for the $5D$ problems.

Figures 1d to 1f analogously visualize differences in rankings on the HPO problems split for the dimensionality. Friedman tests indicated overall significant differences in rankings ($2D$: $\chi^2(4) = 36.32, p < 0.001$, $3D$: $\chi^2(4) = 34.32, p < 0.001$, $5D$: $\chi^2(4) = 34.80, p < 0.001$). Again, MBO and CMAES perform well throughout all three dimensionalities. Notably, GENSA shows lacklustre performance regardless of the dimensionality, failing to outperform Grid or Random. Similarly as on the BBOB problems, Grid tends to fall behind Random for the higher-dimensional problems. We do want to note that critical difference plots for the HPO problems are somewhat underpowered when compared to the BBOB problems due to the difference in the number of benchmark problem which results in larger critical distances, as seen in the figures.

In Fig. 2, we visualize the anytime performance of optimizers by the mean normalized regret averaged over replications split for the dimensionality of problems. The normalized regret is defined for an optimizer trace on a benchmark problem as the distance of the current best solution to the overall best solution found across all optimizers and replications, scaled by the overall range of empirical solution values for this benchmark problem. We choose this metric due to the theoretical optimal solutions being unknown for HPO problems, and apply it to both BBOB and HPO problems to enable performance comparisons. We

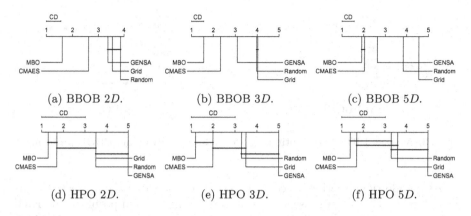

Fig. 1. Critical differences plots for mean ranks of optimizers on BBOB and HPO problems split with respect to the dimensionality.

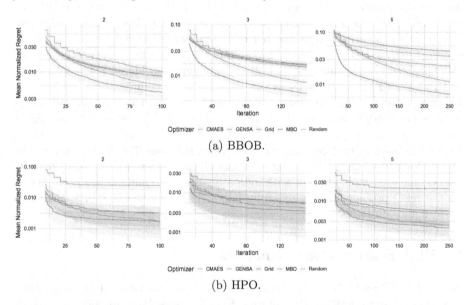

Fig. 2. Anytime mean normalized regret of optimizers on BBOB and HPO problems averaged over replications split for the dimensionality of problems. Ribbons represent standard errors. The x-axis starts after 8% of the optimization budget has been used (initial MBO design).

observe strong anytime performance of MBO and CMAES on both BBOB and HPO problems regardless their dimensionality. GENSA shows good performance on the 5D BBOB problems but shows poor anytime performance on HPO problems in general. Differences in anytime performance are less pronounced on the HPO problems, although we do want to note that the width of the standard error ribbons is strongly influenced by the number of benchmark problems.

(a) BBOB.   (b) HPO.

Fig. 3. Average ERT ratios (optimizers to `Random`) for HPO and BBOB problems.

As an additional performance evaluation, we calculated the Expected Running Time (ERT) [11]. In essence, for a given algorithm and problem, the ERT is defined as $\text{ERT} = \frac{1}{n}\sum_{i=1}^{10} \text{FE}_i$, where n is the number of repetitions which are able to reach a specific target, i refers to an individual repetition, and FE_i denotes the number of function evaluations used. We investigated the ERT of optimizers with the target given as the median of the best `Random` solutions (using $50D$ evaluations) over the ten replications per benchmark problem. We choose this (for BBOB unusual) target due to (1) the theoretical optimum of HPO problems being unknown and (2) `Random` being considered a strong baseline in HPO [2]. To bring all ERTs on the same scale, we computed the ERT ratios between optimizers and `Random` per benchmark problem which further allows us to aggregate these ratios over benchmark problems[2]. We visualize these aggregated ERT ratios separately for the dimensionality of benchmark problems in Fig. 3. We observe that average ERT ratios of `MBO` and `CMAES` are comparably similar for BBOB and HPO problems although the tendency that these optimizers become even more efficient with increasing dimensionality is less pronounced on the HPO problems. `Grid` generally falls behind and `GENSA` shows lacklustre performance on HPO.

5   ELA Feature Space Analysis

For each HPO and BBOB problem, we use $50D$ points sampled by LHS (Min-Max) as an initial design for computing ELA features. We normalize the search space to the unit cube and standardize objective function values per benchmark problem $((y - \hat{\mu})/\hat{\sigma})$ prior to calculating ELA features. This is done to counter potential artefacts that could be seen in ELA features solely due to different value ranges in decision and, in particular, in objective space. We calculate the feature sets `ela_meta`, `ic`, `ela_distr`, `nbc` and `disp`, which were introduced in Sect. 2, using the `flacco` R package [18].

To answer the question whether ELA can be used to distinguish HPO from BBOB problems, we construct a binary classification task using ELA features to predict the label "HPO" vs. "BBOB". We use a decision tree and estimate

---

[2] Following [17], optimizers that did not meet the target in any run were assigned an ERT of the worst ERT on a benchmark problem multiplied by a factor of 10.

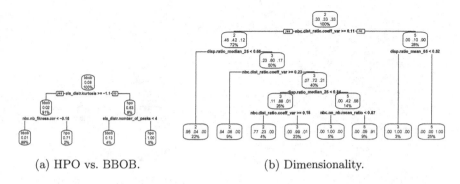

(a) HPO vs. BBOB.   (b) Dimensionality.

Fig. 4. Decision trees for classifying benchmark problems into HPO or BBOB problems (left) and classifying the dimensionality of BBOB problems (right).

the generalization error via 10 times repeated 10-fold cross-validation (stratified for the target). We obtain an estimated classification error of 3.54%. Figure 4a illustrates the decision tree obtained after training on all data. We observe that only few ELA features are needed to correctly classify problems: HPO problems tend to exhibit a lower `ela_distr.kurtosis` combined with more `ela_distr.number_of_peaks` or show a higher `nbc.nb_fitness.cor` than BBOB problems if the first split with respect to the kurtosis has not been affirmed. This finding is supported by visualizations of the $2D$ HPO problems, which we present in our online appendix, i.e., most $2D$ HPO problems have large plateaus resulting in negative kurtosis.

To answer the question whether dimensionality is a different concept for HPO compared to BBOB problems[3] we perform the following analysis: We construct a classification task using ELA features to predict the dimensionality of the problem but only use the BBOB subset for the training of a decision tree. We estimate the generalization error via 10 times repeated 10-fold cross-validation (stratified for the target) and obtain an estimated classification error of 7.39%. We then train the decision tree on all BBOB problems (illustrated in Fig. 4b) and determine the holdout performance on the HPO problems and obtain a classification error of 10%. Only few ELA features of the `disp` and `nbc` group are needed to predict the dimensionality of problems with high accuracy. Intuitively, this is sensible, due to `nbc` features involving the calculation of distance metrics (which themselves should be affected by the dimensionality) and both `nbc` and `disp` features being sensible to the multimodality of problems [16,22] which should also be affected by the dimensionality. Based on the reasonable good hold-out performance of the classifier on the HPO problems, we conclude that "dimensionality" is a similar concept for BBOB and HPO problems.

---

[3] For HPO problems, it is a priori often unclear whether a change in a parameter value also results in relevant objective function changes, i.e., the intrinsic dimensionality of a HPO problem may be lower than the number of hyperparameter suggests.

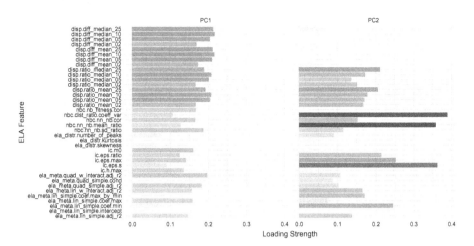

Fig. 5. Factor loadings of ELA features on the first two principle components. Blue indicates a positive loading, whereas red indicates a negative loading.

To gain insight on a meta-level, we performed a PCA on the scaled and centered ELA features of both the HPO and BBOB problems. To ease further interpretation, we select a two component solution that explains roughly 60% of the variance. Figure 5 summarizes factor loadings of ELA features on the first two principle components. Most `disp` features show a medium positive loading on PC1, whereas some `nbc` show medium negative loadings. `ela_meta` features, including R^2 measures of linear and quadratic models, also exhibit medium negative loadings on PC1. We therefore summarize PC1 as a latent dimension that mostly reflects multimodality of problems. Regarding PC2, three features stand out with strong loadings: `nbc.dist_ratio.coeff_var`, `nbc.nn_nb.mean_ratio` and `ic.eps.s`. Moreover, `disp.ratio_*` features generally have a medium negative loading. We observe that all features used by the decision tree in Fig. 4b also have comparably large loadings on PC2. Therefore, we summarize PC2 as an indicator of the dimensionality of problems.

We then performed k-means clustering on the two scaled and centered principal component scores. A silhouette analysis suggested the selection of three clusters. In Fig. 6, we visualize the assignment of HPO and BBOB problems to these three clusters. Labels represent IDs of BBOB and HPO problems. We observe that the dimensionality of problems is almost perfectly reflected in the PC2 alignment. Cluster 2 and 3 can be mostly distinguished along PC2 (cluster 3 contains low dimensional problems and cluster 2 contains higher dimensional problems) whereas cluster 1 contains problems with large PC1 values. HPO problems are exclusively assigned to cluster 2 or 3, exhibiting low variance with respect to their PC1 score, with the PC1 values indicating low multimodality.

As a final analysis we determined the nearest BBOB neighbors of the HPO problems (in ELA feature space based on the cluster analysis, i.e., minimizing the Euclidean distance over the first two principal component scores). For a

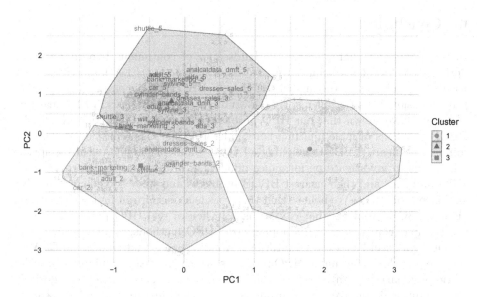

Fig. 6. Cluster analysis of BBOB and HPO problems on the first two principle component scores in ELA feature space.

complete list, see our online appendix. We again computed optimizer rankings based on the average final performance of the optimizers (over the replications), but this time for all HPO problems (regardless their dimensionality) and the subset of BBOB problems that are closest to the HPO problems in ELA feature space (see Fig. 7). Friedman tests indicated overall significant differences in rankings for both HPO ($\chi^2(4) = 104.99, p < 0.001$) and nearest BBOB ($\chi^2(4) = 61.01, p < 0.001$) problems. We observe similar optimizer rankings, with MBO and CMAES outperforming Random or Grid, indicating that closeness in ELA feature space somewhat translates to optimizer performance. Nevertheless, we do have to note that GENSA exhibits poor performance on the HPO problems compared to the nearest BBOB problems. We hypothesize that this may be caused by the performance of GENSA being strongly influenced by its hyperparameter configuration itself and provide an initial investigation in our online appendix.

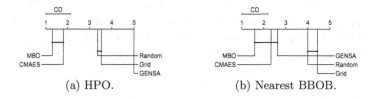

(a) HPO.   (b) Nearest BBOB.

Fig. 7. Critical differences plots for mean ranks of optimizers on all HPO problems (left) and the subset of nearest BBOB problems.

6   Conclusion

In this paper, we characterized the landscapes of continuous hyperparameter optimization problems using ELA. We have shown that ELA features can be used to (1) accurately distinguish HPO from BBOB problems and (2) classify the dimensionality of problems. By performing a cluster analysis in ELA feature space, we have shown that our HPO problems mostly position themselves with BBOB problems of little multimodality, mirroring the results of [30, 32]. Determining the nearest BBOB neighbor of HPO problems in ELA feature space allowed us to investigate performance differences of optimizers with respect to HPO problems and their nearest BBOB problems and we observed comparably similar performance. We believe that this work is an important first step in identifying BBOB problems that can be used in lieu of real HPO problems when, for example, configuring or developing novel HPO methods.

Our work still has several limitations. A major one is that traditional ELA is only applicable to continuous HPO problems, which constitute a minority of real-world problems. In many practical applications, search spaces include categorical and conditionally active hyperparameters – so-called hierarchical, mixed search spaces [34]. In such scenarios, measures such as the number of local optima, fitness-distance correlation or auto-correlation of fitness along a path of a random walk [10, 13] can be used to gain insight into the fitness landscape. Another limitation is that our studied HPO problems all stem from tuning XGBoost, with little variety of comparably low dimensional search spaces, which limits the generalizability of our results.

In future work, we would like to extend our experiments to cover a broader range of HPO settings, in particular different learners and search spaces, but also data sets. We also want to reiterate that HPO is generally noisy and expensive. In our benchmark experiments, costly 10-fold cross-validation with a fixed instantiating per data set was employed to reduce noise to a minimal level. Future work should explore the effect of the variance of the estimated generalization error on the calculation and usage of ELA features which poses a serious challenge for ELA applied to HPO in practice. Besides, we used logloss as a performance metric which by definition is rather "smooth" compared to other metrics such as the classification accuracy (but the concrete choice of performance metric typically depends on the concrete application at hand). Moreover, ELA requires the evaluation of an initial design, which is very costly in the context of HPO. In general, HPO often can be performed with evaluations on multiple fidelity levels, i.e., by reducing the size of training data, and plenty of HPO methods make use of this resulting in significant speed-up [21]. Future work could explore the possibility of using low fidelity evaluations for the initial design required by ELA and how multiple fidelity levels of HPO affect ELA features.

We consider our work as pioneer work and hope to ignite the research interest in studying the landscape properties of HPO problems going beyond fitness measures. We envision that, by improved understanding of HPO landscapes and identifying relevant landscape properties, better optimizers may be designed, and eventually instance-specific algorithm selection and configuration for HPO may be enabled.

Acknowledgement. This work was supported by the German Federal Ministry of Education and Research (BMBF) under Grant No. 01IS18036A. H. Trautmann, R. P. Prager and P. Kerschke acknowledge support by the European Research Center for Information Systems (ERCIS). Further, L. Schäpermeier and P. Kerschke acknowledge support by the Center for Scalable Data Analytics and Artificial Intelligence (ScaDS.AI) Dresden/Leipzig. L. Schneider is supported by the Bavarian Ministry of Economic Affairs, Regional Development and Energy through the Center for Analytics - Data - Applications (ADACenter) within the framework of BAYERN DIGITAL II (20-3410-2-9-8).

References

1. Belkhir, N., Dréo, J., Savéant, P., Schoenauer, M.: Per instance algorithm configuration of CMA-ES with limited budget. In: Proceedings of the Genetic and Evolutionary Computation Conference, pp. 681–688 (2017)
2. Bergstra, J., Bardenet, R., Bengio, Y., Kégl, B.: Algorithms for hyper-parameter optimization. In: Proceedings of the 24th International Conference on Neural Information Processing Systems, pp. 2546–2554 (2011)
3. Bischl, B., et al.: Hyperparameter optimization: Foundations, algorithms, best practices and open challenges. arXiv:2107.05847 [cs, stat] (2021)
4. Bischl, B., et al.: OpenML benchmarking suites. In: Vanschoren, J., Yeung, S. (eds.) Proceedings of the Neural Information Processing Systems Track on Datasets and Benchmarks, vol. 1 (2021)
5. Bischl, B., Mersmann, O., Trautmann, H., Preuß, M.: Algorithm selection based on exploratory landscape analysis and cost-sensitive learning. In: Proceedings of the 14th Annual Conference on Genetic and Evolutionary Computation, pp. 313–320 (2012)
6. Bosman, A.S., Engelbrecht, A.P., Helbig, M.: Progressive gradient walk for neural network fitness landscape analysis. In: Proceedings of the Genetic and Evolutionary Computation Conference Companion, pp. 1473–1480 (2018)
7. Bossek, J.: smoof: Single- and multi-objective optimization test functions. R J. **9**(1), 103–113 (2017). https://journal.r-project.org/archive/2017/RJ-2017-004/index.html
8. Chen, T., Guestrin, C.: XGBoost: A scalable tree boosting system. In: Proceedings of the 22nd ACM SIGKDD International Conference on Knowledge Discovery and Data Mining, pp. 785–794 (2016)
9. Doerr, C., Dreo, J., Kerschke, P.: Making a case for (hyper-)parameter tuning as benchmark problems. In: Proceedings of the Genetic and Evolutionary Computation Conference Companion, pp. 1755–1764 (2019)

10. Hains, D.R., Whitley, L.D., Howe, A.E.: Revisiting the big valley search space structure in the TSP. J. Oper. Res. Soc. **62**(2), 305–312 (2011). https://doi.org/10.1057/jors.2010.116

11. Hansen, N., Auger, A., Finck, S., Ros, R.: Real-parameter black-box optimization benchmarking 2010: Experimental setup. Research Report RR-7215, Inria (2010). https://hal.inria.fr/inria-00462481

12. Hansen, N., Finck, S., Ros, R., Auger, A.: Real-parameter black-box optimization benchmarking 2009: Noiseless functions definitions. Technical report RR-6829, Inria (2009). https://hal.inria.fr/inria-00362633/document

13. Hernando, L., Mendiburu, A., Lozano, J.A.: An evaluation of methods for estimating the number of local optima in combinatorial optimization problems. Evol. Comput. **21**(4), 625–658 (2013)

14. Hutter, F., Kotthoff, L., Vanschoren, J. (eds.): Automated Machine Learning. TSSCML, Springer, Cham (2019). https://doi.org/10.1007/978-3-030-05318-5

15. Jones, D.R., Schonlau, M., Welch, W.J.: Efficient global optimization of expensive black-box functions. J. Global Optim. **13**(4), 455–492 (1998). https://doi.org/10.1023/A:1008306431147

16. Kerschke, P., Preuss, M., Wessing, S., Trautmann, H.: Detecting funnel structures by means of exploratory landscape analysis. In: Proceedings of the 2015 Annual Conference on Genetic and Evolutionary Computation, pp. 265–272 (2015)

17. Kerschke, P., Trautmann, H.: Automated algorithm selection on continuous black-box problems by combining exploratory landscape analysis and machine learning. Evol. Comput. **27**(1), 99–127 (2019)

18. Kerschke, P., Trautmann, H.: Comprehensive feature-based landscape analysis of continuous and constrained optimization problems using the R-package Flacco. In: Bauer, N., Ickstadt, K., Lübke, K., Szepannek, G., Trautmann, H., Vichi, M. (eds.) Applications in Statistical Computing. SCDAKO, pp. 93–123. Springer, Cham (2019). https://doi.org/10.1007/978-3-030-25147-5_7

19. Lang, M., et al.: mlr3: a modern object-oriented machine learning framework in R. J. Open Source Softw. **4**(44), 1903 (2019)

20. Le Riche, R., Picheny, V.: Revisiting Bayesian optimization in the light of the COCO benchmark. Struct. Multidiscip. Optim. **64**(5), 3063–3087 (2021). https://doi.org/10.1007/s00158-021-02977-1

21. Li, L., Jamieson, K., DeSalvo, G., Rostamizadeh, A., Talwalkar, A.: Hyperband: A novel bandit-based approach to hyperparameter optimization. J. Mach. Learn. Res. **18**(1), 6765–6816 (2017)

22. Lunacek, M., Whitley, D.: The dispersion metric and the CMA evolution strategy. In: Proceedings of the 8th Annual Conference on Genetic and Evolutionary Computation, pp. 477–484 (2006)

23. Mersmann, O., Bischl, B., Trautmann, H., Preuss, M., Weihs, C., Rudolph, G.: Exploratory landscape analysis. In: Proceedings of the 13th Annual Conference on Genetic and Evolutionary Computation, pp. 829–836 (2011)

24. Mersmann, O., Preuss, M., Trautmann, H., Bischl, B., Weihs, C.: Analyzing the BBOB results by means of benchmarking concepts. Evol. Comput. **23**(1), 161–185 (2015)

25. Muñoz, M.A., Kirley, M., Halgamuge, S.K.: Exploratory landscape analysis of continuous space optimization problems using information content. IEEE Trans. Evol. Comput. **19**(1), 74–87 (2014)

26. Pimenta, C.G., de Sá, A.G.C., Ochoa, G., Pappa, G.L.: Fitness landscape analysis of automated machine learning search spaces. In: Paquete, L., Zarges, C. (eds.) EvoCOP 2020. LNCS, vol. 12102, pp. 114–130. Springer, Cham (2020). https://doi.org/10.1007/978-3-030-43680-3_8

27. Prager, R.P.: pflacco: A Python Interface of the R Package Flacco, April 2022. https://github.com/Reiyan/pflacco

28. Prager, R.P., Trautmann, H., Wang, H., Bäck, T.H.W., Kerschke, P.: Per-instance configuration of the modularized CMA-ES by means of classifier chains and exploratory landscape analysis. In: 2020 IEEE Symposium Series on Computational Intelligence (SSCI), pp. 996–1003. IEEE (2020)

29. Probst, P., Boulesteix, A.L., Bischl, B.: Tunability: Importance of hyperparameters of machine learning algorithms. J. Mach. Learn. Res. **20**(53), 1–32 (2019)

30. Pushak, Y., Hoos, H.: Algorithm configuration landscapes: In: Auger, A., Fonseca, C.M., Lourenço, N., Machado, P., Paquete, L., Whitley, D. (eds.) PPSN 2018, Part II. LNCS, vol. 11102, pp. 271–283. Springer, Cham (2018). https://doi.org/10.1007/978-3-319-99259-4_22

31. Pushak, Y., Hoos, H.H.: Golden parameter search: Exploiting structure to quickly configure parameters in parallel. In: Proceedings of the 2020 Genetic and Evolutionary Computation Conference, pp. 245–253 (2020)

32. Pushak, Y., Hoos, H.H.: AutoML landscapes. ACM Trans. Evol. Learn. Optim. (TELO) (2022, in print). https://www.cs.ubc.ca/labs/algorithms/Projects/ACLandscapes/PusHoo22a.pdf

33. R Core Team: R: A Language and Environment for Statistical Computing. R Foundation for Statistical Computing, Vienna, Austria (2021). https://www.R-project.org/

34. Thornton, C., Hutter, F., Hoos, H.H., Leyton-Brown, K.: Auto-WEKA: Combined selection and hyperparameter optimization of classification algorithms. In: Proceedings of the 19th ACM SIGKDD International Conference on Knowledge Discovery and Data Mining, pp. 847–855 (2013)

35. Traoré, K.R., Camero, A., Zhu, X.X.: Fitness landscape footprint: A framework to compare neural architecture search problems. arXiv:2111.01584 [cs] (2021)

36. Vanschoren, J., van Rijn, J.N., Bischl, B., Torgo, L.: OpenML: Networked science in machine learning. SIGKDD Explor. **15**(2), 49–60 (2013)

37. Xiang, Y., Gubian, S., Suomela, B., Hoeng, J.: Generalized simulated annealing for global optimization: The GenSA package. R J. **5**(1), 13–28 (2013)

Increasing the Diversity of Benchmark Function Sets Through Affine Recombination

Konstantin Dietrich$^{(\boxtimes)}$ (ID) and Olaf Mersmann (ID)

Cologne University of Applied Sciences, Gummersbach, Germany
{konstantin.dietrich,olaf.mersmann}@th-koeln.de

Abstract. The Black Box Optimization Benchmarking (BBOB) set provides a diverse problem set for continuous optimization benchmarking. At its core lie 24 functions, which are randomly transformed to generate an infinite set of instances. We think this has two benefits: it discourages over adaptation to the benchmark by generating some diversity and it encourages algorithm designs that are invariant to transformations. Using Exploratory Landscape Analysis (ELA) features, one can show that the BBOB functions are not representative of all possible functions. Muñoz and Smith-Miles [15] show that one can generate space-filling test functions using genetic programming. Here we propose a different approach that, while not generating a space-filling function set, is much cheaper. We take affine recombinations of pairs of BBOB functions and use these as additional benchmark functions. This has the advantage that it is trivial to implement, and many of the properties of the resulting functions can easily be derived. Using dimensionality reduction techniques, we show that these new functions "fill the gaps" between the original benchmark functions in the ELA feature space. We therefore believe this is a useful tool since it allows one to span the desired ELA-region from a few well-chosen prototype functions.

Keywords: Exploratory landscape analysis · Benchmarking · Instance generator · Black box continuous optimization

1 Introduction

Playing the benchmarking game when developing or choosing an optimization algorithm is a tricky endeavor. On the one hand, the choice of benchmark functions is crucially important in guiding the development in the right direction by posing challenging and representative problems. On the other hand, there is no guarantee that there is a benchmark function that is similar to "my" real world

Supported by German Federal Ministry of Education and Research in the funding program Forschung an Fachhochschulen under the grant number 13FH007IB6 and German Federal Ministry for Economic Affairs and Climate Action in the funding program Zentrales Innovationsprogramm Mittelstand under the grant number KK5074401BM0.

G. Rudolph et al. (Eds.): PPSN 2022, LNCS 13398, pp. 590–602, 2022.
https://doi.org/10.1007/978-3-031-14714-2_41

problem. A popular choice when faced with this challenge are curated sets of benchmark functions, such as those contained in the BBOB suite [5]. These sets have proven so useful, that they are often used for other purposes as well. In particular, they are often used to generate datasets for a multitude of machine learning tasks related to algorithm selection or algorithm configuration. While the original 24 noiseless BBOB functions were designed to cover many properties of real world problems, they are far from unbiased. Therefore, any dataset generated using only these functions is also biased. For example, Liao et al. [9] showed that the choice of a benchmark set has a non-negligible influence on the benchmark result. This is not really surprising, given that each benchmark suite is designed for a specific purpose.

While we think it is almost impossible to obtain a completely unbiased function set, we believe it is possible to generate a function set which is more uniformly distributed w.r.t. the set of all problems. Why would we want such a set? For one thing, we could be more certain, that we haven't missed any cases in our empirical studies. Furthermore, and possibly more important, there is an ever-increasing effort to use machine learning methods to choose or configure optimization algorithms. As mentioned above, the datasets used to train these models are currently heavily biased by the choice of benchmark suite used to generate them.

With exploratory landscape analysis (ELA) [12] there is a method to empirically assess problem properties. We therefore have a tool to check how uniform our function set is with regard to some ELA features. Here we have to be careful that we cannot assume, that true uniformity in the statistical sense is what we ultimately want since little is known about the scaling of the different classes of ELA features. The other tool we need is some way to ideally arithmetically define new candidate benchmark functions for our set.

One way to approach this problem is proposed by Muñoz and Smith-Miles [15]. They represent every function by its eight dimensional ELA feature vector. Then, using principal component analysis, this space is projected down to two components. This allows them to visualize a model of the instance space and to identify points in the instance space and by extension ELA feature combinations, which are not yet covered. The identified gaps are then filled using genetic programming to find functions with the missing ELA feature combinations. This approach is successful and shows promising results by filling the gaps and even expanding the boundaries of the instance space. However, there are some drawbacks. For example, in contrast to the BBOB function sets, the global optima of the generated functions are not known. But we think the most significant disadvantage is the computational overhead associated with the approach. Ideally, a large number of new benchmark functions are needed to ensure diversity. Randomly generated benchmark functions also tend to not resemble real world problems [6]. We believe genetic programming is especially susceptible here. Once it has found a building block that will reliably exhibit some ELA feature or combination of ELA features, the algorithm will exploit that building block as much as it can. Therefore, the internal structure of many of the generated instances will likely be the same.

An entirely different approach is implemented within the COCO framework [4]. There is an infinite number of instances of each benchmark function included in the COCO code. These are obtained by scaling, rotating, and shearing the parameter and objective space. This scales well and creates just enough diversity to ensure that the functions are not immediately recognized by the optimization algorithms. Still the ELA features, and therefore the properties of the instances barely change as has previously been shown and can also be seen in the plots later in this paper.

Thus, the need to construct novel benchmarking functions remains, and in this article we propose a new approach which tries to address some of the previously mentioned concerns. Our method can be described as something between the genetic programming approach used in Muñoz and Smith-Miles [15] and the simple instance generation mechanism used by COCO to vary the benchmark functions. Our idea is to use well established benchmark functions that were designed by experts as the "basis vectors" from which we construct new functions. Here, the term basis vector is obviously not meant in the literal sense - these functions are certainly not orthogonal! We think the analogy is still helpful because our new functions are convex combinations of pairs of existing benchmark functions. Why restrict ourselves to convex combinations and pairs? We chose convex combinations of pairs of functions because we wanted to stay "in between" the two functions for visualization purposes. To make the most of the advantage, which is to start from easily transformable functions, we shift one function of the recombined pair in such a way that the global optima of both functions are aligned. This way, our results are new benchmark functions with known global optima. We determine the properties of the resulting functions by employing ELA. The resulting data is then used to construct an instance space using principal component analysis similar to the procedure described in Muñoz and Smith-Miles [15]. In the following, the experimental methods will be described in detail.

2   Experimental Methods

2.1   ELA Feature Selection

Research related to ELA has flourished in recent years. This has resulted in an immense number of landscape features which posses different strengths and weaknesses. Many of these are fairly similar and in fact, when calculated for the BBOB functions, are highly correlated. We therefore choose to use only the following 14 features from the literature:

ela_level.lda_mmce_25 Mean cross-validation accuracy (MCVA) of a linear discriminant analysis (LDA) for the 25% level set [12].

ela_level.lda_qda_25 The ratio between the MCVA of a LDA at 25% and the MCVA of a quadratic discriminant analysis (QDA) for the 25% level set [12].

ela_meta.lin_simple.intercept The intercept of the linear regression model approximating the function [12].

ela_meta.lin_simple.coef.max The largest coefficient of the linear regression model approximating the function (excluding the intercept) [12].

ela_meta.lin_simple.adj_r2 Adjusted coefficient of determination of the linear regression model [12].

ela_meta.quad_simple.adj_r2 Adjusted coefficient of determination of the quadratic regression model without any interactions [12].

ela_meta.quad_simple.coef.min_by_max Ratio between the minimum and the maximum absolute values of the quadratic term coefficients in the quadratic regression model without any interactions [12].

ela_distr.skewness Skewness of the cost distribution [12].

disp.ratio_mean_02 Ratio of the pairwise distances of the points having the best 2% fitness values with the pairwise distances of all points in the design [10].

ic.eps.ratio The half partial information sensitivity [13].

ic.eps.s The settling sensitivity [13].

nbc.nb_fitness.cor The correlation between the fitness values of the search points and their indegree in the nearest-better point graph [7].

entropy.y Entropic significance of first order [11].

entropy.sig_dth_order Entropic significance of d-th order [20].

For an exact description of the features, we refer the reader to the above cited literature and the code contained in the supplementary material. Our choice of features stems from the two sets of features selected by Muñoz and Smith-Miles [15] and Renau et al. [18]. Muñoz and Smith-Miles [15] select eight features based on the co-linearity of 33 candidate features. Renau et al. [18] use ten candidate features to determine even smaller sets of feature combinations, that are able to characterize BBOB functions sufficiently for different classification algorithms. Both selection approaches seem valid to us, and the combined set of features is sufficiently small for our purposes. We therefore skip a rigorous selection of features and opt to use the combination of both suggested feature sets with the exception of two. Firstly, we omit the number of peaks of the cost distribution (ela_distr.n_peaks [12]). This is due to the fact that we deem it numerically unstable, meaning it has too many hyperparameters, which in practice changes the estimated number of peaks drastically. Secondly, we leave out the feature pca.expl_var_PC1.cov_init [8]. It only captures information about the sampling strategy and since we don't vary that here, we can safely ignore it. With two overlapping features in both lists, we arrive at the 14 features listed above. It is worth mentioning that Renau et al. [18] findings showed that, during variation of sample size and search space dimension, in no case more than six features are necessary to correctly classify the BBOB functions with a target accuracy of 98%. But since the feature combinations vary w.r.t. dimension and sample size, we include all of their candidate features such that we have a single set of features regardless of the dimensionality of the underling instance space.

Sampling Method. For every function the features are calculated on the basis of $250 \cdot d$ sampled points. We choose this sample size since it is the smallest sample size that resulted in correct classification of the functions for every tested dimension in the previously mentioned paper by Renau et al. [18]. The choice of sampling method has a large influence on the resulting ELA features. In fact, Renau et al. [17] showed that for different sampling strategies, the ELA features do not necessarily converge to the same values.

As recommended by Santner et al. [19] and further discussed by Renau et al. [17], we choose to use Sobol sequences [21] as our sampling strategy. For every function, the feature values are calculated 30 times with samples obtained from different Sobol sequences. We use the scrambling procedure described by Owen [16] to obtain 30 different Sobol sequences.

2.2 Function Generation

As stated before, in order to generate new benchmark functions, we recombine already known and proven benchmark functions. We do this in such a manner that the known global optima of the combined functions are aligned by shifting one of the functions. To perform such an affine recombination of two functions, the pair must be compatible.

Given two real valued functions f_0 and f_1, with

$$x^* := \arg\min f_0(x) = \arg\min f_1(x),$$

we define the family of functions

$$f^{(\alpha)}(x) = (1 - \alpha) \cdot f_0(x) + \alpha \cdot f_1(x) \qquad \alpha \in [0, 1]$$

as the convex recombinations of f_0 and f_1. The domain of $f^{(\alpha)}$ is given by $\operatorname{dom} f_0 \cap \operatorname{dom} f_1$ and, by definition, contains x^*.

Note that given any two functions $\hat{f}_0$ and $\hat{f}_1$, you can always shift $\hat{f}_1$ such that the above conditions are met by defining

$$\tilde{f}_1(x) = \hat{f}_1(x - (x_0^* - x_1^*))$$

where $x_0^* = \arg\min \hat{f}_0(x)$ and $x_1^* = \arg\min \hat{f}_1(x)$ and the domain of $\tilde{f}$ is given by shifting the domain of $\hat{f}_1$ by $x_0^* - x_1^*$.

Like already mentioned, we only use functions from the set of 24 noiseless, single-objective and scalable BBOB functions. These functions are designed to be evaluated over $\mathbb{R}^d$ and their search space is given by $[-5, 5]^d$. Thus, all the functions have their global minimum inside the search space. This means each function from this set is per design compatible to be combined with every other function from the set. This results in $(24^2 - 24)/2 = 276$ possible pairs.

In order to have the new functions once with their optima locations being equal to the optimum location of f_0 and once being equal to the optimum location of f_1, we calculate every pair twice. Once with shifted f_0 and once with

shifted f_1. Thus, this makes a total of 552 pairs. As mentioned before, the transformations performed by COCO to generate different function instances include
translations. Per design, this approach wants to keep algorithms from learning
the optimum location and how to get there by hard while preserving the functions properties (ELA features). Thus, shifting a function by a relatively small
amount does not change its ELA feature values significantly. In line with this,
we do not observe significant differences between pairs with shifted f_0 and pairs
with shifted f_1. For the recombinations we generally only use the first instances
of the BBOB functions. For this proof of concept we generate recombinations
for the dimensions $d \in \{2, 3, 5, 10, 20\}$.

Selection of α Values. Varying α would theoretically allow generating an
infinite number of new functions from only one function recombination. But since
small changes in α cannot be expected to cause huge changes in the resulting
functions properties, it is not sensible to generate too many functions from only
one pair. We also need to take into account that the function value ranges do
not always have comparable scales. This means the values need to be chosen in a
manner that circumvents one function dominating for all α values between zero
and one. To avoid this scenario, we use an algorithmic approach based on the
entropy of the cost distribution (`entropy.y`). Our goal is to choose $n = 20$ values
for α such, that the corresponding values for the entropy of the cost distribution
are spaced equidistantly between the entropy for $\alpha = 0$ (f_0) and $\alpha = 1$ (f_1).
That is, we choose E_1 as the entropy of the cost distribution of f_0 and E_n as
the cost distribution of f_1. Thus, this will generate 18 new entropy values E_2 to
E_{19}. To find the corresponding α values, we iteratively minimize the quadratic
distance of the linearly spaced entropy and the current calculated entropy w.r.t.
α_i

$$\underset{\alpha_{i-1} < \alpha_i < 1}{\arg\min} \; (E(f^{\alpha_i}) - E_i)^2$$

with $i \in \{2, 3, 4, ..., n - 1\}$. This is done in a bounded setting where α_i must
always be bigger than α_{i-1} and smaller than 1 (α_n). Subtracting the α values
that correspond to the original functions (α_1 & α_n) we generate $552 \cdot 18 = 9936$
new functions. Of course, half of these functions are expected to have similar
properties to the first half (shifting either f_0 or f_1). We provide the resulting
ELA feature dataset under an open source license for further use [2].

2.3 Principal Component Analysis

To visualize the diversity of the new functions, we follow the approach outlined
in Muñoz and Smith-Miles [15]. While the method to reduce the dimensions
remains the same (PCA), there are several things that set our visualization
approach apart.

First, we use a different set of features as listed above. From this list, we
omit the two entropy features to avoid any artifacts from our generation of α
values in the visualization. Second, we normalize all the feature values individually to be between 0 and 1 by subtracting the minimum and dividing by the

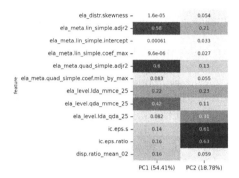

Fig. 1. Absolute values of the PCA transformation matrix.

range. Last and most importantly, we determine our transformation matrix on independently estimated feature vectors of the first 15 instances of all BBOB functions. These feature vectors are again calculated 30 times, and thus we have $24 \times 15 \times 30 = 10\,800$ 12-dimensional feature vectors per dimension. The absolute values of the coefficients in the resulting transformation matrix are shown in Fig. 1. With the first two principal components, it is possible to explain 73.19% of the variance in the instance data set. The first principal component is dominated by the features `ela_meta.quad_simple.adjr2`, `ela_meta.lin_simple.adjr2` and `ela_level.qda_mmce_25`, while the second principal component mostly explains the variance of `ic.eps.ratio`, `ic.eps.s` and `ela_level.lda_qda_25`.

3   Results

The resulting PCA scatter plots are shown in Fig. 2. All five selected dimensions are visualized independently (in SubFig. 2(a) to (e)) even though the same transformation matrix is used in every case. Subfigure 2(f) shows the shared legend for all the plots. The colored points in all the graphs represent the first 15 instances of all BBOB functions. These are the actual instances that we use to determine the PCA transformation matrix. Every instance is evaluated on 30 differently scrambled sobol samples, therefore in every graph there are 15 (instances) $\times$ 30 (sobol samples) points of the same color. As specified in the legend, we use a different color gradient for every one of the five function groups contained by BBOB. The first principal component is always plotted on the x-axis, while the second principal component is plotted on the y-axis. Looking at the color distribution of the plots, we observe that *Separable Functions* and *Functions with low or moderate conditioning* are generally more distributed towards the middle and left of the occupied area in the instance space. *Functions with high conditioning and unimodal* are settled at the top, while *Multi-modal functions with adequate global structure* and *Multi-modal functions with weak global structure* can be found more towards the bottom and right. This observation holds for all dimensions and lets us assume that the employed feature

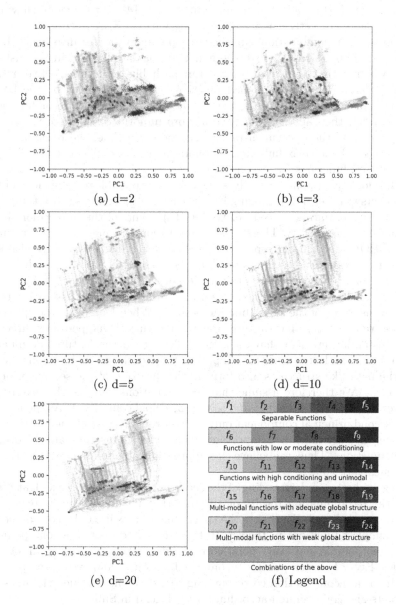

(a) d=2

(b) d=3

(c) d=5

(d) d=10

(e) d=20

(f) Legend

Fig. 2. Instance spaces spanned by the first two principal components that resulted from applying PCA to the ELA feature vector representations of the first 15 instances of all BBOB functions.For every BBOB function there are 15 (instances) × 30 (sobol samples) points of the respective color given in the legend. In gray the 30 repetitions of our 9936 recombinations are plotted. (Color figure online)

set is largely invariant w.r.t. dimension. These results are in accordance with Mersmann et al. [12] and Muñoz and Smith-Miles [14], as they again show that it is possible to correctly classify BBOB functions via ELA features.

In gray, the recombinations are shown as points in the instance space. These give a good impression of where the new functions are located in the instance space. We find that the gray points form path like structures between their building blocks (the two combined functions). This way, a lot by the BBOB instances unoccupied instance space is now occupied by the recombinations. Thus, we "fill the gaps" and achieve a more uniform coverage of the convex hull spanned by the original instances. At the same time, the recombinations rarely exceed this convex hull and, therefore, do not resemble a space-filling set of benchmark functions.

While looking at points with the same color, it is possible to get an idea of how much diversity in terms of changing ELA features is added to the BBOB function set by the instantiating procedure. We see that points of the same color cluster together in clusters of 30. These clusters each represent one of the 15 instances of the BBOB function corresponding to the color. This is especially visible in case of the green, blue, purple and dark red colored functions. As clusters of the same color spread over a certain region of the instance space, they remain rather regional. Therefore, as expected, the variance of ELA features between different instances of the same function is small. When looking at the spreading clusters, we also notice that the direction of spread mostly happens in x-direction. Thus, instantiating might have a larger influence on the features dominating the first principal component. Comparing this to the recombinations we see that the path like structures also form in the direction of the second principal component. We therefore assume that recombinations increase the diversity of features dominating in the first and second principal component. As the features of the second principal component only explain roughly 19% of the total original variance, they are much less important when trying to discriminate the original BBOB functions. Thus, we believe this approach is able to increase the diversity of a benchmark suite.

To get a better understanding of the recombination process, we look at Fig. 3. Here, the recombination results of all pairs that include the unshifted sphere function are depicted. We choose the sphere function as it can be considered one of the simplest functions in the set. In this case, we do not show one point for every sobol sample, but every point corresponds to the average over all 30 sobol samples. Every function pairing is connected by a line, and every gray point on that line is a recombination corresponding to a different α value. The lines and endpoints are again colored according to the legend in SubFig. 3(f).

The figures clearly show paths that interpolate the instance space between the paired up functions. We think it is interesting that most of the points are spaced almost equidistantly along the paths, even though we did not use the entropy feature, used to determine the α values, to construct the instance space. Some paths show abrupt direction changes which, we think, could be attributed to something akin to a phase transition in one or several features. For the two-

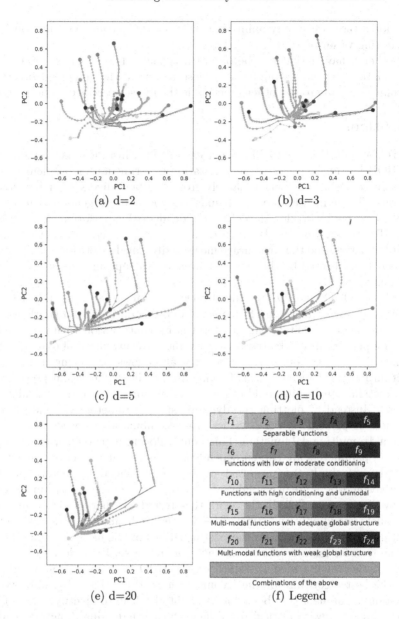

(a) d=2

(b) d=3

(c) d=5

(d) d=10

(e) d=20

(f) Legend

Fig. 3. Instance spaces spanned by the first two principal components that resulted from applying PCA to the ELA feature vector representations of the first 15 instances of all BBOB functions. Every point is the result from the average over the 30 sobol samples.For every dimension all recombinations (gray) that include the sphere function (star) are plotted and connected by a line.The other original BBOB functions are visualized as colored circles according to the legend. (Color figure online)

dimensional case, we supply animations that show the changing function landscape during variation of α [1].

Another behavior that can be observed is that, in some cases, most of the recombinations lie on a straight path close to one of the combined functions. The connecting line to the other function is then best described to be jumping.

4   Outlook

Since this approach is able to fill the convex hull of the instance space, spanned by the BBOB instances, we think generalization experiments for problem specific algorithm selection can benefit largely from the new diversity of benchmark functions. The immense amount of functions that can be generated using the described method, will allow deeper insight on algorithm performance vs. feature values. Testing known algorithms on the recombinations can uncover overfitted hyperparameters, and thus encourage more evenly tuned algorithms. This testing is especially facilitated by the fact that we know the optimum location of every generated function.

Besides COCO, the IOH profiler by Doerr et al. [3] is another benchmark platform for evaluating the performance of iterative optimization heuristics. We believe the main benefit of this platform is its contribution to understanding algorithm performance. The IOH analyzer tracks the evolution of dynamic state parameters during the optimization process, and allows visualizing them vs. the algorithm budget. Currently, this is done during the optimization process of a fixed function. Our results could introduce a new dimension to this visualization approach. Instead of plotting the dynamic state parameters vs. budget, they could now be plotted vs. the α values of a recombination between, e.g., a unimodal and a multimodal function. This would allow insights on to which internal algorithm parameters are chosen dependent on a changing function landscape.

Lastly, we think the proposed method to artificially generate problem instances could be a great candidate for further testing the algorithm selection done by Škvorc et al. [22]. In this work they tested whether an algorithm selection model trained only on artificially generated problems can correctly provide algorithm recommendations for the existing BBOB problems. The results showed that the transfer learning model was not able to do so. It would be interesting to see whether such a model, trained on the BBOB instances, could correctly select the best algorithm for the recombinations. If so, this could improve the understanding of algorithm behavior. It might also help to identify "phase transitions", and therewith critical α_C values for each individual function pairing. These values would mark the degree of mixture between two functions, at which the optimal algorithm changes.

For future work we plan to look into a different algorithmic generation of α values that explicitly considers function ranges. This might be beneficial while pairing functions with ranges that differ by several orders of magnitude. Another interesting direction for the future could be to pair the BBOB functions with Benchmark functions from other Benchmark sets. By doing this, we hope to "expand the borders" of the uniformly covered instance space.

Lastly, we would like to run Benchmark experiments along the recombination paths. This is another approach of finding the critical α_C values. Determining and then marking them for every path through the instance space could create borders between regions of dominating optimization algorithms.

References

1. Dietrich, K., Mersmann, O.: Changing function landscape of two dimensional recombinations (2022). https://doi.org/10.5281/zenodo.6456367
2. Dietrich, K., Mersmann, O.: Exploratory Landscape Analysis Feature Data of Recombinations and noiseless BBOB Instances, pp. 1–15 (2022). https://doi.org/10.5281/zenodo.6456361
3. Doerr, C., Wang, H., Ye, F., van Rijn, S., Bäck, T.: IOHprofiler: A benchmarking and profiling tool for iterative optimization heuristics arxiv.org/abs/1810.05281, October 2018
4. Hansen, N., Auger, A., Ros, R., Mersmann, O., Tušar, T., Brockhoff, D.: COCO: a platform for comparing continuous optimizers in a black-box setting. Optim. Methods Softw. **36**(1), 114–144 (2021). https://doi.org/10.1080/10556788.2020.1808977
5. Hansen, N., Finck, S., Ros, R., Auger, A.: Real-Parameter Black-Box Optimization Benchmarking 2009: Noiseless Functions Definitions. Research Report RR-6829, Inria (2009). https://hal.inria.fr/inria-00362633
6. Hooker, J.: Testing heuristics: we have it all wrong. J. Heuristics **1**(1), 33–42 (1995). https://doi.org/10.1007/BF02430364
7. Kerschke, P., Preuss, M., Wessing, S., Trautmann, H.: Detecting funnel structures by means of exploratory landscape analysis. In: GECCO 2015 - Proceedings of the 2015 Genetic and Evolutionary Computation Conference, pp. 265–272 (2015). https://doi.org/10.1145/2739480.2754642
8. Kerschke, P., Trautmann, H.: Comprehensive feature-based landscape analysis of continuous and constrained optimization problems using the R-package Flacco. In: Bauer, N., Ickstadt, K., Lübke, K., Szepannek, G., Trautmann, H., Vichi, M. (eds.) Applications in Statistical Computing. SCDAKO, pp. 93–123. Springer, Cham (2019). https://doi.org/10.1007/978-3-030-25147-5_7
9. Liao, T., Molina, D., Stützle, T.: Performance evaluation of automatically tuned continuous optimizers on different benchmark sets. Appl. Soft Comput. J. **27**, 490–503 (2015). https://doi.org/10.1016/J.ASOC.2014.11.006
10. Lunacek, M., Whitley, D.: The dispersion metric and the CMA evolution strategy. In: GECCO 2006 - Genetic and Evolutionary Computation Conference, vol. 1, pp. 477–484 (2006). https://doi.org/10.1145/1143997.1144085
11. Marín, J.: How landscape ruggedness influences the performance of real-coded algorithms: a comparative study. Soft Comput. **16**, 683–698 (2012). https://doi.org/10.1007/s00500-011-0781-5
12. Mersmann, O., Bischl, B., Trautmann, H., Preuss, M., Weihs, C., Rudolph, G.: Exploratory landscape analysis. In: Genetic and Evolutionary Computation Conference, GECCO 2011, pp. 829–836 (2011). https://doi.org/10.1145/2001576.2001690
13. Muñoz, M.A., Kirley, M., Halgamuge, S.K.: Exploratory landscape analysis of continuous space optimization problems using information content. IEEE Trans. Evol. Comput. **19**, 74–87 (2015). https://doi.org/10.1109/TEVC.2014.2302006

14. Muñoz, M.A., Smith-Miles, K.: Performance analysis of continuous black-box optimization algorithms via footprints in instance space. Evol. Comput. **25**, 529–554 (2017). https://doi.org/10.1162/EVCO_a_00194

15. Muñoz, M.A., Smith-Miles, K.: Generating new space-filling test instances for continuous black-box optimization. Evol. Comput. **28**, 379–404 (2020). https://doi.org/10.1162/EVCO_A_00262

16. Owen, A.B.: Scrambling sobol' and Niederreiter-Xing points. J. Complex. **14**(4), 466–489 (1998). https://doi.org/10.1006/jcom.1998.0487

17. Renau, Q., Doerr, C., Dreo, J., Doerr, B.: Exploratory landscape analysis is strongly sensitive to the sampling strategy. In: Bäck, T., et al. (eds.) PPSN 2020, Part II. LNCS, vol. 12270, pp. 139–153. Springer, Cham (2020). https://doi.org/10.1007/978-3-030-58115-2_10

18. Renau, Q., Dreo, J., Doerr, C., Doerr, B.: Towards explainable exploratory landscape analysis: extreme feature selection for classifying BBOB functions. In: Castillo, P.A., Jiménez Laredo, J.L. (eds.) EvoApplications 2021. LNCS, vol. 12694, pp. 17–33. Springer, Cham (2021). https://doi.org/10.1007/978-3-030-72699-7_2

19. Santner, T.J., Williams, B.J., Notz, W.I.: The Design and Analysis of Computer Experiments. Springer, New York (2003). https://doi.org/10.1007/978-1-4757-3799-8

20. Seo, D.I., Moon, B.R.: An information-theoretic analysis on the interactions of variables in combinatorial optimization problems. Evol. Comput. **15**, 169–198 (2007). https://doi.org/10.1162/EVCO.2007.15.2.169

21. Sobol', I.M.: On the distribution of points in a cube and the approximate evaluation of integrals. USSR Comput. Math. Math. Phys. **7**, 86–112 (1967). https://doi.org/10.1016/0041-5553(67)90144-9

22. Škvorc, U., Eftimov, T., Korošec, P.: Transfer learning analysis of multi-class classification for landscape-aware algorithm selection. Mathematics **10**(3) (2022). https://doi.org/10.3390/math10030432

Neural Architecture Search: A Visual Analysis

Gabriela Ochoa[1]([⊠]) [ID] and Nadarajen Veerapen[2] [ID]

[1] University of Stirling, Scotland, UK
`gabriela.ochoa@stir.ac.uk`
[2] Univ. Lille, CNRS, Centrale Lille, UMR 9189 CRIStAL, 59000 Lille, France
`nadarajen.veerapen@univ-lille.fr`

Abstract. Neural architecture search (NAS) refers to the use of search heuristics to optimise the topology of deep neural networks. NAS algorithms have produced topologies that outperform human-designed ones. However, contrasting alternative NAS methods is difficult. To address this, several tabular NAS benchmarks have been proposed that exhaustively evaluate all architectures in a given search space. We conduct a thorough fitness landscape analysis of a popular tabular, cell-based NAS benchmark. Our results indicate that NAS landscapes are multi-modal, but have a relatively low number of local optima, from which it is not hard to escape. We confirm that reducing the noise in estimating performance reduces the number of local optima. We hypothesise that local-search based NAS methods are likely to be competitive, which we confirm by implementing a landscape-aware iterated local search algorithm that can outperform more elaborate evolutionary and reinforcement learning NAS methods.

Keywords: Neural architecture search · Fitness landscapes · Local optima networks · Neuroevolution · Neural networks · Deep learning

1 Introduction

Neural architecture search (NAS) is a fast growing topic within automated machine learning (AutoML). The idea is to use search methods to automatically design the architecture (or topology) of deep neural networks. NAS has produced neural network models that surpass the performance of human-designed ones in image recognition [1,2] and natural language processing [1,3]. NAS is a relatively recent term, coined in 2017 by Zoph and Le [1], but the subject of research overlaps with earlier topics such as hyper-parameter optimisation, meta-learning and neuroevolution.

Neuroevolution, the use of evolutionary algorithms to design neural networks, has a long tradition in evolutionary computation with roots in the late 1980s and early 1990s [4]. Most neuroevolution systems optimise both the neural network topology and its weights. However, when scaling up to contemporary deep models

G. Rudolph et al. (Eds.): PPSN 2022, LNCS 13398, pp. 603–615, 2022.
https://doi.org/10.1007/978-3-031-14714-2_42

with millions of weights for supervised learning tasks, gradient-based weight optimisation generally outperforms evolutionary methods. In consequence, many recent neuroevolution systems use gradient-based weight optimisation and only evolve the topology [2,5]. Other approaches to NAS include random search [6, 7], hill-climbing [7], reinforcement learning [1], Bayesian optimisation [8], and gradient-based optimisation [3].

NAS is generally formulated as a discrete optimisation problem $max_{a \in A} f(a)$, where A denotes a set of architectures (the search space) and $f(a)$ denotes the objective function to be maximised[1], often set to the validation accuracy after training with a fixed set of hyper-parameters. Several search spaces have been studied [9], including chain-structured networks, which encode a sequence of layers; multi-branch networks, which incorporate skip connections; and networks consisting of repeated motifs also called *cells* or *blocks*. These cell-based architectures are designed by combining repeated cells in a predefined arrangement. Despite the underlying complexity of deep neural network architectures, many NAS spaces can be encoded as fixed-length strings of symbols of a given alphabet, where symbols represent predefined operations. This is the case of the search space considered in our study (see Sect. 2 for details).

The performance of NAS algorithms crucially depends on the search space structure. However, very little work has been devoted to analysing NAS fitness landscapes [7,10]. Our work is inspired by White et al. [7] findings when studying NAS loss landscapes, which the authors summarise as follows: "...we show that (1) the simplest hill-climbing algorithm is a powerful baseline for NAS, and (2), when the noise in popular NAS benchmark datasets is reduced to a minimum, hill-climbing outperforms many popular state-of-the-art algorithms". Our contributions are to:

- Analyse the fitness landscapes of a established NAS benchmark, using three landscape analysis techniques not previously used in this setting: (i) density-of-states [11], fitness-distance correlation [12], and local optima networks [13]. These techniques explore the landscape global structure and have a strong visual component.
- Explore the impact of reducing the noise in estimating the fitness function (validation accuracy) on the NAS landscape structure.
- Propose a local search-based NAS method informed by the structure of NAS landscapes.

2   The Selected NAS Benchmark

It is challenging to provide fair and statistically sound comparisons among NAS methods due to the different search spaces and training setups, as well as the high computational costs [6,9]. In response to this challenge, several tabular NAS benchmarks have been proposed, which exhaustively evaluate all architectures in a given search space, and store a wealth of training, evaluation and testing

---

[1] NAS can also be formulated as a minimisation problem (minimising validation loss).

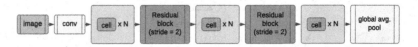

Fig. 1. The macro skeleton of candidate architectures in the search space. The skeleton is shared by all architectures and only the configuration of the cell (visualised in red) is subject to change. (Color figure online)

metrics in queryable look-up tables [14–16]. This facilitates the reproducibility of NAS experiments, drastically reduces the computational costs, and fosters a wider uptake of this topic. Our experiments use one of these tabular benchmarks, specifically, the cell-based topology search space S_t in NATS-Bench [16], also called NAS-Bench-201 [15].

The NATS-Bench Topology Search Space was inspired by the successful cell-based NAS algorithms [1–3], it consists of a predefined macro skeleton where modular (searchable) cells are stacked. Figure 1 illustrates the macro skeleton. The architecture starts with one 3×3 convolution layer with 16 output channels and a batch normalisation layer. The main body contains three stacks of cells, connected by a residual block. Each cell is stacked $N = 5$ times. The architecture ends with a global average pooling layer that flattens the feature map in to a vector. A fully connected layer with softmax is used for the final classification [16]. The macro skeleton remains fixed for all architectures, what changes is the configuration of the red cell in Fig. 1. For a given architecture, all the cells in the macro skeleton will have the same structure. Therefore, searching for a suitable architecture is reduced to searching for a suitable cell.

A cell is represented as a dense directed acyclic graph (DAG), as illustrated in Fig. 2a. Each edge in this DAG is associated with an operation that transforms the feature map from the start to the end node. Operations are selected from a predefined set of five: (A) zeroize, (B) skip connection, (C) 1×1 convolution, (D) 3×3 convolution, and (E) 3×3 average pooling layer. The *zeroise* operation simply drops the associated edge. Therefore, the cell topology is not restricted to densely connected DAGs. The nodes represent the sum of the feature maps from the incident edges. The DAG has $V = 4$ nodes. This number was chosen to allow the encoding of basic residual block-like cells, which require 4 nodes. A complete graph with 4 nodes has combinations of 4 in 2, $\binom{4}{2} = 6$ edges. Since each edge can be one of 5 operations, the search space contains $5^6 = 15,625$ unique neural architectures. Each architecture was trained three times using different random seeds on three popular image classification datasets: CIFAR-10, CIFAR-100, and ImageNet-16-120. The training pipeline and hyper-parameters is the same for all architectures. NATS-Bench [16] provides training, validation, and test loss and accuracy metrics for all architectures that can be queried via an API[2] with negligible computational costs.

---

[2] https://github.com/D-X-Y/NATS-Bench.

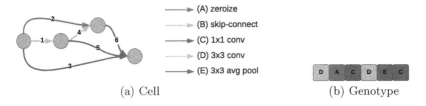

(a) Cell   (b) Genotype

Fig. 2. Encoding of an example architecture showing the mapping from a cell to the corresponding linear genotype. (a) A cell is represented as DAG with six edges representing operations taken from a fixed set of five operations (A – E) as indicated in the legend. (b) A candidate solution (genotype) is encoded as a string of six symbols, each representing the operation associated with the numbered edge in the DAG.

3   Fitness Landscape Analysis

A fitness landscape [17] is a triplet (S, N, f) where S is a set of admissible solutions i.e., a search space, $N : S \longrightarrow 2^S$, is a neighbourhood structure, a function that assigns a set of neighbours $N(s)$ to every solution $s \in S$, and $f : S \longrightarrow \mathbb{R}$ is a fitness function that measures the quality of the corresponding solutions. We define below these three components for our NAS formulation.

Search Space. The search space consists of strings of length $n = 6$ (the number of edges in the DAG representing the cell, Fig. 2a) in the alphabet $\Sigma = \{A, B, C, D, E\}$, where each symbol represents a predefined operation. An example genotype is given in Fig. 2b, where the symbol at position i corresponds to the operation associated to edge i in the DAG. The size of the search space is $|Sigma|^n$, that is, $5^6 = 15,625$ as indicated in Sect. 2.

Neighbourhood Structure. We use the standard Hamming distance 1 neighbourhood (1-change operator). The Hamming distance between two strings is the number of positions in which they differ. Therefore, the neighbourhood $N(s)$ of solution s includes the set of all solutions at a maximum Hamming distance 1 from s. The size of the neighbourhood is $n \times (|\Sigma| - 1)$, that is, $6 \times 4 = 24$.

Fitness Function. To measure the performance of each cell we consider the validation accuracy metric, to be maximised. In NATS-Bench, every architecture (cell) was independently trained three times using three different random seeds. Therefore, there are three sets of metrics for each image dataset. Since we are interested in exploring the effect of noise in the fitness landscape, we follow the approach in [7], where two ways to draw the validation metric were considered: (i) using a single value, and (ii) using the average of the three values to obtain a less noisy estimate. We therefore consider two fitness functions that we call f_{sng} and f_{avg}, to refer to using a single validation accuracy or the average of the three available values, respectively.

3.1   Density of States

The density of states (DOS) [11], plots the number of solutions in the search space with a certain fitness value. Normally, this plot requires sampling the search space, but since we have access to the whole space, we do not need a sample and instead use the complete set of solutions. The density of states gives an indication of the performance of random search or a random initialisation of metaheuristics, as it gives the probability of having a given fitness value when a solution is randomly chosen. Moreover, the right tail of the distribution near optimal fitness values gives a measure of the difficulty of a maximisation problem, the faster the decay, the harder the problem.

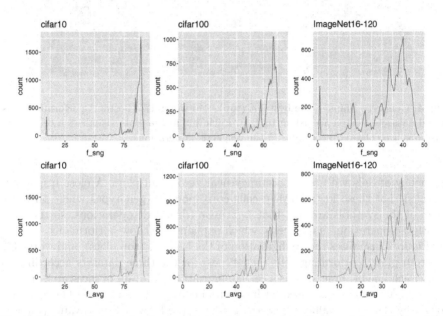

Fig. 3. Density of states for the two fitness functions f_{sng} (top) and f_{avg} (bottom) on all the datasets. The x-axis shows the whole range of validation accuracy values for each dataset, grouped in bins of width 0.5 in order to draw the frequency polygons.

In order to visualise the distribution of fitness function values, Fig. 3, shows frequency polygon plots contrasting the distribution across the two fitness functions, f_{sng} (top) and f_{avg} (bottom), for the three image datasets. There is no clear visual difference between the distributions of the two fitness functions. Notice that the range of accuracy values (x-axis) is different for each image dataset, which is consistent with the difficulty of the respective classification task. The DOS curves show a faster decay towards near optimal fitness in all cases, indicating that NAS landscapes are not completely smooth. For ImageNet, there is wider range of accuracy values with high frequency of solutions, indicating a more complex landscape. Another interesting observation is the high

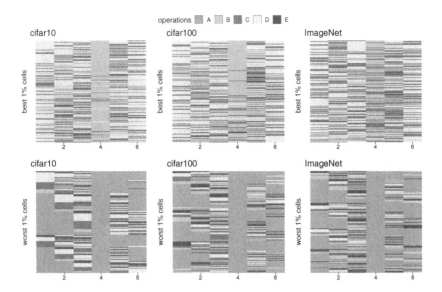

Fig. 4. Genotype maps of the best 1% (top), and worst 1% (bottom) performing cells for all datasets, sorted according to f_{avg}. Each line in the plots visualises a cell where positions are coloured according to the respective operation.

frequency of cells with a low accuracy near zero in all plots. A close inspection revealed that these low accuracy cells correspond to genotypes where three or more of the symbols are 'A', that is *zeroise* (dropped) operations, so they are mostly no-operation, empty cells, which explains their low performance.

Figure 4 visualises the configuration of the best 1% (top plots), and worst 1% (bottom plots) performing cells (genotypes) for the three datasets according to f_{avg}. Each line in these plots is a cell configuration (solution in the search space), where positions are visualised with colours identifying operations. The rows are sorted by their f_{avg} value, where the cell with the highest fitness value (highest average accuracy) in the set is the top line of each plot. We can clearly see that the low performing cells (bottom plots) are those containing a majority of 'A' (zeroise, or drop) operations, thus they are mostly empty cells. Specially the 4th positions is always an 'A' in the worst performing configurations. The best performing configurations (top plots) also show a visible pattern, with the most common operations being 'C' (green) and 'D' (orange), corresponding to 1×1 and 3×3 convolutions, respectively. The exception is the 4th position, which for all datasets is mostly a 'B' (light blue, skip connection) in the best performing cells. The plots suggest that the choice of operation for the 4th position (4th edge in the DAG cell, Fig. 2a) has more impact in performance than the other positions. An analysis of the frequency of operations in the top 1% performing cell across the 3 datasets revealed that they rank as follows: D, C, B, A, E with frequency percentages of: 45.7, 24.8, 19.2, 6.1, 4.1, respectively. We argue that this information can be used to design informed mutation operations that can improve the performance of search heuristics in this domain, and we set to do that in Sect. 4.

3.2   Fitness Distance Correlation

Since the whole search space is available, and thus the optimal cell is known, we can compute the distances from all cells in the search space to the global optimum. Specifically, for each cell i we have a pair (f_i, d_i), where f_i is the validation accuracy (either f_{sng} or f_{sng}) of cell i and d_i is the Hamming distance to the cell with the highest validation accuracy (global optimal cell). The FDC is calculated as the (Spearman) correlation coefficient of this set of (accuracy, distance) pairs.

Figure 5 shows the FDC plots, as well as the Spearman correlation coefficients (R) with significance level (p-values) for each image classification dataset. The top plots show the measurements with a single training seed (f_{sing}), while the bottom plots show the average of the 3 training seeds available (f_{avg}). The regression lines with confidence regions (95%) are also shown. The horizontal axes show the Hamming distance between all architectures and the global optimal architecture, while the vertical axes show the validation accuracy of each architecture. From these plots we can observe that there is a moderate negative correlation (ranging from -0.33 to -0.46) between distance and fitness (validation accuracy), suggesting a gradient towards the global optimum. However, for all studied scenarios, some configurations that differ in 3 or 4 operations from the global optimum reveal a low accuracy value, these are the cells with a high number of *zeroise* operations. For the three datasets, the correlations coefficients are higher when the less noisy estimation of fitness f_{avg} is considered, supporting the insight from [7] indicating that reducing noise in the estimation of fitness can improve search. The range of possible values for R is $[-1, 1]$ where, for a maxisimisation problem, high negative correlations would be regarded as easier for a hill climber. When FDC was proposed [12], $-0.15 \leq R \leq 0.15$ was classified as hard, and $R \leq -0.15$ was considered as misleading for a minimisation problem. Using these criteria, none of the problems used in this study is regarded as hard or misleading. The ImageNet dataset reveals the lowest correlation coefficient, which supports that this is hardest of the three instances.

3.3   Local Optima Networks

To further understand the landscapes' global structure, we extract and analyse local optima networks (LONs)[13]. LONs are graph-based model of landscapes where nodes are local optima and edges are transitions among optima with a given search operator.

Definitions. The relevant definitions, and the procedure to construct the LON modes, are given below.

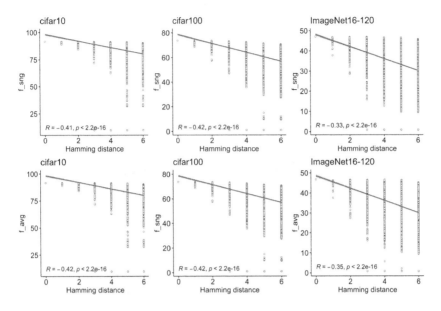

Fig. 5. FDC plots for all datasets. The horizontal axes show the Hamming distance to the global optimum, using the f_{sng} fitness values (top plots), and the less noisy f_{avg} values (bottom plots). The Spearman correlation coefficients with p-value are also shown.

Local Optima. A local optimum, which in our NAS formulation is a maximum, is a solution l such that $\forall s \in N(l)$, $f(l) > f(s)$. Local optima are identified with a best-improvement hill-climbing heuristic using the 1-change (Hamming distance 1) neighbourhood. The set of local optima, denoted by L, corresponds to the nodes in LON model.

Edges. Edges are directed and based on the perturbation operator 2-change. There is an edge from local optimum l_1 to local optimum l_2, if l_2 can be obtained after applying a random perturbation (changing at random 2 locations in the genotype) to l_1 followed by local search. Edges are weighted with estimated frequencies of transition in a sampling process. The weight is the number of times a transition between two local optima occurred when constructing the LON models as detailed below. The set of edges is denoted by E.

LON. The LON is the directed graph $LON = (L, E)$, with node set L, and edge set E as defined above.

LON Sampling and Construction. To construct the LON models for each dataset and fitness function, a sampling process is conducted. It consists in running an iterated local search algorithm (ILS) [18], where the stopping condition is set to $t = 100$ iterations without any improvement. This serves the purpose of empirically estimating the global optimum or the end of a *funnel*, i.e., a solution at the end of an ILS trajectory, where escaping is difficult, if not impossible. While running ILS, we store in a set L all the unique optima obtained after the local search stage, and in a set E all the unique edges obtained after a perturbation followed by local search. To construct the LONs for each image dataset and fitness function, these sets of nodes and edges are aggregated over 1 000 runs, started from different random configurations.

Network Visualisation. One advantage of network models is that they can be visualised, bringing useful insight into their structure. Figure 6 illustrates the LONs for all datasets and fitness functions. The networks capture the whole set of sampled nodes and edges in each case. In the plots, each node is a local optimum and edges are perturbation transitions. Plots were produced using *force-directed* layout methods as implemented in the igraph R library [19]. The global optimum, which was unique in all cases, is highlighted in red. The other local optima are painted in grey. The edges' colour indicate whether they end in a node with better fitness (dark gray), worse fitness (orange) or equal fitness (blue). The size of nodes is proportional to their incoming weighed degree, so larger nodes indicate attractors in the search process.

The networks in Fig. 6 indicate that for all datasets and fitness functions, there is a connected component of nodes that can reach the global optimum (red node) in a few search steps. This is indicative of a multi-modal landscape, but where it is not too hard to reach the global optimum. For all datasets, there are fewer local optima when the less noisy fitness function f_{avg} is used (Figs. 6d, 6e and 6f), which indicates that improving the fitness estimate facilitates the search process, as the search paths to the global optimum become shorter.

The LON analysis indicates that the edges considered (changing 2 locations in the genotype), allow escaping local optima in most cases. This suggests that a local search based method, coupled with a escape mechanism can be a suitable NAS search strategy. This is empirically explored in the next section.

4   Search Performance Analysis

4.1   Competing Algorithms

We propose and implement a local search based algorithm, specifically an iterated local search (ILS) method [18]. ILS is a simple yet powerful metaheuristic that alternates a local search stage with a perturbation stage. We use a first improvement local search with a 1-change neighbourhood, and a perturbation operator that changes 2 positions in the incumbent solution. Only improving moves are accepted. We consider two versions of the ILS: ILS-shuffle where the values for the 1-change operator are explored in random order and ILS-order where the 1-change operator uses insights from the landscape analysis. Specifically, following the frequency profile observed in the 1% best-performing cells (Sect. 3.1 and Fig. 3) we systematically explore neighbours using the following ordering of the operations: D, C, B, A, E.

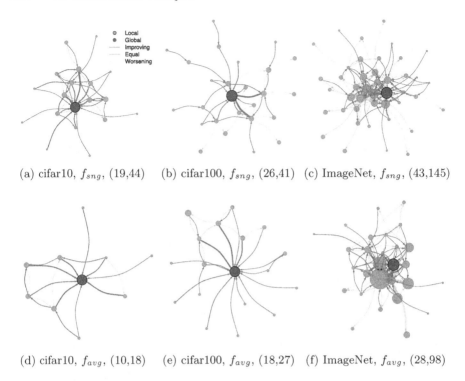

(a) cifar10, f_{sng}, (19,44)   (b) cifar100, f_{sng}, (26,41)   (c) ImageNet, f_{sng}, (43,145)

(d) cifar10, f_{avg}, (10,18)   (e) cifar100, f_{avg}, (18,27)   (f) ImageNet, f_{avg}, (28,98)

Fig. 6. LONs for all datasets and the two fitness functions. For each model, the number of nodes n and edges e are indicated as (n, e). (Color figure online)

We contrast our proposed ILS against the following NAS methods, as implemented in [16].

- **Random search** (RANDOM) [6]. This serves as the baseline. It draws cells at random and returns the best found.
- **Regularised evolution** (REA) [2]. This is a mutation only evolutionary algorithm that uses tournament selection and introduces the notion of age to the individuals. The replacement strategy removes the oldest individual in the population, thus favouring newer cells. This serves as a mechanism to handle the noisy performance estimation.
- **Reinforcement learning** (REINFORCE) [1]. This approach frames NAS as a reinforcement learning problem. The generation of a neural cell correspond to the agent's actions, with the action space identical to the search space. The agent's reward is based on an estimate of the cell performance on unseen data.

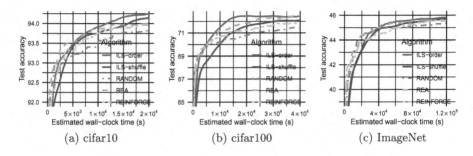

Fig. 7. Evolution of average test accuracy across the three datasets.

4.2 Empirical Setup

Our experiments follow the protocol suggested by NATS-Bench [16]. The benchmark provides performance data on each neural architecture for two scenarios: one with 12 epochs, the other 200. The epoch indicates the number of times the entire training dataset is used while building the model.

We replicate the NATS-Bench experiments by training the models over 12 epochs and using the accuracy calculated on the validation set as feedback to direct the search. This is meant to simulate a faster but less accurate training step. The configurations obtained are then evaluated against the test set of the 200 epoch scenario. The best solution found using 12 epochs is therefore not necessarily the best for 200 epochs. The training time budgets considered for cifar10, cifar100 and ImageNet datasets are 20 000, 40 000 and 120 000 seconds respectively. Each algorithm is executed 30 times.

4.3 Results

The average test accuracy is presented on Fig. 7. The different methods have fairly similar behaviour. ILS, for its part, initially converges slightly slower than the rest since it is costly to evaluate multiple neighbours before accepting a new solution. However, on average, it manages to get ahead of the other approaches within the time budget on the cifar10 and ImageNet datasets. Using `ILS-order` improves convergence speed and the final result. As was previously noted [16], despite its simplicity, random search performs comparatively well, even if it comes in last.

In order to better grasp the overall performance of the different algorithms, Fig. 8 presents boxplots of the test accuracy calculated on the cell configurations found at the end of each run. The `ILS-order` boxplots are fairly tight, indicating that there is little spread in the quality of configurations obtained. In contrast, the results for `ILS-shuffle` on cifar10 and cifar100 are much more spread out. This is because its slower convergence means it hasn't yet reached the very bottom of the landscape. This is not the case on the ImageNet dataset where both ILS versions converge to similar solutions within the allotted budget.

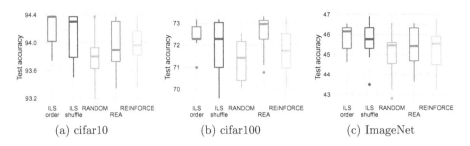

(a) cifar10   (b) cifar100   (c) ImageNet

Fig. 8. Test accuracy distribution for configurations found at the end of 30 runs.

On this problem, the challenge for optimisation methods compared to classic optimisation problems is that the function used to evaluate the end result (test accuracy) is not the same as the objective function (validation accuracy). We know from sampling the landscape and LON analysis that ILS is able to reach the global validation accuracy optima on the benchmarks when there is no time limit, however this is not a guarantee that the same solution will be the best for test accuracy.

Overall, ILS proves to be a viable and competitive approach for optimising neural network topology, especially if appropriate design choices are implemented. Despite its conceptual simplicity, ILS is able to match and even outperform more sophisticated approaches within the time budget in this NAS topology benchmark.

5   Conclusions

We analysed the fitness landscape of a popular tabular, cell-based NAS benchmark for image classification. Our analysis revealed that the landscapes are not trivial to search, they are rugged (multi-modal), however they have a relatively low number of local optima, from which it is not difficult to escape with a simple perturbation operation. Our analysis of the best-performing cells indicated that some operations of the available set appear more frequently than others. We used this information to design a conceptually simple, yet high-performing local search based NAS method. On the studied benchmark, our iterated local search (ILS) implementation outperforms both a reinforcement learning method on the 3 available image datasets, and the state-of-the-art evolutionary method on 2 of the 3 image datasets. Future work will analyse the landscapes of other available NAS-benchmarks, as well as test the performance of the proposed ILS on them. We will also incorporate noise-handling mechanisms into our ILS approach since our landscape analysis reported a smoother landscape when noise is reduced.

References

1. Zoph, B., Le, Q.V.: Neural architecture search with reinforcement learning. In: Conference on Learning Representations, ICLR (2017)

2. Real, E., Aggarwal, A., Huang, Y., Le, Q.V.: Regularized evolution for image classifier architecture search. In: AAAI Conference on Artificial Intelligence, AAAI, pp. 4780–4789. AAAI Press (2019)
3. Liu, H., Simonyan, K., Yang, Y.: DARTS: differentiable architecture search. In: Conference on Learning Representations, ICLR (2019)
4. Schaffer, J., Whitley, D., Eshelman, L.: Combinations of genetic algorithms and neural networks: a survey of the state of the art. In: International Workshop on Combinations of Genetic Algorithms and Neural Networks, pp. 1–37 (1992)
5. Lu, Z., et al.: NSGA-Net: neural architecture search using multi-objective genetic algorithm. In: Genetic and Evolutionary Computation Conference (GECCO), pp. 419–427. ACM, New York, NY, USA (2019)
6. Yu, K., Sciuto, C., Jaggi, M., Musat, C., Salzmann, M.: Evaluating the search phase of neural architecture search. In: Conference on Learning Representations, ICLR (2020)
7. White, C., Nolen, S., Savani, Y.: Exploring the loss landscape in neural architecture search. In: Conference on Uncertainty in Artificial Intelligence, UAI. Proceedings of Machine Learning Research, vol. 161, pp. 654–664. AUAI Press (2021)
8. Kandasamy, K., Neiswanger, W., Schneider, J., Póczos, B., Xing, E.P.: Neural architecture search with Bayesian optimisation and optimal transport. In: Advances in Neural Information Processing Systems, NeurIPS 2018, pp. 2020–2029 (2018)
9. Elsken, T., Metzen, J.H., Hutter, F.: Neural architecture search: a survey. J. Mach. Learn. Res. **20**, 55:1–55:21 (2019)
10. Rodrigues, N.M., Silva, S., Vanneschi, L.: A study of generalization and fitness landscapes for neuroevolution. IEEE Access **8**, 108216–108234 (2020)
11. Rosé, H., Ebeling, W., Asselmeyer, T.: The density of states—a measure of the difficulty of optimisation problems. In: Voigt, H.-M., Ebeling, W., Rechenberg, I., Schwefel, H.-P. (eds.) PPSN 1996. LNCS, vol. 1141, pp. 208–217. Springer, Heidelberg (1996). https://doi.org/10.1007/3-540-61723-X_985
12. Jones, T., Forrest, S.: Fitness distance correlation as a measure of problem difficulty for genetic algorithms. In: International Conference on Genetic Algorithms, pp. 184–192. Morgan Kaufmann (1995)
13. Ochoa, G., Tomassini, M., Verel, S., Darabos, C.: A study of NK landscapes' basins and local optima networks. In: Genetic and Evolutionary Computation Conference, GECCO, pp. 555–562. ACM (2008)
14. Ying, C., Klein, A., Christiansen, E., Real, E., Murphy, K., Hutter, F.: NAS-Bench-101: towards reproducible neural architecture search. In: International Conference on Machine Learning, ICML, vol. 97, pp. 7105–7114. PMLR (2019)
15. Dong, X., Yang, Y.: NAS-Bench-201: extending the scope of reproducible neural architecture search. In: Conference on Learning Representations, ICLR (2020)
16. Dong, X., Liu, L., Musial, K., Gabrys, B.: NATS-Bench: benchmarking NAS algorithms for architecture topology and size. IEEE Trans. Pattern Anal. Mach. Intell. **7**(2022), 3634–3646 (2021)
17. Stadler, P.F.: Fitness landscapes. Appl. Math. Comput. **117**, 187–207 (2002)
18. Lourenço, H.R., Martin, O.C., Stützle, T.: Iterated Local Search: Framework and Applications, pp. 363–397. Springer, US, Boston, MA (2010). https://doi.org/10.1007/978-1-4419-1665-5_12
19. Csardi, G., Nepusz, T.: The igraph software package for complex network research. Int. J. Complex Syst. **1695** (2006)

Author Index

Printed in the United States
by Baker & Taylor Publisher Services